COMPUTER AND PHYSICAL MODELLING IN GEOTECHNICAL ENGINEERING

PROCEEDINGS OF THE INTERNATIONAL SYMPOSIUM ON COMPUTER AND PHYSICAL MODELLING IN GEOTECHNICAL ENGINEERING / BANGKOK 3-6 DECEMBER 1986

Computer and Physical Modelling in Geotechnical Engineering

Edited by
A.S.BALASUBRAMANIAM, G.RANTUCCI, S.CHANDRA, D.T.BERGADO, NOPPADOL PHIENWEJA, S.SARAN & PRINYA NUTALAYA
Asian Institute of Technology, Bangkok

CRC Press
Taylor & Francis Group
Boca Raton London New York

CRC Press is an imprint of the
Taylor & Francis Group, an **informa** business
A BALKEMA BOOK

Published by:
CRC Press/Balkema
P.O.Box 447, 2300 AK Leiden, The Netherlands
e-mail: Pub. NL@tandf.co uk
www.balkema.nl, www.tandf.co.uk, www.crcpress.com

ISBN 13: 978-90-6191-864-6 (hbk)

Visit the Taylor & Francis Web site at
http://www.taylorandfrancis.com

and the CRC Press Web site at
http://www.crcpress.com

The texts of the various papers in this volume were set individually by typists under the supervision of each of the authors concerned.

Computer and Physical Modelling in Geotechnical Engineering, Balasubramaniam et al. (eds)
© 1989 Balkema, Rotterdam. ISBN 90 6191 864 2

Contents

3. *Underground openings and excavations*

4. *Computer controlled testing and investigation of soils*

Computer and Physical Modelling in Geotechnical Engineering, Balasubramaniam et al. (eds)
© 1989 Balkema, Rotterdam. ISBN 90 6191 864 2

Dedications

This volume of proceedings, which completes a decade of successful symposia and activities, is dedicated to Mr R.E.Ward, on the occasion of his retirement, and to the Cambridge Soil Mechanics Group, (currently led by Prof. Andrew Schofield). Prof. Balasubramaniam, one of the authors, was a student of Prof. A.Thurairajah at the University of Ceylon, Colombo, and later his studies were supervised by Dr R.G.James at Cambridge University.

The Cambridge Group of Soil Mechanics was initiated and nurtured by the late Prof. Kenneth Harry Roscoe. Dr Peter Smart, one of his successful students, described Roscoe's philosophy as: (1) Ken Roscoe was a one-hundred percent type of man, determined both to improve the quality of life and to see good research done. (2) It was paradoxical that this determination led him into battles and disagreements; but it was typical of him that these were short-lived. (3) His approach to research was simple. Money was needed, money would be obtained. Space was needed, space would be acquired. Assistance was needed, assistants would be found. Since a large group is more efficient, colleagues and students would be recruited. Since quality depends on criticism, criticism would be provided. Since innovation depends on stimulation, distinguished engineers would be persuaded to visit the Group. Since significant progress depends on it, work would be concentrated on a few fundamental themes. The whole success of the Cambridge Soil Mechanics Group depended on the vigorous execution of this programme. (4) The Roscoe approach to research was even simpler. They were taught to punt correctly from the 'wrong' end, fed a diet of parties, steered through their vicissitudes, and supported in their troubles. From the moment they reached Cambridge they and their wives were spontaneously and warmly adopted as members of the family, by both him and Janet. Perhaps this policy had been consciously evaluated by both of them. (5) People were important to him; quality was important to him; but perhaps Ken Roscoe's greatest contribution was to remind us of the virtue of trying.

It is difficult to describe the successful and bright career of all of Roscoe's students, but for a brief period the Group was subsequently led by Prof. C.P.Wroth, and then by Prof. Andrew Schofield. Some of the other senior students in chronological order are Prof. H.B.Poorooshasb, Prof. A.Thurairajah, Dr R.G.James, Dr J.R.F.Arthur, Dr J.B.Burland, Dr D.L.Coumulos, Dr W.H.Ting and, Prof. Balasubramaniam's classmates, Drs E.C.Hambly, P.J.Avgherinos and N.K.Tovey.

A brief biographical data of Mr R.E.Ward who saw so many bright individuals at Cambridge, is as follows: 'Ralph Ward joined the Cambridge University Engineering Department in 1954. Prior to this he was an instrument-making apprentice and had served in the Armed Forces from 1943 to 1947. He worked in the department's Instrument Shop until 1958, then moved to the Soil Mechanics Laboratory at the request of Mr (later Professor) K.H.Roscoe, head of the Group, which at that time was comprised of two lecturers and three research students. At present there are 20 research students/workers and 9 technicians, now led by Professor A.N.Schofield. Ralph has helped countless budding soil mechanics to pass through the department and has witnessed many improvements and complexities in technique and equipment – the building and development of the large geotechnical centrifuge, and recently the smaller drum centrifuge, in a laboratory sited at Madingley, which is the major expansion of the Group. He is married to Margaret and has two sons and three grandsons.' We wish Ralph and Margaret a happy and prosperous future.

A.S.Balasubramaniam

Mr R.E.Ward

Late Prof. Kenneth Harry Roscoe

1. Stability of natural and man made slopes

Computer and Physical Modelling in Geotechnical Engineering, Balasubramaniam et al. (eds)
© 1989 Balkema, Rotterdam. ISBN 90 6191 864 2

Rapid design assessments from slope stability calculation results

Edward N.Bromhead & Andrew J.Harris
Kingston Polytechnic, Kingston upon Thames, UK

ABSTRACT: With the widespread use of computers, and in particular desk top computers, in geotechnical design offices, slope stability computations have become routine, and simple. There are, however, instances where the application of a few simple principles utilising information routinely supplied by slope stability analysis programs (and equally as routinely ignored in the design office!) can lead to rapid design decisions being taken. These can rule out much repetitive "trial & error" work. These simple principles are described in detail in the paper, with examples, and include:

* Assessment of stress levels in slopes for correlation with laboratory test data (or indeed, for specification of test data at appropriate stress magnitudes)
* Deciding what shear strength parameters give "F=1"
* Assessing whether drainage is a feasible measure
* Approximate sizing of counterweight embankments
* Approximate calculation of anchor loads
* Evaluating forces on walls in landslides
* Location of cuts and fills on an unstable slope
* Prediction of critical seismic accelerations
* Deciding where to concentrate an initial search for a critical slip surface

These techniques are not intended to replace proper and systematic analyses, but to act in conjunction with them to speed up the design process.

1 INTRODUCTION

Few procedures in geotechnical engineering, or for that matter, other branches of civil engineering, are genuinely design methods. Engineering analysis is usually a matter of testing the feasibility or economy of a design concept. The evolution of a design is not therefore the result directly of the computational process, but is the result of a trial-and-error "think of a concept and test it" procedure. Nowhere is this better shown than in slope stability analysis. By and large, slope stability analysis is undertaken with analytical computer software of a variety of levels of sophistication, but virtually all of this software postulates the existence of a "design" or actual cross section, for which it is desired to obtain some index of stability. This index might be the conventional factor of safety, or it might be the alternative critical seismic acceleration - see for instance, Sarma, 1973. Generally, the index of stability is evaluated for a range of trial surfaces to find the most critical: or in cases where the position of an actual failure surface is known, a more restricted range of surfaces in the vicinity of the known failure surface are investigated. However, in no case does the method indicate what design slopes might be acceptable.

Even stability charts, which summarise the results of a number of slope stability analyses, are usually set up to yield a factor of safety: for example Bishop & Morgenstern's charts (1960). To evolve a design section requires several attempts until a satisfactory slope is obtained. The exception is in the case of the simple stability charts by Taylor (1937), from which a design slope can be obtained directly. However, any attempt to extend from a total stress analysis to one invoking effective stress behaviour (e.g. Spencer, 1967) by necessity adopts a recursive procedure since the behaviour of the seepage regime when the slope geometry is changed is not a linearly dependent function.

In this paper a number of simple but nonetheless useful techniques are presented, which speed up the process of finding the desired solution by eliminating many of the fruitless searching analyses. No claim is made that these eliminate the need for a final check analysis, nor that they are rigorous solutions (although some of them are). Their purpose is to assist in the evolution of the final design.

2 EVALUATION OF AVERAGE SHEAR AND NORMAL EFFECTIVE STRESSES

The first step in many of the procedures outlined below is a stress analysis of the slide. In most methods of slope stability analysis, some stress analysis is done, but the details of this are usually lost on production of a "factor of safety", although it is commonplace in non-circular slip analysis to be able to recover at least part of the stress analysis from the output. This takes the form of a table of "interslice" forces, in many cases offered as a means by which the analyst may make a more-or-less subjective assessment of the "reasonableness" of the solution in respect of two principle criteria:

* is the location of the line of thrust within the section at all, or is it within the "middle third" (both sorts of "no-tension" criteria)?
* what is the mobilisation of shear strength between the slices?

In the most straightforward procedure, the program will compute the shear and normal interslice force components explicitly (e.g. Morgenstern & Price's 1965 procedure), from which the equilibrium forces on the slip surface may be obtained. In some software these values form part of the printed output, in others, merely a summary, or occasionally, nothing at all, is produced. The latter result is commonest with "slip-circle" programs, where the output is extensive as a result usually of the analysis of a significant number of individual slip surfaces. Non-circular slip analysis programs usually have some output for the reasons discussed above. With commercial software, it is unlikely that the source code will be available for modification, and whatever output is obtained has to suffice.

If sufficient information is included in the output, then the levels of shear and normal stress acting on the slip surface can be derived from the summed force resultants and the length of slip surface affected. This simple expedient may not be a practical approach, however, and the following procedure can be used in its stead.

To obtain the average shear stress acting along the slip surface, perform an analysis using a single cohesive shear strength parameter acting along the whole length of slip surface. Suppose this to be c^*. A corresponding factor of safety is obtained from the analysis, F^*. It will be found that the cohesion or shear strength (s) required to be available for a factor of safety of 1.0 is:

$$s = c^* / F^*$$

and this is also the average shear stress acting around the slip surface. Indeed, it can easily be shown to be so for a slip circle, although it is less clear in the case of a non-circular slip surface, and there may be a 2 to 3 % difference between the "average" shear stress computed by means of this rule, and that obtained by summing the shear forces and dividing by the appropriate length of slip surface.

In some slip circle software, an overturning (M_o) or resisting (M_r) moment is output. The average shear stress can be obtained from this without an analysis with c^*, since:

$$M_r = F M_o = s L R$$

in which L is the length of slip surface, and R is its radius.

However the average shear stress is obtained, it is thereafter used in the estimation of the average normal effective stress σ_n' by means of an analysis involving an angle of shearing resistance ϕ' (with or without a corresponding c') to produce a new factor of safety F^+, and then:

$$F^+ = (c' + \sigma_n' \tan \phi') / s$$

so that σ_n' may be found from:

$$\sigma_n' = (s F^+ - c') / \tan \phi'$$

This gives the "working" average stresses. One further parameter, the average total normal stress can be obtained from the above equation for σ_n' by first re-analysing the section with zero pore water pressures. It will be found that a minimum of data manipulation is required, if within the data submitted to the program used it is possible to:

a. set all r_u values to zero, or

b. take the unit weight of water to be zero

The average pore water pressure ratio r_u for the slip surface is

$$r_u = (1 - (\sigma_n - \sigma_n') / \sigma_n)$$

3 FINDING SHEAR STRENGTH PARAMETERS FOR "F = 1"

These may be obtained in two ways: from the stress analysis outlined above, or by manipulations of the trial values of c' and Ø' and corresponding factor of safety. Note that there are an infinity of possible c'/Ø' combinations to yield the same factor of safety, and it is commonplace (indeed correct in many cases) to assume c'=0. In this situation, Ø' for F=1, often styled the mobilised angle of shearing resistance, $Ø'_m$, is simply

$$Ø'_m = \tan^{-1} (s / \sigma_n')$$

it is also simple to show that in an analysis using Ø' alone to yield a non-unity factor of safety F^+, that

$$Ø'_m = \tan^{-1} ([\tan Ø'] / F^+)$$

Were there a cohesion present in this analysis, one would also predict a mobilised cohesion c'_m equal to c' / F^+.

4 IS DRAINAGE FEASIBLE?

The feasibility in principle of drainage as a remedial measure is indicated by the magnitude of r_u. Where r_u is high, drainage can produce a large increase in stability, but where low, drainage is of proportionately less use. To estimate the maximum effect of drainage, that is when all pore water pressures have been removed, compute F' from the original factor of safety F_o, where:

$$F' = \frac{s + \sigma_n \tan Ø'}{s + \sigma_n' \tan Ø'} \times F_o$$

or

$$F' = F_o / (1 - r_u)$$

In the case of a non-cohesive soil:

$$F' = F_o \sigma_n / \sigma_n'$$

Obviously, if F' is not as high as the desired factor of safety, F_r, drainage on its own is not feasible.

Suppose that F' exceeds the required factor of safety? What average drainage is required?

$$F_r = \frac{s + \sigma_{nr}' \tan Ø'}{s + \sigma_n' \tan Ø'} \times F_o$$

from which we find the required average normal effective stress σ_{nr}' Since the average drawdown of the piezometric line must be

$$(\sigma_{nr}' - \sigma_n') / (\gamma_w)$$

Is this obtainable? More to the point, since a uniform drawdown is not usually achievable, with only small or perhaps even no drawdown at each end of the slip surface, so that the drawdown is distributed linearly or parabolically (as shown diagrammatically in Figure 1), can twice or one-and-a-half times this drawdown be obtained as a maximum?

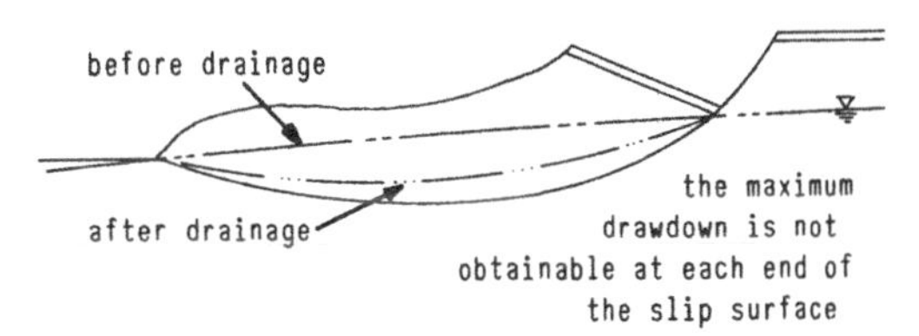

Fig. 1 Likely distribution of draw-down across a rotational slip, following drainage as a remedial measure

5 COUNTERWEIGHT EMBANKMENTS

The optimum location of a counterweight embankment is indicated by the "neutral line" theory (Hutchinson, 1976). The "neutral line" is the trace, on a plan of a particular landslide, of a series of "neutral points". Each of these "neutral points" is defined for a cross section as the location on the ground surface (or on the slip surface perpendicularly beneath it) at which the imposition of a load has no influence on the factor of safety. Loads placed on each side of the neutral point will have a beneficial or detrimental effect on the factor of safety.

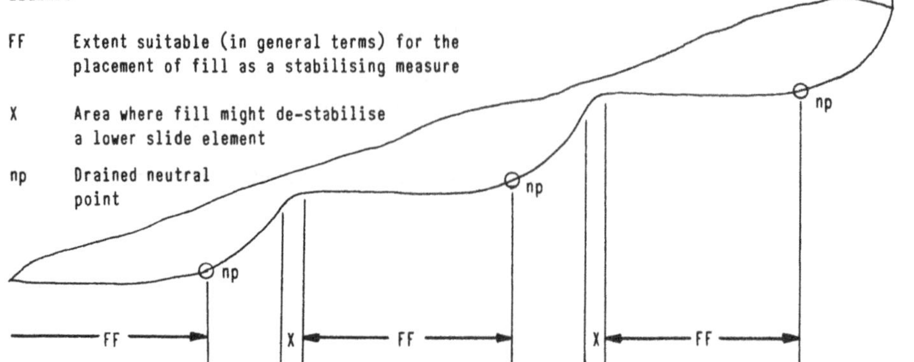

Fig. 2 Diagrammatic representation of the use of the Neutral Point concept in deciding the location of stabilising fills on a complex, "multi-storey" landslide

A counterweight embankment may operate in one of three ways:

* as a dead load on the toe of the slide, so that although its weight decreases the nett overturning force or moment, no contribution to the resisting forces is made. This is typical of loads applied without consolidation drainage of the underlying soil, as in the "short term". (Type 1)
* as a dead load but with some effect on the effective stresses underneath it, as in the "long term" or fully drained condition. (Type 2).
* as a passive resistance at the toe of the slide, which slip movements must displace if further movements of the slide are to take place. (Type 3).

Type 1 & 2 counterweights also represent two extreme cases in terms of the generation and dissipation of pore water pressures under the applied loading. Hutchinson (1976) terms these respectively fully undrained and fully drained loading. An intermediate case, where either the amount of pore pressure generated is less than the amount of the total stress change (i.e. $B < 1.0$), or where some consolidation has taken place from an initially higher generated pore water pressure can be envisaged.

Should the increase of pore water pressure in this intermediate case be represented by a parameter B^* times the increase in vertical stress, the location of neutral points may be found from positions on the slip surface which satisfy the following equation:

$$\tan \alpha = (1 - B^* \sec^2 \alpha) \tan \phi' / F$$

in which is the angle of inclination from the horizontal of the slip surface. There must be neutral points for drained loading ($B^*=0$), undrained loading ($B^*=1$) and intermediate points.

The first step is to locate the neutral points. If more than one cross section through the landslide has been investigated, we can draw the neutral line. This delineates feasible locations for the placement of fills, or for the excavation of cuts in the slope. (Figure 2 shows this for a fairly complex slide. Note that a number of Neutral Points may appear on a section like this).

If it is not possible to place fills where the neutral line delineated fill areas are shown, and at the same time it is not possible either to make cuts where the method indicates potential cut areas, then cut and fill solutions are not possible. However, when the neutral point theory indicates that cuts and/or fills might be permissible, there is the problem of deciding their approximate size. The following is a possible approach:

* on the section, divide the slide into "active" and "passive" zones by a vertical through the appropriate neutral point. Measure areas in the active and passive zones (A_a and A_p respectively in Figure 3).
* express the ratio of these R_{oa} as:

$$R_{oa} = A_p / A_a$$

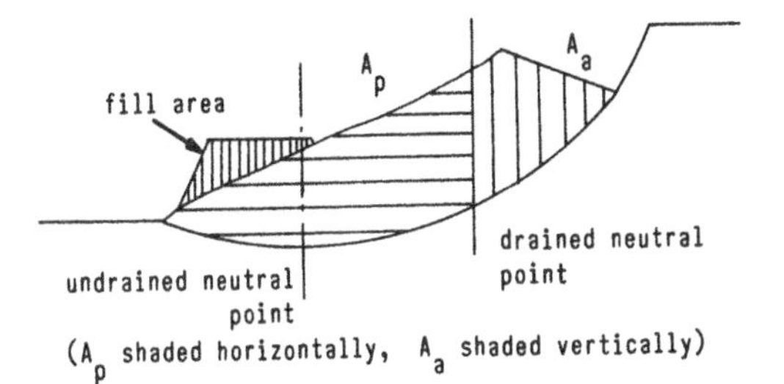

Fig. 3 Assessment of desirable size of a counterweight embankment in terms of slide geometry

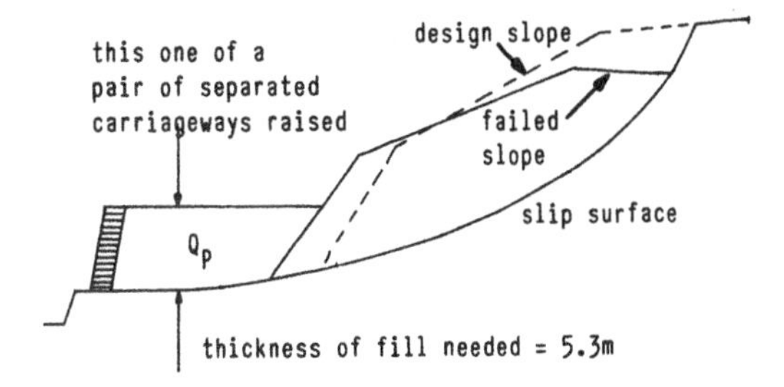

Fig. 4 An entirely passive toe fill: contributes nothing to the shear strength along the slip surface

* following a stabilising muckshift, the new area ratio R_{1a} (defined in the same way as the original area ratio R_{oa}) should be increased in approximately the same proportion as the factors of safety, viz.

$$R_{1a} \;/\; R_{oa} \;=\; F_1 \;/\; F_o$$

Naturally, in the commonest approach, which is to use toe loading without corresponding slide head cuts, the ratios can be evaluated in terms of the passive areas alone, since the active area is a common factor. Partial sums of slice weights can be substituted for cross sectional areas, especially where these form part of the initial stability analysis output, without loss of "accuracy", since this is only a rule of thumb. It should be noted that this does not take into account the greater unit efficiency of fill placed at the toe compared to fill placed nearer the neutral point, and so the likely sign of the error in using this approach can be deduced. Furthermore, the method is really only applicable where the earth moving operation produces a small effect on the overall slope section, rather than inflicting gross changes on the slope geometry.

Loading placed in response to the predictions of the neutral line theory may fall into classes 1 or 2 of the introduction to this section, or indeed, somewhere in between. Class 3 is a different matter. In this case, the toe load contributes nothing to stability through an increase in shearing resistance, but is largely or wholly off the slip, and acts as an additional passive external force. An example of this is explained by Figure 4.

In principle, the calculation required is as follows:

* decide the required passive force, assuming a passive anchor load at the toe of the slope - in the example, 900 kN per metre of slide
* compute the necessary depth of fill to provide this resistance, using Rankine (or similar earth presssure theory) e.g. in the example, with fill of unit weight=20 kN/m^3, and $\phi'=30^o$, $K_p=3$ so:

$$900 = 0.5 \times 3 \times 20 \times d^2$$
$$\text{or } d = \text{approx } 5.3 \text{ m}$$

* the width of such an embankment is decided by external constraints, in the example it had to be at least one road carriageway wide, for instance, and this may be a more rigorous constraint than sliding along its base.

The designer of such a passive fill should be cautioned, however, that the strains necessary to mobilise this resistance can represent considerable further movement of the slide, and this constraint may dictate the use of some other expedient for stabilisation.

With all cut and/or fill solutions, the possibility of alternate failure mechanisms ("over-riding" and "under-riding" slips in the sense of Figure 5) must be considered.

Under-riding slip surfaces are unlikely to prove critical, at least not as shown in the Figure; yet deeper slip surfaces not found or explored in the site investigation can, of course, prove to be a serious problem, although the paper does not seek to explore the competence of the analyst (or his luck!). However, under-riding slip surfaces will almost certainly have lower factors of safety than those computed assuming rising slip surface

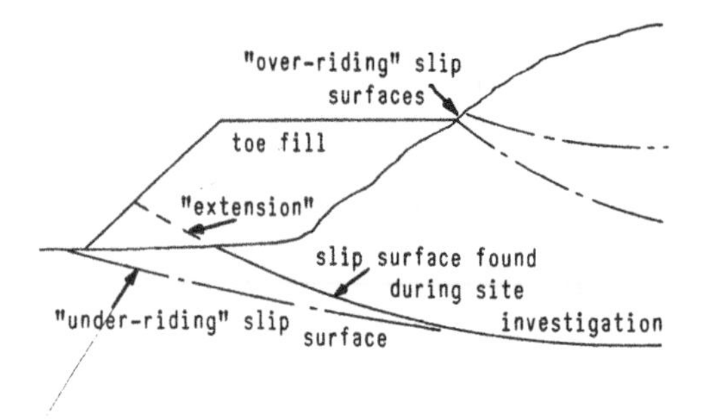

Fig. 5 Detail at the toe of a landslide with "over-riding" and "under-riding" slip surfaces

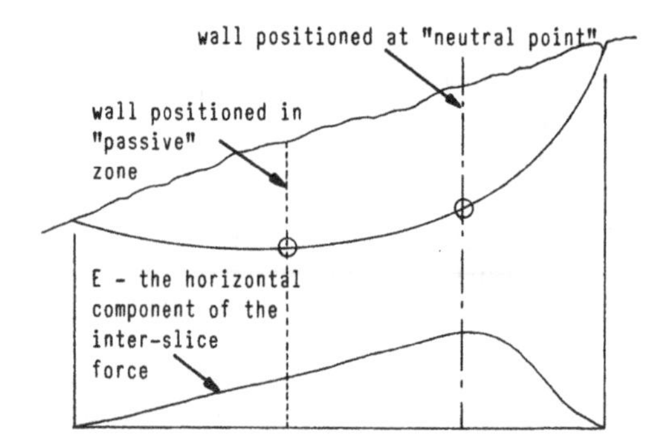

Fig. 6 Estimation of forces on walls in landslides

segments through such a toe fill (particularly when this is rockfill with a high Ø' value). Attention has been drawn to this problem elsewhere, for example, Bromhead (1986), not only in the context of misleading analysis results (if a result is obtainable at all - see for example, Whitman & Bailey, 1967), but also in the disproportionate effect that small removals of material from such a toe load can later have on the real, as distinct from calculated, factor of safety.

6 FORCES ON WALLS IN LANDSLIDES

A retaining wall in a landslide may have one of several roles:

* It retains fill, which in turn acts to resist movement: the wall has no structural loading specifically arising from its retaining action on the slide. (Type 1).
* It is a continuous wall, positioned so as to resist the action of the slide. The loading on the wall arises from movement of the slide. (Type 2).
* It is an intermittent structure, like a row of piers of a viaduct penetrating the slide mass, and acted on by forces as the slide moves past the structure. The structure may not prevent movement of the slide as a whole, but will impede its action locally. (Type 3).

A Type 1 wall can be designed using ordinary earth pressure theories in most cases, or it may act as a wave protection structure if it retains the toe fill stabilising a coastal landslide.

Type 2 walls are more difficult. Essentially, they are passive structures, and can only be stressed up by some further movement of the slide. The key element in analysis and design is to estimate the forces on such a wall.

As a first step, consider the forces between the slices at the intended wall position. Obviously these increase from zero at the toe of the slide to a maximum, (which is found at the 'neutral point'), and then fall to zero at the head of the slide. If a wall at a certain position (Figure 6) is capable of resisting a force equal to the interslice force, then to all intents and purposes the soil below the wall could be removed, without in any way adversely affecting the stability of the given slide (but see below for other effects when the soil is removed!). A wall capable of resisting X times the computed force would have a factor of safety of X against its own failure. It does not raise the factor of safety of the slope, as conventionally calculated, but does increase its reserve of safety against being destabilised by some other means, for example an unloading at the toe, or a rise in the groundwater table. This approach leads to the design of walls much lower in the slide mass, (where the forces are lower), and should be used with caution if the possibility of "over-riding" slip surfaces, as shown in Figure 5, is likely.

An alternative approach is to examine the "shortfall" in resisting force, in the case of a slip circle by examining the relation:

$$F = M_r / M_o$$

and then taking the difference $M_o - M_r$ as the moment to be resisted by the wall. Force equilibrium methods will obviously yield a shortfall in force from an equation analogous to this one. The provision

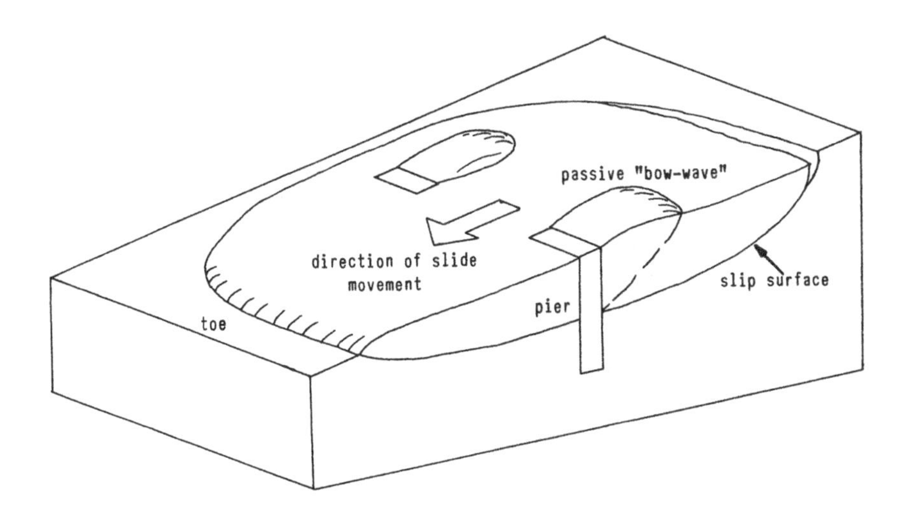

Fig. 7 Isolated piers in a slide. As the slide moves past these, there is a maximum to the force that the pier can be required to sustain

of a retaining wall capable of supplying the shortfall, increases the apparent factor of safety, and can be seen to improve the safety factor to any desired level by increasing M_T in the equation. The optimum position for wall designed on this basis is at, or slightly below, the neutral point.

Non-continuous (type 3) structures pose yet another class of problem. It is possible to view them as rigid obstacles to the progress of the landslide, so that it slides or flows around them. The forces on the piers then amount to the maximum force that the slide can exert. Naturally, this cannot be more than the inter-slice force at the appropriate slide section. An alternative viewpoint is to see the piers as moving upslope through the slide, carving out a swathe of soil in a passive "bow-wave" as they do so. The forces on this passive bow wave can then be calculated from (passive) earth pressure theory, including the side friction on the passive wedges. Naturally, the sum total of these forces, averaged so as to apply to one metre width of the slide, cannot exceed the interlice force: it is perhaps easier to see why with this viewpoint - if they were to do so, the upper part of the slide mass would be carried uphill! An obvious impossibility, more so since the forces exerted on the slide by the piers are reactions to the loads which the slide, under the effects of gravity, applies to the piers. (Figure 7).

7 ANCHOR LOADS

Ground anchors used in slope stabilisation fall into two categories:

* PASSIVE ANCHORS. Passive anchors carry no load initially, but are stressed as the slide continues to move. The maximum load that the slide can apply to a passive anchor is the load which in turn applied to the slide via the anchor reaction, raises the factor of safety to 1. A passive anchor may well have a total resistance of in excess of the load the anchor is called upon to carry: this defines the factor of safety of the anchor itself, not of the slope.
* ACTIVE ANCHORS. These are prestressed anchors: the full anchor force is applied to the slope and may raise the factor of safety of the slope to in excess of 1. Naturally, the anchor itself needs a pull out resistance in excess of the load it is required to carry. Settlement of the anchor pads is not normally a problem, except insofar as it causes a loss of prestress in the tendons.

Passive anchors cannot affect the factor of safety of an already stable slope ($F > 1$) since there is no movement or strain to load them up - or if there is, it will occur at a slow rate. Nevertheless, they do affect the reserve of safety which applies: some destabilising influence will trigger movement which stresses up the anchors until equilibrium is restored. It is difficult to design such anchors, as their resistance is only mobilised up to the re-equilibrium level.

Active anchors are in principle easier to deal with, although in practice, some obstacles arise. First and foremost of these is that some methods of stability analysis make it difficult to incorporate arbitrarily inclined external force components. This is because the methods fail to handle both the effect of the anchor on the shearing resistance of the slip surface through modification of normal stresses AND the change in shear stress, simultaneously. It should be noted that this is fundamental to the correct application of active anchors, and texts which state that you can do one or the other, but not both, are quite wrong. However, for the purpose of approximate selection of anchor loads, we must do one or the other, and then perform check analyses using a complete theory. The anchor calculations differ for slip circles and non-circular slides as follows:

7.1 Slip circles.

As a result of the preliminary stability analysis, and associated stress analysis, the initial factor of safety, F_o, and overturning and restoring moments (M_o & M_r) are known:

$$F_o = M_r / M_o$$

For an improved factor of safety, F_1, assuming no change in M_r, the changed overturning moment M_{1o} can be computed. The change is equivalent to the tangential comp-onent of anchor force times its radius from the slip circle centre and hence the desired anchor force will be obtained. In essence, with anchor force Q at inclination to the slip surface, (Figure 8):

$$Q \; R \; \cos \alpha = M_o - M_{1o}$$

This will be an overestimate of the desired anchor force.

7.2 Non-circular slip surfaces.

The procedure is more subtle for non-circular slip surfaces which lack the convenience of a centre of rotation. Firstly, perform the initial back analysis, and related stress analysis of the slide. Assuming pure "frictional" strength on the slip surface, we can

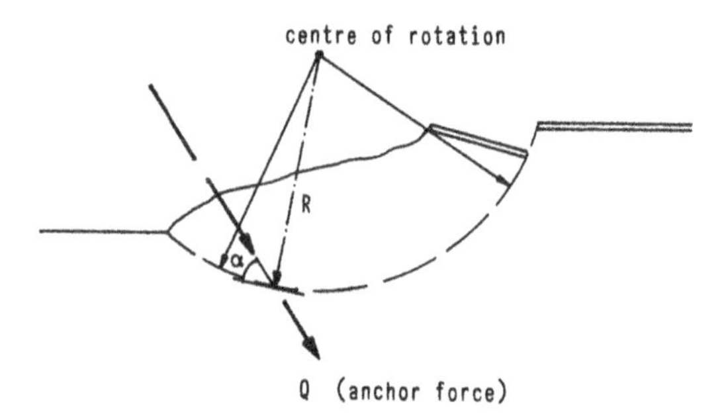

Fig. 8 Anchor force on a slip circle

compute an average increase in the normal effective stress to yield the desired improvement in factor of safety. Multiplying by the total length of slip surface turns this into a normal force which the anchor would need to apply if its entire effect was through the increase of normal stress. As the anchor is most effective if it is inclined to the slip surface, such a load would exceed that actually required, or be an overestimate. However, experience has shown that other factors, for instance, the availability or otherwise of a suitable anchorage zone at depth, can affect the inclination of anchors more than consideration of their optimum inclination, and the figure so obtained is a useful preliminary working estimate. Use of this method for a slip circle gives a different answer to the first method - both are gross approximations.

8 ASSESSMENT OF THE CRITICAL SEISMIC ACCELERATION FROM A STATIC FACTOR OF SAFETY AND VICE VERSA

The slip mass has an average normal stress, σ_n', and an average shear stress, s. This in turn represents an average mobilised angle of shearing resistance of ϕ'_m. If the actual angle of shearing resistance is ϕ' then available "unused friction coefficient" is tan $(\phi' - \phi'_m)$ (and by rights, tan ϕ' should be the same as F tan ϕ'_m, so that even if the slip is through a number of dissimilar soil zones, some with cohesion etc., it should still be possible to find an average ϕ'). Now this is exactly analogous to the case of a block on an inclined plane with a coefficient of friction tan ϕ' and the plane inclined at an angle of ϕ'_m to the horizontal. What further tilt makes the block slide down the plane? This is an angle of $\phi' - \phi'_m$ of course. That additional "tilt" is equivalent to changing the

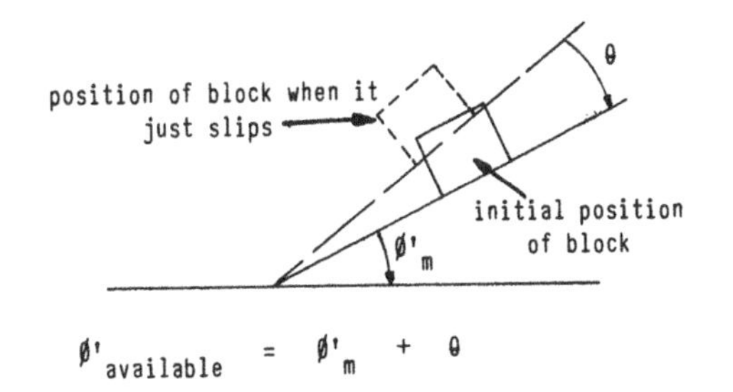

Fig. 9 The relationship between k_c and ϕ'_m, and factor of safety, diagrammatically explained

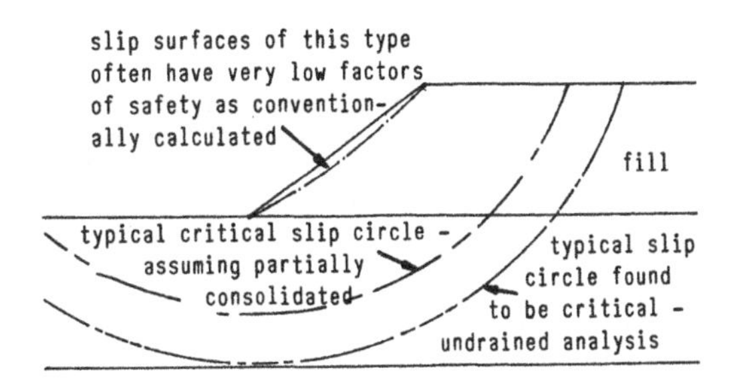

Fig. 10 Location of typical critical slip surfaces

orientation of the nett gravitational acceleration: in effect, of putting on an additional seismic acceleration laterally. (Figure 9 shows this).

A seismic acceleration k_c of approximately tan ($\phi' - \phi'_m$) will bring the block to the verge of failure: this must be an approximate value for the critical seismic acceleration.

The inverse procedure can be followed to obtain the static factor of safety (approximately) from a known critical horizontal seismic acceleration coefficient.

9 FINDING CRITICAL SLIP CIRCLES

Two important cases spring easily to mind: the slope of non-cohesive material, and the embankment on a soft cohesive foundation. In the first case, unless the slope is underlain by a weaker soil, or contains a zone of relatively high pore water pressure within itself, it is often futile doing a slip circle search. The critical surface in such a case is usually a shallow slide parallel to the slope surface. The centre for this is on the perpendicular bisector of the slope face, an "infinite" distance from it. Many software programs get into difficulty numerically at this point, but their factors of safety tend to approach the limit:

$$F = (1 - r_u) \tan \phi' / \tan \alpha$$

wherein is the slope angle. Also, this limit assumes that r_u is evaluated correctly, using pore water pressures and taking into account the direction of seepage. Often it is evaluated carelessly (and incorrectly) from the water table elevation as:

$$"r_u" = \frac{(\text{depth below water table} \times \gamma_w)}{(\text{depth below ground level} \times \gamma)}$$

In this case, the r_u in the formula for factor of safety should be multiplied by $\cos^2 \alpha$ to correct for the error introduced. The equation above derives from the method presented by Haefeli (1948) now known as the "infinite slope method".

Where the slip circles penetrate into a cohesive foundation soil, it is often found that the deepest surfaces possible provide the lowest factor of safety. This results from the inevitable minimum slip circle surface area to overturning moment ratio that a slip circle provides. This observation is "well-known". However, it should be used with care when selecting a small range of slip circles for analysis in cases where:

* The strength increases with depth
* The embankment geometry is adverse (prevents the rapid increase of overturning moment noted above)
* Embankment shear strength plays an important role is yielding a factor of safety

Furthermore, purely cohesive foundations are something of a convenient fiction, and it is more often the case that "softness" is the result of a pore pressure effect. Slip circles will be biased towards the centre of such a deposit, where the pore pressures are retained longest during consolidation. (Figure 10).

10 CONCLUSIONS

It will be seen that a number of important design decisions can be made on the basis of quite preliminary calculations. This does not mean that final stability calculations encompassing all of the design decisions mentioned herein do not need to be made, rather that time-wasting intermediate steps may be safely omitted.

REFERENCES

Bishop, A.W. 1955. The use of the slip circle in the stability analysis of earth slopes. Geotechnique, 5, 7-17.

Bishop, A.W. & Morgenstern, N. 1960. Stability coefficients for earth slopes. Geotechnique, 10, 129-150.

Bromhead, E.N. 1986. The stability of slopes. Publ. The Surrey University Press.

Bromhead, E.N. 1984. Stability of slopes & embankments. Chapter 3 in "Ground Movements & their effect on structures". Eds. R.K. Taylor & P.B. Attewell, Publ. Blackies.

Haefeli, R. 1948. The stability of slopes acted upon by parallel seepage. Proceedings 2nd International Conf on Soil Mechanics, Rotterdam. 1, 57-62.

Hutchinson, J.N. 1977. Assessment of the effectiveness of corrective measures in relation to Geological conditions and types of slope movement. Bull. International Assoc. of Engineering Geology, 16, 131-155.

Morgenstern, N.R. and Price, V.E. 1965. The analysis of the stability of general slip surfaces. Geotechnique, 15, 79-93.

Sarma, S.K. 1973. Stability analysis of embankments and slopes. Geotechnique, 23, 423-433.

Spencer, E.E. 1967. A method of the analysis of the stability of embankments assuming parallel inter-slice forces. Geotechnique, 17, 11-26.

Taylor, D.W. 1937. Stability of earth slopes. Journal Boston Soc. Civil Engineers, 24, 197-246.

Whitman, R.V. & Bailey, W.A. 1967. Use of computers for slope stability analysis. Proceedings of the American Society of Civil Engineers, Journal Soil Mechanics Div., 93, 475-498.

Computer and Physical Modelling in Geotechnical Engineering, Balasubramaniam et al. (eds)
© 1989 Balkema, Rotterdam. ISBN 90 6191 864 2

The deformation and stress responses and stability analysis of reinforced soil structures

G.E.Bauer & A.O.Abd-El Halim
Department of Civil Engineering, Carleton University, Ottawa, Canada

ABSTRACT: In this paper a non-linear F.E. technique is used to study the behaviour of reinforced soil retaining walls and/or embankments. The analysis was supplemented by a rigorous testing program to provide the parameters for the finite element analysis. An extensive limit equilibrium analysis was also carried out to determine the overall stability of such structures. The results from these two analyses are used to evaluate the global behaviour of reinforced soil structures.

1 INTRODUCTION

Over the past ten years the principal author was involved in full-scale and large-scale tests of bridge abutment footings located in compacted granular fills (Bauer et al. 1977; Bauer, 1982). It is advantageous to locate the footing as close as possible to the edge of the embankment and to make the slope as steep as possible in order to keep the bridge span to a minimum length. At this location, of course, the footing has the least factor of safety against a bearing failure and also will experience maximum deformation. The stability can be improved considerably and the settlement can be reduced by reinforcing the embankment. In addition, the face of the slope can be steepened which in turn brings about a reduction in the bridge span.

Most of the reinforced earth structures (Figure 1) built so far use steel strips or ribbed steel elements as reinforcing elements. The common method of expressing the mobilized friction between the metal and the soil is by using a frictional coefficient which is generally determined experimentally.

The behaviour of such structures depends on many factors, such as height of wall or embankment, length and spacing of reinforcing elements, soil properties and applied stresses. A non-linear finite element program, which was especially developed for reinforced earth structures (Hermann, 1977; 1978), was used in this investigation. The study was aimed to analyze the deformation behaviour of the reinforced structure and to determine the tensile stresses in the reinforcing elements. The finite element program was calibrated against the performance of several large-scale walls and prototype structures (Bauer and Mowafy, 1986). This program can compute stresses and deformations only, but cannot calculate a factor of safety against collapse of the structure. In order to investigate the overall stability of reinforced embankments, a limit equilibrium analysis was carried out using a bi-linear failure surface. This analysis yielded critical heights for given soil parameters and geometries of reinforcing elements.

The results of both analyses were combined and are presented in this paper as normalized graphs. These graphs can be used in the preliminary design of reinforced soil structures.

2 NON-LINEAR FINITE ELEMENT ANALYSIS

There are two general approaches to the analysis of reinforced soil systems, namely, discrete and composite representations of the constituents. The two approaches are described in detail by Herrman and Al-Yassin (1978). The results presented herein are based on a 2-D finite element analysis utilizing the composite approach. The programme used in this investigation is called "REA" (Reinforced Earth Analysis) and was developed by Hermann (1978). In a composite representation, for the purpose of analysis, the reinforced system is modelled as a locally homogeneous, orthotropic material termed the "composite material" (e.g., see Romstad et al., 1976). When performing a finite element analysis of a composite representation of a reinforced material, the body is represented by an array of continuum elements whose boundaries need bear no spatial relationship to the geometric arrangement of the reinforcing members. The "composite" material properties assigned to the continuum elements reflect the properties of the matrix material and the reinforcing members, and their composite interaction. The interaction of the reinforcement and the soil is highly non-linear and inelastic. In a direction normal to the reinforcement, the soil and reinforcement experience the same displacement, however, the tangential displacements differ by a relative movement ϕ. In the case when the attainable bond stress has been fully mobilized, the relative movement, ϕ, is the resulting slippage of the reinforcement. The presence of a flexible facing place results in a nonuniform displacement and stress distribution along the face of a reinforced earth wall due to this edge effect. There is a local displacement (in addition to possible slippage) of the reinforcement relative to the average displacement of the composite system. By an appropriate selection of the spring coefficient, it is possible to model this relative movement between soil and wall facing.

The composite approach was used in this program. In order to take into account the nonlinear inelastic soil behaviour, and the slippage and yielding of the reinforcement, an incremental iterative analysis was used in the program. In order to model the nonlinear behaviour of the soil in the program a hyperbolic stress function was employed (Duncan and Chang, 1970). Based on this, the stress-strain behaviour of the soil may be described at all times by instantaneous values of Young's modulus, E_s, and Poisson's ratio, ν, as follows:

$$E_s = KP_a \left(\frac{\sigma_3}{Pa}\right)n \left[1-\frac{R_f(\sigma_1-\sigma_3)(1-\sin)}{2\sigma_3\sin + 2C\cos\phi}\right]^2 \quad (1)$$

$$\nu = \frac{G-F\log\left(\frac{\sigma_3}{Pa}\right)}{(1-A)^2} \quad (2)$$

$$A = \frac{d}{E_s}(\sigma_1-\sigma_3)$$

where Pa = the atmospheric pressure (expressed in the same units as E_s and σ_3);

R_f = failure ratio, $\frac{(\sigma_1-\sigma_3)}{(\sigma_1-\sigma_3)f}$;

K = modulus number;

n = exponent determining the rate of variation of E_s and σ_3; and

d, F and G are Poisson's ratio parameters to be determined experimentally.

The program assumes that the reinforcing elements and the skin element (wall face) behave in an elastic-plastic fashion with isotropic linear strain hardening. The properties needed to describe the distributed reinforcement are the elastic modulus, E, the plastic modulus, E_p, the yield stress, f_y, the cross-sectional area of the reinforcement per unit cross-section area of the reinforced soil, ρ, the surface area of reinforcement per unit volume of reinforced soil, P, and the inclination of reinforcement, θ. For the skin element besides knowing, E, E_p, f_y, one must also specify the absolute value of the ultimate strain, ε_{ult}, at which the material ruptures.

The tensile resistance of the reinforced earth structure is due to the bond strength between reinforcement and soil. The main components of bond strength are the friction, and the cohesion. In the case of granular soils, the only component is friction. The frictional resistance is considered to be small if sheet reinforcement is used. To have a realistic representation of a reinforced earth structure, the slippage between reinforcement and soil must be included in this model. This is done by using a spring coefficient, K', which is chosen such that the relative displacement between reinforcement and soil is represented accurately. For the case where the reinforcing members are strips the analysis yields:

$$K' = \frac{8}{P\,S'^2\left[\ln\frac{4}{ps'} - \frac{3}{4}\right]} \quad (3)$$

where S' = average spacing defined as $S'=2\sqrt{\frac{B_H B_V}{\pi}}$ where B_H and B_V are the horizontal and vertical spacings of the strips, respectively.

When the reinforcement is in the form of a sheet or a mesh with vertical spacing B_V, the analysis yields:

$$K' = \frac{6}{B_V} \quad (4)$$

The interaction between the existing soil and the reinforced block of soil and the foundation material upon which the reinforced earth wall rests, are represented in the F.E. analysis by three different methods. The first one is to represent all or part of this material by finite elements. The second method is to model its constraints by boundary springs. The last and third procedure is to model the interface by a frictional-cohesional law.

The resistance provided by the skin element to movement of the strips relative to the soil can be modelled by boundary springs, K', placed between the soil and the facing plate.

3 LIMIT EQUILIBRIUM ANALYSIS

As mentioned above, the F.E. analysis computes stresses and deformations only; it does not provide a factor of safety against collapse of the structure. In order to make a complete investigation of the the state of a reinforced earth structure, a limit equilibrium analysis is also needed. A new approach for estimating the failure height and the shape of the failure surface for reinforced earth structures is given in this paper. The limit equilibrium analysis presented here is based on the investigation conducted in reference [10], and is extended here to include surcharge effects and mesh reinforcing elements. This extended analysis is capable of defining the failure height of such structures under the effect of external loading and it can also include different soil types for the reinforced soil block and the backfill material. It assumes that the failure pattern can be approximated by two intersecting straight surfaces with the transition occuring at the back edge of the reinforcement, as shown in Figure 2. This assumption was verified by an intensive nonlinear finite element analysis carried out by Romstad et al. (1978). In order to study the effect of surface loads and various soil parameters on the performance of the reinforced wall, several surcharge intensities were considered. Also two different soil densities for the reinforced soil block and the adjacent backfill material were considered as indicated in Figure 2. Two different cases of failure wedges were studied. Firstly, the plane of failure passed through both the reinforced block and the backfill ($H/L>\tan\theta_1$). In the second case, the plane of failure passed only through the reinforced earth block ($H/L<\tan\theta_1$). In the following paragraphs each case will be discussed.

3.1 Case 1: ($H/L>\tan\theta_1$)

Figure 2(a) shows the failure surface which consists of two straight lines. Each of them has a different inclination, θ_1 and θ_2. The point of intersection of the two planes occurs at the back edge of the reinforcement. In order to calculate the normal and shear forces along the failure surface, the mass is divided into rectangles and triangles as shown in Figure 2(b). The equilibrium of the mass bounded by this failure surface is used to calculate the maximum shear forces S_2 and S_3. The total frictional force developed along both failure surfaces (CD and DE, Figure 2) is S_2+S_3, which can be expressed as:

$$\begin{aligned} S_2+S_3=\gamma_1 L^2[(\tfrac{H}{L}-\tan\theta_1)^2(-\tfrac{1}{2}rmk\sin\theta_1 \\ +\tfrac{1}{2}mK\cos\theta_1+\tfrac{1}{2}m\cot\theta_2\sin\theta_2 \\ +\tfrac{1}{2}rmK\sin\theta_2-\tfrac{1}{2}mK\cos\theta_2) \\ +(\tfrac{H}{L}-\tan\theta_1)(\sin\theta_1-rKC_3\sin\theta_1 \\ +kC_3\cos\theta_1+C_3\cot\theta_1\sin\theta_2 \\ +rKC_3\sin\theta_2-KC_3\cos\theta_2) \\ +(C_3\sin\theta_1+\tfrac{1}{2}\tan\theta_1\sin\theta_1)] \qquad (5) \end{aligned}$$

The mobilized shear forces are resisted by the frictional resistance of the soil and the bonding forces of the reinforcing elements to the soil. Assuming the strip resultants are horizontal at failure the total resistance capacity S_r is:

$$\begin{aligned} S_r=\gamma_1 L^2[(\tfrac{H}{L}-\tan\theta_1)^2\{-\tfrac{1}{2}rmK\cos\theta_1 \\ \tan\theta_1-\tfrac{1}{2}mK\sin\theta_1\tan\theta_1 \\ +\tfrac{1}{2}m\cot\theta_2\cos\theta_2\tan(n\theta_1) \\ +\tfrac{1}{2}rmK\cos\theta_2\tan(n\theta_1)+\tfrac{1}{2}mK\sin\theta_2 \\ \tan(n\theta_1)\}+(\tfrac{H}{L}-\tan\theta_1) \\ \{\cos\theta_1\tan\theta_1-rKC_3\cos\theta_1\tan\theta_1 \\ -KC_3\sin\theta_1\tan\theta_1 \end{aligned}$$

$$+ C_3 \cot\theta_2 \cos\theta_2 \tan(n\theta_1) + rKC_3 \cos\theta_2 \tan(n\theta_1) + KC_3 \sin\theta_2 \tan(n\theta_1)\}$$

$$+ (C_3 \cos\theta_1 \tan\theta_1 + \frac{1}{2} \tan\theta_1 \cos\theta_1 \tan\theta_1) + \frac{\Sigma_i T_{imax}}{\gamma L_2} (\sin\theta_1 \tan\theta_1 + \cos\theta_1)] \quad (6)$$

Failure of the structure is assumed to occur when the maximum mobilized shear force along the failure surfaces equals the shear resistance, i.e.,

$$S_2 + S_3 = S_r$$

or

$$S_2 + S_3 - S_r = 0$$

Substituting the values for S_2, S_3 and S_r yields a quadratic equation of the following form:

$$[m \cot\theta_2\{\sin\theta_2 - \cos\theta_2 \tan(\phi_1)\} - rmK \{\sin\theta_1 - \sin\theta_2 - \cos\theta_1 \tan\phi_1 + \cos\theta_2 \tan(n\phi_1)\}$$

$$+ mK\{\cos\theta_1 - \cos\theta_2 + \sin\theta_1 \tan\phi_1 - \sin\theta_2 \tan(n\phi_1)\}](\frac{H}{L} - \tan\theta_1)^2 + 2\{(\sin\theta_1 - \cos\theta_1 \tan\phi_1) + C_3 \cot\theta_2 (\sin\theta_2 - \cos\theta_2 \tan(n\phi_1)$$

$$+ C_3 K[\cos\theta_1 - \cos\theta_2 + \sin\theta_1 \tan\phi_1 - \sin\theta_2 \tan(n\phi_1)] - C_3 KR[\sin\theta_1 - \sin\theta_2 - \cos\theta_1 \tan\phi_1 + \cos\theta_2 \tan(n\phi_1)]\}(\frac{H}{L} - \tan\theta_1)$$

$$+ \tan\theta_1 (\sin\theta_1 - \cos\theta_1 \tan\phi_1) + 2C_3 (\sin\theta_1 - \cos\theta_1 \tan\phi_1) - \frac{2\Sigma_i T_{imax}}{\gamma L^2} (\cos\theta_1 + \sin\theta_1 \tan\phi_1) = 0 \quad (7)$$

where:

$\Sigma_i T_{imax}$: the total resisting capacity of the reinforcing element.

m: the ratio between the unit weight of the reinforced earth mass, γ_1, and the unit weight of the backfill material, γ_2.

n: the ratio between the angle of internal friction of the reinforced earth mass, ϕ_1, and the angle of internal friction of the backfill material, ϕ_2.

r: backfill shear coefficient and can be expressed as:

$$r = - \frac{\cot\theta_2}{K} + \frac{\cos\theta_2 + \sin\theta_2 \tan(n\phi_1)}{\sin\theta_2 - \cos\theta_2 \tan(n\phi_1)}$$

K: coefficient of active earth pressure.

C_1, C_2 and C_3: design parameters.

The total resistance capacity of the reinforcement element $\Sigma_i T_{imax}$ can be expressed as:

$$\frac{\Sigma_i T_{imax}}{L^2} = C_1 \frac{H}{L} \tan\theta_1 [\frac{1}{2} + (C_2 \frac{L}{2H} - 1) \frac{X_y}{X_n} + \frac{L}{H} \tan\theta_1 (-1 + \frac{1}{2} \frac{X_y^2}{X_n^2} - \frac{1}{3} \frac{X_y^3}{X_n^3}) + C_3 \frac{L}{H} (\frac{1}{2} - \frac{X_y}{X_n} + \frac{1}{2} \frac{X_y^2}{X_n^2}) + \frac{1}{2} \frac{X_y^2}{X_n^2}] \quad (8)$$

where: C_1, C_2 and C_3 are design parameters and are expressed as follows:

$$C_1 = \frac{2bs \cdot f \cdot L}{B_H B_V}$$

$$C_2 = \frac{A_s f_y}{2bsf\gamma L^2}$$

$$C_3 = \frac{q0}{\gamma L}$$

f = coefficient of friction between soil and reinforcement

B_H and B_V = the horizontal and the vertical spacing between the reinforcement elements respectively

A_s = cross-sectional area of reinforcement

f_y = yield stress of reinforcement

$\frac{X_y}{X_n}$ = the ratio between the total height, X_n, of the reinforced block cut by the failure plane and the height, X_y, where the reinforcement yields at failure.

The ratio X_y/X_n is given by the following expression:

$$\frac{X_y}{X_n} = \frac{1}{2}\,[(1+\frac{H/L}{\tan\theta_1}+\frac{C_3}{\tan\theta_1}-(1+\frac{H/L}{\tan\theta_1}+\frac{C_3}{\tan\theta_1})^2-\frac{4H/L}{\tan\theta_1}(1+\frac{C_3}{H/L}-\frac{C_2}{H/L})]\qquad(9)$$

3.2 Case 2: $H/L \leq \tan\theta_1$

Similar expressions as equations (7), (8) and (9) can be derived for the case where $H/L \leq \tan\theta_1$. These corresponding expressions will take the following forms:

$$(\frac{H}{L})^2+2C_3(\frac{H}{L}) = \frac{2\tan\theta_1\ \Sigma_i T_{imax}(\cos\theta_1+\sin\theta_1\ \tan\phi_1)}{\gamma L^2\ (\sin\theta_1-\cos\theta_1\ \tan\phi_1)}\qquad(10)$$

$$\Sigma_i T_{imax} = c_1(\frac{H}{L})^2[\frac{1}{2}+(\frac{C_2}{H/L}-1)\frac{X_y}{H}+\frac{H/L}{\tan\phi_1}\{-\frac{1}{6}+\frac{1}{2}(\frac{X_y^2}{H})-\frac{1}{3}(\frac{X_y}{H})^3\}+\frac{1}{2}(\frac{X_y}{H})^2+\frac{C_3}{H/L}\{1-\frac{H/L}{2\tan\theta_1}-\frac{X_y}{H}+\frac{H/L}{2\tan\theta_1}(\frac{X_y}{H})^2\}]\qquad(11)$$

where

$$\frac{X_y}{H} = \frac{1}{2}\,[1+\frac{\tan\theta_1}{H/L}+\frac{C_3}{H/L}-(1+\frac{\tan\theta_1}{H/L}+\frac{C_3}{H/L})^2-\frac{4\tan\theta_1}{H/L}(1+\frac{C_3}{H/L}-\frac{C_2}{H/L})]\qquad(12)$$

Equations (7) and (10) are functions of the unknowns θ_1, θ_2 and H/L. For given values of θ_1 and θ_2, these two equations become quadratic in form for H/L. Their solution is achieved by assuming various values for θ_1 and θ_2 and seeking the minimum value for H/L.

In both the limit equilibrium and F.E. analyses the properties of the georigid SS2 and the granular material were kept constant. Therefore the three design parameters C_1, C_2 and C_3 are only functions of the reinforcement length, L, the vertical spacing, B_V and the surcharge intensity q_0.

The most suitable form to present the results of this investigation is by normalizing the relationships of the ratios H/B_V and L/B_V. These curves were obtained by assuming various values for B_V, and L, and hence calculating the parameters C_1 and C_2. Three values for B_V were taken as 1,000, 1,500 and 2,000 mm. From the limit equilibrium analysis the initial ratios of H/L were obtained. Figure 3 shows the normalized failure ratio curves for reinforced earth walls in the form of H/B_V versus L/B_V for zero surcharge intensity, i.e., C_3=0, and for the three values of vertical mesh spacing. Figures 4 to 6 show similar curves for the five different surcharge intensities, as expressed by C_3=2,4,6,8,10 and for three vertical reinforcement spacings. These figures show that for each curve there is a maximum or critical wall height which cannot be exceeded regardless of the length of the reinforcement mesh specified. At this critical height all of the reinforcement elements have reached the yield stress. As expected, when the value of the surcharge intensity is increased, the critical height of the wall decreases correspondingly. As mentioned before these graphs, Figures 3 to 6, are drawn for failure conditions. Failure is defined by either the overall stability of the wall being not satisfied, or by the stresses within the wall having reached their respective limits. In order to use these curves for preliminary design purposes, a factor of safety and/or a load reduction factor should be applied in order to guarantee adequate overall stability and to limit the stresses within the soil and the steel reinforcing elements to the acceptable limits.

4 REINFORCED SOIL WALL

The U.S. Army Engineer Waterways Experiments Station, Soil and Pavement Laboratory (WES), constructed, instrumented, and loaded to failure two reinforced earth walls [4,10]. A composite finite element analysis was performed of one of the walls and the results compared to the experimental measurements. The same wall was used in the study described here to include the effect of wall inclination, β, load location, b; H/S ratio (where H is the height of the wall and S is the vertical spacing between the reinforcement element); H/B ratio (B is the width of the footing); K_b/K_s ratio (K_b and K_s are the base and embankment modulus numbers, respectively); load inclination, and cohesion effect.

The wall was reinforced with galvanized steel strips, each strip was 0.635 mm thick, 101.6 mm wide, and 3.1 m long. The reinforcing strips were placed 0.77 and 0.61 mm apart in the horizontal and vertical directions, respectively. The wall was constructed in twelve one-foot (305 mm) lifts.

The grid used in the finite element analysis is shown in Figure 7. Elements 1 to 18 are continuum elements representing the existing soil foundation. Elements 19 to 60 are composite elements representing the reinforced earth and elements 61 to 66 are one-dimensional elements representing the face plate. Each of the six layers of elements in the wall were placed in a separate construction increment. Because the wall was constructed against an existing vertical soil face, it was assumed that a well-defined interface existed at this location. This interface was modelled by introducing a frictional-cohesive interface between elements 11 to 60 and an assumed rigid surface. This rigid surface is introduced into the analysis by cantilevering a relatively rigid bending member upwards, from bottom right-hand node; the frictional-cohesional surface is placed between this fictitious bending member and the back of the reinforced earth; values of cohesion and friction of $c=0.35$ kN/m^2 and $\tan\phi=0.67$ were used. The values for the soil properties and reinforcement elements which were needed for the computer program are given by Romstad et al. (1980).

The most pertinent information which resulted from this investigation is discussed in the following sections.

5 EFFECT OF SLOPE INCLINATION (Cotβ)

Four different slopes inclination were used, Cotβ=0, 0.5, 1.0 and 2.0. The tensile stress in the reinforcing strips decreased with a decrease in slope inclination. For example, the maximum tensile stress in the strip at height 1.5 metre above the wall base, which occurs at a distance of 0.9 m from the wall face, is 130 MPa for cotβ=0, 100 MPA for cotβ=0.5, 80 MPa for cotβ=1 and 65 MPa for cotβ=2.

Figure 8 shows that the lateral movement of the wall face decreases as the wall inclination decreases. It was found that the lateral movement decreased from 31 mm for cotβ=0 to 3mm for cotβ=1.0. The settlement of the backfill decreased as the slope of the wall decreased. For example, the value of settlement decreased by 60% for a slope of 2 horizontal to 1 vertical compared to a vertical wall. The maximum settlement, of course, was obtained at the upper port of the wall and close to the wall face.

6 EFFECT OF FOOTING LOCATION

Five footing locations as given by the ratios b/B and D/B were investigated. Typical results for the stress distribution on a vertical section 0.3 m from the crest of the wall were obtained. It was found that the maximum stress in the reinforcing strip depended on the footing location. The maximum stress occurred for the footing location of b/B=3.0 and D/B=2.0 at a section of 2.7 metres from the wall crest. This stress was 124.5 MPa.

7 EFFECT OF H/B Ratio

The effect of the ratio of embankment height to footing width (H/B), on the traction stress of the reinforcing strip, on lateral movement of the wall and on the settlement of the backfill were also investigated. The following observations were made:

1. The maximum tensile stress increased as the H/B ratio decreased. For instance, the maximum tensile stress was 110 MPa at H/B=2 for height 1.5m, 90 MPa for H/B=3 and 80 MPa for H/B=4 and 70 MPa for H/B=5 and 65 MPa for H/B=6.
2. In general, the H/B ratio had little effect on the lateral wall movement.
3. A maximum settlement of over 21 mm was observed at a height of 2.7 m for H/B =2 compared to 14.5 mm of settlement for H/B=6 at the same height. The settlements were found to be smallest for the higher H/B ratios at all heights.

8 EFFECT OF H/S RATIO

In order to study the effect of H/S ratio (height of embankment to spacing of reinforcing strips) on the stress distribution along the reinforcing strips, the lateral movement of the wall and on the settlement of the back fill, four different ratios of H/S (5, 10, 15 and 20) were used. The most pertinent information which resulted from this investigation is summarized as follows. An increase in the H/S ratio will result in a decrease in tensile stress in the reinforcing strips. For instance, the maximum tensile stress on the tie at a height of 1.5 m decreased from 175 MPa to 65 MPa as the H/S ratio increased from 5 to 20. Figure 9 shows the relationship between the lateral movement of the wall face and the wall height for different H/S ratios. In recent years a survey was carried out in the United States and Canada to investigate the performance of over 200 bridges whose abutments or piers were supported on either spread footings, friction pile or end-bearing piles (Bozozuk, 1978; Mouton et al. 1980). Based on the data from the above mentioned survey the following classification of bridge foundation movement was proposed (Bozozuk, 1978):

Tolerable or acceptable:

$S_V < 50$ mm

$S_H < 25$ mm

Harmfull but tolerable:

$50 \text{ mm} \leq S_V \leq 100 \text{ mm}$

$25 \text{ mm} \leq S_H \leq 50 \text{ mm}$

Not tolerable:

$S_V > 100$ mm

$S_H > 50$ mm

where S_V = vertical movement (settlement or heave)

S_H = horizontal movement

Using this classification and the results given in Figure 14, the lateral movement of the wall can be considered tolerable for a H/S ratio of 15 and greater, harmful but tolerable for $10 \leq H/S < 15$, and intolerable if H/S become smaller than 10. For example, in the case of H/S=15, the maximum lateral movement was 24 mm compared to 31 mm at H/S=10 and 101 mm at H/S=5.

9 EFFECT OF LOAD INCLINATION (α)

The effect of load inclination, α, on the wall deformation and the reinforcing stresses are shown in Figures 16 to 18. The angle is measured from the vertical as shown in Figure 1. From these results, the following observations can be made:

1. The direction of the load inclination has a great effect on the wall deformation (lateral movement and settlement). The lateral movement and the settlement increase dramatically as the load inclination, α, is changed from -15° to +15°. For a surcharge load of 47.9 kN/m^2 and H/S=10, the maximum lateral movement was 13 mm for α=-15° compared to 53 mm for α=+15°. Also the settlement increased from 20 mm for α=-15° to 102 mm for α=+15° for a 81.4 kN/m^2 surcharge and H/S=10.
2. The deformation behaviour of the embankment can be minimized by increasing the H/S ratio (Figure 18). For instance by increasing H/S ratio from 10 to 15 for α=+15° and a 47.9 kNm/^{2} surcharge, the maximum lateral movement decreased from 53 mm to 15 mm, and at the same time the vertical settlement was reduced from 50 mm to 10 mm.
3. The maximum tensile stress of the reinforcing strips for a given spacing can be reduced by changing the angle of load inclination (Figure 1), from a positive to a less positive or a negative inclination.
4. For a given load inclination the stresses in the reinforcing elements can be reduced by increasing the H/S ratios.

10 EFFECT OF COHESION

The effect of cohesion in the backfill material on the vertical and lateral movements of the wall were also investigated. The properties of the granular backfill was kept the same as in the previous computer runs, except a unit cohesion of 27.6 kN/m^2 was introduced. This reduced the lateral displacement by about 25% and the settlement by as much as 50% compared to corresponding values for a purely granular material in the case where b/B=D/B=0, and a vertical load application.

11 EFFECT OF BASE RIGIDITY

The results discussed above were all for cases where the underlying foundation material was assumed to be rigid, that is $K_b/K_s=\infty$. In order to investigate the effect of a compressible foundation on the stress and deformation behaviour of the wall system, preliminary computer runs with a K_b/K_s value of 0.6 were performed. These results have shown that both lateral and vertical displacements increase as well as the stresses in the reinforcing strips. The most critical conditions resulting from the variation of the aforementioned parameters were combined and re-analyzed. A summary of results is represented graphically in Figure 10. The wall was vertical (cotβ=0) and had a height of 4.5 m. The reinforcing elements had a length of 3m (H/L=1.5) and a vertical spacing of 1.5 m (H/S=3). The maximum surcharge intensity was 660 kN/m^2 (C_3=10). The wall was surcharged in eleven increments of each 60 kN/m^2. Figure 10 shows the vertical settlement, lateral wall

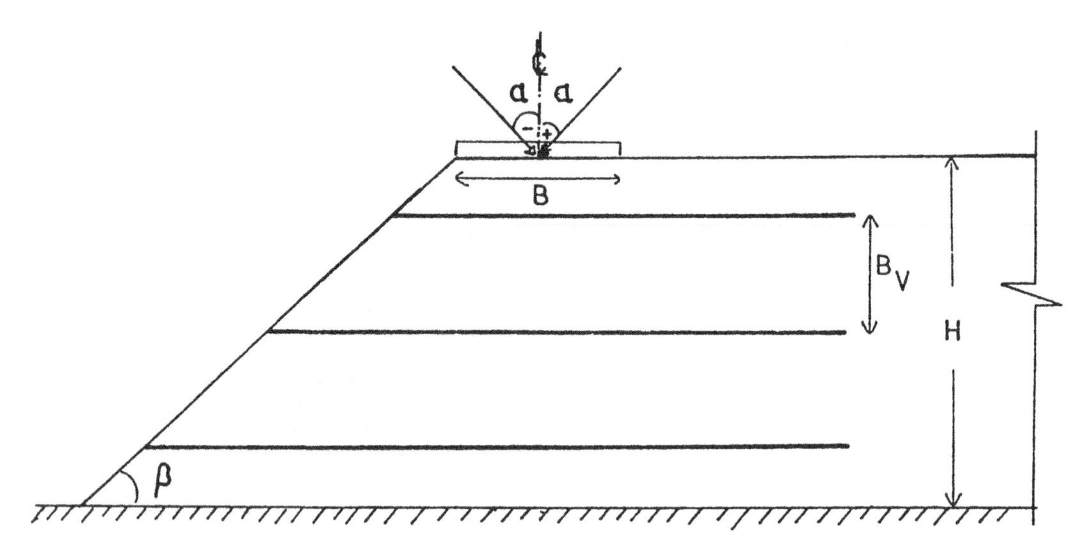

Figure 1: Variable Parameters in Reinforced Earth Structure Analysis

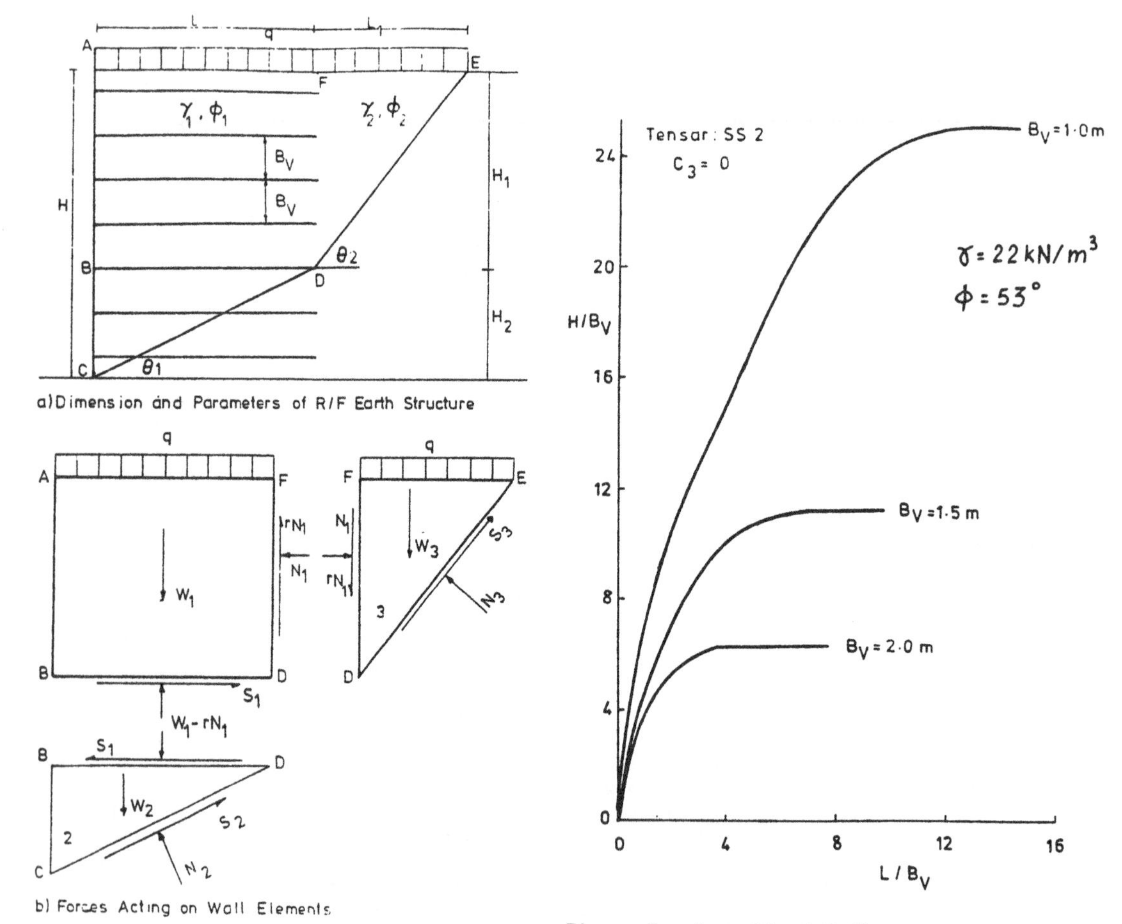

Figure 2: Limit Equilibrium Analysis

Figure 3: Normalized Failure Ratios (No Surcharge)

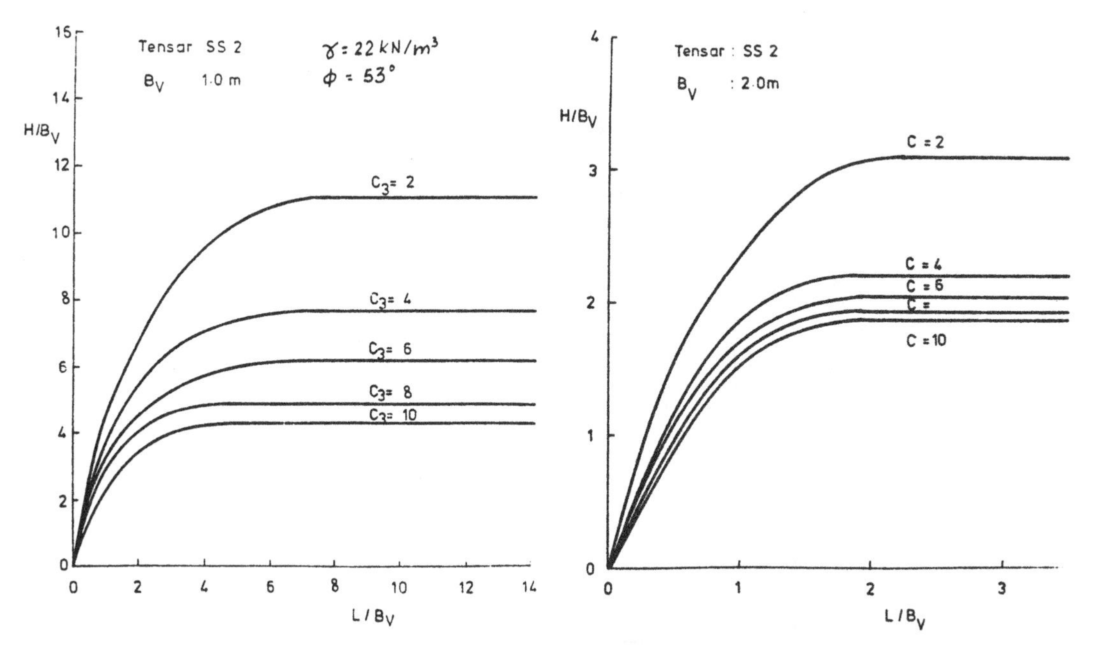

Figure 4: Normalized Failure Ratios (With Surcharge)

Figure 6: Normalized Failure Ratios (With Surcharge)

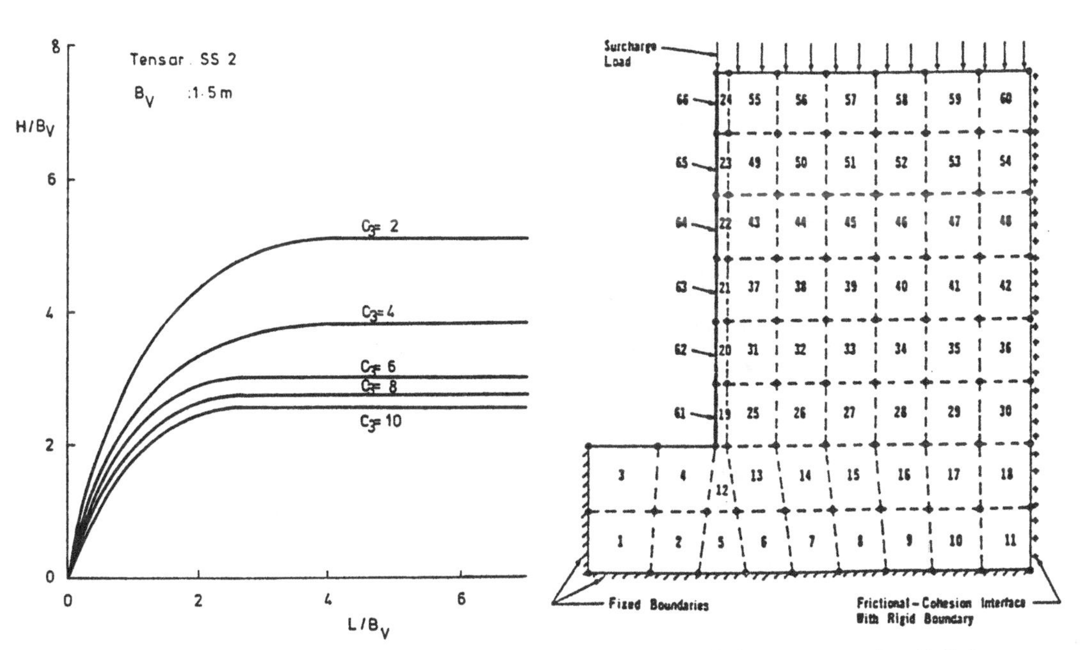

Figure 5: Normalized Failure Ratios (With Surcharge)

Figure 7: Finite Element For W.E.S. Wall

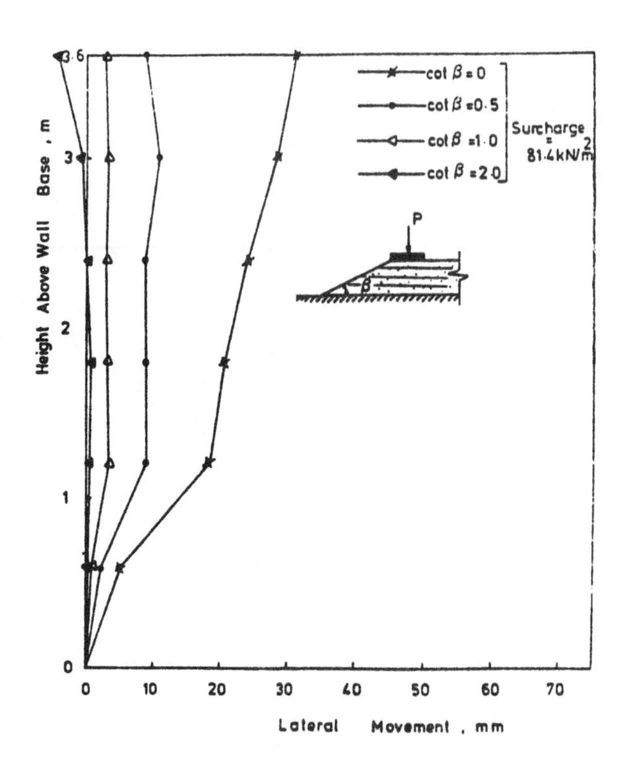

Figure 8: Lateral Movement of Wall Face

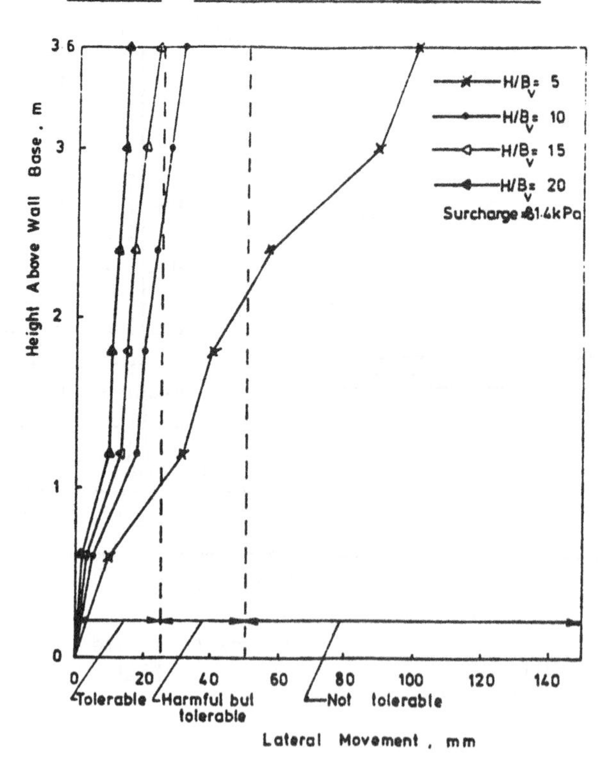

Figure 9: Lateral Movement of Vertical Wall

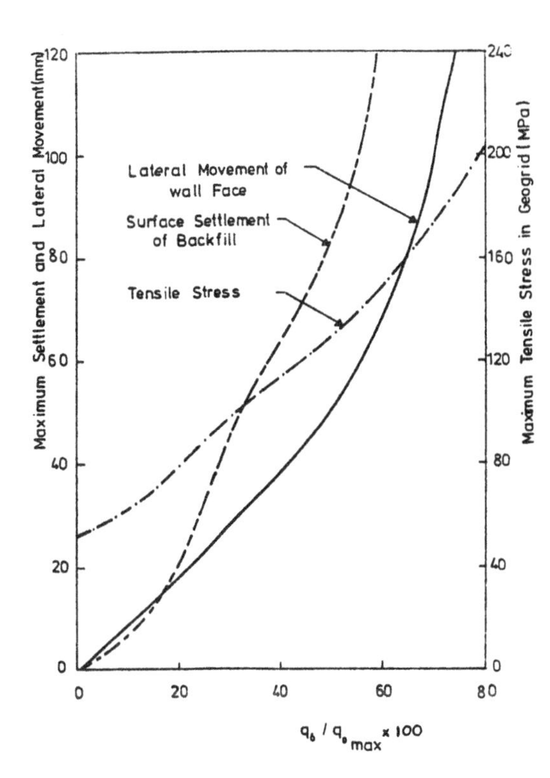

Figure 10: Maximum Settlement, Lateral Movement and Tensile Stress Versus Surcharge Ratio

movement and the maximum, tensile stresses in reinforcing elements plotted against the surcharge ratio, q_0/q_0(max). If a lateral wall movement of 50 mm were taken as a design criterion, the corresponding vertical settlement would be 82 mm, the tensile stress 132 MPa and the surcharge intensity 330 kN/m^2 (q_0/q_0(max)=0.5). These values would be within the acceptable limits for such a structure (Laba et al. 1984).

12 CONCLUSIONS

The finite element method remains a powerful tool in the analysis of deformation and stresses in embankments. It should be kept in mind that the stress-strain relationships or more generally, the constitutive laws used to describe the soil behaviour are very important ingredients in the entire analysis.

It has been shown in this paper that a non-linear finite element program which was especially developed for reinforced embankments, could be used to calculate the deformation and stress responses of various embankments subjected to several loading conditions. The results from the nonlinear finite element analysis were complemented by the results from an extended limit equilibrium analysis to examine the overall stability of reinforced embankment structures.

The limit equilibrium analysis provided the governing geometries for the wall and reinforcements in order for the structure to be stable internally and externally. The finite element analysis provided the additional information with regard to lateral and vertical displacements of the structure and the soil and reinforcement stresses.

The normalized design curves which were presented can, of course, also be used to dimension the reinforced soil structure. But this must be followed by a rigorous analysis as outlined in this paper.

ACKNOWLEDGEMENTS

Financial assistance for this study was provided in part by the first author's NSERC Operating Grant. This assistance was greatly appreciated.

REFERENCES

Bauer, G.E., Shields, D.H. and Scott, J.D., 1977. The bearing capacity of footings on compacted approach fills. Final report to the Ministry of Transportation and Communications on Ontario, Canada.

Bauer, G.E., 1982. Finite element analysis of footings in granular approach fills. Fourth International Conference on Numerical Methods in Geomechanics. Edmonton, Canada, Vol.2, pp.741-747.

Herrmann, L.R. and Al-Yassin, Z., 1978. Numerical analysis of reinforced earth systems. Proc. of the ASCE Symposium on Reinforced Earth, Pittsburg.

Herrmann, L.R., 1978. User's manual for REA (general two-dimensional soils and reinforced earth analysis program). Department of Civil Engineering Report, University of California, Davis.

Romstad, K.M., Herrmann, L.R. and Shen, C.K., 1976. Integrated study of reinforced early - 1: theoretical formulation. Journal of Geotechnical Engineering Division, ASCE, Vol.102, No.GTS.

Bauer, G.E. and Mowafy, Y.M., 1985. A non-linear finite element analysis of reinforced embankments. Fifth International Conference on Numerical Methods in Geomechanics, Nagoya, April, 905-918.

Duncan, J.M., and Chang, C.Y., 1970. Nonlinear analysis of stress and strain in soils. ASCE Journal of the Soil Mechanics and Foundations Division, 96 (SMS), pp.1629-1653.

Romstad, K.M., Al-Yassin, Z., Herrmann, L.R., and Shen, C.K., 1978. Stability analyses of reinforced earth retaining structures. ASCE Symposium on Earth Reinforcement, Pittsburgh.

Bozozuk, M., 1978. Bridge foundation move. tolerable movement of bridge foundations, sand drains, K_0-tests, slopes and culverts, TRR 678, Washington, D.C., pp.17-21.

Mouton, L.K., Gangarao, It.V.S., Tadros, M.K. and Halvoren, G.T., 1980. Tolerable movement criteria for highway bridges - analytical studies. ASCE Convention Florida.

Laba, J.T., Kennedy, J.B. and Seymour, P.H., 1984. Reinforced earth retaining wall under vertical and horizontal strip loading. Can. Geotech. J. 21, pp.407-418.

Computer and Physical Modelling in Geotechnical Engineering, Balasubramaniam et al. (eds)
© 1989 Balkema, Rotterdam. ISBN 90 6191 864 2

Application of an inter-active computer-aided probabilistic slope stability assessment for embankments on soft Bangkok clay

D.T.Bergado
Geotechnical and Transportation Engineering Division, Asian Institute of Technology, Bangkok, Thailand

J.C.Chang
Asian Institute of Technology & RSEA Inc., Taipei, Taiwan

ABSTRACT: This paper is mainly concerned with the application of a 3-D probabilistic model as contained in computer program, PROBISH, based on variance reduction techniques. The results were compared with a previous Monte Carlo analysis. The trial embankment at Nong Ngoo Hao was used. The only random variable considered was the undrained strength in which a deterministic trend with depth was observed. The soft Bangkok clay domain was found to be stationary. Sample autocorrelation functions were computed in both vertical and horizontal directions after detrending the data. The Bjerrum's correction factor on vane shear test results were successfully applied to explain failure conditions at safety factors near unity. The results showed that using a design criteria of probability failure of 10^{-3}, a design safety factor of 1.12 and 1.28 were recommended for 3-D and 2-D analysis, respectively. In the 2-D analysis, the end area resistances and the variance reduction along the axis of the embankment (Z-direction) was neglected. Comparing the results with an earlier work using Monte Carlo techniques, it was found that the variance reduction techniques yielded lower probabilities of failure.

1 INTRODUCTION

Four statistical parameters are needed to describe the variability of soil properties, namely: mean, variance, probability density function (pdf), and autocorrelation function (ACF). In the probabilistic approach to slope stability analysis, the input parameters are considered to be random variables with associated probability density functions at each point of the soil domain. The output variables lead to probability distributions that reflect the uncertainty in the model prediction arising from the heterogeneity of the soil. The conventional approach to accommodate the variability of soils is to use some average properties and assume a homogeneous material for design. Such an approach is called deterministic.

The probabilistic analyses which take into account the variability of the soil build upon the deterministic methods and should be viewed as extensions, not replacements of such methods. Analyses accounting for variability have been used for design of soil slopes, foundations, quality assurance of earthworks for roads and embankments, and for design and interpretation of site investigations.

The purpose of this study is to formulate the probabilistic soil parameters that are used in the application of an interactive computer-aided probabilistic slope stability model. This model use three-dimensional failure surface and is based on variance reduction techniques. The random variable consisted of the undrained shear strength from field vane shear tests. The total stress analysis and Bishop's simplified method of slices were utilized. The results were compared with the corresponding outcome of the Monte Carlo simulation and the results of the trial embankment construction. The site of the case study is the Nong Ngoo Hao trial embankments on the future site of the Second Bangkok International Airport (AIT, 1973 a, b). The Monte Carlo simulation was done in an earlier work by SIVANDRAN (1979).

2 THE SUBSOIL PROFILE IN BANGKOK

The Chao Phraya Plain on which Bangkok City is situated is covered by a surficial

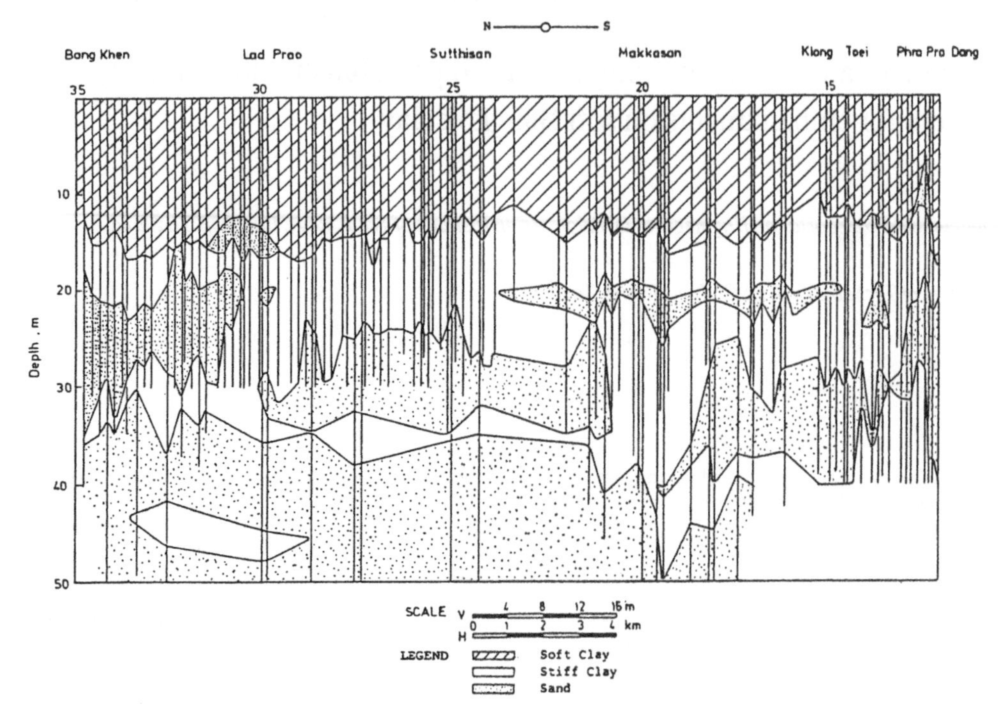

Fig. 1 Profile of subsoils from north to south of Bangkok area.

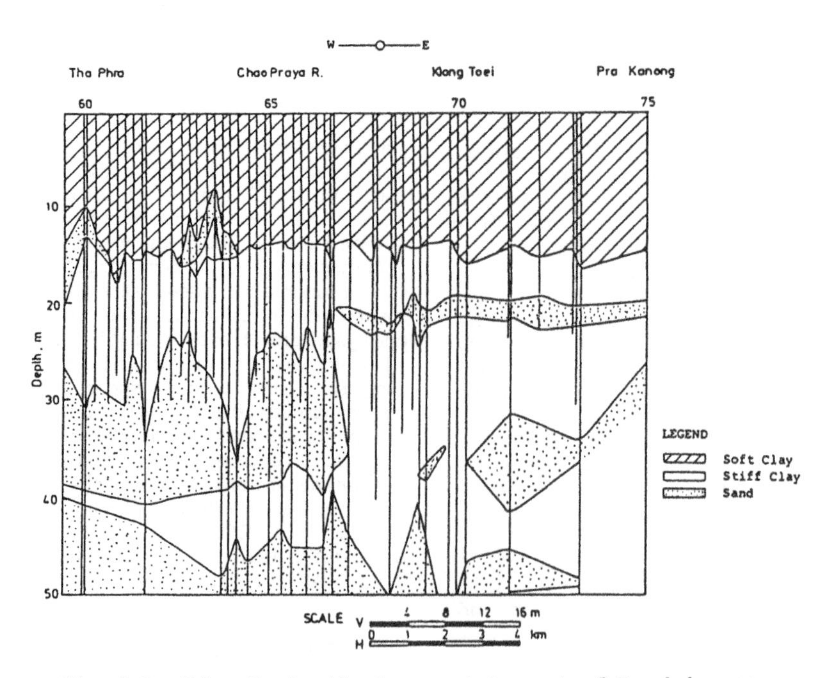

Fig. 2 Profile of subsoils from west to east of Bangkok area.

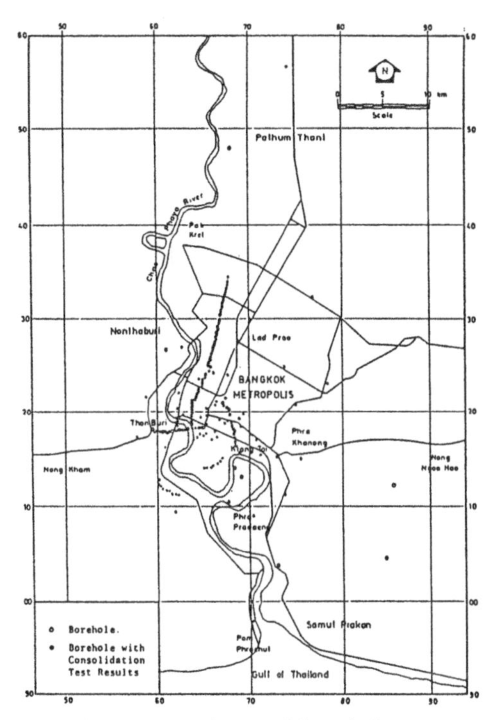

Fig. 3 Locations of boreholes.

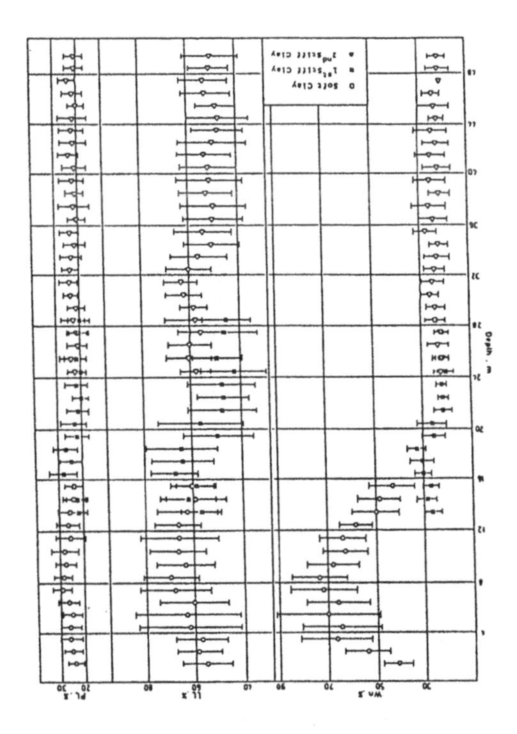

Fig. 4 Variations Wn, LL and PL of Bangkok clay with depth (After TASNEENART, 1984).

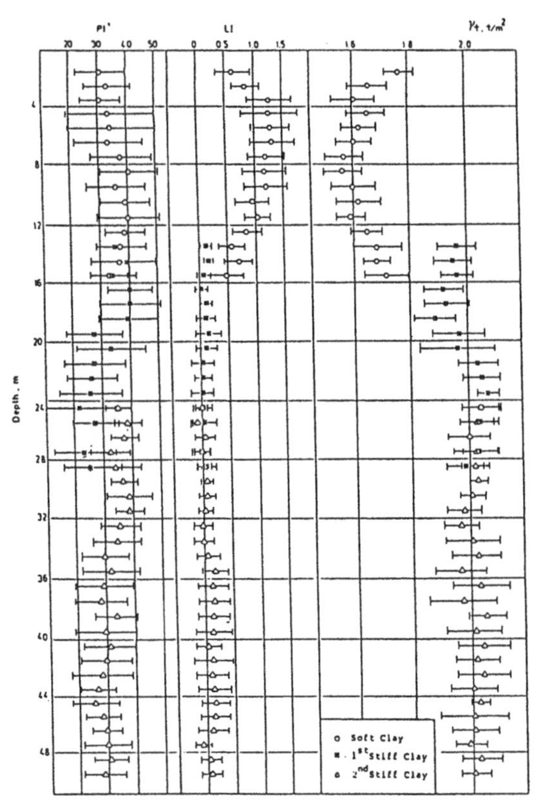

Fig. 5 Variation of PI, LI and γt of Bangkok clay with depth (After TASNEENART, 1984).

deposit of marine clay overlying alternate layers of stiff clay and dense sand. The north-south and east-west soil profiles were plotted by KERDSUWAN (1983) as shown in Figs. 1 and 2, respectively, based on the borehole locations in Fig. 3. According to the strength and compressibility characteristics, the uppermost clay layer is divided into three zones as follows (MOH et al, 1969):

1) An upper zone of weathered clay to a depth of about 2.5 m below the ground surface which behaves as an apparently overconsolidated clay.
2) A highly compressible soft clay underlying weathered clay to a depth of about 15 m which behaves as lightly overconsolidated clay.
3) A stiff clay layer is found underlying the soft clay layer down to a depth of about 20 m overlying sand layer. This layer behaves as overconsolidated clay.

The basic soil properties such as natural water content (ω), Atterberg limits (LL, PL, PI) and total unit weights (γ_t) are plotted in Figs. 4 and 5

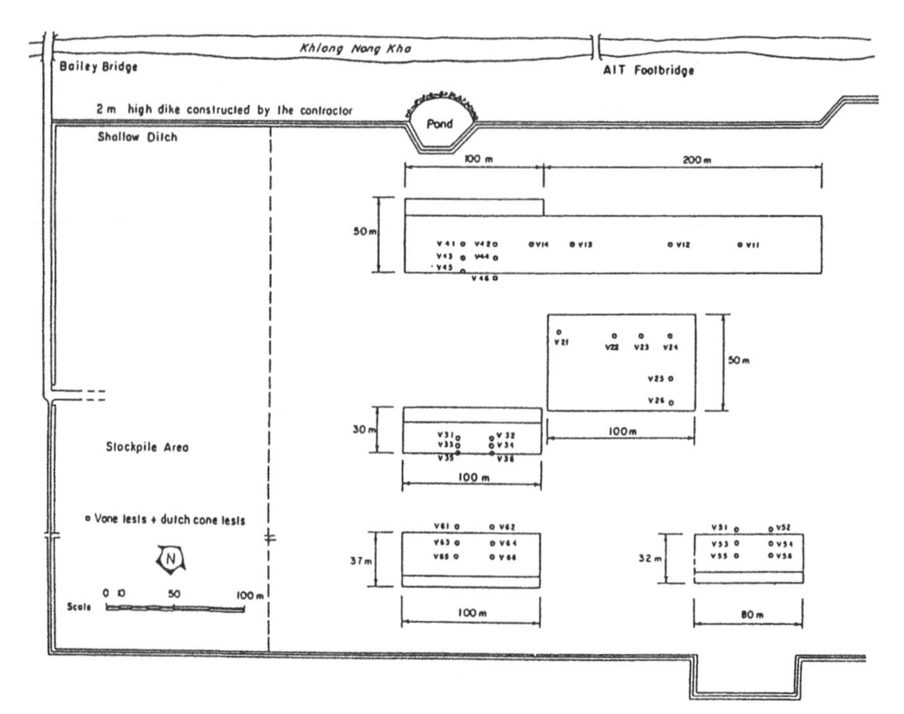

Fig. 6 In-situ testing in Nong Ngoo Hao site (After AIT, 1973a, b).

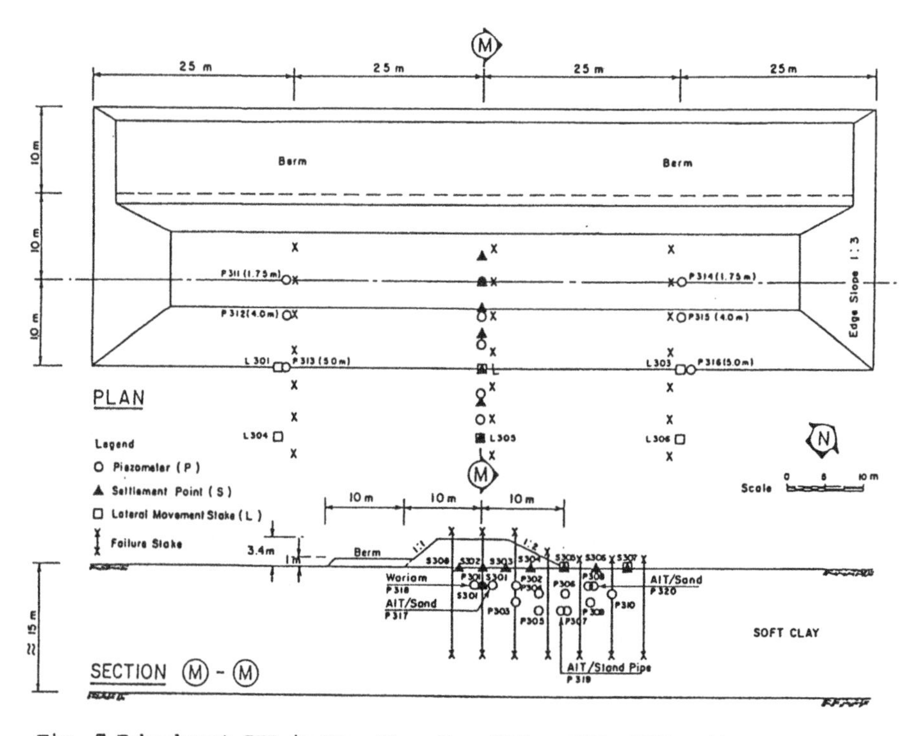

Fig. 7 Embankment III in Nong Ngoo Hao (After AIT, 1973a, b).

including the mean and the range (KERDSUWAN, 1983). These results were obtained through the analysis of the collection of large amounts of data from laboratory tests and borehole records from past investigations.

3 THE NONG NGOO HAO TEST EMBANKMENT

AIT (1973 a, b) studied and built test embankments at Nong Ngoo Hao to evaluate the subsoil as future site of the Second Bangkok International Airport. The site is located about 28 km southeast of Bangkok City. The soil investigation program undertaken at the site included field and laboratory tests as well as construction of full scale trial embankments and excavations. The plan area and test locations at the site are given in Fig. 6. This site is chosen for the case study in this paper due to the availability of soil test data obtained from the comprehensive series of field and laboratory tests. The subsoil profile at the site is essentially similar as those presented in Figs. 1 and 2. The general soil properties are summarized in Table 1.

Full scale embankments were constructed at Nong Ngoo Hao to study the in-situ behavior of the foundation. The fill consisted of sand. Embankment No. III as shown in Fig. 7 was built rapidly to failure. It has plan dimensions of 30 m x 100 m with side slope of 1V : 2H. A berm of 1 m wide was constructed on one side to ensure that failure will occur on the other side which was fully instrumented. Tension cracks were observed after 31 days of construction when the height of the embankment was 3.4 m as shown in Fig. 8. The observed failure surfaces which were close to the assumed failure surfaces are shown in Fig. 9. Other data on lateral movements, settlements, pore pressures, and failure modes were also documented.

Table 1 Mean soil properties at Nong Ngoo Hao.

Property	Weathered Clay	Soft Clay
Natural water content (%)	133 ± 5	112 - 130
Natural Voids ratio	3.86 ± 0.15	3.11 - 3.64
Degree of saturation (%)	95 ± 5	79.8 - 108.0
Specific Gravity	2.73	2.75 ± 0.01
Liquid Limit (%)	123 ± 2	118 ± 1
Plastic Limits (%)	41 ± 2	43 ± 0.3
Dry Density (lb/ft^3)	36 ± 2	40.5
Soluable Salt Content (gm/l)	7.5	13.1
Organic matter (%)	4.0	3.6
pH	6.5	-
Grain Size Distribution		
Sand (%)	7.5	4.0
Silt (%)	23.5	31.7
Clay (%)	69.0	64.3

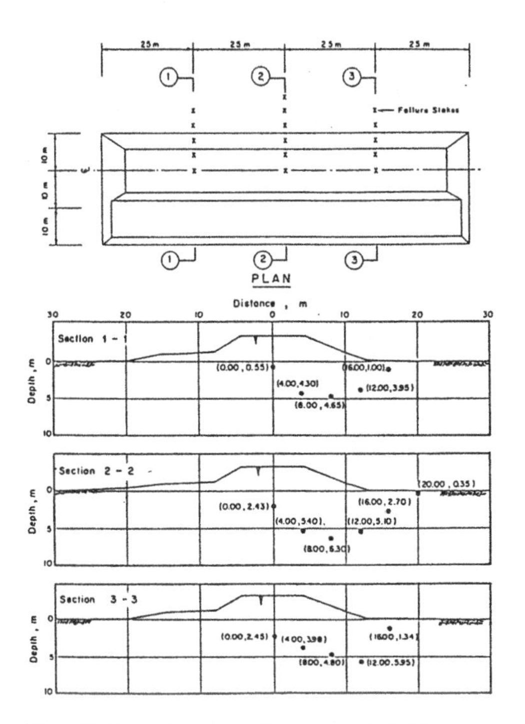

Fig. 8 Coordinates of observed failure surfaces (After AIT, 1973a, b).

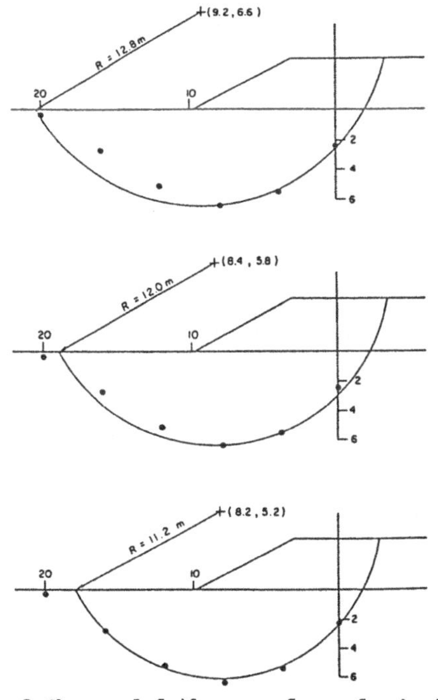

Fig. 9 Observed failure surface of embankment III (After AIT, 1973a, b).

From the previous research on Nong Ngoo Hao trial embankment, BOONSINSUK (1974) concluded that total stress analysis using the method of Fellenius with vane shear strength appeared to be most suitable yielding theoretical safety factors at failure of 1.20 to 1.35 depending on the mode of failure. In this study, the total stress analysis was used with simplified Bishop method of slices. Two modes of failure were considered, namely: (1) assuming that the strength of the sand fill was fully mobilized with friction angle, ϕ_s, of 37.5 degrees, and (2) neglecting the strength of the fill (ϕ_s equals zero). In both modes, the total stress analysis was used in the foundation clay with friction angle, ϕ, of zero.

4 THE SOIL PARAMETERS

The unit weight values of the Bangkok clay are quite uniform for the top 8 m of the subsoil as plotted in Fig. 10. The unit weight of the sand fill also show small variation as shown in Fig. 11. SIVANDRAN (1979) obtained empirical distributions of unit weight measurements in the sand fill and the uppermost 8 m of the subsoil and found that their probability density function (pdf) were both Gaussian. The coefficient of variation (COV) of both the unit weights in the clay foundation and sand fill was about 6% in the average. The variability of the unit weights for every 1 m interval was used in the analysis.

From direct shear tests, it was found that the compacted sand had zero cohesion intercept and friction angle values ranging from 36 to 40 degrees with weighted average of 37.5 degrees. The undrained shear strength of the foundation clay was measured by both Dutch cone tests and field vane tests (AIT, 1973 a, b; WIROJANAGUD, 1974; MEMON, 1974). For the probabilistic analysis, the large amount of data on undrained strength from vane shear and Dutch cone test obtained from previous studies (AIT, 1973 a, b) were utilized. These data were all determined approximately under similar conditions with regard to human, environmental and experimental factors. The spatial variability of the undrained strength of Bangkok clay was analyzed and incorporated in the probabilistic slope stability analysis in this study. In this study, undrained conditions were also assumed in the total stress analysis. The pore pressures, accordingly, are not included in the analysis.

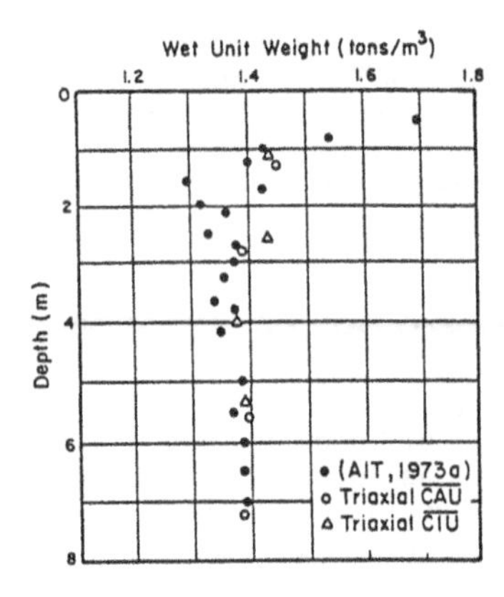

Fig. 10 Wet unit weight of clay.

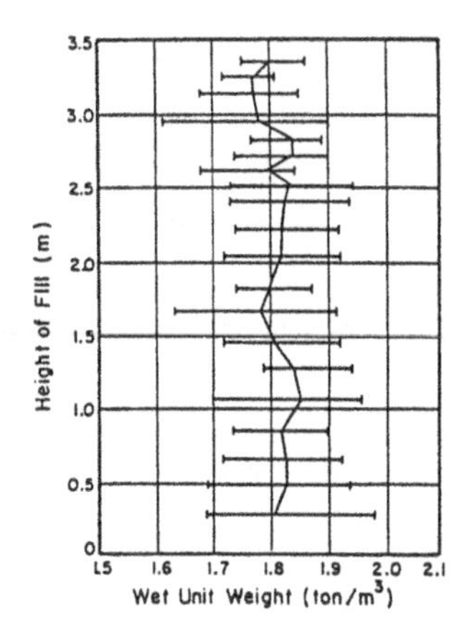

Fig. 11 Wet unit weight of sand.

5 RANDOM DATA

Any observed data representing a physical phenomenon can be broadly classified as being either deterministic or random. In practical terms, the decision as to whether or not physical data are deterministic or random is usually based on the ability to reproduce the data by controlled experiments. If an experiment producing specific data of interest can be repeated many times with identical results (within the limit of experimental error), then the data can generally be considered deterministic. If an experiment cannot be designed which will produce identical results when the experiments are repeated, then the data must usually be considered random in nature. From the above definition, the observed variability of soil properties suggests that the soil properties can be modelled as random data. The collection of all possible sample functions (also called ensemble) which the random phenomenon might have produced in time and/or space according to the probability law is called random or stochastic process.

5.1 Stationary and Ergodic Random Process

A stochastic process is stationary if the same probability law holds at every point in space. The less restrictive assumption of second order stationarity requires that the stochastic process has the same expected value at each point and that the covariance between the random variables at any two points must depend only upon the vector separating them and not on their absolute position. For the ergodic random process, the time-averaged mean and autocorrelation values are equal to the corresponding ensemble averaged values. Underlying this hypothesis is that time plays no role in determining probabilities. Note that only stationary random processes can be ergodic.

Although SIVANDRAN (1979) verified empirically that the probability density function of Su at Nong Ngoo Hao was best fitted to a 2 parameter Beta distribution, the Su can be fitted to a normal distribution if lower levels of significance is used (α = 1%, for instance). Moreover, the normal distribution can be used according to the Central Limit Theorem and for simplicity.

6 STATISTICAL EVALUATION OF THE UNDRAINED SHEAR STRENGTH

Three statistical parameters were needed in the probabilistic analysis, namely: mean, standard deviation and correlation distance. The statistical homogeneity (stationary) was assumed. Thus, the mean, standard deviation and autocorrelation function were the same for the stationary domain. In this study, the domain were the embankment sand fill and the foundation soft Bangkok clay.

6.1 Mean Trends and Regression Analysis

From the field vane tests in 28 locations, it was observed that the vane shear strength increased with depth. To eliminate the effect of the deterministic trends, detrending was done before the vertical autocorrelation was calculated. Regression analysis was used to determined the deterministic trends. It was observed that a second order regression equation was satisfactory with multiple correlation coefficient of about 90-98%. The second order polynomial regression model for the 28 vane test locations is expressed as:

$$Su = 1.889 - 0.171d + 0.023d^2 \quad (1)$$

where d is the depth. This equation was obtained from average undrained strength (Su) at each 1 m interval pooled from 28 test locations. The mean and standard deviation for each 1 m interval for Su is given in Table 2 and plotted in Fig. 12.

Table 2 Mean and standard deviation of field vane shear strength for NNH clay.

Depth (m)	Mean S_u (t/m²)	Standard Deviation	Coeff. of Variation (%)
0 - 1	1.82	0.28	15.4
1 - 2	1.90	0.26	13.7
2 - 3	1.52	0.12	7.9
3 - 4	1.36	0.09	6.6
4 - 5	1.38	0.12	8.7
5 - 6	1.48	0.13	8.6
6 - 7	1.74	0.14	8.0
7 - 8	2.09	0.19	9.1
8 - 9	2.50	0.29	11.6
9 - 10	2.65	0.33	12.5
10 - 11	3.01	0.27	9.0
11 - 12	3.15	0.32	10.2
12 - 13	3.48	0.34	9.8
13 - 14	3.88	0.31	8.0

7 SPATIAL VARIABILITY OF SHEAR STRENGTH

7.1 Sources of Uncertainties

VANMARCKE (1977 a) suggested three major sources of uncertainties of soil properties for the purpose of describing the spatial variation, namely: natural heterogeneity, limited information about subsurface conditions, and measurement errors. The natural heterogeneity of the material is responsible for the spatial variation of shear strength and must be evaluated to properly evaluate the probability of slope failure. The accuracy of the spatial variation estimate depends on the extent of the subsurface investigation and the accuracy of the measurements (the second and third error). Some judgment must be used in planning the subsurface investigation and a compromise between the location and number of soundings and the values of obtaining the information. Even when using the best possible techniques to sample and measure the soil parameter some measurement errors will be present and this could include a measurement bias in the measured values.

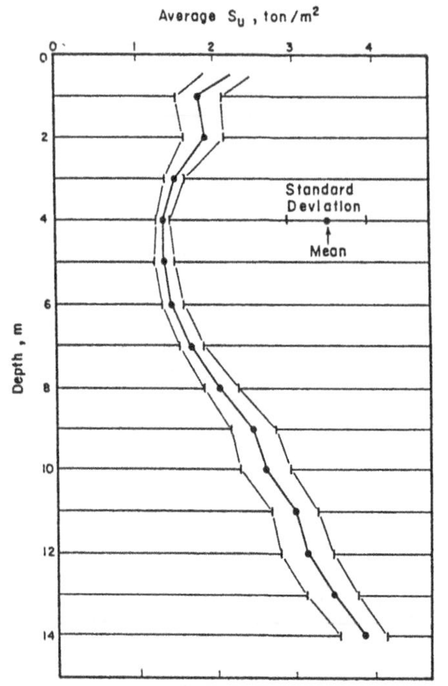

Fig. 12 Undrained shear strength profile of NNH clay from field vane.

7.2 Model Error

Model uncertainty depends on the method of deterministic slope stability analysis and the assumptions that must be made to make the analysis statically determinant. WHITMAN & BAILEY (1967) suggested that the plane strain safety factor computed from Bishop's simplified method may be in error by a much as 5 percent. One of the difficulties in error analysis is in separating bias factors from variability. It is generally accepted that all of the slope stability methods make conservative assumptions. The magnitude of the bias introduced by this conservatism varies for each individual case, depending on the particular slope geometry, failure surface, soil conditions, and pore pressure conditions. Thus, this bias may be considered to be a random variable which can be described by a probability density function (pdf). However, it is doubtful that the pdf of the bias factor is Gaussian. A Beta distribution having a positive skew seems to be a better assumption. CORNELL (1971) has suggested the variance of the safety factor which includes model uncertainty, maybe computed through simulation of variance due to model error, and the variance due to spatial variability as:

$$\tilde{F}^2 = (V_F\bar{F})^2 + \tilde{F}_s^{\,2} \qquad (2)$$

where V_F is the coefficient of variation of the plane strain safety factor, F; $\tilde{F}_s^{\,2}$ is the variance of the safety factor which includes the spatial variability of shear strength; $\tilde{F}^2$ is the total variance of the safety factor, and $\bar{F}$ is the mean plane strain safety factor.

7.3 Variance Reduction of Shear Strength

Variations in shear strength occur naturally in a soil mass. However, the stability of an embankment dam is not effected by very small areas of weakness because these are compensated for by the strength of adjacent areas. Thus, local weaknesses tend to be "averaged" out when the strength of larger areas are considered, even though the point to point variation in the shear strength can be high. There may be several places in the soil mass where the strength is low or high but only for short distances. If the average strength over an averaging length, b, is calculated as the length b is moved along the axis of the embankment, the moving average of Su is much less variable than the point strength values. As a result, the standard deviation of the average values, $\tilde{S}ub$, is less than the standard deviation of the point values, $\tilde{S}up$, as shown in Fig. 13 a. As the averaging length, b, is increased the standard deviation of the averaged shear strength decreases (Fig. 13 b). The standard deviation of shear strength, $\tilde{S}ub$, averaged over the surface area of the failure mass, A, is related to the point standard deviation, $\tilde{S}up$, by:

$$\tilde{S}ub = \Gamma_{s,a}(A)\ \tilde{S}up \qquad (3)$$

where $\Gamma_{s,a}(A)$ is the shear strength reduction function along the surface area of the failure mass. Following ANDERSON et al (1981, 1984) based on the work of SHARP (1982) as applied earlier by BERGADO & ANDERSON (1985), the expression for the variance of Su for a soil type j over a failure surface of length L and width b is given as;

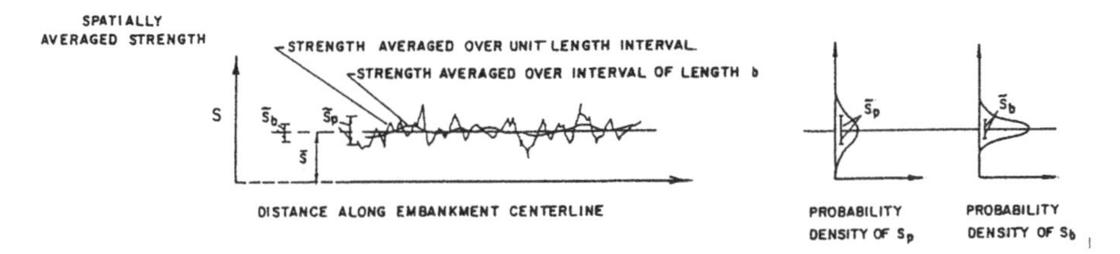

Fig. 13a Spatial averages of undrained shear strength for different averaging lengths (After Anderson et al, 1981).

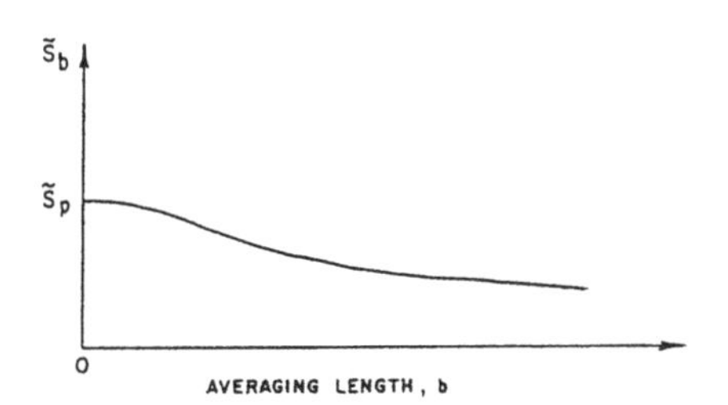

Fig. 13b Dependence of standard deviation on averaging length (After Anderson et al, 1981).

$$\tilde{S}ubj^2 = \Gamma^2{}_{c,z}(b)\,\Gamma^2{}_{c,\ell}(L)\,\tilde{S}upj^2 \qquad (4)$$

where $\tilde{S}upj$ is the point variance for soil type j, $\Gamma^2{}_{c,z}$ and $\Gamma^2{}_{c,\ell}$ are the variance reduction functions along the axis of the embankment and along the arc length, respectively.

7.4 Scale of Fluctuation

There are many variance reduction functions that maybe used to model variance decay. VANMARCKE (1977 a, 1979) described an approximate form using a parameter called the scale of fluctuation, δ, which is a measure of the rate of fluctuation of a soil property about its mean value along a line of the embankment. The scale of fluctuation is also a function of the mean crossing distance of a variable about its mean, $\bar{d}$. The relationship between δ and $\bar{d}$ is given by VANMARCKE (1977 a) as:

$$\delta = \bar{d}/\sqrt{\pi/2} \qquad (5)$$

Therefore, the scale of fluctuation is related to the average distance through which the random variable is either above or below the mean. The variance reduction function in terms of the scale of fluctuation is:

$$\Gamma^2{}_{s,z}(Z) = \delta/Z, \quad Z > \delta \qquad (6)$$

$$\Gamma^2{}_{s,z}(Z) = 1, \quad Z < \delta \qquad (7)$$

where the shear strength is averaged over a distance, Z.

7.5 Autocorrelation Function (ACF)

Other forms of variance reduction functions make use of the autocorrelation functions (ACF) which measure the linear dependency of strength at two points separated by a given distance. VANMARCKE (1977 a, 1979) showed graphical comparison of the various forms of variance reduction functions. The development of the model presented in this report utilizes autocorrelation functions rather than the scale of fluctuation. This method appears to be more versatile when using the method of slices slope stability analysis. Furthermore, the use of autocorrelation function allows more flexibility in evaluating field data.

There are several ways in which the autocorrelation function of shear strength may be described. VANMARCKE (1979) and ALONSO & KRIZEK (1979) have suggested various functional forms for the autocorrelation structure. Autoregressive models utilized simply decaying exponential functions to describe the decay of correlation in space. The type of function and its coefficients are usually estimated from curve fitting techniques. Previous studies have shown the following expressions for describing the autocorrelation of geotechnical parameters.

$$\rho_x\ (\rho x) = \exp\ (-\ (\Delta x/Kx)^2) \quad (8)$$

$$\rho_y\ (\rho y) = \exp\ (-\ (\Delta y/Ky)^2) \quad (9)$$

$$\rho_z\ (\rho z) = \exp\ (-\ (\Delta z/Kz)^2) \quad (10)$$

where ρ_x, ρ_y and ρ_z are lag distances in each direction. The values of Kx, Ky and Kz determine the rate of decay of autocorrelation in each direction and must be determined for each soil type through field programs and data analysis techniques.

The ACF may also be a product of more than one random process and the resulting functional form may not necessarily be simply decaying exponential. For instance, ANDERSON et al (1981) showed that the exponential decay did not adequately described the autocorrelation of tan $\emptyset$. Thus, in order to accommodate various autocorrelation structure, an alternate way to describe the AFC is to specify a series of sample estimates for autocorrelation values at various lag distances and fitting these data points to obtain an empirical autocorrelograms.

7.6 Horizontal Autocorrelation

The horizontal autocorrelations was determined in both the x and z directions which are perpendicular and parallel to the embankment axis, respectively. The x and z autocorrelograms were obtained by considering the values of shear strength along a line at constant elevation using the following expression:

$$\rho_k(\Delta x) = \sum_{i-1}^{n-k} \frac{(Y(X_{i+k})-\bar{Y})\ \ (Y(X_i)-\bar{Y})}{\tilde{Y}^2} \quad (11)$$

where $\bar{Y}$ and $\tilde{Y}^2$ are the mean and variance of the data. The shear strength values were averaged at every 1 m interval which constitute data points at each elevation. The data were pooled by averaging the ACF values over all elevation sets to obtain the overall ACF. There were 10 elevation sets corresponding to 10 m depth.

Due to the irregularly spaced field vane test locations, the following procedure proposed by AGTERBERG (1970) was adopted:

1) Regardless of the positions of the control points in the original data, consider all possible pairs of values, Yi and Yj.
2) Make a histogram of the variable distance Di between pairs (Yi, Yj) using a constant class interval, Δ. Let N(D') represent the number of pairs with $(K-1)\ \Delta < D < (K)\ \Delta$, $K = 1, 2 \ldots, m$ and $D' = K - 1/2$. Note that the constant class interval Δ is estimated by the average sampling interval, $(\sum_{i=1}^{n} Di/n)$.
3) Calculate autocorrelation coefficient, $r^*(D')$, to estimate the function $\rho^*(D')$ from all pairs (Yi, Yj) with interval $(D' - 1/2)\ \Delta < D < (D' + 1/2)\ \Delta$.

The calculated ACF, $\rho^*(D')$ which was average from 10 elevations for x and z directions are plotted in Fig. 14. The results showed that the Z - direction ACF exhibited negative terms even at the smallest lag distance of 19.5 m while the ACF in X - direction exhibited weak correlation at the smallest lag distance of 16.2 m. In view of this difficulty, it was assumed that the horizontal ACF is isotropic and the exponential decay idealized ACF was fitted (refer to Equas. 8,9 and 10). The resulting idealized ACF is plotted in Fig. 14 with K = 11.74.

7.7 Vertical Autocorrelation

There were more available data for estimating the vertical ACF than the horizontal ACF. Moreover, since the data were equally-spaced, there was no need for the Agterberg procedure. The field vane data with equal spacing of 1 m were used. The deterministic trend of the data was first deleted before calculation of ACF using the overall mean trend equation (Equa. 1). The overall ACF was obtained by averaging the ACF from each test location at each lag.

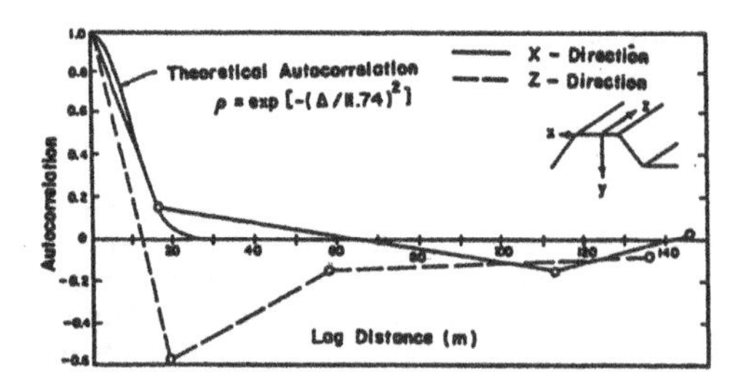

Fig. 14 Autocorrelogram of Horizontal directions.

Six vane shear test locations were considered to calculate the vertical ACF. There was negligible correlation as shown in Table 3. It was necessary to reduce the intervals of data points. The Dutch cone test results obtained at 20 cm intervals provided another alternative. The data were first detrended before the calculation of the vertical ACF. The results are plotted in Fig. 15. The results show a rapidly decaying ACF which if fitted to an exponential function yielded K = 0.184 (see Equas. **8 to 10**).

Table 3 Autocorrelation coefficient for vertical (Y) direction, using overall mean trend.

A.C.F Coeff. / Lag (m) / Bore-hole	Y - Direction 0	1	2	3	4
V11	1	-0.23	-0.54	0.22	-0.19
V12	1	0.29	-0.15	-0.16	-0.33
V13	1	-0.08	-0.15	-0.30	-0.29
V43	1	-0.03	-0.02	-0.35	-0.47
V46	1	-0.05	-0.12	-0.48	-0.05
V47	1	0.51	-0.04	-0.42	-0.79
Average	1	0.07	-0.17	-0.25	-0.35

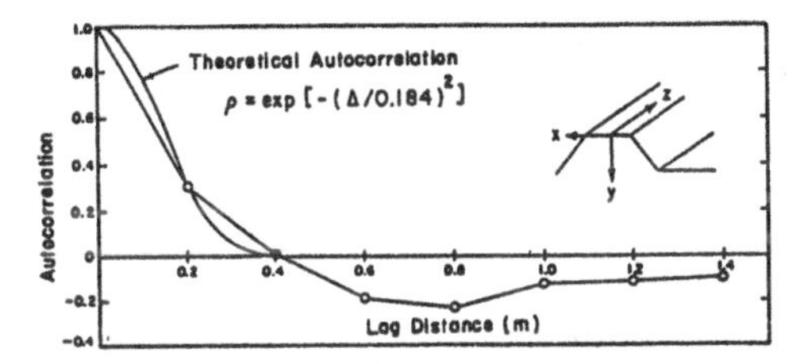

Fig. 15 **Autocorrelogram of vertical direction.**

7.8 Stationarity of the Soil Mass

A distribution-free test in non-parametric statistics which is valuable in data evaluation is the run test. Consider a sequence of N observed values of a random variable X where each observation is classified into one of two mutually exclusive categories which may be identified simply by plus (+) or minus (-). An example might be a sequence of measured values Xi, i = 1, 2, 3 ... N, with a mean value, $\bar{X}$, where each observation $X_i > \bar{X}$ is positive or $X_i < \bar{X}$ is negative. The sequence of plus and minus observations might be as follows:

(++) (-) (++) (-) (+++) (-) (+) (--) (+) (--) (+) (---)
1 2 3 4 5 6 7 8 9 10 11 12

A run (R) is defined as a sequence of identical observations that is followed and preceeded by a different observation on no observation at all. In the above example, there are 12 runs (R = 12) in the sequence of 20 observations (N = 20). The number of runs which occur in a sequence of observations gives an indication as to whether or not the results are independent random observations of the same random variable. The sampling distribution of N independent observations is a random variable R with mean and variance as follows:

$$\bar{R} = \frac{2N_1N_2}{N} + 1 \qquad (12)$$

$$\sigma_R^2 = \frac{2N_1N_2\,(2N_1N_2-N)}{N^2\,(N-1)} \qquad (13)$$

where N_1 is the number of positive observations and N_2 is the number of negative observations. The most direct application of run test to data evaluation involves the testing of a single sequence of observations for homogeneity or stationarity.

Table 4 Run test for mean of vane shear strength.

Depth	Mean (t/m²)	Deviation to Mean Trend	Run Numbers R	Remarks
0 - 1	1.82	0.08		*n = N = N = N/2 = 6
1 - 2	1.90	0.26		
2 - 3	1.52	-0.06		*Acceptance Region
3 - 4	1.36	-0.21		3 < R ≤ 10
4 - 5	1.38	-0.23		
5 - 6	1.48	-0.21		
6 - 7	1.74	-0.08	4	
7 - 8	2.09	0.09		
8 - 9	2.50	0.28		
9 - 10	2.65	0.16		
10 - 11	3.01	0.20		
11 - 12	3.15	-0.02		
12 - 13	3.48	-0.10		
13 - 14	3.88	-0.15		

Table 5 Run test for standard deviation of vane shear strength.

Depth	Standard Deviation (t/m²)	Deviation to Mean Trend	Run Numbers R	Remarks
0 - 1	0.28	+0.07		*n = N = N = N/2 = 6
1 - 2	0.26	+0.05		
2 - 3	0.12	-0.09		*Acceptance Region
3 - 4	0.09	-0.12		3 < R ≤ 10 (α = 5%)
4 - 5	0.12	-0.09		2 < R ≤ 11 (α = 1%)
5 - 6	0.13	-0.08		
6 - 7	0.14	-0.07	3	
7 - 8	0.19	-0.02		
8 - 9	0.29	0.08		
9 - 10	0.33	0.12		
10 - 11	0.27	0.16		
11 - 12	0.32	0.11		
12 - 13	0.34	0.13		
13 - 14	0.31	0.10		

The run test was used to check the stationarity of the mean and variance of undrained strength, Su. Table 4 shows the result of the run test for the mean of Su at each 1 m interval. Before detrending, it was found that $\bar{R}$ equals 2 and the hypothesis of stationarity is rejected. After detrending by the second order polynomial regression equation, $\bar{R}$ equals 4 which meant stationary data at level of significance, α, of 0.05. Table 5 shows the run test results for the standard deviation of Su at 1 m interval. R value was obtained to be 3 which fell at the rejection edge at 5% significance level. However, at 1% significance level, the data is stationary.

8 PROBABILISTIC SLOPE STABILITY ANALYSIS

8.1 Previous Investigations

The procedure of probabilistic analysis and design of earth slopes has been investigated by various researchers. These procedures include the development of mathematical models that:

1) account for the spatial variation and correlation structure of soil properties in both two-and three-dimensional analysis (VANMARCKE, 1975, 1977 a, b, 1979, 1980; ANDERSON et al, 1981; SHARP et al, 1981; BERGADO & ANDERSON, 1985; BERGADO & JU, 1986);
2) integrate the components of uncertainties into overall uncertainty, for example, through multiplicative models of correction factor accounting for each source of error contributing to the discrepancies between laboratory and in-situ strength values (TANG, et al, 1976; YUCEMEN et al, 1975; HOEG & MURARKA, 1970; WU & KRAFT, 1970);
3) combine systematically the relevant information available to assess the statistics of the parameters involves (WU & KRAFT, 1967, 1970; YUCEMEN, 1973);
4) compute the probability of failure for a certain cross-section of slope with a fairly homogenous soil stratum using prescribed probability distribution for resisting and applied moments (ALONSO, 1976; MATSUO & KURODA, 1974; LANGEJAN, 1965);
5) examine the contribution of additional probabilities of failure of sliding surfaces other than the critical one with the largest failure probability (MORLA - CATALAN & CORNELL, 1976);
6) use of straight forward Monte Carlo simulation technique (KRAFT & MUKHOPADHYAY, 1977; SIVANDRAN, 1979);
7) consider the progressive failure of slope when applying probabilistic analysis (CHOWDHURY & A-GRIVAS, 1982).

8.2 Slope Stability Analysis

In conventional slope stability analysis, it is assumed that the width of the failure mass to be infinite such that the analysis is two-dimensional and the safety factor is given as:

$$F = \frac{\text{resisting moment (Mr)}}{\text{driving moment (Md)}} \tag{14}$$

In three-dimensional analysis, the end effect of the failure mass is considered. The cylindrical mass is assumed to have resistance at both ends as shown in Fig. 16 a, b. Then, the factor of safety for length b, F_b, is given as:

$$F_b = \frac{Mr,b}{Mo,b} = \frac{Su\ Lrb + Re}{Wab} \tag{15}$$

where the terms are defined in Fig. 16a, and b.

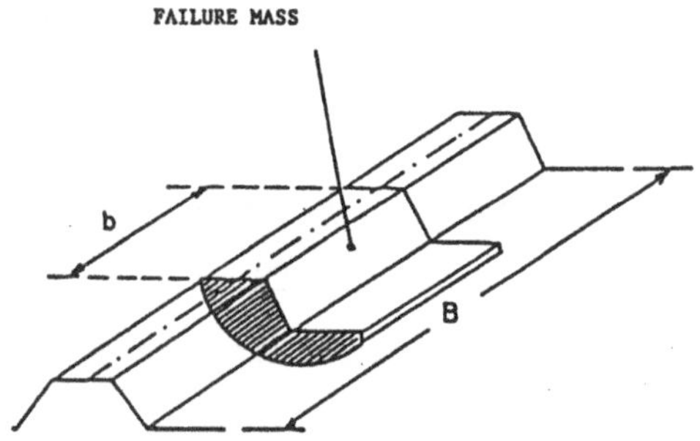

Fig. 16a Typical cylindrical failure mass of an earth embankment (After VANMARCKE, 1977b).

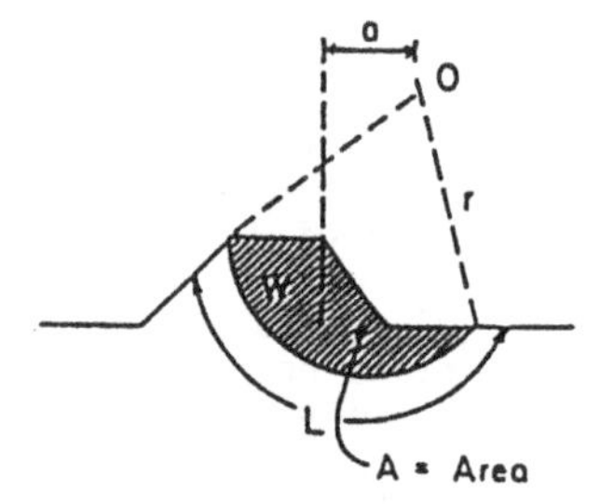

Fig. 16b Typical cross section of failure mass (After VANMARCKE, 1977b).

The mean safety factor, $\bar{F}_b$, for a cylindrical failure mass of width b can be expressed as;

$$\bar{F}_b = \frac{\bar{M}rb + Re}{Mob} = \frac{\bar{S}ub\ Lr + Re}{Mob} \qquad (16)$$

in which Mr is the mean resisting moment and Mo is the driving moment. The variance of the safety factor, $\tilde{F}_b^2$, may be pressed as:

$$\tilde{F}_b^2 = \frac{\tilde{M}^2 r,b}{Mo^2,b} = \frac{\tilde{S}ub^2\ L^2\ r^2}{(Wab)^2} \qquad (17)$$

Thus, the variance of the safety factor depends on the variance of undrained strength. A number of shear strength tests would produce variable results that may be represented by a histogram such as shown in Fig. 17 a. If normal distribution is assumed (Fig. 17 b), then the distribution of the safety factor is also normal.

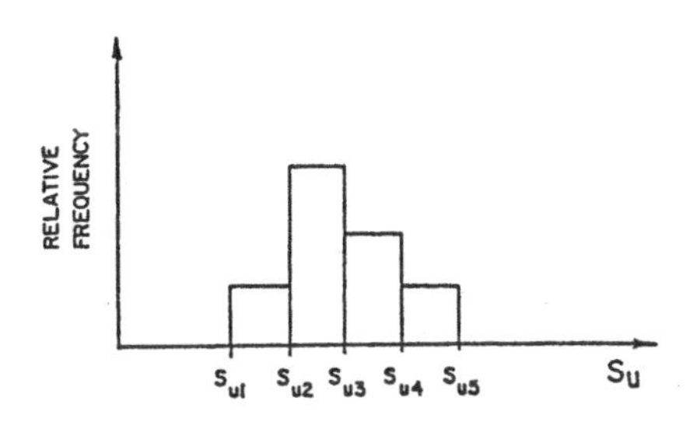

Fig. 17a Frequency histogram of undrained shear strength, S_u.

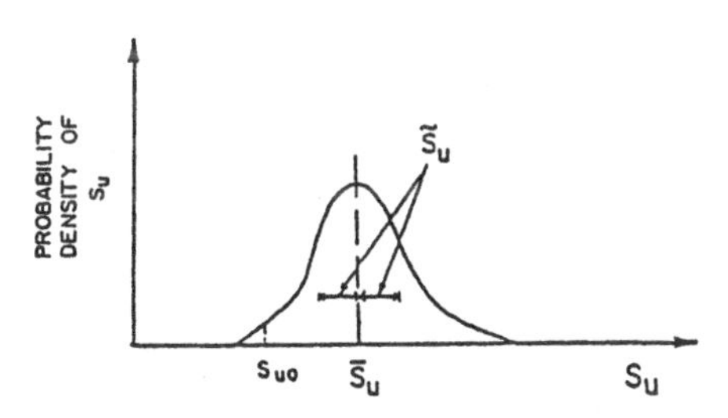

Fig. 17b Probability density function of undrained shear strength, S_u.

The variance reduction, denoted by Γ^2, is simply the ratio of $\tilde{S}^2_{ub}$ to $\tilde{S}^2_{up}$ where $\tilde{S}^2_{ub}$ is the variance of the undrained shear strength average over a length of b and $\tilde{S}^2_{up}$ is the variance of undrained shear strength of the unit length values as discussed earlier (see Fig. 13 a). Thus, the product of the variance reduction and the variance of point values give the variance of the shear strength average over a distance b. The effect of the averaging length, b on $\tilde{S}_{ub}$ as b increases is shown in Fig. 13 b which was based on the original concepts of VANMARCKE (1979a, b; 1980) and detailed in ANDERSON et al (1981).

In performing slope stability analysis for embankment, it is convenient to use a method of slices (Fig. 18) such as Bishop's simplified method (BISHOP, 1955). Using total stress analysis the mean safety factor for a failure mass of width b by the method of slices is given as:

$$\bar{F}_{bs} = \frac{br\ \Sigma\ \bar{S}_{bj}\ L_j + Re}{Mo,b} \qquad (18)$$

where: $\bar{S}_{bj}$ = mean shear strength of the jth soil type averaged along failure plane of width, b; L_j = length of the failure plane passing through the jth soil type; $M_{o,b}$ = overturning moment of failure mass of width b taken as deterministic quantity; Re = contribution of end sections of the failure mass to the resisting moment; and r = radius of failure plane.

The average undrained shear strength, $\bar{S}_{bj}$, for the clay foundation can be expressed in terms of the Mohr-Coulomb strength theory in terms of total stresses as:

$$\bar{S}_{bj} = \left|\frac{\Sigma(S_i \Delta \ell)}{\Sigma \Delta \ell_i}\right|_j = \frac{[\Sigma(S_i \Delta \ell_i)]_j}{L_j} \qquad (19)$$

where L_j is length of the failure surface passing through the jth soil type and $\tilde{S}_{bj}$ is the averaged undrained shear strength of the jth soil type averaged along the failure plane of width b.

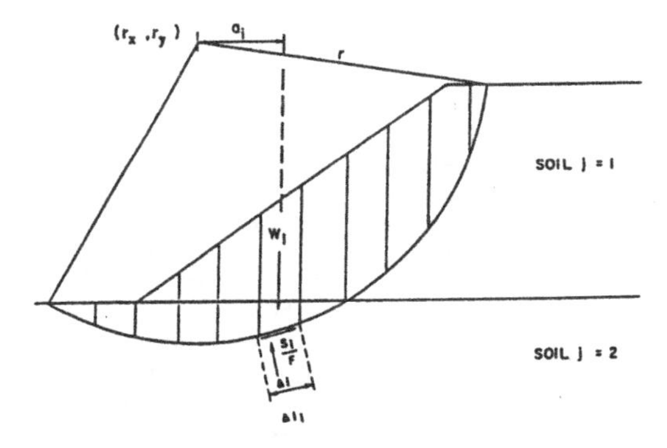

Fig. 18 Slice geometry in method of slices analysis (After Anderson et al, 1981).

Assuming the strength parameters between soil types to be independent, the standard deviation of the factor of safety, $\tilde{F}_{bs}$, due to the spatial variability of shear strength can be expressed as:

$$\tilde{F}_{bs} = \frac{r\sqrt{\Sigma \tilde{S}^2_{bj} L_j^2}}{M_{o,b}} \qquad (20)$$

In Eqs. (19) and (20), the factor of safety appear in the term $M_{o,b}$ which is taken as deterministic quantity. The difference of the vertical interslice forces is assumed to be zero and the factor of safety is determined by trial. Thus, it is the value of shear strength averaged over the failure surface and not the point values that is used in determining the safety factor.

Once the mean and standard deviation of the safety factor have been found, the probability of failure, P_f (b), of a failure mass of width b is defined as the probability that the safety factor is less than one.

$$P_f(b) = P(F_b < 1) \qquad (21)$$

VANMARCKE (1977 b) showed that the probability of failure can be conveniently expressed in terms of a standardized random variable, U, and the reliability index, β_b, as:

$$P_f(b) = P(U < \beta_b) \qquad (22)$$

where:

$$U = \frac{F_b - \bar{F}_b}{\tilde{F}_b} \qquad (23)$$

and

$$\beta b = \frac{\bar{F}_b - 1}{\tilde{F}_b} \qquad (24)$$

Figures 19 and 20 illustrate the effects of the mean and variance of shear strength, Su, for 3 soil types on the probability of failure. Soils 1 and 2 have the same mean but different variances, as shown in Fig. 19. Soil 3 has greater mean and variance than soil 1 as shown in Fig. 20. Figure 19 demonstrates that soil 2 has greater probability of failure than soil 1 even though they have similar mean due to the higher variance of soil 2. Furthermore, soil 3 has greater probability or failure as shown in Fig. 20 than soil 1 even though soil 3 has higher mean because the latter has higher variance than the former.

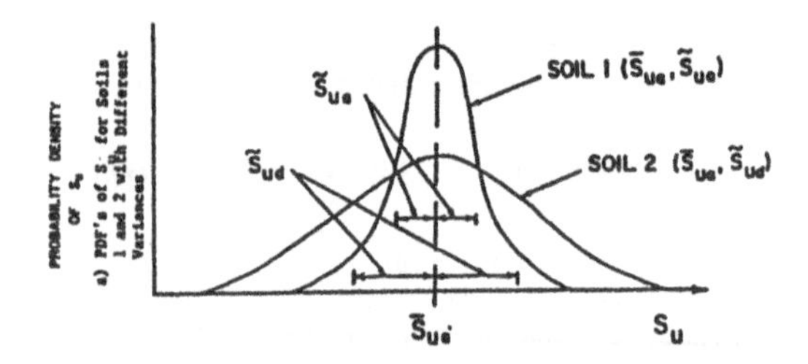

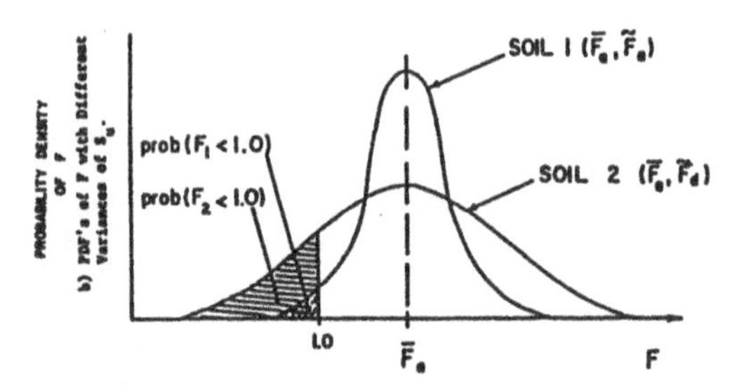

Fig. 19 Effect of different variances in undrained shear strength on probability of failure, Prob (F < 1.0).

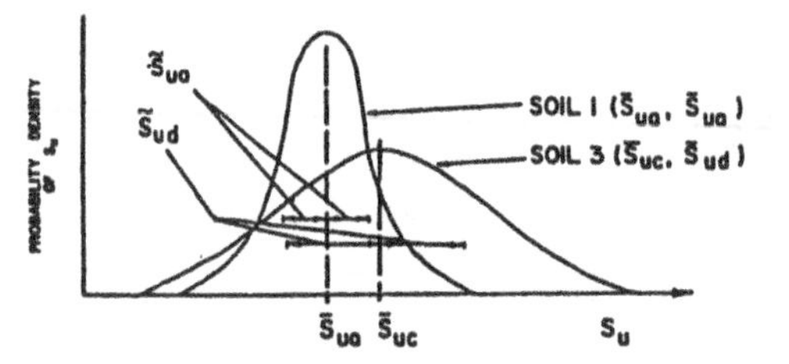

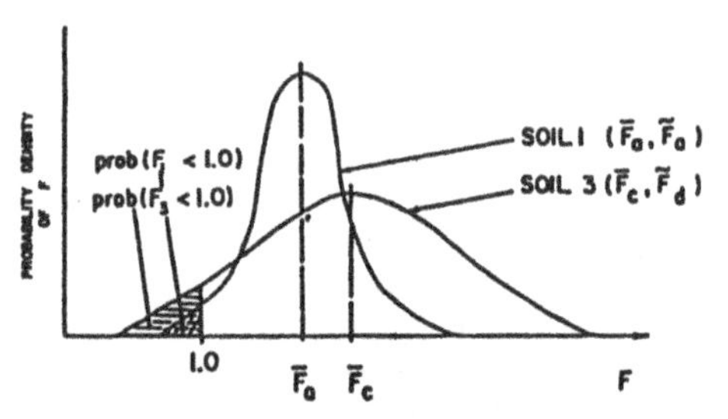

Fig. 20 Effect of different means and variances in undrained shear strength on probability of failure, Prob (F < 1.0).

Unlike conventional slope stability analysis, the safety factor depends on the width, b, of the failure mass. As the width of the failure mass increases the contribution of the end resistance becomes negligible. In the limit as b approaches infinity the value of $\bar{F}_b$ approaches the plane strain safety factor, $\bar{F}_o$, as shown in Fig. 21. Figure 13 b illustrated the effect of the averaging length on the standard deviation of shear strength, $\tilde{S}_{ub}$. The effect of the averaging length on the standard deviation of the safety factor, $\tilde{F}_b$, is similar, that is $\tilde{F}_b$ decreases as b increases. Thus, as b increases $\bar{F}_b$ and $\tilde{F}_b$ decrease. Figure 22 shows several probability density functions of F_b as b increases. Due to the changing rates of $\bar{F}_b$ versus b and $\tilde{F}_b$ versus b, there is a critical width, b_c, at which the probability of failure is maximum. Figure 23 shows a plot of probability of failure versus width of failure mass.

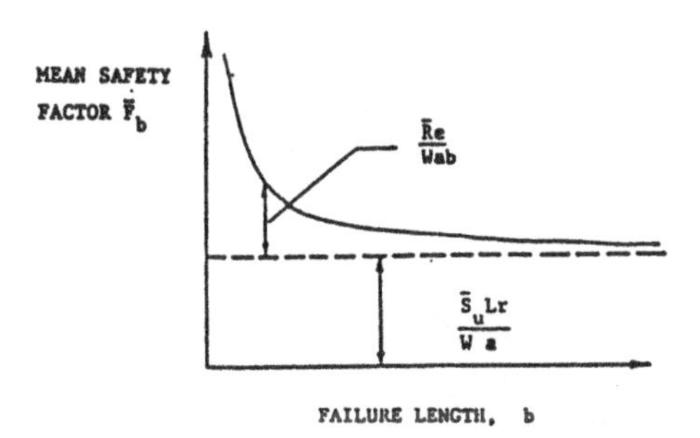

Fig. 21 Influence of failure length, b on $\bar{F}_b$.

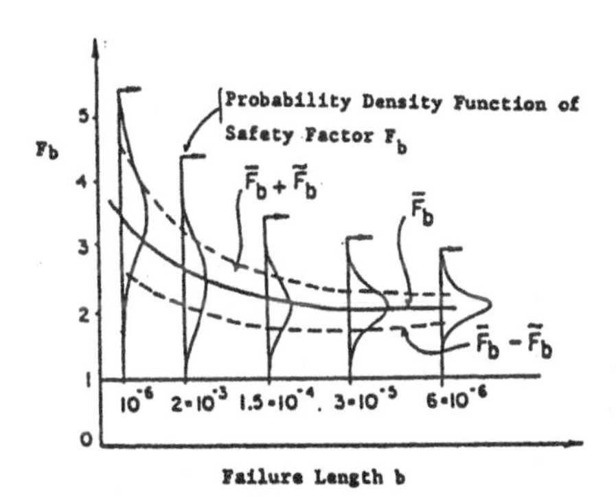

Fig. 22 Probability density function of factor of safety for different failure lengths (for each value of b).

VANMARCKE (1977 b) developed an expression for the critical failure width, b_c, by evaluation of the width at which the reliability index is minimized. The value of b_c is a function only of the end resistance, Re, the mean resisting moment, M_r, are the overturning moment, M_o.

$$b_c = \frac{Re}{\bar{M}r - M_o} \qquad (25)$$

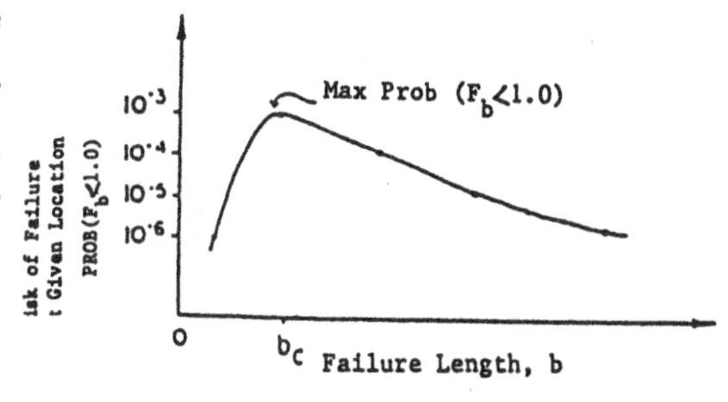

Fig. 23 Probability of failure versus length of failure mass.

This value, b_c, should be used to evaluate the mean and standard deviation of the safety factor, $\bar{F}_b$, and $\tilde{F}_b$. It is interesting to note that the critical width, b_c, is not a function of the variance properties.

For an embankment with an overall lenght, B, less than the critical width, b_c, the probable failure mass includes the entire embankment. However, when the overall embankment width exceeds the critical width, there are many possible placements of the critical width along the embankment. Thus, the probability of failure of the embankment increases as the total embankment width increases as a result of the multiplicity of locations of the critical width. VANMARCKE (1977 b) has developed a method to adjust the probability of failure depending on the total embankment length.

9 APPLICATION OF THE PROBABILISTIC MODEL

In the application of the probabilistic model, the following assumptions were made:

1) The variance of the end resistance, Re, was neglected. VANMARCKE (1977 b) gave detailed explanation for this assumption and showed that the inclusion of the end area variance has negligible effect on the probability of failure.
2) The density of the sand fill and the embankment geometry were treated as deterministic parameters.
3) Pore pressure uncertainties were not included in the probabilistic analysis. The critical condition was assumed in the total stress analysis.
4) The probability density function of the safety factor is assumed to be normal because the undrained shear

strength is assumed to be normal. This simplifying assumption is justified by the Central Limit Theorem.

The applied computed model is based on the program (PROBISH) developed at Utah State University. A flowchart illustrating the probabilistic 3-D model is given in Fig. 24. Utilization of the computer model requires data input for a conventional deterministic analysis as well as the statistical parameters required for the probabilistic analysis. The data required for the conventional analysis includes:

1) Cross-section geometry.
2) Soil parameters, including unit weight, cohesion and friction angle (mean values for probabilistic analysis).
3) Pore pressure data (if effective stress method are applied).
4) Specifications for the location of the failure surface.

The parameters required for the probabilistic analysis include:

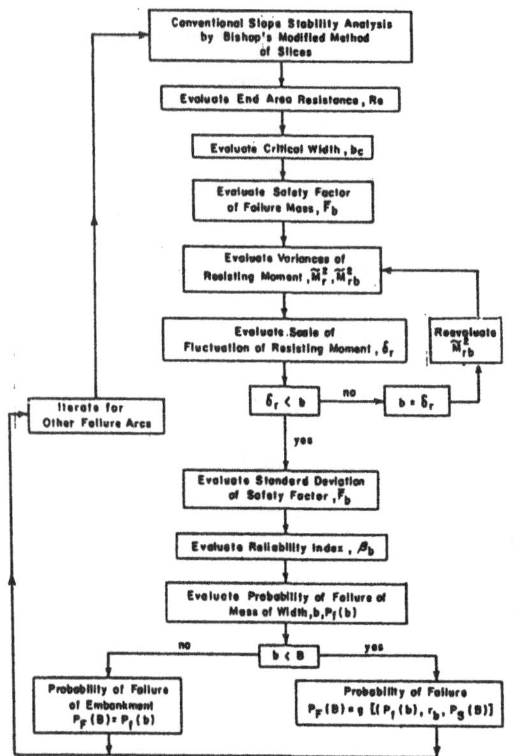

Fig. 24 Computer model flow chart for probability analysis of embankment stability (After Sharp et al, 1981).

1) Standard deviation of the strength parameters for each soil type.
2) Variance decay parameters for shear strength of each soil type.
3) Total length of the embankment.

PROBISH is an interactive computer program for performing deterministic and probabilistic slope stability analysis, it is limited to circular failure surfaces analysis. In deterministic analysis, safety factors are computed by Bishop's simplified method of slices. The probabilistic analysis used Bishop's simplified method for the computation of the safety factor, normal forces, driving moment and other quantities required for computing the probability of failure. Both spatial variability for the shear strength parameters and model error are considered in evaluating the probability of failure. A sample run of PROBISH Program is illustrated in the Appendix.

9.1 Probability of Failure

The probability of failure was analyzed using the typical embankment section a shown in Fig. 25. Two modes of failure were considered, namely: neglecting the strength of the sand fill ($\phi s = 0^{\circ}$) and considering the sand fill ($\phi s = 37.50^{\circ}$). The spatial variability of the undrained shear strength as well as 5% model error were included.

Table 6 shows the values of the mean, variance and autocorrelation estimates of Su that were used in the analysis. The autocorrelation function was assumed to be isotropic in horizontal directions. The foundation clay was divided into 8 layers down to 8 m below the ground which considered to be the maximum depth the critical failure surface will reach.

The vane shear strength correction factor suggested by BJERRUM (1972) was also applied to get more precise safety factor for slope stability analysis.

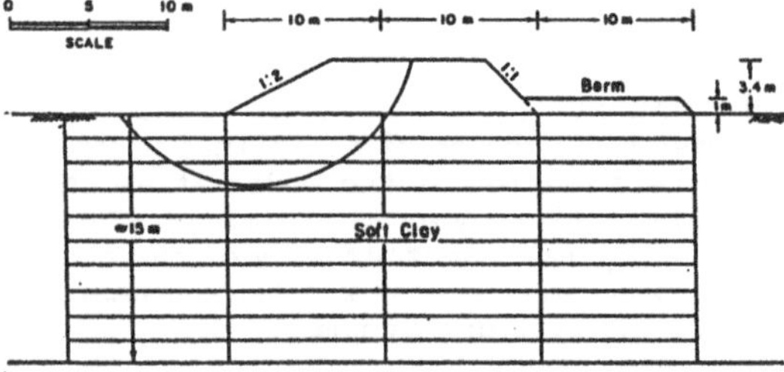

Fig. 25 Typical section of NNH Embankment III.

Table 6 Soil parameters used in NNH Embankment** analysis.

Parameter	Embankment fill	Foundation Clay
Unit weight	1.85 t/m^3	*
tan Ø	0.767	0
tan Ø	0	0
$\bar{c}$	0	$1.889-0.171d+0.023d^2$
$\tilde{c}$	0	1.74 t/m^2
δ_x	0	20.80 m
δ_y	0	0.68 m
δ_z	0	20.80 m

* it varies with depth, the mean value of each 1 m layer is used, please refer to Table 5.3

** Embankment length = 100 m

Table 7 Results of NNH slope analysis without BJERRUM's correction factor, $Ø_s = 37.5^o$ 3-D model.

Type of Errors Included	Uncertainty (critical failure surface)						
	Embankment height (m)	F_o	$\bar{F}_b$	$\tilde{F}_b$	P_f	R.I.	B_c (m)
Spatial variation only	2.0	2.33	3.67	0.099	0.0	26.9	6.5
	2.5	1.88	2.76	0.061	0.0	28.8	9.8
	2.8	1.68	2.37	0.053	0.0	25.8	11.5
	3.0	1.58	2.15	0.060	0.0	19.3	14.4
	3.2	1.49	1.99	0.054	$1.0(10^{-72})$	18.4	16.5
	3.4	1.41	1.82	0.050	$1.0(10^{-58})$	16.4	18.6
	3.7	1.30	1.60	0.040	$6.1(10^{-40})$	15.0	24.6
	4.0	1.21	1.42	0.036	$2.3(10^{-31})$	11.8	33.5
	4.3	1.15	1.30	0.029	$6.6(10^{-25})$	10.5	48.8
	4.5	1.10	1.20	0.022	$1.7(10^{-19})$	9.2	64.7
	5.0	1.02	1.08	0.017	$6.0(10^{-7})$	4.9	100.0
Spatial variation & 5% model error	2.0	2.33	3.67	0.208	$9.3(10^{-36})$	12.8	6.5
	2.5	1.18	2.76	0.151	$6.5(10^{-30})$	11.7	9.8
	2.8	1.68	2.37	0.130	$1.5(10^{-24})$	10.5	11.5
	3.0	1.58	2.15	0.123	$1.5(10^{-19})$	9.4	14.4
	3.2	1.49	1.99	0.113	$3.3(10^{-17})$	8.7	16.5
	3.4	1.41	1.82	0.104	$3.9(10^{-14})$	7.9	18.6
	3.7	1.30	1.60	0.089	$1.6(10^{-10})$	6.7	24.6
	4.0	1.21	1.42	0.080	$4.9(10^{-7})$	5.3	33.5
	4.3	1.15	1.30	0.071	$7.0(10^{-5})$	4.2	48.8
	4.5	1.10	1.20	0.064	$2.2(10^{-3})$	3.2	64.7
	5.0	1.02	1.08	0.057	$7.0(10^{-2})$	1.5	100.0

Table 8 Results of NNH slope analysis with BJERRUM correction factor, $Ø_s = 37.5^o$.

Type of Errors Included	Uncertainty (critical failure surface)						
	Embankment height (m)	F_o	$\bar{F}_b$	$\tilde{F}_b$	P_f	R.I.	B_c (m)
Spatial variation only	2.0	1.66	2.32	0.067	0.0	19.5	9.7
	2.5	1.33	1.67	0.038	$2.2(10^{-67})$	17.5	19.4
	2.8	1.19	1.39	0.029	$1.1(10^{-34})$	13.4	30.6
	3.0	1.12	1.23	0.027	$3.6(10^{-17})$	8.6	53.8
	3.2	1.06	1.13	0.019	$1.4(10^{-11})$	6.4	96.1
	3.4	1.00	1.06	0.018	$7.1(10^{-4})$	3.2	100.0
	3.7	0.92	1.00	0.018	$5.7(10^{-1})$	0.0	75.7
	4.0	0.86	1.00	0.024	$7.1(10^{-1})$	0.0	38.9
	4.3	0.82	1.00	0.024	$7.5(10^{-1})$	0.0	31.7
	4.5	0.79	1.00	0.023	$7.9(10^{-1})$	0.0	24.7
	5.0	0.72	1.00	0.023	$8.3(10^{-1})$	0.0	19.6
Spatial variation & 5% model error	2.0	1.66	2.32	0.134	$2.7(10^{-21})$	9.8	9.7
	2.5	1.33	1.67	0.092	$3.9(10^{-12})$	7.3	19.4
	2.8	1.19	1.39	0.075	$1.2(10^{-6})$	5.2	30.6
	3.0	1.12	1.23	0.067	$1.1(10^{-3})$	3.5	53.8
	3.2	1.06	1.13	0.060	$1.7(10^{-2})$	2.2	96.1
	3.4	1.00	1.06	0.056	$1.5(10^{-1})$	1.03	100.0
	3.7	0.92	1.00	0.053	$5.7(10^{-1})$	0.0	75.7
	4.0	0.86	1.00	0.055	$7.1(10^{-1})$	0.0	38.9
	4.3	0.82	1.00	0.055	$7.5(10^{-1})$	0.0	31.7
	4.5	0.79	1.00	0.055	$7.9(10^{-1})$	0.0	24.7
	5.0	0.72	1.00	0.055	$8.3(10^{-1})$	0.0	19.6

Table 7 and Table 8 show the results of analysis without and with, respectively, application of BJERRUM correction factor and $Ø_s = 37.5^o$. Both Table 7 and Table 8 used the predetermined critical surface for various embankment height. The different embankment height used are 2.0m, 2.5m, 2.8m, 3.0m, 3.2m, 3.4m, 3.7m, 4.0m, 4.3m, 4.5m, 5.0m, respectively, in which at 3.4m the failure occurred in the field.

Table 7 shows that the probability of failure at 3.4m height increased from 10^{-58} to 10^{-14} considering spatial variation of undrained shear strength and 5% model error. $\tilde{F}_b$ ranged from 0.050 to 0.104. The plane strain factor of safety for this condition was 1.41. $\bar{F}_b$ for the critical failure width was 1.82. These indicate the relative small influence of spatial variability and variance reduction on design criteria, and 5% model error exerted much influence on the P_f. Because the high safety factor (F_o = 1.41, $\bar{F}_b$ = 1.82) and low probability of failure (P_f = 10^{-14}) can not explain the failure occurred at embankment height of 3.4 m satisfactorily, it is suspected that the reason was over-estimation of undrained shear strength in field vane. A BJERRUM correction factor of 0.7 (PI = 70%) was applied to the undrained shear strength. The results of analysis (Table 8) shows that the plane strain safety factor was 1.00 and $\bar{F}_b$ equals 1.06 and P_f with model error excluded is 7 x 10^{-4}. P_f value considering 5% model error was 1.5 x 10^{-1}. These could explain the failure condition in embankment height 3.4 m.

The P_f of observed failure surface for embankment height of 3.4 m was found to be 2.5 x 10^{-2} if 5% model error was considered and the correction factor was applied to the vane shear strength. The results are shown on Table 9. Figure 26 shows the variation of the conventional plane strain safety factor with height of embankment. Figure 27 shows the relationship derived between the

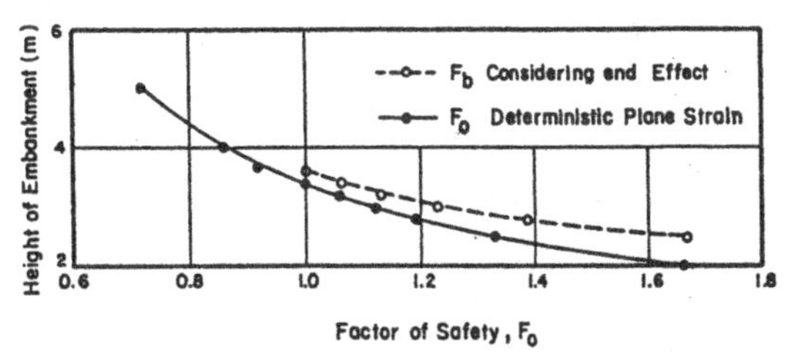

Fig. 26 Factors of safety vs height of embankment, considering the BJERRUM's correction factor.

Table 9 Results of NNH slope analysis with BJERRUM correction factor, $\phi_s = 37.5^o$, 3-D model.

Type of Errors Included	Actual Failure Surface (X , Y)	R	Uncertainty (observed failure surface) F_o	$\bar{F}_b$	$\tilde{F}_b$	P_f*	P_f**	N.L.	B_c (m)
Spatial variation only	(9.2 , 6.6)	12.8	1.10	1.20	0.015	$1.1(10^{-37})$	$6.8(10^{-37})$	12.8	74.3
	(8.4 , 5.8)	12.0	1.09	1.17	0.014	$2.7(10^{-33})$	$1.1(10^{-33})$	12.0	81.8
	(8.2 , 5.2)	11.2	1.05	1.11	0.014	$6.1(10^{-16})$	$6.1(10^{-16})$	8.0	100.0
Spatial variation & 5% model/ error	(9.2 , 6.6)	12.8	1.10	1.20	0.062	$8.0(10^{-4})$	$1.9(10^{-3})$	3.2	74.3
	(8.4 , 5.8)	12.0	1.09	1.17	0.060	$2.3(10^{-3})$	$2.3(10^{-3})$	2.8	81.8
	(8.2 , 5.2)	11.2	1.05	1.11	0.057	$2.5(10^{-2})$	$2.5(10^{-2})$	2.0	100.0

Remark: * P_f for embankment length B
** P_f for full embankment length

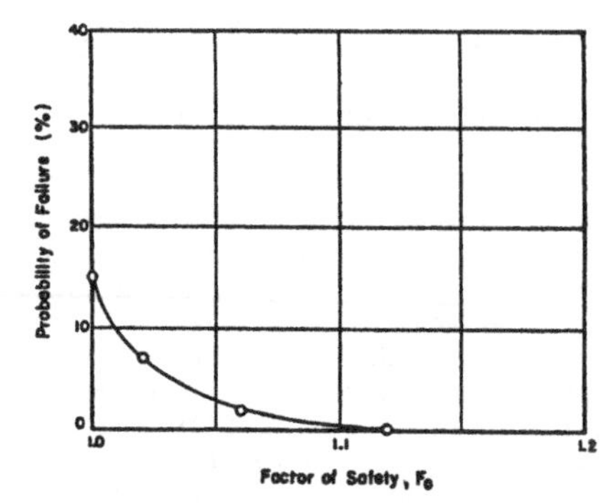

Fig. 27 Probability of failure P_f vs plane strain safety factor F_o, 3-D model at COV of 10%.

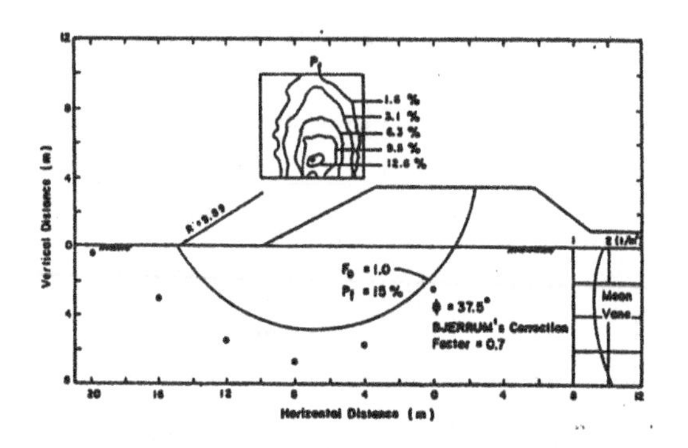

Fig. 28 Critical surface of probabilistic analysis.

probability of failure and the conventional plane strain safety factor with applying the BJERRUM's correction factor with ϕs = 37.5 degrees. The P_f used here was derived from considering both the spatial variability of Su and 5% model error. The P_f decreases asymptotically with safety factor. The P_f reaches values less than 10^{-3} for a corrected safety factor of 1.12. Figure 27 can be used as a design criteria. If the P_f for design is of the order of 10^{-3} and less, a corrected safety factor 1.12 or larger should be selected.

The critical surface of probabilistic slope stability analysis for Nong Ngoo Hao embankment is shown on Fig. 28 in which the contour map of probability of failure for square grid is also plotted. The suggested safety factor of 1.12 for design was derived considering only the shear failure in the embankment and its foundation. It must be checked with associated settlement to assure that the resulted settlement will fall in the range of allowable value when applied to design.

9.2 Sensitivity Analysis

The magnitude of the probability of failure of an embankment depends on the mean safety factor which is a function of the embankment geometry, mean shear strength and mean soil properties, the variance of the shear strength parameters, the variance reduction factors which depend on the autocorrelation functions and the type of uncertainty that is included in the analysis (i.e. model error and measurement error). A sensitivity study was performed to observe the influence of the soil properties and the

Table 10 Influence of mode of failure and failure surface, 3-D model.

Surface	Mode of Failure	Location of Center	Radius (m)	F_s	$\overline{F}_b$	$\widetilde{F}_b$	P_f*	P_f**	R.I.	B_c (m)
Critical Surface	$\phi_s = 0$	(7.0 , 5.0)	9.89	0.91	1.00	0.056	$5.0(10^{-1})$	$6.6(10^{-1})$	0.0	48.9
	$\phi_s = 37.5°$	(7.0 , 5.0)	9.89	1.00	1.06	0.056	$1.5(10^{-1})$	$1.5(10^{-1})$	1.03	100.0
Observed Surface	$\phi_s = 0$	(9.2 , 6.6)	12.8	0.98	1.04	0.054	$2.3(10^{-1})$	$2.3(10^{-1})$	0.8	100.0
		(8.4 , 5.8)	12.0	0.93	1.00	0.052	$5.0(10^{-1})$	$5.6(10^{-1})$	0.0	77.7
		(8.2 , 5.2)	11.2	0.95	1.04	0.052	$4.7(10^{-1})$	$4.7(10^{-1})$	0.07	100.0
	$\phi_s = 37.5°$	(9.2 , 6.6)	12.8	1.10	1.20	0.062	$8.0(10^{-4})$	$1.9(10^{-3})$	3.2	74.3
		(8.4 , 5.8)	12.0	1.09	1.17	0.060	$2.3(10^{-3})$	$4.2(10^{-3})$	2.8	61.8
		(8.2 , 5.2)	11.2	1.05	1.11	0.057	$2.5(10^{-2})$	$2.5(10^{-2})$	2.0	100.0

Remarks : BJERRUM Correction factor was applied and 5% model error are considered

* P_f for embankment length B
** P_f for full embankment length

Table 11 Influence of variance of Su, ϕ = 37.5°, 3-D model.

Standard Deviation of S	Uncertainty (critical failure surface)						
	Embankment height (m)	F_o	$\overline{F}_b$	$\widetilde{F}_b$	P_f	R.I.	B_c (m)
2.45 t/m² C.O.V = 15%	2.0	1.66	2.32	0.184	$1.7(10^{-11})$	7.2	9.7
	2.5	1.33	1.67	0.116	$1.3(10^{-8})$	5.8	19.4
	2.8	1.19	1.39	0.093	$1.2(10^{-4})$	4.2	30.6
	3.0	1.12	1.23	0.084	$1.0(10^{-2})$	2.8	53.8
	3.2	1.06	1.13	0.070	$3.5(10^{-1})$	1.9	96.1
	3.4	1.00	1.06	0.059	$1.7(10^{-1})$	1.0	100.0
	3.7	0.92	1.00	0.057	$5.7(10^{-1})$	0.0	75.7
	4.0	0.86	1.00	0.071	$7.1(10^{-1})$	0.0	38.9
	4.3	0.82	1.00	0.071	$7.5(10^{-1})$	0.0	31.7
	4.5	0.79	1.00	0.070	$7.9(10^{-1})$	0.0	24.7
	5.0	0.72	1.00	0.070	$8.3(10^{-1})$	0.0	19.6
6.10 t/m² C.O.V = 35%	2.0	1.66	2.32	0.358	$2.7(10^{-3})$	3.7	9.7
	2.5	1.33	1.67	0.208	$7.5(10^{-3})$	3.2	19.4
	2.8	1.19	1.39	0.162	$4.6(10^{-2})$	2.4	30.6
	3.0	1.12	1.23	0.150	$1.5(10^{-1})$	1.6	53.8
	3.2	1.06	1.13	0.113	$1.3(10^{-1})$	1.2	96.1
	3.4	1.00	1.06	0.105	$2.9(10^{-1})$	0.5	100.0
	3.7	0.92	1.00	0.102	$5.7(10^{-1})$	0.0	75.7
	4.0	0.86	1.00	0.129	$7.1(10^{-1})$	0.0	38.9
	4.3	0.82	1.00	0.129	$7.5(10^{-1})$	0.0	31.7
	4.5	0.79	1.00	0.125	$7.9(10^{-1})$	0.0	24.7
	5.0	0.72	1.00	0.125	$8.3(10^{-1})$	0.0	19.6

Remarks : BJERRUM Correction factor and 5% model error was considered

probabilistic types of uncertainties on the probability of failure.

a) Influence of Failure Model - The comparison between mode of failure using ϕ_s equals 0 and ϕ_s equals 37.5° on both critical and observed failure surfaces for field embankment height of 3.4 m are shown in Table 10. The failure mode of ϕ_s = 0° make the design conservative.

b) Influence of the Variance of Su - The results given in Table 11 shows the dependence of the probability of failure, P_f, for a critical width b on the standard deviation of undrained shear strength S_u. The autocorrelation coefficient were held constant while the S_u were increased from 1.74 t/m². The P_f predictably increased as shown in Table 11. The relationship between P_f and the corrected conventional safety factor for S_u = 2.56 t/m² and S_u = 6.10 t/m² are presented in Fig. 29.

Table 12 Influence of autocorrelation coefficient, $\emptyset = 37.5^{o}$, 3-D model.

Soil properties : see Table 6.1
Center of rotation : X = 9m, Y = 5m
Radius : 9.89m
Embankment length : 100m
Embankment height : 3.4m
BJERRUM Correction factor not included, 5% midel included

Critical width : 18.6m
F_o = 1.41
$\bar{F}_b$ = 1.82

Trial	$\tilde{S}_u$ (t/m²)	C.O.V (%)	K_x, K_z (m)	K_y (m)	$\tilde{F}_b$	R.1	P_f
1	1.74	10	11.74	0.39	0.104	7.9	3.9(10⁻¹⁴)
2	2.56	15	11.74	0.39	0.117	7.0	3.3(10⁻¹¹)
3	5.22	30	11.74	0.39	0.176	4.6	2.6(10⁻⁵)
4	6.10	35	11.74	0.39	0.198	4.1	2.6(10⁻⁹)
5	1.74	10	10,000	0.39	0.109	7.5	7.0(10⁻¹³)
6	1.74	10	50.00	0.39	0.109	7.5	5.7(10⁻¹³)
7	1.74	10	20.00	0.39	0.107	7.7	2.1(10⁻¹³)
8	1.74	10	5.00	0.39	0.097	8.4	5.8(10⁻¹⁶)
9	1.74	10	11.74	10.00	0.105	7.8	6.6(10⁻¹⁴)
10	1.74	10	11.74	5.00	0.105	7.8	6.4(10⁻¹⁴)
11	1.74	10	11.74	2.00	0.104	7.8	5.1(10⁻¹⁴)
12	1.74	10	11.74	0.20	0.096	8.5	2.8(10⁻¹⁶)

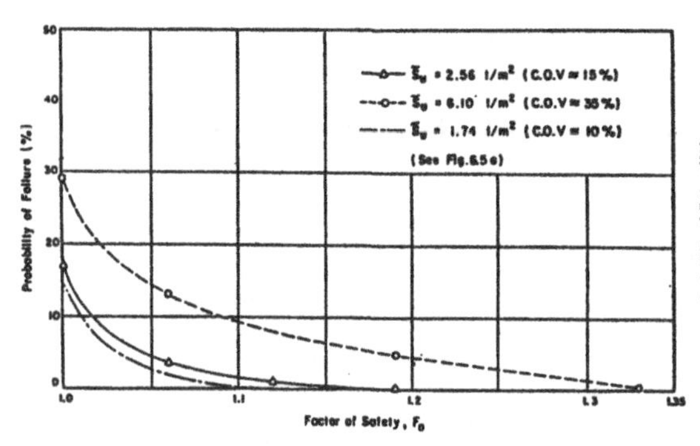

Fig. 29 P_f vs F_o using various standard deviation of undrained shear strength.

Fig. 30 Comparison of P_f vs F_o curve, 2-D model.

c) Influence of the Variance Reduction Coefficient - The sets of trials in Table 12 demonstrate the effect of the variance reduction coefficient. The results indicated that as the variance decay constant increased, $\tilde{F}_b$ increased, thus, increasing the probability of failure. It can be seen that the value of variance decay constant which is dependent on the ACF plays an important role in the magnitude of the probability of failure.

d) Influence of the Measurement Bias - As desribed earlier, the resulted high F_o and low P_f from uncorrected field vane shear strength for embankment height of 3.4 m cannot explain the actual failure condition. This may be due to the measurement bias of undrained shear strength.

The BJERRUM's correction factor can be applied to adjust the possible measurement bias. The effect of adjustment is shown on Table 7 and Table 8. It is noted that applying the correction factor lead to lower mean of S_u, but the COV of S_u remained constant.

9.3 Comparison with SIVANDRAN's Result

SIVANDRAN (1979) stated that the P_f was less than 0.01 for safety factor of 1.5 and the P_f of 10^{-4} was obtained for safety factor of 2.2. To provide for a basis for comparison, the variance reduction in Z-direction (along the axis of the embankment) was neglected in the 3-D probabilistic model. Furthermore, the analysis was done neglecting the strength of sand fill ($\phi_s = 0^o$) due to the presence of tension cracks. The comparison for P_f values versus the plane strain factor of safety, F_o, is given in Fig. 30. It should be noted also that Sivandran (1979) used a coefficient of variation of the unit weight of the soil of 6% and an empirical Beta distribution for the undrained strength.

10 CONCLUSIONS

From the results of this study, the following conclusions can be made:

1) For the description of soil properties as random data, the parameter such as mean, variance, probability density function (pdf), autocorrelation function (ACF) must be considered.
2) From the application of statistical data and probabilistic method on the Nong Ngoo Hao embankment, the safety factor of 1.12 is recommended for (3-D) slope design if the criterion for P_f is 10^{-3} for sand fill strength, ϕ_s, of 37.5 degrees. For more conservative criterion, if the failure mode ϕ_s of zero is considered and variance reduction in Z direction was neglected (2-D), the safety factor of 1.28 is recommended for slope design for equal to
3) The recommendations of items 2 and 3 are only suitable for Bangkok clay of which the COV of undrained shear strength is approximately equal to 10%. If higher COV of Su is observed, the safety factor used for design should be larger. Furthermore, the recommended safety factor for design was derived considering undrained shear failure only, it must be verified with associated settlement when applied to design. If the resulted settlement exceed the allowable value, then the recommended design value must be revised. The design curve for P_f versus plane strain safety factor can be constructed for various COV of undrained shear strength.
4) The BJERRUM's correction factor was successfully applied to explain the actual failure condition satisfactorily. For Nong Ngoo Hao site, the BJERRUM's correction factor has been successfully applied.
5) From 3-D probabilistic slope stability analysis, it was indicated that the spatial variability of soil properties contributed to the probability of slope stability failures. However, it was observed that the model error contributed more to the probability of failure. When the 5% model error was included, the resulting P_f value was able to explain the actual failure condition.
6) Performing a probabilistic slope stability analysis requires sufficient data to statistically characterize the soil properties for each soil zone within the soil structure to be analyzed. Field vane data can be used in embankment slope stability analysis. The widely used Dutch cone test can also be used for probabilistic analysis.
7) Comparing the results of the variance reduction technique in this paper with the previous results of SIVANDRAN (1979) using Monte Carlo techniques on two-dimensional probabilistic slope stability analysis, it was found that the variance reduction lowered the probabilities of failure.

11 REFERENCES

AGTERBERG, F. (1970), "Autocorrelation Functions in Geology", Geostatistics, Merrian Ed., Plenum Press, New York, pp. 113-142.

AIT (1973 a), "Performance Study of Test Sections for New Bangkok Airport at Nong Ngoo Hao", Progress Report Submitted-to Northrop Airport Development Corporation, AIT, Bangkok.

AIT (1973 b), "Performance Study of Test Sections for New Bangkok Airport at Nong Ngoo Hao", Final Report Submitted-to Northrop Airport Development Corporation, AIT, Bangkok.

ALONSO, E.E. and KRIZEK, R.J. (1975), "Stochastic Formulation of Soil Properties", Proc. 2nd ICASP, Aachen, Germany, pp. 9-32.

ALONSO, E.E. (1976), "Risk Analysis of Slopes and It's Application to Slopes in Canadian Sensitive Clays", Geotechnique, Vol 26, No. 3, pp. 453-472.

ANDERSON, L.R., BOWLES, D.S., CANFIELD, R.V. and SHARP K.D. (1981), "Probabilistic Modeling of Tailings Embankment Designs", U.S. Bureau of

Mines-Research Report, Contrac No. J0295 029, Utah State University, Logan, Utah.

ANDERSON, L.R., SHARP, K.V., BOWLES, D.S. and CANFIELD, R.V. (1984), "Application of Methods of Probabilistic Characterization of Soil Properties". Proc. Symp. Probabilistic Characterization of Soil Properties". ASCE, Atlanta, Georgia, U.S.A., D.S. Bowles and H.Y. Ko (eds).

BERGADO, D.T. (1982), "Probabilistic Assessment of the Safety of Earth Slopes Using Pore Water Pressure as a Random Variable", Ph.D. Thesis, Utah State University, Logan, Utah, U.S.A.

BERGADO, D.T. and ANDERSON, L.R. (1985), "Stochastic Analysis of Pore Pressure Uncertainty for the Probabilistic Assesbment of Earth Slopes", Soils and Foundations, Vol. 25, No. 2, pp. 53-76.

BERGADO, D.T. and JU, Y.C. (1986), "Probabilistic Computer Modeling of a Rockfill Dam - A Case of Khao Laem Dam", Soils and Foundations, Vol. 26, No. 4, pp. 183-202.

BISHOP, A.W. (1955), "The Use of the Slip Circle in the Stability Analysis of Slopes", Geotechnique, Vol. 5, No. 1, pp. 7-17.

BJERRUM, L. (1972), "Embankments on Soft Ground", State of the Art Report, ASCE Spec. Conf. on Performance of Earth and Earth-Supported Structures, Lafayette, Vol. 2, pp. 753-769.

BOONSINSUK, P. (1974), "Stability Analysis of a Test Embankment on Nong Ngoo Hao Clay, M. Eng. Thesis No. 696, AIT, Bangkok.

CHANG, J.C. (1985), "Probabilistic Characterization of Bangkok Subsoils", M. Eng. Thesis No. GT-84-8, AIT, Bangkok.

CHOWDHURY, R.N. and GRIVAS, D.A. (1982), "Probabililistic Model of Progressive Failure of Slopes", J. of the Geotechnical Engineering Division, ASCE, Vol. 108, No. GT8, pp. 803-819.

CORNELL, C.A. (1971), "First-Order Uncertainty Analysis of Soils Deformation and Stability", 1st ICASP, Hong Kong, pp. 130-144.

HOEG, K. and MURARKA, R.P. (1974), "Probabilistic Analysis and Design of a Retaining Wall", J. of the Geotechnical Engineering Division, ASCE, Vol. 100, No. GT3, pp. 349-366.

KERDSUWAN, T. (1983), "Basic Properties and Compressibility Characteristics of the First and Second Clay Layer of Bangkok Subsoils", M. Eng. Thesis No. GT-83-35, AIT, Bangkok.

KRAFT, L. and MUKHOPADHYAY, T. (1977), "Probabilitic Analysis of Excavated Earth Slopes", Proc. 9th ICSMFE, Tokyo, Japan.

LANGEJAN, A. (1965), "Some Aspects of Safety Factors in Soil Mechanics Considered as a Problem of Probability", Proc. 6th ICSMFE, Montreal, Canada.

MATSUO, M. and KURODA, (1974), "Probabilistic Approach to Design of Embankments", Soils and Foundations, Vol. 14, No. 2.

MEMON, A. (1974), "Effect of Anisotropy on Shear Strength of Nong Ngoo Hao Clay", M. Eng. Thesis, AIT, Bangkok.

MOH, Z.C., NELSON, J.D. and BRAND, E.W. (1969), "Strength and Deformation Characteristics of Bangkok Clay, Proc. 7th ICSMFE, Mexico, Vol. 1, pp. 287-296.

MORLA-CATALAN, J. and CORNELL, C.A. (1976), "Earth Slope Reliability by Level Crossing Method", J. of the Geotechnical Engineering Division, ASCE, Vol. 102, No. GT6.

SHARP, K.V. (1982), "Probabilistic Modeling of Tailings Embankment Design", Ph.D. Thesis, Utah State University, Logan, Utah, U.S.A.

SHARP, K.D., ANDERSON, L.R., BOWLES, D.S. and CANFIELD, R.V. (1981), "A Model for Assessing Slope Reliability", Proc. of the 1981 Annual Meeting of the Transportation Research Board, Washington, D.C.

SIVANDRAN, C. (1979), "Probabilistic Analysis of Stability and Settlement of Structures on Soft Bangkok Clay. Ph.D. Eng., Dissertation No. GT-79-1, AIT, Bangkok.

TANG, W.H., YUCEMEN, M.S. and H-S ANG A. (1976), "Probability-Based Short Term Design of Soil Slopes", Canadian Geotech. J., Vol. 13, pp. 210-215.

VANMARCKE, E.H. (1975), "Probabilistic Prediction of Levee Settlements", Proc. 2nd ICASP, Aachen, Germany.

VANMARCKE, E.H. (1977 a), "Reliability of Earth Slopes", J. of the Geotechnical Engineering Division, ASCE, Vol. 103, No. GT11, pp. 1247-1265.

VANMARCKE, E.H. (1977 b), "Probabilistic Modeling of Soil Profiles", J. of the Geotechnical Engineering Division, ASCE, Vol. 103, No. GT11, pp. 1237-1246.

VANMARCKE, E.H. (1979), "On the Scale of Fluctuation of Random Functions", M.I.T. Department of Civil Engineering, Research Report No. R79 17, M.I.T., Cambridge, Mass., U.S.A.

VANMARCKE, E.H. (1980), "Probabilistic Stability Analysis of Earth Slopes", Engineering Geology, Vol. 16, pp. 29-50.

WHITMAN, R.V. and BAILEY, W.A. (1967), "Use of Computers for Slope Stability Analysis", J. of the Soil Mech. and Found. Eng. Division, ASCE,. Vol. 93, No. SM6, pp. 1987-2006.

WU, T.H. and KRAFT L.M. (1970), "Safety Analysis of Slopes", J. of Soil Mechanics and Foundation Engineering Division, ASCE, Vol. 96, No. SM2.

YUCEMEN, M.S. (1973), "A Probabilistic Study of Safety and Design of Earth Slopes. Ph.D. Thesis, University of Illinois, at Urbana, U.S.A.

YUCEMEN, M.S., TANG, W.H. and ANG, A.H-S, (1975), "Long Term Stability of Soil Slopes - A Reliability Approach", Proc. 2nd ICASP, Aachen, Germany.

12 APPENDIX

Example problem A1 involves the probabilistic slope stability analysis of an embankment fill on a soft foundation. The embankment geometry is shown on Figure A1. The autocorrelation properties of the two soil types are defined using the theoretical autocorrelation constants, K. Two runs were made in the analysis for 1 trial center. The first run neglected model error. In the second run the model error was included by using option 7 after completion of the first run. The DO ONLY command for specification of a single trial center was used in both cases. Probabilistic analysis requires much more computer time and thus it is often more convenient to use the DO ONLY command in probabilistic analysis once the critical trial center based on safety factor has been located.

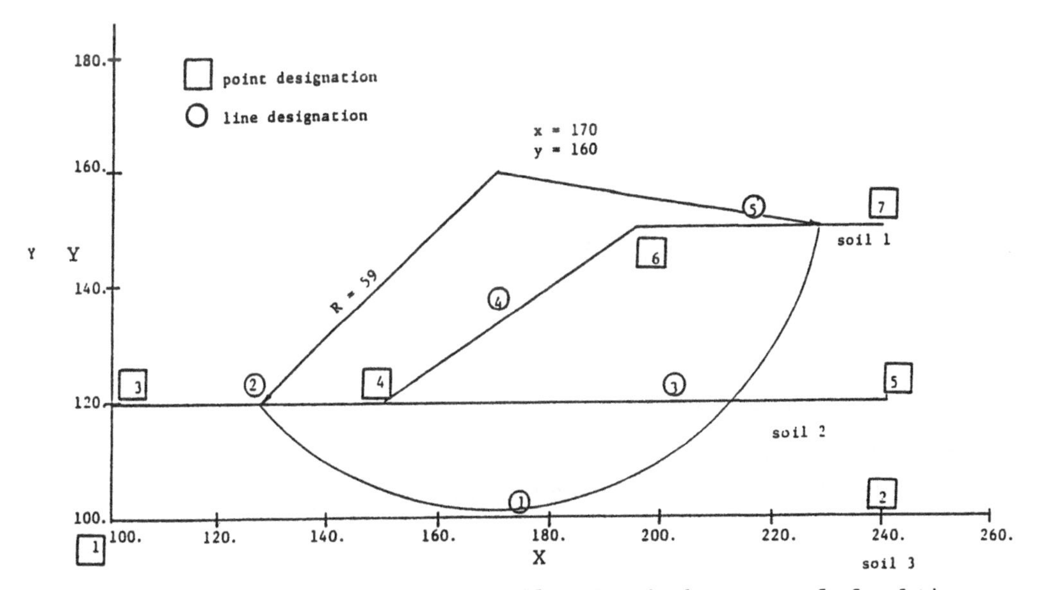

Fig. A1 Geometry used in example problem A, embankment on soft foundation.

```
E PROBISH
#RUNNING 4060

TYPE IN POINT DATA: POINT NO.,X-COORD.,Y-COORD. ON ONE LINE.  USE 100
POINTS OR LESS.  TYPE 999,0,0 AFTER LAST POINT.
N      X      Y

#?
1,100,100
2,240,100
3,100,120
4,150,120
5,240,120
6,195,150
7,240,150
999,,,

NO. X-COORD Y-COORD
 1  100.00  100.00
 2  240.00  100.00
 3  100.00  120.00
 4  150.00  120.00
 5  240.00  120.00
 6  195.00  150.00
 7  240.00  150.00
POINT DATA ARE LISTED ABOVE. ARE THEY ALL CORRECT....
Y

TYPE IN LINE DATA: LINE NO.,TWO POINT NOS., AND ONE SOIL NO. PER LINE.
USE 100 LINES OR LESS. TYPE 999,0,0,0 AS LAST NO. IN FIRST COLUMN.
NO.  P  P  S
1,1,2,3
2,3,4,2
3,4,5,2
4,4,6,1
5,6,7,1
999,,,,

NO.   P   P   S
  1   1   2   3
  2   3   4   2
  3   4   5   2
  4   4   6   1
  5   6   7   1
LINE DATA ARE LISTED ABOVE. ARE THEY ALL CORRECT....
Y

ARE UNIT WEIGHTS SPECIFIED IN ENGLISH OR SI SYSTEM?
E

GAMMA WATER = 62.4 POUNDS PER CUBIC FOOT.

TYPE IN SOIL DATA: SIX ITEMS OF DATA PER LINE. SOIL NO.,GAMMATOTAL,
COHESION,PHI(DEGREES),RU,RC. USE 10 SOILS OR LESS.
TYPE 999,0,0,0,0,0 AS LAST NO. IN SOIL COLUMN.
N  G   C   PHI   RU   RC
1,130,730,12,0,0
2,115,900,0,0,0
3,135,0,45,0,0
999,,,,,,

NO.  GAMMA   CBAR     PHI     RU   RC

 1  130.00  730.00  12.000   0.00 0.00
 2  115.00  900.00   0.000   0.00 0.00
 3  135.00    0.00  45.000   0.00 0.00
SOIL DATA ARE LISTED ABOVE. ARE THEY ALL CORRECT....
Y

DO YOU WISH TO USE PROBABILISTIC LOGIC?
Y

TYPE IN LENGTH OF EMBANKMENT.
750

TOTAL LENGTH OF EMBANKMENT =      750.00

DO YOU WISH TO INCLUDE EFFECTS OF MODEL ERROR?
N
```

```
TYPE IN STANDARD DEVIATIONS OF STRENGTH PARAMETERS FOR
EACH SOILTYPE. TYPE SOIL NUMBER, COHESION STANDARD DEVIATION,
AND TANPHI STANDARD DEVIATION. TYPE 99 AS LAST NUMBER IN SOIL
COLUMN.
 SOIL   CSD   TPSD
1,200,0035
2,375,0
999,,,,

 SOIL      CSD        TPSD

   1      200.000    0.0350
   2      375.000    0.0000
   3        0.000    0.0000
STANDARD DEVIATIONS FOR SOIL STRENGTH COMPONENTS ARE LISTED ABOVE
ARE THEY CORRECT?
Y

 TYPE IN THE CODE TO SPECIFY THE TYPE OF AUTOCORRELATION
 STRUCTURE TO BE USED IN VARIANCE REDUCTION ANALYSIS:

  1    NO VARIANCE REDUCTION
  2    EMPIRICAL AUTOCORRELATION (SET OF DATA POINTS SPECIFYING
       THE AUTOCORRELATION FUNCTION FROM DATA ANALYSIS.)
  3    THEORETICAL AUTOCORRELATION FUNCTION (EXPONENTIAL WITH
       ARGUMENT SQUARED, AND DEFINED BY A CONSTANT IN THE ARGUMENT)

  TYPE THE CODE IN FOR EACH SOILTYPE - COHESION: X,Y,Z AND
  TANPHI: X,Y,Z. TYPE SOIL NUMBER, 3 COHESION CODES, AND
  3 TANPHI CODES. TYPE 99,0,0,0,0,0,0 AFTER LAST SOILTYPE

SOIL CX CY CZ TX TY TZ
1,3,3,3,3,3,3
2,3,3,3,0,0,0
999,,,,,,,

SOIL CX CY CZ TX TY TZ

  1   3  3  3  3  3  3
  2   3  3  3  0  0  0
  3   0  0  0  0  0  0
AUTOCORRELATION CODES ARE LISTED ABOVE,
ARE THEY CORRECT?
Y

 TYPE IN AUTOCORRELATION DATA CORRESPONDING TO PREVIOUS
 INPUTS FOR EACH SOILTYPE. FOR EMPIRICAL DEFINITIONS TYPE IN
 NUMBER OF DATA POINTS, AND THE DATA SETS: AUTOCORRELATION
 VALUE AND LAG DISTANCE FOR EACH POINT.
 FOR EACH EMPIRICAL FUNCTION MAKE SURE THAT THE LAST
 TWO POINTS ARE SPECIFIED SO THAT ANY VALUES EXTRAPOLATED FROM
 THEM YIELDS ACCEPTABLE AUTOCORRELATION VALUES. USE 20 PTS OR
 LESS PER FUNCTION.
 FOR THEORETICAL DEFINITIONS TYPE IN ONLY THE AUTOCORRELATION
 CONSTANT DEFINING THE FUNCTION IN THE SPECIFIED DIRECTION.
 WHEN ALL ENTRIES ARE COMPLETED TYPE 99 AS SOIL NUMBER.

TYPE IN SOIL NUMBER
1

SOIL NUMBER:  1
TYPE IN AUTOCORRELATION CONSTANT FOR SOIL NUMBER  1  CX DIRECTION
70

TYPE IN AUTOCORRELATION CONSTANT FOR SOIL NUMBER  1  CY DIRECTION
20

TYPE IN AUTOCORRELATION CONSTANT FOR SOIL NUMBER  1  CZ DIRECTION
70

TYPE IN AUTOCORRELATION CONSTANT FOR SOIL NUMBER  1  TX DIRECTION
70

TYPE IN AUTOCORRELATION CONSTANT FOR SOIL NUMBER  1  TY DIRECTION
20

TYPE IN AUTOCORRELATION CONSTANT FOR SOIL NUMBER  1  TZ DIRECTION
70

TYPE IN SOIL NUMBER
2
```

Theoretical autocorrelations are used

```
SOIL NUMBER:  2
TYPE IN AUTOCORRELATION CONSTANT FOR SOIL NUMBER  2  CX DIFECTION
100

TYPE IN AUTOCORRELATION CONSTANT FOR SOIL NUMBER  2  CY DIRECTION
10

TYPE IN AUTOCORRELATION CONSTANT FOR SOIL NUMBER  2  CZ DIRECTION
100

TYPE IN SOIL NUMBER          Information for all soil types has been input
99

SOIL      CX         CY         CZ         TX         TY         TZ

  1      70.00      20.00      70.00      70.00      20.00      70.00
  2     100.00      10.00     100.00       0.00       0.00       0.00
  3       0.00       0.00       0.00       0.00       0.00       0.00
THE AUTOCORRELATION  CONSTANTS ARE  PRINTED ABOVE, ARE THEY CORRECT?
Y

TYPE IN PIEZOMETRIC DATA:
TYPE IN PIEZOMETRIC SURFACE NO., POINT NO.,X AND Y COORDS ON SAME LINE.
USE PIEZOMETRIC SURFACE NO. 1 ONLY AS PHREATIC SURFACE.
USE 10 SURFACES OR LESS AND 10 POINTS OR LESS PER SURFACE.
TYPE 999,0,0,0 AFTER LAST SURFACE COORDINATE.
PIEZ.   NO.   X    Y
999,,,,

THE FOLLOWING IS A PRINTOUT OF THE LINE ARRAY.  THE INITIAL  3 LINES
MUST BE THE SURFACE OF THE SLOPE GOING FROM LEFT TO RIGHT. THERE MUST
BE NO VERTICAL LINES AFTER NO.  3.
NO.  X-LEFT    Y-LEFT    X-RGHT    Y-RGHT     SLOPE   SOIL
 1   100.00    120.00    150.00    120.00    0.0000   2
 2   150.00    120.00    195.00    150.00    0.6667   1
 3   195.00    150.00    240.00    150.00    0.0000   1
 4   100.00    100.00    240.00    100.00    0.0000   3
 5   150.00    120.00    240.00    120.00    0.0000   2
TYPE A NUMBER 100 OR LESS. HOW MANY SLICES....
100

TYPE THE LOWEST ELEVATION THAT SHOULD OCCUR ALONG ANY
TRIAL FAILURE SURFACE (YMIN).
100

TYPE THE MINIMUM VALUE FOR THE GREATEST DEPTH OF THE
SLIDING MASS (DMIN).
0

DO YOU WANT FS OUTPUT FOR EACH TRIAL X,Y,R....
Y

TYPE A NUMBER.
 1  TO USE THE AUTOMATIC SEARCH  SUBROUTINE
 2  TO COMPUTE FS ACROSS A PRESCRIBED CONTRL GRID.
 3  TO COMPUTE FS AT A SINGLE TRIAL CENTER AND RADIUS.
3

TYPE THE X AND Y LOCATION FOR THE TRIAL CENTER.
170,160

TYPE IN DESIRED RADIUS.
59

CENTER OF ROTATION AT X =  170.00  Y =  160.00
RADIUS =    59.00
NUMBER OF SLICES USED =   37
PLANE STRAIN SAFETY FACTOR BY BISHOP'S METHOD =    1.35
PLANE STRAIN SAFETY FACTOR BY FELLINIUS' METHOD =   1.34
PROBABLE WIDTH OF FAILURE, BC =   121.096
THREE DIMENSIONAL SAFETY FACTOR FOR WIDTH BC =   1.705
STANDARD DEVIATION OF SAFETY FACTOR, BC =  .297E+00
RELIABILITY INDEX  (BETA - BC) =    2.373

****************************************************************

PROBABILITY OF FAILURE FOR EMBANKMENT LENGTH BC    =  .882E-02
PROBABILITY OF FAILURE FOR FULL EMBANKMENT LENGTH =  .831E-01

****************************************************************

DO YOU WISH TO SPECIFY ANOTHER RADIUS?
N
```

```
DO YOU WISH TO USE A DIFFERENT TRIAL CENTER?
N

CALCULATIONS HAVE BEEN COMPLETED.  TYPE A NUMBER,
1  TO REWORK WITH A DIFFERENT LOCATION FOR THE FS GRID
2  TO REWORK USING A DIFFERENT CENTER FOR THE INITIAL TRIAL CENTER
3  TO REWORK AT A DIFFERENT TRIAL CENTER AND RADIUS (DO ONLY)
4  TO CHANGE MEAN SOIL PROPERTIES (SAME SLOPE GEOMETRY)
5  TO CHANGE PIEZOMETRIC SURFACE(S)
6  TO CHANGE BOTH MEAN SOIL PROPERTIES AND PIEZOMETRIC SURFACES
7  TO CHANGE PROBABILISTIC SOIL PROPERTIES
8  TO CHANGE MEAN AND PROBABILISTIC SOIL PROPERTIES.
9  TO CHANGE MEAN AND PROBABILISTIC SOIL PROPERTIES AND PIEZOMETRIC SURFACES.
10 TO CHANGE PROBABILISTIC SOIL PROPERTIES AND PIEZOMETRIC SURFACES
11 TO CHANGE SLOPE GEOMETRY(WHOLE NEW PROBLEM)
12 TO TERMINATE
7

DO YOU WISH TO USE PROBABILISTIC LOGIC?
Y

TYPE IN LENGTH OF EMBANKMENT.
750

TOTAL LENGTH OF EMBANKMENT =        750.00

DO YOU WISH TO INCLUDE EFFECTS OF MODEL ERROR?
Y

TYPE IN MODEL ERROR IN PERCENT
5
```

Model error is now included

```
TYPE IN STANDARD DEVIATIONS OF STRENGTH PARAMETERS FOR
EACH SOILTYPE. TYPE SOIL NUMBER, COHESION STANDARD DEVIATION,
AND TANPHI STANDARD DEVIATION. TYPE 99 AS LAST NUMBER IN SOIL
COLUMN.
 SOIL   CSD   TPSD
999,,,
```

Standard deviations are not changed from previous inputs

```
 SOIL     CSD       TPSD

   1     200.000    0.0350
   2     375.000    0.0000
   3       0.000    0.0000
STANDARD DEVIATIONS FOR SOIL STRENGTH COMPONENTS ARE LISTED ABOVE
ARE THEY CORRECT?
Y

 TYPE IN THE CODE TO SPECIFY THE TYPE OF AUTOCORRELATION
 STRUCTURE TO BE USED IN VARIANCE REDUCTION ANALYSIS:

  1   NO VARIANCE REDUCTION
  2   EMPIRICAL AUTOCORRELATION (SET OF DATA POINTS SPECIFYING
      THE AUTOCORRELATION FUNCTION FROM DATA ANALYSIS.)
  3   THEORETICAL AUTOCORRELATION FUNCTION (EXPONENTIAL WITH
      ARGUMENT SQUARED, AND DEFINED BY A CONSTANT IN THE ARGUMENT)

  TYPE THE CODE IN FOR EACH SOILTYPE - COHESION: X,Y,Z AND
  TANPHI: X,Y,Z. TYPE SOIL NUMBER, 3 COHESION CODES, AND
  3 TANPHI CODES. TYPE 99,0,0,0,0,0,0 AFTER LAST SOILTYPE

 SOIL CX CY CZ TX TY TZ
 999,,,,,,,
```

Type of autocorrelation definition is not changed from previous input

```
 SOIL CX CY CZ TX TY TZ

  1   3  3  3  3  3  3
  2   3  3  3  0  0  0
  3   0  0  0  0  0  0
 AUTOCORRELATION CODES ARE LISTED ABOVE,
 ARE THEY CORRECT?
 Y

  TYPE IN AUTOCORRELATION DATA CORRESPONDING TO PREVIOUS
  INPUTS FOR EACH SOILTYPE. FOR EMPIRICAL DEFINITIONS TYPE IN
  NUMBER OF DATA POINTS, AND THE DATA SETS: AUTOCORRELATION
  VALUE AND LAG DISTANCE FOR EACH POINT.
  FOR EACH EMPIRICAL FUNCTION MAKE SURE THAT THE LAST
  TWO POINTS ARE SPECIFIED SO THAT ANY VALUES EXTRAPOLATED FROM
  THEM YIELDS ACCEPTABLE AUTOCORRELATION VALUES. USE 20 PTS OR
  LESS PER FUNCTION.
  FOR THEORETICAL DEFINITIONS TYPE IN ONLY THE AUTOCORRELATION
  CONSTANT DEFINING THE FUNCTION IN THE SPECIFIED DIRECTION.
  WHEN ALL ENTRIES ARE COMPLETED TYPE 99 AS SOIL NUMBER.
```

```
TYPE IN SOIL NUMBER                 Autocorrelation constants not changed
99                                  from previous input

SOIL      CX          CY          CZ          TX          TY          TZ

  1       70.00       20.00       70.00       70.00       20.00       70.00
  2      100.00       10.00      100.00        0.00        0.00        0.00
  3        0.00        0.00        0.00        0.00        0.00        0.00
THE AUTOCORRELATION  CONSTANTS ARE  PRINTED ABOVE, ARE THEY CORRECT?
Y

TYPE THE LOWEST ELEVATION THAT SHOULD OCCUR ALONG ANY
TRIAL FAILURE SURFACE (YMIN).
100

TYPE THE MINIMUM VALUE FOR THE GREATEST DEPTH OF THE
SLIDING MASS (DMIN).
0

DO YOU WANT FS OUTPUT FOR EACH TRIAL X,Y,R....
Y

TYPE A NUMBER.
 1  TO USE THE AUTOMATIC SEARCH  SUBROUTINE
 2  TO COMPUTE FS ACROSS A PRESCRIBED CONTRL GRID.
 3  TO COMPUTE FS AT A SINGLE TRIAL CENTER AND RADIUS.
3

TYPE THE X AND Y LOCATION FOR THE TRIAL CENTER.
170,160

TYPE IN DESIRED RADIUS.
59

CENTER OF ROTATION AT X =  170.00   Y =  160.00
RADIUS =    59.00
NUMBER OF SLICES USED =    37
PLANE STRAIN SAFETY FACTOR BY BISHOP'S METHOD =    1.35
PLANE STRAIN SAFETY FACTOR BY FELLINIUS' METHOD =    1.34
PROBABLE WIDTH OF FAILURE, BC =    121.096
THREE DIMENSIONAL SAFETY FACTOR FOR WIDTH BC =    1.705
STANDARD DEVIATION OF SAFETY FACTOR, BC =   .309E+00
RELIABILITY INDEX   (BETA - BC) =     2.281

***************************************************************

PROBABILITY OF FAILURE FOR EMBANKMENT LENGTH BC     =  .113E-01
PROBABILITY OF FAILURE FOR FULL EMBANKMENT LENGTH =  .102E+00

***************************************************************

DO YOU WISH TO SPECIFY ANOTHER RADIUS?
N

DO YOU WISH TO USE A DIFFERENT TRIAL CENTER?
N

CALCULATIONS HAVE BEEN COMPLETED.  TYPE A NUMBER,
1   TO REWORK WITH A DIFFERENT LOCATION FOR THE FS GRID
2   TO REWORK USING A DIFFERENT CENTER FOR THE INITIAL TRIAL CENTER
3   TO REWORK AT A DIFFERENT TRIAL CENTER AND RADIUS (DO ONLY)
4   TO CHANGE MEAN SOIL PROPERTIES (SAME SLOPE GEOMETRY)
5   TO CHANGE PIEZOMETRIC SURFACE(S)
6   TO CHANGE BOTH MEAN SOIL PROPERTIES AND PIEZOMETRIC SURFACES
7   TO CHANGE PROBABILISTIC SOIL PROPERTIES
8   TO CHANGE MEAN AND PROBABILISTIC SOIL PROPERTIES.
9   TO CHANGE MEAN AND PROBABILISTIC SOIL PROPERTIES AND PIEZOMETRIC SURFACES.
10  TO CHANGE PROBABILISTIC SOIL PROPERTIES AND PIEZOMETRIC SURFACES
11  TO CHANGE SLOPE GEOMETRY(WHOLE NEW PROBLEM)
12  TO TERMINATE
12

#ET=14:47.4 PT=4.5 IO=1.2
```

Computer and Physical Modelling in Geotechnical Engineering, Balasubramaniam et al. (eds)
© 1989 Balkema, Rotterdam. ISBN 90 6191 864 2

Back analysis of average strength parameters for critical slip surfaces

Takuo Yamagami & Yasuhiro Ueta
University of Tokushima, Tokushima, Japan

ABSTRACT: This paper is concerned with a simple technique for the back calculation of strength parameters c and ϕ along a given critical slip surface. The method provides unique and reliable solutions because it fulfills the condition that the given slip surface must have a minimum factor of safety as well as satisfying the so called c-tanϕ relationship. In this approach, however, slopes need to be assumed homogeneous. Therefore, the strength parameters obtained should be regarded as average for the critical slip surface.

Since somewhat different treatments are necessary according to the factor of safety equation employed, detailed procedure is given to the back analysis for each of the ordinary method of slices, the Bishop method, and the simplified Janbu method, together with example problems.

1 INTRODUCTION

A technique has been developed for the back analysis of average strength parameters along the given critical slip surface in a homogeneous slope. Back calculations of the strength parameters of soils in a failed slope are often required in landslide control works. Some techniques therefore have already been presented for the back analysis of slope failures. However, much of the previous work does not meet both the essential requirements (see next section) which any satisfactory back analysis method should allow for. This shortcoming will lead to the multiplicity of solutions and/or unreliable results. To the best of the authors' knowledge, only Saito (1980) has proposed a back analysis procedure satisfying the essential requirements; it is, thus, considered to be the most logical of all the conventional methods. His mathematical formulation, however, is not so lucid because of the concept of envelopes he used. This means that it is not always applicable to every factor of safety equation.

In this paper, in contrast to Saito's method, the simple technique is given for back analysis of strength parameters; this will provide unique and reliable solutions, though slopes are assumed to be homogeneous. The proposed method needs different treatments according to the factor of safety equation employed. Therefore we shall first describe the fundamental concepts of the proposed method. These are then followed by the illustration of back calculation procedures when the ordinary method of slices, the Bishop method, and the Janbu method are employed as the factor of safety equation.

2 FUNDAMENTAL CONCEPTS

2.1 Essential requirements which back analysis method should satisfy

First of all, we assume that slopes are homogeneous and isotropic with respect to the strength parameters, c and ϕ. Obviously, natural slopes show very complicated phases and never have uniform strength parameters. However, there is an idea at the base of this study that if we represent such non-uniformly distributed strength parameters by means of a unique pair of c and ϕ, then what values should we employ?

In general, any factor of safety equation may be summarized as follows:

$$F = f(c,\phi,M,F) \tag{1}$$

where F denotes a value of the factor of safety for an assumed slip surface, M signifies the self-weight of the sliding mass, external forces acting on the surface of the ground, the effect of pore water pressure, and so forth. The above equation indicates that the safety factor function f consists of c, ϕ, M and the factor of safety itself.

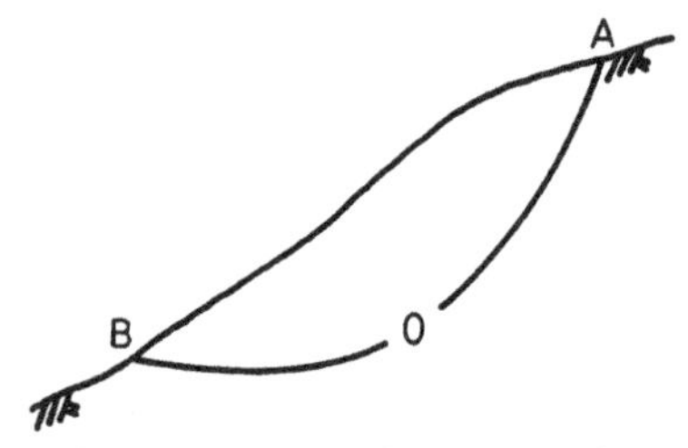

Fig.1 Given critical slip surface

Now, suppose that in Fig.1 the curve AOB is the given critical slip surface (or the failure surface in the case of a failed slope), the value of the factor of safety for the slip surface AOB being the smallest of all the possible adjacent slip surfaces. So, in order to back analyze the strength parameters we must first estimate the value of the factor of safety F_0 for the critical slip surface AOB. In the case of a landslide, F_0 is usually assigned a value of between 1.0 and 0.9 according to the situation. Once the value of F_0 has been determined, we can then write the following expression from Eq.(1):

$$F_0 = f(c,\phi,M,F_0) \qquad (2)$$

The only unknowns are c and ϕ when the above equation is applied to the given critical slip surface. Thus, we can obtain the so called c-tanϕ relationship as shown schematically in Fig.2 where c_{max}

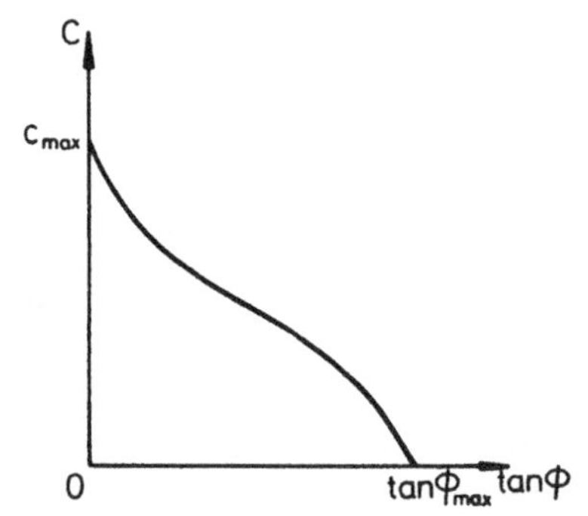

Fig.2 c-tanϕ relationship

and ϕ_{max} denote the maxima of all the respective values of c and ϕ that satisfy Eq.(2). In other words, the strength parameters c_0 and ϕ_0 to be back analyzed must lie somewhere on the c-tanϕ relationship.

To summarize the above discussion, any satisfactory back analysis method should simultaneously meet the following two essential requirements:

a) The strength parameters c_0 and ϕ_0 to be obtained must satisfy the c-tanϕ relationship.

b) The factor of safety F_0 for the critical slip surface is the minimum value of the slope under consideration.

As far as the authors are aware, none of the previous back analysis methods have successfully incorporated these two essential points, thereby leading to the multiplicity of solutions and/or unreliable results as mentioned before.

2.2 The substance of the proposed method (Yamagami and Ueta 1984a)

The central idea of the present method which meets both the essential requirements above is described herein on the basis of Figs.3 and 4. Fig.3 represents a three dimensional space with the factor of safety and the strength parameters. The curve PIQ in Fig.3 schematically illustrates a possible existing range of c and tanϕ corresponding to the factor of safety F_0 for the critical slip surface AOB given in Fig.4. The strength parameters c_0 and tanϕ_0 to be back analyzed should therefore bear on the curve PIQ (see Requirement a) above). Fig.4 shows the critical slip surface and an appropriate number of trial slip surfaces chosen above and below it. Here we will develop the back analysis procedure fixing the end points A and B of the slip surfaces for computational convenience.

Let the factor of safety F be expressed as follows for a trial slip surface other than the critical one:

$$F = f(c,\phi,\underline{M},F) \qquad (3)$$

where the underlined notation $\underline{M}$ indicates the amount evaluated from the trial slip surface. It is anticipated that the distribution of F becomes a curve such as RIS in Fig.3 when c and tanϕ in Eq.(3) are varied along the curve PIQ, i.e. varied such that Requirement a) above is satisfied. Fig.5 illustrates this idea in two dimensional F-c and F-tanϕ co-ordinate systems. However, according to Requirement b) we must always satisfy the following inequality:

$$F \geqq F_0 \tag{4}$$

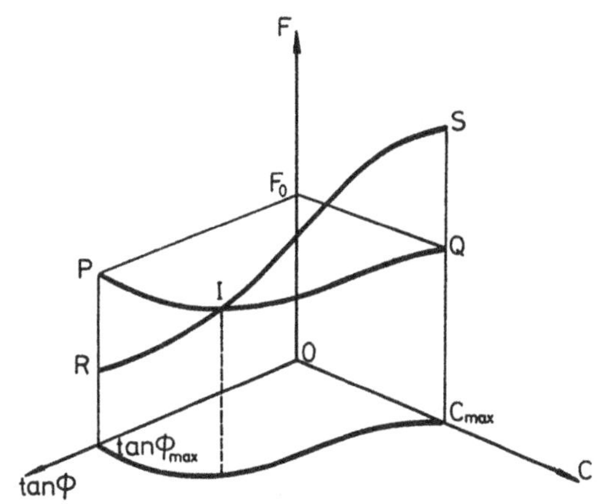

Fig.3 (F,c,tanϕ) three-dimensional space

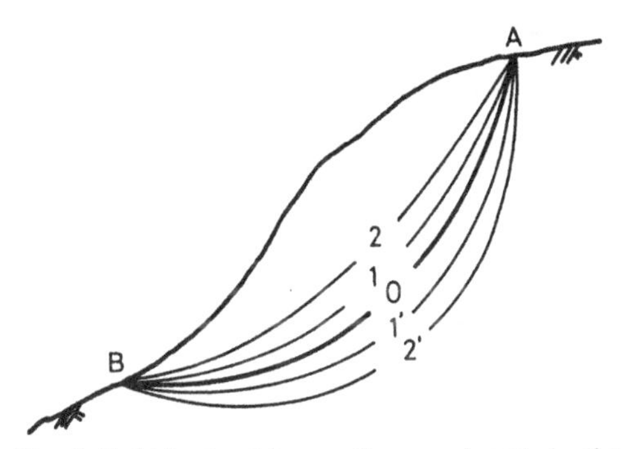

Fig.4 Critical slip surface and trial slip surfaces

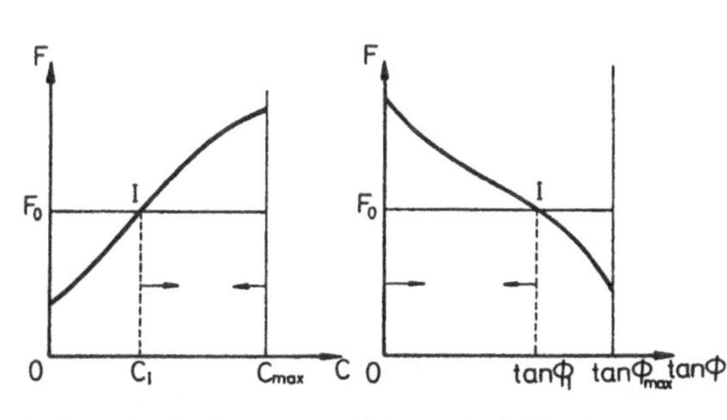

(a) restriction based on F-c relationship

(b) restriction based on F-tanϕ relationship

Fig.5 Restriction of ranges within which the strength parameters should exist

The meaning of the above equation is that with reference to Fig.5 the correct parameters c_0 and $\tan\phi_0$ should exist within the following ranges, respectively:

$$\left.\begin{array}{l} c_I \leqq c_0 \leqq c_{max} \\ 0 \leqq \tan\phi_0 \leqq \tan\phi_I \end{array}\right\} \tag{5}$$

If the inclination of the curve representing the distribution of F is the reverse to that of Fig.3 or 5, then the ranges within which c_0 and $\tan\phi_0$ should bear will become

$$\left.\begin{array}{l} 0 \leqq c_0 \leqq c_I \\ \tan\phi_I \leqq \tan\phi_0 \leqq \tan\phi_{max} \end{array}\right\} \tag{6}$$

In any case it is expected that from one trial slip surface, the existing ranges of c_0 and $\tan\phi_0$ will be considerably restricted by virtue of Eq.(4). Hence, ranges within which c_0 and $\tan\phi_0$ must exist will be extremely restricted by assuming several trial slip surfaces above and below the critical one adjacent to it as shown in Fig.4, and then by applying Eq.(5) or (6) to each of the slip surfaces; thereby enabling us to identify the correct strength parameters c_0 and $\tan\phi_0$.

This is the basic idea of the proposed back analysis method. Although somewhat different treatments are needed according to the factor of safety equation employed for Eq.(1), in every case it has been ascertained that the back analysis is successfully performed based on the fundamental concepts above.

3 APPLICATION TO THE ORDINARY METHOD OF SLICES (Yamagami and Ueta 1984b)

The equation corresponding to Eq.(2), i.e. the c-tanϕ relationship in the ordinary method of slices (Chowdhury 1978) is given by

$$F_0 = \frac{c\hat{L} + \tan\phi\Sigma(N-ul)}{\Sigma T} \tag{7}$$

where the notations above are sufficiently conventional that we can eliminate their explanation here. This equation represents the straight line as shown in Fig.6 or 7 on which the parameters c_0 and $\tan\phi_0$ to be back analyzed should exist.

Let the factor of safety F be expressed as follows for a trial slip surface other than the critical one in Fig.8:

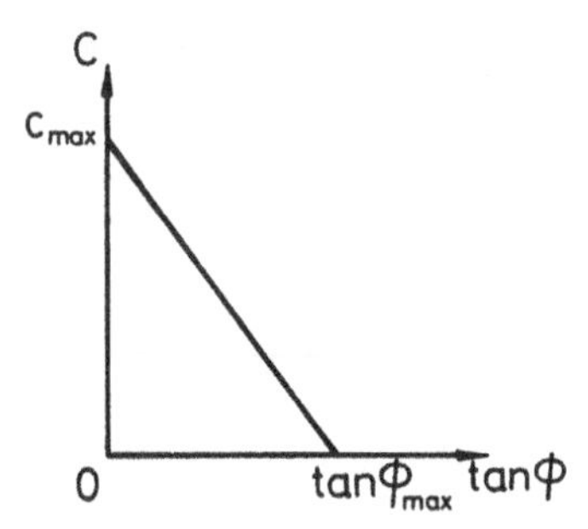

Fig.6 c-tanϕ relationship
(the ordinary method of slices)

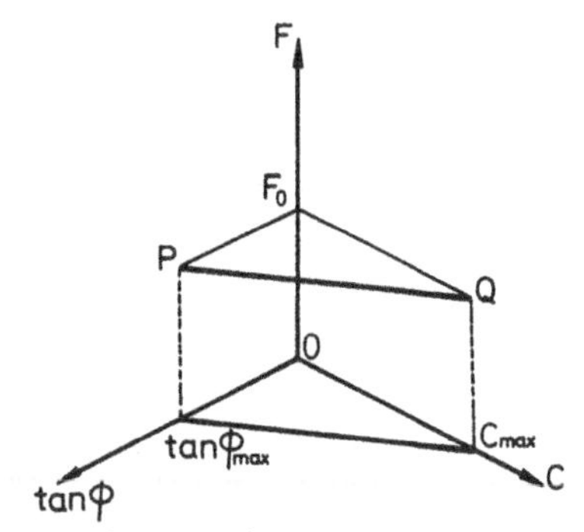

Fig.7 (F,c,tanϕ) space

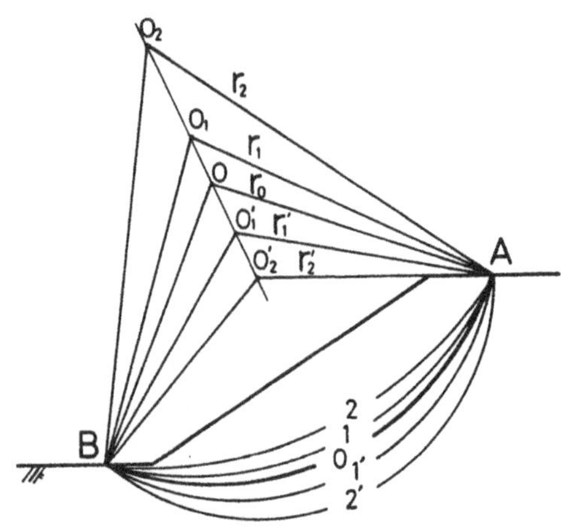

Fig.8 Critical slip circle and trial slip circles

$$F = \frac{c\underline{\hat{L}} + \tan\phi\Sigma(\underline{N}-\underline{u}\underline{l})}{\Sigma\underline{T}} \tag{8}$$

where, as stated earlier, the underlined notations denote the values obtained from a trial slip surface other than the critical one. When c and tanϕ in Eq.(8) are varied along the straight line in Fig.6 or 7, i.e. varied such that Eq.(7) is satisfied (see Requirement a)), F can be expressed as

$$F = \alpha c + \beta \qquad (0 \leqq c \leqq c_{max}) \tag{9}$$

or

$$F = \bar{\alpha}\tan\phi + \bar{\beta} \qquad (0 \leqq \tan\phi \leqq \tan\phi_{max}) \tag{10}$$

where

$$\left.\begin{aligned} \alpha &= \frac{\underline{\hat{L}} - m_0\Sigma(\underline{N}-\underline{u}\underline{l})}{\Sigma\underline{T}} \\ \beta &= \frac{\tan\phi_{max}\Sigma(\underline{N}-\underline{u}\underline{l})}{\Sigma\underline{T}} \end{aligned}\right\} \tag{11}$$

and

$$\left.\begin{aligned} \bar{\alpha} &= \frac{\Sigma(\underline{N}-\underline{u}\underline{l}) - n_0\underline{\hat{L}}}{\Sigma\underline{T}} \\ \bar{\beta} &= \frac{c_{max}\underline{\hat{L}}}{\Sigma\underline{T}} \end{aligned}\right\} \tag{12}$$

where $n_0 = c_{max}/\tan\phi_{max} = 1/m_0$.

Thus, observing Eq.(4), i.e. Requirement b), the cohesion to be obtained must satisfy the following inequalities from Eq.(9):

$$\left.\begin{aligned} c_0 &\geqq (F_0 - \beta)/\alpha \qquad \text{when } \alpha > 0 \\ c_0 &\leqq (F_0 - \beta)/\alpha \qquad \text{when } \alpha < 0 \end{aligned}\right\} \tag{13}$$

Similarly, we get the inequalities for the angle of internal friction as

$$\left.\begin{aligned} \tan\phi_0 &\geqq (F_0 - \bar{\beta})/\bar{\alpha} \qquad \text{when } \bar{\alpha} > 0 \\ \tan\phi_0 &\leqq (F_0 - \bar{\beta})/\bar{\alpha} \qquad \text{when } \bar{\alpha} < 0 \end{aligned}\right\} \tag{14}$$

Fig.9 geometrically illustrates the signification of the above description: First, as shown in Fig.8 we choose an appropriate number of trial slip circles having respectively larger and smaller radii than the radius r_0 of the critical slip circle. Then, by determining Eq.(9) or (10) for each of the trial slip circles, the existing ranges of c_0 and $\tan\phi_0$ will be greatly restricted according to a positive or negative value of α or $\bar{\alpha}$ as shown in Fig.9. As a result, we can approximately identify the correct values.

In order to perform this back analysis procedure more efficiently and systematically, it has been observed that Fig.10 should be used, the vertical axes in Fig.10 representing the radii of trial slip circles r. In general, it has been shown that $\alpha>0$ and $\bar{\alpha}<0$ for the trial slip circles having radii larger than r_0, and inversely $\alpha<0$ and $\bar{\alpha}>0$ for those having radii smaller than r_0. First we plot the values $(F_0-\beta)/\alpha$ (Fig.10(a)) or $(F_0-\bar{\beta})/\bar{\alpha}$ (Fig.10(b)), which are, respectively, the minimum (when $\alpha>0$ or $\bar{\alpha}>0$) or the maximum (when $\alpha<0$ or $\bar{\alpha}<0$) from the possible ranges for c_0 and $\tan\phi_0$ defined by Eq.(13) or (14) against the radius of the corresponding trial slip circle. A smooth curve through these points is drawn in each of Figs.10(a) and (b). The intersection of the curve and the straight line: $r=r_0$ (=const.) will thus provide a single and accurate solution.

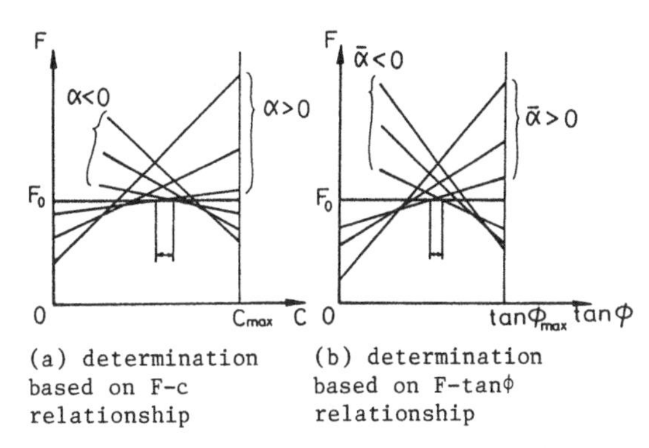

(a) determination based on F-c relationship

(b) determination based on F-tanϕ relationship

Fig.9 Final determination of existing ranges for the strength parameters

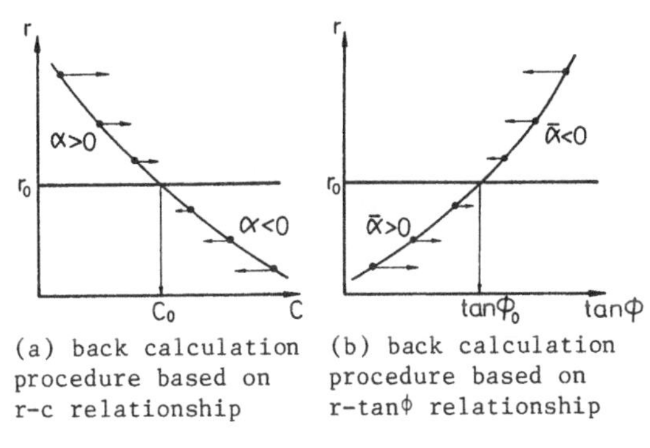

(a) back calculation procedure based on r-c relationship

(b) back calculation procedure based on r-tanϕ relationship

Fig.10 More efficient and systematic back calculation procedure

Example problem — Efficiency of the present method is illustrated here by an extremely simplified example. A slope stability analysis based on the ordinary method of slices in terms of total stresses showed that when $c=1.0 tf/m^2$ and $\phi=10°$, the slope given in Fig.11 has the minimum factor of safety F_0 of 1.282 along the critical slip circle shown in the figure. Thus, we have performed the back analysis of strength parameters, assuming that the location of the critical slip circle and the value of F_0 are known, but the strength parameters are not.

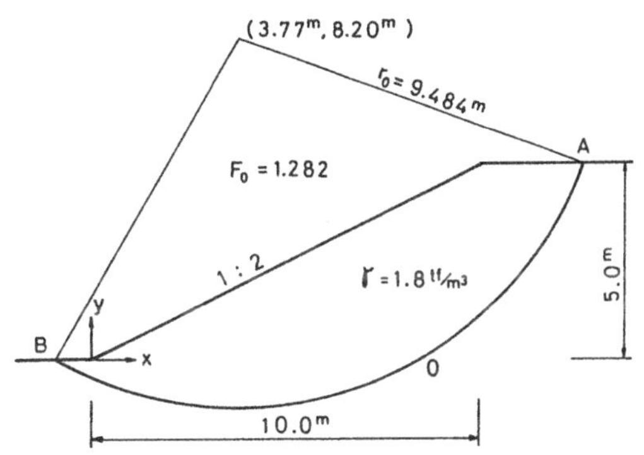

Fig.11 Example problem (total stress analysis)

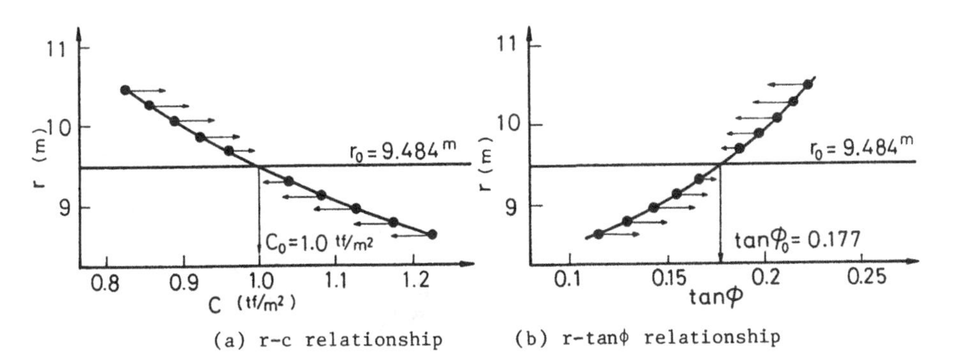

(a) r-c relationship

(b) r-tanϕ relationship

Fig.12 Results of back analysis by the efficient and systematic procedure

Fig.12 shows the computed results in terms of the efficient and systematic procedure explained with Fig.10. As indicated in Fig.12, five trial slip circles were used on each side of the critical slip circle. The back analyzed parameters were $c_0=1.0\text{tf/m}^2$ and $\phi_0=10.04°$ ($\tan\phi_0=0.177$). These results are almost perfect, thereby suggesting the efficiency of the proposed method of analysis.

4 APPLICATION TO THE BISHOP METHOD (Yamagami and Ueta 1985)

The equation indicating the c-tanϕ relationship in the simplified Bishop method (Chowdhury 1978) becomes

$$F_0 = \frac{1}{\Sigma W\sin\alpha}\,\Sigma\left\{\frac{cl\cos\alpha + (W-ul\cos\alpha)\tan\phi}{\cos\alpha + \sin\alpha\tan\phi/F_0}\right\} \qquad (15)$$

with reference to Fig.13. As these notations are also conventional, their definitions will not be elaborated upon. It is clear from Eq.(15) that the c-tanϕ relationship is not linear. Consequently the c-tanϕ curve must be plotted with many discrete points as shown in Fig.14 or 15, where

$$c_{max} = F_0\Sigma W\sin\alpha/\Sigma l \qquad (16)$$

$$\tan\phi_{max} = \frac{F_0\Sigma W\sin\alpha}{\Sigma\left\{\dfrac{W-ul\cos\alpha}{\cos\alpha + \sin\alpha\tan\phi_{max}/F_0}\right\}} \qquad (17)$$

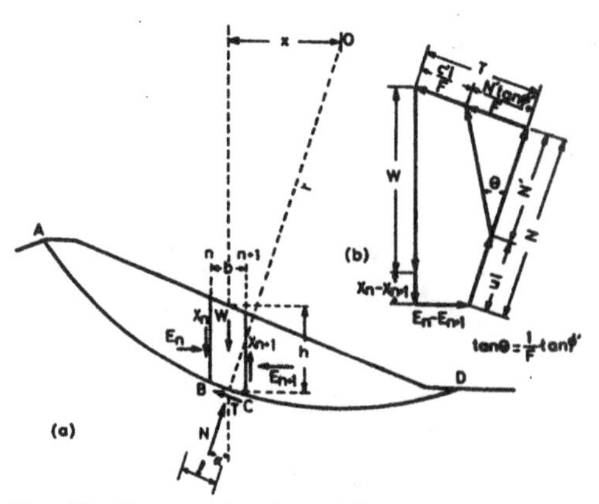

Fig.13 Forces in the Bishop method

Obviously, an iterative procedure, e.g. a Newton-Raphson procedure is necessary to obtain the value of $\tan\phi_{max}$ in Eq.(17).

Suppose that the factor of safety is given by the following expression for a trial slip circle other than the critical one in Fig.16:

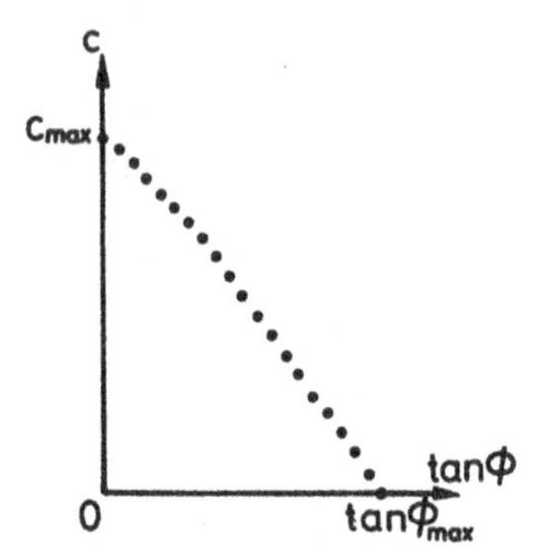

Fig.14 c-tanϕ relationship (the Bishop method)

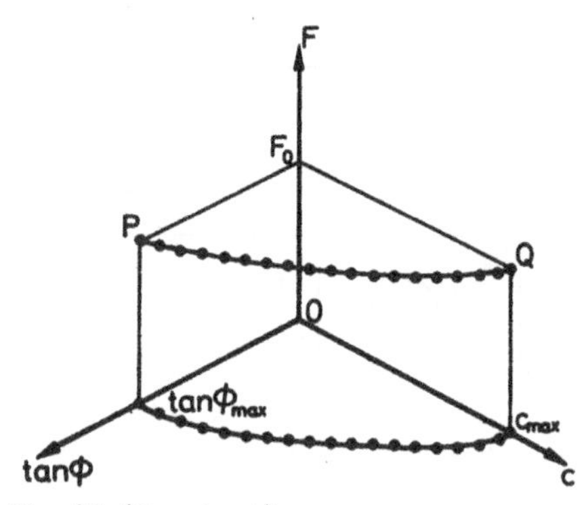

Fig.15 (F,c,tanϕ) space

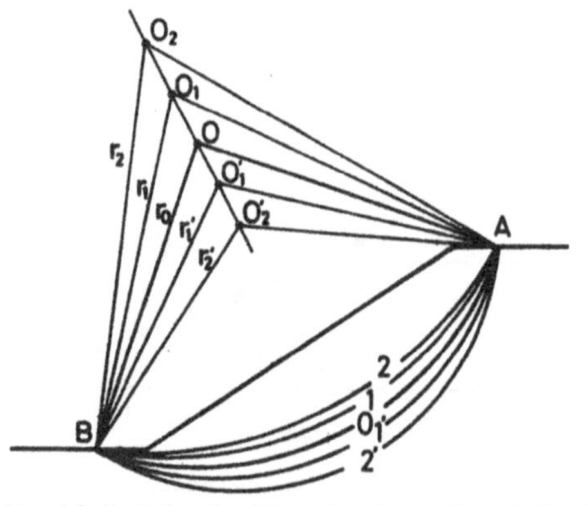

Fig.16 Critical slip circle and trial slip circles

$$F = \frac{1}{\Sigma W \sin\alpha} \Sigma \left\{ \frac{c l \cos\alpha + (W - u l \cos\alpha)\tan\phi}{\cos\alpha + \sin\alpha \tan\phi / F} \right\} \quad (18)$$

As is evident from Eqs.(15) and (18), the variation of F with c or $\tan\phi$ which satisfies the c-$\tan\phi$ relationship cannot be expressed explicitly as it can be in the ordinary method of slices. Thus, in order to get the variation of F with c or $\tan\phi$, we use F versus c or F versus $\tan\phi$ curves as shown in Fig.17; these curves are obtained by substituting the co-ordinates ($\tan\phi$, c) of many points satisfying the c-$\tan\phi$ relationship illustrated in Fig.14 or 15 into Eq.(18), and by determining the corresponding values for F. There is no evidence that the relationships in Fig.17 become linear as in the ordinary method of slices, although many example problems have shown them to be linear. Fig.17 corresponds to the aforementioned Fig.5.

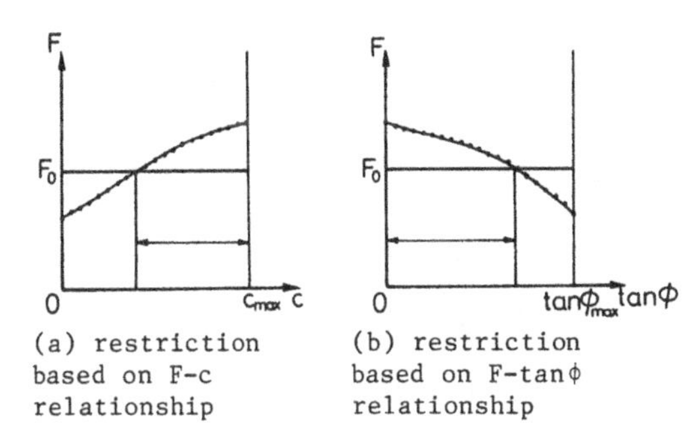

(a) restriction based on F-c relationship

(b) restriction based on F-$\tan\phi$ relationship

Fig.17 Restriction of ranges within which the strength parameters should exist

Fig.18 shows a situation in which the existing ranges of c_0 and $\tan\phi_0$ to be back analyzed have been extremely restricted by applying the above procedure to several trial slip circles.

Use of Fig.19 has been proved to be efficient and systematic to obtain a unique pair of solutions, c_0 and $\tan\phi_0$. The dots • show the relations between the radius of each trial slip circle and the maximum or minimum values of the possible ranges for c_0 and $\tan\phi_0$ corresponding to each trial slip circle. The intersection of a smooth curve through these points and the straight line: $r=r_0$(where r_0 is the radius of the critical slip circle) provides the cohesion (Fig.19(a)) or the coefficient of internal friction (Fig.19(b)). In this case, however, we must read the co-ordinates of the dots in Fig.18 in drawing Fig.19 because we cannot theoretically calculate these values.

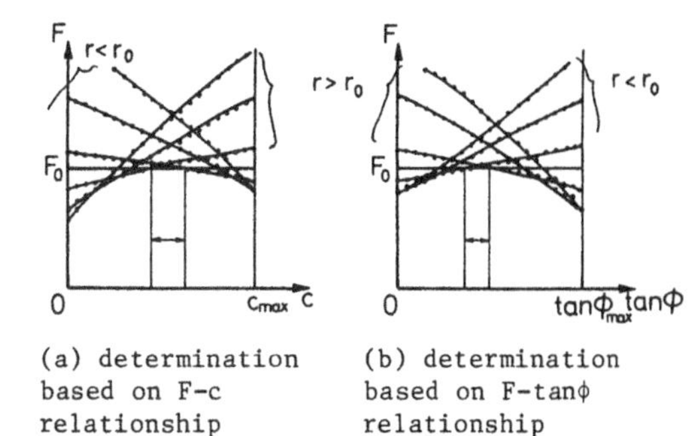

(a) determination based on F-c relationship

(b) determination based on F-$\tan\phi$ relationship

Fig.18 Final determination of existing ranges for the strength parameters

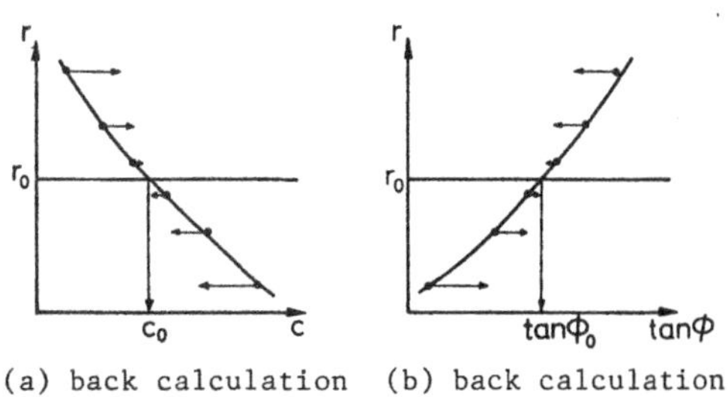

(a) back calculation procedure based on r-c relationship

(b) back calculation procedure based on r-$\tan\phi$ relationship

Fig.19 More efficient and systematic back calculation procedure

Example problems — Here we present two illustrative examples. The first one is shown in Fig.20: a slope stability analysis based on the simplified Bishop method in terms of effective stresses showed that when $c=1.0$tf/m^2 and $\phi=10°$, F_0 equals 1.525 for the critical slip circle of the submerged slope (Fig.20). Thus, we have performed the back analysis of strength parameters, assuming that the location of the critical slip circle and the value of F_0 are known, but the strength parameters are not.

Figs.21 to 23 show a series of computed results. The back analyzed parameters were $c_0=0.946$tf/m^2 and $\tan\phi_0=0.1962$ ($\phi_0=11.10°$), and these are also plotted on the c-$\tan\phi$ relationship together with the correct ones in Fig.21. The results obtained are considered to be sufficient.

Another example is a slide in a clay slope reported by Sevaldson (1956). The slope geometry and observed failure surface are shown in Fig.24. The failure surface configuration is basically circular; therefore we carried out the back analysis of strength parameters by using the simplified Bishop method with $F_0=1.0$ in terms of effective stresses.

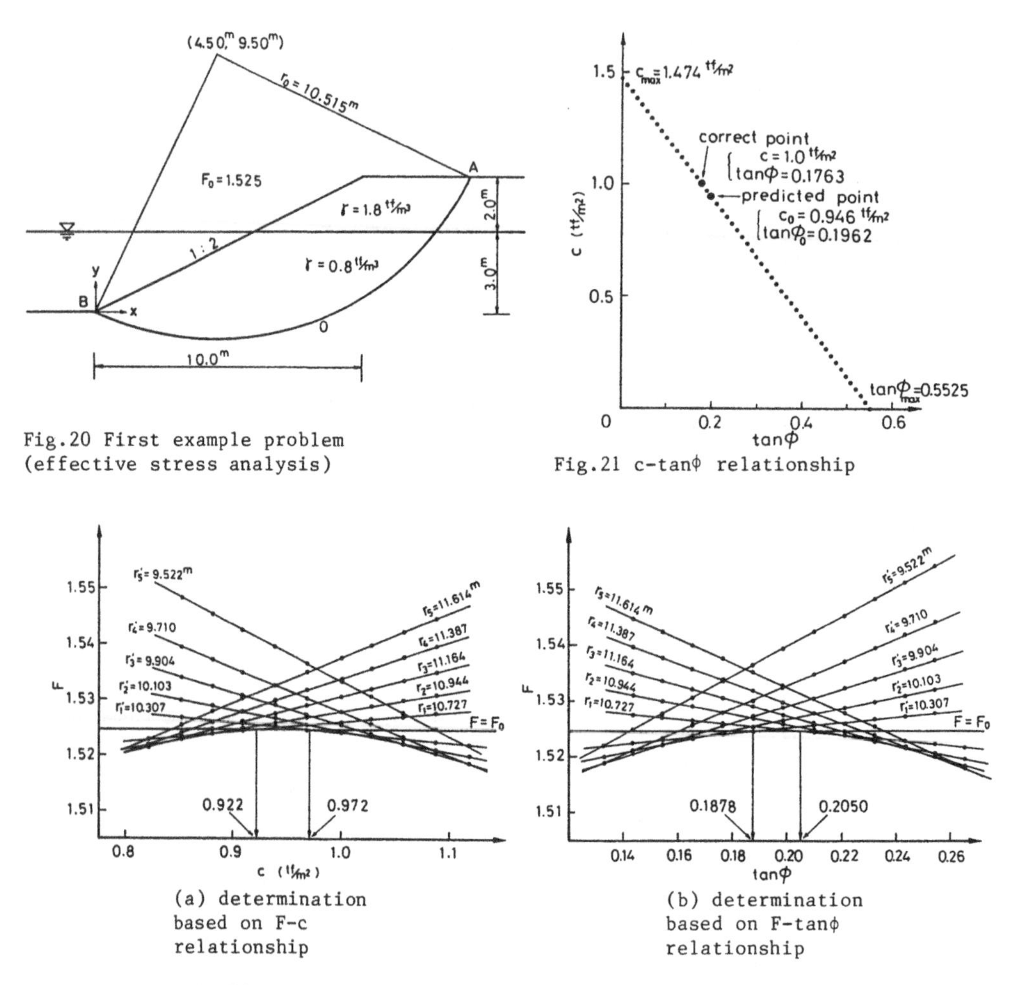

Fig.20 First example problem (effective stress analysis)

Fig.21 c-tanφ relationship

(a) determination based on F-c relationship

(b) determination based on F-tanφ relationship

Fig.22 Determination of existing ranges for the strength parameters

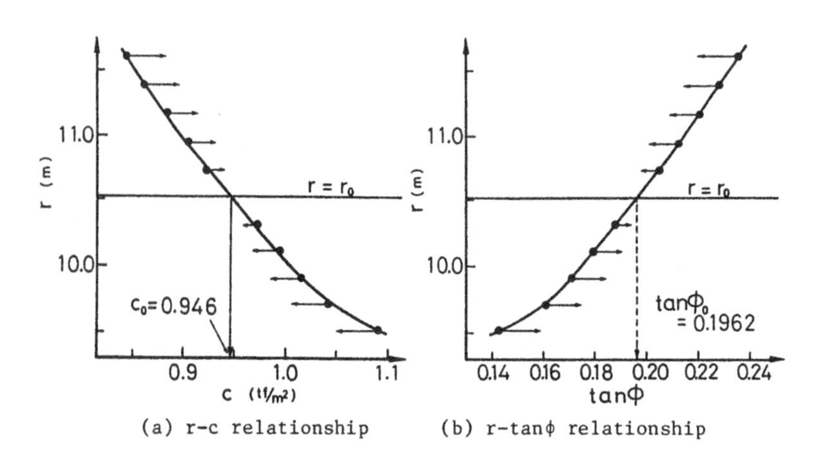

(a) r-c relationship (b) r-tanφ relationship

Fig.23 Results of back analysis by the efficient and systematic procedure

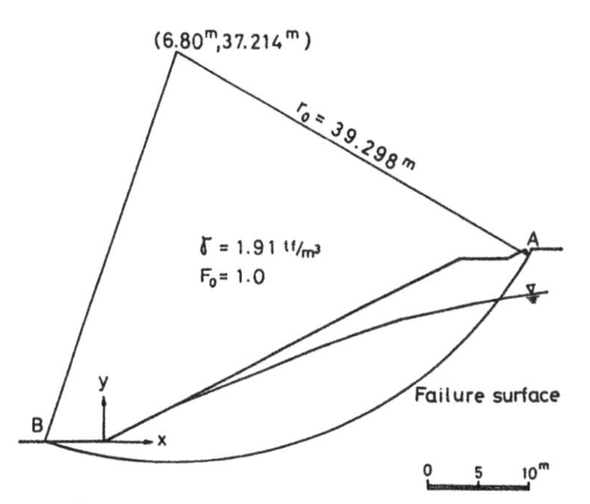

Fig.24 Second example problem
(a slide in a clay slope reported by Sevaldson(1956), effective stress analysis)

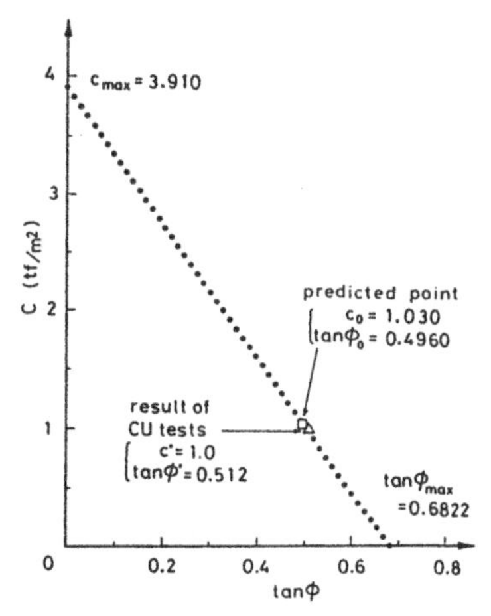

Fig.25 c-tanφ relationship

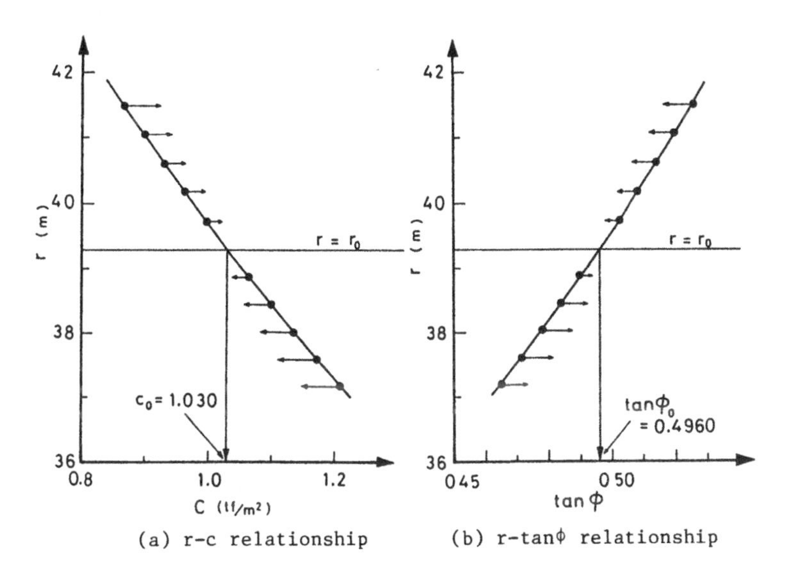

(a) r-c relationship (b) r-tanφ relationship

Fig.26 Results of back analysis by the efficient and systematic procedure

The results are shown in Figs.25 and 26. Due to lack of space greater detail cannot be given here. The strength parameters back analyzed were c_0=1.03tf/m^2 and $\tan\phi_0$=0.496 (ϕ_0=26.38°). According to Sevaldson's triaxial compression tests the average values of c' and φ' were 1.0tf/m^2 and 27.1°(tanφ'=0.512), respectively. It can thus be concluded that the back analyzed values are highly accurate.

5 APPLICATION TO THE JANBU METHOD (Yamagami and Ueta 1984c)

For simplicity, here we describe the back calculation procedure based on the simplified Janbu method. The determination of factors of safety in the rigorous Janbu method inherently involves an iteration process due to indeterminate interslice forces (Janbu 1973). In the case of homogeneous slopes, however, a simple procedure for calculating the factor of safety has been presented, omitting the interslice forces and introducing a correction factor f_0. That is, first, one computes the initial value F_0 for the factor of safety disregarding the interslice forces. Then the final approximate value is given by

$$F = f_0 F_0 \tag{19}$$

A diagram (Janbu 1973) to obtain values for f_0 shows that f_0 may vary from 1.0 up to about 1.13 depending on the curvature ratio of a slip surface and the type of soil. This means that F_0 is always a little smaller than F. If we bear this fact in mind, it is possible to use F_0 - obtained by simplified computation- for landslide control works.

With the above descriptions, the c-tanϕ relationship becomes

$$F_0 = \frac{1}{\Sigma W\tan\alpha}\Sigma\left\{\frac{cl\cos\alpha + (W-ul\cos\alpha)\tan\phi}{\cos^2\alpha(1+\tan\alpha\tan\phi/F_0)}\right\} \tag{20}$$

F_0 being again the factor of safety for the critical slip surface. Thus, the illustration of the relationship must be also based on many discrete points in the same way as the Bishop method (Fig.14 or 15).

A WAY TO DETERMINE TRIAL SLIP SURFACES — Locating trial slip surfaces adjacent to the critical one is not so easy for the Janbu method which is based on non-circular slip surfaces, unlike that for the ordinary method of slices or the Bishop method. This is because there are numerous possible slip surface configurations. Accordingly, we have made the following three assumptions here to locate the trial slip surfaces:

- A slip surface consists of a chain of line segments (see Fig.27).
- The shape of a trial slip surface is quite similar to that of the critical one.
- Trial slip surfaces are created by vertically shifting all the nodes up and down except for the end points on the critical slip surface (see Fig.28).

Under these assumptions the procedure to determine the trial slip surfaces is as follows:

First, we employ a polynomial to express a smooth curve which passes through each node on the critical slip surface:

$$y = a_1x^{n-1} + a_2x^{n-2} + \ldots + a_{n-1}x + a_n \tag{21}$$

where n is the number of nodes including the end points. Note that the polynomial

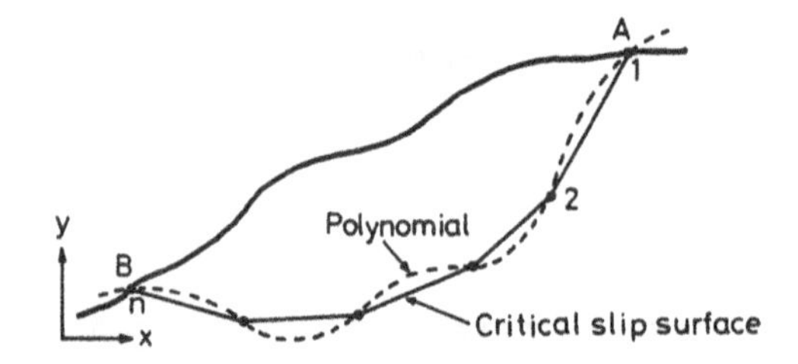

Fig.27 Representation of a curve which passes through the nodes on the critical slip surface by means of a polynomial

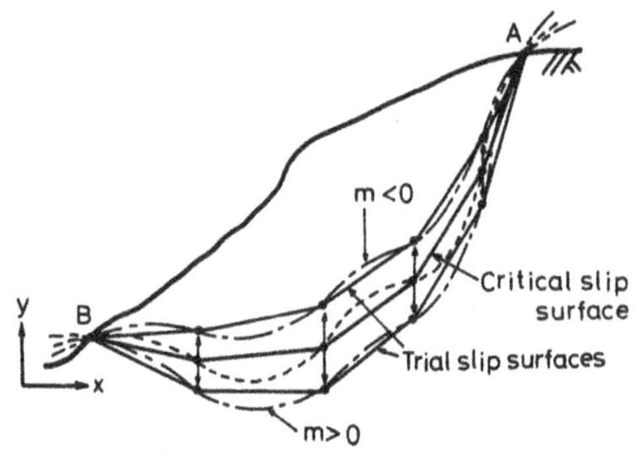

Fig.28 Determination of trial slip surfaces

itself does not express the shape of the critical slip surface as will be briefly mentioned later. Substituting the co-ordinates (x, y) of each node into Eq.(21) and solving the resulting simultaneous equations will yield the values for the coefficients $a_j (j=1 \sim n)$. Next, we can obtain a curve which passes through the end points A and B, and is similar in shape to Eq.(21), by the following equation:

$$y = (a_1 - \frac{1}{m})x^{n-1} + (a_2 - \frac{1}{m})x^{n-2} + \ldots + (a_{n-2} - \frac{1}{m})x^2 + b_1x + b_2 \tag{22}$$

where m is an arbitrary constant($m \neq 0$), and the coefficients b_1 and b_2 are those determined by the condition that the curve given by Eq.(22) passes through the end points A and B. Eq.(22) coincides with Eq.(21) when $m \to \pm\infty$. Given an appropriate value for m, the above equation will produce a curve similar in shape and in close proximity to the original one, Eq.(21), as shown schematically in Fig.28. This curve is located below the original one when m>0 and, conversely, above when m<0. The intersections of the curve determined in this way and the vertical lines passing through each node on the original curve then become the nodes on a trial slip surface. Connecting these n nodes, including the end points, with line segments, we finally obtain one trial slip surface. In this case, a necessary number of trial slip surfaces may be created with some different values for m. Note that the function of Eqs.(21) and (22) is not to express the shapes of slip surfaces. These polynomials pass through each node, but are generally so varied that we cannot employ them as slip surface configurations. It should be emphasized that these polynomials only play an auxiliary role in creating trial slip surfaces.

An iterative procedure used to determine the values for m is briefly introduced here. For convenience, we shall respectively locate the same number of trial slip surfaces above and below the critical one by adopting several pairs of m, where the values of m in each pair are of the same magnitude as their opposite signs. Eq.(22) indicates that the greater the absolute value of m, the nearer the curve given by this equation approaches the original one, Eq.(21). Thus, we first determine the absolute minimum of all the values for m to be adopted, i.e. the values for m corresponding to the outermost trial slip surfaces. To do this, we locate the nodes on the corresponding slip surface in terms of the aforementioned procedure with an appropriately assumed value for m. Subsequently, we calculate the vertical distances between each of the newly located nodes and the corresponding one on the critical slip surface. This is followed by the comparison of the calculated distances with a prescribed tolerance which has been introduced so that trial slip surfaces exist adjacent to the critical one. Values of the prescribed tolerance in our experience were in the order of 2.0 or 3.0 meters. At this stage, if even a vertical distance is larger than the prescribed tolerance, we then increase the present value of m ten times and repeat the above process by using this new value for m. When the convergence is reached the value of m becomes the final, correct value for the outermost trial slip surface. After this, the required number of m's is determined by increasing the absolute values for m two consecutive times.

Now, assume that the factor of safety F is expressed as follows for an arbitrary trial slip surface:

$$F = \frac{1}{\Sigma \underline{W}\tan\underline{\alpha}} \Sigma\left\{\frac{c\underline{l}\cos\underline{\alpha} + (\underline{W}-\underline{u}\underline{l}\cos\underline{\alpha})\tan\phi}{\cos^2\underline{\alpha}(1+\tan\underline{\alpha}\tan\phi/F)}\right\} \quad (23)$$

In this case as well, a similar procedure to that of the Bishop method is needed to examine the variations of F with c or tanϕ which satisfy the c-tanϕ relationship, Eq.(20). Hence, the results lead to the schematic diagrams shown in Fig.29.

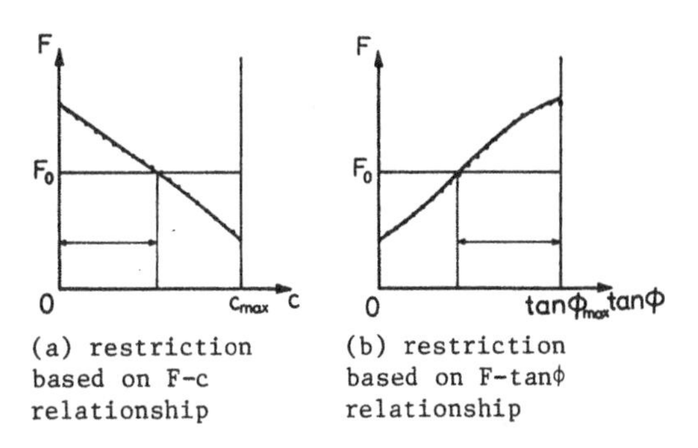

(a) restriction based on F-c relationship
(b) restriction based on F-tanϕ relationship

Fig.29 Restriction of ranges within which the strength parameters should exist

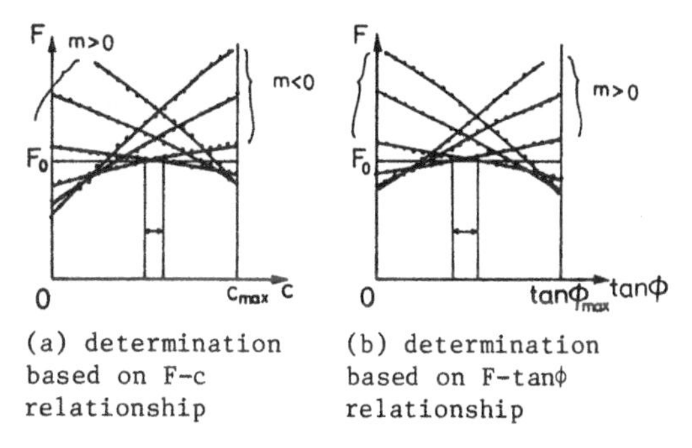

(a) determination based on F-c relationship
(b) determination based on F-tanϕ relationship

Fig.30 Final determination of existing ranges for the strength parameters

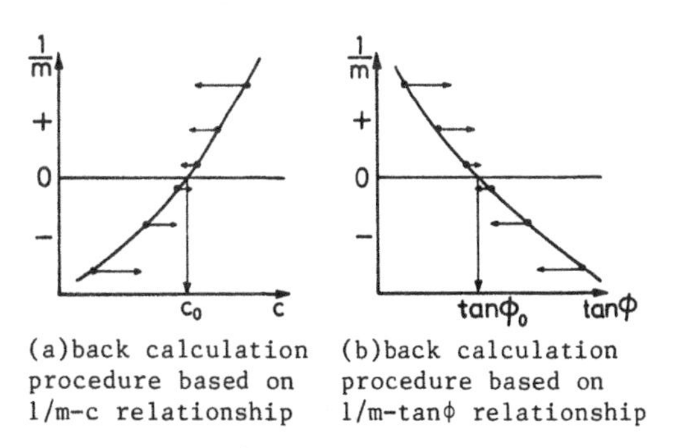

(a)back calculation procedure based on 1/m-c relationship
(b)back calculation procedure based on 1/m-tanϕ relationship

Fig.31 More efficient and systematic back calculation procedure

Fig.30 schematically illustrates a situation in which the existing ranges of the strength parameters to be obtained have been restricted by applying the procedure shown in Fig.29 to several trial slip surfaces adjacent to the critical one.

A more efficient and systematic approach for determining a unique pair of solutions is to utilize Fig.31. The dots • indicate the relationships between the reciprocal of m and the maximum or minimum values of the possible ranges for c_0 and $\tan\phi_0$ corresponding to each trial slip surface (see Fig.30). These maximum or minimum values should be read from Fig.30. Consequently, the abscissa of the intersection of a smooth curve connecting the dots and the straight line: $1/m=0$ will yield the solution as shown in Fig.31.

Example problems — The first example (see Fig.32) considers a simple slope involving homogeneous soil with zero pore water pressure. A computer program (Yamagami and Ueta 1986) based on the simplified Janbu method showed that when $c=4.25\mathrm{tf/m^2}$ and $\phi=15°$, the slope in Fig.32 has the minimum factor of safety F_0 of 1.263 along the critical slip surface. Thus, as with the preceding example problems, we have back analyzed strength parameters under the assumption that these strength parameters are unknown, but that other factors are known.

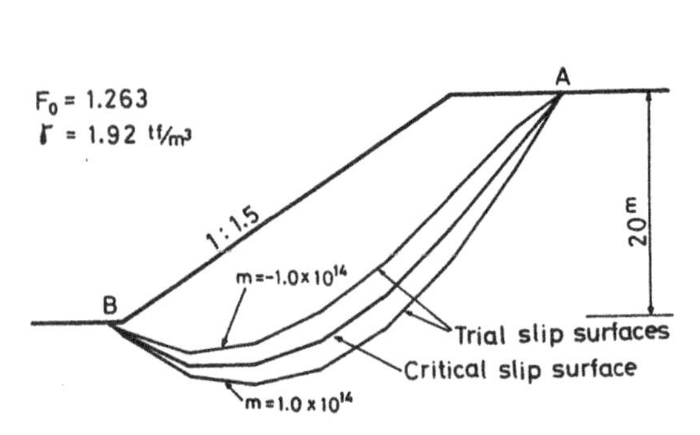

Fig.32 First example problem (given critical slip surface and the computed outermost trial slip surfaces)

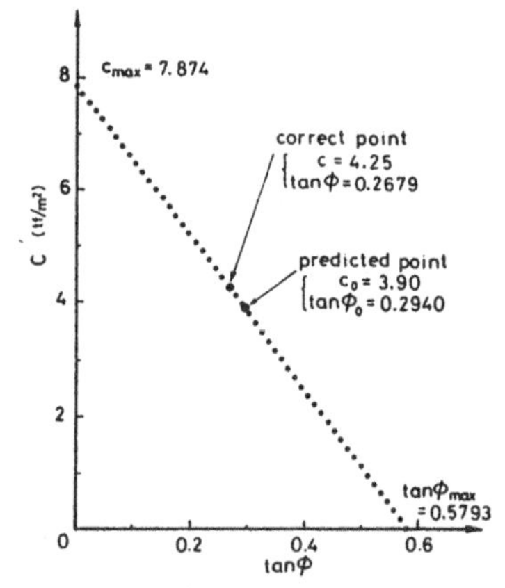

Fig.33 c–tanφ relationship

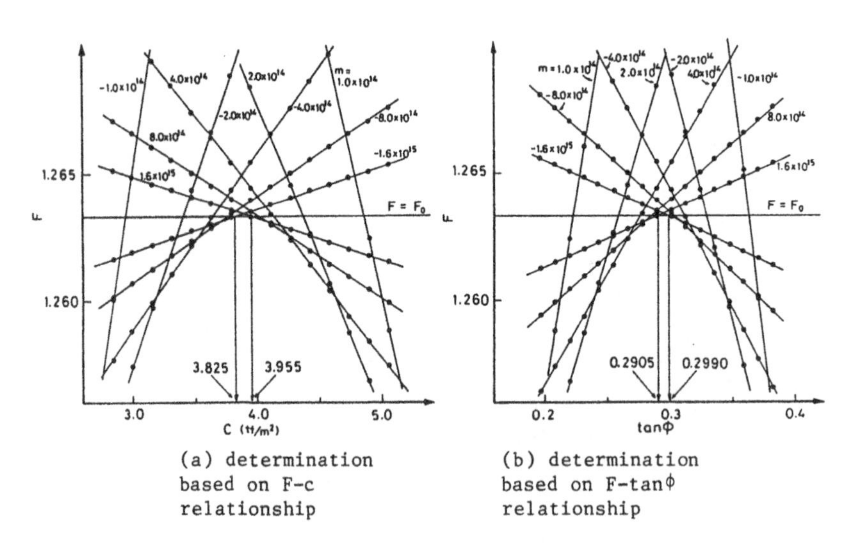

(a) determination based on F-c relationship

(b) determination based on F-tanφ relationship

Fig.34 Determination of existing ranges for the strength parameters

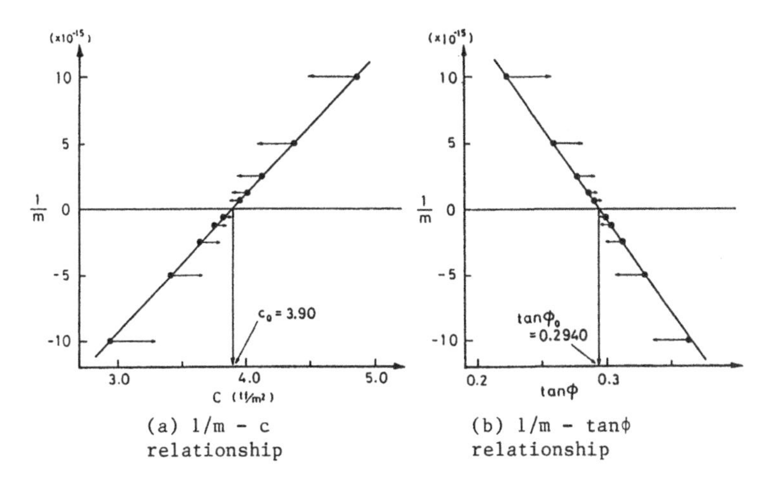

(a) 1/m - c relationship

(b) 1/m - tanϕ relationship

Fig.35 Results of back analysis by the efficient and systematic procedure

The results are shown in Figs.33 to 35. As seen in Fig.35, the parameters obtained were c_0=3.90tf/m^2 and $\tan\phi_0$=0.2940 (ϕ_0=16.38°). These values are also indicated on the c-tanϕ relationship in Fig.33 together with the correct values. The values for m determined automatically by the computer are given in Fig.34; a starting value of m in the iteration process was 1.0x10^7 and a value for the prescribed tolerance was 4 meters. For reference, the outermost trial slip surfaces automatically produced by the computer are also illustrated in Fig.32. There are five trial slip surfaces on each side of the critical one.

The second example is concerned with a failed natural slope shown in Fig.36. This case history was reported by Futakuchi, et al. (1985). Again due to lack of space, only the results are shown in Figs.37 and 38. In this case, although there is no information available on the correct strength parameters, we have considered the back analyzed values, c_0=0.592tf/m^2 and ϕ_0=34.07°($\tan\phi_0$=0.6764), to be adequate. The starting value of m used to determine trial slip surfaces herein was 1.0x10^7 and the value for the prescribed tolerance was 2 meters. Two outer trial slip surfaces created by the computer on each side of the critical one are also shown in Fig.36.

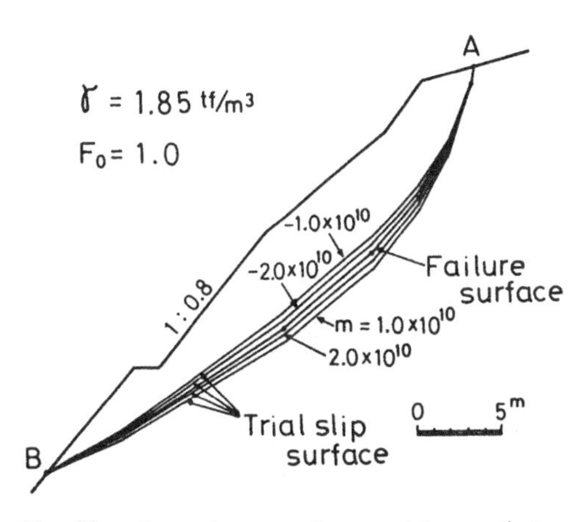

Fig.36 Second example problem (given failure surface and two outer trial slip surfaces created on each side)

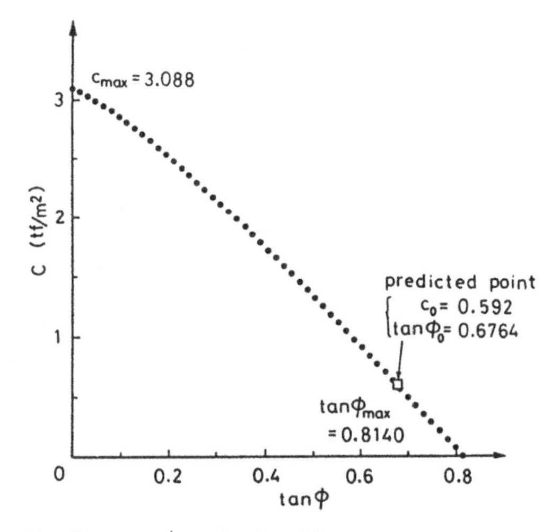

Fig.37 c-tanϕ relationship

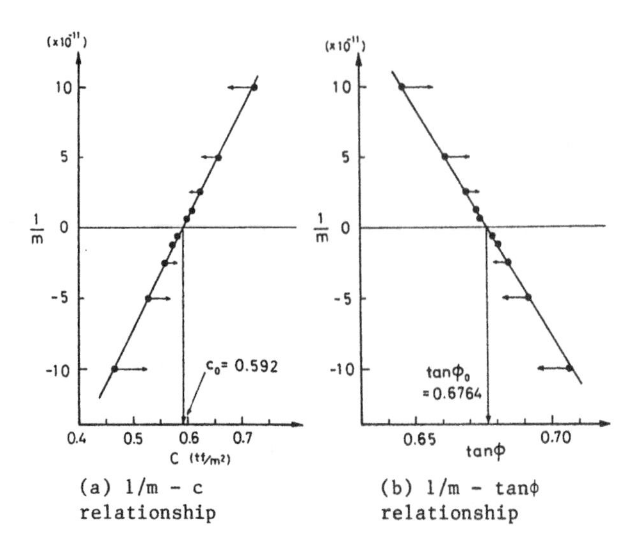

Fig.38 Results of back analysis by the efficient and systematic procedure

It is concluded that from the results obtained by the two example problems above, the back analysis procedure based on the Janbu method is also of much practical use.

6 CONCLUDING REMARKS

The back analysis method has been presented for the average strength parameters along a given critical slip surface, assuming the slopes to be homogeneous. Somewhat different treatments are necessary for the method in accordance with the factor of safety equation to be employed. For this reason, certain fundamental concepts have first been described. These are followed by the back calculation procedures for three different factor of safety equations with one or two example problems for each, namely the ordinary method of slices, the Bishop method, and the Janbu method. As a result, all the solutions obtained were unique and reliable; this is because despite its mathematical formulation being quite simple, the proposed method has met the two essential requirements.

Experiences through example problems have shown that the use of a computer is recommended even in a simple case such as the ordinary method of slices where manual calculation is possible. Otherwise undesirable solutions might result from a scatter of data due to personal errors, e.g. when reading numerical values such as angles or areas from diagrams.

The proposed method can be applied only to homogeneous slopes. However, an alternative approach (Yamagami and Ueta 1987) has been developed based on a nonlinear programming technique which may prove to be a very promising tool for non-homogeneous slopes.

REFERENCES

Chowdhury, R. N. 1978. Slope Analysis, Elsevier.

Futakuchi, T., et al. 1985. Strength Parameters for Stability Analysis of Cut Slope, The 20th Japan National Conference on Soil Mechanics and Foundation Engineering: 1431-1434. (in Japanese)

Janbu, N. 1973. Slope Stability Computations, Embankment-dam Engineering, Casagrande Volume (eds. R. C. Hirschfeld and S. J. Poulos), John Wiley & Sons, : 47-86.

Saito, M. 1980. Reverse Calculation Method to Obtain c & ϕ on a Slip Surface, Proc. Inter. Sympo. on Landslides, New Delhi, Vol.I: 281-284.

Sevaldson, R. A. 1956. The Slide in Lodalen, October 6th, 1954, Geotechnique, Vol.6, No.4: 167-182.

Yamagami, T. and Y. Ueta 1984a. A New Method for Inverse Calculation of c and ϕ on a Slip Surface (Part I) -Fundamental Concept-, Journal of Japan Landslide Society, Vol.21, No.2: 16-21. (in Japanese)

Yamagami, T. and Y. Ueta 1984b. A New Method for Inverse Calculation of c and ϕ on a Slip Surface (Part II) -Inverse Calculation on the Basis of Fellenius Method-, Journal of Japan Landslide Society, Vol.21, No.3: 24-31. (in Japanese)

Yamagami, T. and Y. Ueta 1984c. A Method for Inverse Calculation of c and ϕ on a Slip Surface Based on the Janbu Method, Proc. the 6th Japan Symposium on Rock Mechanics: 299-304. (in Japanese)

Yamagami, T. and Y. Ueta 1985. A New Method for Inverse Calculation of c and ϕ on a Slip Surface (Part III) -Inverse Calculation Based on the Bishop Method-, Journal of Japan Landslide Society, Vol.21, No.4: 10-17. (in Japanese)

Yamagami, T. and Y. Ueta 1986. Noncircular Slip Surface Analysis of the Stability of Slopes -An Application of Dynamic Programming to the Janbu Method-, Journal of Japan Landslide Society, Vol.22, No.4: 8-16.

Yamagami, T. and Y. Ueta 1987. Optimization Technique for Back Analysis of Failed Slopes, 8th Asian Regional Conference on Soil Mechanics and Foundation Engineering. (submitted)

Computer and Physical Modelling in Geotechnical Engineering, Balasubramaniam et al. (eds)
© 1989 Balkema, Rotterdam. ISBN 90 6191 864 2

Implementation of a probabilistic stability analysis method on microcomputers and Markovian approaches

F.Oboni
CSD Colombi Schmutz Dorthe, Engineers and Geologists, Le Mont-sur-Lausanne, Switzerland

F.Russo
Department of Mathematics, Swiss Federal Institute of Technology, Lausanne, Switzerland

ABSTRACT: The implementation of a recent probabilistic slope stability analysis method on a micro-computer is presented. After a short definition of the capabilities of the method and its theoretical aspects, a programming-oriented explanation of the details of the solution is given. The listing of a programm written in BASIC language is shown. Finally, two Markovian models are presented and discussed. An actual landslide in the Swiss Alps is analyzed with the proposed models.

1 INTRODUCTION

Since several years, it has been recognized that the strength parameters of soils are random variables (Singh [32], Schultze [31], Harr [15], Hammit [14], Lumb [19], Recordon [29]).

As a consequence of this fact, one should note the huge effort related to the introduction of the probabilistic analysis, particularly in the field of slope stability in the last decade (Alsonso [1], Grivas [9], Biernatowski [2, 3], Matsuo and Kuroda [21], Vanmarcke [34]).

In the majority of the earlier papers on this topic, the attention has been focused on the introduction of random strength parameters in classical deterministic models, using numerical methods such as Taylor's series expansion or Monte-Carlo simulation.

In the latest years, many papers discussed the development of new models, where the random nature of the parameters is taken into account at the very beginning of the formulation of the problems (Chowdhury and Grivas [7], Oboni and Bourdeau [25], Grivas and Harr [10, 11, 12, 13], Oboni, Bourdeau and Russo [27]).

The aim of this paper is to present the implementation of the Oboni and Bourdeau slope stability analysis method on a microcomputer as well as the latest developments in the Markovian approach. This method has been developed as a part of the DUTI research programm (Detection and Use of Landslide Prone Areas) supported by the EPFL (Swiss Federal Institute of Technology, Lausanne), and has been used on several landslides in the Swiss Alps in good agreement with the observed behaviour (Engel, Noverraz, Oboni [8], Oboni [22], Oboni, Bourdeau and Bonnard [26], Bonnard [6]).

2 CAPABILITIES OF THE METHOD

In slope stability problems, a good engineering analysis should provide answers to such question as :

1. In the whole set of potential slip surfaces, which is the most critical ? [25]
2. How likely is the failure to initiate ? [25]
3. Where will tensile cracks be located ? [25]
4. If a failure occurs, what will be the longitudinal extent of the distressed area ? [27]
5. Where will be the fastest sliding areas in a given slide and what are the ratios of velocity among them ? [27]
6. Where are the best locations for retaining structures and which force has to be introduced in the prestressed anchors in order to reach a given reliability in the slope ? [24]

Global analysis methods cannot help engineers in sloving these kind of problems, because slope failures are not global phenomena : the plastification of the soil along the candidate slip surface begins somewhere in the slope, then, it propagates, but the actual failure is only experimented of the toe of the slope fails. Failure is here introduced in a rigid-plastic definition where the deformations of the body are neglected.

Once the toe has failed, the next section of the slide may or may not slide and so forth, until the failure has progressed to the head of the sliding mass or has stopped after reaching an "a priori" unknown location.

The Oboni and Bourdeau method is a slice method in which the local equilibrium conditions are analyzed in order to define the probabilities of transition of the progressive failure, understood as a stochastic process.

By these means, it can help engineers to answer the questions asked in the beginning of this section.

In the present paper, the attention will first be focused on the development of a micro-computer program which gives the solution to the questions 2. and 3., and subsequently on the discussion of Markovian approaches.

It should be noted that in the Oboni and Bourdeau original paper [25], the solution to the question 1. was given under the form of a maximization scheme : the functional to be maximized was the probability of transition of the first slice, at the toe. As the maximization algorithm needs a bigger computer, we have left it aside.

3 STABILITY OF ONE ELEMENT OF THE SLOPE

3.1 Definitions

In this section, the necessary equations to the solution of a generical slice i are developed.

After defining the slip surface to be analyzed, the potential sliding mass is discretized in n differential (thin) slices of uniform width b_i and base slope α_i. The slices are numbered starting from 1 at the toe to n at the head of the slide.

As the soil stratification is assumed to be known, it is possible to define for each slice the following parameters :

T_{ij}	Specific weight of the $1 \leq j \leq m$ soil strata at the ith slice
h_{ij}	Height of the $1 \leq j \leq m$ soil strata at the axis of the ith slice
$\bar{c}_i$, S_{ci}	Cohesion of the soil strata at the slip surface, at the ith slice (mean value, standard deviation)
$\bar{\phi}_i$, $S_{\phi i}$	Friction of the soil strata at the ith slice (mean value, standard deviation)

It should be noted that the specific weight T_{ij} is considered as a deterministic variable : this is due to the fact that the coefficient of variation (C.O.V.) of T is generally very low when compared to the C.O.V. of the other parameters.

As the skewness of the parameters is very expensive to determine and is therefore often neglected in practical applications, we assume in this paper that all random variables are symmetrically distributed.

If a water table is present in the slope, it is necessary to give for each slice the following data :

$\bar{h}_{wi}$, S_{hwi}	Height of the water table above the slip surface at the ith slice (mean value, standard deviation)
η_i	Slope of the water table at the ith slice
T_w	Specific weight of water

Generally, the S_{hwi} is kept constant all over the slope so that the water table fluctuation is reduced to a global vertical translation : this is not due to a limitation of the model, but to a difficulty of precisely monitoring the behaviour of the water table in the field.

3.2 Equations

The following equations can now be written [25] :

$$W_i = \sum_{j=1}^{m} T_{ij}\, b_i\, h_{ij} \quad \text{(weight of the } i^{th} \text{ slice)}$$

$$W_i' = W_i - T_w\, h_{wi}\, b_i \quad \text{(buoyant weight of the } i^{th} \text{ slice)}$$

$$S_i = \sin \eta_i\, T_w\, h_{wi}\, b_i \quad \text{(seepage force through the } i^{th} \text{ slice)}$$

Before computing the capacity C_i (shear resistance) and the demand D_i (shear driving force), a P_{i+1} force has to be introduced.

P_{i+1} is the result of the deficit of shear resistance accumulated by the slide from the nth slice down to the ith. As the nth slice is located at the head of the slide, P_{n+1} is obviously equal to zero.

Assuming the Mohr-Coulomb failure criterion for the shear strenght of the materials, C_i and D_i are computed as follows :

$$C_i = (W_i' \cos \alpha_i - S\perp_i - P\perp_{i+1}) \cdot \operatorname{tg} \phi_i' + c_i'\, b_i/\cos \alpha_i$$

$$D_i = W_i' \sin \alpha_i + S//_i + P//_{i+1}$$

where the signs // and $\perp$ stand respectively for the parallel and the normal component of the force with respect to the base of the slice :

$S_{//i} = S_i \cos (\alpha_i - \eta_i)$
$S_{\perp i} = S_i \sin (\alpha_i - \eta_i)$
$P_{//i+1} = P_{i+1} \cos (\alpha_i - \alpha_{i+1})$
$P_{\perp i+1} = P_{i+1} \sin (\alpha_i - \alpha_{i+1})$

The safety margin SM_i of the slice is equal to :
$SM_i = C_i - D_i$
and the local probability of failure of the ith slice is :
$p_{fi} = p\ [SM_i \leqq 0]$
It should be noted that p_{fi} is a conditional probability which denotes the probability that the ith slice fails if the $i\text{-}1^{th}$ slice has already failed. The ith element of the slope will in turn transfer to the $i\text{-}1^{th}$ one a force P_i^* equal to :
$P_i^* = D_i - C_i = - SM_i$
As soils are supposed to be unable to carry tensile forces, P_i^* has to be left zero bounded : this can be formulated by the following expression :

$$P_i = P_i^* \ \frac{1 + P_i^*/|P_i^*|}{2}$$

3.3 Moments of random functions

As C_i, D_i, P_i, SM_i are functions of four random variables c_i, ϕ_i, P_{i+1}, h_{wi}, it is necessary to evaluate their mean value as well as their standard deviation. This can be done in a very efficient way by using the Rosenblueth Point Estimate Method, PEM (Rosenblueth [30]).
When dealing with symmetrically distributed uncorrelated parameters, the PEM for a random function F is carried out in the following way :
1. For each parameter X_k entering in the definition of F, two point estimates are computed, namely :
$X_k^+ = \overline{X}_k + S_{xk}$
$X_k^- = \overline{X}_k - S_{xk}$
2. 2^N permutations of the parameters point estimates are written, where N is the number of random parameters : $1 \leqq k \leqq N$. For our case, N = 4, so that the array of permutation is :

k	1	2	3	4 = N
1	+	+	+	+
2	-	+	+	+
3	+	-	+	+
4	-	-	+	+
5	+	+	-	+
6	-	+	-	+
7	+	-	-	+
8	-	-	-	+
9	+	+	+	-
10	-	+	+	-
11	+	-	+	-
12	-	-	+	-
13	+	+	-	-
14	-	+	-	-
15	+	-	-	-
16 = 2^N	-	-	-	-

(con't)

NB : Here, the sign + stands for X_k^+, - for X_k^-

3. For each line (permutation), the point estimate of the random function $F^{jk\ell m}$ is computed, where j,k,ℓ,m = + or - stand for the signs of the point estimates of the parameters.
4. The following sums are then computed :

$$Sum_F = \sum_{t=1}^{2^N} F^{jk\ell m}$$

$$Sum_F{}^2 = \sum_{t=1}^{2^N} (F^{jk\ell m})^2$$

5. Mean and standard deviation of F are then yielded by the following expressions :

$$\overline{F} = \frac{Sum_F}{2^N}$$

$$S_F{}^2 = \frac{Sum_F{}^2}{2^N} - (\overline{F})^2$$

$$S_F = \sqrt{S_F{}^2}$$

In the program given at the end of this paper, the PEM for four random variables is introduced by using a very compact and efficient algorithm for defining the signs of the X_k parameter point estimate line by line : in this way, the construction of the array of permutations can be avoided.
Another particular feature of the program is that the covariance of C_i and D_i is computed by using the following expression (Oboni [23]) :

$$\text{cov}\ (C_i, D_i) = - \frac{\Sigma\ C_i{}^{jk\ell m}\ \ \Sigma\ D_i{}^{jk\ell m}}{(2^N)^2} + \frac{\Sigma\ (C_i{}^{jk\ell m}\ \ D_i{}^{jk\ell m}}{2^N}$$

The correlation coefficient $\rho_{Ci,Di}$ is then found to be :

$$\rho_{C_iD_i} = \frac{\text{cov}\ (C_i,D_i)}{S_{D_i} \cdot S_{C_i}}$$

```
100 DIM AMEAN(4),SD(4),PAR(4)
150 ISHORT=1
200 INPUT "impression courte ? (1=oui) ";ISHORT
250 P=0!:SP=0!:AP1=0!
300 INPUT "nombre de tranches à calculer ? ";N
350 FOR I=1 TO N
375 PRINT " TRANCHE ";I
400 INPUT "   poids                          ";W
450 INPUT "   largeur                        ";B
500 INPUT "   pente de la base               ";A
550 INPUT "   pente de la nappe              ";EW
600 INPUT "   hauteur moyenne de la nappe    ";AMEAN(3)
650 INPUT "   éc.type de ses fluctuations    ";SD(3)
700 INPUT "   moyenne du frottement          ";AMEAN(2)
750 INPUT "   éc. type du frottement         ";SD(2)
800 INPUT "   moyenne de la cohesion         ";AMEAN(1)
850 INPUT "   éc. type de la cohesion        ";SD(1)
855 CLS
860 PRINT " Données enregistrées"
865 PRINT " poids";W;" largeur";B;" pente de la base";A
870 PRINT " pente de la nappe";EW
875 PRINT " cohésion moyenne";AMEAN(1);" éc.type";SD(1)
880 PRINT " frottement moyen";AMEAN(2);" éc.type";SD(2)
885 PRINT " niveau eau moyen";AMEAN(3);" éc.type";SD(3)
900 AMEAN(4)=P:SD(4)=SP
950 A=.01745*A:EW=.01745*EW
1000 AMEAN(2)=.01745*AMEAN(2):SD(2)=.01745*SD(2)
1050 GOSUB 1150
1100 NEXT I
1125 END
1150 CA=COS(A):SA=SIN(A)
1200 DW=A-EW:DP=A-AP1
1250 SSM=0:SSM2=0:SCP=0:SCP2=0:SCD=0:SC=0:SC2=0:SD=0:SD2=0
1300 FOR K=1 TO 16
1350 FOR IC=1 TO 4
1400 L=0
1450 L=L+1
1500 IF L>IC THEN GOTO 2000
1550 AM=(K-L)/2^(IC-1)
1600 AM1=1!*INT(AM)
1650 DEL=ABS(AM-AM1)
1700 IF DEL>.01 THEN GOTO 1450
1750 M=INT(AM) MOD 2
1800 SI=-1!
1850 IF M=0 THEN SI=1!
1900 PAR(IC)=AMEAN(IC)+SI*SD(IC)
1950 GOTO 2050
2000 PRINT "";
2050 NEXT IC
2100 IF PAR(3)<0! THEN PAR(3)=0
2150 IF PAR(4)<0! THEN PAR(4)=0!
2200 IF ISHORT=0 THEN PRINT PAR(1) PAR(2) PAR(3) PAR(4)
2250 W1=W-(10*PAR(3)*B)
2300 IF ISHORT=0 THEN PRINT "   poids apparent ";W;"   poids déjaujé ";W1
2350 SO=PAR(3)*10*B*SIN(EW)
2400 S1=SO*SIN(DW)
2450 S2=SO*COS(DW)
2500 C=(W1*CA-S1-PAR(4)*SIN(DP))*TAN(PAR(2))+PAR(1)*B/CA
2550 D=W1*SA+S2+PAR(4)*COS(DP)
2600 SM=C-D
2650 SC=SC+C
2700 SD=SD+D
2750 SC2=SC2+C^2
2800 SD2=SD2+D^2
2850 SCD=SCD+D*C
2900 IF ISHORT=0 THEN PRINT W1 SO S1 S2 SM C D
2950 SSM=SSM+SM
3000 SSM2=SSM2+SM^2
3050 IF SM<0 THEN SCP=SCP-SM:SCP2=SCP2+SM^2
3100 NEXT K
3150 SMM=SSM/16
3200 SMV=SSM2/16-SMM^2
3250 SMSD=SQR(SMV)
3300 PRINT " marge de sécurité moyenne ";SMM
3350 PRINT " éc. type de la marge de sécurité";SMSD
3400 CM=SC/16:DM=SD/16
3450 CSDV=SQR(SC2/16-SC*SC/256)
3500 DSDV=SQR(SD2/16-SD*SD/256)
3550 PRINT " capacité moyenne ";CM;" éc. type ";CSDV
3600 PRINT " demande moyenne  ";DM;" éc. type ";DSDV
3650 COVCD=SCD/16-SC*SD/256
3700 RHOCD=COVCD/(CSDV*DSDV)
3750 PRINT " covariance (C,D) ";COVCD;" correl. ";RHOCD
3800 P=SCP/16
3850 SP=SCP2/16-P^2
3900 PRINT " force trans. moy.";P; " éc. type ";SP
3950 AP1=A
4000 RETURN
```

NB :		
	Tranche	: slice
	éc.type	: std. deviation
	moyenne	: mean value
	nappe	: phreatic level
	frottement	: friction angle

4 STABILITY OF THE COMPLETE SLOPE

The computation of the stability of the entire sliding body begins with the evaluation of the forces involved in the nth element (top of the slope) and its conditional probability of failure.

After transferring the P_n force to the $n\text{-}1^{th}$ element, the forces of the $n\text{-}1^{th}$ element are computed and the cycle is repeated down to the toe element of the slope.

A good example of this way of doing is a toboggan with several children seated on it, each of them representing a slice in the actual slope.

Moreover, the distribution of the transmitted force P_i can be approximated with an empirical distribution like the Beta (Pearson type I) distribution to compute the retaining force (anchors) to be introduced in the slice, in order to reach a given reliability (Oboni [24]).

5 A FIRST MARKOVIAN MODEL

Although the progression of failure is a tridimensional mechanism, in the Oboni and Bourdeau model, only one profile is analyzed (2-D model). The Markovian model exposed in this section could be applied to a system of parallel and independent longitudinal profiles.

The failure is supposed to move forward with discrete steps, following the same basic rules as given by Chowdhury and Grivas [7] :

1. jumps are forbidden;
2. failure propagates from the toe to the head of the slide.

This phenomenon can be mathematically explained by a Markov chain, the states of which are : j = 0, 1, ... n. j = 0 represents the state before that any failure has taken place, $j \in \{1, \dots n\}$ symbolizes the fact that the slice j has slidden; n stands clearly for the end of the process. One notes by $p_{i-1,i}$ (resp. $p_{i-1,i-1}$) the conditional probability p_{fi} (resp. $1-p_{fi}$) with $1 \leqq i \leqq n$; $p_{01} = p_{f1}$ is the probability of triggering of the mechanism, p_{nn} is always 1.

The formulation of the Oboni and Bourdeau method satisfies the Markov condition which requires that future does not depend on the past, if the present is known. In fact, if the jth slice has failed, the future of the progression is only connected to $p_{i,i+1}$, $i \geqq j$, and is independent of previous steps.

The transition matrix associated to this stochastic process is :

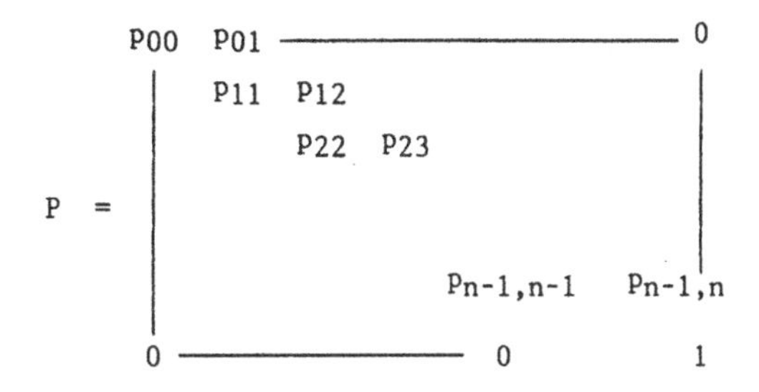

As the p_{ij} are clearly non negative and the sum of the terms of each line is equal to 1, the transition matrix P is a stochastic matrix.

The Nth power of P, which is noted by P^N, is also stochastic matrix and the term $(P^N)_{ij}$, $0 \leqq i, j \leqq n$, is the probability of passage in N fictious time units, from state i to state j. This "time" may be physically understood as a transition from one parallel profile to another. This note opens the door to a more complicated (3-D) model, the study of which could be the aim of a future paper.

6 ABSORPTION OF THE PROGRESSION OF A LOCAL FAILURE

6.1 Definitions

A problem that can be handled with the help of Markovian formalism is the evaluation of the probability $\beta_0{}^j$, i.e. the probability that the phenomenon stops at level j ($0 \leqq j \leqq n$), or, more clearly, after the failure of the jth slice.

The case j = 0 means that the mechanism has not started; when j > 0, the physical result is the failure of a first group of slices belonging to the potential slide. It is also interesting to know the average value M_0, the variance V_0 and the skewness coefficient S_0 of the random number of slices which are included in this first phase of the failure. If the slices are chosen uniformly wide, then M_0 can be understood as the average length of the distressed areas, taking into account the width of the slices.

In the Oboni and Bourdeau original paper (except for the case $p_{ii} = 1$ for any i, $0 \leq i \leq n-1$), it was assumed that after a certain number of steps, the process always reached the "final catastrophe" or the state where all slices had failed. In Markovian language, this implies that the only absorbing state is n. Since we are interested in studying the occurence of intermediate failures, states $0 \leq j \leq n-1$ must become in a certain way absorbing. In order to achieve this goal, the previous model must be modified.

A new Markov chain must be defined on 2n+1 states; at each level j ($0 \leq j \leq n-1$) will correspond two states : the first, noted by +j, symbolizes the fact that the process has just reached level j. The second, noted -j, expresses that the phenomenon has stopped at j and slice j+1 will not be involved in this failure. State n, which will be also noted by -n, stands for the final step of the failure. The new transition matrix is given as follows :

$$P_{+j,-j} = p_{j,j}$$

$$P_{+j,+(j+1)} = p_{j,j+1} \qquad 0 \leq j \leq n-1$$

$$P_{-j,-j} = 1$$

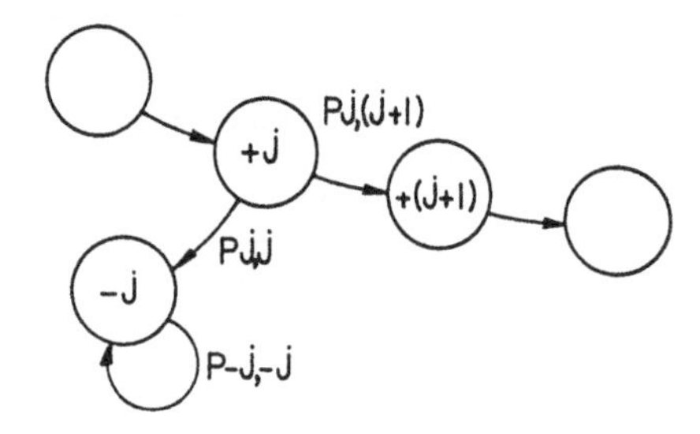

States +0, +1, ..., +(n-1) are transient, -0,-1,...,-(n-1), -n are absorbing.

Assuming that the failure has already progressed until a certain level ℓ, the number of blocks which could still fail is on average $m_\ell - 1$, and the probability of a stop at level j, $j \geq \ell$, is given by $\beta_\ell{}^j$. In Markovian language, m_ℓ stands for the "absorption time" of a chain leaving from ℓ with transition matrix P, and $\beta_\ell{}^j$ is the probability of absorption in -j. These are similar quantities can be computed as shown hereafter.

6.2 Absorption time

Let A be a generical matrix defined on the previous 2n+1 states; we will note by A_T, a restriction of A to transient states. By using this notation, the matrix P_T is given by :

$$(P_T)_{i,i+1} = p_{i,i+1}, \quad 0 \leq i \leq n-1$$

The rest of the matrix is filled by vanishing terms. Let $Q = (I_T - P_T)^{-1}$, where I_T is a n by n identity matrix.

It is not difficult to prove that :

$$Q_{ij} = \prod_{k=1}^{j-1} p_{k,k+1}, \quad 0 \leq i,\ j \leq n-1$$

where the convention is that the product of zero (resp. m < 0) factors is one (resp. zero). If the number of slices is for example n = 4, the Q_{ij} of Q_T terms can be obtained computing the following products :

i \ j	0	1	2	3
0	1	P01	P01P02	P01P12P23
1	0	1	P12	P12P23
2	0	0	1	P23
3	0	0	0	1

According to Kemeny and Snell [17, theorem 3.5.4], the vector of the absorption time is finally yielded by :

$$m = \begin{vmatrix} m_0 \\ m_1 \\ \cdot \\ \cdot \\ m_{n-1} \end{vmatrix} = Q_T \cdot \begin{vmatrix} 1 \\ \cdot \\ \cdot \\ \cdot \\ 1 \end{vmatrix}$$

6.3 Probabilities and state of absorption

Consider the matrix B : n x n+1, defined by : $\beta_{\ell j} = \beta_\ell{}^j$ ($\beta_\ell{}^j = 0$ if $j < \ell$).

A line $\ell \in \{0, 1,\ldots,n-1\}$ represents a transient state + ℓ, a column $j \in \{0, 1,\ldots,n\}$ stands for an absorbing state -j. Hoel, Port and Stone [16] show that B is the product of Q_T with $R = (r_{ij})$, where r_{ij} is the probability that the chain jumps in one step from the transient state i to the absorbing state -j.

In our case, R is a diagonal matrix with term $(p_{0,0},\ldots p_{n-1,n-1})$. Thus, it can be shown that :

$$\beta_{\ell j} = \prod_{k=\ell}^{j-1} p_{k,k+1}\, p_{j,j}, \quad 0 \leq \ell \leq n-1, \quad 0 \leq j \leq n$$

with the same conventions on product as before.

The mean of the state of absorption (understood as a positive number) of a chain starting from ℓ is :

$$M_\ell = \sum_{j=\ell}^{n} \beta_\ell{}^j \cdot j$$

we can easily obtain the intuitively clear relation $M_\ell = m_\ell - 1 + \ell$

The variance V_ℓ of the absorbing state is found to be :

$$V_\ell = \sum_{j=\ell}^{n} \beta_\ell{}^i \, (j - M_\ell)^2$$

M_ℓ and V_ℓ are respectively the first and the second central moment of the distribution of $(\beta_\ell{}^i)_{j=0}^n$; their physical meaning is respectively a location and its associated dispersion of the aborption of the failure within the slope. Another interesting parameter is the skewness coefficient of the same distribution; we will note it by S_ℓ, though Kendall and Stuart [18] refer to it by $\sqrt{\beta_1}$.

S_ℓ is the quotient of the third central moment of $(\beta_\ell{}^j)_{j=0}^n$ and $V_\ell{}^{3/2}$.

$$S_\ell = \frac{\sum_{j=\ell}^{n} \beta_\ell{}^j \, (j - M_\ell)^3}{V_\ell{}^{3/2}}$$

It can be shown that $-1 \leqq S_\ell \leqq 1$ and that S_ℓ is positive (resp. negative) when the "weight" of the considered distribution is on the right (resp. on the left) of M_ℓ.

7 ESTIMATION OF RELATIVE VELOCITY

Markovian approaches can also allow us predictions on relative velocity of the phenomenon. These quantities are useful to compare the behaviour of different parts of the same slope. In this context, one will need again the first Markovian model, presented at section 5, with n+1 strates and P as transition matrix. If the chain has started from i, the "average time of first visit" at state j can be computed : it will be noted by m_{ij}. Knowing that failure has already reached level i ($0 \leqq i \leqq n-1$), m_{ij} stands for the fictious time required by the progression to go from i to j. The relative velocity proposed is :

$$v_{ij} = \frac{j-1}{m_{ij}} \left| \frac{\text{"blocks"}}{\text{"time"}} \right|$$

It can be shown [17, Kemeny and Snell, Theorem 4.4.4.] that

$m_{ij} = (I - P^{ij})^{-1} \cdot 1, \; 0 \leqq i < j \leqq n$

where P^{ij} is the restriction of P to states i,...j-1. The j-1 elements of vector 1 are all equal to 1.

Thus :

$$m_{ij} = \sum_{k=i}^{j-1} \frac{1}{p_{k,k+1}}$$

It should be noted that maximal relative velocity is given when $p_{k,k+1}$, $i \leqq k \leqq j-1$; in this case :

$$m_{ij} = \sum_{k=i}^{j-1} 1 = j - i$$

Therefore, the relative velocity $v_{ij} = 1$

8 CASE STUDY

8.1 Description of the slide and available data

In this chapter the Peillettes slide, located in the swiss Alps, will be analyzed and discussed.

The toe of this slide is at a height of 1050 m while the main scarp is located at an altitude of 1950 m. The width of the sliding mass ranges from 200 to 500 m and its depth has been defined by geological, geophysical and inclinometric surveys. The sliding body is made of extremely decomposed rocks (chlorito sericitic quartzite ₫ phillites) which can be defined, following the USCS standards as GC-CL (28 to 32 % gravel, 42 to 48 % clay ₫ silt).

The following geotechnical parameters were used for the analysis :

$\overline{T} = 20$ KN·m^{-3}

$\overline{\phi} = 35°$ $S\phi = 3°$ C.O.V. $\phi \cong 10$ %

$\overline{c} = 20$ KN·m^{-2} $Sc = 5$ KN·m^{-2} C.O.V. $c \cong 25$ %

The slope have been equipped with 3 piezometric casings so that it was possible to define the water table fluctuations as follows :

average height of the water table measured from the slice base: $\overline{h} = 1 \div 35.5$ m (see fig)

standard deviation of h: sh = 3 m

8.2 Results of the analysis

The results given by the Oboni and Bourdeau method are plotted in the figure as well as the geometry of the slope and the average position of the water table.

The transition probabilities pij and pjj are also given in the table

i	j	pij	pjj
0	1	0.76	0.24
1	2	0.96	0.04
2	3	0.98	0.02
3	4	0.93	0.07
4	5	0.81	0.09
5	6	0.74	0.26
6	7	0.96	0.04
7	8	0.93	0.07
8	9	0.96	0.04
9	10	0.01	0.99
10	11	0.06	0.94
11	12	0.11	0.89
12	13	0.08	0.98
13	14	0.35	0.65
14	15	0.004	0.996
15	16	0.0003	0.9997
16	17	0.15	0.85
17	18	0.94	0.06

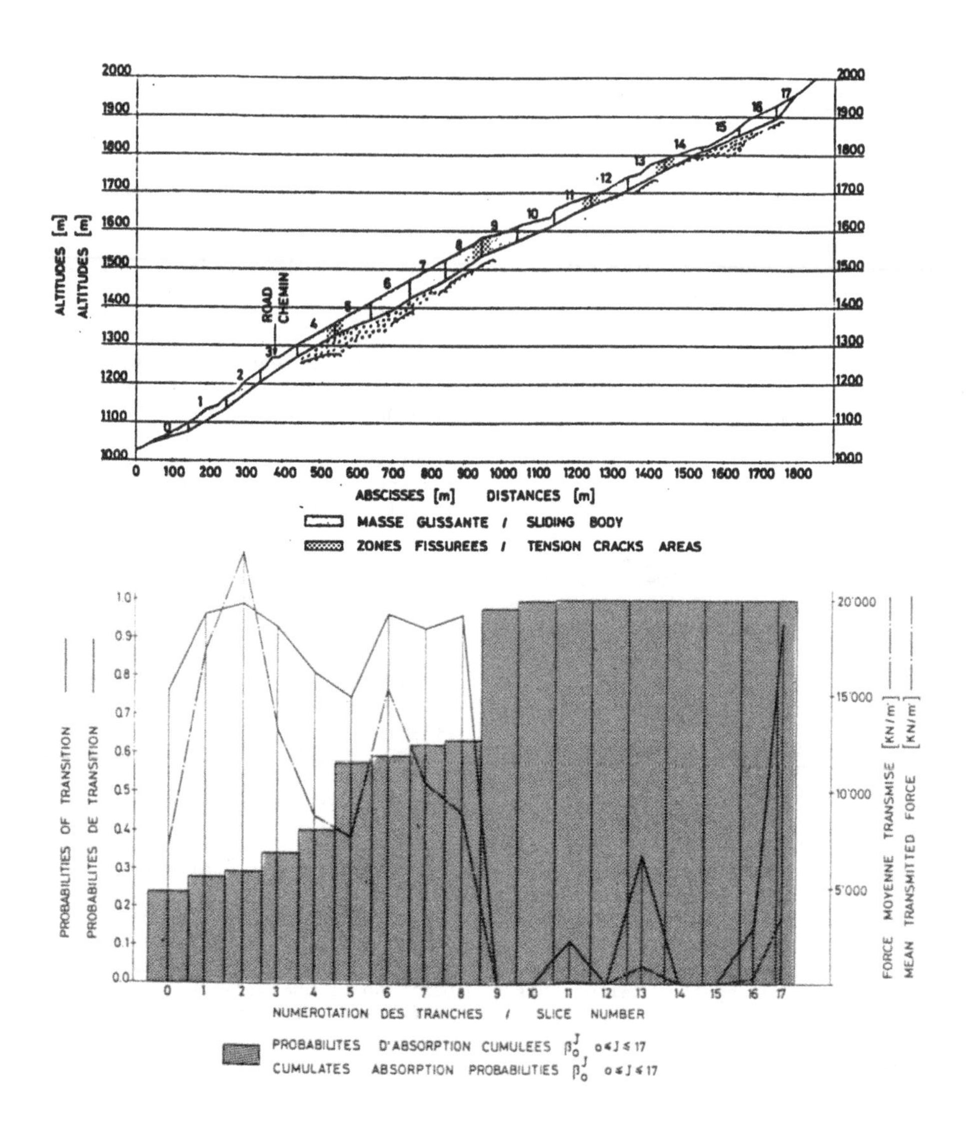
ALTITUDES [m]
ALTITUDES [m]
ROAD
CHEMIN
ABSCISSES [m] DISTANCES [m]
MASSE GLISSANTE / SLIDING BODY
ZONES FISSUREES / TENSION CRACKS AREAS
PROBABILITIES OF TRANSITION
PROBABILITES DE TRANSITION
FORCE MOYENNE TRANSMISE [KN/m²]
MEAN TRANSMITTED FORCE [KN/m²]
NUMEROTATION DES TRANCHES / SLICE NUMBER
PROBABILITES D'ABSORPTION CUMULEES β_o^J $o \leq J \leq 17$
CUMULATES ABSORPTION PROBABILITIES β_o^J $o \leq J \leq 17$

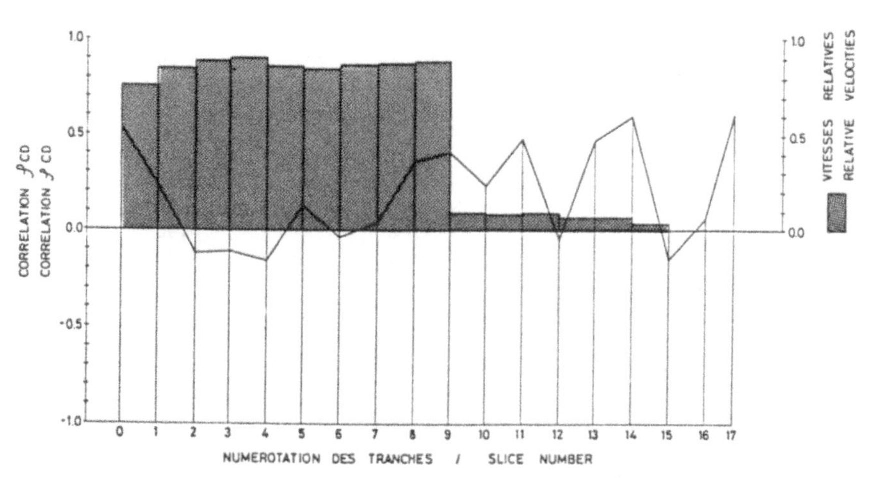
CORRELATION ρ CD
CORRELATION ρ CD
VITESSES RELATIVES
RELATIVE VELOCITIES
NUMEROTATION DES TRANCHES / SLICE NUMBER

First of all the absorption time m_0 was computed and found to be $m_0 = 5.80$

Thus the absorption state M_0 is obviously $M_0 = m_1 + 0-1 = 4.80$

The absorption probabilities β_{1j}, j = 1,...18 were then computed in order to obtain the variance V_0 of the absorption time m_0.

Thus $V_0 = 12.798$ and the standard deviation $\sigma_0 = 3.577$

The skewness So of the distribution of $(\beta_\ell{}^j)_{\ell=0}$ was then found to be $S_0 = -0.066$

The relative velocity analysis was then performed for each slice group 0j with $1 \leq j \leq 18$ so that :

$$
\begin{aligned}
v^{01} &= 0.76 \\
v^{02} &= 0.848 \\
v^{03} &= 0.888 \\
v^{04} &= 0.898 \\
v^{05} &= 0.86 \\
v^{06} &= 0.852 \\
v^{07} &= 0.866 \\
v^{08} &= 0.874 \\
v^{09} &= 0.883 \\
v^{010} &= 0.09 \\
v^{011} &= 0.087 \\
v^{012} &= 0.089 \\
v^{013} &= 0.07 \\
v^{014} &= 0.074 \\
v^{015} &= 0.034 \\
v^{016} &= 0.004 \\
v^{017} &= 0.004 \\
v^{018} &= 0.005
\end{aligned}
$$

As it can be seen, there is a drastic decrease of relative velocity when the 10th and following slices are involved in the failure :

For this reason the slide was finally analyzed in two sections : the downhill section going from slice 0 up to the 9th and the uphill section going from the 10th slice up to the 17th; the results were the following : $v^{09} = 0.801$, $v^{1017} = 0.003$ so that $\frac{v^{09}}{v^{1017}} = 315.$

8.3 Discussion of the results

Simultaneously with the numerical analysis extensive geological and geodetical surveys were carried out by the DUTI project geologist and surveyors; the result of this effort was a map giving the position of the tension cracks and subdividing the slope in areas of homogeneous velocity of slide.

The locations of the cracks defined with the Oboni and Bourdeau method were in good agreement with the observed actual cracks in the slope.

The same can be said on the relative speed estimate : in the matter of fact, geologists and surveyors agreed to define two main sliding bodies, one going from slice 0 to 9, the other from slice 10 to 17.

The measured relative velocity ranged from 230. to 250., the highest absolute speed being so high as 35 m/year.

As it can be seen, the agreement of the numerical results with the field observations was very good although the available data were of poor quality.

The cumulative absorption probability function $F\beta_0{}^j$ $0 < j < n$ has been plotted in the figure : it is interesting to note that the absorption is almost a certainty after the 10th slice.

The mean transmitted force plot can be used to define the best locations for retaining structures (anchors, drainage systems and others) : in the Peillettes slide these locations are X = 400 ÷ 500 and X = 850 ÷ 950 because in these areas the increase of the resistance deficit is very strong.

Moreover the first and second moment of the P force can be used for the probabilistic design of prestressed anchors if the P force distribution is approximated with an empirical distribution (for example Beta or Pearson type I).

The skewness of the distribution of the probabilities of absorption $(\beta_0{}^j)^n_{j=0}$ could be used as an indicator of the hazard involved in a failure of the slope : if the skewness is strongly positive (uphill long tailed, unimodal distribution) the slide will be faster and more hazardous than if the skewness is strongly negative (downhill long tailed unimodal distribution).

Although this is intuitively correct further actual case studies have to be performed in order to confirm this assumption.

The correlation coefficient $\rho CiDi$ has also been plotted.

These coefficients have not yet been used as indicators of the stability conditions of the slope.

They give an idea on how the capacity is related to the demand in a given slice.

In the matter of fact, in the Peillettes slide the correlation is positive when the uphill transmitted force $\bar{P}$ has little mean values, negative when the $\bar{P}$ value is high; so that the ρCD plot could be used, joint to the pij diagram, in defining the tension cracks locations.

Finally it can be shown that if the soils are strongly cohesive the correlation between capacity and demand is very weak; the same trend is observed if the soil exhibits strong friction angles but the effect on the correlation is milder.

9 CONCLUSIONS

The Oboni and Bourdeau stability analysis method has proved to be a powerful tool for solving problems even if the geometry of the solpe is very complicated and many strata of soils are present.

Interesting results concerning the possible behaviour of a slope can be drawn even if the statistics on the radom variables are of poor quality.

The assumed failure criterion for the shear strenght of the materials is for instance the Mohr, Coulomb criterion; all the developments of the method and the Markovian approach are independent from this criterion : if new, more accurate or more economical, rheological models happened to be introduced in the future their implementation in the Oboni and Bourdeau method will be easy to carry out.

Finally it should be noted that the proposed method assumes that in a slope not only the capacity, but also the demand are correlated random variables which have only local meaning i.e. that the probability of failure is not constant all along the slip surface.

REFERENCES

1- Alonso E.E., Risk analysis of slopes and its application to slopes in canadian sensitive clays. Geotechnique 26, No 3, 1976

2- Biernatowski K., Stability of slopes in probabilistic solution. Proc. 7th ICSMFE, Mexico, 1969

3- Biernatowski K., Stability of slopes in variational and probabilistic solution. Proc. 6th ECSMFE, Vienna, 1976

4- Bishop A.W., The strenght of soils as engineering materials. Geotechnique Vol. 16, No 2, 1966

5- Bjerrum L., Progressive failure in slopes of overconsolidated plastic clays and clay holes. 3rd Terzaghi lecture ASCE, Vol. 93, No SM5, 1967

6- Bonnard Ch., Use of probabilistic method of stability analysis for the design of stabilisation works. 3ème Sém. sur les méthodes probabilistes, Ecole Polytechnique Fédérale de Lausanne, 1984

7- Chowdhury R., Grivas D.A., Probabilistic model of progressive failure of slopes. J. of GED, ASCE Vol. 108, FT6, 1982

8- Engel T., Noverraz F., Oboni F., Le glissement de la Chenaula. BTSR No 27, 1984

9- Grivas D.A., Probability theory and reliability of earth slopes. Politechnic Institute, Troy, N.Y., 1977

10- Grivas D.A., Harr M.E., Stochastic propagation of rupture surfaces within slopes. Proc. 2nd ICASP, Aachen, 1975

11- Grivas D.A., Harr M.E., Consolidation - a probabilistic approach. J. of Engg. Mech. Div., June 1978

12- Grivas D.A., Harr M.E., The path flow and its effect on consolidation rates. Proc 3rd ICNMG, Aachen, 1979

13- Grivas D.A., Harr M.E., A reliability approach to the design of soil slopes. VII CEMSTF, Brighton, Vol. 1, 1979

14- Hammit J.M., Statistical analysis of date from comparative laboratory test programs sponsored by ACIL USAEWES Corps of Engineers. Miscellaneous papers No 4-785, 1966

15- Harr M.E., Mechanics of particulate media. McGraw Hill, New York, 1977

16- Hoel P.G., Port S.C., Stone C.J., Introduction to stochastic processes. Houghton Mifflin, Boston, 1972

17- Kemeny J.G., Snell J.L., Finite Markov chains. D. Van Nostrand Company, Amsterdam, 1960

18- Kendall J.M., Stuart A., The advanced theory of statistics. Vol. 1. Charles Griffin and Co. Ltd., London, 1976

19- Lumb P., Precision and accuracy of soil tests. Statistic and Probability in Civil Eng., Hongkong University Press, 1972

20- Markov A.A., Wahrscheinlichkeitsrechnung. Leipzig, 1912

21- Matsuo M., Kuroda K., Probabilistic approach to design of embankments. Soil and Foundations, Vol. 14, No 2, 1974

22- Oboni F., Le concept de probabilité de transition appliqué aux analyses de stabilité des pentes. Colloque "Mouvements de terrain", Université de Caen, 1984

23- Oboni F., Calcul de la covariance entre résistance et sollicitation avec la méthode de Rosenblueth. Note interne, Ecole Polytechnique Fédérale de Lausanne, LMS, 1984

24- Oboni F., Un nuovo metodo per la verifica della stabilità dei pendi. Bollet. Società mineraria subalpina, Torino, 1984

25- Oboni F., Bourdeau P.L., Determination of the critical slip surface in stability problems. Proc. 4th ICASP, Florence, Italy, 1983

26- Oboni F., Bourdeau P.L., Bonnard Ch., Probabilistic analysis of Swiss landslides. IVth Int. Symp. on Landslides, Toronto, 1984

27- Oboni F., Bourdeau P.L., Russo F., Utilisation des processus markoviens dans les analyses de stabilité des pentes. 3ème Sém. sur les méthodes probabilistes, Ecole Polytechnique Fédérale de Lausanne, 1984

28- Peck R.B., Stability of natural slopes. ASCE Vol. 93, SM 4, 1977
29- Recordon E., Despond J.-M., Dispersion des caractéristiques des sols naturels considérés comme homogènes. Congr. int. MSTF, Tokyo, 1977
30- Rosenblueth E., Point estimates for probability moments. Proc. Nat. Acad. Sc. USA, Vol. 72, No 10, 1975
31- Schultze E., Frequency distributions and correlation of soil properties. Proc. 1st ICASP, Hongkong University Press, Hongkong, 1972
32- Singh A., How reliable is the factor of safety in foundation engineering. Proc. 1st ICASP, Hongkong University Press, Hongkong, 1972
33- Terzaghi K., Mechanism of landslides. Geological Soc. of America. Bertrey Vol., pp. 83-102, Harward Soil Mech. Series, 1950
34- Vanmarcke E.H., Reliability of earth slopes. J. of GED., Vol. 103, No GT11

Computer and Physical Modelling in Geotechnical Engineering, Balasubramaniam et al. (eds)
© 1989 Balkema, Rotterdam. ISBN 90 6191 864 2

Automated slope stability analysis using mathematical programming technique

P.K.Basudhar & Yudhbir
Department of Civil Engineering, Indian Institute of Technology, Kanpur, India

ABSTRACT: The study pertains to the application of Sequential Unconstrained Minimization Technique as a tool for autosearching the critical slip surface and evaluation of the corresponding minimum factor of safety for nonhomogeneous dam sections with geologic discontinuities in the foundation.

INTRODUCTION

Slope stability analysis is essentially a problem of optimization (Baker and Garber, 1977; Basudhar, 1976) namely the determination of the slip surface that yields the minimum factor of safety. In most of the analyses the slip surface has been assumed to be of particular geometry e.g. straight line passing through the toe, a circular arc, a log spiral or an arc of a cycloid. The assumptions regarding the shape of the slip surface greatly simplify the computations involved in the analysis but the restriction imposed on the slip surface geometry may lead to the bypassing of the actual critical surface (Talesnick and Baker, 1984).

Many slope stability softwares using the limit equilibrium analysis have been described in literature (Fredlund, 1977); some of the studies have been summarised and critically discussed by Whitman and Bailey (1967), Wright et.al. (1973), Fredlund and Krahn (1976), Zeitlen et. al. (1977), Eisenstein (1977), Fredlund et.al. (1981), Ching and Fredlund (1983). Most of the programs provided an automated version of the existing methods of slope stability analysis. The need for autosearch led to the use of sophisticated optimization algorithms (Krugman and Krizek, 1973; Basudhar et.al., 1979; Narayan and Ramamurthy, 1980); however, in these studies circular slip surfaces were assumed. The methods of slices (Morgenstern and Price, 1965; Spencer, 1973; Janbu, 1973) valid for any arbitrary slip surface may be used in conjunction with optimization techniques for a more generalized analysis.

Successful applications of mathematical programming techniques to the slope stability analysis have been reported (Baker and Garber, 1977; Castillo and Revilla, 1977; Ramamurthy et.al., 1977; Martin, 1982; Munro, 1982; Celestino and Duncan, 1981; Baker, 1980; Nguyen, 1985; Boutrup et.al., 1979; Bhowmik, 1984; Bhowmik and Basudhar, 1986; Babu, 1986; Dhawan, 1986; Yudhbir and Basudhar, 1986). In this paper an attempt has been made to report and highlight some aspects of the studies in applying mathematical programming technique, that has been carried out at I.I.T. Kanpur by the authors and their students.

ANALYSIS

General

The pseudostatic stability analysis of the D/S slope of a dam under steady state seepage and earthquake loading is carried out by using the generalized procedure of slices (GPS) in conjunction with sequential unconstrained minimization technique

(SUMT). The method is capable of locating the critical shear surface corresponding to the minimum factor of safety without putting any prior restriction on the nature of the slip surface e.g. circle, logarithmic spiral etc. In this technique the stability problem is posed as an optimization problem wherein the factor of safety is minimized with respect to the co-ordinates of the slip surface and thus critical surface is located.

To study the effect of earthquake it has been assumed that the quake imposes a horizontal acceleration having an amplitude that is equal to certain given percentage of gravity and the resulting horizontal force acts at the centre of gravity of each slice.

The effect of pore water pressure on the D/S slope stability has been taken care of by considering an average value of pore pressure parameters, r_u, for the entire section.

The computer software developed by Babu (1986) to study the stability of zoned dam has been used to evaluate the stability of the slope. Fig. 1 shows the idealized cross section of an embankment of heterogeneous mass. Given the geometry of the dam section and the soil properties of different zones of the dam, the problem is to determine the shape and location of the shear surface that gives the least factor of safety.

The generalized procedure of slices developed by Janbu (1973) for homogeneous soils is assumed to be valid in analysing the dam with proper choice of the shear strength parameters and vertical stress at the base of each slice depending on its location. For reasons of space and brevity the method is not described here.

Statement of the Problem

Fig. 1 shows the geometry of the slope with a general potential slip surface and with the sliding mass divided into N number of slices.

For the given geometry of the dam section and soil properties the factor of safety is a function of the shape and location of the potential slip surface. The problem is to determine the shape and location of the shear zone that gives the minimum factor of safety.

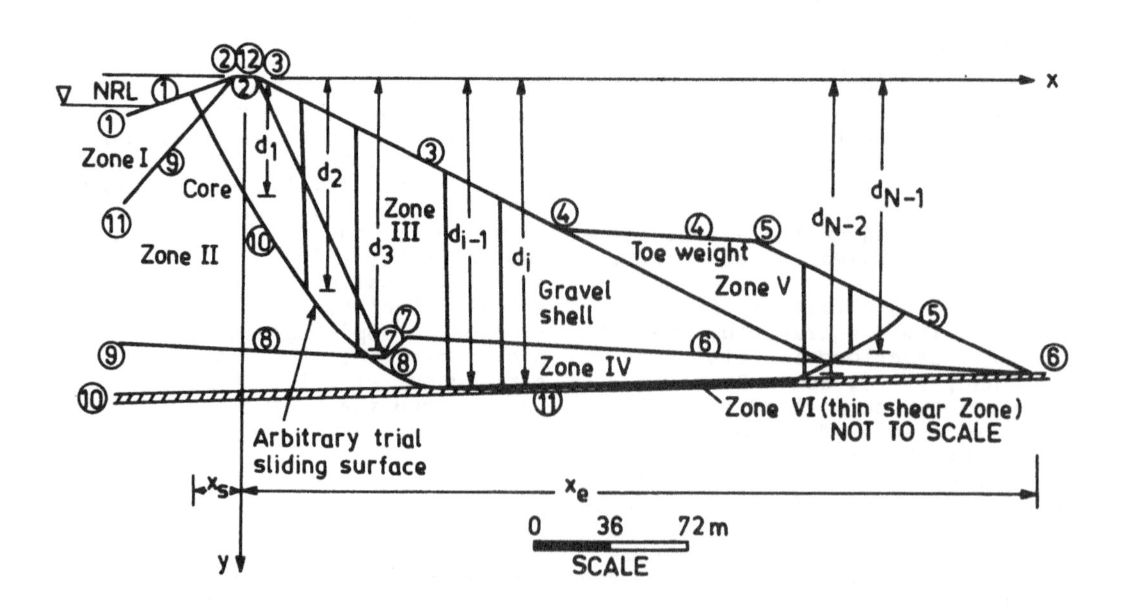

FIG. 1. IDEALIZED SECTION OF A ZONED DAM WITH THE POTENTIAL SLIDING MASS DIVIDED INTO SLICES.

Design Variables and Objective Function

The crest and centre line of the dam section are chosen as X-axis and Y-axis respectively. The co-ordinate system is shown in the Fig. 1. Let x_s and x_e be the x-co-ordinates of the starting and end points of the slip surface respectively and let $d_1, d_2, \ldots, d_{N-1}$ be the y-co-ordinates of the slip surface at the interfaces, which are equispaced on the slip surface. Given the positions of the x-co-ordinates of the starting and end points of the slip surface respective y-co-ordinate can be calculated. The x-co-ordinates of the intermediate points at intersections can be calculated since the slice width and number of slices are known. With this the co-ordinates of the points along the slip surface are completely defined and the factor of safety can be expressed as a function of these co-ordinates. In the adopted procedure the factor of safety is minimized with respect to these y-co-ordinates of the interfaces of the slices and the x-co-ordinate of the starting and end points of the slip surface. Once the optimal design vector is found out,the design vector along with the x-co-ordinates of the slice interfaces will define the actual critical shear surface.

The elements of the design vector $\vec{D}$ are as follows

$$\vec{D}^T = (d_1, d_2, d_3, \ldots, d_{N-1}, x_s, x_e) \quad (1a)$$

writing $d_N = x_s$ and $d_{N+1} = x_e$ one obtains

$$\vec{D}^T = (d_1, d_2, d_3, \ldots, d_{N+1}) \quad (1b)$$

So the problem involves (N+1) variables where N is the number of slices into which the sliding mass is divided.

The objective function is the factor of safety and an expression for the same can be obtained from the original paper of Janbu. The factor of safety can be written in terms of the design vector as

$$F = f(\vec{D}) = f(d_1, d_2, d_3, \ldots, d_{N+1}) \quad (2)$$

Constraints

In order to ensure that the slip surface is physically reasonable and acceptable the following constraints are imposed on the shear surface:

1. The curvature of slip surface should be such that it is always concave upwards. This requires

$$d_{i+1} - 2d_i + d_{i-1} \leq 0 \quad (3)$$

2. The slip surface should be within the cross section of the dam. This requires

$$h_i - d_i \leq 0 \quad (4)$$

where h_i = y co-ordinate of the intersection point of top boundary line of the dam section with the vertical drawn through the point whose y-co-ordinate is d_i; i varies from 1 to N-1.

Adding all the above constraints the total number of constraints (M) is equal to (2N-2). So the problem is one of (N+1) design variables and (2N-2) side constraints.

Slices, Width of Slices and Zoning

For computational purpose in the idealized dam section (Fig. 1) the different zones are numbered. All the intersection points defining the dam geometry are also to be numbered in order. The end points of any straight line whether or not intersected by any other straight line is also treated as intersection point (Fig. 1, points 1, 9, 10 and 11). Similarly all the straight lines are also numbered in order. For clarity numbers of intersection points are encircled and numbers of straight lines are marked in semicircles drawn on respective straight lines.

Once the dam geometry is defined by the co-ordinates of the key intersection points, the coefficients of each straight line defining the dam section can be generated on the computer.

If the number of slices are very large it will pose great difficulty in numerical calculations as the number of design variables become very large and beyond a certain optimal number of slices it does not improve the accuracy of the obtained factor of safety. If the number of slices is too small the obtained results are not reliable. So there should be a trade off study and the calculations should be carried out with the obtained optimal number of slices. A good choice about the width of the slice can be made depending on the length of the dam and the distance between the two nearest intersection points. In the present study for large dams a width of slice of 24 m is found to be quite satisfactory. However, this is problem dependent. During optimization process at any stage the width of slice should not be greater than the width of the slice that is permitted. This may be put as a constraint. Otherwise the calculations of weights of the slices will not be correct and hence the obtained factor of safety will be in error. Correctness of the weight calculations of the slices using the computer program developed by Babu has been ensured, for different possible slip surfaces and position of slices, by comparing the results with the manually computed values. Details are given in Babu (1986) and are not reported here. However, one point to be highlighted is that for slice weight calculations or choosing the appropriate strength parameter along the base of a slice one need to locate the appropriate zone for different critical points. For finding the appropriate zone for any given point in the body of the dam the following procedure was adopted. For the considered point the coordinates are substituted in all the equations surrounding the zone, the inequality condition that are to be satisfied by the equations corresponding to the surrounding straight lines are noted. Hence, for a given point to lie in a particular zone it has to satisfy the inequality conditions of straight lines surrounding that zone. Due to round off errors the starting and end points of slip surface may not exactly satisfy the equations of straight lines on which they are lying. Hence instead of giving a strict equality condition a small margin of 0.0001 is provided. The end points of the slip surface cannot be beyond point 6 (Fig. 1). This is put as an additional side constraints.

Earthquake Force Considerations

Earthquakes may cause the failure of earth embankments which under ordinary conditions would be amply safe. A quake actually imposes displacements rather than forces, and the forces resulting from the displacements are dependent in a complicated way on dynamic stress-strain relationships of the embankment material. An empirical approach, used in regions that are subjected to earthquakes, consists in assuming that the quake imposes a horizontal acceleration with an amplitude that is equal to some given percentage of gravity. It is assumed in the analysis that the horizontal earthquake force acts at the centre of gravity of each slice.

Presence of Thin Shear Plane in the Dam Foundation

If very thin shear zone is present in the foundation a reasonable thickness of the shear zone than the actual one is to be chosen so that during optimization process the intersection points of interfaces with the shear surface along the shear zone will not fall outside the shear zone. In the present study, while the actual thickness of shear zone is only 0.3 m, the value chosen on either side of shear zone is, however, 1 m. Though the thickness chosen is very high it has been observed that the critical shear surface does not move much from the exact shear plane, and, as such, the results are not affected in any way. But the scheme ensures that the shear surface does not lift off from the shear zone.

Optimization Formulation

We are now in a position to state the problem of finding the critical slip surface and the corresponding minimum factor of safety as a mathematical programming problem as follows.

Find the design vector $\vec{D}_m$ such that $F = f(\vec{D}_m)$ is the minimum of $f(\vec{D})$ subject to

$$g_j(\vec{D}_m) \leq 0; \quad j = 1, 2, \ldots, M.$$

where M is the total number of constraints.

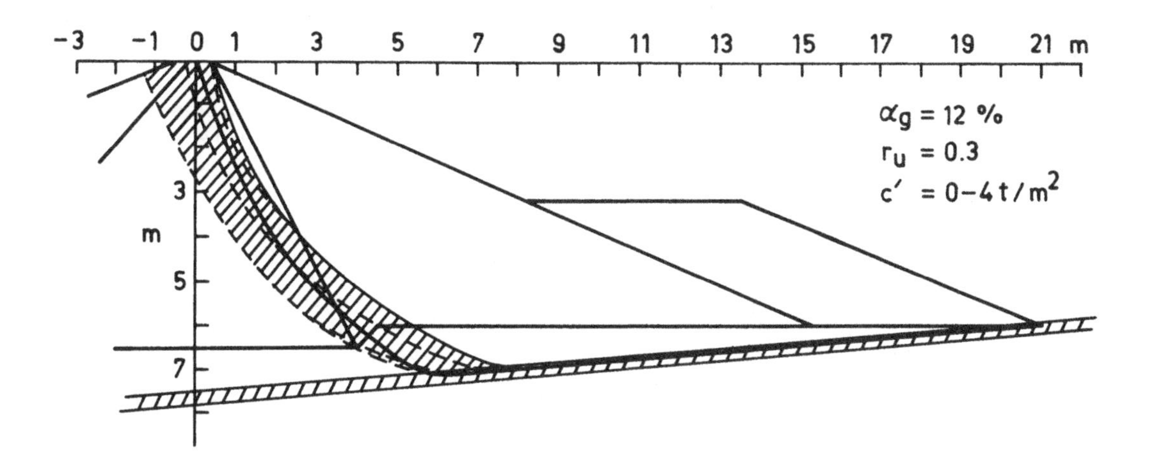

FIG. 2. CRITICAL SHEAR SURFACES.

The best way of deciding the initial design vector is to choose the initial slip surface in such a way that a major portion of it lies along the existing shear plane. If the initial design vector is chosen in any other way then during the optimization process it is very likely that the weak shear plane will be bypassed and the obtained solution would not be the global minimum.

The stability analysis of zoned dams consisting of weak thin shear zones in the foundation needs special consideration while using the present technique. One must be very careful in choosing the penalty parameter (r); it should be chosen initially in a manner such that the objective function and penalty term have equal weightage. If it is not done so there is a possibility of increase in the objective function value during the optimization process even though the penalty function gets minimized. This will result in lifting off the shear surface from the existing shear plane and it will be almost impossible to hit bark the same shear surface again and the solution converges to a local minimum.

In addition to the constraints (Eqs. 3 and 4), Bhowmik (1984) introduced the active earth pressure constraint on the slope of the slip surface at its intersection with the top surface of the slope and if the shear surface had a negative slope at the base he imposed the passive earth pressure condition (see Janbu, 1973).

However, present study revealed that if the active failure condition, which is an equality constraint at the left extreme intersection point had been satisfied, the optimization scheme did not have the freedom to move away from the constraint surface and the search probably continued along the constraint surface imposing a severe limitation on its efficiency resulting in premature termination. As such, it was felt that there is no loss of generality if these constraints are not imposed at the extreme points and full freedom is given to the slip surface to take its own shape within the limitation of concavity only (Babu, 1986). This procedure gave excellent results.

The obtained results for various values of pore pressure parameters, r_u, and horizontal acceleration, αg, revealed that the factor of safety is linearly related to r_u for various values of C' and horizontal acceleration.

Minimization Procedure

The sequential unconstrained minimization technique using the interior penalty function formulation in combination with Powell's multidimensional search and quadratic fit for finding the minimizing steps, has been used. Interior penalty function method needs a feasible point for starting the solution. As in this case it is possible to get a feasible initial design vector, the interior penalty function method has been chosen for the analysis. The basic object of the penalty function method is to convert the original constrained problem into one of unconstrained minimization by blending the constraints into a composite function (Ψ). The detailed background of these methods are available in standard textbooks on optimization (Fox, 1971).

For problems with inequality constraints only, the Ψ-function is defined as:

$$\Psi(\vec{D}, r_k) = F(\vec{D}) - r_k \sum_{j=1}^{M} \frac{1}{g_j(\vec{D})}$$

where F is to be minimized over all $\vec{D}$, satisfying

$$g_j(\vec{D}) \leq 0; \quad j = 1, 2, \ldots, M$$

The penalty parameter r_k is made successively smaller in order to obtain the constrained minimum of F.

Using the ideas contained in this paper Babu (1986) developed the Sequential Unconstrained Minimization Stability Analysis Package (SUMSTAB) whose strength and weakness were further studied by Dhawan (1986) under the authors' supervision.

RESULTS AND DISCUSSIONS

The following design parameters have been used in the analysis:

Unit weight:	Core	2.04 t/m^3
	Shell	2.4 t/m^3
	Toe-weight	2.4 t/m^3
Angle of shearing resistance (ϕ):	Core	26.5°
	Shell and toe-weight	35°
	Shear zone	18°
Effective cohesion (C'):	Core	0, 2.0, 4.0 t/m^2
	Shell and toe-weight	0
	Shear zone	0
Pore pressure ratio (r_u):		0.3, 0.4, 0.5
Horizontal acceleration,		α = 0, 0.05 g, 0.10 g, 0.12 g

Numerical results have been obtained by using the DEC SYSTEM 1090 and the SUMSTAB package. For r_u=0.3 and C' = 0-4 t/m^2 typical results for αg = 12% are presented in Fig. 2.

It is observed from the figure that the critical surfaces crowd in a narrow zone in the core; the major portion of it ofcourse being controlled by the shear zone. These results indicate the location of any possible failure surface during an earthquake for the assumed psuedo-static analysis. The factor of safety for these slip surfaces ranges from 0.9 to 0.92. A typical result showing distribution of interslice forces and normal and shear stresses is shown in Fig. 3. This result is most critical in checking the reasonableness of computations.

It is interesting to observe that the nature of the critical failure surfaces is approximately similar to wedge configuration.

It is well known that initial starting point plays a great role in getting the desired solution within reasonable number of iterations. If the obtained final solution is same irrespective of the starting points the obtained solution is the global minimum. This aspect has been studied for the given problem and has been found that though the final solutions corresponding to different initial design vectors are different the value of the factor of safety does not differ much (about 4 to 5%). It has also been observed that all the final optimal surfaces lie in a narrow band in the core and very near to the core in the foundation and over the rest of the portion they are the same and pass along the existing shear zone. As such, to arrive at the critical shear surface and the corresponding minimum factor of safety it is necessary to have solutions for different initial design vectors.

The Bishop and Morgenstern (1960) linear relationship between factor of safety F and r_u has been found to be equally valid for different values of horizontal earthquake acc-

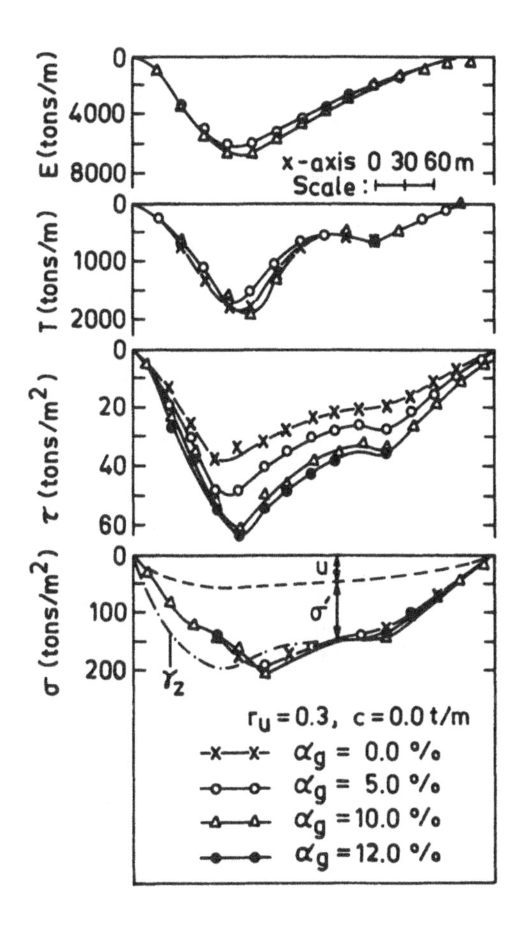

FIG. 3. INTERSLICE FORCE AND STRESSES ALONG A NON-CIRCULAR CRITICAL SHEAR SURFACE.

eleration. The 'm' and 'n' coefficients in their relationship have been correlated with the fraction of horizontal acceleration (Dhawan, 1986; Yudhbir and Basudhar, 1986).

CONCLUSIONS

A software package utilising Sequential Unconstrained Optimization Technique has been formulated to enable autosearch for critical surfaces and corresponding minimum factor of safety using Janbu's GPS method, without imposing a priori assumptions regarding the shape of the potential failure surface. The method has been used for homogeneous and heterogeneous sections of embankments and has been extended to consider the influence of thin shear zones in the foundation materials.

REFERENCES

Babu, N.S. 1986. Optimization techniques in stability analysis of zoned dams. M.Tech. thesis, Indian Institute of Technology, Kanpur.

Baker, R. 1980. Determination of the critical slip surface in slope stability computations. International Journal for Numerical and Analytical Methods in Geomechanics, Vol. 4, pp. 333-359.

Baker, R. and Garber, M. 1977. Variational approach to slope stability. Proc. of the 9th Int. Conf. on Soil Mech. and Found. Engg., Tokyo, Vol. 2, pp. 9-12.

Basudhar, P.K. 1976. Some applications of mathematical programming techniques to stability problems in geotechnical engineering. Ph.D. thesis, Deptt. of Civil Engg., Indian Institute of Technology, Kanpur.

Basudhar, P.K., Valsangkar, A.J. and Madhav, M.R. 1979. Nonlinear programming in automated slope stability analysis. Ind. Geotech. J., Vol. 9, No. 3, pp. 212-219.

Bhowmik, S.K. 1984. Optimization techniques in automated slope stability analysis. M.Tech. thesis, Indian Institute of Technology, Kanpur.

Bhowmik, S.K. and Basudhar, P.K. 1986. Sequential unconstrained minimization technique in slope stability analysis. Communicated.

Bishop, A.W. and Morgenstern, N. 1960. Stability coefficients for earth slopes. Geotechnique, Vol. X, No. 4, pp. 129-150.

Boutrup, E., Lovell, C.W. and Siegel, R.A. 1979. STABL2 - A computer program for general slope stability analysis. Numerical Methods in Geomechanics, Proceedings of the Third International Conference on Numerical Methods in Geomechanics, Aachen, Ed. W. Wittke, Vol. 2, pp. 747-757, A.A. Balkema, Rotterdam.

Castillo, E. and Revilla, J. 1977. The calculus of variations and the stability of slopes. Proc. of the IX Int. Conf. on Soil Mech.

and Found. Engg. (IX ICSMFE), Vol. 2, pp. 25-30, Tokyo.
Celestino, T.B. and Duncan, J.M. 1981. Simplified search for non-circular slip surfaces. X ICSMFE, Vol. 3, pp. 391-394, Stockholm.
Ching, R.K.H. and Fredlund, D.G. 1983. Some difficulties associated with the limit equilibrium method. Can. Geotech. J., Vol. 20, pp. 661-672.
Dhawan, R.K. 1986. Parametric studies in slope stability using SUMSTAB package. M.Tech. thesis, Indian Institute of Technology, Kanpur.
Eisenstein, Z. 1977. Computer analysis in earth dam engineering. Proc. of the Speciality Session on Computers in Soil Mechanics: Present and Future, IX International Conference on Soil Mech. and Found. Engg., Tokyo, July, M.A.A. Publishing Company, pp. 77-122.
Fox, R.L. 1971. Optimization methods for engineering design. Addison-Wesley, Reading, Mass.
Fredlund, D.G. 1977. Slope stability software usage in Canada . Proc. of the Speciality Session in Computers in Soil Mech.: Present and Future, IX Int. Conf. on Soil Mech. and Found. Engg., pp. 289-302.
Fredlund, D.G., Krahn, J. and Pufahl, D.E. 1981. The relationship between limit equilibrium slope stability method. X ICSMFE, pp. 409-416.
Fredlund, D.G. and Krahn, J. 1976. Comparison of slope stability methods of analysis. Proceedings, Twenty-Ninth Canadian Geotech. Conf., Vancouver, pp. 858-874.
Janbu, N. 1973. Slope stability computations. Embankment Dam Engineering, Casagrande Vol., John Wiley and Sons, Inc.
Krugman, P.K. and Krizek, R.J. 1973. Stability charts for inhomogeneous soil condition. Geotechnical Engineering, Vol. 4, pp. 1-13, Journal of South East-Asian Society of Soil Engineering.
Martins, J.B. 1982. Embankments and slopes by mathematical programming. J.B. Martins (Ed.), Numerical Methods in Geomechanics, pp. 305-334, D. Reidel Publishing Company.
Morgenstern, N.R. and Price, W.D. 1965. The analysis of the stability of general slip surfaces. Geotechnique, 15, No. 1, pp. 79-93.
Munro, J. 1982. Plastic analysis in geomechanics by mathematical programming. J.B. Martin (Ed.), Numerical Methods in Geomechanics, pp. 247-272, D. Reidel Publishing Company.
Narayan, C.G.P. and Ramamurthy, T. 1980. Computer algorithm for slip circle analysis. Indian Geotechnical Journal, Vol. 10, No. 2, pp. 164-172.
Nguyen, V.U. 1985. Determination of critical slope failure surfaces. ASCE, Journal of Geotechnical Engineering, Vol. 111, No. 2, Feb., pp. 238-250.
Ramamurthy, T., Narayan, C.G.P. and Bhatkar, U.P. 1977. Variational methods for slope stability analysis. IX ICSMFE, Vol. 2, pp. 139-142.
Spencer, E. 1973. Thurst line criterion in embankment stability analysis, Geotechnique, Vol. 23, No. 1, pp. 85-100.
Talesnick, M. and Baker, R. 1984. Comparison of observed and calculated slip surface in slope stability calculations. Canadian Geotech. J., Vol. 21, No. 4, Nov., pp. 713-719.
Yudhbir and Basudhar, P.K. 1986. Revised stability analysis of down stream slope of Beas dam at Pong. A report submitted to the Beas Dam Design Directorate.
Whitman, R.V. and Bailey, W.A. 1967. Use of computers for slope stability analysis. J. of Soil Mech. and Found. Div., ASCE, Vol. 93, No. SM4, pp. 475-498.
Wright, S.G., Kulhawy, F.H. and Duncan, J.M. 1973. Accuracy of equilibrium slope stability analysis. Journal of the Soil Mech. and Found. Div., ASCE, Vol. 99, No. SM10, pp. 783-791.
Zeitlen, J.G., Wiseman, G., Komornik, A. and Birnbaum, A. 1977. Experiences with computer based techniques in solving geotechnical engineering problems in Israel. IX Int. Conf. on Soil Mech. and Found. Engg., pp. 303-313.

Computer and Physical Modelling in Geotechnical Engineering, Balasubramaniam et al. (eds)
© 1989 Balkema, Rotterdam. ISBN 90 6191 864 2

Dynamic slope stability analysis using the finite element method

N.C.Koutsabeloulis & D.V.Griffiths
University of Manchester, UK

ABSTRACT: Finite element solutions are presented to determine the yield acceleration coefficient for the seismic design of cohesionless slopes. An "explicit" time integration scheme is employed for the dynamic analysis and the dynamic slope stability depends on the static slope stability. Results are compared to those provided by classical theory.

1 INTRODUCTION

Many of the engineering works required to support modern high-density agricultural and industrial societies involve earth cut and fill slopes. Many of these slopes are located in earthquake regions of the world, hence a rational consideration of earthquake action in the design and construction is essential to ensure safe and economical works.

The present work is an attempt to use the finite element method to evaluate the seismic response of slopes and dams to subjected dynamic loading.

Seed and Goodman (1964) have defined a horizontal stability coefficient called the "yield acceleration" which represents the horizontal acceleration at which slippage and large deformations develop. Later Goodman and Seed (1966) tried to incorporate their previous work with Newmark's (1965) in order to measure displacements due to the previously defined "yield acceleration".

Both papers by Seed and Goodman presented results obtained using a shaking table and the input motion was of the sinusoidal type. The only analytical method available to determine this stability coefficient is provided by the classical theory which considers the stability of a block resting on the face of an infinite slope.

Whitman and Bailey (1967) suggested that in order to solve slope stability problems a method must be developed which includes the stress-strain characteristics of the soil. Lysmer (1978) also stated that the ultimate aim of a dynamic analysis is to be able to predict permanent deformations using non-linear material properties. The method described here aims to predict both the seismic coefficient and give an indication of the likely displacements.

Both static and dynamic cases of slope stability are considered using a non-linear elastic perfectly plastic stress strain law. Eight noded rectangular elements with a 2 x 2 integration were used throughout. Plane strain conditions were considered in relation to a non-associated flow rule using Mohr-Coulomb's criterion.

The static analysis used the "viscoplastic" numerical approach while the dynamic one used an "initial stress" type incorporated into an "explicit" dynamic scheme. Drained conditions were considered and the soil had no physical damping.

2 SOIL BEHAVIOUR - NUMERICAL MODELS

It is generally accepted that soils behave in a more complicated manner than a simple elastic theory can predict. One of the main physical features of soil behaviour is the irrecoverability of strains which elastic theory cannot predict. The relationship between stresses and strains is a complicated function and Figure 1 presents typical results that might be obtained from a drained triaxial test on a soil. The stress-strain behaviour is clearly non-linear; both the stiffness and strength are stress dependent. The simplest amongst the non-linear soil models is the elasto-perfectly plastic one shown in Figure 2.

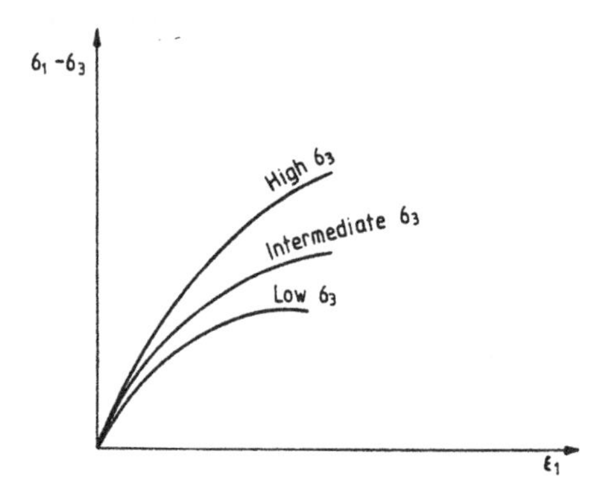

FIG. 1. STRESS STRAIN SOIL BEHAVIOUR

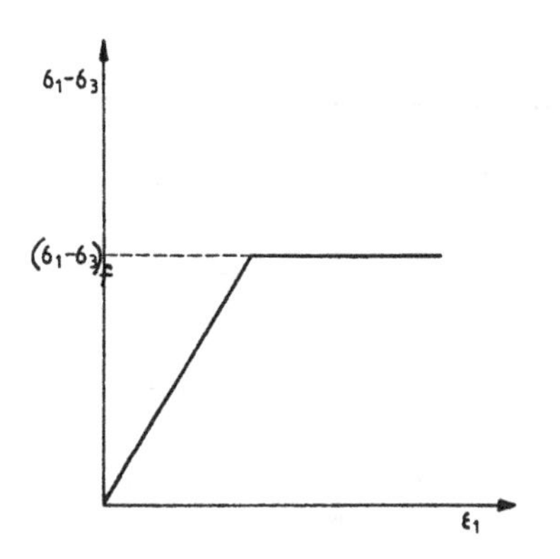

FIG. 2 ELASTO PERFECTLY PLASTIC MODEL

In this theory a 'yield surface' separates stress states which give rise to both elastic and plastic (irrecoverable) strains. This implies that the soil behaves elastically until a failure criterion is violated, at which point plastic behaviour commences.

Numerically the soil behaviour can be determined by different models amongst them the "viscoplastic" and the "initial stress" approaches. The "viscoplastic" approach, was first introduced by Zienkiewicz et al (1972) and it has been shown, Griffiths (1980, 1981) to be an efficient and versatile approach of solving plasticity problems in geomechanics.

This approach which falls into the initial strain of solution techniques, iterates using equivalent elastic solutions until any stresses that originally violated yield have returned to the failure surface within quite strict tolerances. The convergence criterion was implemented by observing the change in the body forces from one iteration to the next. These self-equilibrium body forces are incremented at each iteration by an amount directly related to the magnitude by which the stresses still violate yield. As the stresses return to the yield surface the increment of body-forces diminishes.

This procedure temporarily allows stresses to exist outside yield until they are corrected.

This numerical approach was adopted to perform the static analysis because previous experience, Griffiths (1980), showed its efficiency in a static slope stability problem in which the load due to the self-weight of the soil was applied in one increment.

The differences between "viscoplasticity" and "inital stress" appears in the way they handle body forces due to yielded integration points. In the "initial stress" approach those bodyforces are re-evaluated at the beginning of each iteration instead of being incremented each time. Because convergence applies to loads and not to stresses, the "initial stress" due to the way it tries to achieve this convergence might leave some integration points outside the yield surface. For this reason, correction factors are applied at the end of each load increment to practically eliminate overshoot. The "initial stress" approach as well as the correction of the overshoot was firstly introduced by Nayak et al (1972).

Since it was used in an "explicit" type dynamic algorithm, there was no need for any iterations and convergence in the dynamic analysis was said to be achieved in one increment. At the end of each increment, Nayak's correction factor was applied.

In the static case, convergence was said to have occurred when the change of body forces, non-dimensionalised with respect to the largest absolute value nowhere exceeded a tolerance which had to be defined during the analysis. This was because, for the special case of cohesionless soil properties, the value of the convergence tolerance appeared to have a significant effect on the result.

3 DYNAMIC ANALYSIS

In most numerical solutions of continuum problems the governing partial differential equations are first discretised in space, in order to yield a set of ordinary differential equations in time. If one considers this process in the domain of space-time, the integration then is along lines perpendicular to the space subdomain. There are two basic types of schemes for integrating the ordinary differential equations "explicit" and "implicit".

In "explicit" schemes, differential equations are used that permit the displacement at the next time step to be found in terms of the accelerations and displacements at the previous step, so that the procedure does not involve the solution of any equations.

In "implicit" schemes the differential equations for the displacements at the next time step involve the accelerations at the next time step, so the determination of the displacements involves the solution of a system of equations. For the purposes of the present work the "explicit" scheme was used. The basic equations were given by Bathe and Wilson (1976), but a refinement was made in order to economise space by Smith (1982), so that the displacements in the new time steps were derived from the acceleration and velocities from the previous step and not from the previous two steps as others suggested, i.e. Belytschko (1978), Chang (1979).

The method also can incorporate material and geometry damping. The equations thus were:

$$ {}^{t+\Delta t}\underline{x} = {}^{t}\underline{x} + \Delta t\,{}^{t}\underline{x} + 1/2\Delta t^{2}\,{}^{t}\underline{x} \qquad (1) $$

$$ {}^{t+\Delta t}\underline{x} = {}^{t}\underline{x} + 1/2\Delta t({}^{t}\underline{x} + {}^{t+\Delta t}\underline{x}) \qquad (2) $$

$$ {}^{t+\Delta t}\underline{x} = \Delta\underline{R}/(\underline{M} + 1/2\Delta t\underline{C}) \qquad (3) $$

where $\underline{M}$ is the mass matrix
$\underline{C}$ is the damping matrix
Δt is the time step used in the analysis
${}^{t}\underline{x}$ is the vector of displacements time t
${}^{t}\underline{x}$ is the vector of velocities at time t
${}^{t}\underline{x}$ is the vector of accelerations at time t
and $\Delta\underline{R}$ is the sum of the external and internal loads.

In the present work in order to determine the time step of the analysis the eigenproblem was firstly solved, while the time step was said to be

$$ \Delta t < T_n/4 \qquad (4) $$

where T_n is the period derived from the highest frequency of the system. The above equation is a crude approximation of the formula proposed by Argyris et al (1973), but it kept the "explicit" method stable throughout the analysis because it predicted a lower bound value. The mass matrix ($\underline{M}$) formulation used a technique proposed by Hinton et al (1976). The end result of the procedure gives a "consistent" mass matrix from which the "lumped" matrix can easily be derived if required.

The damping of the system can be due either to the material properties or to the geometry. The geometry damping is due to effects at the boundaries of the system because of multi-reflections of the propagating waves through the soil. These effects were taken into account by using the approach suggested by Lysmer and Kuhlemeyer (1968). Unified boundaries as proposed by White et al (1977) were also incorporated by the authors, but the result was not significantly altered.

For the purposes of the present work no damping effects were included in the analysis and the elasto-plastic work was done by an "initial stress" approach, as mentioned in the previous paragraph. The sequence of the operations, thus, is shown in Figure 3 and is similar to that proposed by Smith (1982).

4 STATIC ANALYSIS

The static problem is the basis of the work which has to be done in dynamic analysis. Stresses due to the weight of the soil itself must be evaluated at the end of the static analysis which must be used as initial stresses at the beginning of the dynamic analysis. It is obvious then that the more thorough the static analysis, the more accurate the dynamic analysis is likely to be.

Firstly, some points arising from the static analyses are discussed:

4.1 Effect of Poisson's ratio, v, and Young's modulus, E

Classical theory provides a straightforward relationship which evaluates the static factor of safety in the case of cohesionless materials, which is

$$F^{st} = \tan\phi/\tan\alpha \qquad (5)$$

where: ϕ is the friction angle
and α is the slope angle.

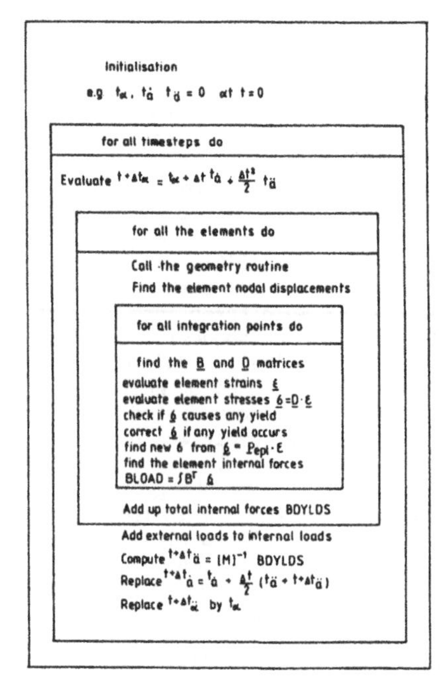

FIG 3. EXPLICIT METHOD USING INITIAL STRESS METHOD

In finite element analyses, using any iterative technique such as "viscoplasticity" or "initial stress", the Young's modulus, E, and the Poisson's ratio, v, are directly involved in the analysis.

Their effect on a static slope stability problem was therefore investigated, analysing three different slopes shown in Figures 4, 5 and 6.

In order to make the tests compatible, all three slopes were brought close to failure with the friction angle slightly higher than the slope angle in each case. The ratio δ_p/δ_e was examined, which specified the ratio of (the vertical plastic displacement at the crest)/(its relative elastic component). Figure 7 shows that for fixed values of Young's modulus, unit weight and convergence tolerance, Poisson's ratio varied almost linearly with the ratio δ_p/δ_e at failure. Figure 8, on the other hand, shows that fixed values of Poisson's ratio, unit weight and the same tolerance as before, the δ_p/δ_e remained constant. Therefore we can conclude that the Young's modulus makes no difference to the end result, while Poisson's ratio affects it in a normalised way.

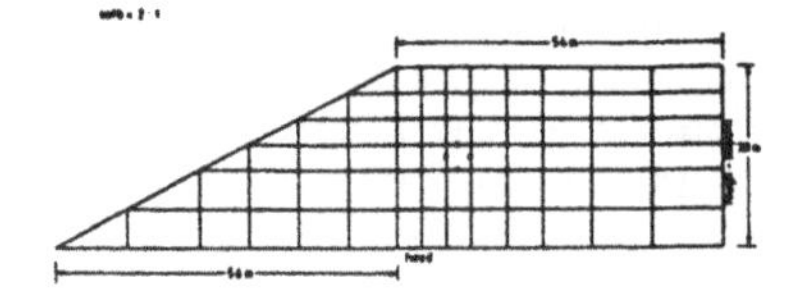

Fig. 4 Finite element discretisation of a slope COTB = 2:1

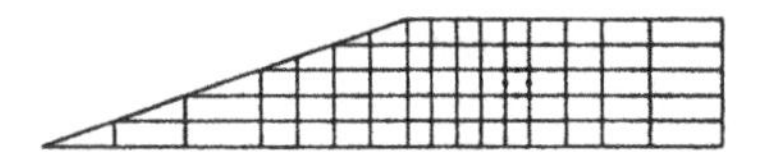

Fig. 5 Finite element discretisation of a slope COTB = 3:1

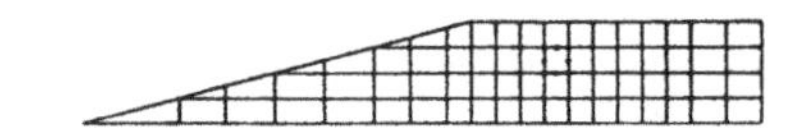

Fig. 6 Finite element discretisation of a slope COTB = 4:1

4.2 Effect of tolerance used in the convergence criterion

A parameter which did have a significant affect on accuracy was the convergence criterion. As mentioned in paragraph 2, convergence is checked against the change in body forces from one iteration to another. For cohesionless materials these changes of the body forces are very sensitive because the shear strength depends only upon the friction angle. The higher the friction angle becomes the more sensitive these changes are and the lower the tolerance should be.

An example is shown in Figures 9 and 10, where the crest and the toe plastic components of the vertical displacements respectively were monitored for different tolerances of the convergence criterion, when the friction angle of $\phi = 32.5°$. They demonstrated that a tolerance can be determined which will eliminate the numerical error and its influence on the end result.

4.3 Evaluation of the factor of safety

Zienkiewicz et al (1975) reported results of static slope stability problems for cohesive soils incorporating both associated and non-associated flow rules using the Mohr-Coulomb criterion. The computed factors of safety were very close to the theoretical ones.

In the present section the same rules applied for cohesionless materials after determining a tolerance for the convergence criterion and fixed values for the Young's modulus $E = 130000\ kN/m^2$ and the Poisson's ratio $v = 0.30$. The slope of Figure 4 was used, while the friction angle was reduced by an amount of $\tan\phi' = \tan\phi'/F$.

The ϕ' at which no convergence could be achieved would determine the friction angle for failure, while the expected factor of safety would be $F = \tan\phi/\tan\phi'$.

It was also useful to monitor the spread of plasticity while the angle was continually reduced.

At $\phi = 40°$ plasticity spots appeared along the face of the slope, while a tension zone was observed at the vertical boundary, see Fig. 11. Similar tension zones were reported by Smith and Hobbs (1974).

As the ϕ angle was reduced the plasticity spread towards the toe from the crest, while the tension zone became deeper. A new plastic zone appeared in the middle of the slope, see Fig 12.

When the factor of safety from equation (5) became almost unity, the plasticity zone was spread throughout the slope, indicating failure conditions, see Fig 13. At the same time the vertical crest displacement was monitored as shown in Figure 14. This indicated a sudden increase in displacements at a friction angle ϕ between 26.5° and 26.3°. The analytical solution obtained by equation (5) gave $\phi = 26.565°$.

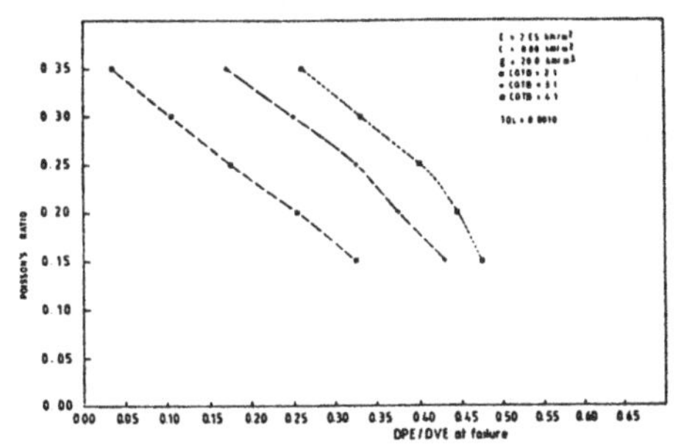

Fig. 7 Influence of poisson's ratio on the ratio DVP/DVE

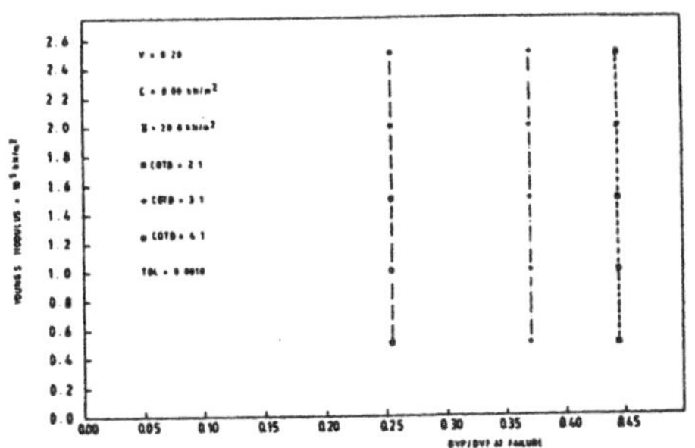

Fig. 8 Influence of young's modulus on the ratio DVP/DVE

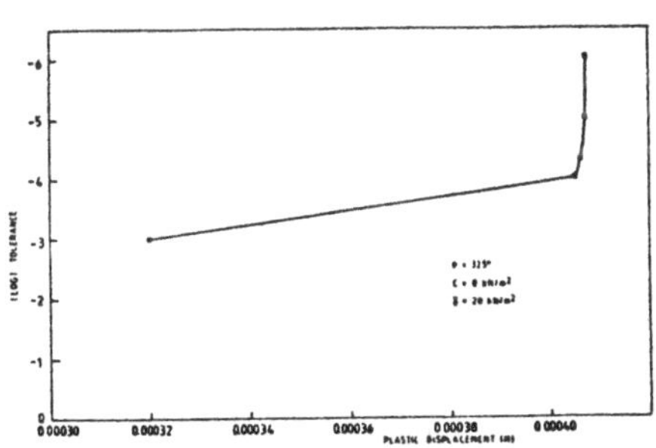

Fig. 9 Variation of vertical crest plastic displacement with convergence tolerance

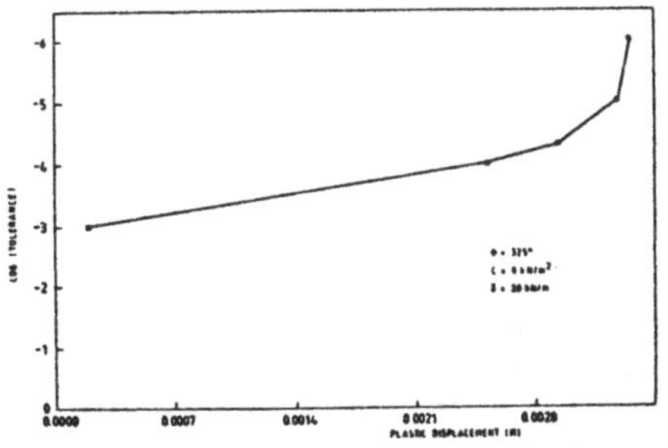

Fig. 10 Variation of toe vertical plastic displacement with convergence tolerance

5 THE DYNAMIC STABILITY COEFFICIENT

The conclusion of such a thorough static analysis was to evaluate the correct stresses which will be used in order to determine the dynamic stability coefficient, the "yield acceleration" as Seed and Goodman (1964) named it, for cohesionless slopes. Classical theory which considers the case of the block resting on a slope determines this stability number as

$$Ky = \tan(\phi - a) \qquad (6)$$

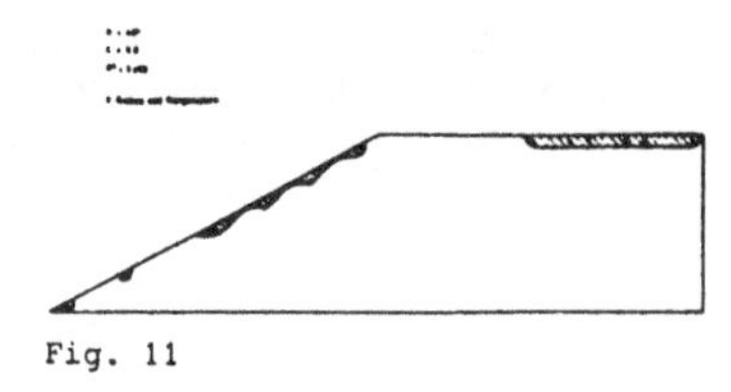

Fig. 11

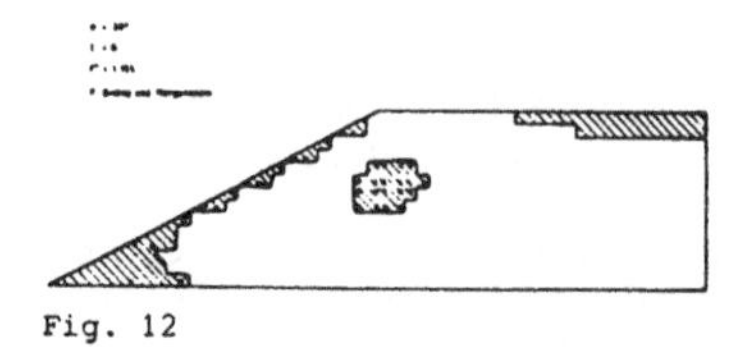

Fig. 12

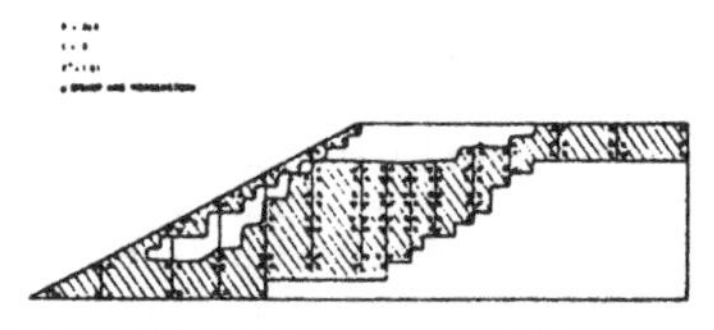

Fig. 13 Yielding zones within the slope produced by F.E. analysis

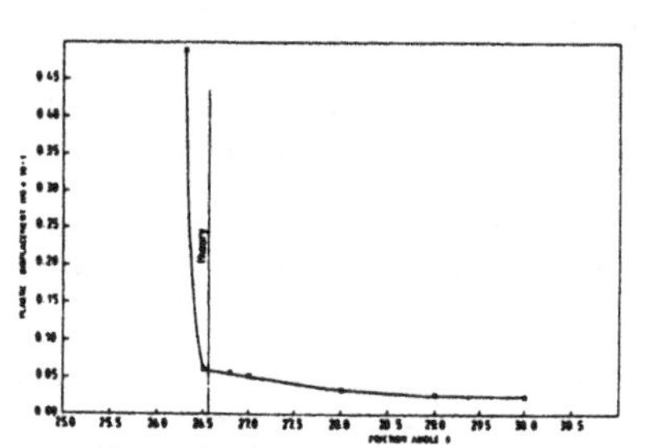

Fig. 14 Vertical crest displacement variation with the friction angle

Seed and Goodman (1964) obtained higher values than those proposed by equation (6) and their explanation was that a shear strength appeared along the face of the slope which they called "shear intercept". (Equivalent to a small cohesion). Morris (1979), reassessed the work of Seed and Goodman using centrifuge tests and noticed differences of up to 24% from those obtained by equation (6). He suggested that as a basis of comparison, the relationship

$$\theta = a + \tan^{-1}(ky) > \phi \qquad (7)$$

could be used where ϕ was the angle needed to cause slippage.

At the end of his work he suggested that some of these differences occurred due to the input horizontal motion which was, as in Seed and Goodman's case, a sinusoidal type whereas equation (6) assumes a steady with time input motion.

In the present work a steady input acceleration was adopted and the behaviour of a slope was monitored for a time span of $t = 4 \sim 5 * T_1$, where T_1 was the fundamental period of the system, obtained from an eigenvalue analysis. Failure was said to have occurred for that Ky input acceleration at which the horizontal movement of the crest of the slope increased steadily with time.

The mesh shown in Figure 4 was adopted and after firstly performing a static analysis the Ky value defining the seismic coefficient was found for a friction and of $\phi = 40°$. Initially a Ky = 0.09g acceleration was input and the crest and toe behaviour plotted in Figure 15. The oscillations of the crest suggested that some elastic resistance was still present. Figures 16, 17 and 18 show the crest displacement for increasing values of Ky. For Ky = 0.16g, the displacements showed no elastic oscillation but just increased steadily with time and at this point dynamic "failure" was said to have occurred. The corresponding value of from equation (7) was equal to 35.8° compared with the 'exact' value of 40°. Similar analyses were also performed for $\phi = 32.5°$ and $\phi = 27.5°$ and the results summarised in Figure 19. The computed values of Ky were compared to those given by equation (6), with the factor of safety obtained from equation (5).

6 CONCLUSIONS

A method has been presented to determine the seismic stability coefficient for safe design of cohesionless slopes or dams against earthquakes or blast. The numerical modelling was performed using the finite element method. A static analysis was necessary prior to dynamic analysis and some assumptions were necessarily made to determine which Ky value represents the seismic coefficient. Non-linear material behaviour was considered incorporating the "viscoplastic" approach for the static analysis, while an "initial stress" approach was incorporated into an "explicit" time interaction scheme. The results obtained were in good agreement with theory. Eight-noded quadrilaterals were used for both the static and dynamic analyses and a non-associated flow rule was related to the

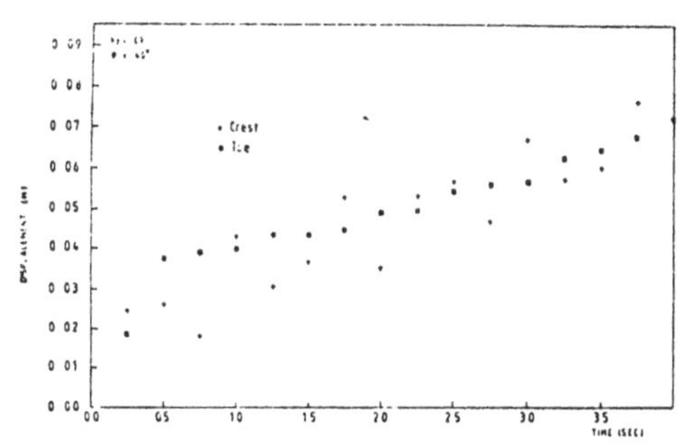

Fig. 15 Toe and crest displacement time histories

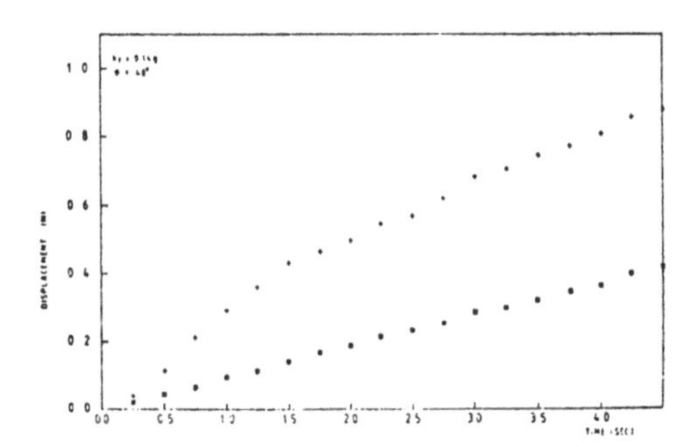

Fig. 16 Toe and crest horizontal displacement time histories

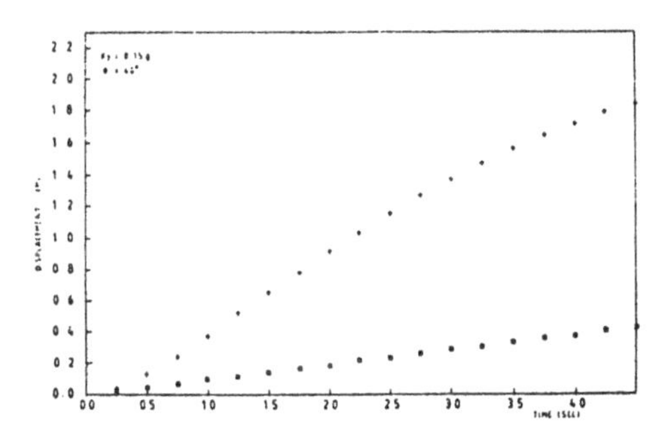

Fig. 17 Toe and crest horizontal displacement time histories

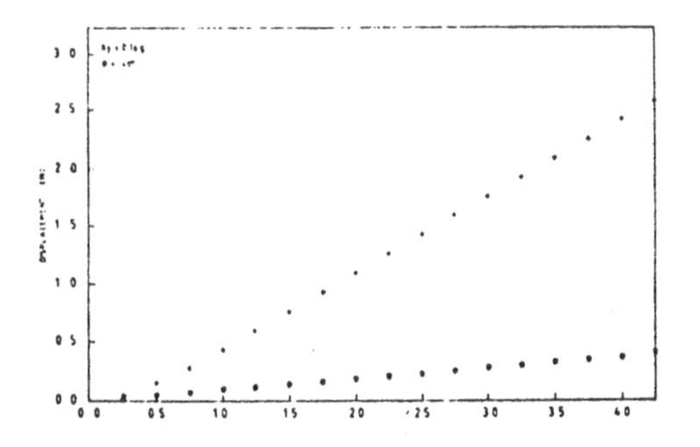

Fig. 18 Toe and crest horizontal displacement time histories

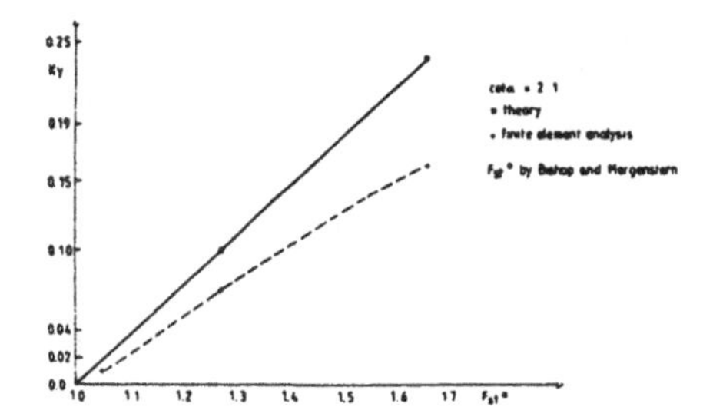

Fig. 19 Comparisons of results obtained using F.E. analysis and theoretical ones

Mohr-Coulomb criterion. The system had no damping either due to soil properties or due to viscous boundaries.

Possible advantages of the method described herein over other numerical approaches (i.e. Vrymoed (1981)) is that both displacements and stresses are considered. As in the methods described by Newmark (1965), displacements and permanent deformations form the main criterion for dynamic "failure".

REFERENCES

Seed, H.B. and Goodman, R.E. "Earthquake stability of slopes of cohesionless soils". ASCE, Journal of the Soil Mechanics and Foundations Division, Vol 90, No.SM6, November 1964.

Goodman, R.E. and Seed, H.B. "Earthquake-induced displacements in sand embankments". ASCE, Journal of the Soil Mechanics and Foundations Division, Vol 92, No.SM12, March 1966.

Newmark, N.M. "Effects of earthquakes on dams and embankments". Geotechnique 15, No.2, 139-160, 1965.

Whitman, R.V., Bailey, W.A. "Use of computers for slope stability analysis". Journal of the Soil Mechanics and Foundations Division, ASCE, Vol 93, No.SM4, 1967, pp 475-498.

Lysmer, J. "Analytical procedures in soil dynamics". Conference on Earthquake Engineering and Soil Dynamics, Passadena, USA, 19-21/6/78.

Griffiths, D.V. "Finite element analyses of walls, footings and slopes". Ph.D. thesis, University of Manchester 1980.

Griffiths, D.V. "Computation of strain softening behaviour". In Symposium on the implementation of computer procedures and stress-strain laws in Geotechnical Engineering, Chicago eds, C.S.Desai and S.K.Saxena pp 591-603, Durham Press 1981.

Zienkiewicz, O.C. and Cormeau, I.C. "Viscoplastic solution by the finite element process". Arch.Mech.24, 1972, pp 873-888.

Nayak, G.C. and Zienkiewicz, O.C. "Elasto-plastic stress analysis. A generalisation for various constitutive relations including strain softening. I.J.N.M.E., Vol 5, pp 113-135 1972.

Bathe, K.J., Wilson, E.L. "Numerical methods in finite element analysis". Prentice Hall Inc., 1976.

Smith, I.M. "Programming the finite element method with application to Geotechnics". Wiley and Sons, 1982.

Belytschko, T. "Explicit time integration of Structure-Mechanical systems". Advanced Structural Dynamics, Seminar held at ISPRA, Italy, October 9-13, 1978.

Chang, C.T., "Non-linear response of earth dams and foundations on earthquakes". Ph.D. thesis, University of Wales, 1979.

Hinton, E., Rock, T., Zienkiewicz, O.C. "A note on mass lumping and related processes in the finite element method". Earthquake Engineering and Structural Dynamics, Vol 4, pp 245-249, 1976.

Lysmer, J., Kuhlemeyer, R.L. "Finite dynamic model for infinite media" ASCE, Journal of the Engineering Hydraulics Division, Vol.95, No.EM4, August 1969.

White, W., Valliappan, S., Lee, I.K. "Unified boundary for finite dynamic models". ASCE, Journal of the Engineering Mechanical Division, Vol 103, No.EM5, October 1977.

Zienkiewicz, O.C., Hympheson, C., Lewis R.W. " Associated and non-associated viscoplasticity and plasticity in soil [illegible]. Geotechnique 25(4), 197[illegible] [illegible]689.

Smit[illegible] .. and Hobbs "Finite element [illegible]ysis of centrifuged and built-up slopes". Geotechnique 24, No.4, 1974, pp 531-559.

Morris, D.V. "The centrifuge modelling of soil-structure interaction and earthquake behaviour". Technical Report, Cambridge University, 1979.

Vrymoed, I. "Dynamic FEM model for Croville Dam". ASCE, Journal of the Geotechnical Engineering Division, Vol 107, No.GT8, August 1981.

2. Design and analysis of foundations

Computer and Physical Modelling in Geotechnical Engineering, Balasubramaniam et al. (eds)
© 1989 Balkema, Rotterdam. ISBN 90 6191 864 2

Computer aided designing, detailing and drafting of transmission line tower foundations

A.R.Santhakumar & S.Arunachalam
Anna University, India

ABSTRACT: The choice of desk top computer in preference to a computer terminal is generally bound by the cost and ease of programming. In certain aspects, a mini-computer has a distinct advantage over much larger systems. One such is the area of computer aided interactive designs where man and machine combine. This paper presents an example of the application of CAD for a Reinforced Concrete Tower Foundation Design. The flow chart of the programme used is included.

1 INTRODUCTION

Programmes have been prepared for structural analysis, detailed design, bar scheduling and costing. For small and medium size design office, the personal computer programme will be quite useful. These programmes enable the final design including costing of tower foundations. This becomes fairly important in view of the fact that the foundation cost is likely to be 20 to 30% of the tower cost depending on the vagaries of the soil.

2 STAGES OF DEVELOPMENT OF PROGRAMME

The development of the general purpose programme was scheduled in the following six stages:

1 Analysis technique and choice of type of foundation.
2 Calculation of area of steel.
3 Bar selection and concrete grade selection.
4 Detailing.
5 Bar scheduling.
6 Costing.

3 FEATURES OF VARIOUS STEPS

The six steps described in the previous section are explained below.

1 The analysis technique and choice of type of foundation depended on

. Properties of the soil
. Tower type
. Soil structure interaction analysis
. Parametric design to satisfy soil condition and
. Structural design.

The major loads that were considered are listed below.

. Uplift
. Down thrust
. Lateral load and
. Overturning moment.

The foundation should satisfy the factors of safety specified by the relevant code of practice (2). Two conditions of loading were considered.

. Normal condition
. Broken wire condition.

The following types of soil were envisaged and the design equations used were based on N value normally obtained from field test (Standard Penetration Test (SPT))

. Sandy soil
. Claey soil
. Clay-silt-sand mixtures (c-Ø soil)
. Rock

The design equations are not presented here but has been described in detail elsewhere (3).

In general, the factor of safety for the various loads considered were obtained and checked against those desired. The factor of safety was adjusted to be not more than 10% of those desired by Do Looping computer technique.

2 Calculation of area of steel

Equations were developed to optimise the area of reinforcement for axial force, shear force and moment for a given effective cross section of concrete and permissible stresses in steel and concrete. The bar selections were made such that while round-

-ing off for the available bar sizes least excess was provided.

3 Bar selection and concrete grade selection

The programme routine was developed to list size, number and layout of reinforcement. This was done in an interactive mode. The details of this, are shown in the form of a flow chart (Fig.1). The cost of the foundation depends on area of concrete and grade of concrete used. Therefore cost of foundation was worked out, for more than one alternative, and the grade of concrete which results in the least cost, is chosen by the programme. Later this grade is adjusted to suit the available material in the field. This way a concious choice of grade of concrete is made with the object of effecting economy.

4 Detailing

The programme itself gives the detailing of reinforcement. A typical output will indicate the bar sizes, the number of bars, the lengths of bars and the overlaps to be provided (Fig.2).

5 Bar scheduling

The programme prints out bar schedule based on identified types of cranking and presents the results in the form of a table (Table 1).

6 Costing

The programme calculates the total quantity of steel and concrete. Based on the prevailing rates which are input, the overall cost of the alternatives chosen are computed.

4 EXAMPLE

Input

- Design forces on tower leg
  - Ultimate compression : 81400 Kg.
  - Ultimate uplift : 58250 Kg.
  - Ultimate shear : 2250 Kg.
- Tower data
  - Base width : 4m
  - Height : 36m
- Soil data
  - Soil type : medium dense sand
  - Site location : Manali (Madras)
  - Average SPT Value : 12

  Insitu density $r=1.79t/m^3$, $r_{sub}=1.0t/m^3$,

  Angle of internal friction $\phi = 32^o$
  Water table : 1.5m below ground level.
- Type of foundation selected
  Programme based on the above data selects pad footing of size 2.5m X 2.5m 2.5m depth.

Analysis

- Check for uplift
  The uplift capacity is assessed as 58.32t using the equation,

$$T_u = CBD + s_f BrD^2 K_u \tan\phi + W \quad (1)$$

Where C = cohesion
B = base width and
D = depth of foundation
W = weight of soil above foundation base
K_u = earth pressure coefficient
s_f = shape factor

- Check for bearing capacity
  The bearing capacity of soil for this foundation is assessed as 94.8t using the equation,

$$Q_{ub} = CN_c s_c d_c i + q(N_q - 1) s_q d_q i_q + 1/2 \; B_r N_r s_r d_r i_r W' \quad (2)$$

Where N_c, N_q, N_r are bearing capacity factors
s,d,i are shape, depth and inclination factors respectively
W' = water table correction factor
q = surcharge

- Check for settlement
  The settlement of the foundation is computed as 0.31cm using the equation,

$$S_i = \frac{IqB(1-\mu^2)}{E_s} \quad (3)$$

Where I = influence factor
q = soil pressure
E and μ = Youngs' modulus and poissons ratio of soil respectively.

- Check for lateral capacity
  The maximum lateral deflection at service load is computed as 0.31cm using the equation,

$$Y_{max} = \frac{A_y H T^3}{EI} + \frac{B_y M_t T^2}{EI} \quad (4)$$

Where A_y and B_y are deflection coefficients.
H = shear force
M_t = bending moment

T = stiffness factor and
EI = flexural rigidity of foundation.

The maximum bending moment is computed as 154856 Kgcm using the equation,

$M_{max} = A_m TH + B_m M_t$

Where A_m and B_m are moment coefficients

Structural Design

- Design of shaft

  Input data

  Deflection at ultimate load = 0.62cm

  Total moment : 205324 Kgcm

  Compression : 81400 Kg

- Selection of grade of concrete and steel

  Programme for the given data selects M_{15} concrete and Fe 415 steel and their design stresses are

  f_{ck}=150Kg/cm^2 and f_y=4150Kg/cm^2 respectively.

  The area of steel and concrete dimension are obtained as

  As : 15.9cm^2 and 30cm dia.

  The details given by the programme satisfy IS 456/1978 (4) code requirements are listed below.

  Longitudinal reinforcement : 16mm dia RTS 8 nos.

  Lateral reinforcement : Circular hoops of 8 mm dia at 15cmc/c.

  Anchorage length : 112cm.

  Design of base

  The base dimension is obtained as d_1 = 75cm and d_2 = 20 cm and the bar details as 10mm dia at 20 cmc/c.

The various alternatives that can be got in the computer aided design of foundations for a particular type of foundation is shown in table (1).

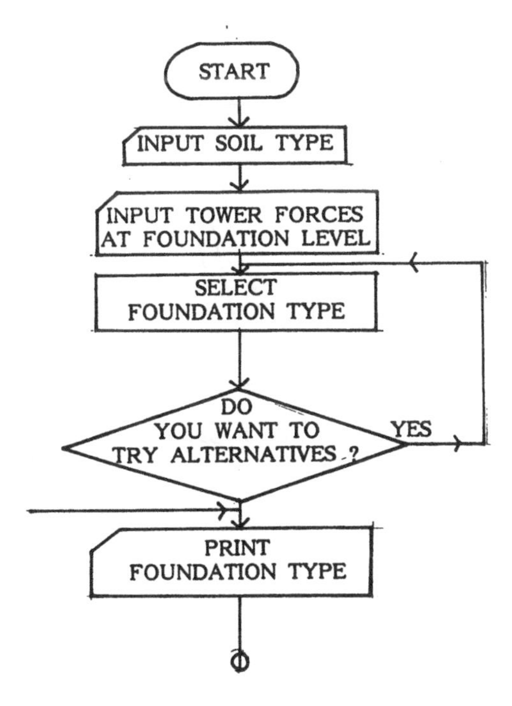

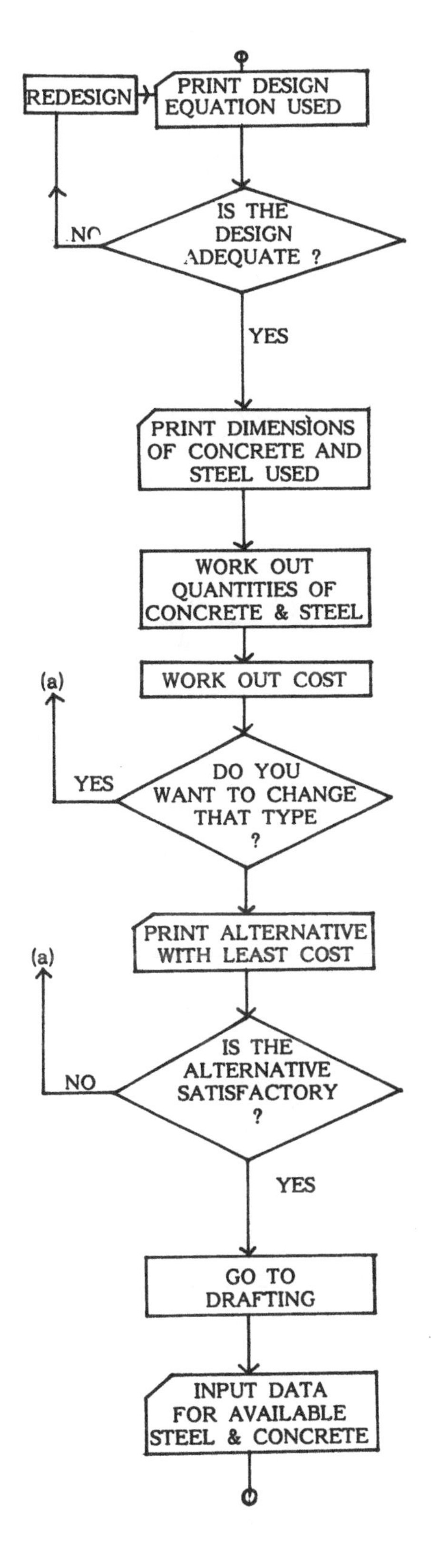

Table 1. Foundation detailing and scheduling

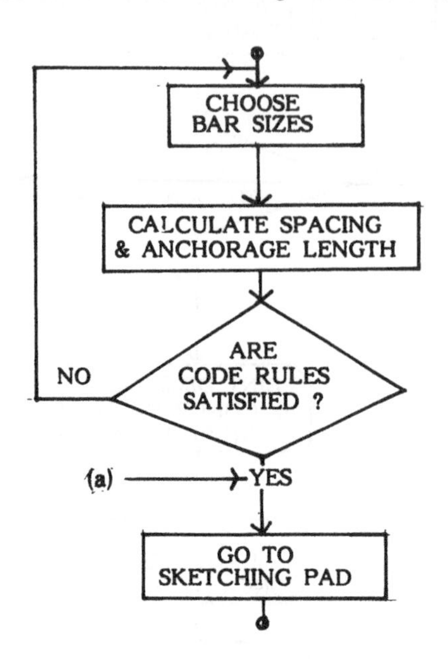

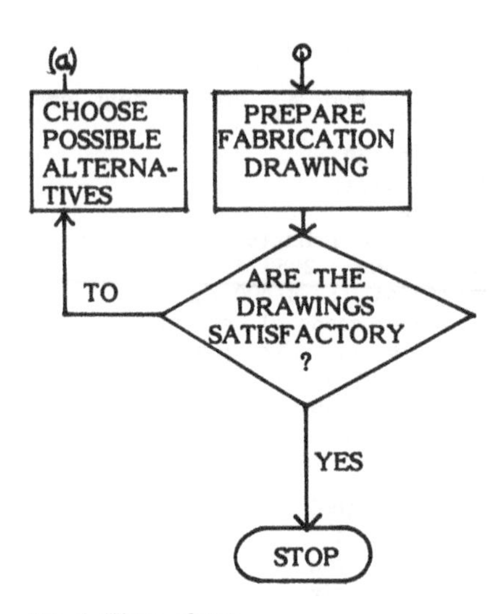

Fig.1 Flow chart

TOWER TYPE A
NORMAL FOUNDATION

DIRECT BEARING

TOTAL COMPRESSION PER LEG ...	KG.	81400
OVER LOAD OF CONCRETE	KG.	5468
TOTAL THRUST	KG.	86868
BASE WIDTH PROVIDED	M.	2.500
ULT. BEARING STRENGTH	KG.	94832
FACTOR OF SAFETY		2.18
CHECK FOR UPLIFT		
(ANGLE OF REPOSE Ø = 32.00)		
EFFECTIVE DEPTH BELOW G.L. .. H..	M.	2.500
CONC. VOLUME BELOW G.L.		
(EXCLUDING PAD)	CU.M.	1.028
VOLUME OF EARTH	CU. M.	13.350
NET WEIGHT OF EARTH	KG.	20871
CONC. VOLUME INCLUDING PAD ...	CU.M.	2.278
WEIGHT OF CONCRETE	KG.	5467
TOTAL ANCHORAGES........................	KG.	58320
NET UPLIFT.......................................	KG.	58250
FACTOR OF SAFETY		2.00
VOLUME OF CONCRETE PER		
TOWER	CU.M	9.112
MAX. VOLUME OF EXCAVATION		
PER TOWER	CU.M.	15.625

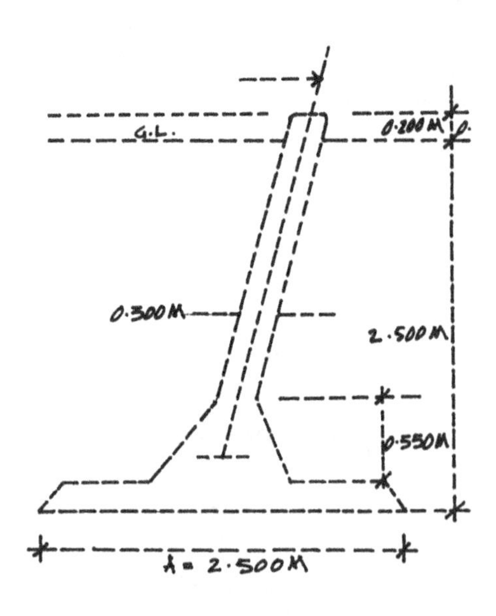

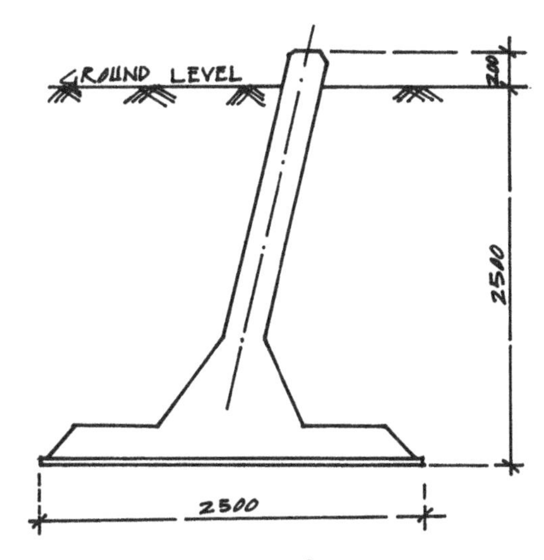

Fig.2 Footing outline for 0° - angle tower

5 CONCLUSION

The time required to prepare detailed drawing, bare schedule and detailed estimate for tower foundation design is remarkably reduced by the use of computers.

The use of computers can save at least 10% on the overall cost of foundations.

The cost of the foundation including the excavation cost can be correctly assessed by the programme. Hence, the programme developed can be quite useful for biding purposes and for choosing alternative bids.

Any type of unusual soil conditions can be taken care of with ease.

The programme eliminates the need for a draftsman and increases the design office efficiency.

Detailed design drawing can be prepared using the programme.

Final fabrication for the reinforcement grill can be prepared using the programme.

REFERENCES

Santhakumar, A.R. 1983. Mini computer aided interactive designing, detailing and costing of reinforced concrete multi storeyed buildings. First National Seminar on Computers and Civil Engineering, Madras.

IS 4091-1979. Code of practice for design and construction of foundations for transmission line towers, New Delhi.

Arunachalam, S., Boominathan, S. and Santhakumar, A.R. 1986. Analysis and design of tower foundations, Annual Seminar on Transmission Line Materials, CPRI, Bangalore.

IS 456 - 1978. Code of practice for plain and reinforced concrete, New Delhi.

Computer and Physical Modelling in Geotechnical Engineering, Balasubramaniam et al. (eds)
© 1989 Balkema, Rotterdam. ISBN 90 6191 864 2

Microcomputer aided design of mat foundation using rigid and flexible footing procedures

Chaim J.Poran
Civil Engineering Department, Polytechnic University, Brooklyn, New York, N.Y., USA

Fabio L.E.Liscidini
Polytechnic University, Brooklyn, New York, N.Y., USA

ABSTRACT: An interactive graphics microcomputer BASIC code for the structural analysis of mat footings is presented. The code incorporates two conventional methods; the rigid footing and the approximate flexible footing procedures. Based on the theory of plates on elastic foundations the flexible footing procedure is especially useful for analyzing mat foundations in general condition where a flexible mat supports columns at random locations with varying intensities of loads. The code is user friendly and does not require any input data files. The output includes plots of moment and shear distributions to be used for the detailed structural design. A design example is presented illustrating the many features of the code and the results are compared with the literature indicating excellent match.

INTRODUCTION

A mat or raft foundation is a combined footing that may cover part or the entire area under structure supporting a number of columns and walls. In most cases mat foundations are used for soils that have low bearing capacity where often spread footings may cover more than half of the building area. By combining all individual footings into one mat the pressure applied on the supporting soil is reduced, the bearing capacity is often increased, and it may also be more economical.

The mat considered in this paper is a common rectangular flat concrete slab with uniform thickness. This type of mat is suitable where the column loads are relatively small to moderate and the column spacing is relatively uniform and small. For this type of mats the structural design can be carried out by the two conventional methods used here: the rigid and the flexible footing procedures.

Other analytical procedures using finite element and finite difference methods are available. These methods are more computationally complicated and more suitable for mat footings with variable cross section and soils with variable coefficient of subgrade reaction, and are not discussed in this paper.

A manual analysis of a mat footing supporting a large number of column may become quite complex and time consuming. MAT - the BASIC interactive graphics code presented herein is a very useful analytical tool, simple to use and, if necessary, may be easily modified, due to its modular structure. The user is expected to select the appropriate design method (rigid or flexible) as prompted by the code. The two design methods are described as follows.

DESIGN METHODS

Conventional Rigid Procedure: In this method the mat is assumed infinitely rigid and therefore its infinitely small flexural deflections does not alter the distribution of the contact pressure with the soil. The contact pressure is therefore assumed to have a planar distribution with a centroid which coincides with the line of action of the resultant of all column loads supported by the mat. A mat with relatively uniform column spacing and loads (which do not vary more than 20% between adjacent columns) is considered rigid when the minimum column spacing is less than $1.75/\lambda$ or when the mat supports a rigid superstructure (according to ACI, 1966) where λ(the characteristic coefficient) is defined by Hetenyi (1946):

$$\lambda = \sqrt[4]{\frac{k\ b}{4E_c I}} \qquad (1)$$

where

k = coefficient of subgrade reaction
b = width of a strip of mat
E_c= modulus of elasticity of concrete
I = moment of intertia of the strip of width b

Based onthese assumptions the mat is analyzed by statics. The mat thickness is designed according to the governing building to resist the punching shear under the critical column.

As shown in Figure 1, the whole mat is analyzed in strips in each of the two perpendicular directions assuming that the strips act independently. By ignoring the shear transfer between adjoining strips each strip may be analyzed as an independent beam subject to column loads and contact pressure. Often this simplified analysis does not satisfy statics but is acceptable as long as the mat is sufficiently rigid as defined by ACI (1966). The dead weight of the mat concrete does not cause any flexural stress and therefore is excluded from this analysis.

The step by step analysis of a rigid mat is outlined by Das (1984) as follows:

1) The resultant force acting on the mat is calculated as the sum of all column loads supported by it.

2) The coordinates of the resultant force about the centroid of the mat area (the resultant eccentricities)are calculated and the contact pressure on the soil **q** may be determined for any point on the mat as

$$q = \frac{Q}{A} \pm Q \frac{e_y}{x} y \pm Q \frac{e_x}{y} x \quad (2)$$

where

Q = total load on the mat
A = total area of the mat
x,y = coordinates of any given point on the mat with respect to the x and y axes passing through the centroid of the mat area
e_x, e_y = coordinates of the resultant forces

The contact pressure is then determined at the column locations.

3) The maximum contact pressure q_{max} calculated by Eq.2 is compared with the allowable soil pressure to verify that

$$q_{max} \leq q_{all(net)}$$

4) The mat is divided into strips in the x and y directions (shown in Fig. 1) between the centers of the columns bays.

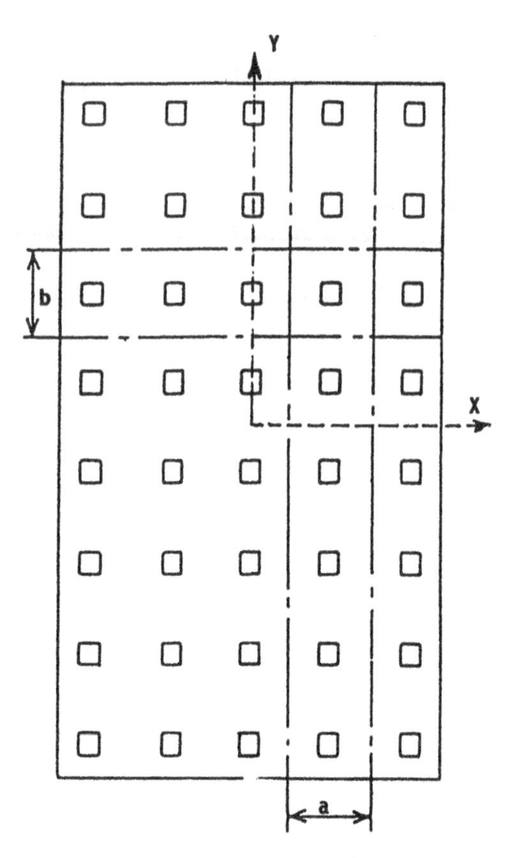

Figure 1. Rigid mat design where a and b are typical strips.

5) An average contact pressure for each strip is calculated. Based on the average contact pressure the total average load is determined and the modified average contact pressure and the column load modification factors are calculated.

6) The modified values are used for the calculation of the shear and moment diagram for the strips. The procedure (steps 5 and 6) is repeated for all strips in the x and y direction.

Based on the governing building codes the structural design of the mat may now be completed by calculating the thickness and the required reinforcement based on the shear maximum and minimum moment per unit width obtained from step 6.

Flexible Footing Procedure: ACI (1966) suggested to use the theory of plates on elastic foundations (Hetenyi, 1946) as a basis for flexible mat foundation design.

According to ACI (1966) the effect of a concentrated load on a typical mat has been found to be damped out quite rapidly. It is possible, therefore, to consider the mat as a plate and determine the effect of a column load in the area surrounding the load. By superimposing all the column loads within the zone of influence, the total effect of all the column loads at any point is determined. This zone of influence is generally not large and it will not be necessary to consider columns further than two bays in all directions to determine stresses at a particular point in most problems. Since the effect of each load is transmitted through the mat in a radial direction, polar coordinates are used in the procedure as follows:

1) The mat thickness (h) is calculated to resist the punching shear of the maximum column load according to the local building codes.

2) The flexible rigidity R is determined as

$$R = \frac{E_c h^3}{12(1 - \mu_c^2)} \qquad (3)$$

where E_c = Young's modulus of mat concrete
μ_c = Poisson's ration of mat concrete

3) The radius of effective stiffness L is determined as

$$L = \sqrt[4]{\frac{R}{k}} \qquad (4)$$

where k = coefficient of subgrade reaction

The zone of influence of any column load will be in the order of 3 to 4L.

4) As shown in Figure 2 the moment caused by a column load is determined at a point in polar coordinate system as

$$M_r = -\frac{Q}{4}\left[Z_4\left(\frac{r}{L}\right) - (1-\mu)\frac{Z_3'\left(\frac{r}{L}\right)}{\frac{r}{L}}\right] \qquad (5)$$

$$M_t = -\frac{Q}{4}\left[\mu Z_4\left(\frac{r}{L}\right) + (1-\mu)\frac{Z_3'\left(\frac{r}{L}\right)}{\frac{r}{L}}\right] \qquad (6)$$

where

r = distance of point under investigation from point load along radius
L = radius of effective stiffness
M_r, M_t = radial and tangential moments (polar coordinates) for a unit width of mat
Q = column load
Z_3, Z_3', Z_4, Z_4' = functions which can be found in Table III. Beams on Elastic Foundations Hetenyi (1946). Also shown in Figure 3.

and the deflection ρ at the point is

$$\rho = \frac{QL^2}{4D} Z_3\left(\frac{r}{L}\right) \qquad (7)$$

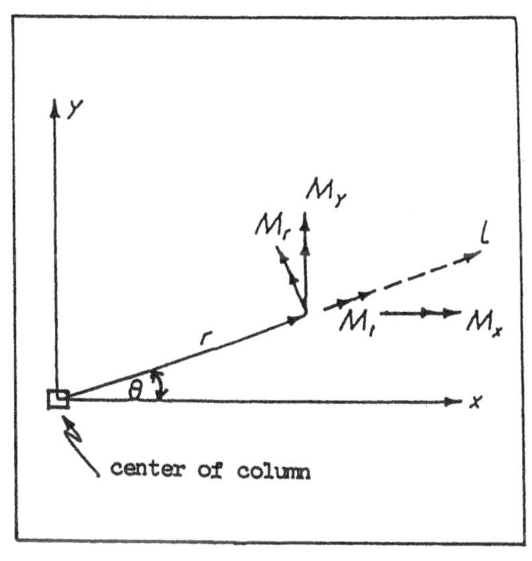

Figure 2. Coordinate system for flexible mat design.

5) The radial and tangential moments are converted to rectangular coordinates as

$$M_x = M_r \cos^2 + M_t \sin^2 \qquad (8)$$

$$M_y = M_r \sin^2 + M_t \cos^2 \qquad (9)$$

6) The shear force per unit width of mat V is determined as

$$V = -\frac{Q}{4L} Z_4' \frac{r}{L} \qquad (10)$$

7) The effects of a stiff foundation wall at the edge of a mat, and the case when the edge of the mat is located within the radius of influence of a column load under analysis, may be incorporated into the procedure according to ACI (1966).

These effects are not included in the current version of MAT.

8) Finally, the structural design may be completed based on the maximum values of shear and moment per unit width of mat according to the governing building codes.

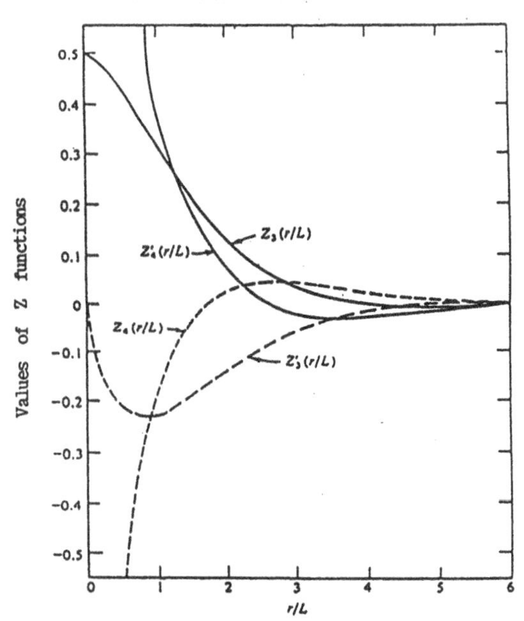

Figure 3. Z functions for moment, shear and deflection for flexible mat design (after Hetenyi, 1946)

THE COMPUTER CODE

MAT is a user friendly interactive graphics computer code. The code is programmed in GW-BASIC version 2.02 by Microsoft using the AT&T model PC6300 microcomputer using the MS-DOS version 2.11 operating system by Microsoft. This modular code is fully compatible with IBM XT and AT personal computers and may be easily modified for use with other personal computers which have a minimum of 128 kilobytes RAM.

The flow chart of MAT is shown in Figures 4a and 4b. The code output includes printer plots of the moment and shear diagrams to be used for the detailed reinforcement design.

The user selects between the rigid and flexible mat design procedures which are described in the flow charts shown in Figures 4a and 4b respectively.

The menues of MAT are described in Figure 5. The main program menue includes four menues; mat geometry definition, rigid footing and flexible footing procedures, and exit. Secondary menues are included in the first three menues as well as an additional multi-function secondary menue which is included in the geometry definition menue in order to create the geometry file. A few printer plots of screen images from this secondary menue are shown in Figure 6 to demonstrate the simplicity of this interactive code. SI or English units may be selected for the input data as appropriate.

EXAMPLE PROBLEM

Rigid Footing Procedure: Design example #4.4 from DAS (1984) is analyzed to illustrate the application of MAT. The problem definition, including column spacing and loads, is shown in the printed screen image in Figure 7a. The eccentricity of the resultant of the column loads and the final soil pressure values calculated at the columns locations are also shown in printed screen images (Figures 7a and 7b respectively). The code output includes printed values and printer plots of moment and shear diagrams for every strip in the x and y directions of the mat. For example, the printed and plotted moment and shear results for strip #3 (the upper strip in the x direction in Figure 7a) are shown in Figure 8. All the results from MAT match almost exactly the results from Das.

Flexible Footing Examples: According to ACI (1966) a mat may be considered flexible where column spacing is more than $1.75/\lambda$.

A number of flexible mat analyses were performed. In general it is observed that the magnitude of the maximum moments and shear forces per unit width tend to decrease as the value of subgrade reaction coefficient increases (denser soil) where all other parameter are unchanged. Due to the absence of a solved example of flexible mat in the literature reviewed here, no comparison is available at this time and the analyses results are therefore not included. However, a parametric study is currently being conducted to check the sensitivity of the flexible mat procedure to geometric parameters, columns loads and coefficient of subgrade reaction.

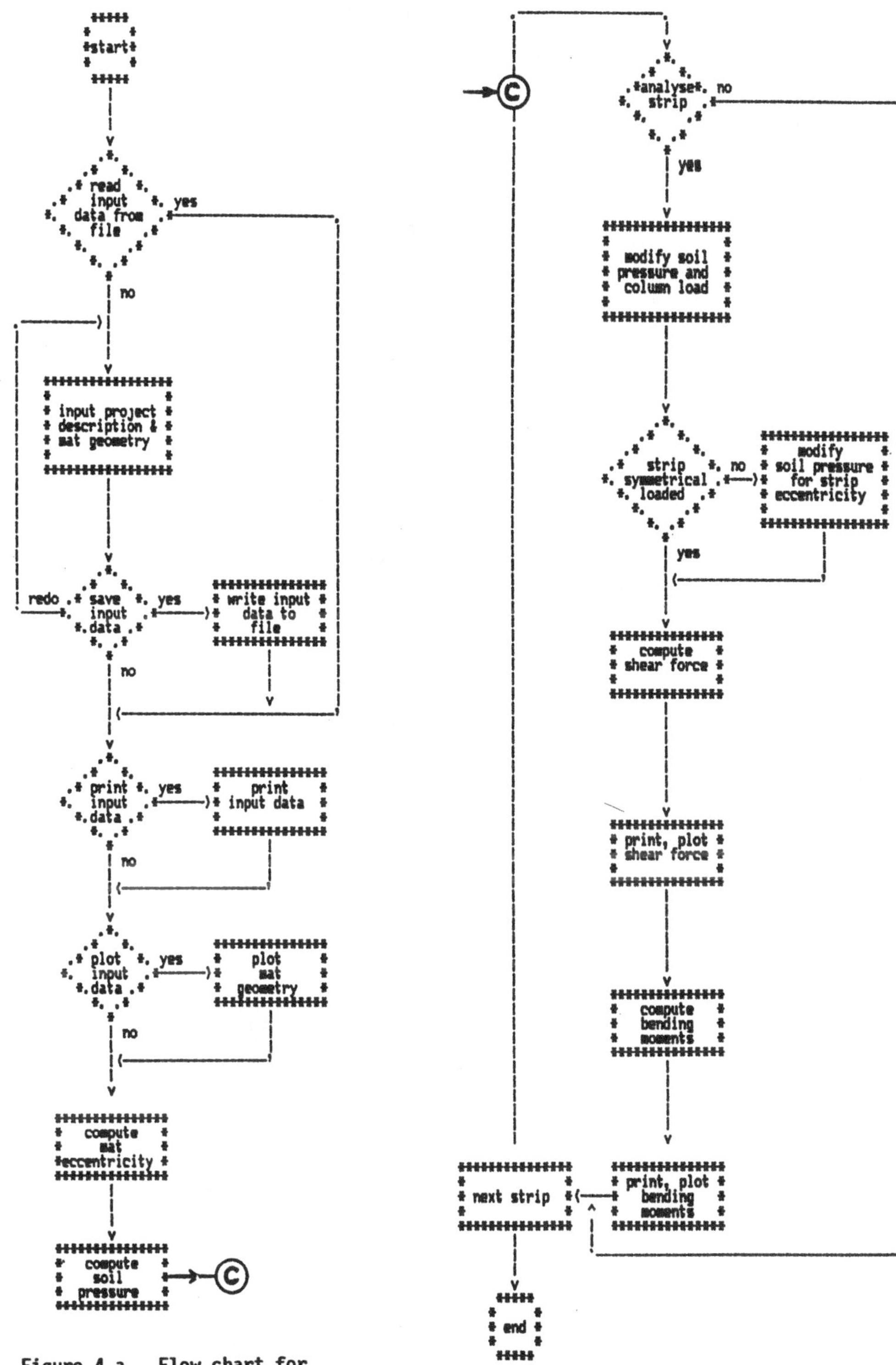

Figure 4.a. Flow chart for rigid mat design.

Figure 4.a. continued...

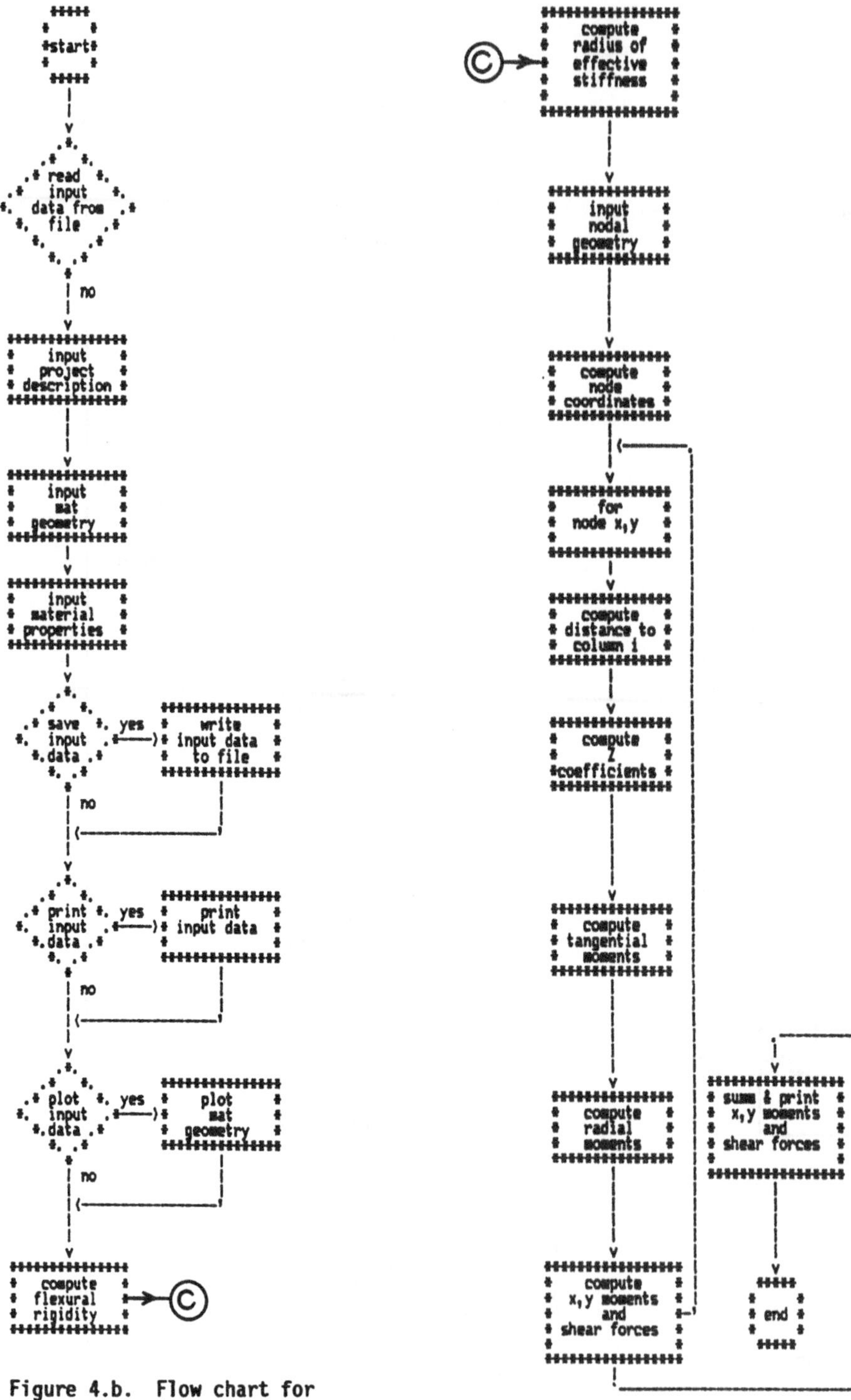

Figure 4.b. Flow chart for flexible mat design.

Figure 4.b. Continued...

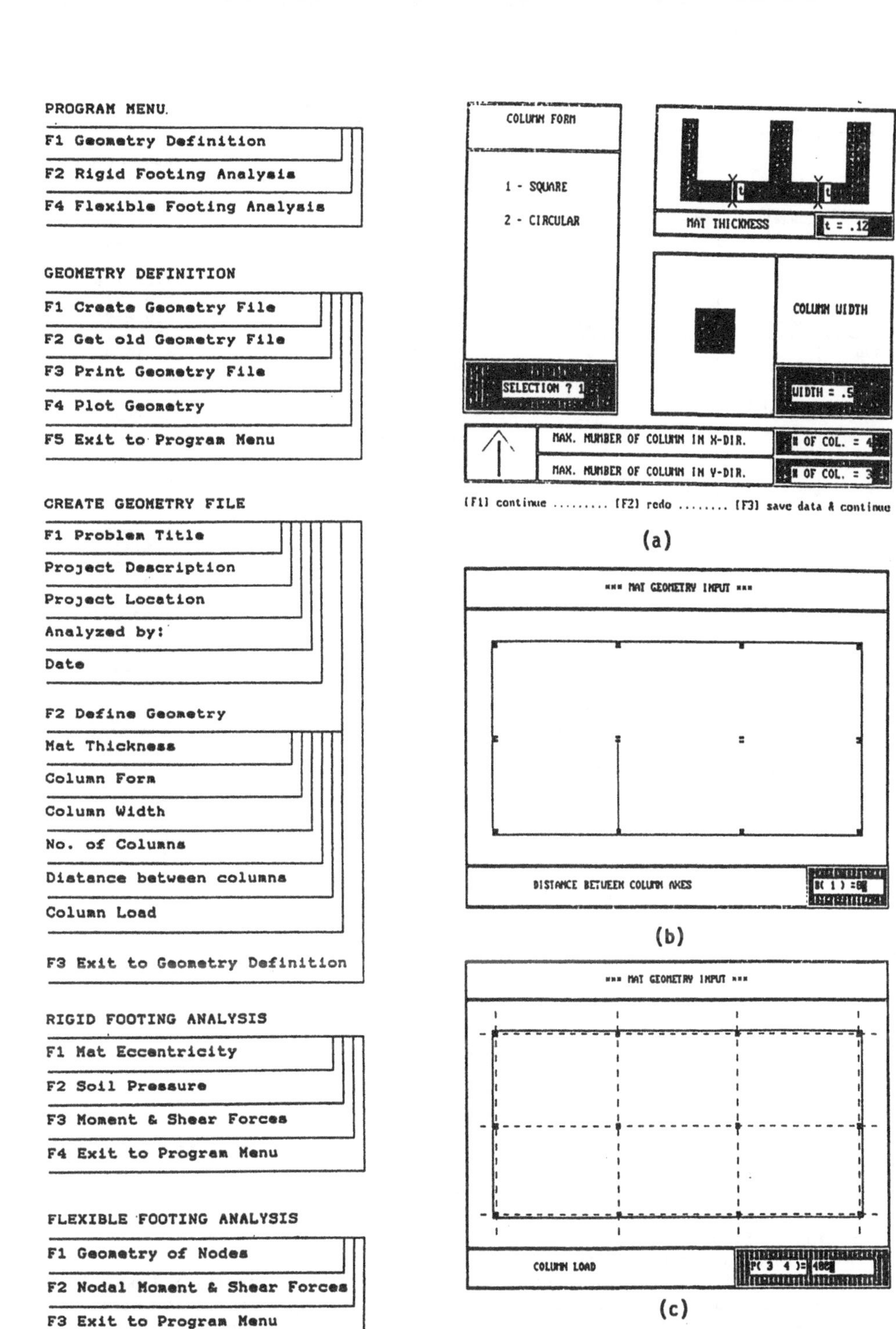

Figure 5. Menues of MAT

Figure 6. Printed screen images of input menues of MAT: a) mat thickness & column data, b) mat geometry data, and c) column loads.

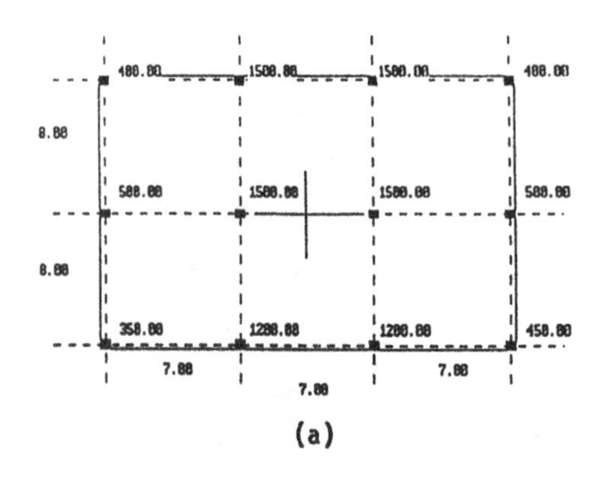

(a)

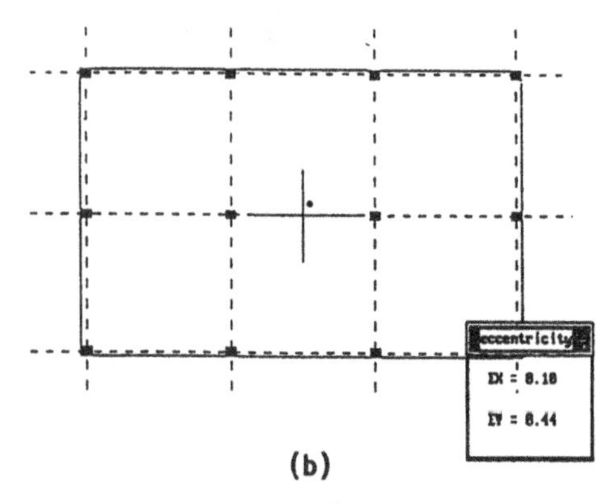

(b)

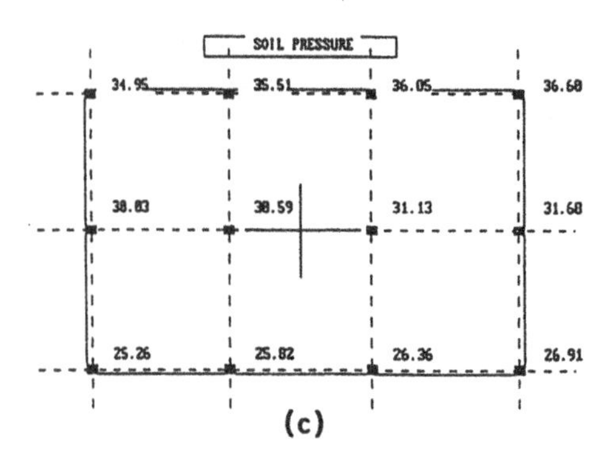

(c)

Figure 7. Printed screen images of output results for rigid mat design example: a) problem definition in SI units (column loads in KN and spacings in m.) b) eccentricity of column loads resultant in m., and c) final soil pressures in KN/$_m$2.

****

RESULTS STRIP 3

****

X-COORDINATE	SHEAR-FORCE
0	0
.25	-41.10056
.25	330.9677
7.25	-819.848
7.25	575.4078
14.25	-575.4078
14.25	819.8481
21.25	-330.9673
21.25	41.10083
21.5	0

X-COORDINATE	MOMENT
0	0
.25	5.13757
1.25	-243.629
2.25	-327.9932
3.25	-247.9553
4.25	-3.515137
5.25	405.3274
6.25	978.572
7.25	1716.219
8.25	1223.012
9.25	894.208
10.25	729.8042
11.25	729.8057
12.25	894.2061
13.25	1223.011
14.25	1716.219
15.25	978.5703
16.25	405.3262
17.25	-3.517578
18.25	-247.9551
19.25	-327.9942
20.25	-243.6328
21.25	5.136719
21.5	0

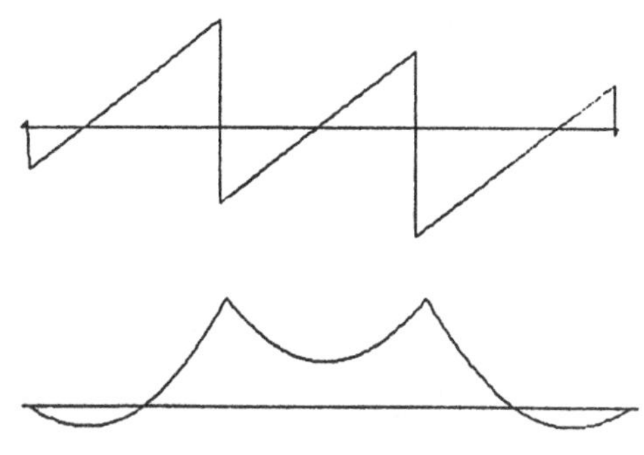

MAX. POS. SHEAR FORCE = 819.85
MAX. NEG. SHEAR FORCE = 819.85
MAX. NEG. MOMENT = 327.99
MAX. POS. MOMENT = 1716.22

Figure 8. Printed results and moment and shear plots for strip # 3 (the upper strip in the x direction) of the rigid mat design example.

CONCLUSIONS

The computer code MAT is a useful aid in designing mat footing using rigid or flexible procedures. The interactive graphics BASIC microcomputer code is user friendly and may be easily modified for different microcomputers. The moment and shear force results match well when compared with a published example. The results of the flexible footing procedure are sensitive to the coefficient of subgrade reaction and generally the maximum moments and shear forces tend to decrease as the coefficient of subgrade reaction increases.

REFERENCES

American Concrete Institute Committee 436 (1966). "Suggested Design Procedures for Combined Footings and Mats," Journal of the American Concrete Institute, Vol.63, No. 10, pp. 1041-1057.

AT&T. (1985). Programmer's Guide, AT&T Personal Computer 6300, GW BASIC by Microsoft, AT&T, U.S.A.

Cheung, Y.K., and Nag, D.K. (1968). Plates and beams on elastic foundations - Linear and nonlinear behavior, Geotechnique, June.

Das, B.M., (1984). Principles of Foundation Engineering, Brooks/Cole Engineering Division, Monterey, California, pp.176-206.

Hetenyi, M. (1946). Beams on Elastic Foundations, University of Michigan Press, Ann Arbor.

Teng, Wayne C. (1949). A study of contract pressure against a large raft foundation, Geotechnique, I. 4, pp.222.

Teng, Wayne C. (1962). Foundation Design, Prentice-Hall, Inc. Englewood Cliffs, N.J., pp. 174-191.

Terzaghi, K. (1955). Evaluation of the Coefficient of Subgrade Reactions, Geotechnique, Institute of Engineers, London, Vol. 5, No. 4, pp. 197-326.

Vesic, A.S. (1961). Bending of Beams Resting on Isotropic Elastic Solid, Journal of the Engineering Mechanics Division, American Society of Civil Engineers, Vol. 87, No. EM2, pp. 35-53.

Computer and Physical Modelling in Geotechnical Engineering, Balasubramaniam et al. (eds)
© 1989 Balkema, Rotterdam. ISBN 90 6191 864 2

Predicting the relationship between time and negative skin frictions

Paul M.Reed
Dames & Moore International, Jakarta, Indonesia

ABSTRACT: The development of negative skin friction or downdrag on piles is time related. However, because of the complexity of including the time dimension in an analysis, only the maximum downdrag is usually calculated and applied to the pile. This paper presents a method for analyzing the relationship between time and negative skin friction. The approach is to consider the build up of effective lateral geostatic stresses with time. Then using a transfer function to relate downdrag to these stresses, the increase in downdrag can be studied. A case where actual downdrag forces have been measured in the field is used to test the proposed model. The model is then applied to a real design problem where it is found that downdrag forces at the time of installation are substantially less than the maximum occuring over the life of the piles. Finally, the paper suggests that the magnitude of movement under axial compressive loads is not enough to significantly reduce negative skin friction.

1 INTRODUCTION

A pile driven through soil that is consolidating will, at some time in its history, experience loads which will increase the axial force in the pile. Moreover, these downdrag forces caused by negative skin friction will vary with time, although classic soil mechanics addresses only the determination of the maximum downdrag at whatever time. However, predicting the relationship between time and negative skin friction may be an important aspect of the design of a pile foundation, not simply the maximum downdrag. This time variation of forces will depend largely upon the rate of settlement of the soil surrounding the pile, the time of application of any axial loads and the load transfer relationship between the pile and soil.

Most of the literature discusses methods of computing the maximum negative skin friction and determining the point on the pile where the negative friction becomes positive. This is due, in part, because actual field data on negative skin friction are scarce and because analyzing the development of downdrag forces with time is extremely complex. However, it is not always the case that the maximum downdrag will develop during the life of the structure. Thus, the pile may be overdesigned if the maximum downdrag is used.

This paper presents a method for predicting the relationship between time and negative skin friction. The transfer functions for both positive and negative skin friction are briefly described, and some of the recent work on time-downdrag relationships are discussed. A computer program which uses the results from a finite difference solution to the consolidation process in conjuction with a finite difference solution for load transfer is proposed. Finally, the program is used to predict the relationship between time and negative skin friction for a case where these values have been measured in the field and for the design of pile foundations for a major structure.

2 BACKGROUND

Textbooks on foundation engineering suggest methods of computing the maximum negative skin friction developed by a consolidating soil surrounding a pile. In general, they do not discuss how the downdrag forces may change with time and how the time of pile loading relates to the negative stresses that are generated. Methods for determining maximum downdrag usually involve some form of the following equations:

$$f_{ns} = \bar{\sigma}_H \tan\delta = k_o \bar{\sigma}_V \tan\delta = \alpha\bar{\sigma}_V$$

where

σ_H = effective lateral geostatic stress
σ_V = effective vertical geostatic stress
K_o = effective earth pressure coefficient
δ = friction angle between the pile material and contact soil.

Studies by Lee (5), Baligh (2), Fellenius (7) and others have suggested that the value of the product K_o and $\tan\delta$ lies between approximately 0.20 and 0.50. The lower value is recommended for soft clays and the upper value for dense sands.

In summary, the procedure for calculating the maximum negative skin friction consists of the following steps:

1. Determine the depth to the bottom of the deepest consolidating soil layer above the pile tip.

2. Assume that the soils to the bottom of this layer or to some point just above the bottom will create downdrag (negative skin friction stresses) along the sides of the pile.

3. Calculate the negative skin friction as :

$$P_{NSF} = \sum_{i=1}^{NL} C \times L_i \times \alpha \qquad (1)$$

where

P_{NSF} = total maximum downdrag force over the pile length

C = pile perimeter

L_i = thickness of the i-th consolidating layer (the last layer may be considered slightly less than the total thickness of the layer).

NL = number of consolidating layers

Most authors (3, 7) have introduced the concept of the neutral point or neutral axis into the calculations. This is considered to be the point on the pile where there is no relative moment between the pile and the soil. Above this point the forces tend to drag the pile down whereas below the neutral axis the forces tend to prop the pile up. Thus, the neutral point can be said to be the location on the pile where neither negative nor positive forces act. Locating this point is complex because it requires a knowledge of the settlement of the individual soil layers and a knowledge of the settlement of the pile along the axis. Suggested methods of determining the neutral axis can be found in References 3 and 7.

The Japanese building code, for example, recommends placing the neutral axis at a depth of $L/\sqrt{2}$ below the top of the existing ground surface.

Recent computer-oriented solutions (1) have addressed the time-downdrag relationships and attempted to locate analytically the location of the neutral axis.

3 RECENT DEVELOPMENTS

One recent development in analyzing the relationship between negative skin friction and time for end bearing piles was proposed by Poulos and Davis (10) who used an elastic model in their work. Initially, they assumed that the soil was perfectly elastic and that there was no slip between the pile and the soil. The analysis was later extended to include the effect of slip and to include the effect of pile crushing.

Alonso (1), et. al., used a stress transfer approach at the interface to solve in closed form the development of negative skin friction with time and pile settlements. The transfer function had the same form as equation 1.

Finally, Fellenius (7) discusses qualitatively the build-up of downdrag forces with time and elaborates on the concept of a neutral axis. In another paper, Fellenius (6) describes the results of measurements of downdrag forces recorded for over 43 months and the implication of time on the forces is apparent.

Morever, while the relationship between time and negative skin friction has been recognized, only two papers to the author's knowledge have attempted to analyze the problem. The method of analysis presented in the following section is meant, by no means, to be complete, but rather to focus on the complexity of the problem and suggest a systematic, rational approach for predicting the relationship between time and negative skin friction.

4 PROPOSED METHOD OF ANALYSIS

Fig. 1 shows the general problem of a pile located in soil layers that are undergoing consolidation. An axial load may or may not be applied immediately, but with the application of the axial load the relative moment between a point on the pile and a point in the adjacent soil will change. The first step in determining the magnitude of the downdrag forces as a function of time is to compute the time rate of increase of the effective stresses in the soil. In the present paper, these were determined using a finite difference solution to the Terzaghi-Rendulic consolidation equation (12). The solution is perfectly general and allows a time rate of surcharge loading, the installation of drains and a stress-dependent coefficient of consolidation.

From an examination of the layer settlement, a first estimate of the depth to the bottom of the consolidating layer can be made. Hence, it can be assumed that all shear stresses above the bottom of the consolidating layer are negative while those below are positive. If the pile is permeable, the location of the pile corresponds to a drain point and no excess pore pressures will build up adjacent to the pile.

The effective vertical stress as a function of time in each consolidating layer (or at each point on the side of the pile) can be used to determine the downdrag forces using a transfer function such as equation 1. The downdrag forces are then applied to the sides of the pile at points along the pile above the neutral axis.

The response of the pile to downdrag forces and any axial load that may be applied at some time during the life of the pile can be analyzed using a procedure suggested by Coyle and Reese (4). The procedure, however, requires that load-soil displacement curves for each layer or sublayer be supplied. This information may not be readily available from the laboratory results, but some estimate can be made using methods discussed by Reese (11), Coyle (5) and Matlock (8).

The depth where the deflection of the pile due to axial and downdrag forces equals the deflection of the consolidating soil is the new neutral axis. The location of the neutral axis will change with the time change of deflections. Points on the pile that were previously acted upon by negative skin friction above the neutral axis may now be supporting the pile through positive skin friction since the points are below the neutral axis. Fig. 2 shows the suggested process for determining the downdrag forces, the neutral axis and the revised downdrag forces. The procedure is repeated until the change in the neutral axis is less than some specified tolerance.

5 APPLICATIONS

Field measurements by Lee and Lumb (9) provide an opportunity to assess the validity of the procedure. Little more than a description of the soils, a log of the soil boring and the time-rate of settlement was provided in the paper. Using representative published values of the unit weight and consolidation properties, the time-settlement curve shown in Fig. 3 was simulated using a finite difference solution. The soil properties were adjusted and a refined curve produced. The procedure was continued until the simulated curve was close to the actual field curve.

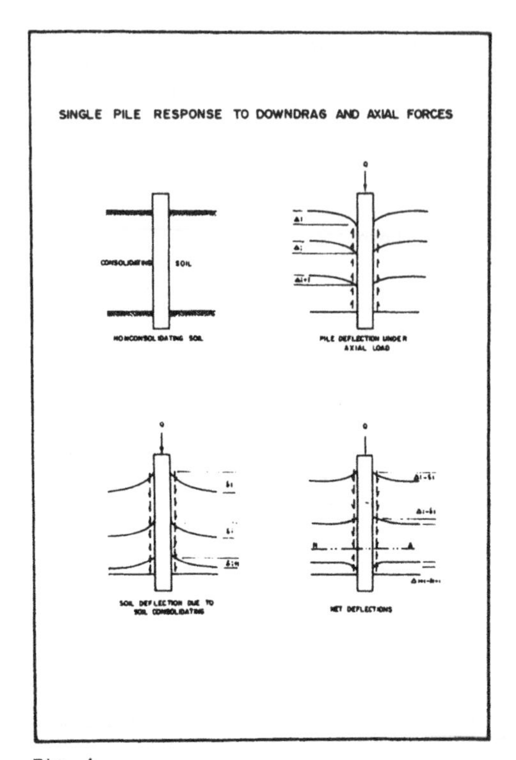

Fig. 1

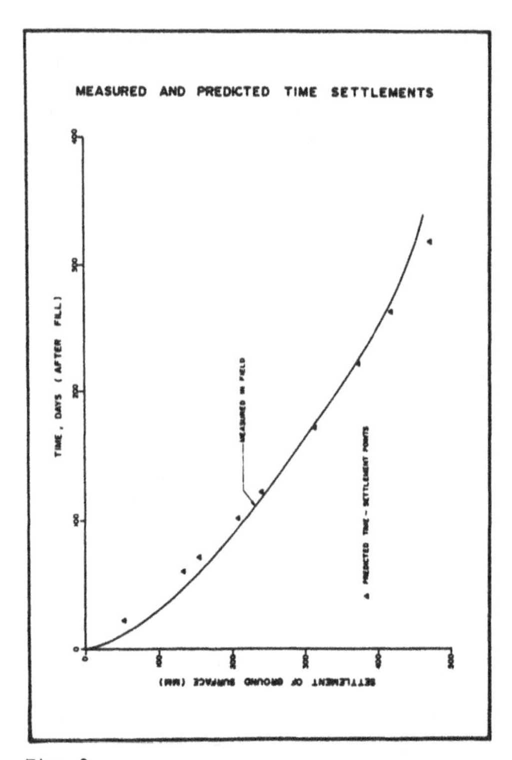

Fig. 3

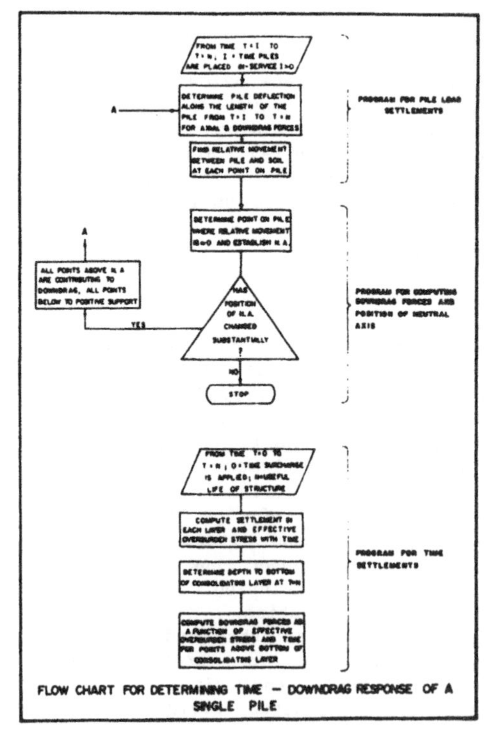

FLOW CHART FOR DETERMINING TIME — DOWNDRAG RESPONSE OF A SINGLE PILE

Fig. 2

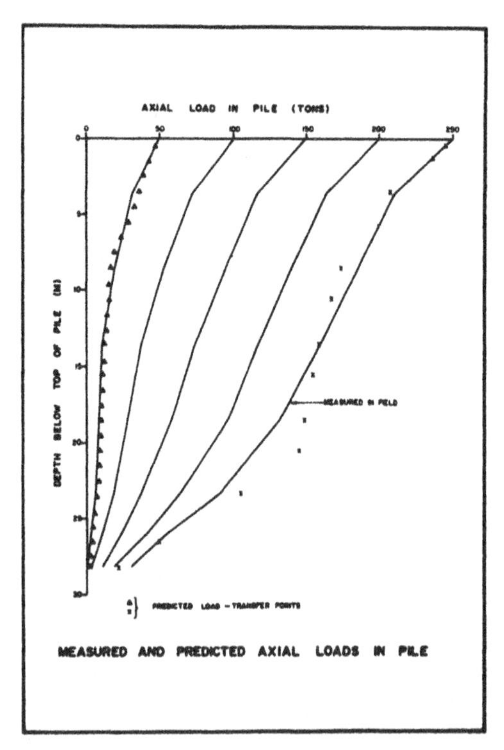

MEASURED AND PREDICTED AXIAL LOADS IN PILE

Fig. 4

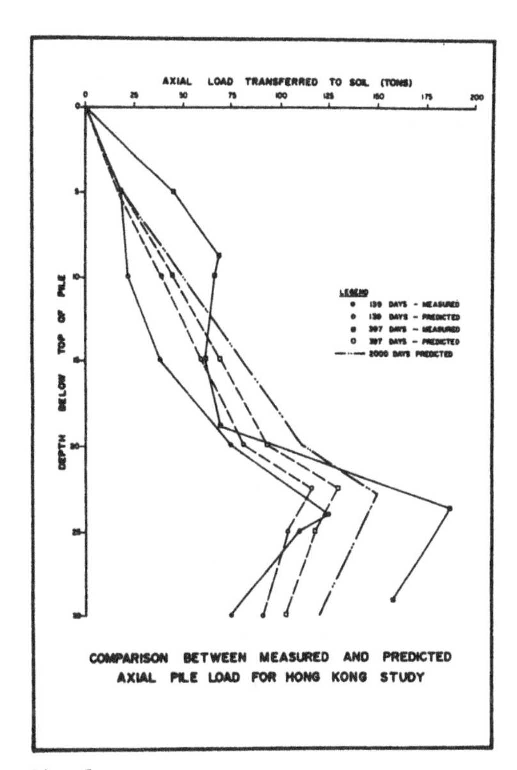

Fig. 5

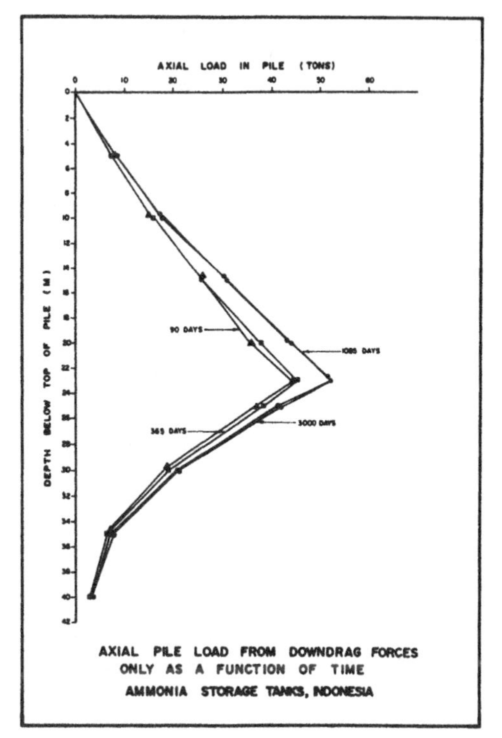

Fig. 7

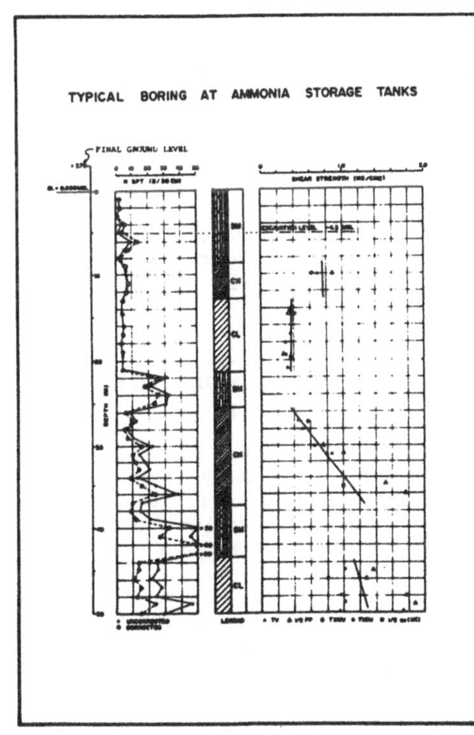

Fig. 6

Fig. 4 shows the load transferred to the soil during pile load testing. Again, since specific soil properties were not reported in the Lee and Lumb paper, representative values of shear strength had to be assumed. Typical load transfer values were taken from Coyle and an attempt to match the curves shown in Fig. 4 was made. This proved to be a little more arduous than the effort to duplicate the time-settlement curve. However, reasonably close curves as shown in the Fig. 4 were produced by adjusting the load-transfer soil properties after each run.

Thus, given reasonable values for the soil classification, consolidation and shear strength properties, it was felt that the time-downdrag forces could be predicted using the procedure of the previous section. The results of the computer analysis are shown in Fig. 5 where measured load-transfer curves from field measurements for two times -- 139 and 397 days -- were selected for comparison with predicted solutions.

At 139 days the predicted solution

both in magnitude and shape is very close to that actually measured in the field. For 397 days the predicted values are slightly less than the measured values although the load-transfer curve is similar. It may be possible to improve the predicted response by adjusting the load transfer curves and perhaps the α - values. However, the present solution gives confidence to the method although additional comparisons to field measurements are necessary.

The second application is the design of 762 mm diameter pipe piles to be used to support ammonia storage tanks at a fertilizer complex in Indonesia. The construction sequence is one of filling approximately 3.7 m above grade, driving piles after four months when all the fill is in place, removing 1.0 m of fill to provide ventilation under the tanks then placing an axial load up to 140 tons on each pile. A log of the typical soil conditions at the tank site is shown in Fig. 6.

The calculation of downdrag forces and the location of the neutral axis with time followed the sequence shown in Fig. 5. The time-history of load transfer is shown in Fig. 7. It is apparent that there is a build-up of forces to about 1095 days, then the forces appear to remain constant at that point. This indicates that if the piles are loaded approximately three years after they are driven, then virtually no downdrag forces would have to be taken into account in the design. Further, in these studies it was found that the deflection of the soils is of an order several times larger than the deflection of the pile under axial loading. Therefore, the position of the neutral axis remained relatively constant.

6 CONCLUSIONS

The results of the study are not particularly surprising, but they do provide some clarification on the relationship between time and downdrag forces. The conclusions that can be drawn from this investigation for the soil and loading conditions described in this paper are summarized as follows:

1. The magnitude of downdrag forces will vary with time and these forces appear to be a function of the effective overburden stresses and the relative movement between the pile and the adjacent soil.

2. The maximum downdrag force may occur when there is no axial load on the pile, and dissipate prior to the pile being placed in-service. Therefore, it may not be prudent to always assume that maximum downdrag occurs throughout the life of the pile.

3. The two case studies indicate that pile movement under axial load is not large enough to shift the neutral axis higher.

4. A time-settlement analysis of the soil to determine the build-up of downdrag forces and a load-transfer analysis to determine the distribution of axial forces in the pile appears to provide a reasonable estimate of the time-downdrag relationship of a pile in a soil which is still consolidating.

REFERENCES

Alonso, E.E., et. al., (1984), "Negative Skin Friction on Piles: A Simplified Analysis and Prediction Procedure," Geotechnique 34, No. 3, pp 341-357

Baligh, m.m., et al (1978) "Downdrag on Bitumen Coated Piles," Journal of the Geotechnical Engineering Division, ASCE, Vo 104, GT 11, November, pp 1355-1363

Bowles, J.E. (1982), Foundation Analysis and Design, McGraw Hill Book Company, New York, 816 pp

Coyle, H.M., and L.C, Reese (1966), "Load Transfer of Axially Loaded Piles in Clay," Journal of the Soil Mechanics and Foundation Division, ASCE, Vol 92, SM 2 March, pp 1 - 26

Coyle, H.M. and I.H. Sulaiman (1967), "Skin Friction for Steel Piles in Sand," Journal of the Soil Mechanics and Foundation Division, ASCE, Vol 93, SM6, pp 261 - 278

Felleniu, B.H. (1972), "Downdrag on Piles in Clay Due to Negative Skin

Friction," Canadian Geotechnical Journal, Vol 9, No. 4, November, pp 323 - 337

Felleniu, B.H. (1984), "Negative Skin Friction and Settlement on Pile," 2nd Int. Geotechnical Seminar on Pile Foundation, Singapore, pp 28 - 30

Matlock, H., et. al., "AXCOL3: A Program for Discrete - Element Solution of Axially Loaded Members with Linear or Nonlinear Supports," Report to the American Petroleum Institute, March 1976, 66 pp

Lee, P.K.K. and P. Lumb (1982), "Field Measurements of Negative Skin Friction on Steel Tube Piles in Hong Kong," Proc. of the 7th Southeast Asian Geotechnical Conference, Hong Kong, pp 363 - 374

Poulos, H.G. and E.H. Davis (1972), "The Development of Negative Friction with Time in End Bearing Piles," The Australian Geomechanics Journal, Vol. G2, No. 1, pp 11 - 20

Reese, L.C., et. al. (1969), "An Investigation of the Interaction Between Bored Piles and Soil," Proc. of the 7th Int. Conference on Soil Mechanics and Foundation Engineering, Vol. 2, pp 211-215

SD3, User's Manual (1982) "Finite Difference Analysis of Consolidation Problems Involving Either One-Dimensional Vertical Flow or Two-Dimensional Flow with Drainage Wicks," Univ. of Texas

Computer and Physical Modelling in Geotechnical Engineering, Balasubramaniam et al. (eds)
© 1989 Balkema, Rotterdam. ISBN 90 6191 864 2

Flexural response of tapered beams on bilinear foundation

Sarvesh Chandra & Phanuwat Suriyachat
Division of Geotechnical & Transportation Engineering, Asian Institute of Technology, Bangkok, Thailand

ABSTRACT: The response of finite beams with variable moment of inertia resting on an elastic foundation has been studied because of its applicability to many design problems in Civil and Aeronautical Engineering. In the present study, an analysis is carried out of a circular tapered beam resting on a nonlinear subgrade, where the response can be idealised by a bilinear representation. Closed form solutions are obtained for a tapered beam with free-free end conditions and for two different loaings. The results are presented in the form of nondimensional charts.

I INTRODUCTION

The problem of finite nonuniform beams resting on an elastic foundtion has received considerable attention because of its wide applicability to many design problems in Civil Engineering and to other engineering structures. A comprehensive review of the various solutions to this class of problems on the assumption of validity of Winkler's model can be found in Hetenyi [1]. Attempts have been made to improve on the Winkler model and a comprehensive review of these is presented by Kerr [2]. Rhines [3] used an elasto-plastic model to simulate the intensive shear in the soil.

The pressure versus settlement response as obtained by a load test is highly nonlinear in most soils and particularly so in weak soils. A typical result of a field test (plate load test) underlining the nonlinear nature of the soil is presented in Fig. 1. However, one difficulty associated with the assumption of nonlinear response of soil is that of rigorous mathematical analysis. One way to overcome this difficulty and at the same time obtain better results would be to approximate the foundation response by a bilinear curve. Fig. 2 shows the load versus settlement response of the soils as idealised by a bilinear representation. With this idealisation of the soil medium, a tapered beam of finite length with free-free end conditions resting on a bilinear foundation is analysed and its flexural response studied. The beam is shown in Fig. 3(a), with Zones I and II corresponding to the portions AB and BC of Fig. 2 respectively. The point load is applied at one end and the other end is free. In Case II as represented in Fig. 3(b), the beam is subjected to a couple instead of the point load at one end.

The governing differential equation of a beam with varying cross-section resting on an elastic foundation is a fourth order differential equation with variable coefficient. In general, the solution can be obtained by using finite differences or finite elements, or by employing series solutions. Kosko [4], showed that closed form solutions can be obtained for a particular variation of the section of the beam i.e. a tapered beam of circular cross-section resting on a Winkler medium. Subsequently, Valsangkar and Basudhar [5] presented solutions for a tapered beam with circular corss-section resting on an elasto-plastic foundation.

The present work is an extension of previous work and incorporates the nonlinear response of a soil which can be idealised by a bilinear representation, as shown in Fig. 2. The results are presented in nondimensional form and the effect of various parameters on the flexural behaviour of the beam emerges clearly in the two different loading conditions.

II THEORETICAL ANALYSIS

The flexural rigidity of the beam is assumed to vary as the fourth power of the abcissae from the origin, as shown in Fig. 3(a). In the analysis presented here, it is assumed that the foundation reaction up to length L_1, is represented by the portion BC and for $x > L_1$, the foundation reaction is represented by the portion AB of Fig. 3(a).

The governing differential equation for the beam of variable cross-section resting on an elastic foundation is:

$$\frac{d^2}{dx^2}\left[EI(x)\frac{d^2y}{dx^2}\right] + p = 0 \qquad (1)$$

where,

- E - Young's modulus of beam material
- I(x) - variable moment of inertial $= I_o(1+bx)^4$
- y - settlement of the beam
- x - coordinate measured along the axis of beam
- p - the soil reaction $= (K_o-K_1) y_o+K_1y$ for zone I $= K_oy$ for zone II
- b - $(c-1/L)$

and

- c - ratio of diameter at x = L to the diameter at x = 0.

Substituting for I (x) and p in Eq. (1) leads to the following differential equations governing the flexural behaviour in Zone I and Zone II respectively:

$$\frac{d^2}{dx^2}\left[EI_o(1+bx)^4\frac{d^2y}{dx^2}\right] + (K_o-K_1)y_o + K_1y = 0; \; 0 < x < L_1 \qquad (2)$$

and

$$\frac{d^2}{dx^2}\left[EI_o(1+bx)^4\frac{d^2y}{dx^2}\right] + (K_oy) = 0; \; L_1 < x < L \qquad (3)$$

In terms of the following nondimensional parameters:

$$Z = x/R, \quad Z' = L/R, \quad Z_1 = L_1/R$$

and

$$A_y = \frac{yEI_o}{K_oY_oR^4} \quad \text{where} \quad R = 4\sqrt{\frac{EI}{K_o}} \qquad (4)$$

Equations (2) and (3) can be rewritten as:

$$\frac{d^2}{dz^2}\left[(1+d'Z)^4\frac{d^2A_y}{dz^2}\right] + r_1 + r_oA'_y = 0; \; 0 < Z < Z_1 \qquad (5)$$

and

$$\frac{d^2}{dz^2}\left[(1+d'Z)^4\frac{d^2A_y}{dz^2}\right] + A_y = 0; \; Z_1 < Z < Z' \qquad (6)$$

where $d' = c - 1/Z'$

Substituting $(1+d'Z) = n$ and $n = e^v$ in equations (5) and (6), we get:

$$\frac{d^4A'_y}{dv^4} + 2\frac{d^3A'_y}{dv^3} - \frac{d^2A'_y}{dv^2} + 2\frac{dA'_y}{dv} + \psi A'_y = 0; \; 0 < v \leq v_1 \qquad (7)$$

$$\frac{D^4A_y}{dv^4} + 2\frac{d^3A_y}{dv^3} - \frac{d^2A_y}{dv^2} + 2\frac{dA_y}{dv} + \psi' A_y$$

$$= 0; \quad v_1 < v \leq v' \qquad (8)$$

where $A'_y = (1 + \frac{r_o}{r_1} A_y)$

Let $\psi = \frac{r_o}{(d')^4}$, $v_1 = 1_n (1 + d'Z_1)$

and $\psi' = (\frac{1}{d'})^4$, $v' = 1_n (1 + d'Z')$

The equations (7) and (8) can be solved in closed form and the solution for Ay is obtained as:

$$A_y = (r_1/r_o)(1+d'Z)^{-0.5}$$
$$[C_1 \cos h(\sigma l_n(1+d'Z)) \cos(\tau l_n(1+d'Z)) + C_2 \cos h(\sigma l_n + d'Z) \sin(\tau l_n(1+d'Z)) + C_3 \sin h(\sigma l_n(1+d'Z)(\cos(\tau l_n + d'Z) + C_4 \sin h(\tau l_n(1+d'Z) - 1]$$
$$0 < Z < Z_1 \qquad (9)$$

and

$$A_y = (1+d'Z)^{-0.5}[C_5 \cos h(\sigma l_n(1+d'Z) \cos(\tau l_n(1+d'Z)) + C_6 \cos h(\sigma l_n(1+d'Z)) \sin(\tau l_n(1+d'Z)) + C_7 \sin h(\sigma l_n(1+d'Z) \cos(\tau l_n(1+d'Z)) + C_8 \sin h(\sigma l_n (1+d'Z) \sin(\tau l_n(1+d'Z))$$
$$Z_1 < Z < Z' \qquad (10)$$

where

$\sigma = (\rho + 1.25)^{0.5}/\sqrt{2}$

$\tau = (\rho - 1.25)^{0.5} \sqrt{2}$

$\rho = (\psi + 9/16)^{0.5}$

$\psi = r_o/(d')^4$

$\sigma' = (\rho' + 1.25)^{0.5}/\sqrt{2}$

$\tau' = (\rho' - 1.25)^{0.5}/\sqrt{2}$

$\rho' = (\psi' + 9/16)^{0.5}$

$\psi' = (1/d')^4$

and

$C_1, C_2, \ldots C_8$ are constants of integration.

The eight constants of integration along with the unknown load P_1 which is necessary for the development of the bilinear zone with K_1 as foundation modulus up to a length L_1 are obtained from the boundary conditions at x = 0 and x = L and from the conditions of continuity at the interface between the two elastic zones. These are:

(a) Case 1 Free-Free tapered beam with a concentrated load at the tip:

$$EI_o \frac{d^2y}{dx^2} = 0 \text{ at } x = 0 \text{ i.e.}$$

$$\text{at } Z = 0, \frac{d^2A_y}{dz^2} = 0 \qquad (11)$$

and

$$\frac{d}{dx}[EI(x)\frac{d^2y}{dx^2}] = P_t$$

at x = 0, i.e. at Z = 0

$$(1+d'Z)^4 \frac{d^3A_y}{dZ^3} = \frac{P_t}{K_oY_oR_o} = U \qquad (12)$$

$$\text{At } Z = Z', \frac{d^2A_y}{dz^2} = 0 \qquad (13)$$

and

$$\frac{d^3A_y}{dZ^3} = 0 \qquad (14)$$

Conditions of continuity at interface are:

$$(A_y) \text{ Zone I} = (A_y) \text{ Zone II} \qquad (15)$$

$$(\frac{dA_y}{dZ}) \text{ Zone I} = (\frac{dA_y}{dZ}) \text{ Zone II} \qquad (16)$$

$$(\frac{d^2A_y}{dZ^2}) \text{ Zone I} = (\frac{d^2A_y}{dZ^2}) \text{ Zone II} \qquad (17)$$

$$(\frac{d^3A_y}{dZ^3}) \text{ Zone I} = (\frac{d^3A_y}{dZ}) \text{ Zone II} \qquad (18)$$

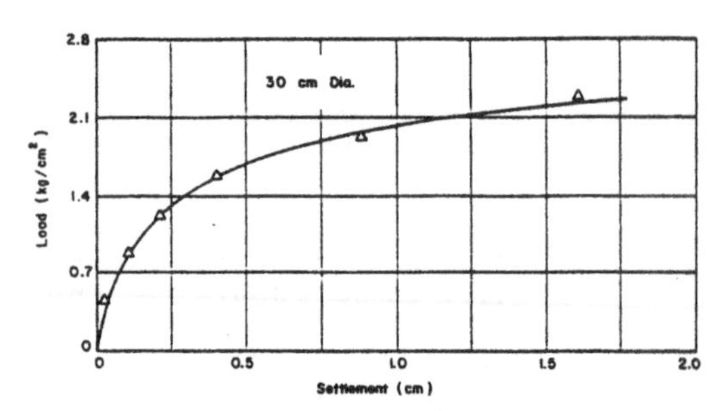

Fig. 1 A Typical Plate Load Test Result

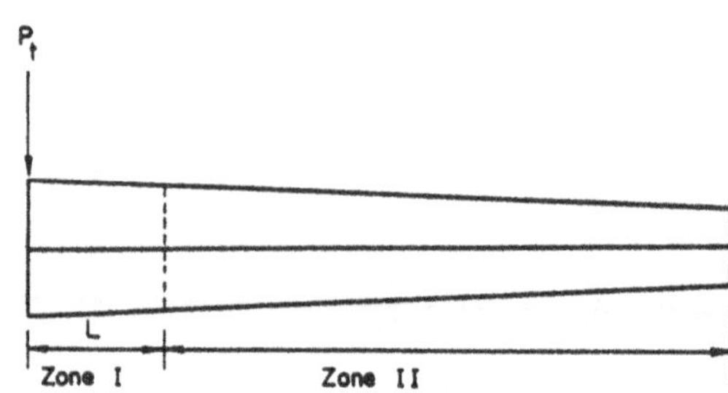

(Fig. 3(a) Free-Free Tapered Beam with Load at free end.

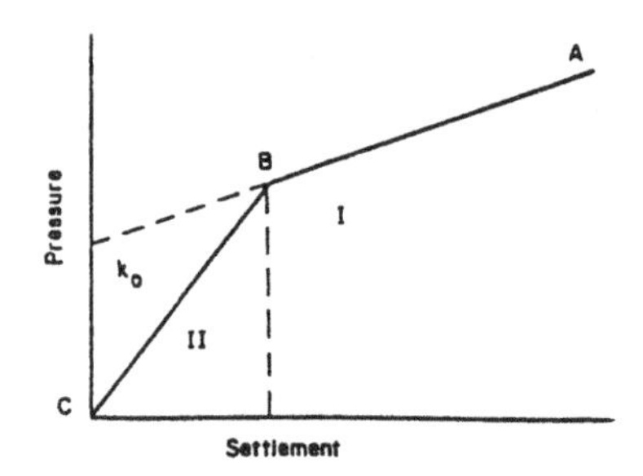

Fig. 2 Bilinear Representation

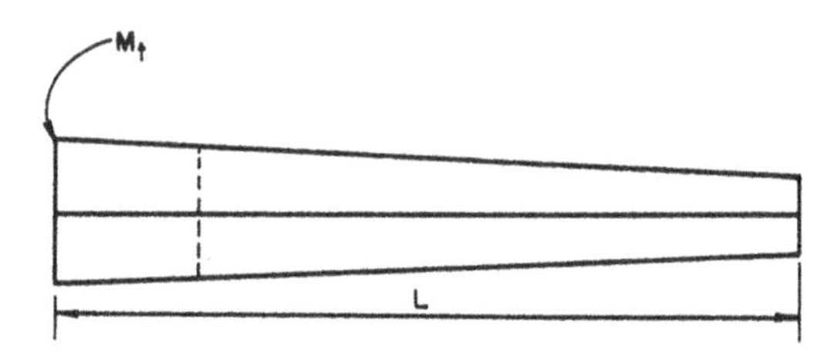

Fig. 3(b) Free-Free Tapered Beam with Moment at free end.

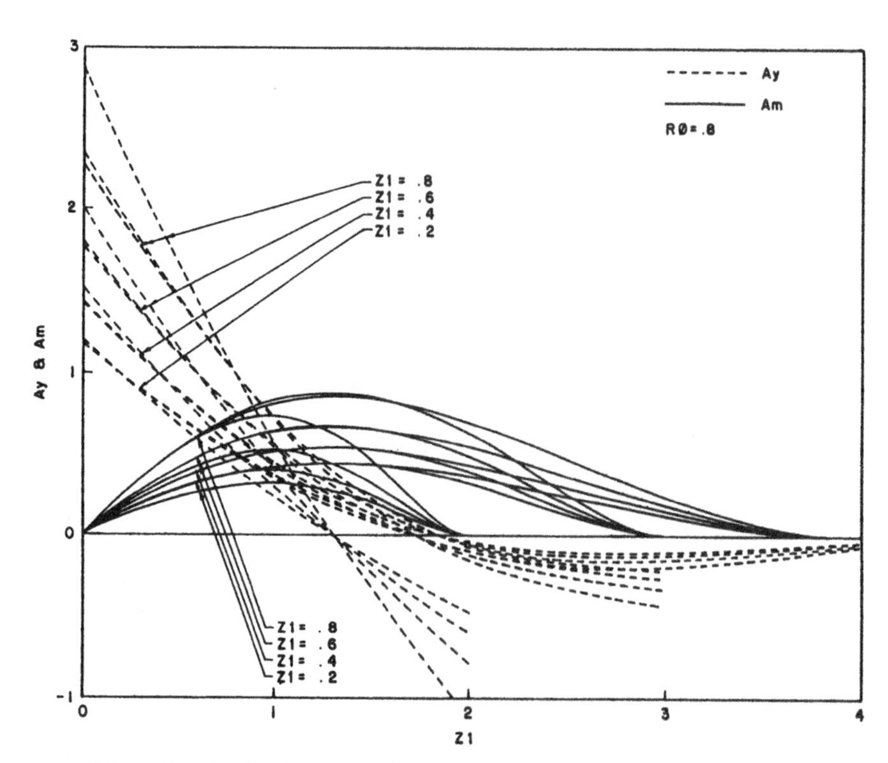

Fig. 4 Variation of Moment and Settlement along T (Free-Free Beam with Load at free end).

Also, from the pressure continuity:

$(r_1 + r_0 A_y)$ Zone I = (A_y) Zone II

(b) Case 2 Free-Free tapered beam with a moment at the tip:

The governing differential equation and the general solution remain the same as in the previous problem. However, the boundary conditions differ and are as follows:

$$EI_o \frac{d^2y}{dx^2} = M_t \text{ at } x = 0, \text{ i.e.}$$

$$\text{at } Z = 0 \qquad \frac{d^2A_y}{dZ^2} = M_t \qquad (20)$$

$$\frac{d}{dx}[EI(x) \frac{d^2y}{dx^2}] = 0$$

$$\text{at } x = 0, \text{ i.e. at } Z = Z'$$

$$4d' \frac{d^2A_y}{dZ^2} + \frac{d^3A_y}{dZ^3} = 0 \qquad (21)$$

The nine simultaneous equations obtained from the boundary conditions and conditions of continuity at the interface are solved by matrix inversion using IBM personal computer; the eight constants of integration and the unknown load factor are obtained.

III RESULTS AND DISCUSSIONS

The solution procedure presented above is used to obtain the numerical results for a tapered beam with free ends and subjected to a concentrated load at the end with the larger diameter, and also for a further case of a tapered beam subjected to a concentrated moment instead of the load. The values of RO, ratio of foundation modulli in Zones I and II are taken as 0.2, 0.4, 0.6 and 0.8. The ratio of diameters at two ends, C, is taken as 0.6 in all cases, as eariler studies by Valsangkar [6] indicate that the variation in the value of C has no marked influence on the flexural behaviour of beams for C ranging from 1.0 to 0.6.

In this analysis, three values of Z' = 2, 3 and 4 are used, in which Z' = 2 corresponds to the short beam whereas Z' = 4 corresponds to the long beam. The results of the two cases considered are discussed below.

(a) Free-Free Tapered Beam with Load at Free End

Typical results of the analysis are presented in Fig. 4 through 6 for the concentrated load case. Fig. 4 presents the variation of nondimensional settlement (Ay) and nondimensional moment (Am) for three different values: Z' = 2, 3 and 4 for value of RO = 0.8. From the plots for values of RO ranging between 0.2 and 0.8, it was observed that RO (ratio of the foundation moduli, K1/K0) has no significant effect on the response. This substantiates further the observations made by Chandra [7] for a beam with uniform cross-section resting on a bilinear foundation. The maximum settlement is observed at Z = 0, where the load is applied, whereas the maximum moment is somewhere close to the mid point of the beam. The short beam shows a tendency to lift at the other end to a greater extent than is observed with the long beam. Further, maximum settlement occurs in the case of the short beam and progressively reduces as Z increases and the settlement increases with the increase in Z1.

The variations of maximum settlement and moment with Z1 are presented in Fig. 5 for RO = 0.8. The maximum settlements increase with the increasing value of Z1 and with decreasing value of Z'. The results obtained for RO ranging between 0.2 and 0.8 are more or less similar, indicating further that RO is not a very sensitive parameter change in maximum settlement or moment values. The maximum moment increases with the increase in Z1 values whereas the maximum moment is obtained for Z = 3.

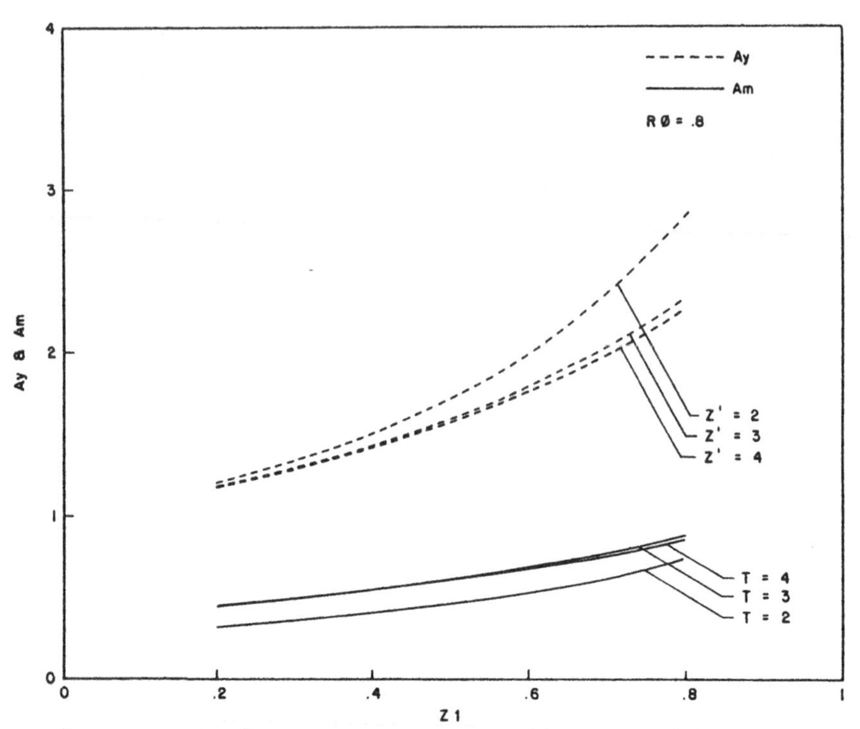

Fig. 5 Variation of Maximum Moment and Settlement with T 1
(Free-Free Beam with Load at free end).

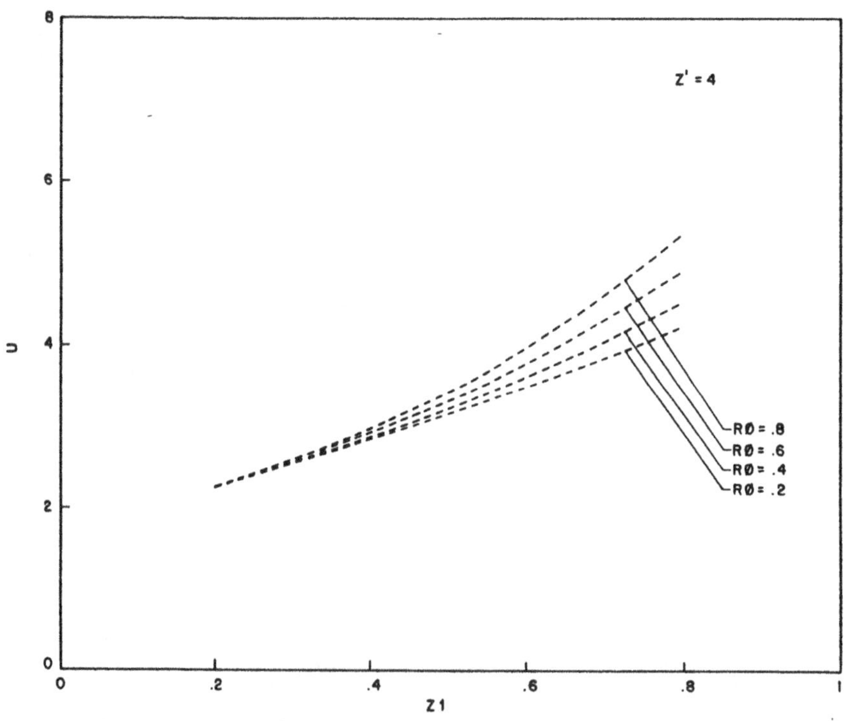

Fig. 6 Variation of Load Factor with Z 1
(Free-Free Beam with Moment at free end).

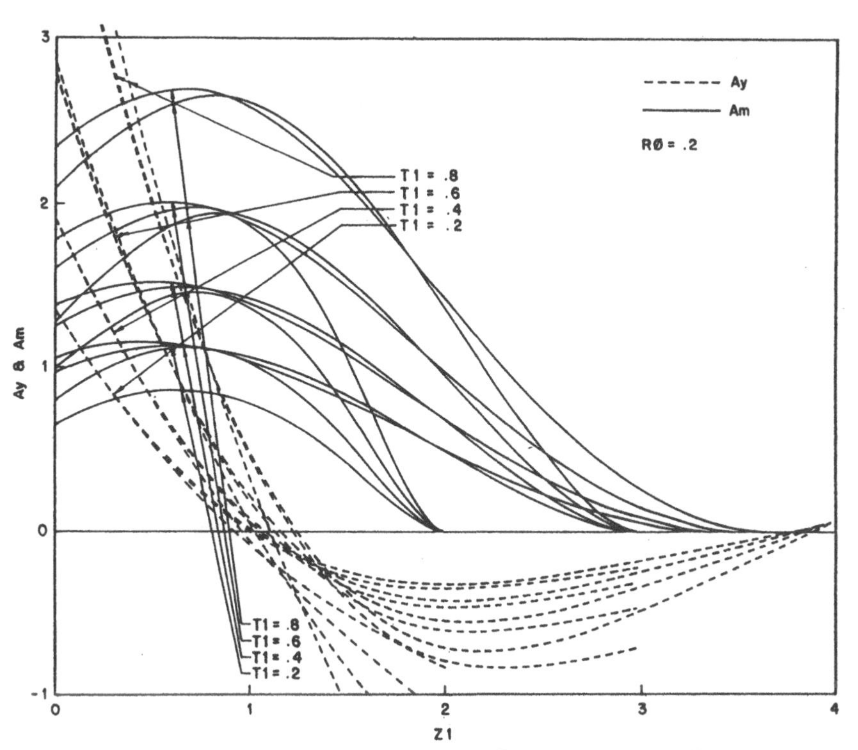

Fig. 7 Variation of Moment and Settlement along T (Free-Free Beam with Moment at free end).

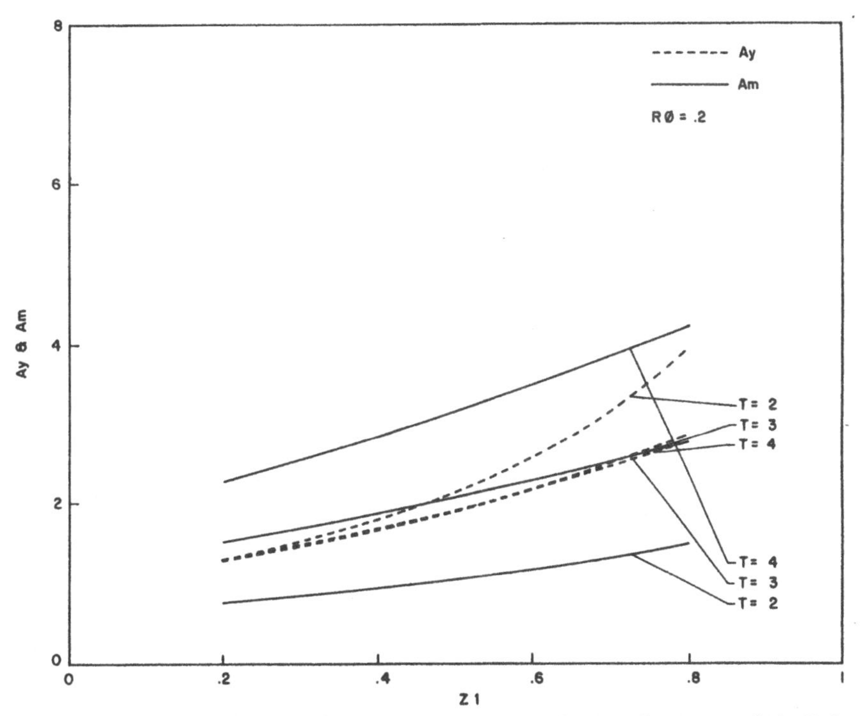

Fig. 8 Variation of Maximum Moment and Settlement with T 1 (Free-Free Beam with Moment at free end).

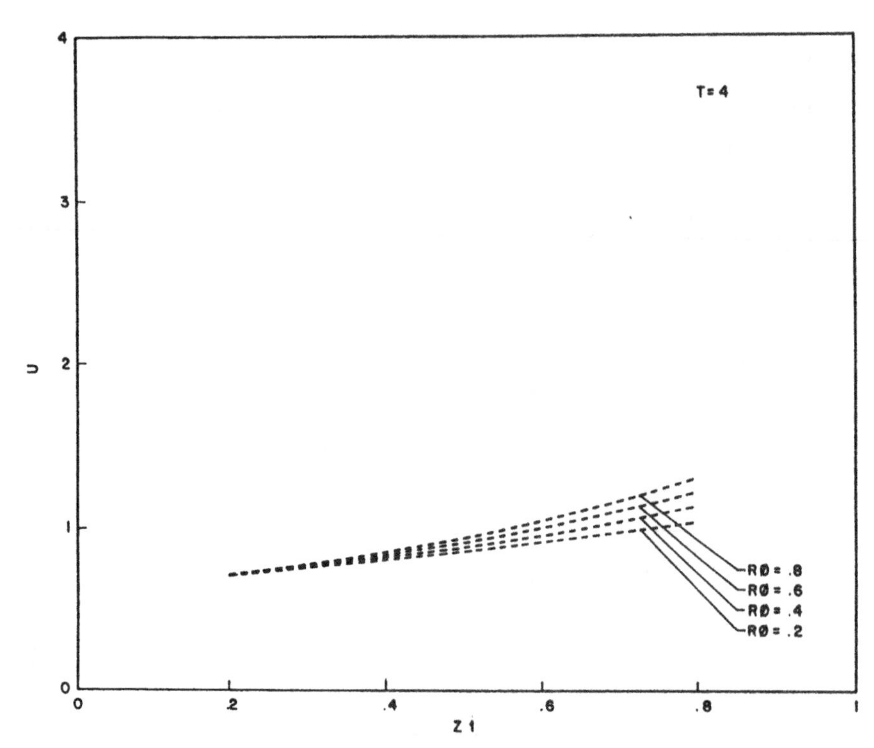

Fig. 9 Variation of Load Factor with T1 (Free-Free Beam with Load at free end).

Fig. 6 shows the relationship between load factor (U) versus Z1 for 0.8 value of RO. For the given value of RO with increasing values of Z1, the load factor U increases, thus implying that with increasing load at the free end, the extent of Zone I increases. It is also seen that the relationship between U and Z1 is essentially nonlinear for all the values of Z'. This is due to the assumed nature of the load settlement curve for the foundation. With increasing values of RO (which implies greater resistance of foundation medium in Zone I) for the same value of Z1, greater loads need to be applied at Z = 0, especially in the range of Z1 = 0 and above. However, for values of Z1 less than 0.4, parameter RO has no appreciable effect.

(b) Free-Free Tapered Beam with Moment at Free End

Figs. 7 through 9 present the results of the analysis of a tapered beam subjected to concentrated moment at the tip. Fig. 7 shows the variation of nondimensional settlement (Ay) and nondimensional moment (Am) for three different values of Z': 2, 3 and 4 for a typical value of RO = 0.2. In this case also it has been observed that RO has no significant effect on the response of the beam. The maximum settlement is observed at Z = 0 where the moment is applied, whereas the maximum moment is somewhere close to Z = 1. This implies that shape of the beam influences the position of maximum bending moment. The short beam shows a tendency to lift at the other end to a greater extent than is observed in the long beam. Further, maximum settlement occurs in the case of the short beam and progressively decreases as Z' increases and the settlement increases with increase in Z1.

The variations of maximum settlement and moment with Z1 are presented in Fig. 8 for RO = 0.2. The maximum settlements increase with increasing value of Z1 and with decreasing value of Z', whereas the maximum moment increases with increasing values of Z1 and Z'.

Fig. 9 shows the relationship between load factor (U) versus Z1 for values of RO = 0.2. It is seen that the value of U increases with increase in Z1, RO and Z' values. This trend can be explained as increasing values of Z1 implying large extent of Zone I and increasing values of RO implying relatively stiffer foundation medium in Zone 1 requiring larger values of U.

IV CONCLUSIONS

In this investigation, results have been obtained for settlement and moments of a nonuniform beam for two different loading conditions, resting on a foundation whose response has been approximated by a billinear model. A particular case of non-uniform beam is considered which results in a closed form solution. However, in practice the situation may be different and closed form solutions always cannot be expected. This solution shows only the trend of behaviour; other suitable techniques may be employed to solve more complicated problems.

NOTATION

A_y nondimensional settlement at any point

A_m non-dimensional moment at any point

b c-1/L

c ratio of diameter at x = L to the diameter at x = 0

C_1-C_8 constants of integration

d' c-1/Z'

E Young's modulus of beam

I_o moment of inertia at x - 0

I_w moment of inertia at any point x

K_o modulus of subgrade reaction in zone II

I_1 modulus of subgrade reaction in zone I

L length of beam

L_1 length beyond which the slope of soil response changes

n (1 + d'Z)

r_o or R_o ratio of foundation moduli in zone I and zone II

r_1 $(1 - r_o)$

U nondimensional load factor

Um nondimensional moment factor

V $\log_e n$

V1 l_n (1 + d'Z)

V' l_n (1 + d'Z)

w surface displacement

x coordinates measured along the axis of the beam

y_o settlement at interface of zone I and II

y settlement of any point

Z x/R

Z_1 L_1/R

Z_2 L/R

ν Poison's ratio

σ $(\rho + 1.25)^{0.5}/\sqrt{2}$

σ' $(\rho' + 1.25)^{0.5}/\sqrt{2}$

ρ $(\psi + 9/16)^{0.5}$

ψ $r_o/(d')^4$

τ $(\rho - 1.25)^{0.5}/\sqrt{2}$

τ' $(\rho' - 1.25)^{0.5}/\sqrt{2}$

ρ' $(\psi' + 9/16)^{0.5}$

ψ' $(1/d')^4$

REFERENCES

M. Hetenyi, Beams on Elastic Foundation, University of Michigan Press, Ann Arbor, 1946.

A.D. Keer, A Study of a new foundation model, Acta Mechanica, I(2): 135-147 (1965).

W.J. Rhines, Elastic plastic foundation model for punch shear failure, J. Soil Mech. Foundation Engg. ASCE, 95, No. 3: 819-819 (1969).

E. Kosko, On a class of tapered beams on elastic foundations, National Research Council of Canada, Aeronantical Report, 1R-425, 1965.

A.J. Valsangkar and P.K. Basudhar, Tapered beam on elastic-plastic foundation, J. Aeronantical Soc. India, 24(4): 422-422 (1972).

A.J. Valsangkar, Flexural and buckling behaviour of individual piles and two dimensional analysis of group of piles, Ph. D. Thesis, I.I. Sc. Bangalore, India, 1969.

S. Chandra, Flexural response of beams on bilinear Foundation, M. Tech Thesis, I.I.T. Kanpur, India.

Computer and Physical Modelling in Geotechnical Engineering, Balasubramaniam et al. (eds)
© 1989 Balkema, Rotterdam. ISBN 90 6191 864 2

Acquisition by microcomputer of some ground vibration data during pile driving

A.R.Selby
Durham University, UK

ABSTRACT: The action of driving piles causes transient vibrations to be set up in the ground. These vibrations are potentially damaging to adjacent buried pipelines and structures. A portable microprocessor-based recording system has been designed and built for studying in detail the magnitude and form of such vibrations both at and below ground surface.

1. INTRODUCTION

In order to assess and control the risk of piling-induced damage to adjacent buried pipelines and structures it is necessary to measure ground vibrations and to compare these with recommended limits for the appropriate structure. It is common commercial practice to record at one station radial, transverse and vertical ground surface velocities (eg Fig. 1) using a single set of three geophones.

In the UK a local authority has the power, under the Control of Pollution Act (1974) to serve notice that the party committing the nuisance must remove the cause, or appeal against the writ on the plea that the levels of vibration are not so severe as to be considered a nuisance. Under the same Act, a preferable procedure (Attewell, 1986) is for the contractor to apply to the local authority for a Consent to Work Agreement, outlining the proposed method of pile installation. If the authority decides to approve the application it will be subject to limits on vibrations appropriate to the neighbourhood and to the construction. No limits on tolerable vibrations are given in the Act, but guidance can be obtained from several sources (Attewell, 1983, 1986, Skipp, 1984, Steffans, 1974).

In the planning stages of a piling contract, however, it would be desirable to have a predictive method for likely maximum ground vibrations and, towards this end, a number of case records have been assembled by Palmer (1983), and an empirical transmission law was proposed some years ago by Attewell and Farmer (1973), which was more recently given some theoretical justification (Attewell, 1985).

However, there is wide scatter among reported measurements, partly caused by limited recording facilities. In addition, the traditional hammers (winched drop hammer, air hammer, and diesel hammer) are rapidly being replaced by modern hydraulic hammers and vibrodrivers because of their improved performance and controlability.

In consequence, detailed multi-channel records of vibrations are being taken on construction sites in the north of England to provide data on ground vibrations caused by modern hammers, operating in a range of conditions of ground, pile-type and penetration, and of course at different stand-off distances from the pile. Currently, measurements are being taken in free ground, with vibrations generally unmodified by the presence of adjacent structures. Nor are structural vibrations yet being measured. Future work may include such situations.

2. MEASUREMENT OF GROUND VIBRATIONS

In order to measure a component of ground vibration, a transducer is required to convert the transient ground vibration into an electrical signal, which may require amplification. The signal is then recorded and usually displayed.

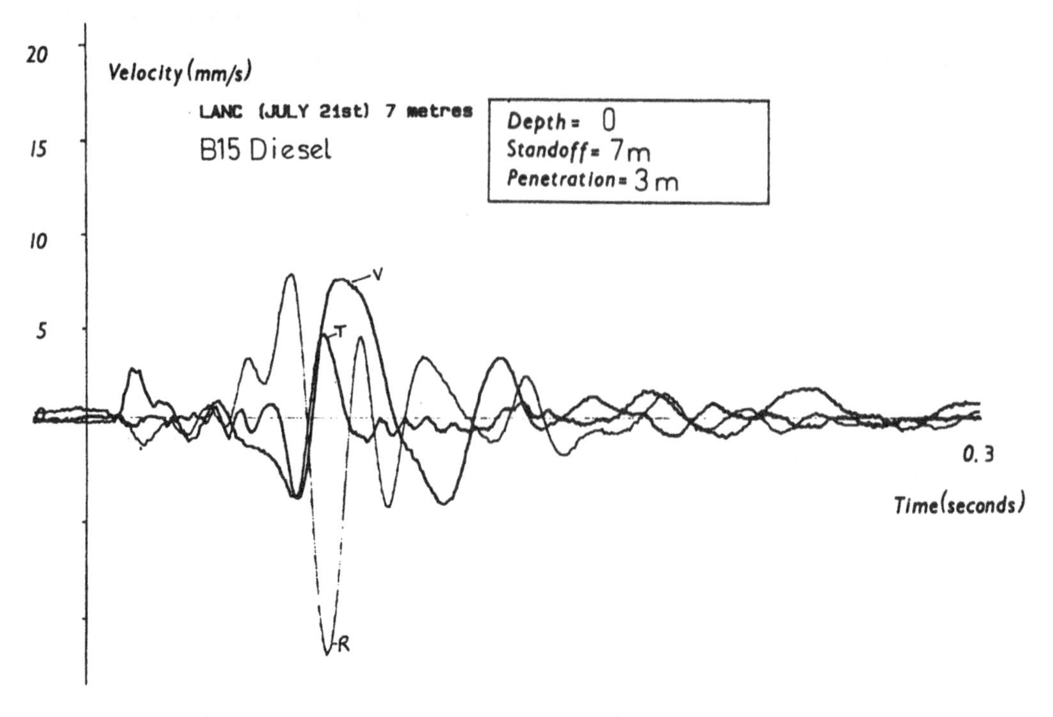

Figure 1. Typical ground vibrations caused by a diesel hammer.

2.1 Transducers

Various transducers are available to respond to transient velocity or acceleration, converting the mechanical disturbances into electrical signals. Every transducer has a limited range of frequency in which a uniform or flat response to a vibration level is obtained. It is vital that a chosen transducer responds uniformly and accurately over the frequency range of the vibrations. In the case of driven piling a very high percentage of energy transmitted in vibration form is in the range 6-80Hz.

A velocity transducer (geophone), which measures the transient velocity of a sprung, damped mass by a moving coil in a permanent magnetic field, gives a large output signal proportional to transient velocity, and requires no excitation (driving) voltage supply; consequently a pre-amplifier is unnecessary. The geophone must be correctly orientated, vertically or horizontally, but shows little cross-sensitivity. When ordering, the buyer must select both low-frequency limit (related to spring/mass resonance) and damping factor; suitable values would be 4Hz or less, and about 0.6 of critical damping (see Figure 2). Unfortunately the price of such an instrument tends to increase significantly with its lower frequency capability. The upper end frequency is less well-defined and is a function of construction detail.

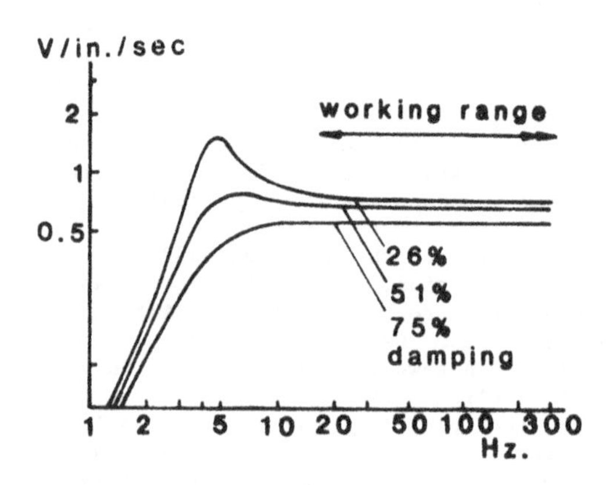

Figure 2. Geophone response characteristic (from Sensor, Geosource).

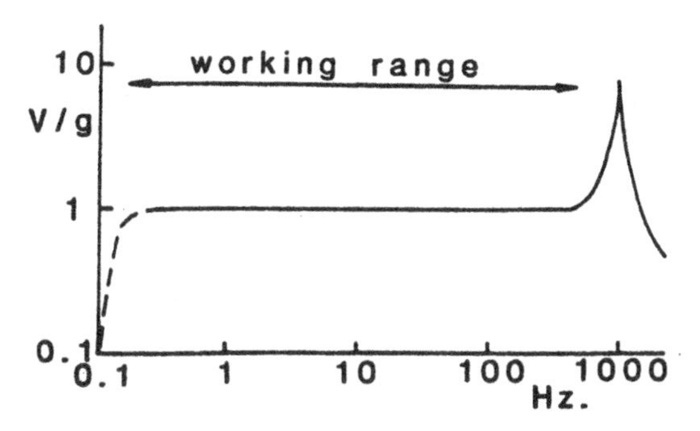

Figure 3. Accelerometer and charge amplifier response.

An acceleration transducer (accelerometer) would consist typically of a small mass mounted on a piezoelectric crystal, which generates a small electrostatic charge proportional to strain caused by acceleration of the mass. A charge preamplifier is required to convert the small charge output into a more substantial voltage signal. In this type of transducer, resonance of the mass on the crystal causes an upper frequency limit typically of about 1 to 2kHz, and the lower limit to frequency is a function of the preamplifier and cable noise. Careful design of a transducer with integral charge preamplifier can yield a limit below 0.1Hz (Figure 3). Accelerometers are robust, and can be mounted in any orientation, but the charge preamplifier, which must be incorporated, requires a power source (batteries or mains supply).

2.2 Recording devices

The simplest recording medium is a trace on paper, by pen on to ordinary paper, or by means of heat-sensitive, light-sensitive, or ultra-violet paper. Obviously, the 'record' is also the display, which is available immediately in the field.

Magnetic storage can be achieved by reel-to-reel tape recorder, cassette recorder, or on video-format cassette or small 'floppy disc'. Storage by amplitude modulation (AM) or direct recording is unsuitable for the frequency range in question and either frequency modulation (FM) or digital recording must be used.

Frequency modulation is widely used in scientific recording. A high frequency carrier wave is frequency-modified by an input signal, and the modified carrier is stored on broad tape, video tape or cassette tape. Replay of the modified wave through demodulation gives an output voltage which is a scaled version of the input signal. Some 14 channels can be stored on a 25mm wide tape, and several signals may be stored on one channel by 'multiplexing'.

The modern approach, aligned with rapid advances in electronics and microcomputer technology, is digital storage. In essence, an input signal is sampled rapidly, at say 500-1000 times per second, and the magnitudes of the samples are stored on cassette or floppy disc. The sample rate must be sufficiently rapid to define accurately the character of the signal.

2.3 Signal display and processing

Display of a transient signal is desirable so that the form of the wave can be appreciated. A permanent trace can be made on paper, or the waveform may be viewed, temporarily, on the screen of a microcomputer.

The primary information to be obtained from the signals (Figure 1) is:

1. The maximum value of a signal, usually as peak particle velocity (ppv) in mm/sec, of the radial, tranverse or vertical component.

2. The maximum vector v_{res} of ppv at a station, at any instant in time, where

$v_{res} = \sqrt{v_r^2 + v_t^2 + v_v^2}$. (Note that this is not the same as the root of the sum of the squares of individual ppv's since these do not occur at the same time.)

3. Attenuation with stand-off distance from the source (i.e. the pile), of directional ppv's and of the vectorial ppv.

4. The dominant frequencies of the signals, normally by fast Fourier transform (FFT) analysis.

2.4 Choice of system

The choice between geophone and accelerometer is not obvious, because both types of transducer can fulfill the requirements of ground vibration sensing. However, the geophone is in more common use because it self-generates a strong voltage signal, and also because ground vibration limits are usually defined by peak particle velocities: structural damage to buildings is a function of strain, which can be shown to be related to ground velocity. Within selection of a record/display system, a paper trace produced in the field has some advantage in terms of instant information, but it may suffer some of the following disadvantages.

(a) There may be a limit on the number of channels displayed simultaneously. Six or eight channels are common maxima, although some chart or u/v recorders may offer up to 14; the width of trace is often limited in multichannel traces. Nine channels is a useful facility to cover three sets of three orthogonal nests of geophones at different stand-off.

(b) Manual extraction of numerical values of velocities is tedious and inaccurate, even when using a 'mouse' cursor for computer digitization. Rapid pen movement often results in a trace of poor definition. Purely manual analysis of traces is particularly tedious in the estimation of the maximum vector of ppv, and frequency analysis can be attempted only a superficial level.

(c) Detailed signal analysis can only be undertaken sensibly on a computer, after inaccurate manual digitization.

(d) Compensation for variable geophone calibrations may need to be made manually.

The second approach to data recording, by F.M. on broad, video or cassette tapes, may suffer from fluctuation of tape speed between record and replay. Replayed signals can be plotted by air-jet pen for visual examination, and may also be fed to a digital storage oscilloscope for single channel analysis for ppv and dominant frequencies, but without automatic facility for vector ppv, or signal comparisons for attenuation of wave velocities. An alternative approach is to work from the air-jet pen traces, using a computer digitizer, with the inherent disadvantages discussed previously. The over-riding disadvantage of F.M. recording of multi-channel signals is the cost of the tape or cassette recorder; a 14 channel model is likely to cost in the order of £10 000 at 1986/87 prices.

The modern approach, offering significant advantages in power of data collection and versatility in processing, is by digital acquisition. Commercial data loggers are available, but a dedicated system is now described, designed in collaboration with the Microprocessor Centre, Durham University, where the system was built. This type of equipment is likely to become the obvious future choice of the discerning engineer who is involved in ground vibration monitoring and interpretation.

2.5 A Portable digital recorder (PDR)

The complete system comprises five sets of three low-frequency geophones with coaxial cables connected through a manifold to the portable digital recorder (PDR). The PDR consists of a sixteen channel multiplexer and twelve-bit analog-to-digital converter, a computer board built around a Motorola 68000 processor chip, and 5.25in floppy disc drive, a liquid crystal display, and a sixteen character key pad. Data are captured and dumped to a floppy disc which is then transferred to an SBC Duet 16 desk top microcomputer, with colour plotter, for data processing and display.

The procedure for capturing data in the field is initiated by inserting into the drive a floppy disc which carries a short start-up program to be read by the microprocessor board. The operator must then specify a range of parameters which control data capture. The main parameters are trigger levels, sampling rate, duration of the sample, whether pre-trigger data are required, and the number and identifications of active channels. Commands to commence data capure are then input, and when the next 'event' occurs, of vibration due to a pile-hammer blow, the PDR samples the signals from the geophones very rapidly, and stores the signal amplitudes in RAM, from which the data are transferred to floppy disc.

The digital information is then processed on the Duet 16, which consists of a keyboard and microcomputer unit, colour monitor, dual disc drive and a separate plotter. The recorded signals from the geophones can be displayed on the monitor using seven colours for easy discrimination, and a hard copy can be obtained on the plotter. A calibration coefficient for each geophone is incorporated, so that the traces are consistent, and in sensible units of particle velocities in mm/sec. The program offers a range of optional facilities including the following:

(a) Presentation of any chosen combination of geophone signals, eg. radial, transverse and vertical at one station, or the vertical traces from five stations.

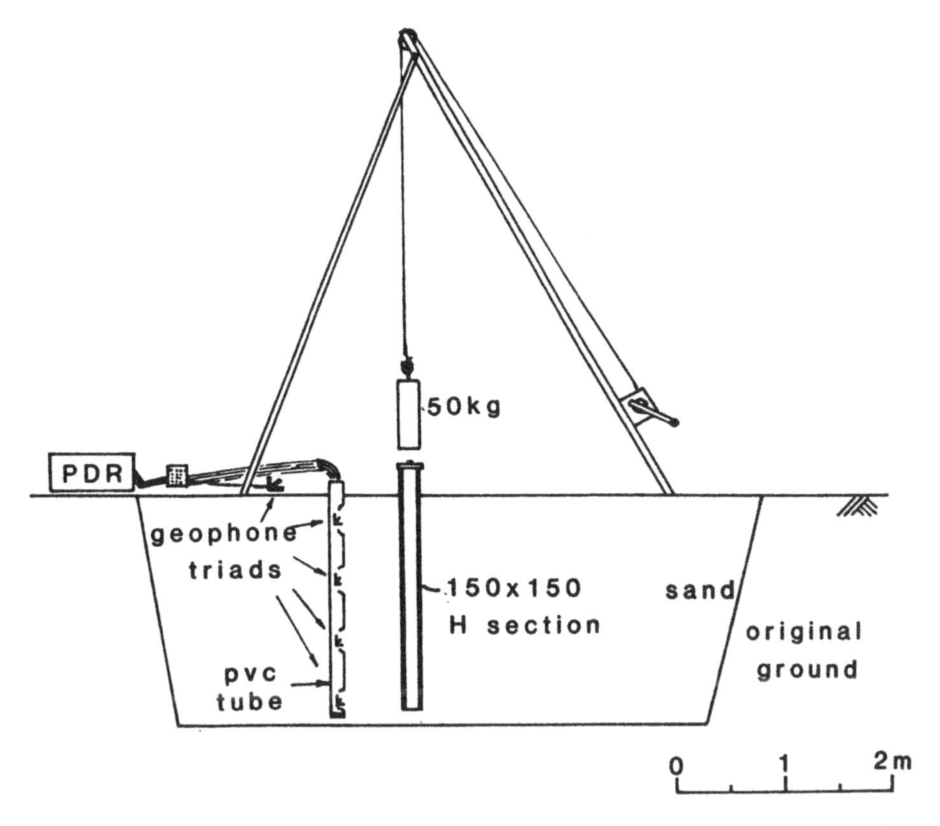

Figure 4. Section showing field tests for sub-surface ground vibrations.

(b) Vector resolution of the orthogonal (RTV) velocities from a station, for all the sample times. The vector waveform can be displayed, or the maximum value can be computed.

(c) The dominant frequencies of a trace or of a vector trace can be assessed using a fast Fourier transform subroutine and a plot of signal density against frequency can be displayed with either a natural or a log base. The trace to be analysed by FFT can be 'windowed', so that the part of the signal of most relevance to the event is chosen.

(d) From a chosen combination of geophone signals, estimation of wave velocities can be made.

(e) Simple options such as identification of a plot by a title, choice of colours for combinations of traces, and style of display of combinations (with common velocity axis, or with displaced axes) are incorporated.

A system with these capabilities enables reliable measurements to be made of ground vibrations caused by pile driving on construction sites, or by drop-weight piling under more controlled field test conditions. Further details of the system are given by Paterson (1985).

3. GROUND VIBRATIONS CAUSED BY PILE DRIVING

In an attempt to provide improved predictive methods of piling-induced ground vibrations two lines of approach are being addressed. The first is a mathematical description of the ground waves generated by shaft friction, toe penetration, and other less obvious sources, in an attempt to synthesise waveforms equivalent to those actually recorded on site. This approach requires some knowledge of energy transfer levels due to the various sources, and controlled field tests are being undertaken in Durham. Useful model test results have been reported by Heerema

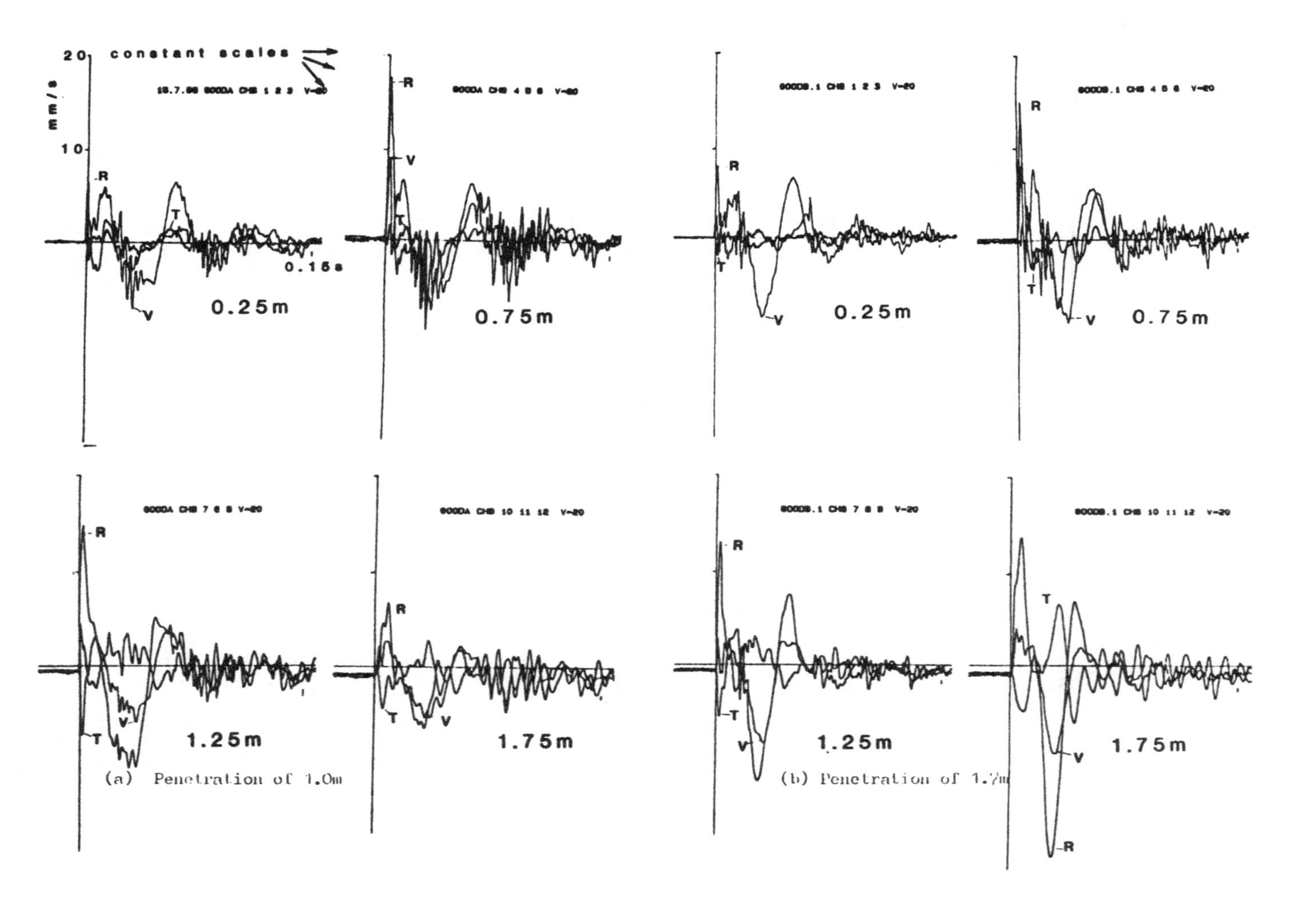

Figure 5, Below ground vibrations 0.5m stand-off.

(1979, 1981). The second line of investigation is empirical. Site piling operations in the North of England are being monitored to allow assessment of the effects of the main variables of hammer type, pile type, penetration/stick-up, soil types and particular site conditions.

Results of some field tests and of site case-studies will now be discussed.

3.1 Controlled field (model) tests

A pit 5m long by 2m wide by 2m deep, was filled with uniformly graded yellow sand (N_{10} = 0.12mm. N_{60} = 0.32mm) having a void ratio of 0.64. A number of windowed vertical pipes allowed triads of orthogonally-orientated geophones to be buried at depths of 0.25m, 0.75m, 1.25m, and to be easily recovered after each test. A 2m length of 150 x 150mm 'H' section steel pile was driven into the sand by a hand-winched 50kg drop hammer, at 0.2m, 0.5m, 1.0m, or 1.7m stand-off distance (Figure 4).

Two example sets of traces are shown in Figure 5. The first set relates to the condition of a short penetration of 1.0m and while vertical shear is dominant at the 0.25m shallow station the waves at the deeper stations comprise substantial vertical and radial components attributable to P waves, emanating from the pile toe.

The second set of traces relates to a pile embedded to 1.7m depth. A reasonably uniform set of vertical shear, SV, waves caused by shaft friction, is evident, although both radial P waves and horizontal shear SH waves cannot be ignored. From a number of traces, contour plots of vectored ppvs have been derived, and an example is included in Figure 6a. This gives an indication of the sources of energy causing the ground waves. Additional tests were conducted using deliberately eccentric blows to the pile head as a crude model of lateral upper-pile motion due to whip, Poskitt and Yip (1987). The resultant modified contour plots are shown in Figure 6b. Further examples and details are given by Cahm (1986).

In addition, conventional log-log plots of resolved ppv's (v_{res}) against ($\sqrt{W_o}/r$) were prepared, e.g. Figure 7, where W_o is notional input energy and r is stand-off. Linear regression lines were produced for each station, with the following results:

Data recorded at 0.25m depth

$$v_{res} = 0.12 \cdot \left[\frac{\sqrt{W_o}}{r}\right]^{1.44}$$

Data from 0.75m depth

$$v_{res} = 0.34 \cdot \left[\frac{\sqrt{W_o}}{r}\right]^{1.13}$$

Data from 1.25m depth

$$v_{res} = 0.55 \cdot \left[\frac{\sqrt{W_o}}{r}\right]^{0.93}$$

Data from 7.75m depth

$$v_{res} = 0.41 \cdot \left[\frac{\sqrt{W_o}}{r}\right]^{1.08}$$

These sub-surface ground vibrations can be compared with Attewell and Farmer's (1973) median line estimate of

$$v = 0.75 \cdot \left[\frac{\sqrt{W_o}}{r}\right]^{0.87}$$

and upper bound of

$$v = 1.5 \cdot \frac{\sqrt{W_o}}{r}$$

The observed data of ppv's are all considerably lower than these expressions. Attenuation rates, defined by the power of the term in parentheses, are broadly similar, although the shallowest station shows more rapid fall-off. This may be influenced partly by inclusion of the results from eccentric hammer blows.

Initial conclusions from these test results are that the major source of energy causing ground waves is toe penetration, while shaft friction generates lower energy levels. The ground waves resulting from the toe penetration comprise significant components of both expansive P, waves, and shear waves, but the waves emanating from shaft friction are predominantly in vertical shear, SV. The consequences of an eccentric hammer blow are most noticeable near to ground surface, where additional P waves occur, primarily in the direction of eccentricity.

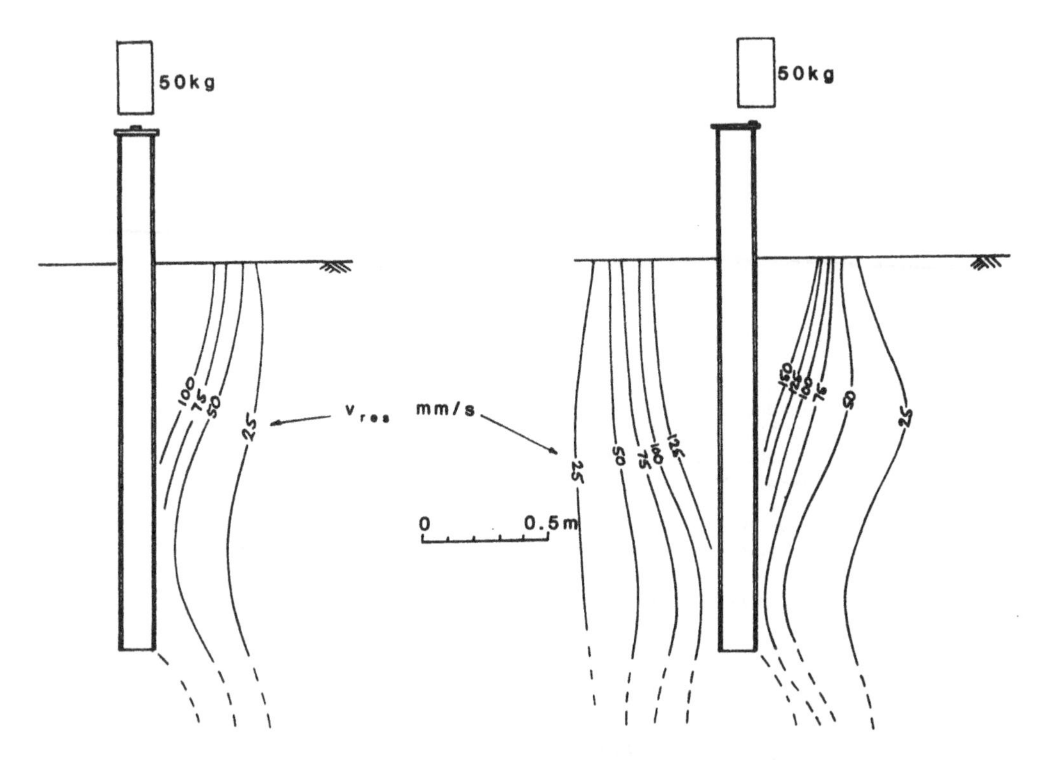

Figure 6. Contours of resolved sub-surface ppv's, axial and eccentric blows

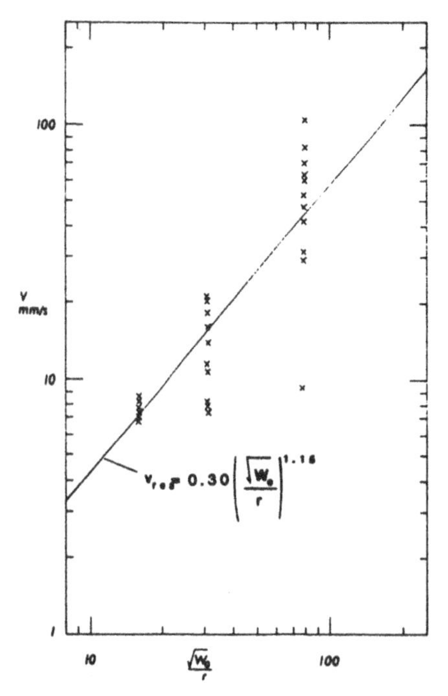

Figure 7. Regression line for resolved ppv's for all sub-surface axial blow records.

Attenuation is approximately proportional to the inverse of distance, so that fall-off in ppv is very rapid close in to the pile axis. Spectral analysis showed frequencies of the order of 20-30Hz carrying most of the transmitted energy.

3.2 Sheet piling for riverbank protection, Lancaster

As part of a larger programme of riverbank improvement on the Lune estuary in North West England, a stretch of the south bank within the City of Lancaster is to be protected by a line of 25W Larssen sheet piles driven on the outside of a deteriorated existing sheet pile quay wall, Figure 8. The new wall is of cantilever design, and relies upon suitable penetration into the gravel beds which underlie the river bed silts.

At one end of the contract a viaduct crosses the river, with a relatively recent steel deck resting on Victorian masonry piers. There was some concern over possible vibration damage by piling operations close to one of the piers, so an initial survey of vibration levels was undertaken across the quayside surface

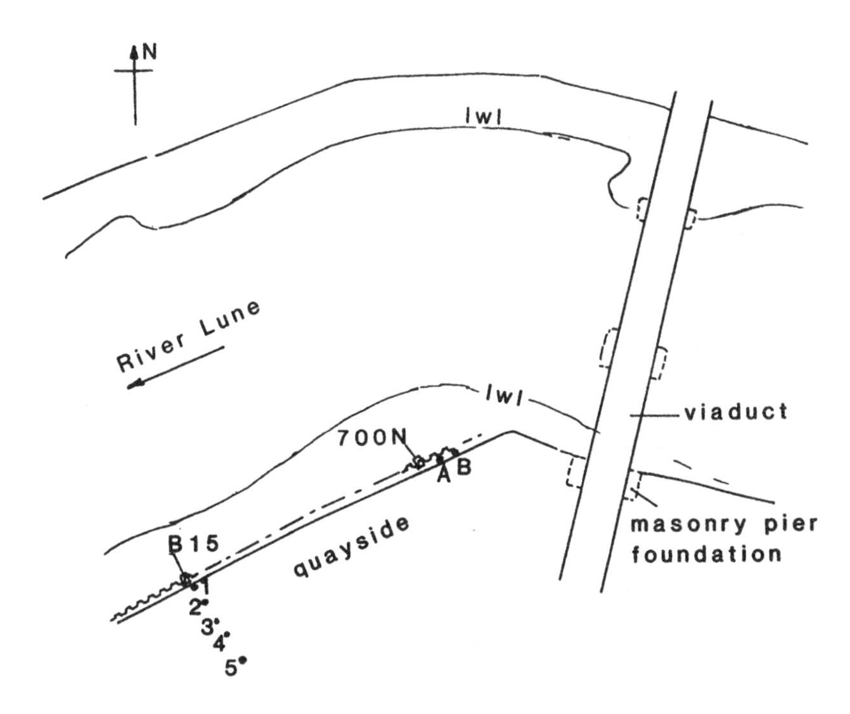

Figure 8a. Sheet piling location plan showing geophone positions

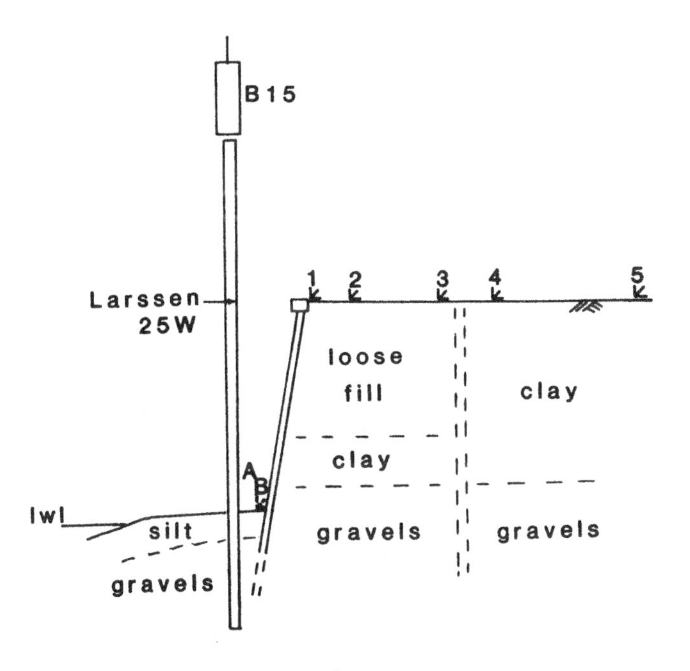

Figure 8b. Section through quayside.

(Figure 8b). The hammer in use was a BSP B15 diesel, with quoted energy rating of 3630kg m. The results were initially rather surprising, in that the ppv's of Station 1 were relatively small, with a maximum of 5.5mm/sec. The largest ppv's (11.8mm/s (R) or 13.9mm/s (res)) were recorded at Station 4 (see Table 1 and Figure 1) 7m horizontally away from the driven pile.

Table 1. Quayside ground surface vibration, BSP B15 hammer.

Station	Stand-off	ppv's (mm.sec)			
		v_r	v_t	v_v	v_{res}
1	2m	5.5	2.9	-	5.5
2	3m	5.7	2.9	4.2	6.3
3	5.5m	6.2	6.9	6.0	9.4
4	7m	11.8	4.2	6.8	13.9
5	11m	3.4	1.6	2.6	3.5

Observation of Figure 8b reveals the reason for this: there is no straight clear path from the embedded pile length to Station 1, while a clear direct path runs to Station 4, through stiffer soil. The picture is further confused by the presence of the old timber wharf, and the variability of the soils, in particular the fill. However, there was a strong possibility of ppv's in excess of 20mm/sec being induced in a masonry pier founded on bedrock should piling be continued to within 2 to 3m.

The contractor then changed to a BSP 700N air hammer having a nominal energy of 650kg m. Vibrations on the quay top were reduced to levels of less than 1mm/sec. Geophones were then placed in the mud close to the existing wall, at 2m and 3.5m along the wall from the driven pile, at low tide. The consequent records are shown in Figure 9. The ppv, s of around 7mm/sec, or some 32 vibrars (see below) were considered to be tolerable for the masonry pier.

The investigation highlighted the value of multichannel recording, by affording confidence in apparently unexpected results. The air hammer signals were also processed by FFT, and showed quite dominant frequencies of around 12Hz. This information was of value for converting the data into vibrars, where strength in vibrars

$= 10 \log_{10} 1.6\pi^4 a^2 f^3$,

with a = peak displacement and f =

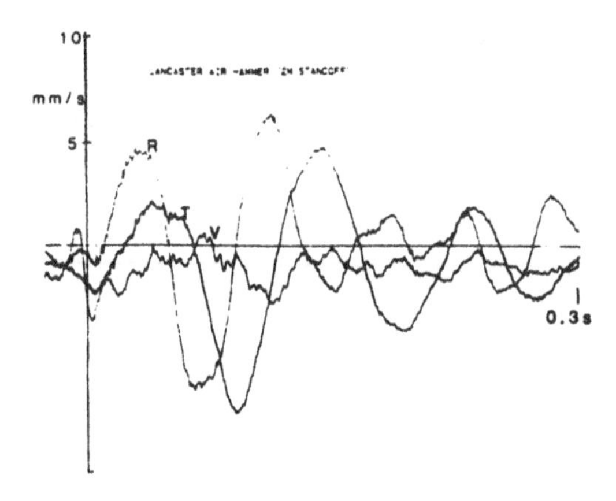

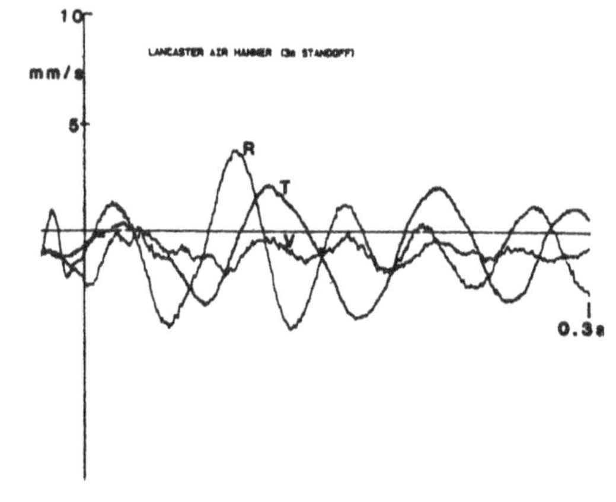

Figure 9. Mudline ground vibrations, air hammer, Lancaster.

frequency in Hz . This was the unit preferred by the viaduct owners.

3.3 Field data with some modern hammers

Additional field data were collected on several visits to a site during construction of a contract by Fairclough Civil Engineering Ltd., in Yorkshire. A range of piling operations was conducted to provide bearing foundations and small retaining walls.

a) Bearing piles. Foundations consisted of steel 'H' piles, 356 x 368 section and 32m long. They were driven through very soft alluvial clay down to the bedrock consisting of the Carboniferous sandstones of the Upper Palaeozoic measures. The piles were driven in lengths of 16m, with a butt weld connection. A BSP 357 hydraulic hammer was used, with a hammer weight of

3 tonnes falling through 1.4m. Traces of ground surface vibrations are shown in Figure 10, with the traces now plotted individually for clarity. Table 2 summarises the largest ppv's. Clean signals are apparent (except for the transverse trace at 2m stand-off) and this adds confirmation to the claim that the 357 is an efficient hammer.

Table 2. Summary of ppv's, hydraulic hammer on H piles.

Disc No.DATA19 H-Steel Pile Hydraulic-Hammer

Geophones Code			A	B	C	D	E
Stand-off [r] (m)			2.0	5.0	9.0	13.0	18.0
(1 / r)			0.5	0.2	0.11	0.08	0.06
KHH1	Max. Velocity mm/s D=16.0m	Radial	9.90	7.59	9.37	17.00	10.30
		Transvers	11.52	5.34	3.48	8.12	1.79
		Vertical	7.14	3.59	4.89	4.78	2.73
		Vres(t)	12.47	8.58	9.40	18.78	10.64
KHH2	Max. Velocity mm/s D=17.0m	Radial	8.73	6.22	10.55	14.75	10.02
		Transvers	6.32	5.05	2.67	5.81	1.79
		Vertical	7.34	3.32	3.95	4.69	2.43
		Vres(t)	9.41	7.32	10.68	16.01	10.29
KHH3	Max. Velocity mm/s D=18.5m	Radial	5.49	5.58	6.56	9.97	8.30
		Transvers	4.27	5.72	3.19	4.71	1.79
		Vertical	4.30	1.98	2.83	3.78	1.89
		V res(t)	6.32	5.99	7.16	10.95	8.43
KHH4	Max. Velocity mm/s D=20.0m	Radial	6.21	6.12	7.25	9.43	9.19
		Transvers	3.81	5.15	4.00	3.60	2.05
		Vertical	3.81	2.07	2.71	4.14	1.59
		Vres(t)	6.82	6.19	7.74	10.22	9.25
KHH5	Max.s Velocity mm/s D=21.0m	Radial	10.08	7.04	7.33	10.51	9.93
		Transvers	7.71	5.43	4.08	3.42	2.14
		Vertical	6.85	2.60	2.78	4.96	1.69
		Vres(t)	10.74	7.19	7.82	11.26	10.22

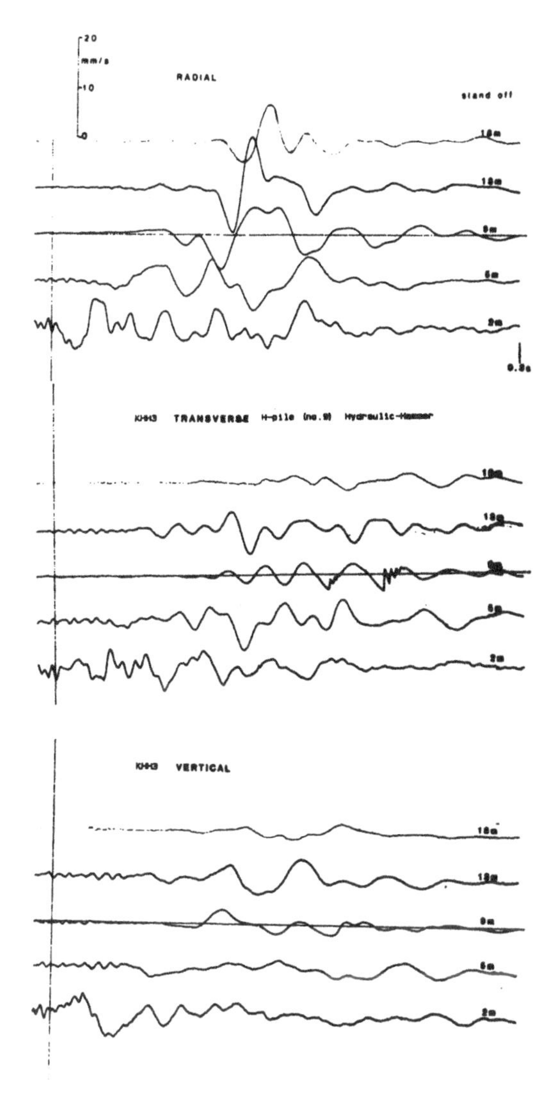

Figure 10. Ground surface vibrations, hydraulic hammer on H piles.

The traces and ppv magnitudes also reveal some intresting indications of the main sources of energy transfer from the pile into the ground. The rather ragged traces of particle velocities at 2m stand-off may have been caused by 'clatter' or minor impacts between the pile and the guide frame and by lateral 'whip' (Poskitt & Yip, 1987) due to strut-type deformation of the pile length above ground. These random, higher frequency, signals are quickly attenuated or filtered out by the ground.

The signals recorded further away from the pile are generally 'cleaner', and show one main cycle; broadly sinusoidal, with the radial component being dominant.

It is particularly interesting that the radial ppv's do not show uniform attenuation with stand-off as has been previously proposed (see Table 2 and Figure 11). In fact, they tend to show a maximum at some 15m away from the pile.

This unexpected behaviour was observed several times, on this and on other sites. One possible explanation might be suggested, on the assumption that the major energy transfer from pile to soil occurs at the toe, and that ground waves (as a mixture of compressive P and shear

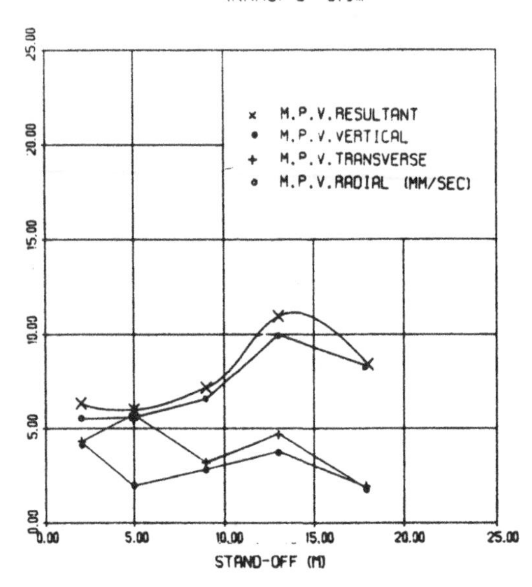

Figure 11. Variation in ppv's with stand-off, hydraulic hammer.

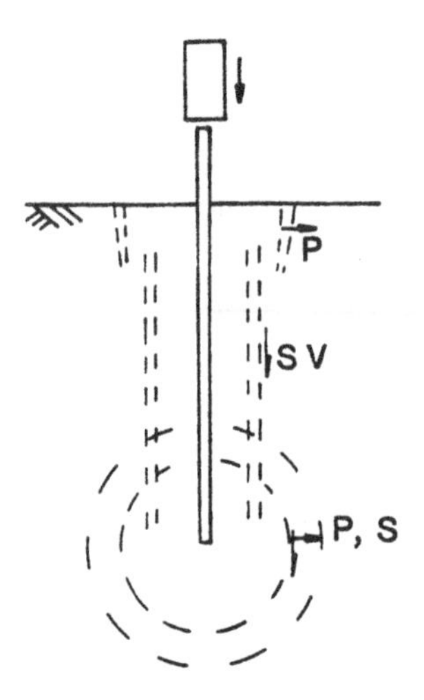

Figure 12a. Possible wave types.

S) are propagated spherically outwards from the toe (Attewell & Farmer, 1973), Figure 12a. The wave energy, however, is likely to be non-uniform around the expanding spherical wave front, with a maximum energy content vertically below the toe, and a very low energy level vertically above the toe (Figure 12b). When geometrical divergence and material attenuation is added, then a possible explanation can be made, by reference to Figure 12c.

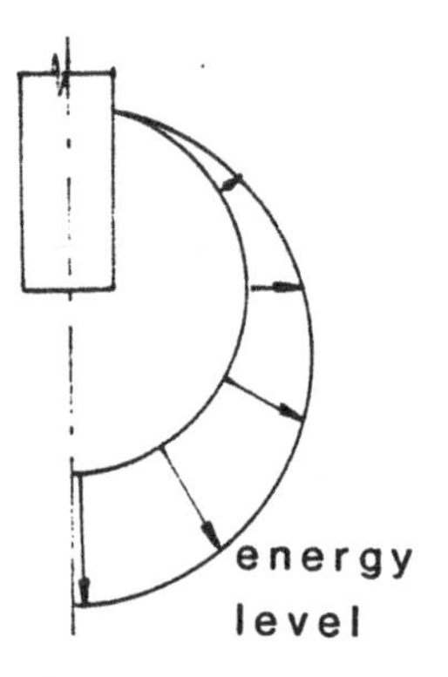

Figure 12b. Possible toe energy transfer.

At any point A, the observed surface ppv, v_a might be derived as

$$v_a = v_r \, \alpha_a . \exp[-\beta r/2] . \frac{\sqrt{4\pi \, r_o^2}}{\sqrt{4\pi \, r^2}}$$

where it is assumed that

$\alpha = \sin^{1/2}(\phi/2)$, $r_o = 1\text{m}$, and $\beta = 0.1$.

Then $v_5 = v_r \times 0.314 \times 0.281 \times 1/25.4$
$= v_r \times 0.00347$

For $v_5 \simeq 7\text{mm/sec}$, take $v_r = 2000$,

so that $v_5 = 6.9\text{mm/sec}$.

Hence

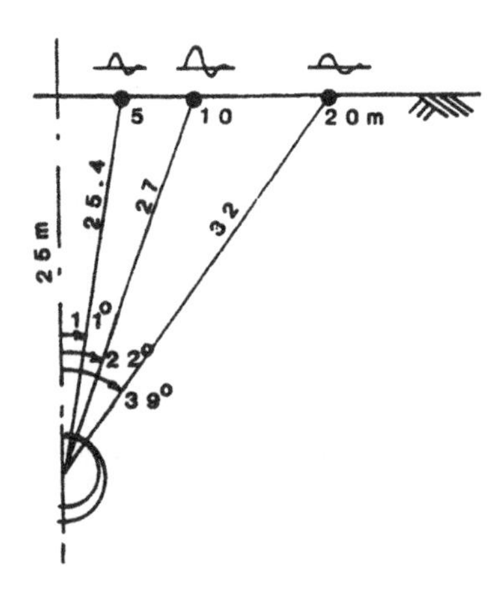

Figure 12c. Possible propagation of toe waves.

$$v_{10} = 2000 \times 0.435 \times 0.253/26.9$$
$$= 8.17\text{mm/s}$$

$$v_{15} = 2000 \times 0.517 \times 0.233/29.2$$
$$= 8.26\text{mm/s}$$

and

$$v_{20} = 2000 \times 0.575 \times 0.20/32.0$$
$$= 7.18\text{mm/s}$$

This example calculation should not be treated as a serious technique for estimation of ppv,s, but rather as a demonstration of a possible explanation of the observed phenomenon of a peak in the plot of ppv against stand-off.

It must be emphasised that such observations are possible only with multichannel recording equipment, because the variability from one blow to the next precludes such a deduction when using a single mobile triad of geophones.

In general terms, the resolved ppv's of 16 to 18mm/s at 13m stand-off show some potential risk of damage to sensitive structures, but the signals of around 10mm/s at 18m stand-off give less cause for concern.

b) Sheet piles. The ground vibrations generated by a small vibrodriver in driving short lengths of Frodingham sheet piling, type 3N and 9m long, are entirely different in character. The traces in Figure 13, are highly periodic in nature, with a frequency of 22.5Hz, which corresponds to the driving frequency of the vibrodriver. Close in to the pile, the three components (R, T and V) are of broadly equal magnitude and typically between 15 and 20mm/s (see Table 3).

The second point to note is that the time-resolved vector ppv's reduce with stand-off in the more 'expected' manner (Figure 14). A log plot (Figure 15) of v_{res} against ($\sqrt{W}/r$), where W is the nominal energy per cycle of the vibrodriver gives an expression for record KVS7, of

$$v_{res} = 1.48 \left[\frac{\sqrt{W_o}}{r}\right]^{0.98}$$

This is almost coincident with the Attewell and Farmer upper bound equation previously mentioned. It might be deduced that the energy transfer from a vibro-driven, short, sheet pile is substantially different from the transfer mechanism of an H pile driven by impact.

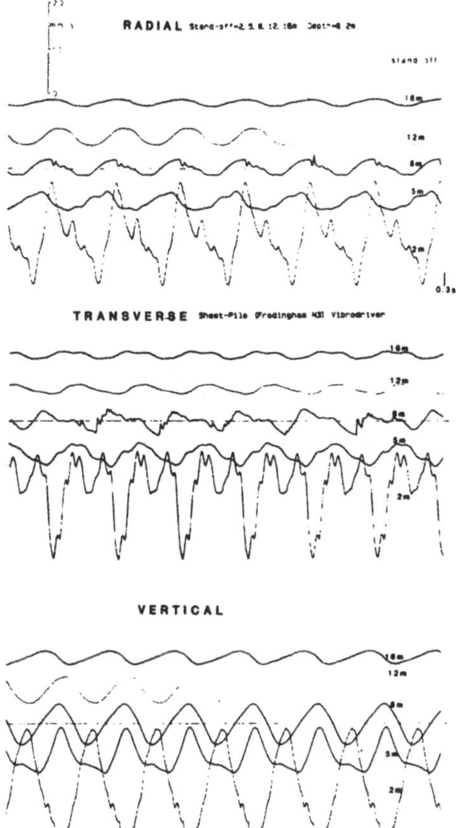

Figure 13. Ground surface vibrations, vibrodriver on sheet piles.

Table 3. Summary of ppv's vibrodriver on sheet piles.

Disc No.DATA18 Steel Sheet Pile

Geophones Code			A	B	C	D	E
Stand-off [r] (m)			2.0	5.0	8.0	12.0	16.0
(1/r)			0.5	0.2	0.12	0.08	0.06
KVS 6	Max. Velocity mm/s D=6.0m	Radial	7.02	2.65	6.44	2.80	1.15
		Transvers	6.87	3.14	4.32	1.75	1.37
		Vertical	14.47	6.02	4.98	3.51	2.02
		Vres(t)	16.18	6.47	8.04	4.53	2.22
KVS 7	Max. Velocity mm/s D=6.2m	Radial	14.58	3.02	3.81	2.80	1.34
		Transvers	19.23	3.34	3.85	1.85	1.28
		Vertical	16.82	8.08	5.75	3.60	2.32
		Vres(t)	24.45	8.38	6.85	4.48	2.47
KVS 8	Max. Velocity mm/s D=6.5m	Radial	15.84	4.02	2.36	3.52	2.44
		Transvers	19.51	6.29	6.07	1.75	1.72
		Vertical	17.11	9.70	7.85	4.97	2.52
		Vres(t)	25.18	11.09	8.22	5.99	2.77

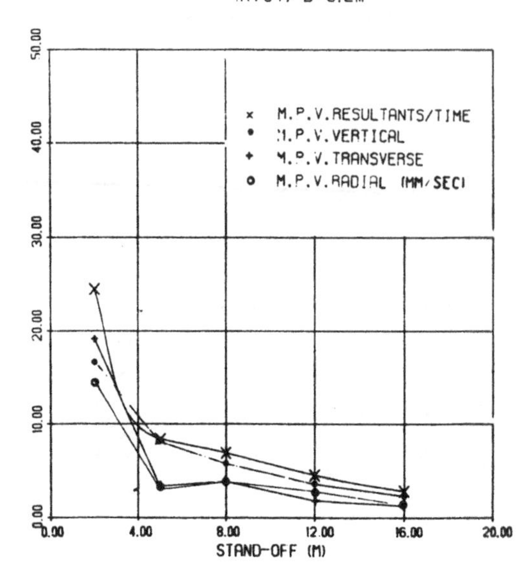

Figure 14. Variation in ppv's with stand-off, vibrodriver.

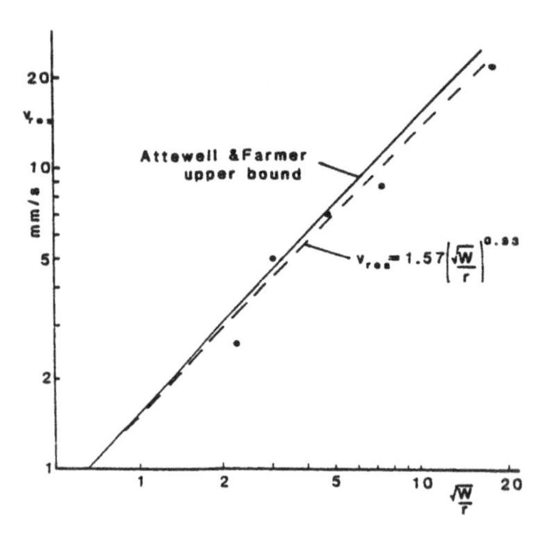

Figure 15. Attenuation of vibrodriver ppv's.

In this case study, the sheet piles were short, driven into an accessible free-ground surface, and well-supported by a substantial gate, with a carefully aligned vibrodriver on the pile head. Consequently, the level of ground vibrations dropped very quickly to an acceptable level. Observations at other sites, with less-controlled vibro-driving, have shown substantially larger ppv's at considerable stand-off distances.

4. CONCLUSIONS

The purpose-built multi-channel recording equipment described in this paper has shown its flexible capabilities in a variety of situations related to measurement of transient vibrations caused by pile driving. The only minor (and rare) loss of reliability arose from contamination of the disc drive by dust and dirt during use on site. The importance of multi-channel observations in a range of situations is clearly demonstrated.

Some below-ground vibrations observed very close to a small pile are reported, with very high values of ppv's, especially when the driving blow is eccentric.

When piles are to be driven near to structures sensitive to vibrations, the risk of damage can be reduced by selection of hammer type and of driving procedure.

The trend in pile driving equipment is towards more controllable hydraulic hammers and vibrodrivers. The hydraulic hammer causes a clean ground vibration trace, and when driving long piles it may show a peak in particle velocities at some 10-15m out from the pile. A tentative explanation for this behaviour has been suggested.

Vibrodriving which is commonly used on sheet piling, causes a periodic form of ground vibrations, with a relatively high value of ppv, for the energy per cycle.

No conclusion can yet be drawn concerning the effects of soil type upon propagation of ground waves, but as further data is obtained some trends may emerge.

The collaboration with Bullen & Partners, UK Consulting Engineers, over the Lancaster case-study is gratefully acknowledged. Thanks are expressed to Fairclough Civil Engineering Ltd, for permission to visit the site in Yorkshire.

REFERENCES

Attewell, P.B. & Farmer, I.W. 1973. Attenuation of ground vibrations from pile driving, Ground Engineering, 6, 4, 26-29.

Attewell, P.B. 1983. Assessing the effects of explosive detonations and mechanical impact upon adjacent buried pipelines, WRC Ext. Rep. No. 98E.
Attewell, P.B. 1985. Estimation of ground vibration caused by driven piling in soil, Proc. XII Int. Conf. SM & F Engrg., San Francisco, Ca. Ed. D. Resendes & M.P. Romo, 1-10.
Attewell, P.B. 1986. Noise and vibration in civil engineering, Mun. Eng. 3. June 139-158.
Cahm, J.H. 1986. Pile driving and ground vibration, Dissertation MSc Advanced Course in Engineering Geology, University of Durham.
The Control of Pollution Act 1974. HMSO.
Heerema, E.P. 1979. Relationships between wall friction, displacement velocity and horizontal stress in clay and in sand, for pile drivability analysis, Ground Engineering, 12, 6, 30-37.
Heerema, E.P. 1981. Dynamic point resistance in sand and in clay for pile drivability analysis, Ground Engineering, 14, 6, 30-37.
Palmer, D.J. 1983. Ground borne vibrations arising from piling, draft CIRIA Report 106/2.
Paterson, S. 1985. A portable digital seismic data recorder for measurement of vibrations during piling, Dissertation: MSc Advanced Course in Engineering Geology, University of Durham.
Poskitt, T.J. & Yip, K.L. 1987. Monitoring and analysis of pile driving during installation, Proc. of Int. Conf. on Foundations and Tunnels, Vol. 1. Eng. Tech. Press, 200-204.
Skipp, B.O. 1984. Dynamic ground movements - man-made vibrations, 'Ground movements and their effects on structures' Ed. Attewell & Taylor, Surrey University Press.
Steffans, R.J. 1974. Structural vibration and damage, BRE report, DoE.

Computer and Physical Modelling in Geotechnical Engineering, Balasubramaniam et al. (eds)
© 1989 Balkema, Rotterdam. ISBN 90 6191 864 2

Prediction of horizontal behavior of pier foundations

Yukio Yoshii
Tokyo Electric Power Company Inc., Japan

Kazuma Uto
Tokai University, Japan

Mikio Takeuchi & Yasuhide Seno
Okumura Corp., Japan

1 INTRODUCTION

Approximately 70% of Japan is covered by mountainous terrain.

In recent years, transmission towers and highways have been constructed on the mountainside. Therefore, these bases often employed pier foundations.

However, the problems of the horizontal resistance and displacement of pier foundations on a slope have not yet been made clear.

In these circumstances, the authors conducted a series of in-situ full-scale horizontal loading tests to the pier foundations on a slope in order to clarify their behavior when they are subjected to horizontal loads.

The authors intend to state the result of the observation during the full-scale horizontal loading tests aided by micro-computers, the method of prediction of the horizontal behavior of pier foundations and the result of numerical calculations for the pier foundations.

2 FULL-SCALE HORIZONTAL LOADING TEST[1)2)3)]

2.1 Outline of the tests

In Fig.1 and Table 1 show the outline of the conditions of ground at three different locations where the loading test was carried out for each pier specimen. The positions of measuring instruments for each conditions and the outline of testing apparatus are shown in Fig.1. In Fig.1(c), the pier C' is a reaction pier against to the pier C. Also, the structure of the monitoring system using the micro-computer is shown in Table 2.

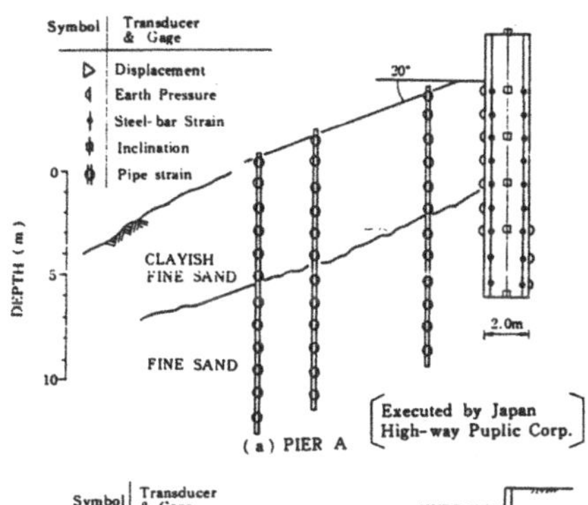

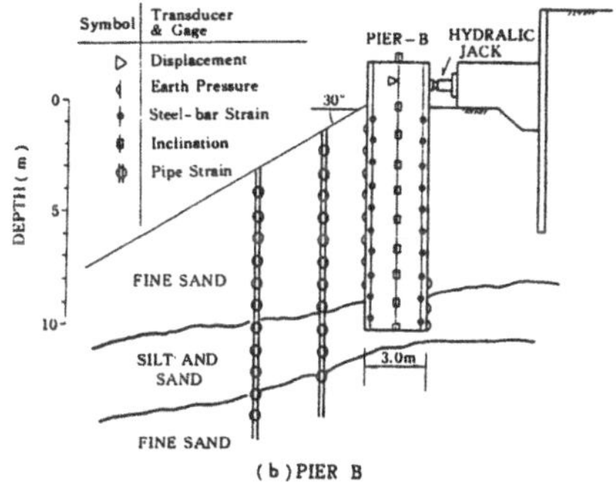

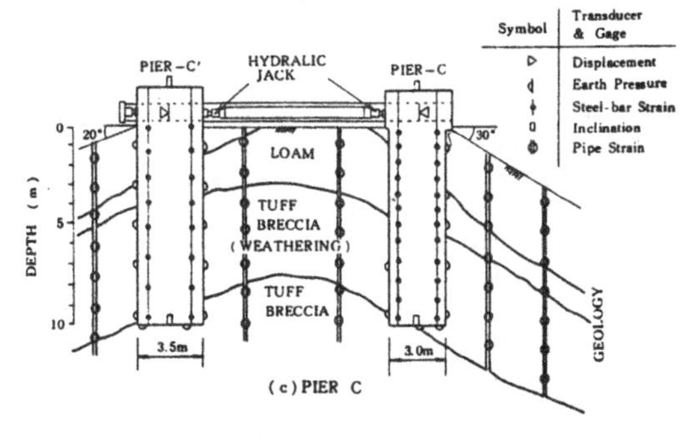

Figure 1. Testing apparatus and arrengement of instruments

Table 1. Site geology

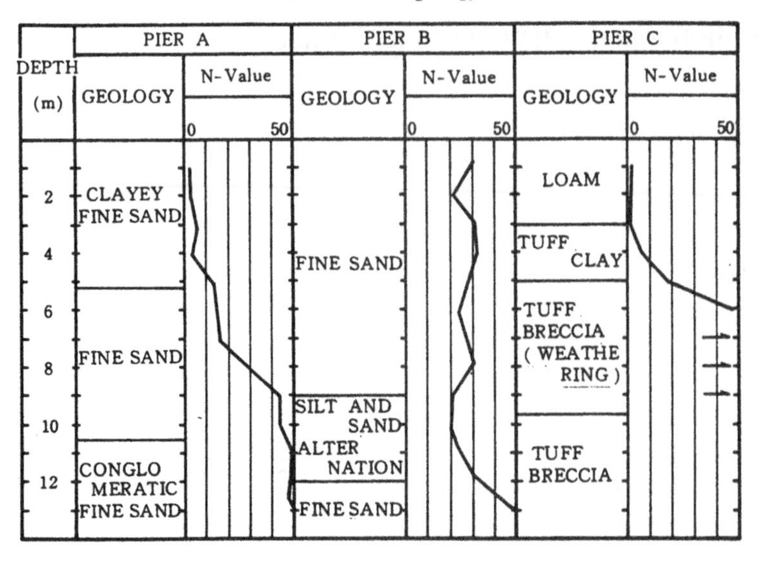

2.2 Test results

The load - displacement curves at the each pier head are shown in Fig.2.

When the load reached 5.89 MN for the pier A, 4.66 MN for the pier B and 9.81 MN for the pier C, the displacement at the pier head increased suddenly. Strain data for reinforcing bars showed that at the same loading levels, each pier had not reached the ultimate resistance.

Therefore, most of the ground in front of each pier seemed to have reached a plastic condition.

In Fig.3, the distribution of the ground reaction measured by the earth pressure gauges that are buried in front of the pier B and pier C is shown. It appears that the intensity of the ground reaction above the center of rotation (in case of the pier B, at around a depth of 8.0 m, and in case of the pier C, at around a depth of 8.0 m~9.0 m) tends to increase toward the direction of depth, even if the ground still remains within the elastic range. As shown in Fig.3, the zone of ground in front of the pier that has turned plastic is gradually expanding downward from the top.

In Fig.4, the conditions of the progress of cracks developed in the ground around the pier C are shown. The crack ① that started at the side surface of the pier proceed toward the direction of 45° to the direction of loading. After the test, the test ground was excavated and the

Table 2. Structure of the monitoring System

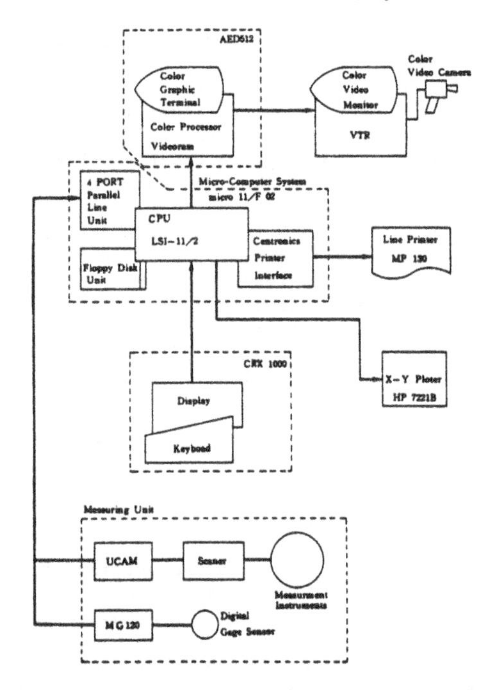

conditions of cracks were observed. It was found that the crack ① had run downward making an angle of 45° with the horizontal ground surface as shown in Fig.4(b). The size of the zone of cracks was about three

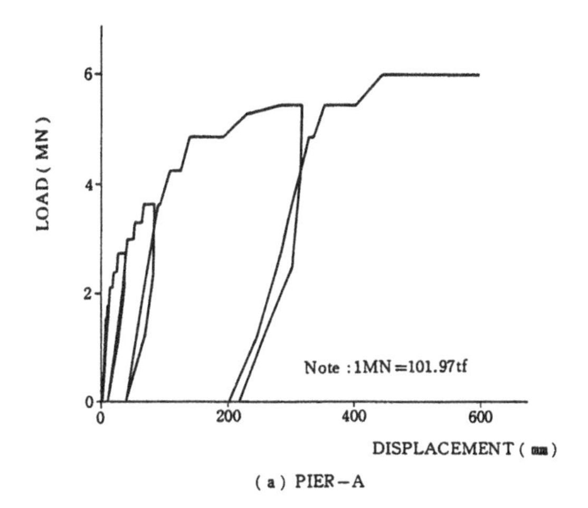

(a) PIER-A

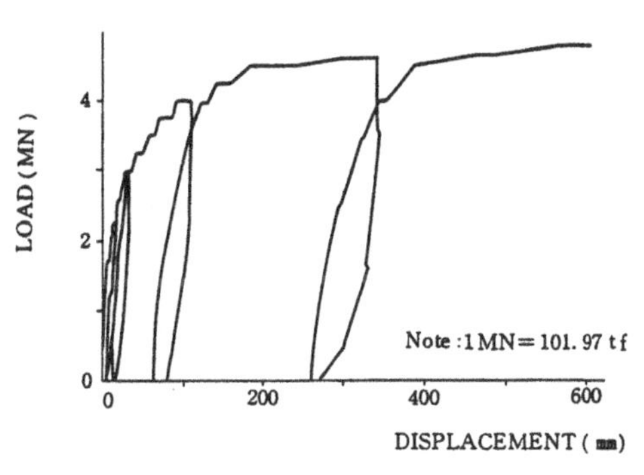

(b) PIER-B

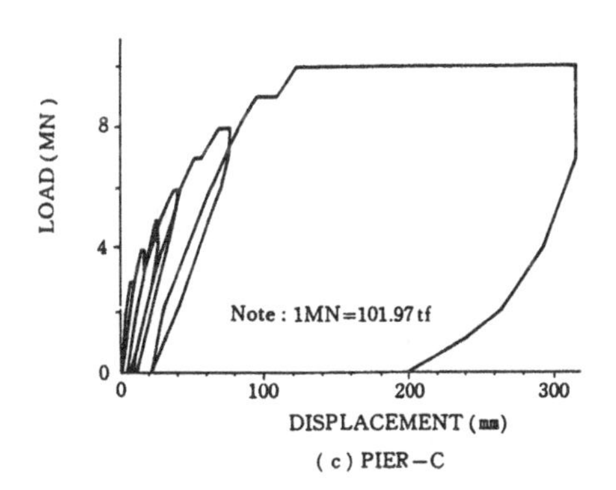

(c) PIER-C

Figure 2. Load-displacement curve at the pier head

times as large as the pier diameter. The crack ③ made about 30° to the normal to the slope as shown in Fig.4(c). If the angle ϕ of internal friction of soil is introduced, this angle approximately corresponds $(45° - \phi/2)$.

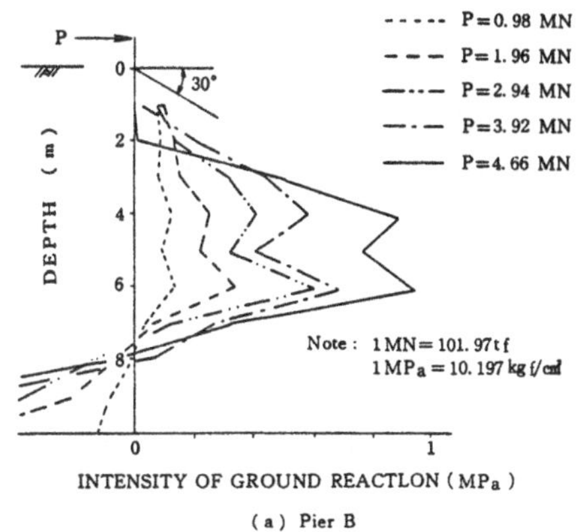

(a) Pier B

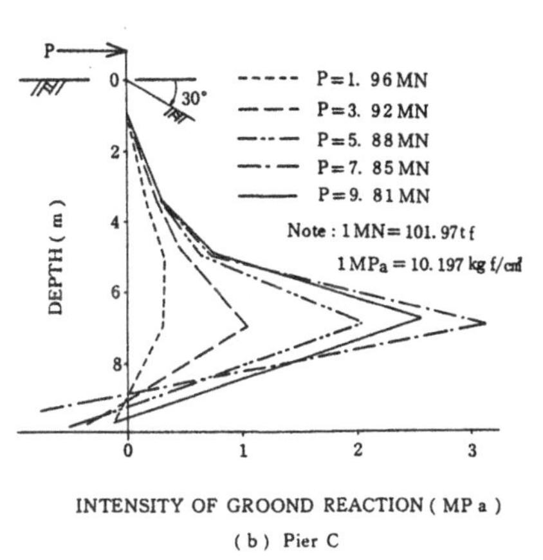

(b) Pier C

Figure 3. Distribution of ground reaction

Similar phenomena to the above were observed with respect to both of the pier A and B as well.

3 ULTIMATE GROUND BEARING CAPACITY[5)]

3.1 Sliding Model of the Ground

Observations of the development of cracks in the soil-pier system in the test led to a assumption that the mass of earth as shown in Fig.5(a) was thrust upwards. The extent of the lateral spread of the thrust of the soil mass is limited to three times the diameter of the pier. It was further assumed that the force Fi, acts on each sliding surface ① of the earth mass in accordance with Coulomb's law of shear resistance.

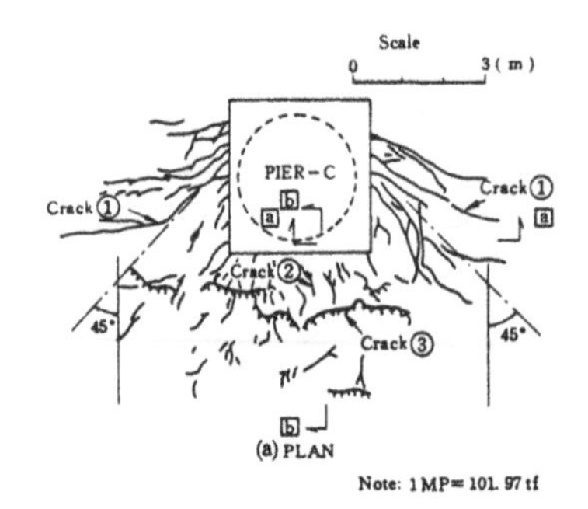

NUMBER	LOAD(MN)	APPEARANCE SITUATION OF CRACKS
①	3.43	The tensile cracks appeared from the side of the pier.
②	4.41～7.85	Radial cracks appeared towards the front of the pier.
③	8.83～	The front of the pier was apparently thrust forwards. Simultaneously, cracks occurred in a direction perpendicular to radial cracks ②, and subsequently formed a circular Shape.

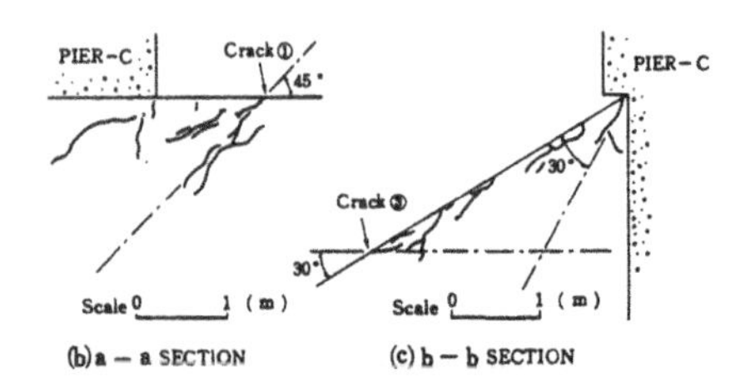

Figure 4. Crack occurrence in ground (Pier C)

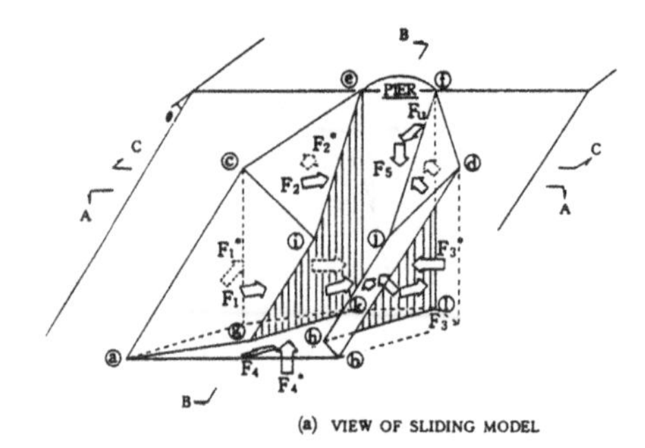

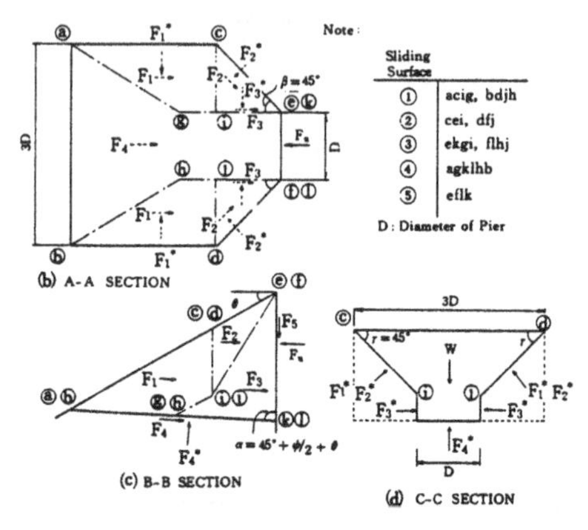

Figure 5. Sliding model of ground

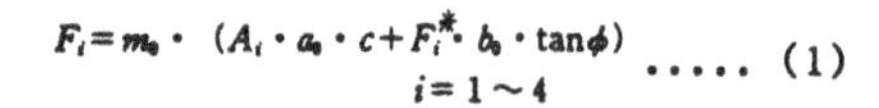

$$F_i = m_0 \cdot (A_i \cdot a_0 \cdot c + F_i^* \cdot b_0 \cdot \tan\phi) \quad i = 1 \sim 4 \qquad \cdots\cdots (1)$$

Where, Fi* is the force normal to each surface and assumed to the represented by the following formulas:

$$F_i^* = \int_A (K_0 \cdot \gamma_t \cdot x \cdot \sin\gamma + \gamma_t \cdot x \cdot \cos\gamma)\, dA_i \quad i = 1,\ 2$$

$$F_3^* = \int_A K_0 \cdot \gamma_t \cdot x dA_3 \qquad \cdots\cdots (2)$$

F_4^*: determined for equilibrium of the mass of earth

where,

C: cohesion

ϕ: angle of internal friction

γ_t: unit weight

x: depth

Ai: area of sliding surface ⓘ

γ: angle of crack along the direction of depth in C-C section (45°)

K_0: coefficient of earth pressure at rest (0.5)

a_0: coefficient of strength reduction factor of soil by collapse $0 \leqq a_0 \leqq 1$

b_0: coefficient of shear transfer factor across crack $0 \leqq b_0 \leqq 1$

m_0: degree of mobilization $0 \leqq m_0 \leqq 1$

Along the surface ⑤, considering only the ultimate bearing capacity Fu and ignoring the shear resistance F_5, Fu is derived from the balance of formula (1) and the weight of the mass of earth W.

$$F_u = \{W' - 2 \cdot \sum_{i=1}^{2} F_i^* \cdot \cos\gamma + (2 \sum_{i=1}^{3} F_i + m_0 \cdot A_4 \cdot c) \cdot \cos\alpha\} \cdot \frac{\cos\alpha + m_0 \cdot \sin\alpha \cdot \tan\phi}{\sin\alpha - m_0 \cdot \cos\alpha \cdot \tan\phi} - 2 \cdot F_2^* \cdot \sin\gamma \cdot \sin\beta + (2 \cdot \sum_{i=1}^{3} F_i + m_0 \cdot A_4 \cdot c) \cdot \sin\alpha$$

where,

α : angle between the bottom surface and the pier ($45° + \phi/2 + \theta$)

θ : inclination of the ground surface

β : angle of cracks to the lateral direction (45°)

The intensity of ultimate bearing capacity Pu at every depth is obtained by differentiating Fu with respect to the depth and dividing by the pier diameter.

$$Pu = \frac{1}{D} \cdot \frac{dFu}{dx} \qquad \cdots\cdots (4)$$

3.2 Simplification of the sliding model of the ground

In order to simplify formula (3), the sliding surfaces formed by broken lines as shown Fig.5 are assumed. The shearing force is assumed to work along the sliding surface ④ only at the bottom and the resisting forces on the other surfaces are disregarded.

Accordingly, formula (3) can be simplified as follows:

$$F_u = \frac{W \cdot (\cos\alpha + \sin\alpha \cdot \tan\phi) + c \cdot A}{\sin\alpha - \cos\alpha \cdot \tan\phi} \quad \cdots\cdots (5)$$

where,

$\alpha = 45° + \phi.2 + \theta$

A: the area of the sliding surface at the bottom.

3.3 Application of formula for ultimate bearing capacity

In applying formula (3), the values of a_0 and b_0 are assumed to be zero along the sliding surface ② and ① along the other surfaces, and the value of m_0 is assumed to be 1 along every surface.

The intensity of ultimate bearing capacity is calculated by the formula (4), using each of the strict formula (3) and by the simplified formula (5) using the values of c, ϕ and γ_t shown in Table 3 of 5.2. In Fig.6, the observed value corrected by equilibrium for loading and calculated value of the intensity of ultimate bearing capacity of the ground of the piers A, B and C are presented. From this result, it is clear that the calculated values are in good agreement with observed values and both calculated values are about the same.

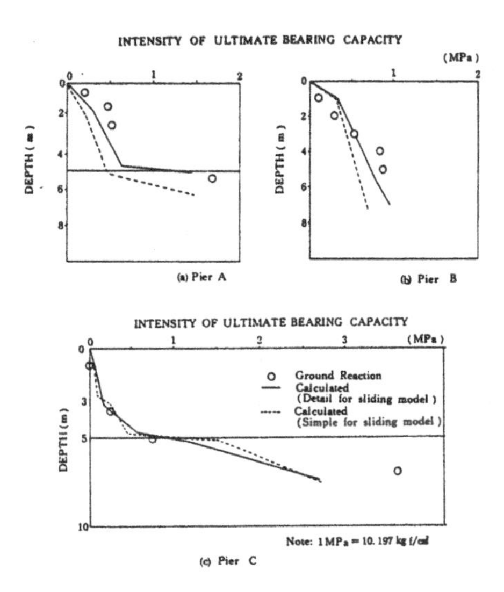

Figure 6. Result of calculations for intensity of ultimate bearing capacity

Therefore, the authors will apply the simplified formula to the analysis in Chapter 5.

4 DIFFERENTIAL EQUATION FOR THE "BEAM ON SHEAR TYPE GROUND" MODEL[6]

4.1 Ground Reaction and shear resistance

The ground reaction per unit length Px is generally represented by formula (6) using Winkler's discrete spring.

$$P_x = k \cdot D \cdot y \qquad \cdots\cdots (6)$$

where,

k: modulus of foundation

D: diameter of pier foundation

y: horizontal displacement

The horizontal displacement y of a pier loaded horizontally at its head becomes larger toward the top layer. If it is assumed that the modulus of foundation is constant along the direction of depth, formula (6) means that the ground reaction becomes larger toward the top layer. However, the result of measurement of the ground reaction in the loading tests (Fig.3) shows that it tends to increase toward the direction of depth when the ground is within the elastic limit. In order to explain such phenomena with the coefficient of the ground reaction, modulus k should be assumed to increase toward the direction of depth. But, as shown Table 1, soil property arround the pier B is homogenious at any depth. It is, therefore, difficult to assume that the modulus k increase linearly along the depth.

When the ground around the pier B is within elastisity, the ground reaction to the upper part of the pier is almost uniform. It means the shear strain of ground is uniform. Here, the pier behavior is analyzed assuming stress-strain relationship in a model of a beam on shear-type ground.

4.2 Introduction of Differential Equation

The differential equation is derived under the following assumptions:

(1) As shown in Fig.7, when the horizontal load begins to act at the pier head, shear strain is caused on the straight sliding surface with an angle generates (Refer to 3.1).

(2) A small slice ABCD in the ground in front of the pier, as illustrated in Fig.7, deforms into A'B C'D' and the shear strain of a small slice is uniform.

(3) The shear strain of this small slice is proportional to the angle of deflection of the pier.

From Fig.7, when the horizontal load acts at top of pier, the angle of inclination of the pier becomes negative.

$$\frac{dy}{dx} < 0 \qquad \text{..... (7)}$$

The shear strain of the small slice is represented as follows:

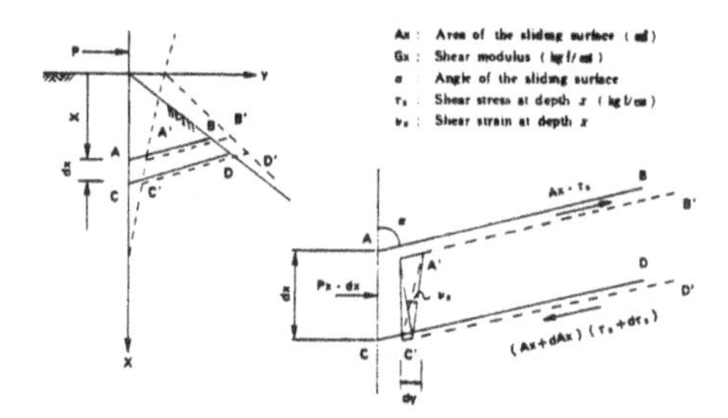

Figure 7. Infinitesimal slice in front of the pier

$$\gamma_x = \frac{dy}{\sin\alpha} \Big/ \frac{dx}{\sin\alpha}$$

$$= \frac{dy}{dx} \cdot \frac{1}{\sin^2\alpha} \qquad \text{..... (8)}$$

Taking the direction of the shear stress τ_x in Fig.7 as positive, τ_x becomes as follows:

$$\tau_x = -Gx \cdot \frac{dy}{dx} \cdot \frac{1}{\sin^2\alpha} \qquad \text{..... (9)}$$

Differentiating τ_x with respect to x, we obtain:

$$\frac{d\tau_x}{dx} = -Gx \cdot \frac{d^2y}{dx^2} \cdot \frac{1}{\sin^2\alpha} \qquad \text{..... (10)}$$

From the condition of equilibrium at the small slice dx, equation (11) is obtained.

$$P_x \cdot dx \cdot \sin\alpha + Ax \cdot \tau_x - (Ax + dA_x)\ (\tau_x + d\tau_x) = 0 \qquad \text{..... (11)}$$

The ground reaction per unit length Px becomes equation (12) by neglecting the second term.

$$Px = \left(Ax \cdot \frac{d\tau x}{dx} + \tau_x \cdot \frac{dAx}{dx}\right) \cdot \frac{1}{\sin\alpha} \qquad \text{..... (12)}$$

Substituting equations (9) and (10) in equation (12), Px becomes:

$$P_x = -\frac{A_x \cdot G_x}{\sin^3\alpha} \cdot \frac{d^2y}{dx^2} - \frac{G_x}{\sin^3\alpha} \cdot \frac{dA_x}{dx} \cdot \frac{dy}{dx} \qquad \text{..... (13)}$$

Accordingly, the differential equation that determines the deflection curve of a pier on shear-type ground is represented as follows:

$$EI \cdot \frac{d^4y}{dx^4} - \frac{A_x \cdot G_x}{\sin^3\alpha} \cdot \frac{d^2y}{dx^2} - \frac{G_x}{\sin^3\alpha} \cdot \frac{dA_x}{dx} \cdot \frac{dy}{dx} = 0 \qquad \text{... (14)}$$

The model that is represented by this differential equation is hence called the "Beam on Shear Type Ground" model.

5 SIMULATION USING THE "BEAM ON SHEAR TYPE GROUND" MODEL

5.1 Model for simulation

The model for simulation is shown in Fig.8. The part of the pier that is displaced toward the direction of slope is assumed to be a beam on the shear type ground and sliding model as illustrated in Chapter 3. For the part of the pier that lies below the center of rotation, its pattern of failure has not been made clear at this moment. Therefore, assumption was made for calculation to follow Winkler's theory. At the bottom surface of the pier, the shear spring k_s and the rotational spring k_r are considered.

5.2 Constants for soil

In Table 3, the constants for soil used in the analysis of the pier A, B and C are shown. With respect to the cohesion and the angle of internal friction, the values obtained by the CD test are used for sandy soil and those obtained by the UU test for cohesive soil and rock.

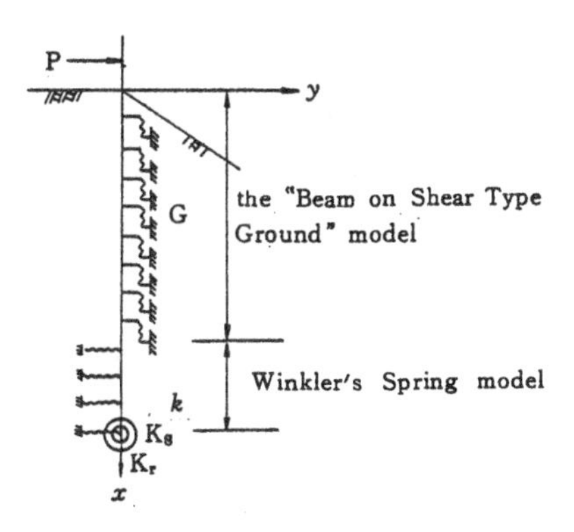

Figure 8. Simulation model

For the shear modulus G, the calculation is made by equation (15) and the upper limit of the shear stress τ_u is calculated by equation (16). From a practical view point, the relation of $\tau - \gamma$ is simplified by the bilinear form.

$$G = \alpha^* \cdot \frac{E_{50}}{2(1+\nu)} \quad \text{..... (15)}$$

$$\tau_u = \frac{1}{A} \cdot F_u \cdot \sin\alpha \quad \text{..... (16)}$$

Table 3. Soil properties

PIER NAME	PARAMETER	FIRST LAYER	SECOND LAYER
PIER A	COHESION C (MPa)	0.03	0.06
	INTERNAL FRICTION ANGLE ϕ (Degree)	32.0	40.0
	UNIT WEIGHT γt (MN/m³)	$1.47*10^{-2}$	$1.57*10^{-2}$
	POISSON'S NUMBER ν	0.40	0.40
	MODULUS OF DEFORMATION E_{50} (MPa)	4.90 (Depth=2.0m) 29.42 (Depth=5.0m)	83.35 (Depth=8.0m)
PIER B	COHESION C (MPa)	0.04	0.05
	INTERNAL FRICTION ANGLE ϕ (Degree)	34.4	36.9
	UNIT WEIGHT γt (MN/m³)	$1.77*10^{-2}$	$1.86*10^{-2}$
	POISSON'S NUMBER ν	0.40	0.40
	MODULUS OF DEFORMATION E_{50} (MPa)	39.91 (Depth=3.0m) 45.11 (Depth=7.0m) 47.07 (Depth=8.5m)	74.14 (Depth=10m)
PIER C	COHESION C (MPa)	0.013	0.26
	INTERNAL FRICTION ANGLE ϕ (Degree)	24.0	50.0
	UNIT WEIGHT γt (MN/m³)	$1.37*10^{-2}$	$2.06*10^{-2}$
	POISSON'S NUMBER ν	0.40	0.40
	MODULUS OF DEFORMATION E_{50} (MPa)	27.46 (Depth=5.0m)	325.1 (Depth=5.5m) 325.1 (Depth=10m)

Note : When the horizontal load is working at the pier head, the ground in front of the pier is passive condision. So E_{50} is gotten from $\sigma \sim \varepsilon$ curve ($\sigma_3 = \gamma t \cdot H$).
where , γt : Unit Weight
H : Hight of Earth Covering

1.0 MPa =10.197 kgf /cm², 1.0 MN =101.97 tf

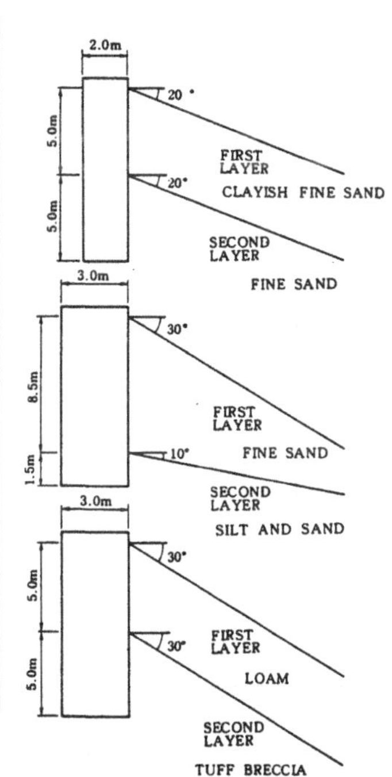

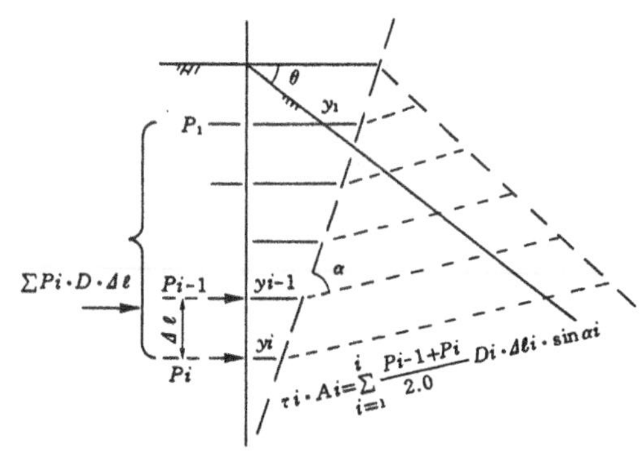

yi : Measured displacement at the point ⓘ (cm)
Pi : Measured ground reaction at the point ⓘ (kgf/cm²)
Di : Diameter of pier (cm)
$\Delta \ell_i$: Setting interval of earth pressure gage (cm)
αi : Angle of shear sliding
Ai : Area of sliding resistance from ⓘ toward the ground surface (cm²)
τi : Equivalent shear stress at the point ⓘ
γi : Equivalent shear strain at the point ⓘ

Figure 9. Explamation of $\tau - \gamma$ relation

where, α*: correction coefficient (Soil: 4.0, Rock: 1.0)

E_{50}: deformation modulus (see the remarks of Table 3)

ν : Poission's ratio

Fu,A, α : see equation (1)

The actual equivalent shear stress τ_i and strain γ_i are obtained by substituting P_i, y_i, A_i and α_i as shown in Fig.9 in equation (17) and (18).

$$\tau_i = \frac{1}{2A_i} \cdot \sum_{j=1}^{i} (P_{j-1} + P_j) \cdot D_i \cdot \Delta \ell_i \quad \cdots\cdots (17)$$

$$\gamma_i = \frac{1}{\sin^2 \alpha} \cdot \frac{(y_{i-1} - y_i)}{\Delta \ell_i} \quad \cdots\cdots (18)$$

Fig.10 shows the τ - γ curve obtained from the measured values and the bilinear τ-γ curve which is used for simulation. The fact that the gradient of the τ - γ relation in both curves fit well each other over the range where γ is small seems to suggest that equation (15) using result of triaxial tests is reasonable. Besides, the upper limit value of the shear stress could well be estimated.

The value of Winkler's spring k_H, shear spring k_s and rotational spring k_r at the bottom are calculated by equation (19), (20) and (21) below.

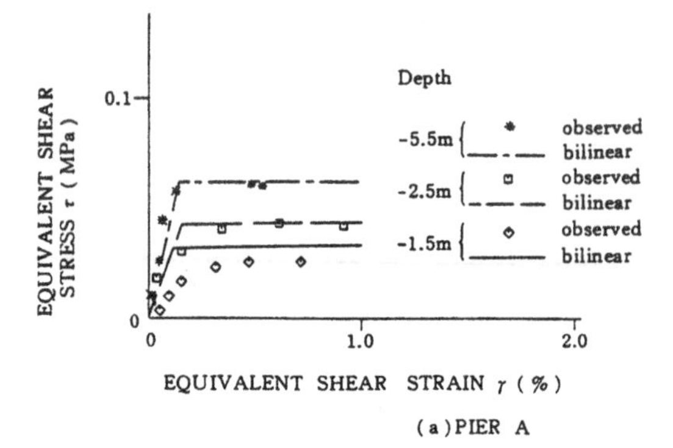

(a) PIER A

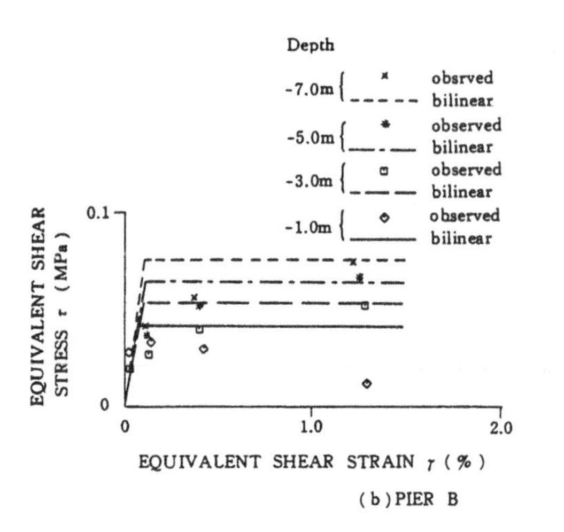

(b) PIER B

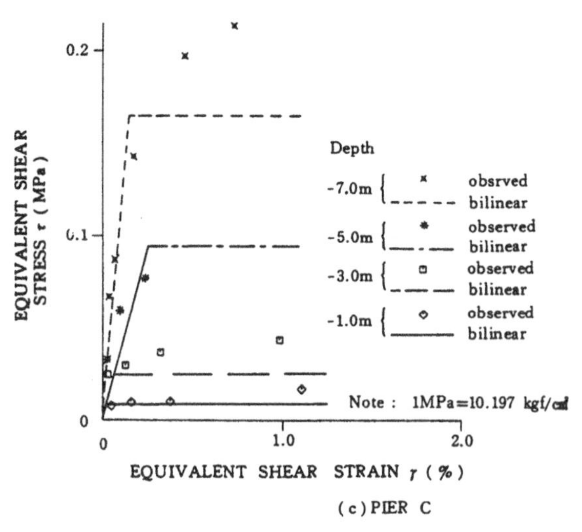

(c) PIER C

Figure 10. $\tau - \gamma$ CURVE

$$k_H = \frac{1}{30}\alpha \cdot E_{50} \cdot \left(\frac{D}{30}\right)^{-3/4} \quad \text{..... (19)}$$

$$k_S = \frac{1}{3} k_H \cdot B \quad \text{..... (20)}$$

$$k_r = k_H \cdot I \quad \text{..... (21)}$$

where, α: correction factor (Soil: 4.0, Rocks: 1.0)

E_{50}: see equation (15)

D: diameter of pier

B: area of pier at bottom

I: moment of inertia of pier at bottom

5.3 Result of Calculation

In this paragraph, the result of calculation made by using the $\tau - \gamma$ relation simplified into the bilinear form in equations (15) and (16) is presented.

Fig.11 shows the result of calculations for the relation between the load-displacement at the pier head and their measured values for each of pier A, B and C. Correlation coefficients of the calculated values for the measured ones are indicated. Though the correlation looks rather poor for the pier B, it may be said that fairly good correlation was established as a whole.

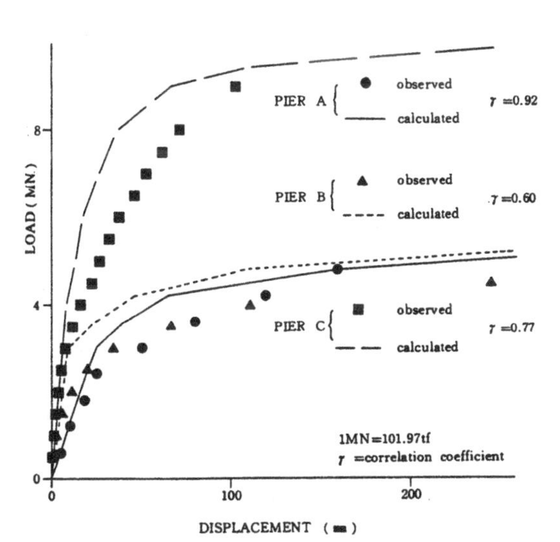

Figure 11. Result of simulation for load-pier head

6 APPLICATION TO FLAT GROUND[4)]

6.1 Model for simulation

The colapse mode of the ground around the long pier is generally assumed to be the following. In the case of the ground where the constrained force is small, the sliding surface develops upwards and the mass of earth thrusts upwards. On the other hand, in the case of the ground where the constrained force is large, the sliding surface turns around the pier horizontally, and then the ground collapses.

Therefore, in Fig.12, the ground consisting of one zone (Region I) where the constrained force is small and another zone (Region II) where the constrained force is small is devided into separate models. In Region I, the differential equation (14) of the "Beam on Shear Type Ground" model and the aforementioned equation (5) for the intensity of the ultimate bearing capacity of the ground are applied. In Region II, the differential equation (22) using Winkler's discrete spring and equations (23), (24) for the intensity of the ultimate capacity of the ground that is obtained as a 2-dimentional solution are applied.

$$EI \cdot \frac{d^4y}{dx^4} - k \cdot D \cdot y = 0 \quad \text{..... (22)}$$

where,

k,D,y: see equation (6)

$$Pu = 9Cu \quad \text{..... (23)}$$

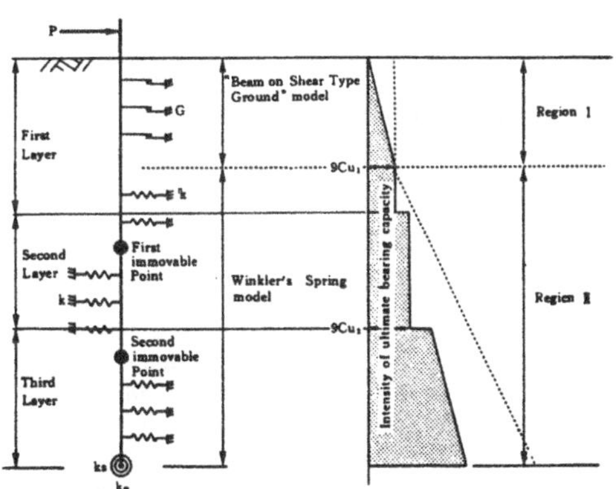

Figure. 12 Simulation model

(Broms' formula, in case of cohesive ground)

$$Pu = C \cdot \cot\phi \cdot \tan^2\left(\frac{\pi}{4}+\frac{\phi}{2}\right) \cdot \exp(\pi \cdot \tan\phi) - 1$$

$$+K_{or} \cdot Z \cdot \tan^2\left(\frac{\pi}{4}+\frac{\phi}{2}\right) \cdot \exp(\pi \cdot \tan\phi) - K_{Ar} \cdot Z \quad \ldots\ldots (24)$$

(modified Prandl's equation, in case of sandy ground)

The boundary between Region I and Region II is where the intensities of the ultimate capacity of the ground for the both regions are the same.

6.2 Constants for soil

The outline of the geology at the test site is shown in Fig.13. The ground at the test site was regarded as having three different layers, and the constants for soil in Table 4 were used for the calculation.

6.3 Result of calculation

The result of the calculation for the relation between load-displacement at the pier head and the observed values are shown in Fig.14. In spite of having used the soil constants obtained from N-values, good agreement was achieved.

7 CONCLUSIONS

Through a series of the in-situ full-scale horizontal loading tests, the following two results were obtained.

(1) The ground in front of the pier in failure forms a sliding mass of earth with the limited dimensions.

(2) The distribution of the ground reaction in front of the pier increases toward the direction of depth.

From the observation result(1), the authors proposed the sliding model of earth mass. Further, it was pointed out that this model could well express the bearing capacity of the ground. From the observation result(2), the "Beam on Shear Type Ground" model was proposed. It was indicated that the behavior of pier foundation could be predicted quite precisely by application of this model. It was further shown that this concept could be also applied to the pier foundation on flat ground.

By applying this concept, it was pointed out that both of bearing capacity and deformation of pier foundation under horizontal loading could be dealt with consistently.

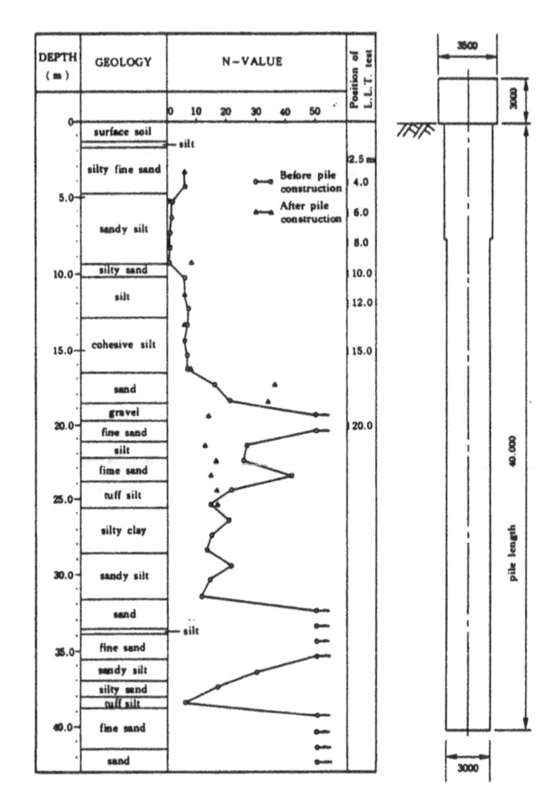

Figure.13 Soil geology

Table 4. Soil Properties

Parameter	First Layer	Second Layer	Third Layer
N-VALUE	4	8	20
COHESION C (kgf/cm²)	0.25*	0.50*	0.00
ANGLE OF INTERNAL FRICTION ϕ (°)	0.00	0.00	30.0**
UNIT WEIGHT γt (kgf/cm³)	1.60	1.60	1.80
POISSON'S RATIO ν	0.40	0.40	0.40
DEFORMATION MODULUS E_L (kgf/cm²)	25.4~ 32.0	214.9~ 245.6	449.9~ 500.0

Diameter =3.0m
Cohesive soil H=10.0m
Second Layer Cohesiue soil H=6.5m
Third Layer Sandy Soil H=23.5m

* $C=N/16$ ** $\phi=15+(12N)^{1/2}$

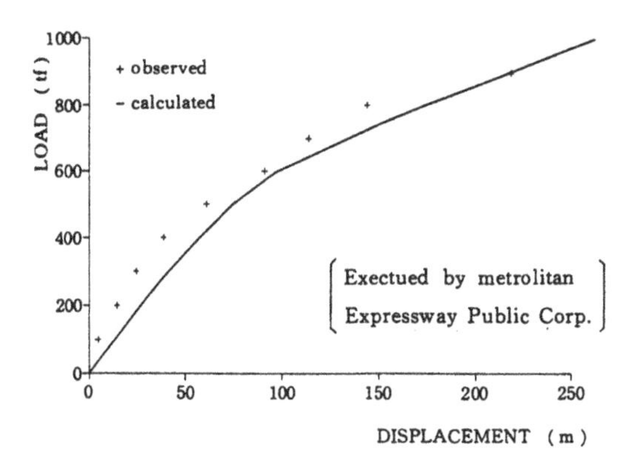

Figure.14 Load–Displacement curve

8 POSTSCRIPT

In this paper, the authors reported the result of a series of the horizontal loading tests of pier foundation monitoring by micro-computer and the prediction method of horizontal behavior of the pier. If there is a chance for another loading test, it will be intended to develop experiment system controlled by A.I. computer which has knowledge base and inference engine. This paper introduced a part of these knowledge base and inference engine.

In ending this paper, the writers wish to thank Professor Kuno of Chuo University (Present Chairman of The Japanese Society of Soil Mechanics and Foundation Engineering) who have given us guidances, and the staff of Japan Highway Public Corporation and Metropolitan Expressway Public Corporation who have given us guidance as well as permission for using the data.

REFERENCES

1) Japan Highway Public Corporation, Kanetsu Highway: Report of Horizontal Loading Tests of Pier Foundations on a Slope, 1981
2) Tokyo Electric Power Company Inc., Report of Full-scale Horizontal Loading Tests of Pier Foundations (No.2), 1982
3) Tokyo Electric Power Company Inc., Report of Full-scale Horizontal Loading Tests of Pier Foundations (No.1), 1981
4) Express Highway Research Foundation of Japan, Report about a New Structure Type of Elevated Bridge for Metropolitan Highway
5) H.Maeda, Horizontal Behavior of Pier Foundation on a Soft Rock Slope, the 5th ISRM, 1983, pp.81～84
6) K.Uto, H.Maeda, Y.Yoshii, M.Takeuchi K.Kinoshita, A.Koga, Horizontal Behavior of Pier Foundations in a Shearing Type Ground Model, the 5th ICONMIG, 1985, pp.781～788

Computer and Physical Modelling in Geotechnical Engineering, Balasubramaniam et al. (eds)
© 1989 Balkema, Rotterdam. ISBN 90 6191 864 2

The SVI system for stress wave measurement and analysis of pile driving using the program SVIDYN 1

N.T.Tien & T.V.Cuong
Institute for Building Science and Technology, Hanoi, Vietnam

B.Berggren & B.Möller
Swedish Geotechnical Institute, Sweden

ABSTRACT: In this paper a system, SVI, for stress wave measurement and analysis is presented. In the analysis, transfer of energy, approximate static and dynamic bearing capacity and determination of pile damage can be evaluated. A computer program, SVIDYN-1, according to Smith's approach and similar to the CAPWAP procedure is presented. Results and stress wave analysis for field tests are presented. In the analysis different soil models, such as nonlinear viscosity and standard viscosity can be used.

INTRODUCTION

The rapid development of electronic components, the need to optimize foundation cost and new driving techniques have altogether led to a better understanding on the static-dynamic behaviour of piles. Static load tests were earlier the only way to confirm the calculated ultimate bearing capacity of piles. As static load tests are expensive, only a limited number of tests can be carried out. However, using stress wave measurement on piles in combination with computerized stress wave theory, it is economically possible to test a large portion of the piles at a given site.

With equipments and computer programs, commercially available, it is possible to evaluate the ultimate bearing capacity in the field for each blow during driving or redriving. Also other parameters can be calculated such as the energy coming into the pile during each blow and evaluation of pile damage.

Several commercial systems are available on the market. The system developed by GRL, Goble, Rausche, Likins and Associates (Cleveland, Ohio, USA) is in use all over the world.

Besides simple determination of the bearing capacity of endbearing piles with the "Case-method", the GRL system contains the CAPWAP and the WEAP programs and others. The TNO in the Netherlands and PiD in Sweden have developed other systems for stress wave measurement and analysis.

A simple system for stress wave measurements in-situ and computer programs for analysis are developed at the Swedish Geotechnical Institute and the Institute for Building Science and Technology in Hanoi, Vietnam. Description of this system, the SVI system, introduction of the computer program and some results of measurements and analysis are presented.

1 THE SVI-SYSTEM FOR STRESS WAVE MEASUREMENT

The SVI-system for stress wave measurement is shown in Fig. 1. The system consists of

- signal unit
- cassette tape recorder
- digital storage oscilloscope
- micro computer, including plotter and printer

During field measurements the signal equipment (signal unit, cassette tape recorder and oscilloscope) is used.

Measured accelerations and strains from the pile top are transformed to an electrical signal, amplified, recorded and displayed by this group. During replay in the office, the tape recorder and the oscilloscope are connected to the computer to transfer the signals into the computer by a transfer computer program.

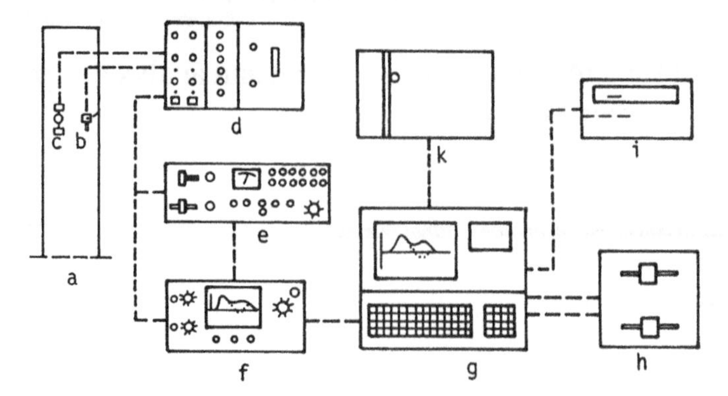

Fig. 1 The SVI system for stress wawe measurement and analysis.
a. pile
b. accelerometer
c. strain gauge
d. signal unit
e. cassette tape recorder
f. oscilloscope
g. computer
h. floppy disk unit
i. printer
k. plotter

2 PROGRAM SVIDYN-1

The program SVIDYN-1 for analysis has been written for the system. The first subroutine, CS 586, evaluates the pile bearing capacity in the same way as the CASE method. The second subroutine, CW 686, is based on Smith's approach (Smith 1960). This subroutine CW 686 is developed using a program of Bowles (Bowles, 1974) and a further development of Litkouchi (1979). However, the hammer and cushion are not simulated by CW 686.

2.1 Subroutine CS 586

The purposes of this subroutine are:

- Integration of measured acceleration to obtain velocity.
- Analysis of bearing capacity of the pile according to the CASE method.
- Computation of the energy transferred to the pile.
- Determination of severity of pile damage.

2.1.1 The CASE method - evaluation of the bearing capacity of the pile

The CASE method was developed at the CASE Western Reserve University by Goble and his coworkers (1973, 1974). The method permits to calculate the resistance of a pile from top measurement of strains and accelerations during pile driving and redriving. Total resistance R_T is defined by

$$R_T = \frac{F(t_1)+F(t_2)}{2} + \frac{Z}{2}\,[v(t_1)-v(t_2)] \quad (1)$$

where $F(t_1)$, $v(t_2)$, $F(t_1)$, $v(t_2)$ = measured forces and velocities at time t_1 and t_2 respectively
$t_2 = t_1 + 2\,L/c$, t_1 = time of maximum force
L = pile length
c = bar velocity
Z = AE/c = impedance of the pile

The static component is separated by

$$R_S = R_T - Jc(2F(t_1)-R_T) \quad (2)$$

Jc, the CASE damping factor, is assumed to be acting at the pile point. Jc is nondimensional and recommended values Jc = 0.05 for sand, Jc = 0.3 for silt and Jc = 1.1 for clay are proposed by Goble (1975).

2.1.2 Calculation of transferred energy

From the force and velocity measured, the energy transferred into the pile from the driving system is determined by:

$$E = \int F(t)\cdot v(t)\,dt \quad (3)$$

The efficiency of the driving system can be evaluated by the ratio between measured energy (Eq 3) and theoretical potential manufacturer's rated or kinetic energy just before impact.

2.1.3 Determination of pile damage

Changes in pile cross section at a depth x below the pile top will cause force and velocity records to diverge at the time 2 x/c after initial impact. The ratio B between the impedance of the upper pile section and the impedance of the lower one indicates how much pile cross section is left.

A classification of pile damage, according to Goble et al (1979) is presented in Table 1.

Table 1. Classification of pile damage.

B	Severity of damage
1.0	Undamaged
0.8-1.0	Slight damage
0.6-0.8	Damage
<0.6	Broken

2.2 Subroutine CW 686

The subroutine CW 686 can be used for calculation of the top force, using measured or calculated velocities at the pile top at boundary conditions. The soil resistance, quake and damping values along the pile shaft and at the pile point can be varied to get a better match between measured and calculated pile top force.

The output of the results is a curve showing force vs time and simulation of the static curve, load-settlement at the pile top and at the pile base.

2.2.1 Basic equations and proceudre of analysis

When a pile driving hammer hits the pile top, a mechanical stress increase is caused and moves down the pile. The pile is supposed to be a slender, elastic and homogeneous pile with Young's modulus E and a constant cross section area A. The one-dimensional wave equation may be derived from consideration of internal forces and motion

$$\frac{\delta^2 u}{\delta t^2} = \frac{E}{\rho}\frac{\delta^2 u}{\delta x^2} = \frac{c^2}{1}\frac{\delta^2 u}{\delta x^2} \qquad (4)$$

where
ρ = density of the pile material
u = displacement
c = wave velocity = $(E/\rho)^{0.5}$
x = length coordinate
t = time

As soil resistances are generated at the pile skin and at the pile tip during the passage of the stress wave, equation 4 becomes

$$\frac{\delta^2 u}{\delta t^2} - \frac{E}{\rho}\frac{\delta^2 u}{\delta x^2} + R_T = 0 \qquad (5)$$

where
R_T = ku + JR_{sv}
k = static soil stiffness
J = damping coefficient
R_S = static soil resistance
v = particle velocity of pile element

This means that all soil resistances generated during the pile driving are composed of a static portion, depending on the displacement u, and a dynamic portion, usually a function of the velocity. The dynamic resistance, depending on the damping process that occurs both at the pile skin and at the pile tip, represents the energy transformation from the pile into the soil.

The method developed by Smith (1960) is a finite difference, in which equation 5 is solved to determine the pile set for a given ultimate pile load. The pile soil system is idealized as shown in Fig. 2.

The rheological model of the soil in Smith's approach is shown in Fig. 3, and the total dynamic force for every pile-soil element is expressed as

$$R_T = R_S\ (1+Jv) \text{ for } u < Q \qquad (6)$$

$$R_T = R_S + k'\ QJv \text{ for } u > Q \qquad (7)$$

R_S = static component = k'u
J = damping coefficient of every element
v = particle velocity of pile element
k' = R_u/Q
R_u = ultimate static soil resistance for every element
Q = quake = maximum elastic displacement for every soil element
u = displacement of pile element

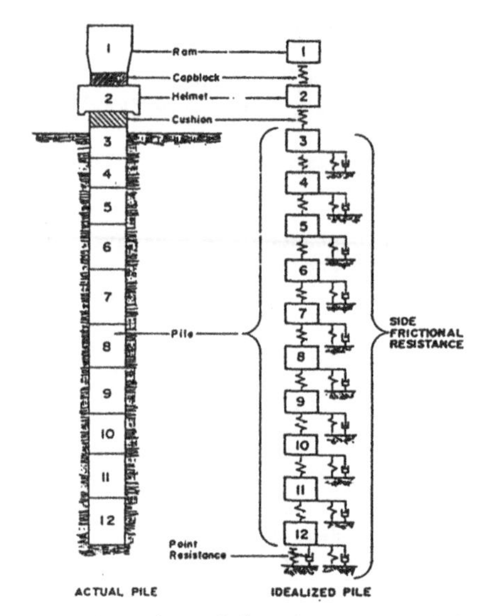

Fig. 2 Pile model and hammer model.

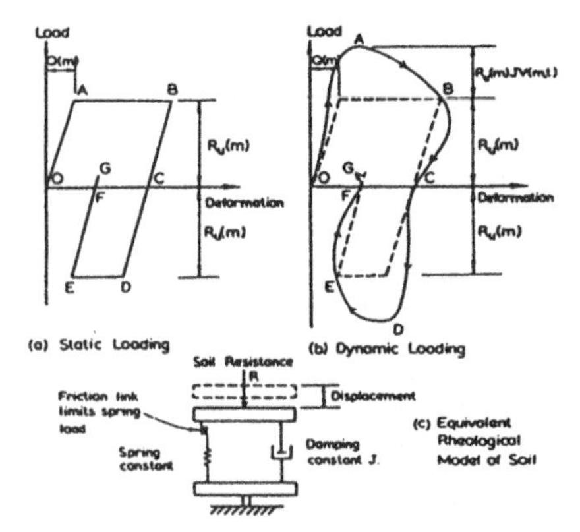

Fig. 3. Load deformation relationships for soil (after Poulos and Davis, 1980)

By using the subroutine CW 686, which has been developed from a program developed by Bowles (1974) and Litkouchi (1979), who used Smith's approach, the force time history near the pile top can be estimated from input data of velocity time history and assumed values of R_u, J and Q for every element and at pile tip. The displacement, velocity and force at every element and also at the pile tip are evaluated at each time of increment. The computed force will be compared with the measured forces. If they don't match well enough a new set of soil resistance will be input and the procedure starts again, and will stop when the agreement between them is good. This procedure of iteration is similar to the procedure in the CAPWAP program as shown in Fig. 4.

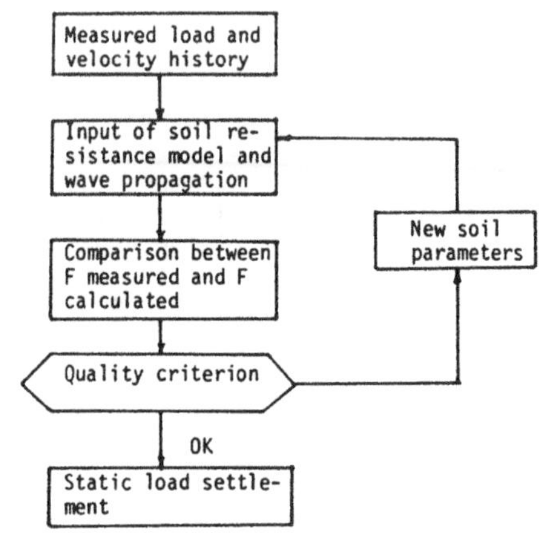

Fig. 4 CW 686 analysis procedure.

2.2.2 Input

- Measured velocity vs time at the pile top can be used as boundary condition. The calculated velocity
- Length of the pile L, Young's modulus of the pile material E.
- Number of pile element NP, NP $\leq$ 40, and internal damping of the pile.
- Soil resistance and quake value for every pile element and at pile tip.
- Different soil models can be used
  - o Soil resistance, quake and damping values for every pile-soil element and at the pile tip can be varied according to the user.
  - o Nonlinear viscosity according to recommendation of Likouchi and Poskitt (1980) and others:

$$R_T = R_S\ (1+Jv^N) \qquad (8)$$

where N can be input by the user, N = 1 corresponds to Smith's soil model.

  - o Standard viscous damping C according to the equation:

$$R_{dyn} = k'u + Cv \qquad (9)$$

where C = kN/m/s
k'= spring stiffness
u = pile displacement

It is noted that the damping factor C is constant in this soil model, while damping factor J in equation 8, 9 and 11 is proportional to the static resistance.

2.2.3 Output of results

- Force-time mesh calculated at the pile top.
- Simulation of static load test curve.
- Displacement of the element at the pile top.
- Displacement and force history of any element in the system.

3 RESULTS OF ANALYSIS

In order to make a control of the reliability of the computer program SVIDYN-1, data from field measurements are analysed by the SVI-system using SVIDYN-1 and CAPWAP programs.

3.1 Measurements from field test No 1

Stress wave measurements were taken on a precast concrete pile, with the following properties:

Area = 0.079 m^2
Length = 23.93 m
Density = 25 kN/m^3

Fig. 5 shows the measured force and velocity.

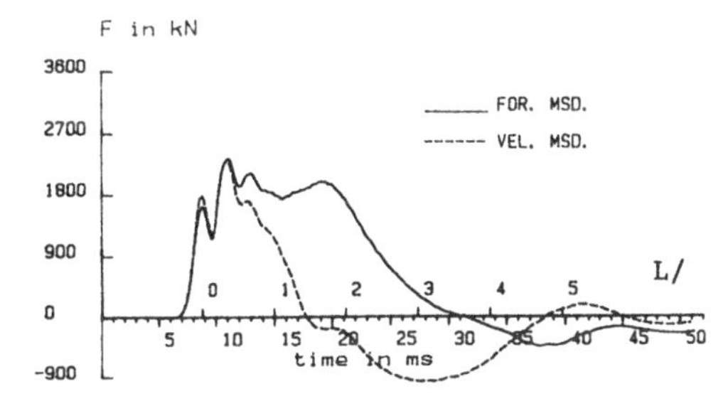

Fig. 5. Measured force and velocity.

Results of analysis by SVIDYN-1 is presented in Fig. 6. Similar soil resistances, damping factors and quake values as in the CAPWAP analysis are used.
The simulation of the static load test by CAPWAP and SVIDYN-1 is shown in Fig. 7. The agreement between the results is good.

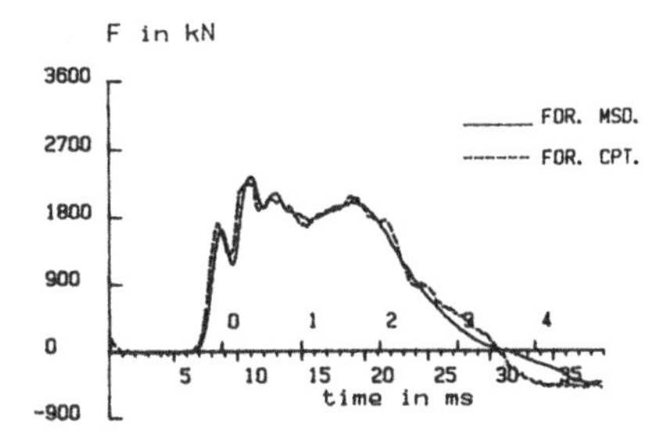

Fig. 6. Results of analysis by SVIDYN-1.

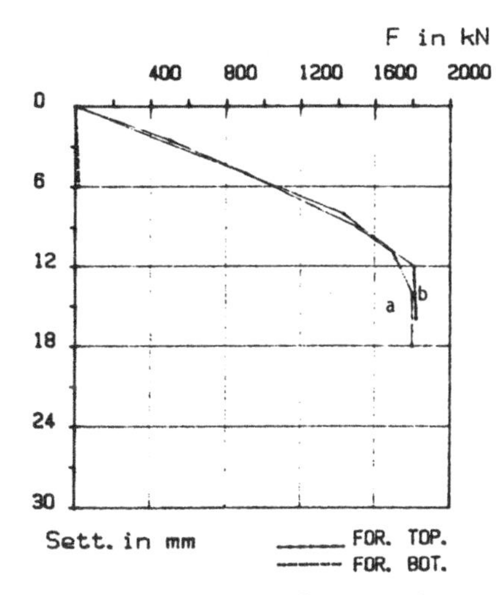

Fig. 7. Simulation of static load test. a. CAPWAP, b. SVIDYN-1.

3.2 Measurements from field test No 2

Stress wave measurements at test site No 2 were carried out on a steel pipe pile with an outer diameter of 101.6 mm and a cross section area of 1689 mm^2. Results of analysis by CAPWAP and SVIDYN-1 are presented in Fig. 8 and Fig. 9. Similar soil conditions (damping, quake) are used in the two analysis, however the point resistances are different. A good agreement between the measured force history and the calculated force history is obtained in both cases.
The simulation of a static load test by SVIDYN-1 is presented in Fig. 10 together with data from a field test and results from a CAPWAP analysis. The agreement between the two analysed forces and the measured force is fairly good.

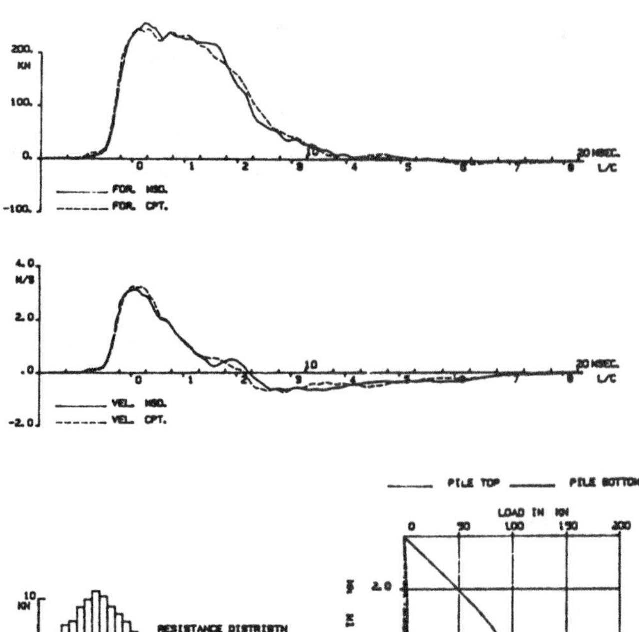

Fig. 8. Results from CAPWAP analysis.

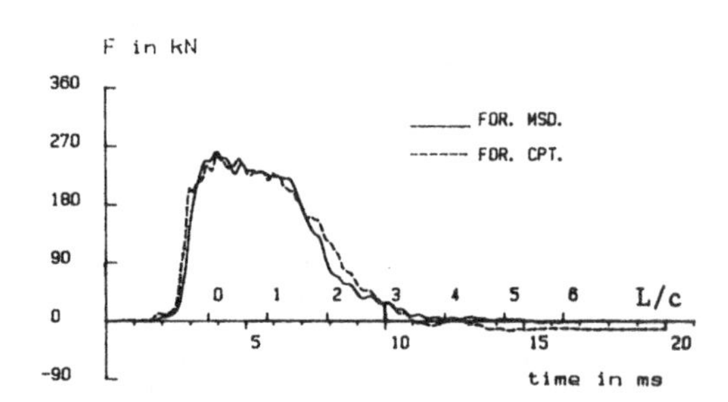

Fig. 9 Results from analysis by SVIDYN-1.

4. CONCLUSIONS AND FURTHER DEVELOPMENT

4.1 A new simple system

A new simple system for stress wave measurement and analysis is presented. Results from analysis and comparison with other computer programs show that the SVI system and SVIDYN-1 computer program are reliable.

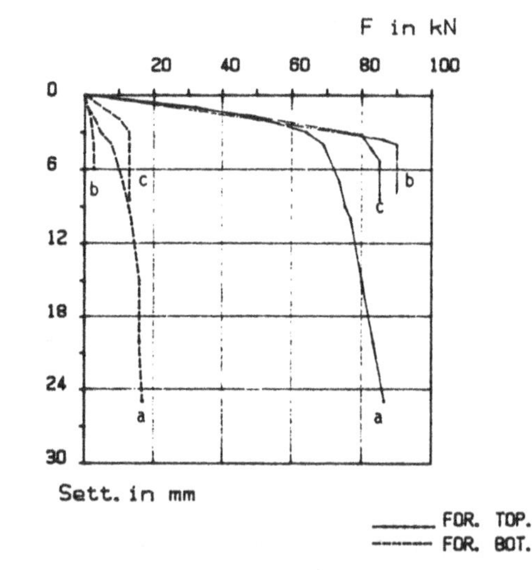

Fig. 10 Static load test.
a. data from static load test
b. CAPWAP analysis
c. SVIDYN-13.3.3

4.2 Soil models

Soil models according to Smith and others, taking into account non-linear viscosity, standard viscosity can be used for stress wave analysis in the SVI system. Comparisons between different soil models and results from field tests will improve the knowledgement on the soil pile behaviour.

5 ACKNOWLEDGEMENT

The authors gratefully acknowledge the support from SAREC, Sweden, and SCTC of Vietnam for the research cooperation between the Swedish Geotechnical Institute (SGI) and the Institute for Building Science and Technology (IBST).

Special thanks to members of SGI and IBST for the contribution to this research work. The first author has been supported by the Swedish Institute during the course of the work and many thanks are due to this organization.

6 REFERENCES

Balthaus, H.G. and Kielbassa, 1986. Numerical modelling of pile driving. Proc. of 2nd International Symposium on Numerical Models in Geomechanics, Ghent.

Bowles, J.E., 1979. Analytical and Computer Methods in foundation engineering. Mc Graw Hill book company.

Fisher, H.C., 1984. Stress wave theory for pile driving application. Proc. of 2nd Int. Conf. on the application of Stress Wave Theory on Piles, Stockholm.

Goble, G.G. & Likins, G.E., 1973. A static and dynamic pile test in west Palm Beach, Florida. Case Institue of Technology.

Goble, G.G. & Likins, G.E., 1974. Predicting the bearing capacity from dynamic measurement, Case Institute of Technology.

Litkouchi, S.,1979. The behaviour of foundation piles during driving. Thesis, University of London.

Litkouchi, S. & Poskitt, 1980. Damping constants for pile driveability calculation. Geotechnique 20, No 1.

Poulos, H.G. & Davis, E.H., 1980. Pile foundation analysis and design, JWS.

Rausche, F. & Goble, 1979. Determination of pile damage by top measurements. Behaviour of deep foundation, ASTM, STP 670.

Smith, E.A., 1960. Pile driving analysis by wave equation, J of SMFD, ASCE, Vol. 86, SM4.

Trinh viet Cuong & Nguyen Truong Tien, 1986. User manual for stress wave measurement.

Computer and Physical Modelling in Geotechnical Engineering, Balasubramaniam et al. (eds)
© 1989 Balkema, Rotterdam. ISBN 90 6191 864 2

Computer design and analysis of pile caps and spread footings

German M.San Juan
San Juan & Associates Co., Ltd, Consulting Engineers, Bangkok, Thailand

ABSTRACT: This program designs and analyze pile caps and spread footing. The footings may be designed for any service load and soil pressure and pile caps for any service load, pile capacity and pile spacing. The program also analyze pile caps with piles driven at off-location. It is written in FORTRAN IV language.

1 INTRODUCTION

This program was developed out of necessity while the author was a structural engineer with a Chicago consulting firm way back in 1976. The problem encountered was routine analysis of hundreds of piles driven out-of location from the original design position. A manual analysis of this work is time-consuming. Other application is the routine design of pile caps and spread footings.

The first generation version of this program was develop in 1974 to produce pile capacity design handbook widely used in the United States today. The program was written in FORTRAN IV language for the GA 1830 computer with 16-k core capacity.

The present program is being modified to be compatible with microcomputer including conversion to metric units and latest American Concrete Institute (ACI-318) design code.

For the purpose of this presentation the original versions of the program are presented which was based on ACI 318-71 building code and in English units. By presenting, the original program, the practicing engineers, students or users have the option to modify the program according to their design code preference.

2 PROGRAM SCOPE AND LIMITS

The program design footings for any service load soil pressure or service load pile capacity and pile spacing. The pile spacing and cap outline may be unsymmetrical. The column may be rectangular and placed off-center on the footings. Load and bending moments in two directions can be applied to the footing from the column. The program also analyzes pile caps for as-driven pile locations.

The program performs exact calculations accounting for all refinements consistent with the (1974) state of the art. The reinforcement selection bars are based on U.S. standards for reinforcing bars such as #2 to #11 bars.

The designs produced conform to ACI 318-71 building code. Two exceptions are that the allowable punching shear at d/2 from the face of the column is modified, and that a calculation for punching shear at the face of the column is made. See "Method of Solution" for explanations. Shear and moment calculations account for partial pile reactions of piles near the critical sections, and for the benefit provided by the caps concrete weight in reducing moment or shear. The solutions given for spread footings loaded only with vertical load are exact solutions. When moment is applied to spread footings, the soil reaction is approximated to be uniform over each "pile's (really soil element's area). The error introduced by this approximation is more severe if the moment is large. Where a high degree of accuracy is desired, the footing may be divided into 10 x 10 (or 100) elements of area.

Pile cap footings are limited to a maximum of 100 piles.

3 METHOD OF SOLUTIONS

The centroid of the piles is found by moment-area principles. The resultant moment in each of two directions is calculated. Then the reaction in each pile is calculated.

Cap depth is determined by either beam shear, punching shear at d/2 from the face of the column, deep beam shear, or punching shear at the column face. The allowable punching shear stress at d/2 from the face of the column, v_u, is limited as follows:

$$v_u \leq 4\sqrt{f'_c} \quad \text{(square columns)}$$

$$v_u \leq 2.5 + 3.0 \left(\frac{\text{short column dimension}}{\text{long column dimension}}\right)$$

$$\leq 4\sqrt{f'_c}$$

Beam shear is checked on all four sides of the column. The allowable shear stress is calculated according to section 11.4.2 of the ACI 318-73 code.

If a pile is located within a distance d from the face of the column, the pile cap is checked for deep beam shear at a critical section midway between the pile and the face of the column. The allowable deep beam shear stress is calculated according to section 11.9.2 of the ACI 318-73 code and then modified by multiplying by d/w and given the limit of $v_c \leq 10\sqrt{f'_c}$

If a pile is located within a distance d/2 from any face of the column, the pile cap is checked for deep beam punching shear at critical section located at the face of the column. The allowable deep beam punching shear stress is calculated as:

$$v_u = 2\sqrt{f'_c}\ (1 + \frac{d}{c})\ \frac{d}{w}$$

$$v_u \leq 32\sqrt{f'_c}$$

Where c is the maximum column dimension and w is the distance between the column face and the nearest pile.

Flexural reinforcement is determined in each of two directions by the maximum moment which occurs on either of two sides of the column. The bar size chosen can be developed in the available embedment length from the face of the column to the bar at the nearest edge of the footing. The minimum amount of reinforcement is equal to or greater than 0.0018 bh. The maximum bar spacing is 18 inches. This may result in the choice of a bar size smaller than the largest bar which can satisfy development length. When the amount of reinforcement provided less than 200/f_y, a number 8 bar size would be printed as " #8 " and when the amount is more than 200/f_y, a number 8 bar size would be printed as " *8 ".

4 PREPARATION OF INPUT DATA

4.1 Batch runs

A single set of input data should be preceded by a " // XEQ PILE " card and end with a "999" card. Several sets of input data may be batch executed in sequence on the computer. The first page of input should use the " // XEQ PILE" card. Following pages should have this card omitted. Each set of data should follow one another without any blank cards. The last set of data should end with a "999" card at the bottom of the form.

4.2 Data card sequencing

Each data card is provided with a two or three letter sequence ID which enables the computer to verify the proper order of input cards, to avoid card publication and omission. The sequence of cards on the input form is the sequence they are read by the program.

The cards which must be input are dependent on the input option used. The heading above each card indicates the options for which it is used. The table below may also be used.

Table I. Card sequencing

OPTION	# OF PILES	NECESSARY CARDS
0	Not entered	P1, HED
1	Any #	P1, HED, P2, P3, PGR
2	2 - 10	P1, HED, P2, P3, PX1, PY1, PP1
	11 - 20	P1, HED, P2, P3, PX1, PY1, PP1, PX2, PY2, PP2
	21 - 30	P1, HED, P2, P3, PX1, PY1, PP1, PX2, PY2, PP2, PX3, PY3, PP3
	31 - 40	P1, HED, P2, P3, PX1, ---, PP4
	41 - 50	P1, HED, P2, P3, PX1, ---, PP5
3	16 - 100	P1, HED, P2, P3

4.3 P1. - card

"# PILES" is the number of piles in a pile cap. This data is ignored for OPTION 0. It may vary from 2 to 50 piles for OPTIONS 1 and 2. It may vary from 4^2 to 10^2 for OPTION 3. The only permissible data entries for OPTION 3 are; 0 (16 used), 16, 25, 36, 49, 64, 81, and 100.

"INPUT OPTIONS" may have the values 0, 1, 2, or 3. OPTION 0 will produce a table of pile caps of 2 piles up to 40 piles except that the maximum total pile group service load capacity is limited to 3000 kips. All pile arrangements and pile cap dimensions are determined internally by the program. The maximum column load is determined by the maximum pile load.

OPTION 1 is used for the design of single pile cap where the pile locations are determined by a grid intersection method of input. Cap dimensions must be given as input data.

OPTION 2 is used for the design of single pile cap where the pile locations are determined by the x and y coordinates of each pile. The cap dimensions must be given as input data.

OPTION 3 is used to design a single spread footing. Caps must be rectangular in shape. The soil under the footing is automatically divided into a 4 x 4, 5 x 5, 6 x 6, 7 x 7, 8 x 8, 9 x 9, or 10 x 10 grid of soil elements. The soil reaction is uniformly distributed over the entire area of each element.

"PILE CAPACITY (TONS)", which is the individual pile service load capacity in tons, must be given for OPTION 0, 1, and 2,

"SOIL CAPACITY (TSF)", which is the service load soil bearing capacity in tons per square foot, must be given for OPTION 3

The balance of the data input is optional input data, as is indicated by a heavy underline. If any of the date is left blank, its value is taken equal to the default value.

"f'_c (KSI) ... (3)" is the design strength of foundation concrete in kips per square inch.

"f'_x (KSI) ... (60)" is the design yield strength of the foundation reinforcement.

"COVER (IN) ... (9) or (3)" is the clear cover from the bottom of the concrete to the bottom bar in the foundation in inches. If left blank, 9 inches is used for pile caps and 3 inches is used for spread footings.

"S = PILE SPACING (FT.) ... (3)" is the pile spacing in feet used for OPTION 0 and 1. The data is ignored for OPTION 2 and 3

"AVG. OVERLOAD FACTOR ... (1.60)" is the average load factor used for strength design. It is used as a multiplier of service load shear and bending moments in design of the pile cap.

"f_b (KSI) AVERAGE BEARING ... (4)"is the average strength design bearing stress in kips per square inch on the gross area at the base of the concrete column. f_b is used to determine the smaller square column size which can sustain the load in the column. If the column size is specified in the P2 card, the cap design is based on the size specified.

"PILE DIAMETER (IN.) ... (12)" is the pile diameter in inches. It is the distance over which the pile reaction is uniformly distributed. If this data is given for OPTION 3, it is ignored in the calculations. OPTION 3, automatically calculates the distance over which the soil pressure is uniformly distributed as the footing length (or width) divided by the number of elements in each direction ($\sqrt{\#PILES}$.)

4.4 Hed - card (title card)

This card must be used with all input options. Heading information must be given in this card to identify job, designer data, etc.

4.5 P2 - card (general input)

This card must be used with OPTION 1, 2, and 3. Pile cap length and width dimension in feet must be given. The subscripts of A, B, C and D indicate the quadrant numbers where the pile cap plan dimension are measured. X and Y are any arbitrary reference axes which are parallel to the sides of the cap as in Fig. 1.

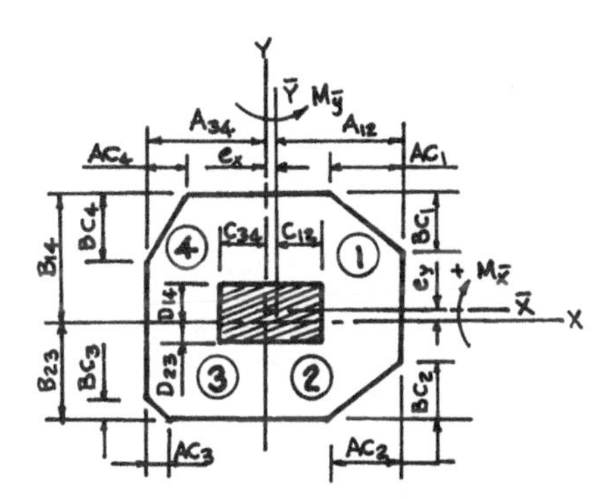

Fig. 1. Pile cap plan

The column size may be left blank in which case the smallest square column will be used. See P1 - CARD ... f_b data input for further details.

The depth of the cap may be left blank, in which case the program calculates and uses the least depth that will satisfy punching and beam shear.

The "depth" of cap is defined as the total out-to-out concrete thickness of the cap.

The user may define the depth of the cap in feet and inches, feet only, or inches only.

4.6 P3 - card (corner cuts (Ft.) & column loads)

This card must be present for OPTION 1, 2, and 3. AC and BC are dimensions in feet of the cut corners of the cap. The subscript refers to the quadrant of the cap. All values are zero for rectangular pile caps.

When OPTION 3 is used, only rectangular spread footings are permitted. All AC and BC values must be zero.

Optional input data for column loads, moments, and eccentricities may be given. If any eccentricity or moment is specified the axial load must be given. X and Y are the centroidal axes of the column and parallel to the X and Y axes respectively of foundation.

"SERVICE LOAD P (KIPS)" is the service axial load in the column above the foundation. When this value is given, the shear and moment capacity of the foundation is obtained to resist the pile reactions resulting from this load, its eccentricity, applied moment (if any), and the cap weight. When the service load is not specified, the pile reaction is based on pile service load capacity given in the P1 - CARD.

"ECCENTRICITY e_x (IN.)" is the X-distance in inches between the Y - axis and the Y-axis.

"SERVICE MOMENT M_x ft-k)" is the service load moment about the $\bar{X}$-axis in foot kips.

4.7 PGR -card (pile input by grid)

This card is used only for OPTION 1. It is used to locate the pile under a cap by specifying the grid patterns.

Up to 6 grid patterns may be superimposed.

The patterns are subjected to the following conditions:

1) The pattern is located symmetrically about both the X and Y axes.

2) The grid lines are spaced evenly in the X direction.

3) The grid lines are spaced evenly in the Y direction.

4) A pile is located at all the intersection of the grid lines.

Each grid pattern is specified by --

"LINES X" - the number of lines crossing the X - axis (a number between 1 and 9)

"LINES Y" - the number of lines crossing the Y - axis (a number between 1 and 9)

"SPACING X" - the spacing between the grid lines crossing the X - axis.

"SPACING Y" - the spacing between the grid lines grossing the Y - axis.

An example of a 13-pile pattern is given in Fig. 2.

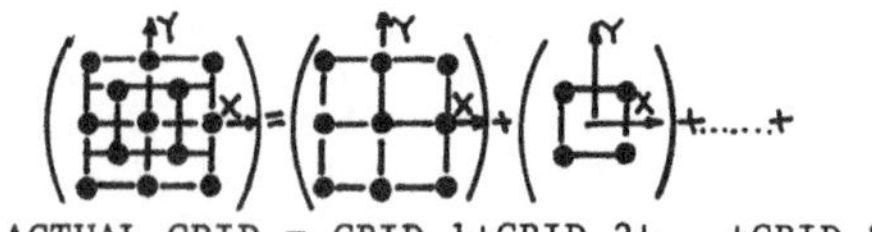

ACTUAL GRID = GRID 1+GRID 2+...+GRID 8

LINES X = NO. CROSSING THE X-AXIS
LINES Y = NO. CROSSING THE Y-AXIS

X-SPACING = $X_1 S\sqrt{1} + X_2 S\sqrt{2} + X_3 S\sqrt{3}$
Y-SPACING = $Y_1 S\sqrt{1} + Y_2 S\sqrt{2} + Y_3 S\sqrt{3}$

Fig. 2. Option 1 by grid

The X - spacing is calculated as:

$$X_1 \; S\sqrt{1} + X_2 \; S\sqrt{2} + X_3 \; S\sqrt{3}$$

Where X_1, X_2, and X_3, are data given in the PGR - CARD, and S is the pile spacing given in the P1 - CARD. X_1, X_2, and X_3 may vary from 0 to 9.

The Y - spacing is calculated as:

$$Y_1 \; S\sqrt{1} + Y_2 \; S\sqrt{2} + Y_3 \; S\sqrt{3}$$

Where Y_1, Y_2, and Y_3, are data given in the PGR - CARD.

4.8 PX, PY, & PP - cards (pile input by coordinates)

These cards are used only with OPTION 2 to specify each pile's location by its coordinates (see Fig.3). 3 cards must be given for each ten or less piles. The X - coordinates in feet are given in the PX- card. The corresponding Y - coordinates in feet are given directly below in the PY - card. The corresponding fraction of each pile which is effective in resisting vertical load is given directly below in the PY - card. A pile may have a vertical load resistance fraction less than 1.0 if it is battered or otherwise not as effective as the pile capacity specified in the P1-card.

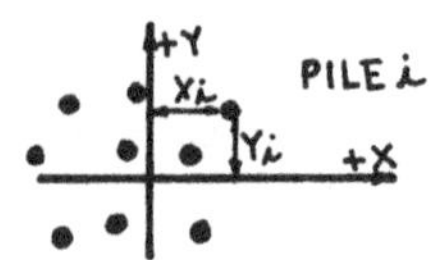

Xi & Yi = COORDINATES (FT.) OF PILE i

Fig. 3 Option 2 by coordinates

SAMPLE PROBLEM NO. 1

FY= 60000. F'C=3750. PSI
PILE SPACING = 3.250 FT
COL. PU / AG =5.000 KSI
PILE DIAMETER = 12.00 IN
REINF. COVER = 8.00 IN

75 T O N P I L E S

NO. OF PILES	COL. LOAD D+L	COL. LOAD 1.60X (D+L)	CAP SIZE A	CAP SIZE B	CAP SIZE T	COL. SIZE	REINF. STEEL LONG BARS	REINF. STEEL SHORT BARS	QUANTITIES CONC C.Y.	QUANTITIES STEEL TON	QUANTITIES FORM S.F	MAX. COL. DOWEL
2	294	470	5- 9	2- 6	33	10	6-# 6	5-# 4	1.4	0.026	45	#10
3	440	705	5- 9	5- 4	31	12	6-# 6	3 WAYS	2.2	0.070	49	# 9
			1- 7	1- 8								
4	585	936	5- 9	5- 9	36	14	8-# 7	8-# 7	3.6	0.085	69	#11
5	726	1163	7- 2	7- 2	36	15	8-# 8	7-# 9	5.7	0.150	86	#11
6	874	1398	9- 0	5- 9	40	17	7-#10	13-# 7	6.3	0.197	98	#11
7	1011	1618	9- 0	8- 2	42	18	6-#10	8-# 9	9.5	0.213	120	#14
8	1161	1858	9- 0	8- 2	42	19	7-#10	11-# 9	9.5	0.271	120	#14
9	1301	2082	9- 0	9- 0	40	20	8-#10	8-#10	12.0	0.292	144	#14
10	1441	2305	12- 3	8- 2	47	21	10-#10	13-# 9	14.5	0.422	159	#14
11	1586	2537	12- 3	8- 2	51	23	9-#11	14-# 9	15.7	0.463	173	#14
12	1726	2763	12- 3	9- 0	53	24	11-#11	14-# 9	18.0	0.545	187	#18
13	1855	2968	13-10	9- 0	61	24	8-#14	18-# 9	23.4	0.668	232	#18
14	2010	3217	12- 3	11- 5	51	25	11-#11	11-#11	22.0	0.662	201	#14
15	2146	3434	13-10	12- 3	54	26	12-#11	12-#11	25.6	0.799	206	#18
			6- 5	8- 0								
16	2296	3674	12- 3	12- 3	55	27	13-#11	13-#11	25.4	0.811	224	#18
17	2436	3898	13-10	12- 3	59	28	13-#11	11-#11	27.9	0.803	225	#18
			6- 5	8- 0								
18	2575	4120	13-10	12- 3	59	29	15-#11	12-#11	30.8	0.905	256	#18
19	2708	4333	14- 8	12- 3	63	29	11-#14	13-#11	34.9	1.001	282	#18
20	2840	4545	15- 6	12- 3	67	30	12-#14	14-#11	39.2	1.125	309	#18

Fig. 4. Output: Sample Problem No. 1

.9 Error messages

he program will determine some errors in nput data, Some of the errors will cause ermination of execution, others will be rinted only as warning messages. When he cap depth is specified by the input lata and if either the shear stress or the unching stress is excessive, a message to hat effect will be printed.

Examples of other error messages are:

Reinforcement is greater than .75 p_b(.75 f balance reinforcement)

Input card out of order.

Pile capacity not specified in input.

Check pile spacing for probable error.

Check load factor for probable error.

Check steel yield for probable error.

Check concrete f_c' for probable error.

Check f_b bearing stress for probable rror.

Check COVER for probable error.

A pile is located outside the concrete cap.

Check pile diameter for a probable error.

5 OUTPUT DATA

OPTION 0 produces a table of pile caps as shown in sample problem 1, Input values are repeated in the heading. The column load (in kips) is given as the service load under the heading "D + L" and as the design load under OLF x (D + L) where OLF is the average overload factor given as input. The "CAP SIZE" in feet and inches measured in the X-direction is given under the "A" heading and in the Y-direction is given under the "B" heading value given in parenthesis are the dimensions of the cap along the side between cut corners. They are not given for rectangular footings. "T" is the total thickness of the cap in inches.

```
---------------------------------------------------------------------------
             PILE 30 REV. SPIESS DECK SAN JUAN 10-5-74
---------------------------------------------------------------------------
                                             FY= 60000. F'C=4000. PSI
                                             PILE SPACING  =****** FT
                                             COL. PU / AG  =4.000 KSI
                                             PILE DIAMETER = 11.00 IN
                                             REINF. COVER  =  3.00 IN
------------------------ 40 T O N   P I L E S -----------------------


      COL. LOAD     CAP SIZE            REINF. STEEL    QUANTITIES
 NO.  --------- -------------           ------------- -------------- MAX.
 OF        1.55X                  COL.  LONG    SHORT CONC STEEL FORM COL.
PILES D+L (D+L)   A     B     T   SIZE  BARS    BARS  C.Y.  TON  S.F  DOWEL


 4   148   229  2- 6  3- 0 .24    4    6 # 6   7 # 6  2.2 0.048  44   # 8
              + 2- 6+ 3- 0     +  4X 7+ 7
              = 5- 0= 6- 0     =  8 X 14
  CORNER  X-DIMENSION  Y-DIMENSION
CUTS          FEET         FEET
QUAD
     1        0.000        0.000
     2        0.000        0.000
     3        0.000        0.000
     4        0.000        0.000
  COLUMN LOAD   =        148.000 KIPS
  X-ECCENTRICITY=          0.000 INCHES
  Y-ECCENTRICITY=          0.000 INCHES
  X-AXIS BENDING=        128.000 FT-KIPS
  Y-AXIS BENDING=         49.000 FT-KIPS
  PILE NO X-DIMENSION  Y-DIMENSION LOAD(ACTUAL)  LOAD(ALLOW)
              FEET         FEET         KIPS         KIPS
     1        1.000        1.419       86.305       80.000      OVERSTRESS
     2        0.750       -2.080       28.536       80.000
     3       -2.000       -1.419       -6.549       80.000      TENSION
     4       -1.250        1.379       48.706       80.000
```

PILE CAP ANALYSIS

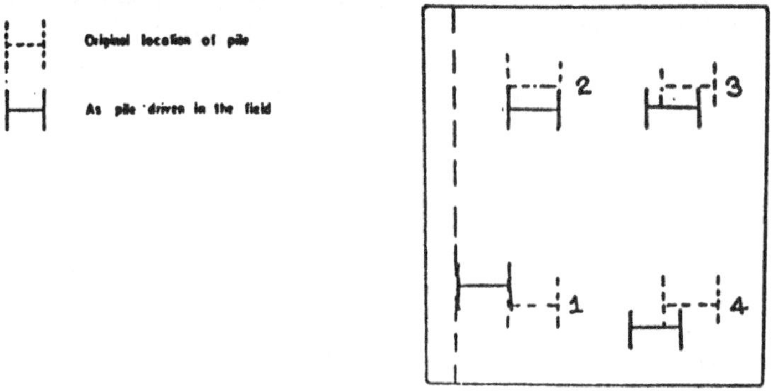

Original location of pile

As pile driven in the field

2 3 1 4

Fig. 5. Output: Sample Problem No. 2

The minimum square column size in inches is given under the heading "COL. SIZE".

The reinforcement is given by the quantity and size of bars for each of two directions of the footing. "LONG BARS" always placed below "SHORT BARS". Variation in flexural depth may cause slight variation between long and short bars used for square foundations.

A "#" sign is used to indicate bar size if the percentage of reinforcement is less than $200/f_y$. A "*" sign is used to indicate bar size if the percentage of reinforcement is greater than $200/f_y$. An "H" sign is used to indicate that the bars must be hooked to develop their strength.

The quantities of material and the units of measure are given towards the right side of the table.

The maximum size of column dowel in compression that can be developed in the depth of the foundation is given under the heading "MAX. COL. DOWEL".

```
           SAMPLE PROBLEM NO. 3
---------------------------------------------------------------------------------
                                                    FY= 60000, F'C=3500, PSI
                                                    PILE SPACING  =****** FT
                                                    COL. PU / AG  =4.000 KSI
                                                    PILE DIAMETER = 10.00 IN
                                                    REINF. COVER  =  4.00 IN
----------------------- 2.50 T S F   S O I L   -----------------------

       COL. LOAD     CAP SIZE          REINF. STEEL    QUANTITIES
  NO.  --------- -------------        ------------ -------------- MAX.
  OF        1.52X                COL.  LONG    SHORT CONC STEEL FORM COL.
 PILES D+L (D+L)    A     B    T SIZE  BARS    BARS  C.Y.  TON  S.F  DOWEL

  100  100   152  6- 0  3- 0 18   7    5 # 7 11 # 6  4.0 0.104  54  # 5
              + 6- 0+ 3- 0     +   7X12+12
              =12- 0= 6- 0     =  15 X 25
 CORNER  X-DIMENSION  Y-DIMENSION
CUTS        FEET         FEET
QUAD
   1        0.000        0.000
   2        0.000        0.000
   3        0.000        0.000
   4        0.000        0.000
COLUMN LOAD    =       100.000 KIPS
X-ECCENTRICITY=          0.000 INCHES
Y-ECCENTRICITY=          0.000 INCHES
X-AXIS BENDING=         80.000 FT-KIPS
Y-AXIS BENDING=          0.000 FT-KIPS
SOIL    X-DIMENSION  Y-DIMENSION STRESS(ACTL)  STRESS(ALLOW)
 AREA      FEET         FEET        KSF           KSF
   1      -5.399       -2.699      0.603         4.999
   2      -4.199       -2.699      0.603         4.999
   3      -2.999       -2.699      0.603         4.999
   4      -1.799       -2.699      0.603         4.999
   5      -0.599       -2.699      0.603         4.999
   6       0.600       -2.699      0.603         4.999
   7       1.800       -2.699      0.603         4.999
   8       3.000       -2.699      0.603         4.999
   9       4.200       -2.699      0.603         4.999
  10       5.399       -2.699      0.603         4.999
  11      -5.399       -2.099      0.828         4.999
  12      -4.199       -2.099      0.828         4.999
  13      -2.999       -2.099      0.828         4.999
  14      -1.799       -2.099      0.828         4.999
  15      -0.599       -2.099      0.828         4.999
  16       0.600       -2.099      0.828         4.999
  17       1.800       -2.099      0.828         4.999
  18       3.000       -2.099      0.828         4.999
  19       4.200       -2.099      0.828         4.999
  20       5.399       -2.099      0.828         4.999
  21      -5.399       -1.499      1.052         4.999
  22      -4.199       -1.499      1.052         4.999
  23      -2.999       -1.499      1.052         4.999
  24      -1.799       -1.499      1.052         4.999
```

Fig. 6. Output: Sample Problem No. 3

OPTION 1, 2 and 3 produce output for a single pile cap or spread footing in a different format. See sample problem 2.

Each component of the cap and column size is given individually which is followed by the overall dimension after the " = " sign. The dimension applicable to quadrant 1 is always given first. Sizes are rounded to the nearest inch which may make the total slightly different than the addition of the components. Internal calculations are preformed using exact values to the nearest 1/2 inch.

The dimensions of the corner cuts in feet are given for each of the four quadrants.

The service column load, eccentricities, and applied moments are given.

FOR OPTION 1 and 2, the pile location coordinates and service loads are given and compared to the allowable service pile loads. TENSION or OVERSTRESS messages are printed when the condition occurs.

For OPTION 3 the centroid of each soil element is given, followed by service load bearing pressure in kip per square foot and is compared to the allowable service load pressure.

The program assumes piles and soil are effective in tension.

If piles cannot be developed in tension or for spread footings where contact with the soil is lost, a manual design or redesign must be made.

6 SAMPLE PROBLEMS

6.1 Example problem no. 1 (design)

A series of design are required for 75 ton piles. Given: fc'= 3750 psi, f_y=60,000 psi, cover = 8", pile spacing = 3'-3",averaged OLF = 1.60, fb (average strength design) column bearing stress=5 ksi, pile diameter=12"

Output: See Fig. 4

6.2 Example problem no. 2 (analysis)

A 4-pile cap has been designed. During construction the piles were driven in the incorrect positions. Check the adequacy of the piles. Given data is in the output sheet.

Output: See Fig. 5

6.3 Example problem no. 3

A spread footing is to be designed for 2.5 TSF soil. Design using OPTION 3. The plan of the footing is 12'-0" by 8'-0" and column size 1'-3" (oriented in the long dimension) by 2'-1". Due to the large moment, divide the footing into 10 x 10 area moments, ie, # piles = 100. Other data: fc' = 3,500 psi, cover = 4", OLF = 1.52, load = 100 kips, moment = 80 ft-kips

Output: See Fig. 6

7 CONCLUSIONS

A rational program for the design and analysis of pile caps and spread footing has been presented. The footings may be designed for any service load soil pressure or service load capacity and pile spacing. The program has wide application also in the analysis of pile caps for as-driven pile location up to a maximum of 100 piles.

The program can design or analyze pile caps whose spacing and cap outline are unsymmetrical or column place off-center on the footing.

For practical purposes, the program is applicable to present design practice even though the program was based on the ACI 318-71 building code.

REFERENCES

ACI 318-71, Building code requirement for reinforced concrete, 1971, American Concrete Institute, Detroit, Michigan, 1971.

CRSI Handbook, 1982, Concrete Reinforcing Steel Institute, Schaumburg, Illinois.

San Juan, G.M., 1984 Computer application in building design systems. Presented at the December, 1984, International Conference on Computer Technology and Its Application, CI-Premier, Bangkok,Thailand.

San Juan, G.M., 1985 Computer in building design. Construction Management Journal, Bangkok and the Philippines.

Computer and Physical Modelling in Geotechnical Engineering, Balasubramaniam et al. (eds)
© 1989 Balkema, Rotterdam. ISBN 90 6191 864 2

Modelling axially loaded piles: Comparisons with pile test data

K.C.Cheung, R.O.Davis & G.Mullenger
University of Canterbury, Christchurch, New Zealand

ABSTRACT: Theoretical models for load diffusion from an axially loaded cylindrical pile are developed using two modern consitutive relations for undrained soil behaviour. The models are used to predict response from two well-documented pile tests. Reasonably good comparisons between theoretical and measured pile behaviour are obtained.

1 INTRODUCTION

As progressively more sophisticated models for soil behaviour become available, it is of interest to employ the models in attempts to predict or reproduce actual field measurements in realistic loading geometries. In this paper, two constitutive models for soil are investigated in this way. One is the bounding surface model of Dafalias (1980) developed from ideas associated with classical theories of plasticity. The second is the rate-type model of Davis and Mullenger (1979) founded upon the theory of hypoelasticity. Both models have been implemented in a simple computer analysis of the axially-loaded elastic pile problem, assuming undrained soil response and perfect bonding between pile and soil. It is possible to calculate the load diffusion from the pile into the soil for non-homogeneous and layered soils, and to predict the load-displacement response of the pile.

Two pile tests previously reported in the literature are used for comparison with the model prediction. The first test was described by O'Neil, et.al.(1982 a,b, 1983). It involved a steel pile placed in over-consolidated clays near Houston. The second test was carried out at Hendon, in London clay, and was reported by Cooke (1979). In both cases relatively complete description of the soils involved are available, and both shear transfer and load-displacement data were obtained during the tests. The Houston soil profile included three distinct strata, whereas the Hendon profile exhibited a more smoothly varying inhomogeneity. Both piles were "floating" in the sense that no significantly stronger strata lay beneath the pile tip.

Both the bounding surface and rate-type models compare favourably with the test data, particularly with regard to load diffusion behaviour and the low stress, load-displacement response. Both models over-estimate the ultimate pile load for the Houston test, but give good results for the Hendon test. The poor result for the Houston test may be related to the assumption of perfect bonding between pile and soil. The rate-type model, which is conceptually simpler than the bounding surface model, appears to perform equally as well in the comparisons. No doubt this is due to the relatively simple loading geometry, but it does emphasize the fact that greater modelling complexity is frequently not warranted in practical applications.

2 SOIL MODELS

Two existing soil models based on the concepts of critical state soil mechanics will be studied. Both models have no distinct elastic region during monotonic loading. One is the bounding surface model (Dafalias, 1980). It is founded on the classical theory of plasticity but employs the concept of a bounding surface in addition to a yield surface. The second model is the rate-type soil model (Davis and Mullenger, 1979) developed from the theory of hypoelasticity. Yielding and plastic flow are incorporated within the model without a formal flow rule or hardening rule such as those found in the theory of plasticity. An elementary formulation for both of the models is given in this section. For further details of either model, the reader is referred to the original reference cited above.

2.1 Bounding surface model

The bounding surface model incorporates a yield surface which lies within a bounding surface of similar shape in stress space. The yield surface may degenerate to a point under certain conditions. Following the conventional theory of plasticity, the strain rate is decomposed into elastic and plastic parts

$$\dot{\epsilon}_{ij} = \dot{\epsilon}^{e}_{ij} + \dot{\epsilon}^{P}_{ij} \tag{1}$$

Using the associated flow rule and the assumption of a bounding surface, Dafalias (1980) arrived at the following equations

$$\dot{\epsilon}^{P}_{ij} = \langle L \rangle n_{ij} \tag{2}$$

where $\langle\ \rangle$ denotes the Heaviside function and L is the a scalar function

$$L = \frac{1}{H} \dot{\sigma}_{kl}\, n_{kl} = \frac{1}{\bar{H}} \dot{\bar{\sigma}}_{kl}\, n_{kl} \tag{3}$$

$$n_{ij} = \frac{1}{g} \frac{\partial F}{\delta \bar{\sigma}_{ij}} \quad , \quad g = \left(\frac{\partial F}{\partial \bar{\sigma}_{ij}} \frac{\partial F}{\partial \bar{\sigma}_{ij}} \right)^{\frac{1}{2}} \tag{4}$$

Here n_{ij} represents the normal vector to the bounding surface which is defined by

$$F = F(\bar{\sigma}_{ij}, \epsilon^{P}_{kk}) = 0 \tag{5}$$

Overbars denote quantities associated with the bounding surface. For example $\bar{\sigma}_{ij}$ is the image of σ_{ij}, the current stress, on the bounding surface (see Figure 1). H denotes the plastic modulus defined below.

Substituting (2) into (1) and assuming undrained behaviour as well as elastic incremental constitutive relations governed by Hooke's law, and finally inverting the resulting relationship we arrive at the following equation

$$\dot{\sigma}_{ij} = 2G\, \dot{\epsilon}_{ij} - [2G(n_{ij} - \frac{1}{3} n_{kk}\delta_{ij}) + B\, n_{kk}\delta_{ij}]\langle L \rangle \tag{6}$$

where G and B denote the elastic shear and bulk modulii and L may be written

$$L = \frac{-2G\, \dot{\epsilon}_{k\ell} n_{k\ell}}{H + Bn_{mm} + 2G\,(1 - \frac{1}{3} n^{2}_{mm})} \tag{7}$$

Equation (6) is a general formulation of the bounding surface model independent of the exact form of the bounding surface itself. For general stress states, it is convenient to introduce the stress invariants

$$J_1 = \sigma_{kk} \ ,\ J_2 = S_{ij}S_{ij} \ ,\ J_3 = \det(S_{ij}) \tag{8}$$

where S_{ij} denotes the deviatoric part of σ_{ij}. The third stress invariant may be replaced by the Lode angle θ, defined by

$$-\frac{\pi}{6} \le \theta = \frac{1}{3} \sin^{-1} \left[3\sqrt{3} \frac{J_3}{J_2^{3/2}} \right] \le \frac{\pi}{6} \tag{9}$$

Then if a radial mapping rule of the form

$$\bar{\sigma}_{ij} = \alpha(\sigma_{kl} \ , \epsilon^{P}_{mm})\, \sigma_{ij} \tag{10}$$

is employed, the stress invariants of the image stress $\bar{\sigma}_{ij}$ are given by $\bar{J}_1 = \alpha J_1$, $\bar{J}_2 = \alpha^2 J_2$, $\bar{J}_3 = \alpha^3 J_3$. The distance factor $\alpha\ [1 \le \alpha < \infty]$ is obtained from equation (5).

In the $\sqrt{J_2}$ versus J_1 invariant space illustrated in Figure 1, the slope of the critical state line will be denoted by N. Specific values of N determined from triaxial compression and extension tests will be denoted N_c and N_e. For other stress states, N may be expressed as a function of θ

$$N(\theta) = \frac{2n\, N_c}{1 + n - (1-n)\sin 3\theta} \tag{11}$$

where $n = N_e/N_c$. Thus the general shape of the bounding surface may depend upon the lode angle.

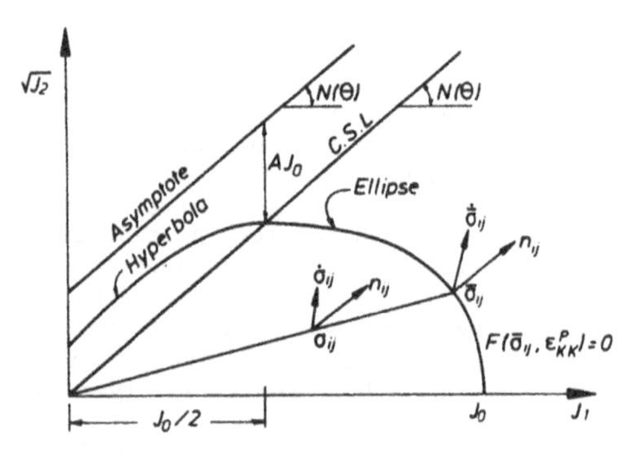

Fig.1. Schematic representation of bounding surface and radial mapping rule

Two alternative shapes for the bounding surface have been defined by Dafalias (1980). The choice of surface depends upon the current dimensionless stress $\eta=\sqrt{J_2}/J_1$. Whenever $0 < \eta \leq N$, the bounding surface is elliptic in shape, defined by

$$F = (\bar{J}_1 - J_o)\bar{J}_1 + \bar{J}_2/N^2 = 0 \qquad (12)$$

in which $J_o = J_o(\epsilon_{kk})$, the hardening parameter, is the intersection of the surface with the J_1 axis. In contrast, if $\eta > N$, a hyperbolic surface is used, defined by

$$F = (\bar{J}_1 - J_o)\bar{J}_1 - \frac{\bar{J}_2}{N^2} - \frac{A_c}{N_c} J_o^2 + J_o(1 + \frac{2A_c}{N_c})\frac{\sqrt{J_2}}{N} \qquad (13)$$

Here A_c is also a material constant, representing the fraction of J_o by which the critical state line lies below the asymptote to the hyperbolic surface for triaxial compression, as illustrated in Figure 1.

Isotropic hardening is governed by the rule

$$\frac{dJ_o}{d\epsilon^p_{kk}} = \left(\frac{1 + e_o}{\lambda - \kappa}\right) J_o \qquad (14)$$

in which e_o is the initial void ratio, and λ and κ denote the slopes of the virgin compression and swelling lines in void ratio-logarithim of pressure space.

Finally, the plastic modulus H is given by

$$\bar{H} = - \frac{\partial F}{\partial \epsilon^P_{kk}} \frac{N_{jj}}{g} \qquad (15)$$

and is related to H by the so-called interpolation rule

$$H = \bar{H} + U (\alpha - 1) \qquad (16)$$

where U is the hardening shape function

$$U = P_a\, u(\theta) \left(1 + \left|\frac{\eta}{N}\right|^{-x}\right) \left\{9 \left(\frac{\partial F}{\partial \bar{J}_1}\right)^2 + \frac{2}{3}\left(\frac{\partial F}{\partial \sqrt{\bar{J}_2}}\right)^2 \right\} \qquad (17)$$

and p_a is atmospheric pressure providing correct dimension. Here $u(\theta)$ has the same form as $N(\theta)$ [equation (11)] with u_c and u_e being parameters measured from triaxial compression and extension tests. The exponent x is a positive number which ensures H becomes singular if η approaches zero, and hence purely elastic loading results.

Using equations (8) through (17) the stress in undrained (constant volume) deformation may be obtained from (6). The bulk and shear modulii may be arbitrary function of the stress invariants, but we will simply take G constant and

$$B = (1 + e_o)\, J_1 \,/\, 3 \qquad (18)$$

The model parameters λ, κ, N_c, N_e, G and e_o can be directly evaluated from conventional experiments. The parameters u_c, u_e, and A_c must be evaluated from curve fitting technique applied to particular test results and x has been taken as 0.2.

2.2 Rate-type model

The rate-type model is based on the theory of hypoelasticity (Truesdell and Noll, 1965). It obeys the general constitutive relation

$$\overset{\circ}{\sigma}_{ij} = \Gamma_{ij}\,(\sigma_{mn}, e, D_{mn}) \qquad (19)$$

Here Γ_{ij} is an isotropic tensor function whose arguments are the stress σ_{mn}, void ratio e, and rate of deformation tensor D_{mn}. The co-rotational stress rate

$$\overset{\circ}{\sigma}_{ij} = \dot{\sigma}_{ij} - \sigma_{kj}W_{ik} + \sigma_{ik}W_{kj} \qquad (20)$$

is used, where W_{ij} is the spin tensor. A special form of (19) has been taken by Davis and Mullenger (1979) as the starting point for their model development. This is

$$\overset{\circ}{\sigma}_{ij} = (a_1 D_{kk} + a_2 \sigma_{km} D_{mk})\,\delta_{ij} + (a_3 D_{kk} + a_4 \sigma_{km} D_{mk})\sigma_{ij} + a_5 D_{ij} \qquad (21)$$

in which a_1, ..., a_5 were assumed to be functions of e only. During monotonic loading, the coefficients were taken to be

$$a_1 = -\frac{2}{3} G$$

$$a_2 = a_3 = \frac{2Gp_c}{M^2} \qquad (22)$$

$$a_4 = -\frac{2G}{M^2}$$

$$a_5 = 2G$$

in which $p_c = p_c(e)$ is the critical state pressure and

$M = 3N_c p_c$

It should be noted here that the critical state pressure p_c for the rate-type model is only a function of void ratio, and therefore remains constant in undrained deformation.

Setting $D_{kk} = 0$ in (21) reduces that equation to undrained conditions. Then use of (22) gives the final, constant volume, constitutive equation

$$\overset{\circ}{\sigma}_{ij} = \frac{2G}{M^2}\left[-(\sigma_{ij} - p_c\delta_{ij})\sigma_{km}D_{mk} + M^2 D_{ij}\right] \quad (23)$$

For these conditions, the strength parameter M is a constant which is found to be equal to $2\sqrt{2}/3\ c_u$, where c_u is the undrained shear strength found from triaxial tests.

3. MODEL PARAMETERS

In order to make use of either of the soil models, we must define the initial stress state, stress history, and material parameters for the soil. We can assume the stress history consists of one-dimensional compression and (possibly) extension processes. If we further assume negligible shear stresses on vertical and horizontal surfaces the initial stress state and stress history can be fully characterized by the vertical stress, σ_{zi}, the coefficient of lateral earth pressure at rest, K, and the overconsolidation ratio, OCR.

The initial overburden stress σ_{zi} may be obtained directly from measurements of soil unit weight and pore water pressure. The value of K and OCR are not independent. For normally consolidated soils, K may be estimated using Jaky's formula

$$K_{nc} = 1 - \sin\phi \quad (24)$$

where ϕ denotes the angle of internal friction. For overconsolidated soils, Parry (1977) has suggested that K may be approximated by

$$K_{oc} = K_{nc}\sqrt{OCR} \quad (25)$$

Slope of the consolidation and swelling lines λ and κ, may be related to more familiar compression and swelling indices C_c and C_s,

$$\lambda = C_c/\ln 10 \quad \text{and} \quad \kappa = C_s/\ln 10 \quad (26)$$

The undrained shear strength in triaxial compression, c_u, is required to characterize the soil's strength. Its value may be predicted from the soil models discussed above. For either the bounding surface or rate type models, c_u is given by

$$c_u = \frac{3}{2}\sqrt{\frac{3}{2}}\,N_c\,p_f \quad (27)$$

Here, p_f is the mean stress at failure. In the case of the bounding surface model, p_f is found from

$$p_f = p_i^{\kappa/\lambda}(p_o/2)^{(1-\kappa/\lambda)} \quad (28)$$

where p_i is the initial mean stress

$$p_i = \frac{1}{3}\sigma_{zi}(1 + 2K) \quad (29)$$

and P_o is related to J_o by

$$p_o = \frac{1}{3}J_o \quad (30)$$

In contrast, p_f in the rate type model is simply

$$p_f = \frac{1}{2}\,p_o \quad (31)$$

Equation (31) results from the simplifying fact associated with the rate type model that the yield surface remains stationary in undrained deformation.

4 LOAD DIFFUSION FROM A FRICTION PILE

It has been demonstrated (Mullenger, et.al., 1984) that the induced shear traction due to pile shaft displacement is relatively independent of initial stress conditions in the surrounding soil. Furthermore, the initial conditions will only seriously affect the stress state within approximately one pile radius of the pile wall. This suggests we consider only homogeneous initial stress states surrounding the pile. We therefore assume a pre-existing cylindrical cavity in the "undisturbed" semi-infinite soil mass into which the elastic pile is installed. Following the usual sign convention of soil mechanics and taking the z-axis pointing downward, we further assume all soil displacements are vertical so that the only non-vanishing strain rate is

$$\epsilon_{rz} = -\frac{1}{2}\frac{\partial u_z}{\partial r} \quad (32)$$

Here u_z is the vertical component of displacement within the soil.

Equilibrium requires that

$$\sigma_{rz} = \frac{a}{r}(\sigma_{rz})_a \quad (33)$$

where a denotes the pile radius, and $(\sigma_{rz})_a$ is the induced shear traction at the pile-soil interface. For either of the constitutive models we may construct a relationship of the following form

$$\epsilon_{rz} = \chi(\sigma_{rz}) \tag{34}$$

To arrive at (34) it is necessary to integrate the constitutive equations for the special case where ϵ_{rz} is the only non-zero strain. This integration is carried out in detail for the rate-type model in the article by Mullenger, et.al(1984). A similar development may be used with the bounding surface model.

Finally, integrating (32) and using (33) and (34), we find the pile shaft displacement is given in terms of the shear stress $(\sigma_{rz})_a$ by

$$\Delta_a = - \int_{r_m}^{a} 2\, \epsilon_{rz}\, dr = \Phi[(\sigma_{rz})_a] \tag{35}$$

where r_m is the radius of influence. Following Randolph(1978), we assume

$$r_m = 1.25\, \ell\, [G(\ell/2)\ /\ G(\ell)] \tag{36}$$

in which ℓ denotes the embedded length of pile, and the shear modulus G is assumed to vary with depth z. An analytic expression for (35) derived for the rate-type model was given by Mullenger, et.al.(1984). For the bounding surface model, numerical integration is required.

After the response function $\Delta_a = \Phi[(\sigma_{rz})_a]$ is obtained, the pile response follows from simultaneous solution of

$$\frac{\partial \Delta_a}{\partial z} = \frac{P(z)}{E_p A_p} \tag{37}$$

and

$$\frac{\partial P}{\partial z} = 2\pi a \Phi^{-1}[\Delta_a] \tag{38}$$

where P = P(z) is the pile shaft force, E_p is Young's modulus, and A_p the pile cross-sectional area. Equations (37) and (38) must be integrated numerically, even in the case where analytic representations for Δ_a is available. Once this integration has been accomplished, the axial force and displacement at all points on the pile shaft are known.

5 COMPARISON WITH PILE TESTS

The theoretical pile model will be compared with results from two well documented pile tests. Both tests were performed in heavily over-consolidated clays and both test piles were closed-end pipe piles. Since the test piles were not bedded on a stiff stratum, we will ignore tip resistance.

The first pile test was performed at Houston by O'Neil, et. al. (1982 a, b, 1983). The pile was a 273 mm diameter, 13 m long steel pipe pile, embedded in three distinct strata. Soil properties are summarized in Table 1, while the relevant model parameters are given in Table 2.

In Table 1, E_u denotes Young's modulus for undrained conditions. The unit weight of the test site soil was 20.1 ± 0.8 kPa and relationships for K_{oc} and OCR were reported as

$$K_{oc} = 3.5\, z^{-0.55}\ , \quad OCR = 16\, z^{-0.55}$$

for depths z in excess of 3 m.

Values for G in Table 2 result from dividing E_u by 3, the appropriate factor for Poisson's ratio equal to 0.5. Values of N_e and u_e were taken as $N_e = 0.8N_c$ and $u_e = u_c$ in all three strata.

The undrained shear strength c_u was calculated using (27) and is directly compared with experimental values reported by O'Neil, et.al (1982 a,b, 1983) in Figure 2. Average values of c_u were used over 1.0m intervals giving the block-like distribution shown in the figure. Measured strengths seemed to agree well with calculated values except in layer A.

The given model parameters and calculated strength profile were then used to estimate the shear transfer along the pile at working load. The radius of influence r_m was taken to be 10.0m. Results of these calculations are shown in Figure 3 for both soil models. Measured shear transfer data are also shown in the figure. Clearly, both soil models show quite good agreement with the measured data. Computed total-load versus pile head displacement is shown in Figure 4. Here it is seen that both theoretical models over-estimate the measured failure load by significant amounts.

It was noted when integrating (37) and (38) to find that load-displacement relationship at the pile head, that very abrupt failure

Table 1. Summary of soil properties at Houston

Stratum	Depth(m)	ϕ	λ	C_u(kPa)	E_u (kPa)
A	0-2.6	22°	0.06-0.144	115	72,000
B	2.6-7.9	23°	0.06-0.144	86	105,000
C	7.9-14.6	27°	0.04-0.057	158	173,000

Table 2. Model parameters for Houston soils

Stratum	G(kPa)	λ	κ	N_c	A_c	u_c
A	24,000	0.13	0.06	0.233	0.08	8.0
B	35,000	0.14	0.06	0.245	0.08	20.0
C	57,500	0.057	0.016	0.292	0.10	100.0

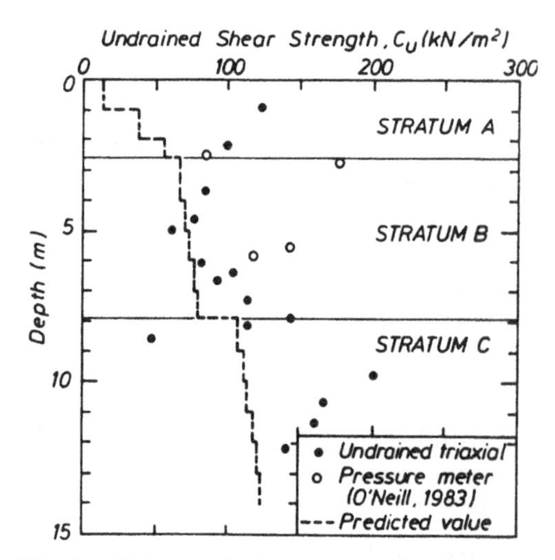

Fig.2. Undrained shear strength, Houston

occured with both soil models. The load-displacement curve remained relatively linear nearly to the ultimate load, and when the failure point was reached, the curve became horizontal almost immediately. The failure points are marked (f) on Figure 4.

The second test pile was located at Hendon and was reported by Cooke (1979). The pile diameter was 168mm. London clay at this test site extends from the ground surface to a depth of 25 m. Properties for this soil are available from several sources (Windle, 1977; Marsland, et.al.,1977). Figures 5 and 6 show measured data for undrained shear strength and shear modulus

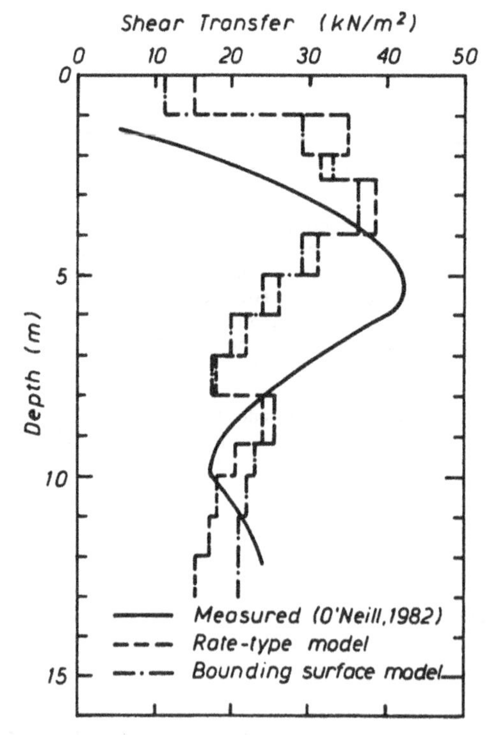

Fig.3. Shear transfer curve at working load, Houston

for various depths at the test site. The selected values of lateral earth pressure at rest are compared with measured data provided by Windle (1977) in Figure 7. The overconsolidation ratio was computed from equation (25) assuming ϕ is equal to 22.5

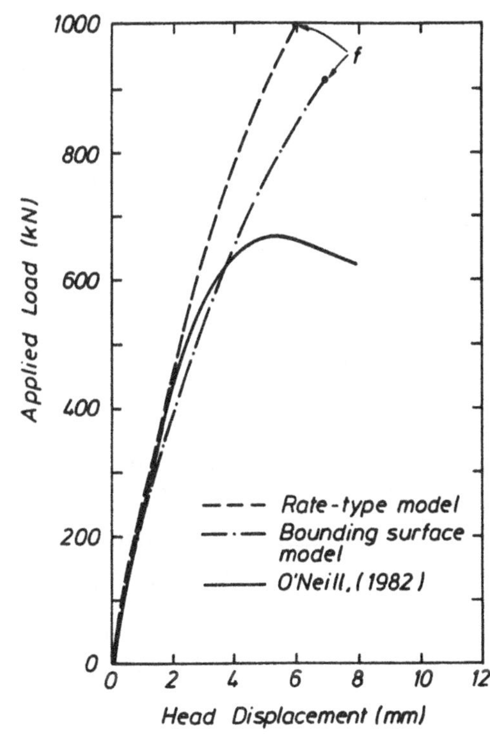

Fig.4. Load versus head displacement, Houston

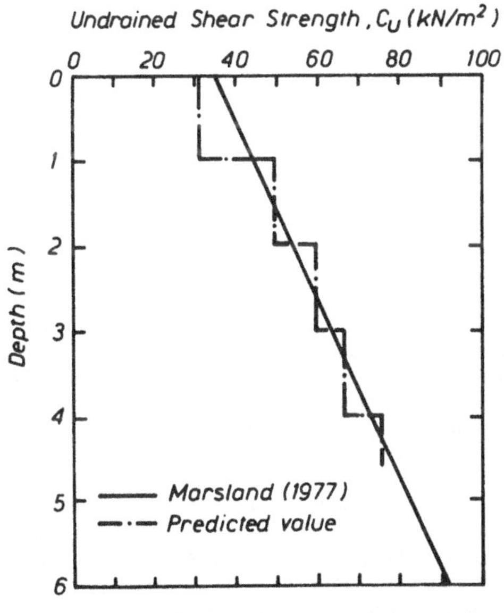

Fig.5. Undrained shear strength, Hendon

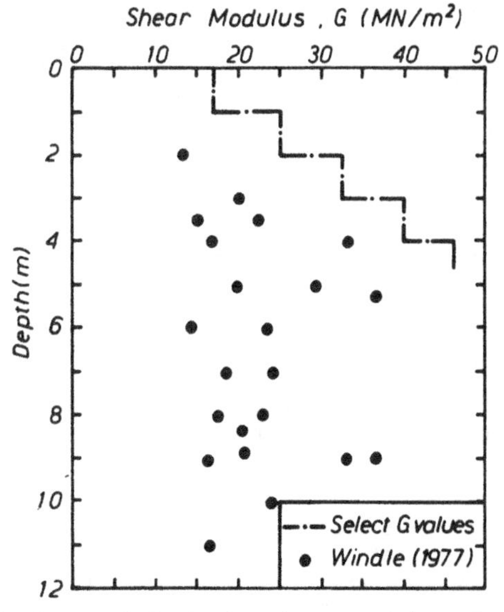

Fig.6. Undrained shear modulii from pressuremeter tests in Londay clay

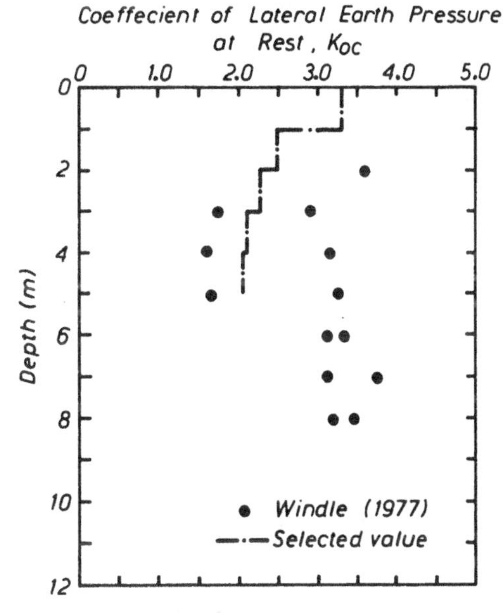

Fig.7. Coefficient of lateral earth pressure versus depth, Hendon

Table 3. Model parameters for London clay

Depth(m)	G(kPa)	U_c
0.5	17,000	100
1.5	25,000	140
2.5	33,000	180
3.5	40,500	240
4.3	46,000	290

degrees. Remaining model parameters are summarized below and in Table 3.

$N_c = 0.24$, $n = 0.81$, $A_c = 0.05$,
$u_e^c/u_c = 1.0$, $\lambda = 0.14$, $\kappa = 0.05$,
$r_m = 4.6\text{m}$

Three test piles were jacked into the clay to a depth of 4.6m with 500 mm centre to centre spacings. Measured shear transfer data for the centre pile are shown in Figure 8 together with the calculated shear transfer using the two soil models. The theoretical solutions are essentially identical in this case and only one line is drawn for both models. The five pairs of curves represent values at five increasing applied loads. Remarkably good agreement between measured and predicted response is found. Finally, the total axial load versus displacement response is illustrated in Figure 9. Load-displacement curves for two of the test piles are shown together

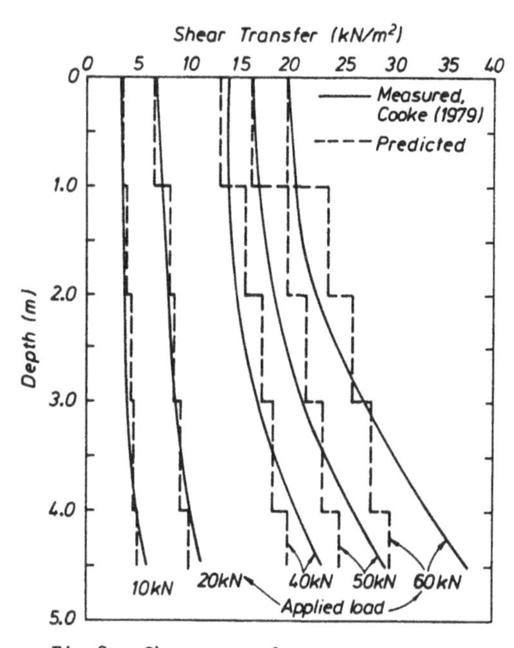

Fig.8. Shear transfer curve, Hendon

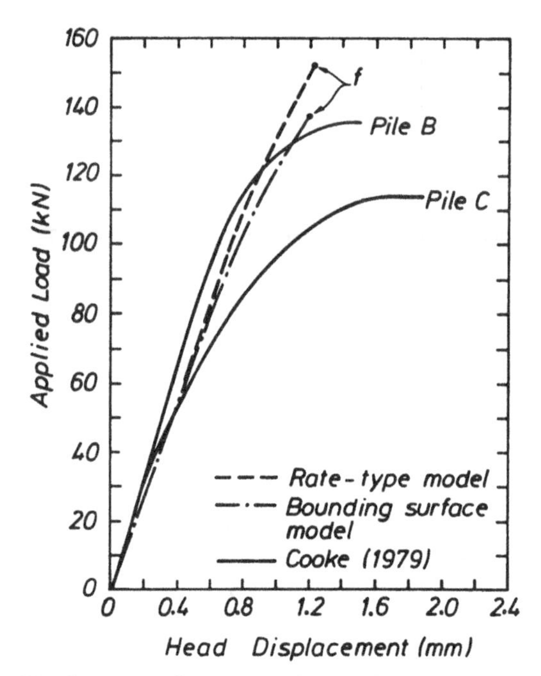

Fig.9. Load versus head displacement, Hendon.

with predicted curves for the two soil models. In this case, good agreement is found.

6 CONCLUSIONS

The essentially one-dimensional pile model which has been used here (in the sense that only one non-zero strain component is allowed) grossly oversimplifies the actual three dimensional soil-pile system. Nevertheless, surprisingly good agreement between theoretical predictions of shear transfer and actual measured data is found for both test piles at working loads. Approaching ultimate load the theoretical solutions have a more mixed success, overestimating the ultimate load in the Houston test but coming quite close in the Hendon test.

The two soil models represent extremes of sophistication. The bounding surface model is quite complex and offers an extremely wide range of possible responses, while the rate-type model is particularly simple but encompasses a far more limited response range. These facts are clearly reflected by the number of parameters required for each model: ten for the bounding surface model versus two for the rate-type model. For the class of problems considered here, however, both models appear to give equally good predictions. This is only slightly suprising in light of the relatively simple but non-homogeneous, deformation field which has been represented; nevertheless it emphasizes the point that use of more complex modelling methods is not always justified.

REFERENCES

Banerjee, P.K., Stipho, A.S.,(1978) "Associated and non-associated constitutive relations for undrained behaviour of isotropic soft clays". Int. J. Num. Anal. Methods Geomech.,Vol.2, pp.35-56

Cooke, R.W., Price,G., Tarr,K., (1979) "Jacked piles in London clay: a study of load transfer and settlement under working conditions". Geotechnique 29, No.2, pp.113-147.

Dafalias, Y.P., Herrman, L.R.,(1980) "A generalised bounding surface constitutive model for clays". Application of Plasticity and Generalized Stress-Strain in Geotechnical Engineering Yong, R.N., Selig, E.T.,(eds) Hollywood, Florida. pp.78-95.

Davis, R.O., Mullenger, G., (1979) "A simple rate-type constitutive representation for granular media". Proc. 3rd. Int. Conf. Num. Methods Geomech. Aachen, 2-6 April, pp.415-421.

Davis, R.O., Scott, R.F., Mullenger, G., (1984) "Rapid expansion of a cylindrical cavity in a rate-type soil". Int. J. Num. Anal. Methods Geomech., Vol.8, pp.125-140

Marsland, A., Randolph, M.F.,(1977) "Comparisons of results from pressuremeter tests and large in situ plate tests in London clay". Geotechnique 27, No.2 pp.217-243.

Mullenger, G., Scott, R.F., Davis, R.O., (1984) "Rapid shearing in a rate-type soil surrounding a cylindrical cavity". Int. J. Num. Anal. Methods Geomech., Vol.8, pp:141-156.

O'Neill, M.W., Havkins, R.A., Audibert, J.M.E., (1982a) "Installation of pile group in over consolidated clay". J. Geot Eng. Div. ASCE, Vol. 108, GT11, pp:1369 -1385.

O'Neill, M.W., Havkins, R.A., Mahar, L.J., (1982b) "Load transfer mechanisms in piles and pile groups". J. Geot. Eng. Div. ASCE, Vol. 108, GT12, pp.1605-1623.

O'Neill, M.W. (1983) "Side load transfer in driven and drilled piles". J. Geot. Eng. Div. ASCE, Vol. 109, No.10 pp: 1259-1266.

Parry, R.H.G., (1977) "A study of skin friction on piles in stiff clay". Ground Eng., Vol. 10, No.8, pp: 33-37.

Randolph, M.F., Wroth, C.P., (1978) "Analysis of deformation for vertically loaded piles". J. Geot. Eng. Div., ASCE, Vol.104, GT12, pp.1465-1488.

Roscoe, K.H., Burland, J.B., (1968) "On the generalized stress-strain behaviour of wet clay". In Heyman, J., Leckie, F.A., (eds) Engineering Plasticity, Cambridge Univ. Press. pp. 535-609.

Schofield, A.N., Wroth, C.P., (1968) "Critical State Soil Mechanics" McGraw-Hill, London.

Truesdell, C., Noll, W., (1965) "The Non-Linear Field Theories of Mechanics". In Flugge, S., (ed) Handbuch der Physik, III/3, Belin, Springer-Verlag.

Windle, D., Worth, C.P., (1977) "In situ measurement of the properties of stiff clays". Proc. 9th. Int. Conf. S.M. & F.E., pp.347-352.

Computer and Physical Modelling in Geotechnical Engineering, Balasubramaniam et al. (eds)
© 1989 Balkema, Rotterdam. ISBN 90 6191 864 2

Evaluation of alternate footing systems using computer aided design

B.V.Ranganatham & D.Bhanu Prasad
Department of Civil Engineering, Indian Institute of Science, Bangalore, India

ABSTRACT : Punching shear, besides being difficult to predict, is catastrophic with little residual strength. As such the rational alternate forms that ensure flexual failure involve the conceptual beam-slab configuration such as uniaxial beam-slab, axial beams-slab and diagonal beams-slab footings. Recent research has demonstrated that all the forms are effective in avoiding punching failure and that the yield line method is adequate to predict their strength. This paper reports the results of their computer aided design, cost estimate and evaluation relative to that of the conventional uniform thick footing. Another alternative form in vogue is footing thickened at the column junction with uniform taper and its cost effectiveness is also evaluated. The input variables are the column load from super structure and the foundation defining parameter either in terms of allowable foundation pressure or compressibility index of the foundation medium. The design outputs are the material quantities (concrete and steel) and the total cost. An in-depth analysis of the design outputs reveals the simple uniaxial beam-slab system with refined structural detailing to be the economic choice.

INTRODUCTION

Square isolated footings of reinforced concrete are the most widely adopted foundations under essentially concentric loads. Traditionally they are of uniform thickness owing to the ease of construction. Such conventional footings have a tendency to fail abruptly by punching which therefore governs the design. Experimental research studies on tapered footings have shown their improved performance against punching (Rengaraju, 1972; Shyam Prasad, 1978). The extensive use of a beam-slab configuration in preference to flat slab construction in the super-structure has motivated the choice of the beam-slab footing as a rational alternative to conventional uniform thick footing. The concentrated column reaction is distributed as line reactions along the beams on to the slab that in turn spreads the line reactions laterally on the foundation medium. The efficacy of this concept has been established by extensive testing (Rengaraju, 1972; Shyam Prasad, 1978; Rajagopalan Nair, 1981). These studies have demonstrated that punching is avoided in all adequately designed alternative forms and that yield line method adequately predicts the strength. However, since there has been no attempt at assessing the material requirement and relative cost of these footing systems, the present study is devoted to fill that identified gap.

FOOTING SYSTEMS - CONFIGURATION AND DESIGN APPROACH

Five feasible configurations are thus adopted for the investigation (conventional uniform thick CF, tapered TF, uniaxial beam-slab UB-S, axial beams-slab AB-S, and diagonal beams-slab DB-S). Figs. 1a and 1b report the general configuration of conventional and tapered footings. Also shown in fig. 1a are critical sections at which checks for wide beam shear and diagonal tension are made as per IS code (IS:456-1978). The IS code specifies a minimum thickness of 150mm and clear cover of 50mm for footings. Fig. 1c indicates the reinforcement pattern. The design is made in accordance with IS code and the so designed flexural capacity coincides with the yield line solution for an assumed contact pressure distribution. Fig. 2a and 2b report the rigid and flexible modes of failure respectively of the uniaxial beam-slab footing. Also shown in these figures are the virtual displacements adopted for the virtual work method of yield line analysis. The yield line solutions for the design of uniaxial beam-slab footing are given below.

$$\frac{W^+}{M} = \frac{8}{(1-\alpha)^2} \quad (1)$$

where W^+ is the ultimate load on the footing, M is the yield moment capacity per unit width of the footing slab and α is the ratio of column side to that of the footing.

$$M_b = m^* M \quad (2)$$

where m^* is the optimal ratio of moment capacity of beam per unit width (M_b) to the yield moment of the slab of the footing.

The optimal ratio m* is for the simultaneous failure of the beam and slab of the footing. The value of m* for uniaxial beam-slab footing is $1/\alpha$. Details of reinforcement are reported in fig. 2c. Designed shear reinforcement in the form of stirrups is provided in the beam.

The uniaxial beam-slab system has the great facility for saving material (concrete and steel) by curtailing steel in both beam and slab and tapering the slab. The detailing to effect the saving of materials is termed the refined design of uniaxial beam-slab footing (fig. 3) keeping the basic design same.

The configuration and the yield line patterns including virtual displacements assumed for the axial beams-slab footing are reported in fig. 4. The yield line solutions for the design of axial beams-slab footing are given below.

$$\frac{W^+}{M} = \frac{24(1+i)}{(1-\alpha)^2} \quad (3)$$

$$m^* = \frac{2+3i}{\alpha} \quad (4)$$

where i is the ratio of negative to positive yield moments of the slab.

The configuration and the governing yield line patterns with associated virtual displacements for the diagonal beams-slab footing are reported in fig.5. The expressions for the design based on yield line solution are given below.

$$\frac{W^+}{M} = \frac{24(2+i)}{(1-\alpha)^2} \quad (5)$$

$$m^* = \frac{2(2+i)(2-3\alpha)(8-6\alpha-3\alpha^2)-8(1-\alpha)^3}{4\alpha(1-\alpha)^2} \quad (6)$$

Owing to the complexities in analysis and the practical fabrication difficulties in the construction of axial and diagonal beams-slab footings, refinements in design detailing to effect material saving have not been attempted.

SUBSURFACE CHARACTERISTICS AND PLAN-DIMENSIONING

The plan dimension of a footing is arrived at by matching the column load with the characteristics of the subsurface medium. The twin design criteria for the determination of allowable bearing pressure of a footing are the bearing capacity (shear strength), and the settlement (compressibility) with the load factor and limiting settlement being stipulated by the code (IS:1904-1978). It is well recognised that the majority of designs is governed by settlement criterion since bearing failure is invariably preceded by plastic and unacceptable level of deformation. Either of the two parameters, coefficient of volume compressibility (m_v) and compression index (C_c), borne out of oedometer test results, is frequently used to define the comprebility characteristics of the soil. Since C_c is a constant in the normally consolidated range, the computation procedure using C_c is simpler than that which makes use of m_v , which varies non-linearly with the stress level even for normally consolidated clay. For computing settlement employing C_c , it is necessary to determine the initial effective stress which is a product of the density of soil and depth below ground surface. The compression index versus dry density relation (Oswald, 1981) is used to arrive at the initial void ratio and the effective stress. In the design of a footing founded at 1.5m below ground level, a tentative plan dimension is chosen to start with, the settlement of such a foundation is then determined for the given C_c-value and the procedure iterated till the computed settlement complies with the maximum value specified by the code and the resulting uniform pressure becomes the allowable pressure appropriate to the column load and the compression index of the soil. The iterative procedure accounts for the actual weight of the footing as against the conventional procedure of assuming it as ten percent of the column load.

COMPUTER-AIDED DESIGN AND RESULTS

For a given column load and soil compressibility characteristics the design is seen to be involved and iterative. The study to be relevant to practice has to cover a useful range of design inputs,viz. column loads (25-600 tons), C_c-values (0.05-0.3) and footing systems (6 in all). The design outputs for a comprehensive evaluation of alternative systems are material quantities (concrete and steel), total cost and cost break-up in terms of those for slab and beam as well as those of concrete and steel. This obviously involves thousands of trial designs which has been accomplished by a computer software developed (flow chart reported in fig. 6) and processed on DEC 1090 system. Since the total cost is to form the basis for identifying the optimal system, it has been computed using the prevailing local prices for M20 grade concrete (Rs. 600/- per cu.m) and HYSD steel (Rs. 5500/- per ton). The rates also include fabrication charges. Fig. 7a reports the variation of q_a (allowable pressure) with C_c (compression index), each curve in it being for one value of total load on the footing inclusive of its self weight. As expected, the allowable pressure decreases with increase in compressibility of soil, the rate of decrease decreasing with increase of C_c-value. The same results are reported in fig. 7b which reports the variation of q_a with total footing load, each curve in it being for one value of C_c. It is interesting to note that q_a decreases with increase in footing load, the rate of decrease decreasing with increase of footing load. The results emphasize the fact that the allowable pressure is influenced both by the compressibility of foundation medium and the total footing load.

Fig. 8a reports the variation of total cost of footing (in thousand rupees) versus C_c-value for a column load of 100 tons, each curve in it being for one footing system. It is interesting to observe that the cost increases quite linearly with C_c since the more compressible the soil, the larger and more expensive is the footing. Fig. 8b reports the variation of cost with column load for one particular value of C_c (=0.10) each curve in it being for one footing system. It is significant to observe the highly nonlinear increase in cost with column load. Although it is realised by researchers that the allowable pressure is a function of the characteristics of the soil deposit and the column load, the design practice still is to assign a value of allowable pressure primarily on the basis of the soil condition. Also there are numerous soil deposits for which allowable pressure is assessed differently from the herein described C_c-based procedure (for instance the plate-load test based procedure for granular deposits). In order to cope with such situations, computer aided designs are carried out in terms of design inputs of allowable pressure and column load treating them as independent variables. Fig. 9a reports the variation of total cost (in units of Rs. 1000/-) with column load for an allowable pressure q_a of 10 tons/m²; each curve in it being for one footing system. The cost evaluation is done by comparing the cost of the alternative footing systems relative to that of the conventional uniform thick footing. The comparison is in terms of either the ratio of the cost of the alternative footing to that of conventional ($F_{x/c}$) or the reduction in cost of alternate footing from that of the conventional footing (F_{c-x}). The variations of $F_{x/c}$ and F_{c-x} with the column load are reported in the inset figures. It is worth noting that the saving in cost increases rapidly with column load. In otherwords, the efficient alternative (uniaxial beam-slab footing of refined design) will cost just about 55% of that of the conventional footing for a column load of 500 tons. Fig. 9b reports the variation of total cost with allowable pressure, each curve in it being for one footing system (full lines being for 100 tons column load and dashed lines for 50 tons column load). As can be expected the cost reduces nonlinearly with an increase in allowable pressure. The relative cost evaluation in terms of $F_{x/c}$ and F_{c-x} are reported in inset figures. Fig.10a reports variation of quantity of concrete (volume in cu.m) with column load for an allowable pressure of 10 tons/m², each curve in it being for one footing system. It is striking to observe the similarity between the variation of concrete quanity with load and the variation of total cost with load from which is inferred that the total cost is primarily affected by the quantity of concrete. As can be expected, the variation of concrete quantity with allowable pressure (q_a) for a column load of 100 tons (fig. 10b) is similar to that of the total cost with q_a. Figs. 11a and 11b report the variation of steel quantity with column load (for a particular allowable pressure) and with allowable pressure (for a particular column load) respectively. Though the nonlinear increase in quantity of steel with load is similar to the cost variation, there is difference in the relative ordering of the footing systems. Whereas the relative cost of the footing systems follows the order CF>UB-S> AB-S> DB-S> TF>UB-S refined, the ordering of the systems with respect to steel quantity is UB-S> AB-S> DB-S>CF>UB-S refined >TF.

The results of beam-slab footing systems are further analysed in terms of the cost component of the beam relative to that of the slab. Fig. 12a reports the variation of the relative beam cost (i.e. cost of beam(s)/cost of slab) with working load and fig. 12b reports the variation of relative cost with allowable pressure. Two bounding curves are reported for each system.It is significant to observe that for the diagonal beams-slab footings the cost of beams is about 200-250% that of the slab and that for axial beams-slab footings is about 70-75%, whereas for the uniaxial beam-slab footing it is a mere 31%. In uniaxial beam-slab footing, both the slab and the beam are designed to resist the same level of bending moment and as such the ratio of their relative cost is an inverse measure of their cost effectiveness (lower cost of a unit to resist the same external moment means higher cost effectiveness). Viewed from this angle the beam is seen to be three times cost effective relative to the slab.

The quantity of concrete besides determining the economic behaviour of the footing system is an essential design input as the slab weight of the footing is additive to the column load to arrive at the plan dimension. Theoretical considerations and a statistical analysis of design outputs yield the following expression for self weight.

$$W_f = \alpha_c \frac{W_c^{1.5}}{q_a} \qquad (7)$$

where W_f is weight of footing, W_c is the working load on column, q_a is allowable pressure, and α_c is an empirical constant.

The ratio of the self weight of footing to the column load will then become

$$\frac{W_f}{W_c} = \alpha_c \frac{\sqrt{W_c}}{q_a} \qquad (8)$$

Variation of the ratio of self weight to the column load with column load (fig. 13a) and that with allowable pressure (fig. 13b) confirm the expected trend of behaviour. Also shown in these figures is the 10% line generally assumed in routine design. The study conclusively demonstrates the limitation of the design practice of assuming the self weight of footing to be 10% of the column load, since the deviations are far too wide (unsafe in some situations and conservative for others).

SUMMARY CONCLUSIONS

The ultimate failure of the conventional uniform thick reinforced concrete footings widely adopted in practice is in punching with little post-peak residual strength and this motivated the development of alternate footing systems that ensure failure in flexure. The present study is devoted to economic evaluation of five alternate footing systems to cover the useful range of design inputs and involves thousands of iterative trial designs. It is accomplished by developing a computer software and processing on DEC 1090 system. The design outputs conclusively confirm the uniaxial beam-slab footing with refined design details (tapering the slab and curtailing the reinforcement) to be the economic choice, the percentage saving in cost relative to that of conventional footing increases with the size of footing as necessiated by larger column load and/or a lower allowable pressure. The cost of uniaxial beam-slab footing with refined design details could be as low as 55% of that of the conventional footing. Since the cost varies as a power of the column load, useful economy is possible by limiting load concentration. For instance 50% saving is a theoretical possibility if one column transmitting 200 tons could be replaced by four independent columns carrying 50 tons each. The results help to quantify the interdependence among (i) column load, compressibility characteristic of foundation soil and allowable pressure, and (ii) column load, allowable pressure and self weight of footing.

In other words these results bring out the gross inadequacies of the design assumptions : (i) the allowable pressure is a function of only soil condition and (ii) the self weight of footing is a constant ratio of column load. It suffices to conclude that rational design judgements are made possible by the effective use of computer-aided design of footing systems.

REFERENCES

IS:456-1978. Indian Standard Code of Practice for Plain and Reinforced Concrete, (Third Revision).

IS:1904-1978. Indian Standard Code of Practice for Structural Safety of Buildings: Shallow Foundations (Second Revision).

Oswald, R.H. (1980). "Universal Compression Index Equation", Journal of Geotechnical Division, ASCE, Vol. 106, No. GT11, Nov. pp 1179-1200.

Rajagopalan Nair, R.P. (1981). "Strength and Structural Behaviour of Alternate Footing Systems", Ph.D Thesis, Indian Institute of Science, Bangalore, India.

Rengaraju, V.R. (1972). "Strength and Structural Behaviour of R.C. Foundations under Concentrated Loads", Ph.D Thesis, Indian Institute of Science, Bangalore, India.

Shyam Prasad, Y. (1981). "Strength and Structural Behaviour of Strip, Isolated and Combined Footings", Ph.D Thesis, Indian Institute of Science, Bangalore, India.

NOTATIONS

a half side of square footing

C_c compression index of soil

D overall depth of slab

D_b overall depth of beam

D_{min} overall minimum depth of footing at edge

d effective depth of slab section

$F_{x/c}$ cost ratio of alternative to conventional footing

F_{c-x} difference in cost of conventional and alternative footing

i ratio of negative to positive yield moment capacity of slab

M yield moment capacity of footing slab per unit width

M_b yield moment capacity of beam per unit width

m_v coefficient of compressibility of soil

m^* optimal value of M_b/M

q_a allowable bearing pressure on soil

W_c working load on column

W_f weight of footing

W^+ ultimate load on footing

α ratio of side of square column to that of square footing

α_c empirical constant

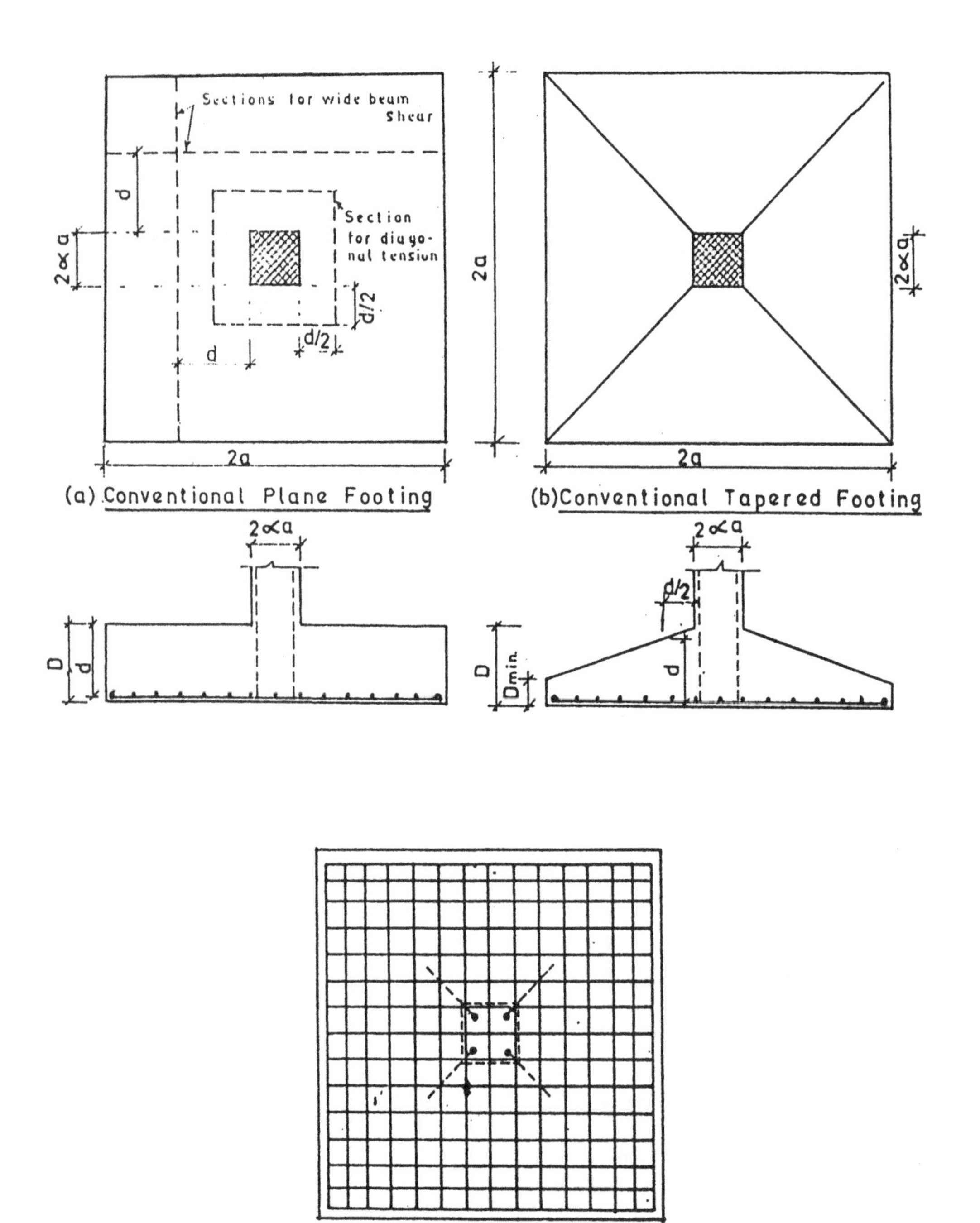

Fig. 1 Conventional and tapered footing

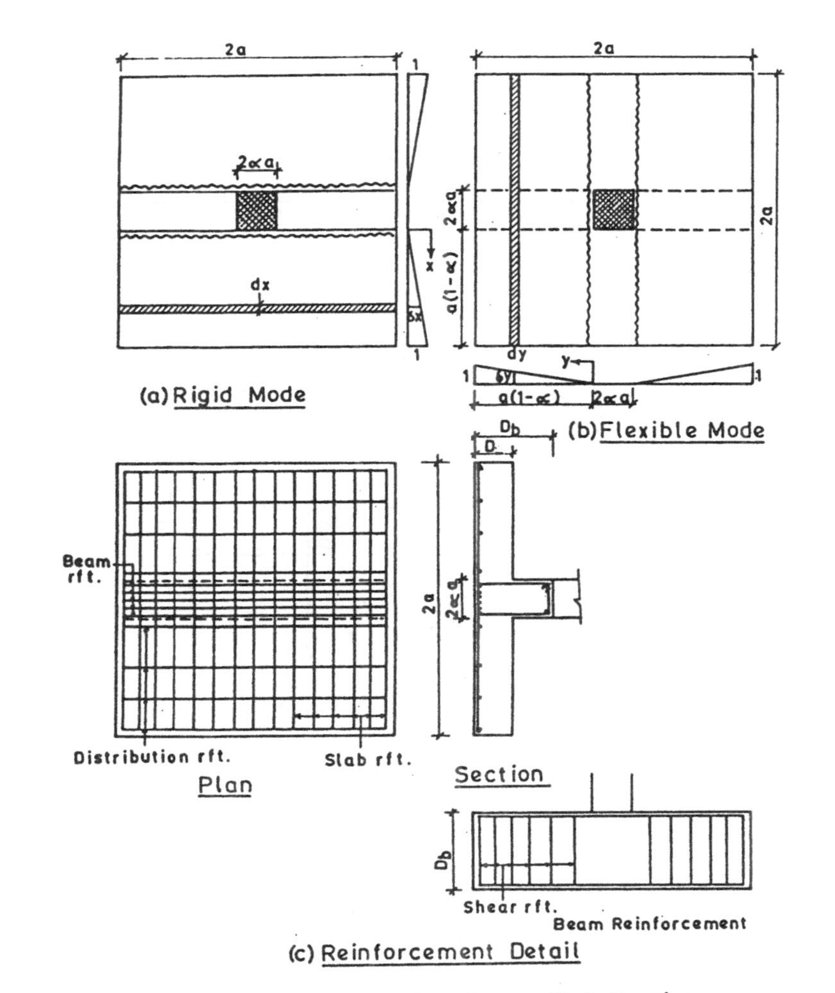

Fig. 2. Uniaxial Beam-Slab Footing

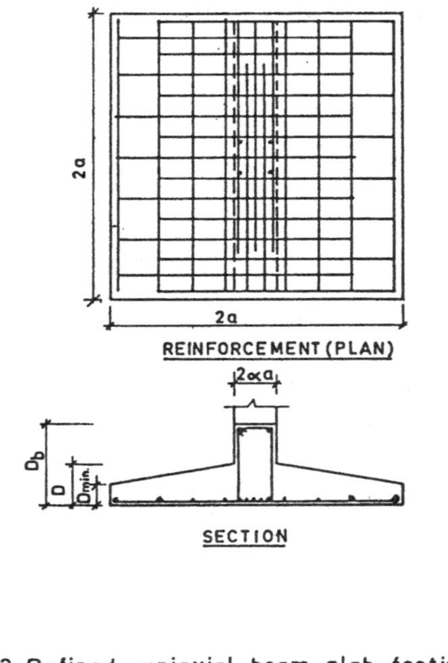

Fig. 3. Refined uniaxial beam-slab footing

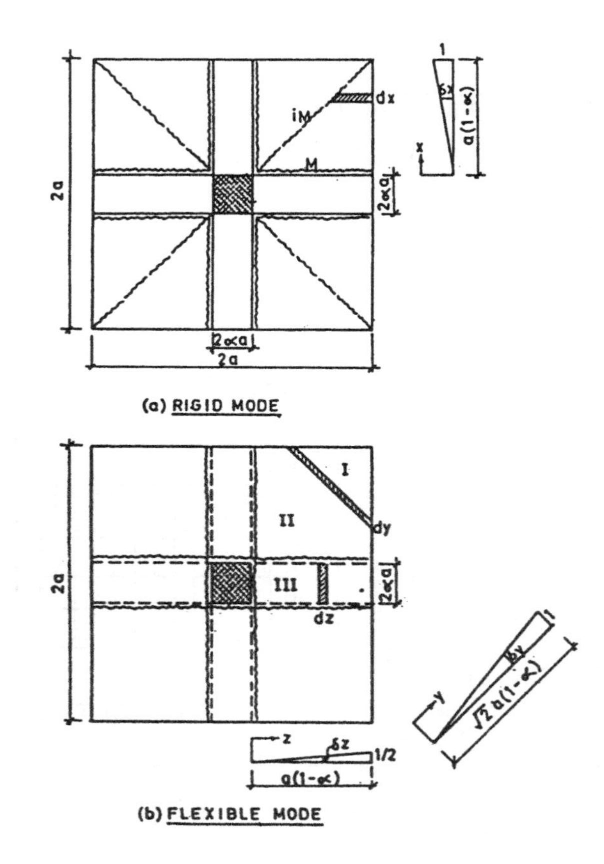

Fig. 4 Axial Beams-Slab Footing

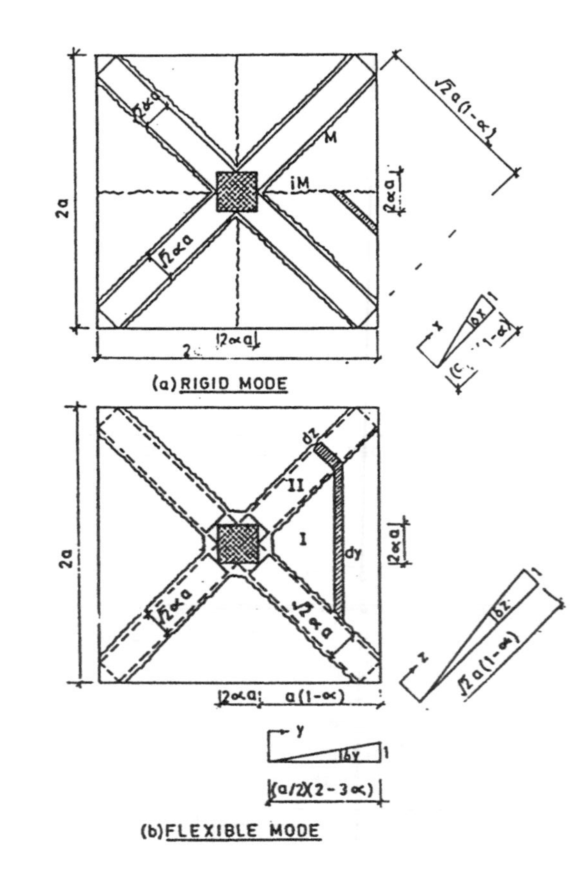

Fig. 5 Diagonal Beams-Slab Footing

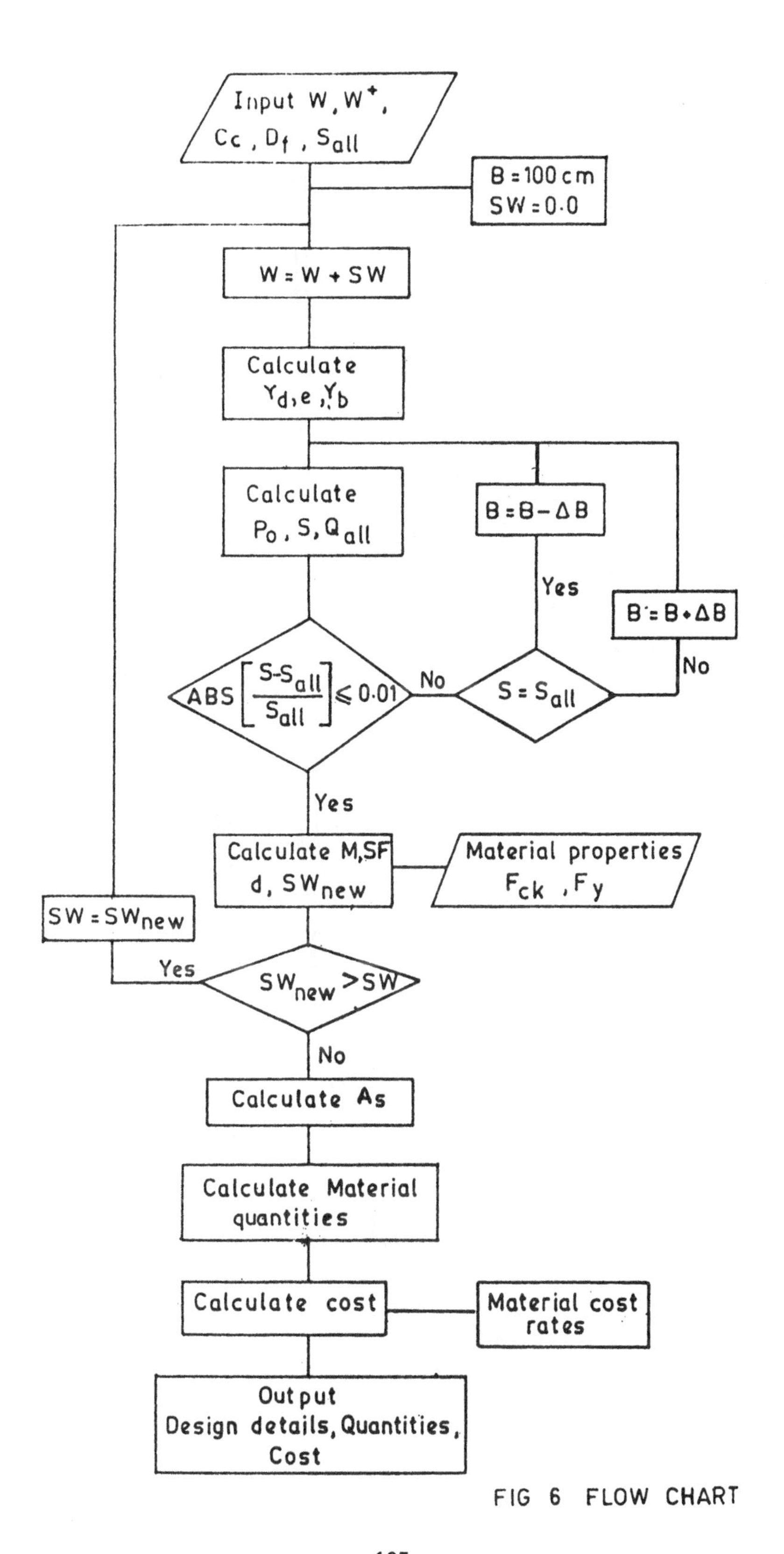

FIG 6 FLOW CHART

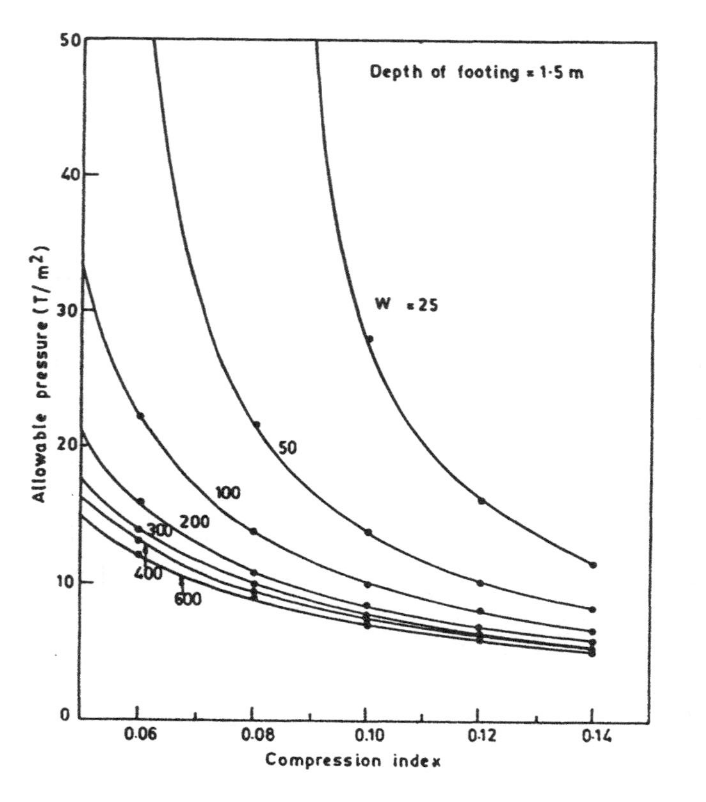

FIG.7a. q_a Vs Cc

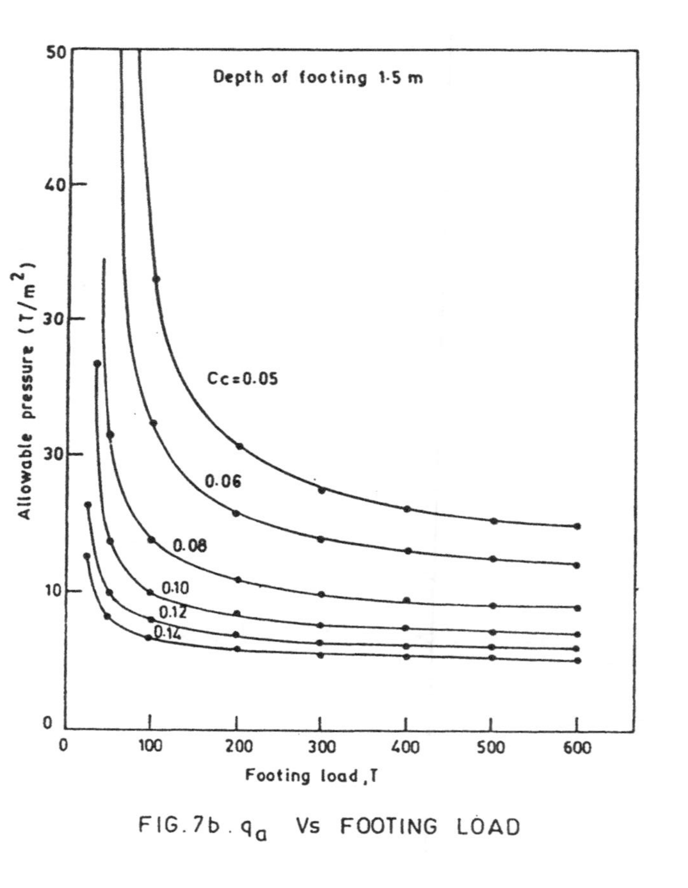

FIG.7b. q_a Vs FOOTING LOAD

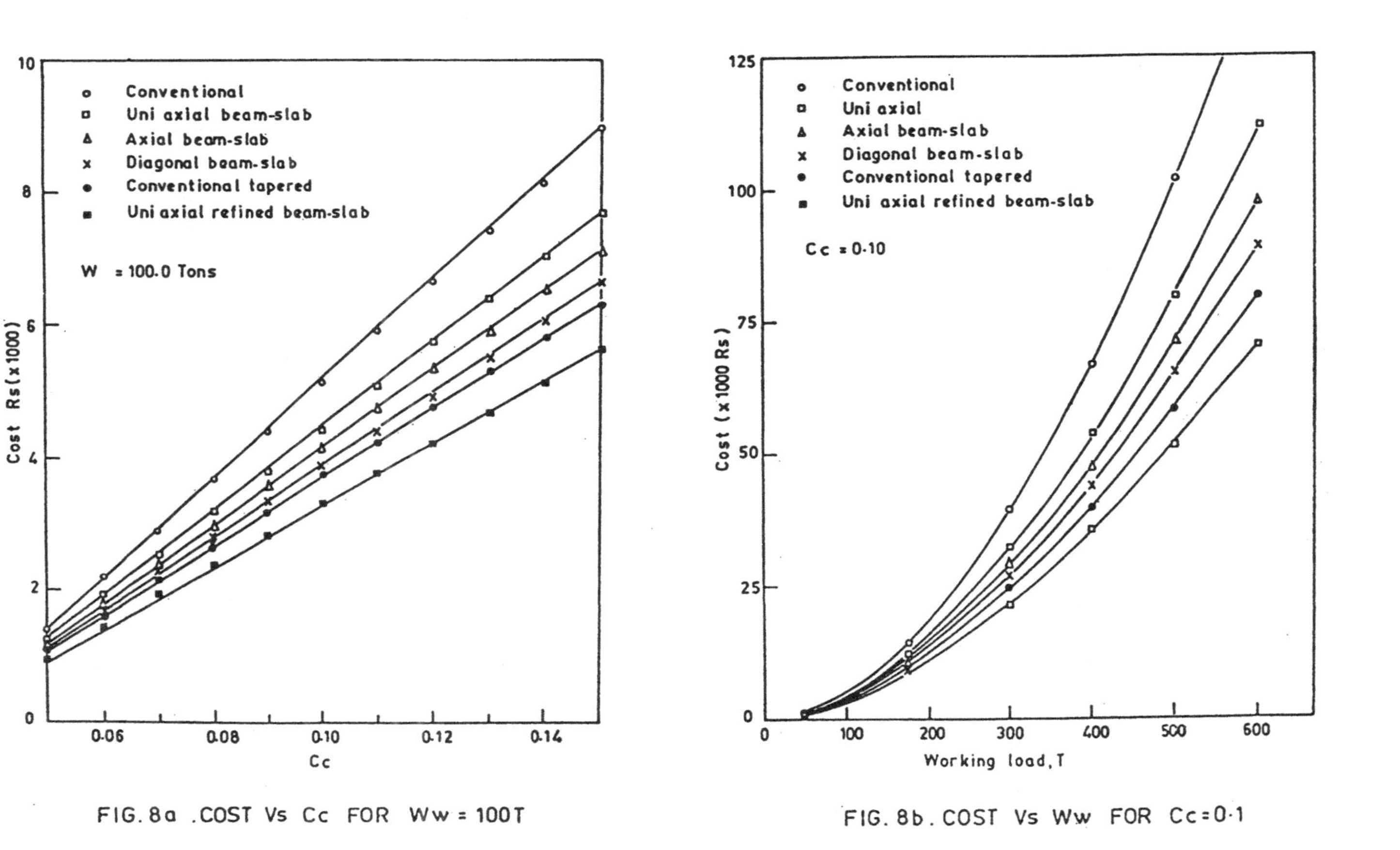

FIG. 8a .COST Vs Cc FOR Ww = 100T

FIG. 8b. COST Vs Ww FOR Cc = 0·1

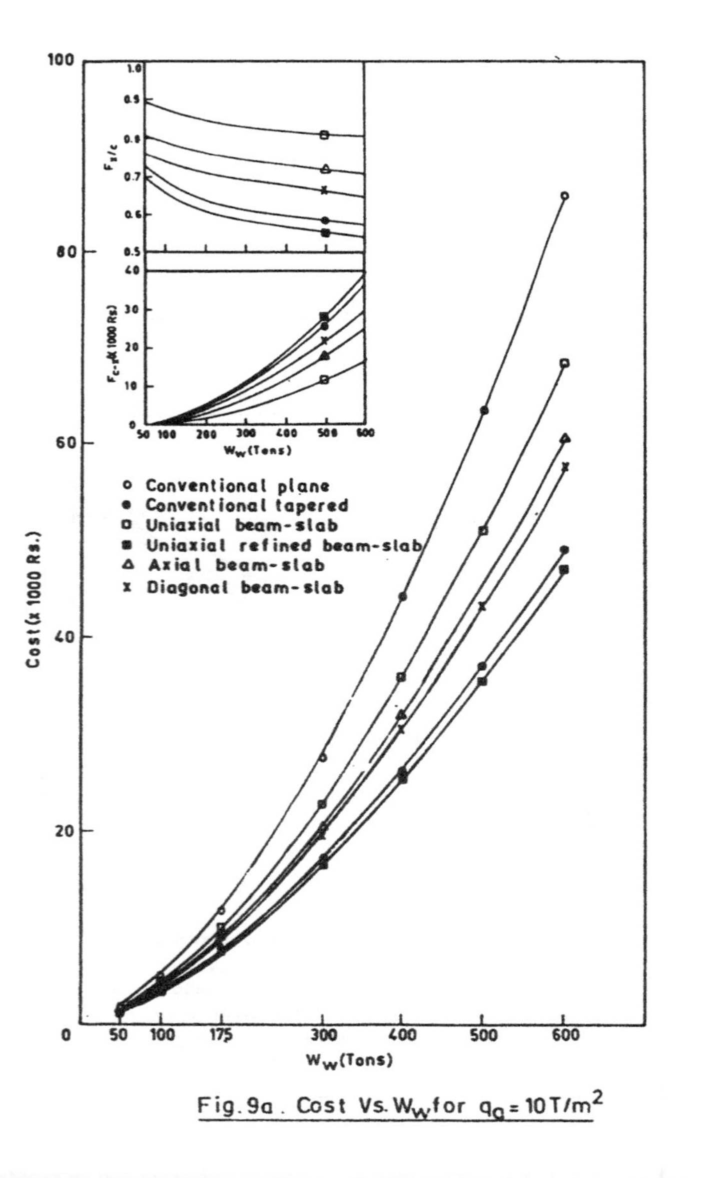

Fig. 9a. Cost Vs. W_W for $q_a = 10 T/m^2$

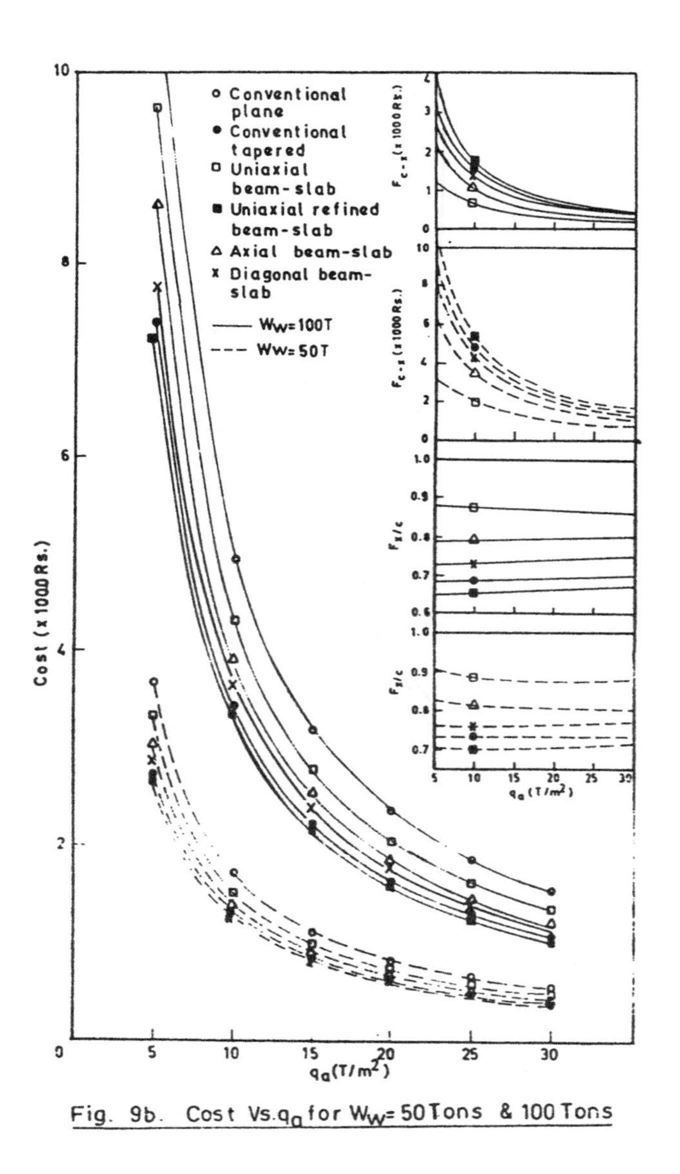

Fig. 9b. Cost Vs. q_a for W_W= 50 Tons & 100 Tons

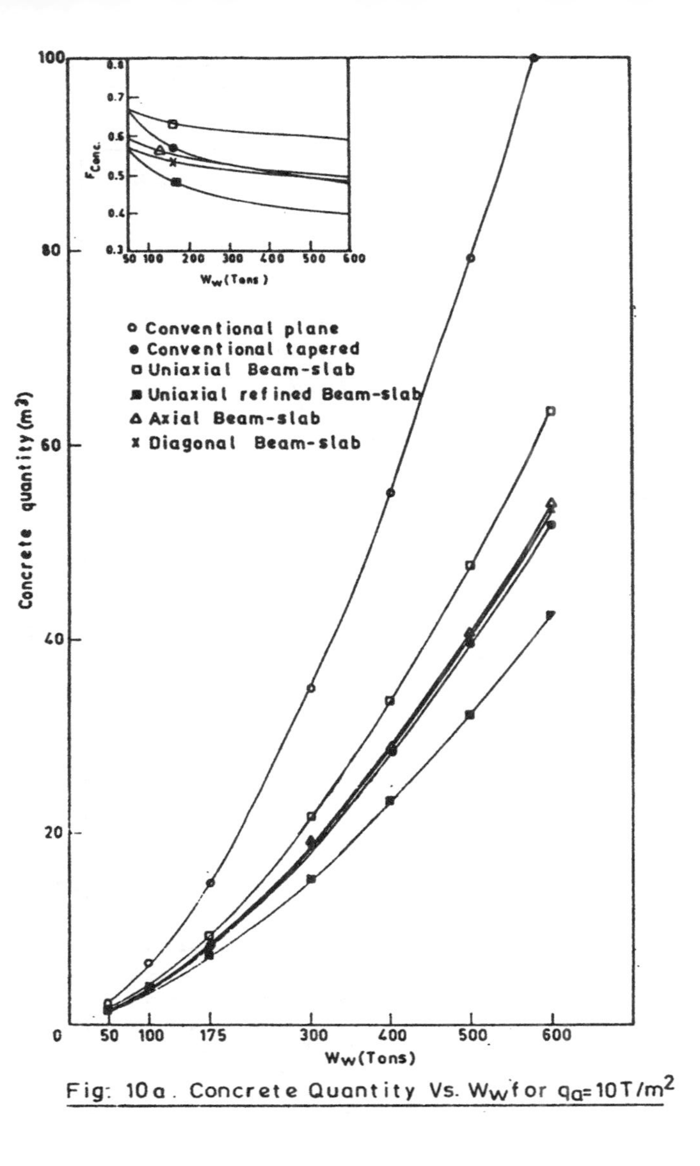

Fig: 10a. Concrete Quantity Vs. W_w for q_a= 10 T/m^2

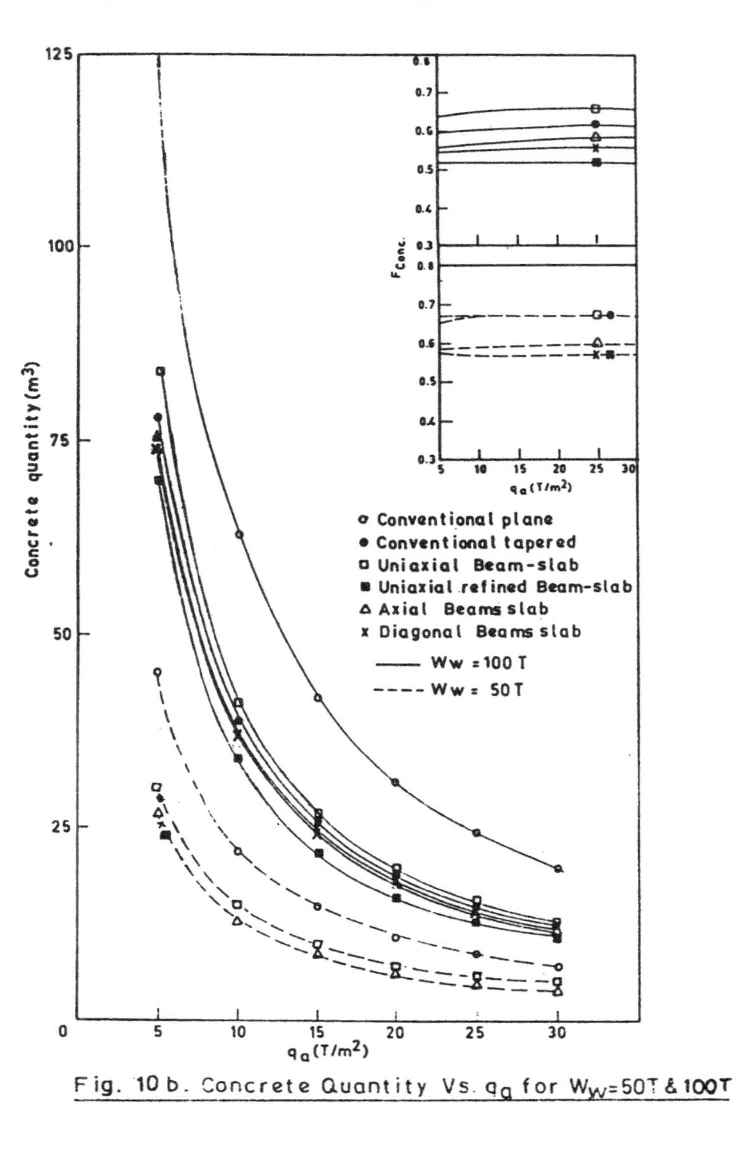

Fig. 10b. Concrete Quantity Vs. q_a for W_w= 50T & 100T

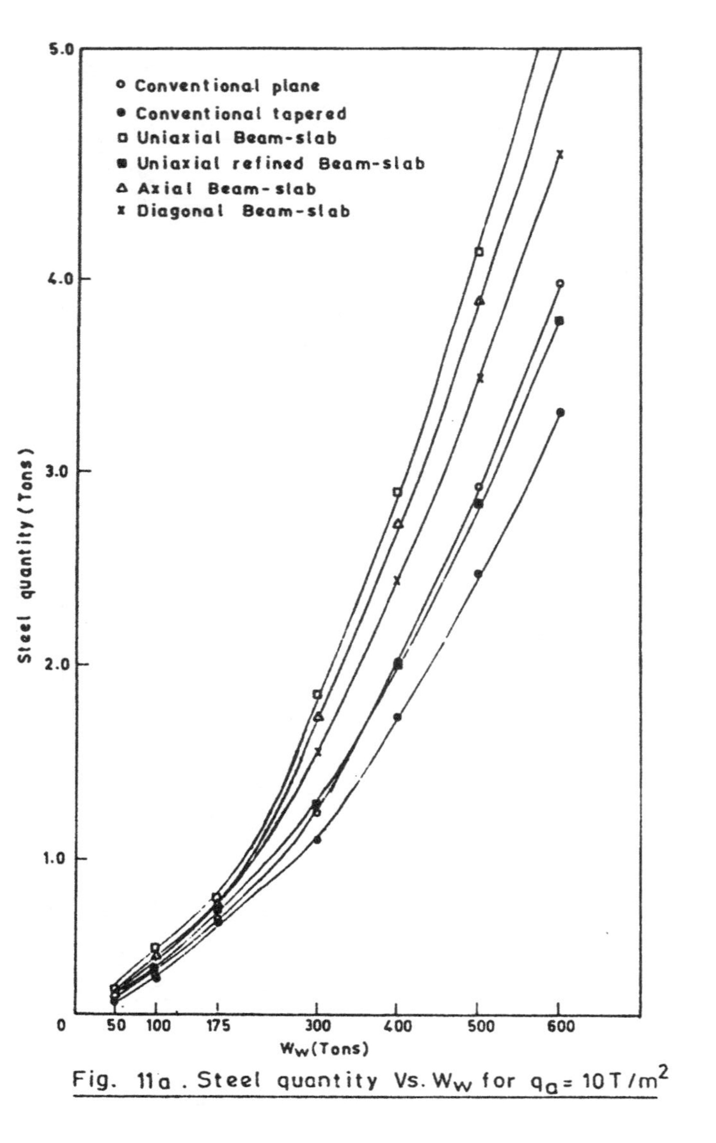

Fig. 11a . Steel quantity Vs. W_w for $q_a = 10\,T/m^2$

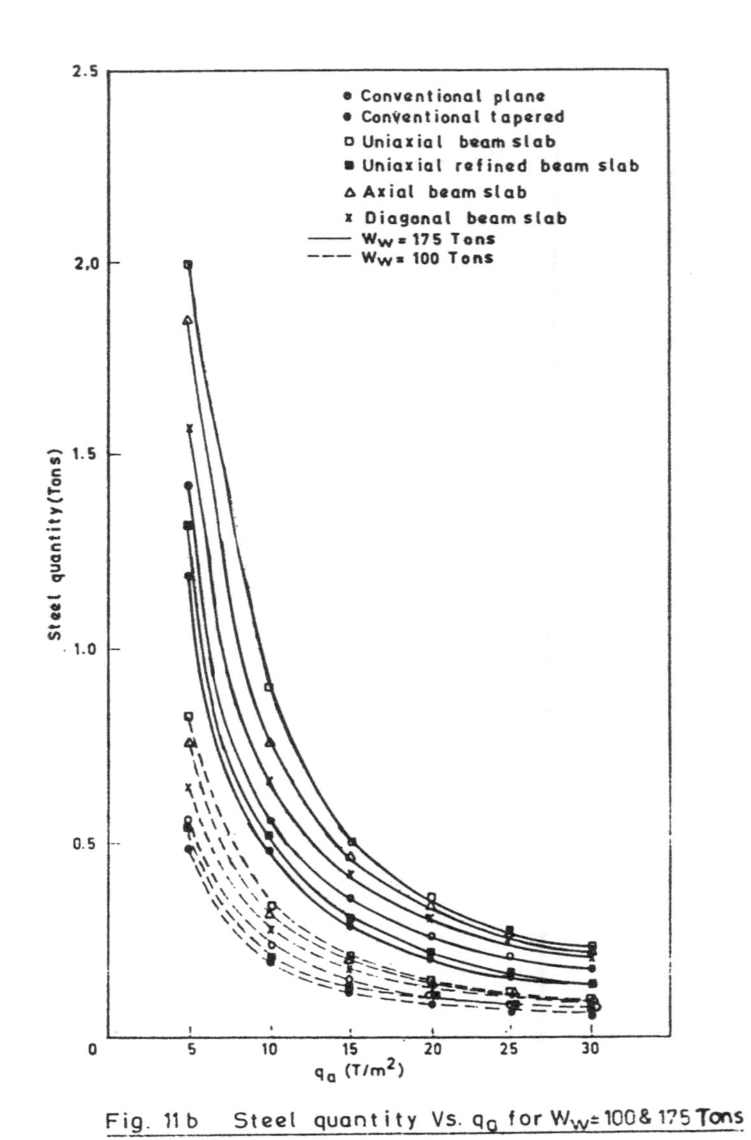

Fig. 11b Steel quantity Vs. q_a for $W_w = 100$ & 175 Tons

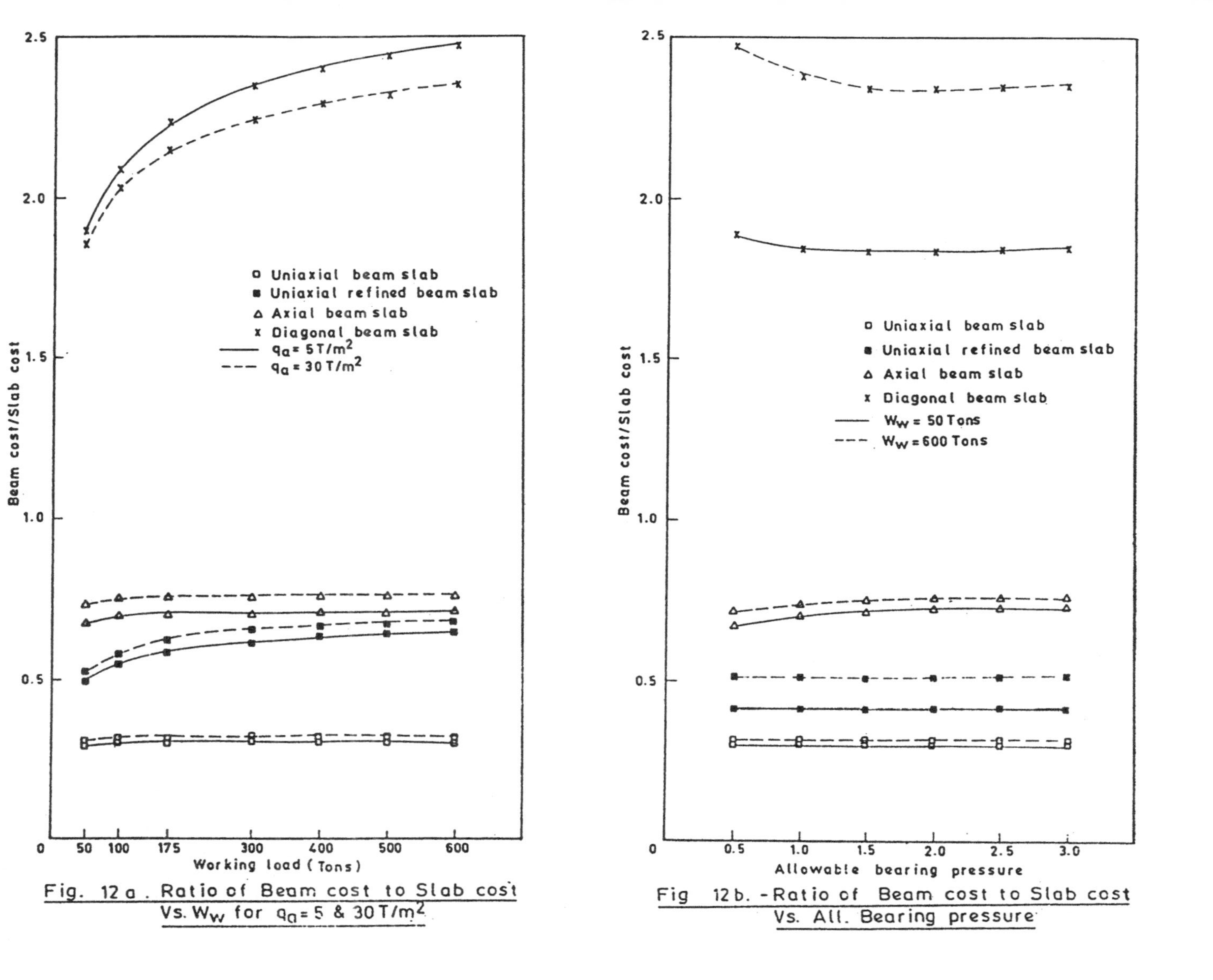

Fig. 12 a . Ratio of Beam cost to Slab cost Vs. W_W for $q_a = 5$ & $30\ T/m^2$

Fig 12 b. - Ratio of Beam cost to Slab cost Vs. All. Bearing pressure

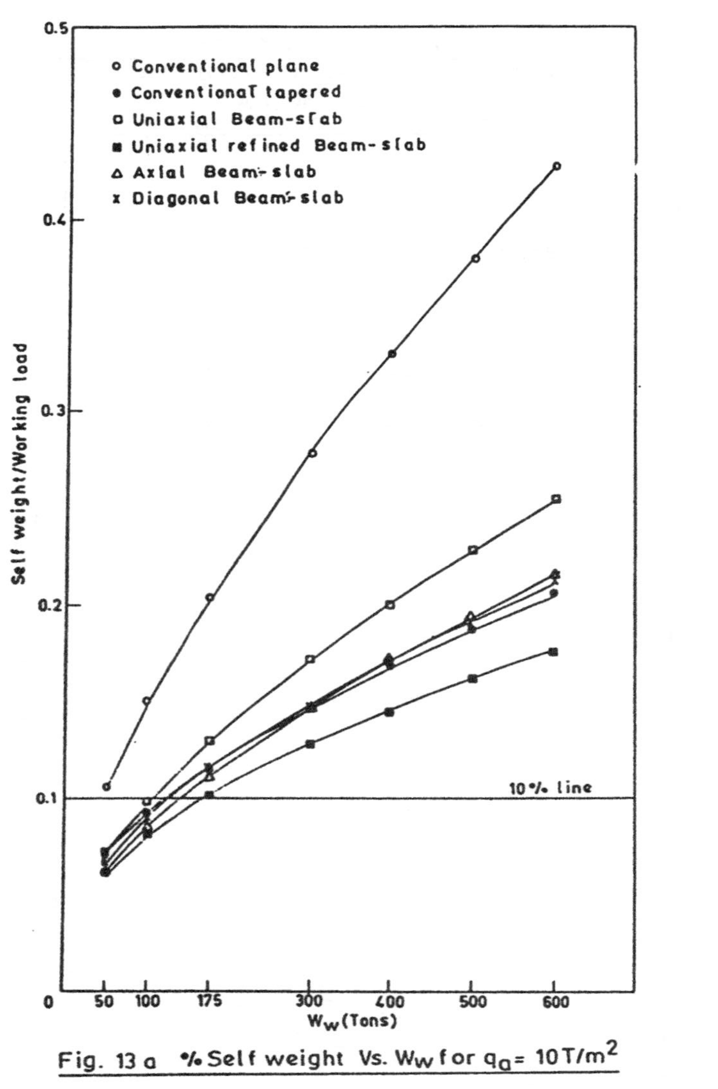

Fig. 13 a % Self weight Vs. W_w for $q_a = 10 T/m^2$

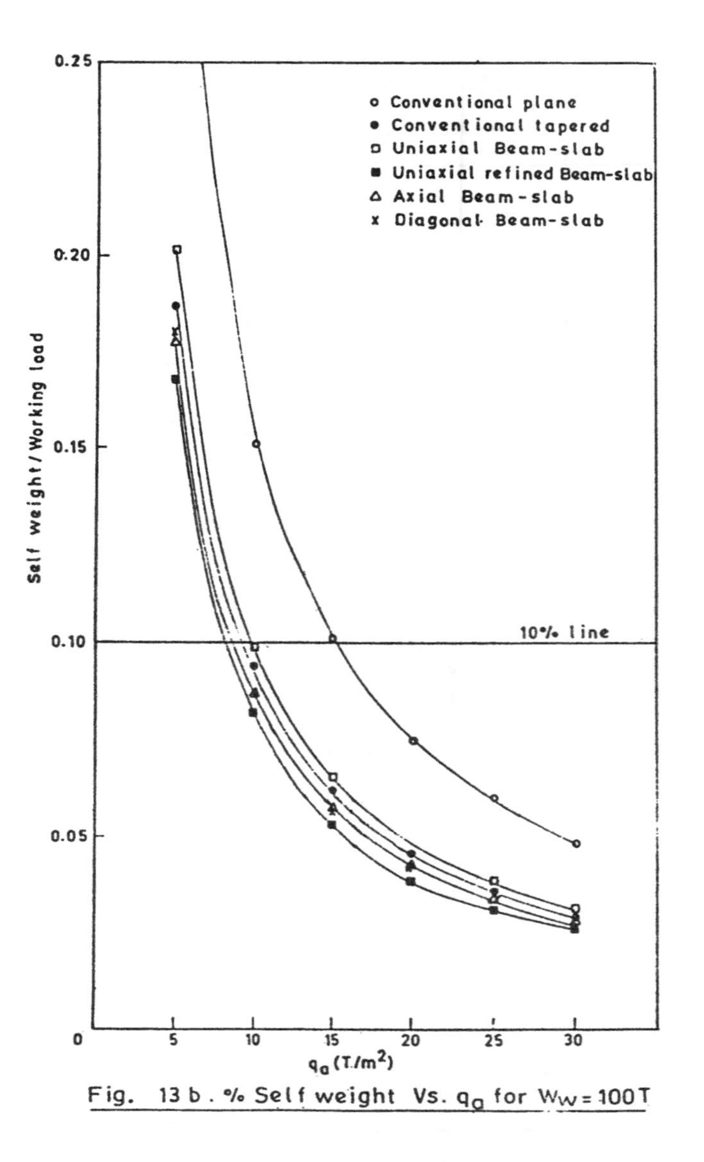

Fig. 13 b. % Self weight Vs. q_a for W_w = 100T

Computer and Physical Modelling in Geotechnical Engineering, Balasubramaniam et al. (eds)
© 1989 Balkema, Rotterdam. ISBN 90 6191 864 2

Analysis of the problems of impact in soil masses using centrifuge tests

P.Lepert & J.F.Corté
Laboratoire Central des Ponts et Chaussées, Nantes, France

ABSTRACT : This paper presents the results of two series of experiments dealing with centrifuge modeling of projectile penetration in soil media. This approach was limited to impacts of large rigid bodies (50 to 500 tons) with low velocity (around 30 m/s), with which are mostly concerned the civil engineering applications. Tests were performed at a 100 g-level on 1/100 scaled models of granular and coherent soils arranged in homogeneous or multi-layers sites. The projectiles as well as the soil masses were instrumented to provide detailed data for subsequent analysis. Apart from the concrete conclusions which were drawn from the experiments, further processing of the results emphasized on the importance of the energy radiated in the soil during the impact. A simple hyperbolic model has been sucessfully used to fit the projectile deceleration versus velocity curves from the different tests. This empirical model appeared to be also suitable to describe the penetration of projectiles in multi-layered sites.

I - Introduction

It is only relatively recently in the long history of soil mechanics that there has been a degree of interest in investigating the impacts of projectiles on soil media. And even here we must distinguish two categories within this family of problems. Military applications amongst others are usually relevant of the first category, whereas the other, as yet hardly explored, is more concerned with some special civil engineering problems. The main difference between the two categories lies in the conditions of impact : high velocity and limited mass in the case of a weapon, low velocity and possibly a very large mass in the case of civil engineering applications. The approach to the problem differs according to category.
For thirty years, investigations of the penetration of projectiles in soil media have been aimed almost solely at the construction of simple theoretical models (Fuchs, 1963) in which the deceleration of the projectile in the target is described by linear or parabolic function of velocity, the coefficients of which are more or less succesfully fitted to the geomechanical parameters of the target. Various experiments, almost always full scale, have been carried out to achieve this fit.

For example, after Chisholm (1962)

$$(1)\quad \frac{dv}{dt} = bv + c$$

after Allen, Mayfield, and Morrison (1957)

$$(2a)\quad \frac{dv}{dt} = av^2 \text{ pour } V_0 > v > V_c$$

$$(2b)\quad \frac{dv}{dt} = bv^2 + c \text{ pour } V_c > v > o$$

and after Hakala (1965) :

$$(3a)\quad \frac{dv}{dt} = av^2 + cx + d \text{ pour } V_0 > v > V_c$$

$$(3b)\quad \frac{dv}{dt} = bv + cx + d \text{ pour } V_0 > v > o$$

where a, b, and c are coefficients, x is the penetration of the projectile at time t, v, its velocity at time t, Vo its initial velocity, and Vc a critical velocity.
Examining the same field, Holsapple and Schmidt (1982) began by considering explosion craters and attempted to extend the results to the analysis of impact-produced craters.

More recently, considering only impact velocities less than 10 m/s, Wang (1971) has approached the problem by formulating an energy balance in the context of a quasi-static approach. The results obtained seem to be borne out by tests conducted at the same time. It should however be noted that these tests involved impacts produced with very small masses that induced rather low stress levels and were therefore not very representative of civil engineering applications.

The experiments carried out by the authors of this article led them to believe that a quasi-static approach was unsuited to this second category of applications. This position will be justified below through the analysis of two actual cases, dealt with in an original manner. The approach taken is based on an elementary observation : the conditions most favourable to the experimental analysis of a phenomenon are satisfied when the phenomenon can be reproduced through a series of tests in a realistic setting, fully and correctly instrumented. This would seem offhand to be inconceivable for civil engineering problems involving projectiles ranging from a few tens to a few hundreds of tons and energies that often exceed a hundred million joules. But experiments on a small-scale models in artificial gravity can get around this difficulty without impairing the validity of the method.

II - Centrifuge experiments

IIa : Principle

The rheology of geomaterials is typically nonlinear. This means that the behaviour of a soil depends in particular on the level of the loadings to which it is subject. Accordingly an experiment on a small-scale model is representative only if it reproduces, at all points in the model, the stress or strain fields found in the full-scale structure. In most cases, this condition can be met by subjecting the small-scale model to artificial gravity, i.e., by carrying out the experiment in a centrifuge.
Thus, when mass-related forces are preponderant, a model to 1/Nth scale must be subjected to a centrifugal acceleration N times as great as natural gravity. The reduction factors to be applied to the other values are deduced from the equations of continuum mechanics. They are listed in table 1.
The materials used in making the model must, strictly speaking, have the same mechanical behaviour as those used in the structure being modelled. As a general rule, then, the modél and the actual structure are made of the same material.

IIb : Centrifuging practice

The experiments described in the balance of this article were carried out using the centrifuge installed at the Laboratoire Central des Ponts et Chaussées at Nantes, France, in 1985. This device (figure 1) consists primarily of a central vertical shaft driving an asymmetrical rotor with a pivoting basket on one end and counterweights on the other (Corté and al, 1986).

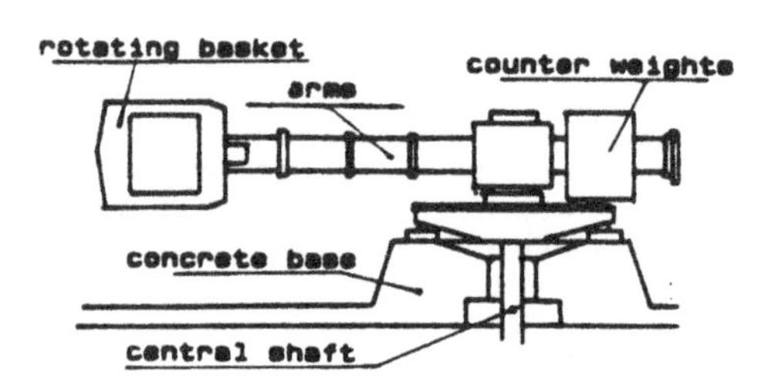

Figure 1. The LCPC's centrifuge

The model tested is built in a container placed on the tray of the basket, which is 5.50 metres from the axis and can therefore, at a speed of 2.1 rps, subject a 2,000-kg model to a mean acceleration of 100g. If the model does not exceed 500 kg, it can even be accelerated to 200 g.
This machine and its environment, designed to provide optimum conditions for experiments in soil mechanics, were therefore suitable for the two experiments described below.

III - Modelling the impact of a rock against an earthfill

On the 1st of May, 1977 a large rock fall occured in the French Alps. When plans were made in 1985 to Build an expressway at the same location, some way had to be found of protecting it from accidents of this type. A geological study was carried out to characterize the risk, i.e., the origin, type, and size of rocks that might be expected to fall. Their trajectory was then investigated using a numerical model of the site. These calculations finally led to a preliminary design (position, height, size) for an earthfill required to stop rocks not exceeding 200 cubic metres, striking it at velocities close to 30 metres per second.

In the absence of an appropriate design method, it was decided that experiments on a small-scale model were the only practical way of testing the effectiveness of this structure.

IIIa : Choice of similarity factor and model

The characteristics of the structure to be modeled are shown in figure 2. It is built on a slope inclined about 30 %. Given its dimensions, and the space available in the basket of the centrifuge, a scale of 1/100th was judged appropriate. This meant (cf. § IIa) that the experiment would be conducted at an artificial gravity of 100 g. Refer to table 1 for the similarity factors to be applied to the other mechanical values under these conditions.

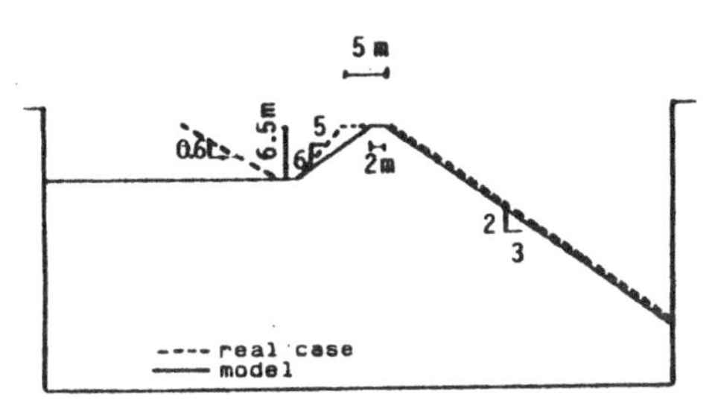

Figure 2. Cross-section and model of site investigated in the first example

Since the material of which the actual structure as to be made had not yet been chosen at the time of the scale-model study, several experiments were carried out on models made of a number materials (coarse or fine sands) having different densities. The slope of the mountainside uphill of the earthfill was not modeled.

IIIb : Experimental arrangements

Free fall would have been the simplest way to achieve the required impact velocity, but could not produce the correct angle of impact. The projectile was accordingly guided in its fall practically to the point of impact. The shape of this projectile was, naturally, one of the unknowns of the problem. It was accordingly decided to use two projectiles, one a hemisphere and the other a cube. Both had a mass of 0.58 kg and a density close to that of the rock.

In practice, the projectile was attached to the top of a curved rail that served to guide it (figure 3). At 100 g, a pneumatic system operated from the control room released it, and it then struck the model.

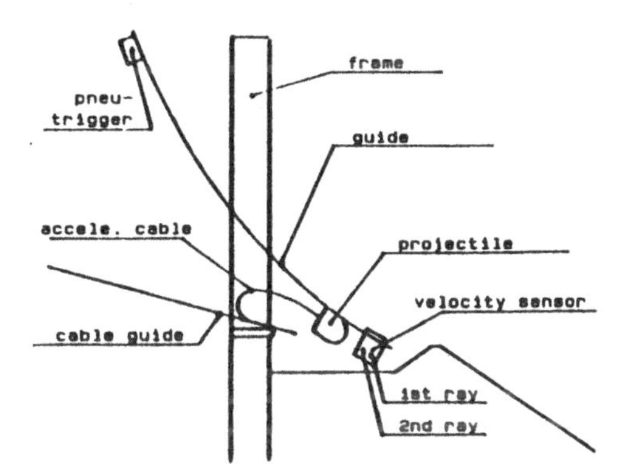

Figure 3. Device for dropping the projectile on the earthwork in the centrifuge

IIIc : Measurements and observations

At the end of its fall, the projectile passed in turn through two infrared beams 30 mm apart, making it possible to check its velocity just before impact. An accelerometer screwed into the body of the projectile, parallel to its centreline, measured its deceleration as it drove into the earthfill. Three accelerometers buried in the slope, below the earthfill, detected the waves engendered by the impact in the soil.

After the machine was stopped, the depth of penetration of the projectile and its final inclination were recorded.

IIId : Processing of measurements

Before any interpretation, the measurements made during the test were smoothed and the zero offsets and electrical drifts of the measurement channels were corrected.

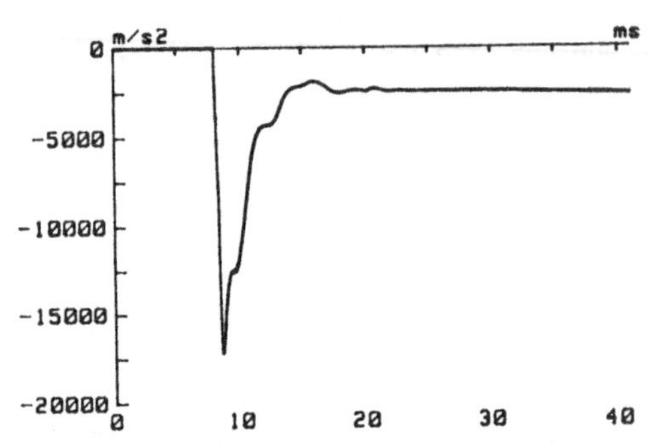

Figure 4. Deceleration of the projectile as it penetrates the earthwork model.

The deceleration signal from the projectile (figure 4) was integrated to give its velocity curve, which was fitted back to the initial value measured by the infrared beams. This velocity was integrated in turn and a similar adjustment was applied to the penetration curve, using the final value recorded after the centrifuge had stopped.
The same processing could be applied to the acceleration curves measured in the soil, since the particle velocity before and after the passage of the shock generated by the impact was zero. Assuming this shock to consist of locally plane waves, formulae (4) and (5) could be used to determine the associated normal stress and the energy conveyed :

(4) $\sigma = Zv$

where σ : normal stress
Z : impedance of soil
v : particle velocity

(5) $E = \int_0^t z\, v^2\, dt$

E : energy conveyed

IIIe : Analysis of results

Several tests carried out on smaller and smaller models showed that the size of the planned structure had to be halved for the projectile to pass completely through it, and even in this case it was stopped immediately after passing through the earthfill.

A more precise analysis of the measurements shows that the stage of deceleration of the projectile, referred to the full-scale situation, lasts about one second (figure 5). The braking force applied to the rock shows a pronounced peak in the first hundred milliseconds. The maximum value of this peak ranges from 70 to 200 MN depending on the type of material of which the model is made and the shape of the rock. Figures 5a to 5d make it possible to compare several deceleration versus velocity curves for the projectile. All of the curves obtained with the hemispherical projectile have the same characteristic shape : from an initial velocity close to 26 m/s, there is first a rapid rise of decele-ration, which goes through a maximum and then decreases in a roughly linear fashion with velocity. The second peak observed during this stage, highly ate-nuated but significant, is perhaps related to a rotation of the projectile during its penetration. On the other hand, the curve obtained with the cubical projectile is quite different, since it includes a rather long plateau after the maximum.

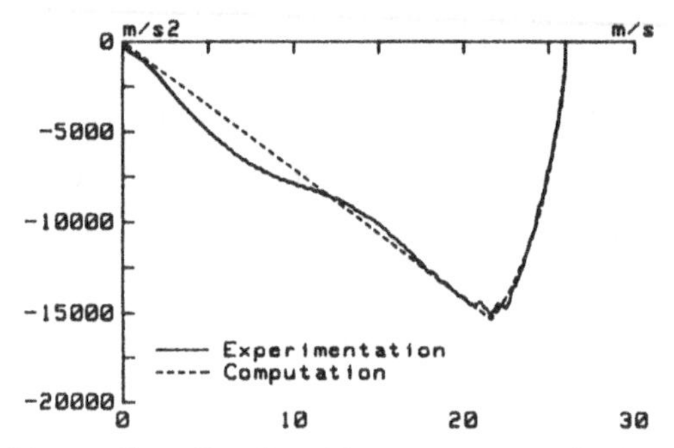

Figure 5a. Deceleration versus velocity for a hemispherical projectile penetrating an earthwork of dense Fontainebleau sand

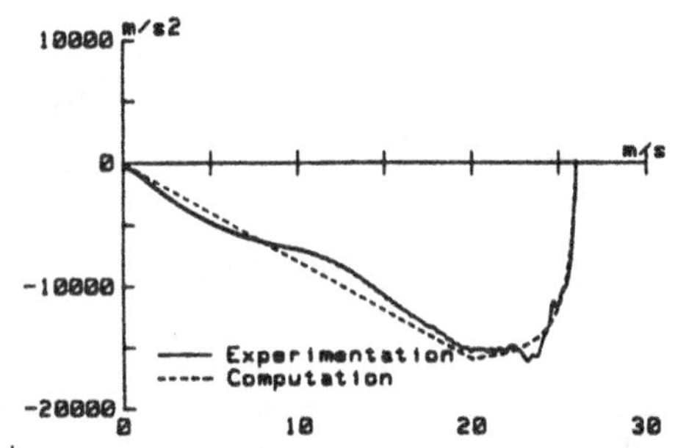

Figure 5b. Deceleration versus velocity for a hemispherical projectile penetrating an earthwork of loose Fontainebleau sand.

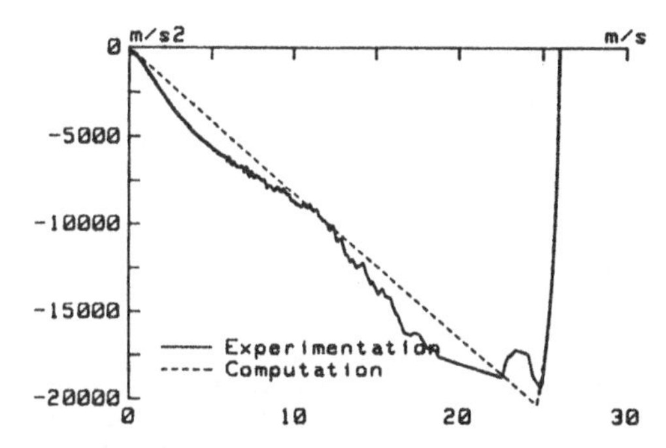

Figure 5c. Deceleration versus velocity for a hemispherical projectile penetrating an earthwork of loose Loire sand.

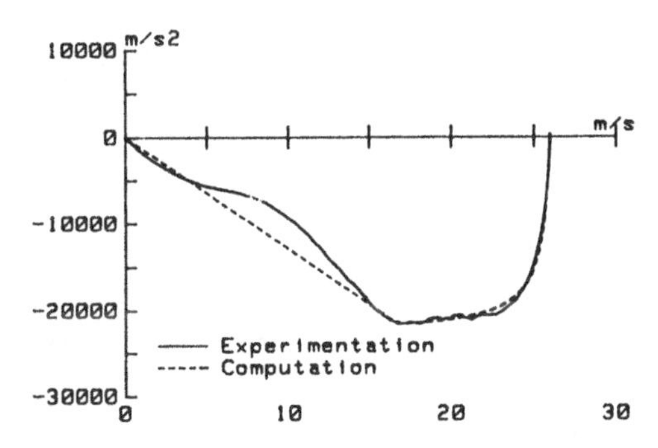

Figure 5d. Deceleration versus velocity for a cubical projectile penetrating an earthwork of loose Fontainebleau sand.

The following hyperbolic model seemed to be the best for describing this movement of the projectile under all the conditions investigated :

$$(6a)\quad \frac{dv}{dt} = a + \frac{b}{c-v} \qquad \text{pour } V_o > v > V_c$$

$$(6b)\quad \frac{dv}{dt} = \left(a + \frac{b}{c-V_c}\right)\frac{v}{V_c} \qquad \text{pour } V_c > v > 0$$

The quality of the fits obtained may be judged on figures 6a to 6d. Moreover, integration of these equations, in the various cases studied, yielded a computed final penetration of the projectile that was rather close to the measured value (difference less than 20 %). Parameter "c" has the dimension of a velocity. Parameters "a" and "b", like characteristic velocity Vc, depend on the type of soil and the shape of the projectile. Other tests will be needed to learn more about these relationships.
Equation (6a) and (6b) are close to the parabolic model proposed by Hakala (cf. equations 3a and 3b) but are simpler to process numerically because they do not explicitly refer to the peneration of the projectile.
It is also observed that, beyond a close in zone characterized by large plastic deformations, the impact engenders a shock consisting of elastic waves. Knowledge of the particle velocities at different points in the soil makes it possible to determine, among other things, the energy this wave conveys (equations 1 and 2). It decreases with increasing distance of the wave from its source, both because of geometrical attenuation and because of the internal damping of the material (figure 6). Thus, 25 metres from the point of impact, in other words, outside the zone of large deformations, but close enough to the source for the wave to be treated as a spherical one, the residual energy is of the order of 25,000 Joules, or about 30 % of the kinetic energy of the projectile at the moment of impact.

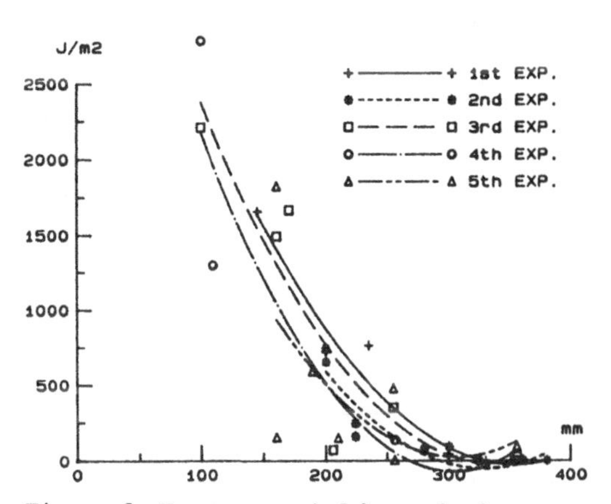

Figure 6. Energy carried by a shock versus distance from source

This last finding leads to the conclusion that a quasi-static analysis that neglects the phenomenon of the radiation of energy is unsuited to deal with the problem. Accordingly, in a first stage, an empirical approach based on experimental considerations, such as the one proposed above, seems preferable.

IIIf : Assessment of the first experiment

In practice, this experiment confirmed that the planned earthfill would be effective against the rock falls anticipated.
From the theoretical standpoint, it demonstrated the need to take the large amount of energy radiated in the soil into account in modelling impacts on soil media.

A simple hyperbolic model seems to be able to describe the movement of the projectile in the soil media.

IV - Impact of a rigid body in a lake site

The next example concerns an object falling on a lake site. The main focus was the waves engendered by the impact and their repercussions on the environment.
The projectile, a cylinder 9 m long and 2.5 m in diameter having a mass of approximately 50 tons fell freely and nearly vertically from a height of 60 metres into a shallow pond. A retaining structure was to be built 15 m from the presumed point of impact. The practical aim of the study was therefore to determine the design method suited to this structure and dynamic loadings to be taken into account.

IVa : Modelling

Two configurations of the site were modelled. The first represented the situation before dredging of the pond : a layer of clay 2.5 m thick overlying a sandstone mass that was weathered near the top, covered by 4.0 metres of water. After dredging, the layer of clay was replaced by an equivalent depth of water.
A 1/100 th scale model (cf. table 1) was built for each of these two configurations. In the first case, a reconstituted kaolinite, compacted to a value close to the Proctor optimum, was used to represent the layer of clay, while a dense sand was used for the substratum. The "sandstone" of the second model was a block of sand treated with 4 % cement.
The projectile was represented by a simple PVC tube 9 cm long and 2.5 cm in diameter, weighing 52 grams. A device similar to the one described in the preceding section served to ensure the desired conditions of impact. It should be noted that in this case the fall of the projectile was not guided, since the Coriolis force naturally produced the desired inclination at the moment of impact. The velocity of 30m/s was reached after a fall of 600 mm.

IVb : Acquisition and processing of measurements

The instrumentation of the projectile and site were the same as in the falling rock study. Altogether, four accelerometers were used, one in the projectile and the others in the soil at various distances from the point of impact, oriented either horizontally or vertically. A special transducer was developped to measure the height of the waves caused by the projectile 's passage throught the water.
The measurements were acquired and processed as in the first example. The observations after the centrifuge had stopped included recording the inclination and depth of peneration of the projectile in the soil constituting the bottom of the model.

IVc : Analysis of results

On the first model, it was observed that the projectile completely perforated the layer of clay and penetrated about 2 metres into the underlying soil. On the other hand, it rebounded from the cement-treated sand of the second model and merely chipped it.
An attempt was made to apply the simple hyperbolic model defined by equations (6a) and (6b) to the deceleration versus velocity curves of the projectiles in the different layers of material. In the case of penetration in the first model, figure 7 clearly shows two successive peaks, representing the impact with the water, then with the clay.

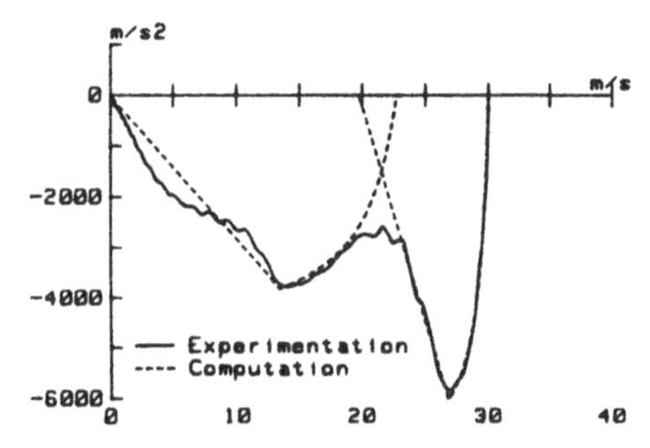

Figure 7. Deceleration versus velocity for a cylindrical projectile penetrating the water/clay/sand model

This curve can in fact be described as the sum of two simple hyperbolic models as defined in § III. The first faithfully reproduces the entry of the projectile into the water, the second its penetration into the clayed layer.

$$(7) \quad \frac{dv}{dt} = \left(\frac{dv}{dt}\right)_1 + \left(\frac{dv}{dt}\right)_2$$

with, for the layer of water :

$$(8a) \quad (\frac{dv}{dt})_1 = a_1 + \frac{b_1}{c_1 - v} \qquad \text{pour } V_{01} > v > V_{c1}$$

$$(8b) \quad (\frac{dv}{dt})_1 = (a_1 + \frac{b_1}{c_1 - V_{c1}})(\frac{v - V_{m1}}{V_{c1} - V_{m1}}) \quad \text{pour } V_{c1} > v > V_{m1}$$

and, for the layer of clay :

$$(9a) \quad (\frac{dv}{dt})_2 = a_2 + \frac{b_2}{c_2 - v} \qquad \text{pour } V_{02} > v > V_{c2}$$

$$(9b) \quad (\frac{dv}{dt})_2 = (a_2 + \frac{b_2}{c_2 - V_{c2}}) \frac{v}{V_{c2}} \qquad \text{pour } V_{c2} > v > 0$$

This analysis was not extended to the final penetration into the substratum, because of the small energy involved. Figure 8 compares the measured and modelled deceleration signals.
In the case of penetration of the second model, it will be noted in figure 9 that entry into the water is correctly modelled, but that the hyperbolic model does not fit the interaction with the underlying cement-treated sand. This agrees with the finding that the projectile rebounded from this material.
When the kinetic energy of the projectile was determined as a function of its depth of penetration in the first model, it was found that 25 to 30 % of the total energy was dissipated in the water and radiated as waves. The balance was shared by the clay (35 to 40 %) and the subs-tratum (30 to 35 %). In the second model, the deeper water absorbed 40 to 50 % of the initial energy and the rest was transmitted to the substratum.
The maximum values of the acceleration curves in the soil were recorded. These values are shown in figures 10 and 11 as a function of distance from the point of

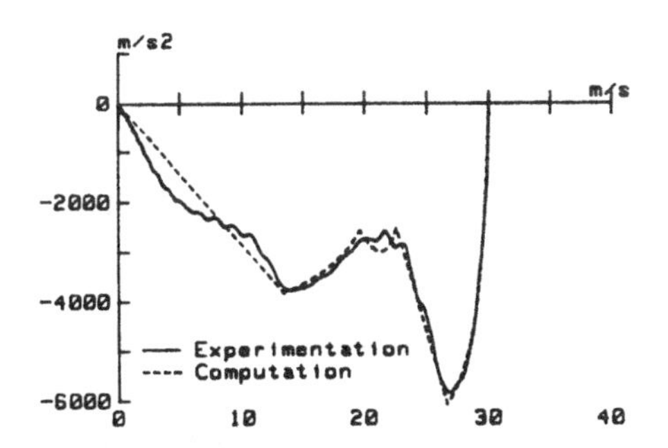

Figure 8. Model of movement of cylindrical projectile penetrating the water/clay/sand model

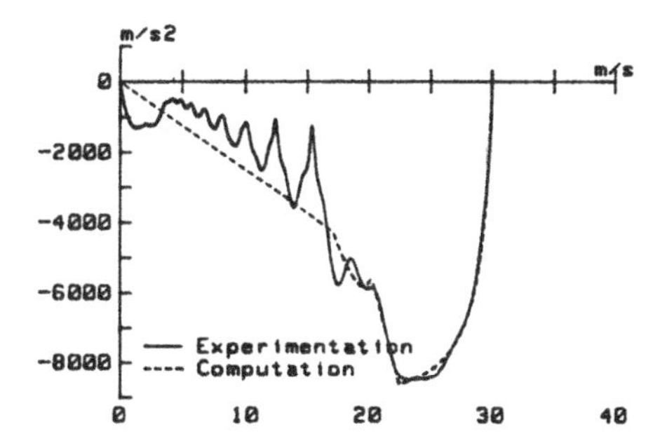

Figure 9. Model of movement of cylindrical projectile penetrating the water/ sandstone model

impact, and grouped into decay curves, according to the type of material and the orientation of the measurement axis. It can be seen that the acceleration levels are significant out to about 15 metres and negligible at greater distances. It can also be seen that the horizontal components decay more rapidly than the vertical components. Finally, the accelerations recorded in the cement-treated sand are larger than those in the clay, but of shorter duration.
The energy conveyed by the waves was computed by treating them as cylindrical surface waves. At 20 metres from the point of impact, they carry about 500,000 Joules in the layer of clay, or 6 to 7 % of the energy released in this layer. At 30 metres, they carry only 1 % of this energy. The fraction of the energy radiated in the substratum of the second model was smaller. It should however be noted that the projectile rebounded from this layer, and that in consequence the fraction of the energy transmitted to the cement-trea-

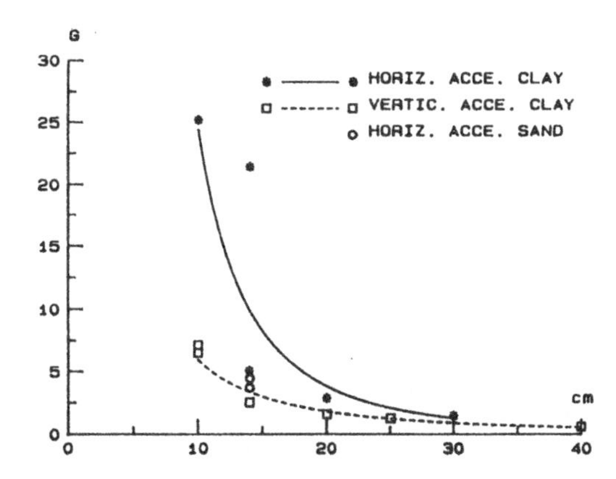

Figure 10. Acceleration in soil versus distance from point of impact for the water/clay/sand model

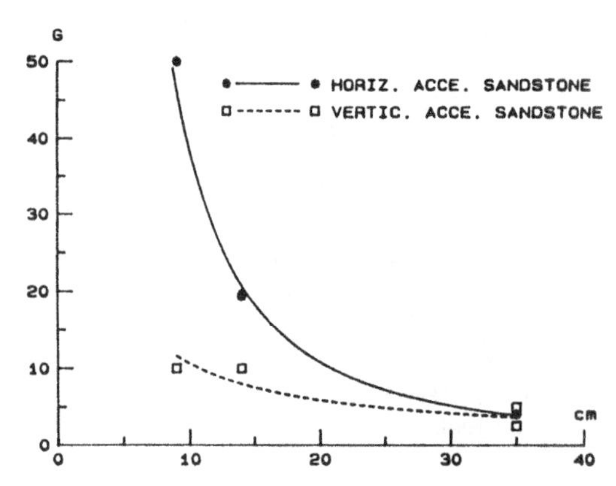

Figure 11. Acceleration in soil versus distance from point of impact for the water/sandstone model

ted sand might be much smaller than the estimate made on the basis of the energy versus penetration curves.

IVd : Assessment of the second experiment

The centrifuge experiment clearly showed that the zone within 15 metres of the presumed point of impact was critical. For any structure there, then, there should be a dynamic response calculation based on the accelerograms recorded during the tests.
Otherwise, this study confirmed that an impact cannot be investigated without taking the energy radiated in the site into account. This conclusion is all the more necessary in that, in this example, a large fraction of the energy was absorbed by the hydrodynamic effects of the impact.
Finally, in the case of a target consisting of several layers of materials, a combination of several hyperbolic models seems to be able to yield a correct description of the movement of the projectile.

V - Overall conclusion

In the absence of other means of analysis, the testing of physical models in a centrifuge can be an effective way of dealing with the problems of impacts in soils as they concern civil engineering. It allows concrete conclusions to be drawn directly and yields quantitative information about the effects of such accidents.
This experimental method may be regarded primarily as a means of investigating these problems in all their generality,

Table n° 1 : Main scale factors

gravity	N	10^{2}
length	1/N	10^{-2}
time	1/N	10^{-2}
velocity	1	1
stress	1	1
mass	$1/N^{3}$	10^{-6}
energy	$1/N^{3}$	10^{-6}

and of arriving at an empirical, but precise, model of the penetration of a projectile in a soil mass. The information yielded by the two experiments presented here may contribute to this. The empirical formulae proposed must be confirmed and completed by the relationships of their coefficients to the properties of the projectile and of the target. The centrifuge seems to be the most suitable means of further research in this direction.

REFERENCES

ALLEN W.A., MAYFIELD E.B., MORRISSON H.L. (1957) "Dynamics of Projectile Penetration Sand" Journal of Applied Physics, Vol 28, N° 3

CHISHOLM P.B., PELEZE R.M., PUGH F.L. (1962) "Earth Trajectory Model for Subsurface Balistic and Thrusted Vehicule" TDR - AFSWC - TDR - 62 - 106

CORTE J.F., GARNIER J. (1986) "Une centrifugeuse pour la recherche en géotechnique" Bulletin de Liaison des Laboratoires des Ponts et Chaussées, n° 146

FUCHS O. (1963) "Impact Phenomena" AIAA Journal, Vol 1, N° 9

HAKALA W.W. (1965) "Resistance of a Granular Medium to Normal Impact of a Rigid Projectile" Thesis presented to the Virginia Polytechnic Institute, Blacburn,

HOLSAPPLE K.A., SCHMIDT R.M. (1982) "On the Scaling of Crater Dimensions - 2 Impact Processes" Journal of Geophysical Research, Vol 87, N. B3

WANG W.L. (1971) "Low Velocity Projectile Penetration" Journal of the Soil Mechanics and Foundations Division, Vol 97, SM 12

3. Underground openings and excavations

Computer and Physical Modelling in Geotechnical Engineering, Balasubramaniam et al. (eds)
© 1989 Balkema, Rotterdam. ISBN 90 6191 864 2

Computer aided bound solutions in a cohesive material

H.N.Seneviratne & M.Uthayakumar
University of Peradeniya, Sri Lanka

ABSTRACT: Some methods of obtaining bound solutions for stability problems in fully cohesive soils using simple computer programs are described here. The effectiveness of some of these techniques, of not only in bracketing the exact solution but also in providing insight into the mechanics of deformation and load transfer is demonstrated by analysing a trial problem.

1 INTRODUCTION

One of the most frequently encountered problem in soil mechanics and foundation design is the determination of failure conditions. The exact solution for such a problem is difficult to obtain except in cases of simple geometry and uniform soil conditions. Numerical techniques such as finite elements can be costly, time consuming and, may not always give the correct answer for this type of problems.

Alternatively, the limit theorems of plasticity can be used to find bounds for the exact solution provided that the pre-collapse deformations do not substantially alter the original geometry of the problem. These bound solutions can sometimes be very simple and need not be refined to an accuracy beyond the design requirements. In addition they may give some insight into the mechanics of deformation and load transfer which may be useful to a designer.

An elastic - perfectly plastic total stress - strain model with a constant undrained shear strength C_u is used here to describe the undrained behaviour of a fully cohesive soil. Upper bound solutions for such cases are not difficult to obtain even accounting for the effects of self weight and non-homogeneity of the material . Complete lower bounds are comparatively difficult to obtain than the upper bounds. In both cases computers can play a useful role not only in calculations but also in optimisation to obtain the best bounds.

This paper describes some well known techniques of obtaining lower and upper bounds for stability problems in fully cohesive soils under undrained conditions. The use of computers for this purpose is illustrated by an example on the stability analysis of a near surface rectangular tunnel under plane strain conditions.

2 BOUND THEOREMS OF PLASTICITY

The limit theorems are applicable to problems in elastic - perfectly plastic materials provided that the pre-collapse deformations are not excessive. According to the theory of plasticity the collapse load of a perfect plastic body is unique. The load is interpreted here as a force which leads to the failure of a body; a supporting force is a negative load in this sense.

The lower bound theorem states that if any stress field which supports the loads and is everywhere in equilibrium without yield criterion being violated, can be found then the body will not collapse or just collapse under such loads. The upper bound theorem states that the loads determined from a work calculation based on a kinematically admissible collapse mechanism which satisfy the boundary displacement conditions, will be higher than or equal to the actual collapse load.

The following corollaries of these theorems can be useful. The collapse load cannot be increased/decreased by removal/addition of weightless material if the positions of the loads are unchanged. It cannot also be increased/decreased by replacing some material of the body by weaker/stronger material. These corollaries help to iron out small non-homogeneities, to ignore the effect of work hardening, and to simplify complex configurations. Chen (1975) is a very useful reference on the use of these theorems in soil mechanics and foundation engineering.

3 UPPER BOUND TECHNIQUES

In the case of plane strain the simplest upper bound method is the use of a mechanism consisting of a number of rectangular or triangular blocks sliding against each other. In many axisymmetric problems complex elements with curved boundaries are needed to satisfy the incompressibility and compatibility requirements. The procedure for determining the collapse load is to derive an expression for it in terms of varying dimensions and to optimise it with respect to these variables. Other than in the cases where the expression for the collapse load is simple the optimisation is best carried out using a computer program. Section 6 gives an example of this technique. It is possible to obtain results close to the exact solution using only two or three rigid blocks provided that the chosen mechanism approximately describe the true mode of failure. Hence, a prior knowledge of the correct mode of failure either from past experience or from model testing is very useful in obtaining good upper bounds.

Improvement of the solution by subdividing the larger blocks while keeping the mode of deformation essentially the same, would not be substantial and diminishes rapidly with increasing subdivision. A substantial improvement of the solution is always associated with a significant change in the assumed mode of deformation. It is also possible to use rotating blocks or zones of simple shear instead of rigid blocks. This is commonly done in the analysis of bearing capacity and slope stability problems. If a computer is used, it is not difficult to incorporate material non-uniformities and self weight forces in this analysis.

If only the collapse load not the mode of failure is required a simple optimisation technique like uni-directional search is adequate as the expression for the collapse load varies very slowly close to the optimum point. If the correct mode of failure is desired more sophisticated optimisation techniques like the method of steepest descent may be preferable. If the mechanism chosen resemble the actual mode of failure the optimum mechanism can be very close to the actual mode of collapse (e.g. see Fig. 1).

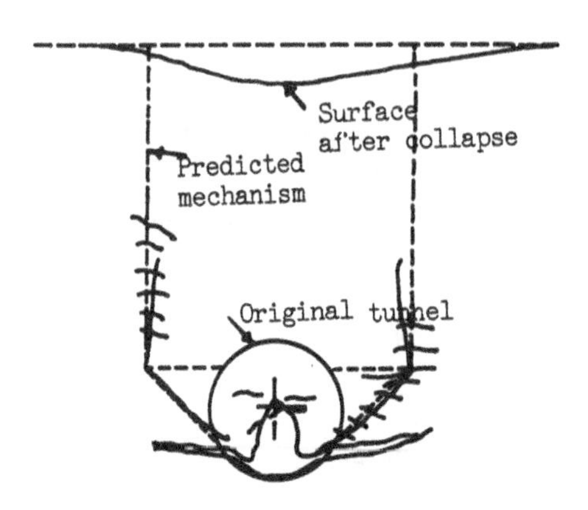

Fig.1 Predicted and Observed Failure of a shallow tunnel. (Seneviratne (1979))

Another upper bound technique popularly used in metal plasticity (see e.g. Johnson and Mellor (1962))is the method of velocity fields. In this method the deformation behaviour with a yield zone is described along the velocity characteristics. The collapse load is estimated from a work calculation as in the previous case. The classical Hill's and Prandtl's solutions for the bearing capacity of a shallow foundation (see Hill (1950)) are derived in this manner. Even though this method requires more expertise than the previous technique, it is possible to get nearly exact solutions using this method. It can also incorporate material non-uniformities and self weight, and a computer is useful for performing numerical computations.

4 LOWER BOUND TECHNIQUES

In general lower bounds are more difficult to obtain than upper bounds.

In some cases it is possible to use an existing solution to a different problem to derive a valid lower bound for a particular problem. As an example, the classical thick cylinder solution is used in section 7 to give a lower bound solution for a tunnelling problem. A problem can be simplified by dividing it into regions of familiar stress distributions like Rankind zones, hydrostatic stress, radial stress zones etc. If the equilibrium conditions across the regions and the stress boundary conditions are fully satisfied then the particular solution is a valid lower bound.

A more elegant method through it requires more expertise and skill is the method of characteristics (Sokolovsky (1965)). A comprehensive account on the use of this method is given by Booker and Davis (1977). In this method the combined yield condition and equilibrium equations is expressed along the characteristic directions. Fig. 2 shows the characteristic field around a near surface circular tunnel in a cohesive soil. This method can incorporate the effects of soil non-uniformity and self weight much more easily than the previous method. It can also easily cope with non-uniform stress boundary conditions. A computer is virtually essential for numerical computations involved with this technique.

A proper lower bound solution should satisfy all stress boundary conditions exactly. For many problems involving a semi-infinite half space the lateral boundaries can assume to be displacement controlled so that a valid lower bound solution does not have to satisfy any lateral boundary conditions. A proper lower bound solution should be extended to satisfy all stress boundary conditions without violating yield.

5 TRIAL PROBLEM

Fig. 3 shows the configuration of the trial problem. The geometry of the problem can be described by the two dimensionless parameters B/D; breadth

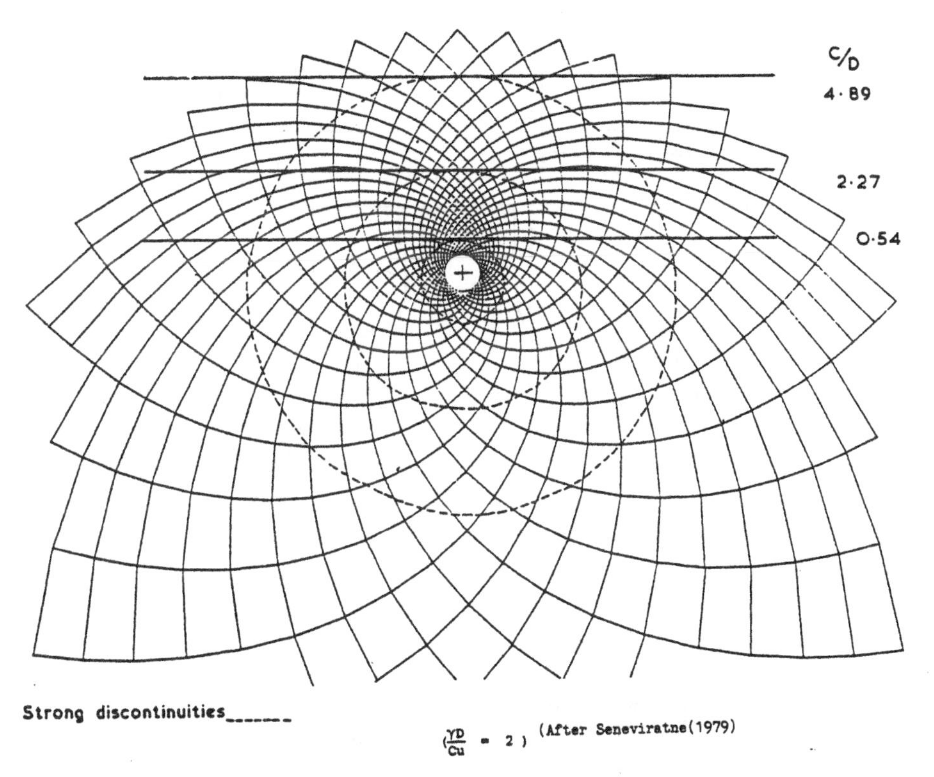

Fig.2 Stress characteristics around a plane strain circular tunnel.

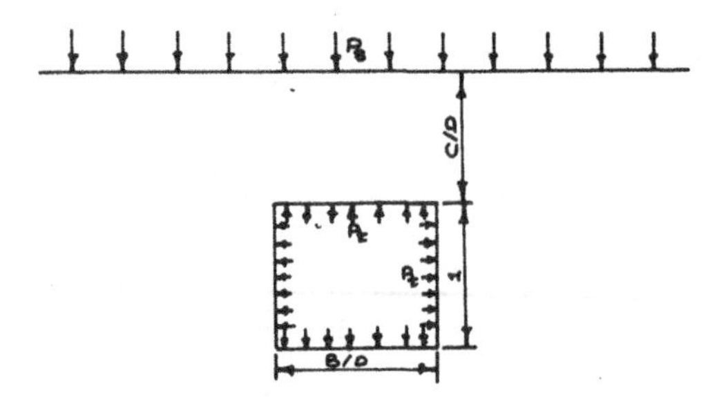

Fig.3 Configuration of the trial problem

to diameter ratio and, C/D; cover to diameter ratio. The tunnel is assumed to be long, hence it is a plane strain problem. The soil is assumed to be having a constant undrained shear strength, C_u. The self weight of the soil is characterised by the dimensionless factor $\gamma D/C_u$. The tunnel is supported by an internal pressure p_t against the surface pressure p_s. It can be shown that the failure is governmed by the difference between p_s and p_t. A dimensionless stability number N defined as;

$$N = (p_s - p_t)/C_u + \gamma D/C_u(C/D+1/2)$$

is used for interpretation of the results. This number was first defined by Broms and Bennermark (1967) who investigated the stability of tunnel headings.

6 UPPER BOUND SOLUTIONS

Fig. 4 shows six mechanisms consisting of rigid blocks chosen for analysis.

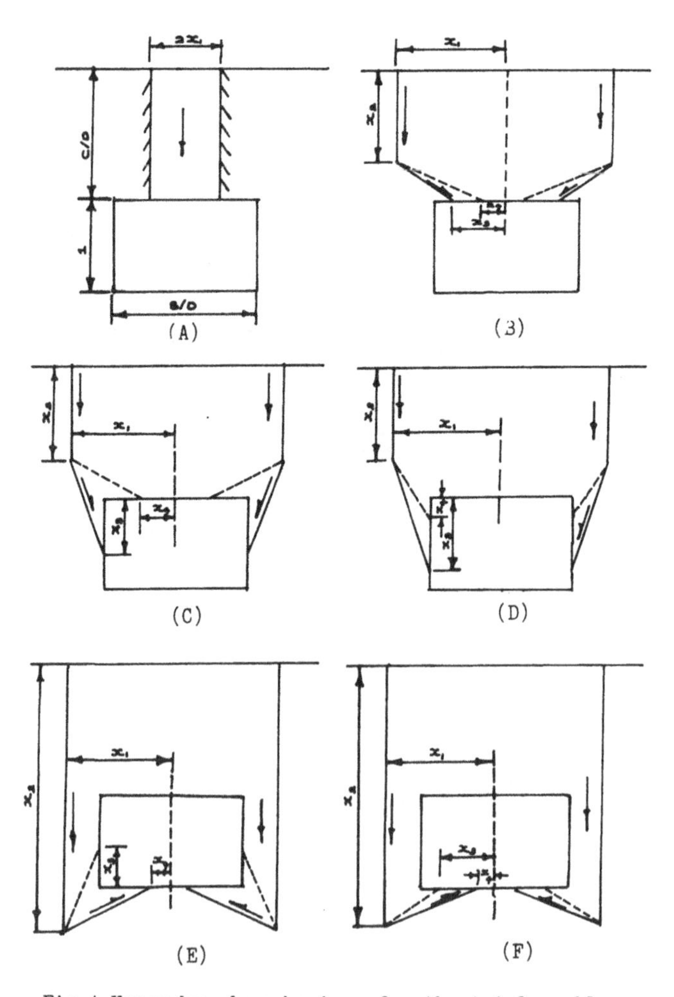

Fig.4 Upper bound mechanisms for the trial problem

The mechanism A is a 'roof' mechanism which can easily be optimised to give,

$$N = 2C/B + \gamma D/2C_u, \; x_1 = B$$

The mechanisms B,C,D,E and F were optimised with respect to the variables x_1, x_2, x_3 and x_4 using a computer program. The optimisation procedure used was as follows;

(i) Choose starting values for the variables to form a feasible mechanism and select step lengths for the four variables.

(ii) Vary x_1 to give the minimum collapse load.

(iii) repeat (ii) but varying x_2,x_3 and x_4 instead of x_1 in each turn.

(iv) Go back to step (ii) with halved step lengths unless sufficient convergence is reached. This uni-directional optimisation was carried out for $\gamma D/C_u$ values of 0,2 and 4. The B/D ratios used were 0.25, 0.5, 1, 2.5, and 5 respectively. The calculations were performed for a range of C/D of 0 to 5.

7 LOWER BOUND SOLUTIONS

It was not attempted to find lower bound solutions for the cases with self weight. However, Fig. 2 shows an example of how this can be carried out. The stability numbers reduce to $(p_s-p_t)/C_u$ when self weight is ignored. The solution A is illustrated in Fig. 5. The radially symmetric stress field within the annular region shown in the figure is given by,

$$\sigma_r = p_t + 2C_u.\ln(r/R_1)$$

$$\sigma_\theta = \sigma_r + 2C_u$$

$$\tau_{r\theta} = 0$$

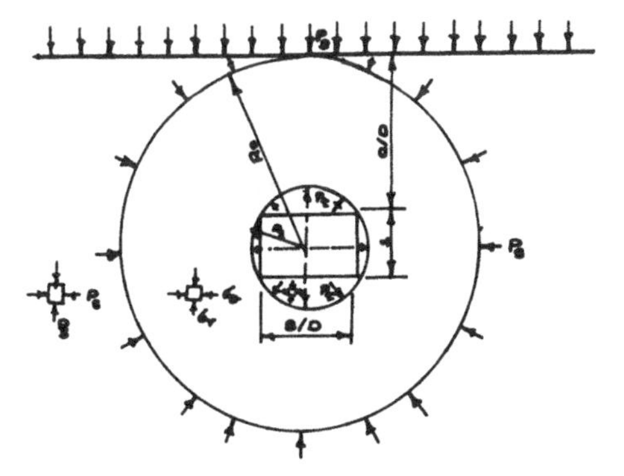

Fig.5 Lower bound solution A

using the normal notation. Then the lower bound solution is;

$$N = 2. \ln (R_2/R_1)$$

This solution was also optimised with respect to R_1 using a computer program.

The solution B was derived using the lower bound solution for a loaded trapezoidal wedge given by Shield (1954) (see Fig. 6). For a given B/D ratio, a lower bound can be found for a specific C/D ratio by combining four trapezoidal wedges as illustrated in the figure. The same solution can be demonstrated to be valid for a tunnel with a larger C/D ratio or a smaller B/D ratio (see Fig.6).

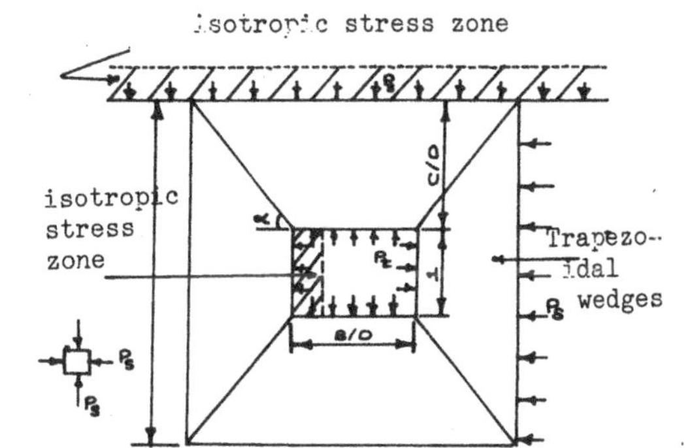

Fig.6 Lower bound solution B

8 RESULTS

Fig. 7 shows the optimum upper and lower bound solutions for the stability number, N at different B/D and C/D ratios for the weightless case. The exact solution is bracketed to within ±15% by the upper and lower bounds except at very high or very low B/D ratios whenC/D is very low. For tunnels with low C/D ratios the mechanism A gave the best upper bound. As C/D increases the mechanism D became critical. The other mechanisms were not found to be critical in the range of B/D and C/D considered in the analysis. The lower bound solution A gave answers for the whole range but the solution B was not possible at low C/D ratios. This solution B was found to be critical only at B/D ratios of 0.25 and 0.5 and for a C/D range of 1 to 1.6.

At low C/D ratios, little arching takes place hence the failure is due to shear failure across the cover of the tunnel. The roof mechanism closely

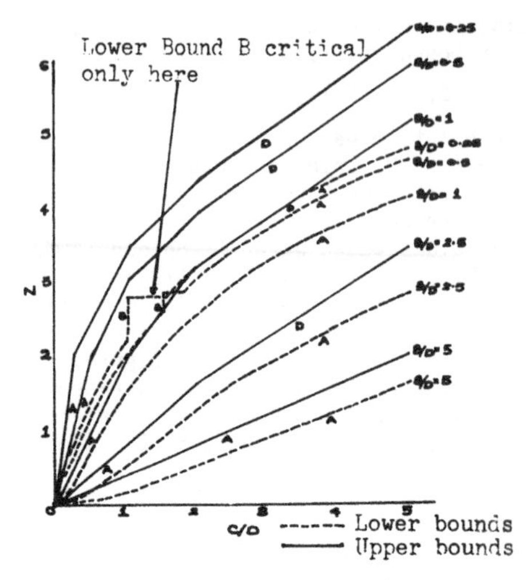

Fig.7 Upper and lower bounds (weightless case)

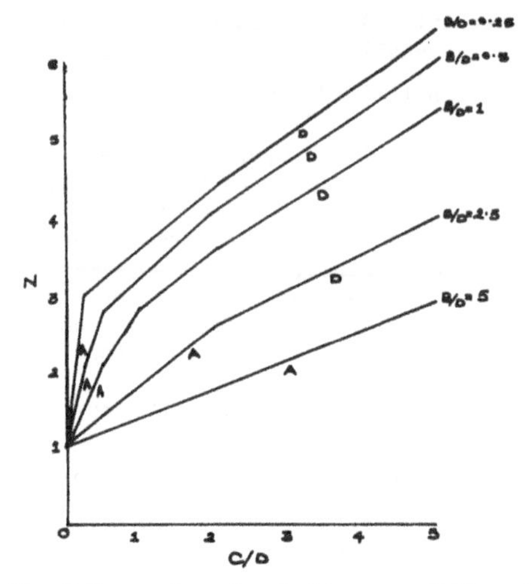

Fig.8 Upper bound solutions : $\gamma D/C_u = 2$

follows this mode of failure. As C/D ratio increases, arching becomes the predominant mode of load transfer and failure occurs due to the collapse of this arch at the end points. The C/D ratio at this point of transition from roof failure to arching mode increases with B/D ratio. This is reasonable as an arch is more difficult to be formed at high B/D ratios. The reason for poor bracketing of the exact solution at very high or very low B/D ratios when C/D ratio is very low, is likely to be due to the inadequacy of the lower bound solutions. A circular arch does not correctly represent load transfer in such cases.

Figs. 8 and 9 show the variation of N with C/D and B/D ratios for the tunnels with self weight as given by optimum upper bound mechanisms. As in weightless case, mechanism A gives the best results at low C/D ratios and mechanism D gives the best results at high C/D ratios; other mechanisms were not found to be critical in the range considered. Even with increasing effect of self weight a bottom heave mechanism like F was not found to be critical at high C/D ratios. Perhaps another type of mechanism with a combination of heave and arching may give better results in this range. This analysis was not extended beyond C/D ratios of 5 as the pre-collapse deformations would have been substantial.

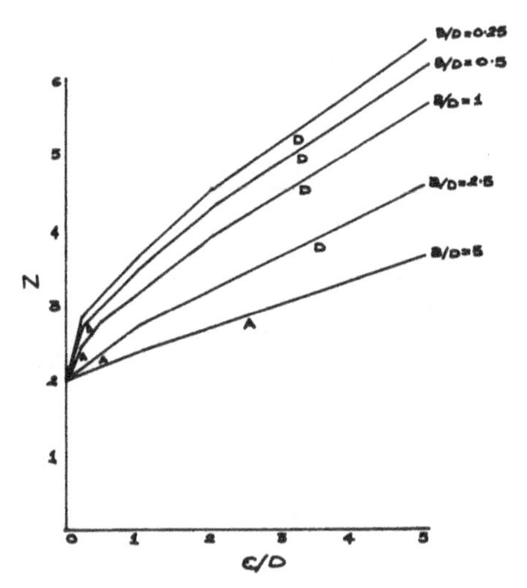

Fig.9 Upper bound solutions : $\gamma D/C_u = 4$

9 CONCLUSIONS

The effectiveness of the use of bound theorems of plasticity in obtaining stability solutions was successfully demonstrated using a trial problem on the stability of a long rectangular tunnel close to the ground surface. The optimisation to obtian the best bounds was carried out using simple computer programs. The solutions derived in this manner, bracketed the exact answers to within ± 15% in most of the cases considered, and gave a good indication of the mechanics of load transfer.

ACKNOWLEDGEMENTS

The authors wish to express their thanks to Mrs. P. Pereira for typing the manuscript and Miss G. Panchalingam for drawing the diagrams.

REFERENCES

Booker, J.R. & Davis, E.H. 1977. Stability analysis by plasticity theory. In C.S. Desai & J.T. Christian (eds.), Numerical methods in geotechnical engineering, Chapter 21, Mc-Graw Hill.

Broms, B.B. & Bennermark, H. 1967. Stability of clay in vertical openings. Proc. ASCE, 193, SM1, 71-94.

Chen, Wai-Fah 1975. Limit analysis and soil plasticity. Amsterdam: Elsevier.

Hill, R. 1950. The mathematical theory of plasticity. Oxford University Press. Amen House. London.

Johnson, W. & Mellor, P. 1962. Plasticity for mechanical engineers. New York: Van Nostrand.

Seneviratne, H.N. 1979. Deformations and pore pressure dissipation around shallow tunnels in soft clay. Ph.D. Thesis. Cambridge University. U.K.

Shield, R.T. 1954. Stress and velocity fields in soil mechanics. J. Maths. Phys. 33 (2): 144-156.

Sokolovsky, V.V. 1965. Statics of granular media. Pergamon Press.

Computer and Physical Modelling in Geotechnical Engineering, Balasubramaniam et al. (eds)
© 1989 Balkema, Rotterdam. ISBN 90 6191 864 2

Strain analysis of jointed rock masses for monitoring the stability of underground openings

S.Sakurai & T.Ine
Department of Civil Engineering, Kobe University, Kobe, Japan

ABSTRACT: In this paper a back analysis method is proposed for determining strain distributions around the underground openings excavated in jointed rock masses. In the proposed method, a simple linear mechanical model is introduced, in which the three different deformational modes can be analyzed by using a single model based on continuum mechanics.

1 INTRODUCTION

The stability of underground openings such as tunnels, underground powerhouses, etc., can be assessed in terms of strain. Sakurai (Sakurai 1981) proposed the Direct Strain Evaluation Technique (DSET), the basic idea of which is the following.

The strain occurring around underground openings is first determined from filed measurements, and is compared with the threshold value called the "critical strain". If the strain becomes greater than the critical strain, the stability is questionable, and additional shotcrete or rock bolls must be installed to stabilize the openings.

When the number of displacement measuring points is sufficiently large, strain can be determined directly from measured displacements using only the kinematic relationship (Sakurai 1981). In this approach, no other information such as initial stress or mechanical constants are needed. In practice, however, the number of extensometers used to measure displacements is not large enough to necessitate limitation of the data of displacement measurements. Therefore, direct determination of strain from measured displacements appears to be difficult.

In order to overcome this difficulty, "back analysis", a reverse calculation of ordinary stress analysis, is extremely useful (Sakurai and Takeuchi 1983). This back analysis can provide initial stress and mechanical constants from measured displacements, which are then used as input data for an ordinary analysis such as FEM or BEM to determine strain distributions.

In this paper a back analysis method is proposed for determining strain distributions around the underground openings excavated in jointed rock masses. The deformational behavior of jointed rock masses may be classified into the following three modes, i.e., (1) spalling of joints, (2) sliding along particular slip surface, and (3) plastic flow. Therefore, the mechanical model used in the back analysis must represent all these deformational modes.

In the proposed back analysis method, a simple liner mechanical model is introduced, in which the three different deformational modes can be analyzed by using a single model based on continuum mechanics.

2 CONSTITUTIVE EQUATIONS

In order to simulate all three modes of deformations, i.e., spalling, sliding,and plastic flow, the following constitutive equation is proposed, which is a further extension of the equation developed for analyzing the deformational behavior of cut slopes (Sakurai et al 1986).

The proposed equation is expressed in terms of x'-y' local coordinates as (see Fig. 1(a)),

$$\{ \sigma' \} = [D']\{ \varepsilon' \} \qquad (1)$$

where

$$[D'] = \frac{E_2}{(1+\nu_1)(1-\nu_1-2n\nu_2^2)} \cdot \begin{bmatrix} n(1-n\nu_2^2) & n\nu_2(1+\nu_1) & 0 \\ n\nu_2(1+\nu_1) & 1-\nu_1^2 & 0 \\ 0 & 0 & m(1+\nu_1)(1-\nu_1-2n\nu_2^2) \end{bmatrix}$$

$$(n = E_1/E_2 \quad , \quad m = G_2/E_2) \tag{2}$$

Hence,it is transformed into the x-y global coordinates as follows,

$$\{\sigma\} = [D]\{\varepsilon\} \tag{3}$$

where

$$[D] = [T][D'][T]^T \tag{4}$$

[T] is a transformation matrix expressed as,

$$[T] = \begin{bmatrix} \cos^2\alpha & \sin^2\alpha & -2\sin\alpha\cos\alpha \\ \sin^2\alpha & \cos^2\alpha & 2\sin\alpha\cos\alpha \\ \sin\alpha\cos\alpha & -\sin\alpha\cos\alpha & \cos^2\alpha-\sin^2\alpha \end{bmatrix} \tag{5}$$

α is the angle between x'- and x- coordinate system.

It should be noted that Eq. (3) can represent all three modes of deformations by changing the material constants, particularly n and m, which are called anisotropic parameters.

2.1 Spalling of joints

The spalling of joints shown in Fig. 1(b) can be represented by increasing the anisotropic parameter n, i.e., by reducing the value of E_2 against E_1. Poisson's ratio ν_2 is taken to be zero, because spalling in the direction of y'-axis makes no movement in x'-axis. In this case, the other anisotropic parameter m must be taken as $m = 1/2(1+\nu_1)$.

2.2 Sliding along joints or slip surface

When sliding occurs along the joints parallel to x'-axis, the anisotropic parameter m can be reduced to a small value ($m < 1/2(1+\nu)$), while $n = 1.0$ and $\nu_1 = \nu_2$.

2.3 Plastic flow

Material under the plastic state tends to slide along the two families of potential sliding planes with an angle of $\pm(45° + \phi/2)$ from the maximum principal stress direction, as shown Fig. 2. ϕ denotes the internal friction angle.

Let us take two different coordinate systems for consideration of the mechanical behavior of the two conjugate slip planes, as shown in Fig. 2. The stress-strain relationship for each family of slip planes is given in the same form as Eqs. (1) and (3) for the local and global coordinate systems, respectively. The total strain is assumed to be expressed as,

$$\{\varepsilon\} = \frac{1}{2}[\{\varepsilon_1\} + \{\varepsilon_2\}] \tag{6}$$

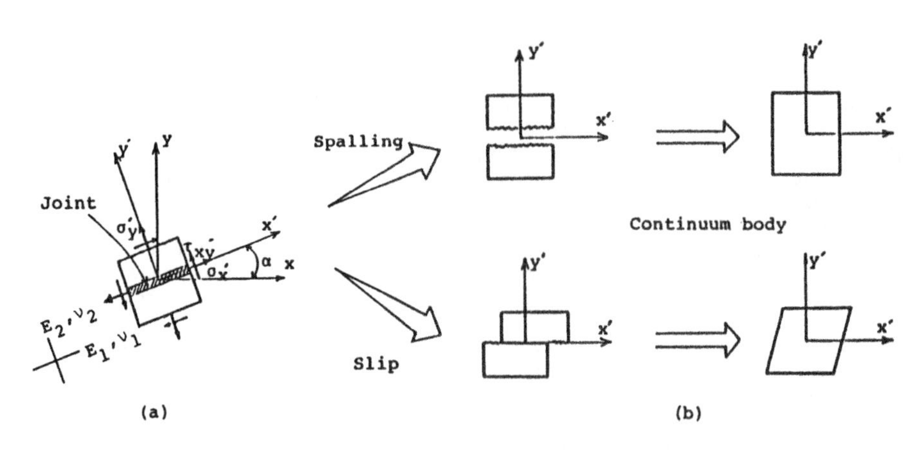

Fig. 1 Modeling for discontinuous deformation in continuum mechanics

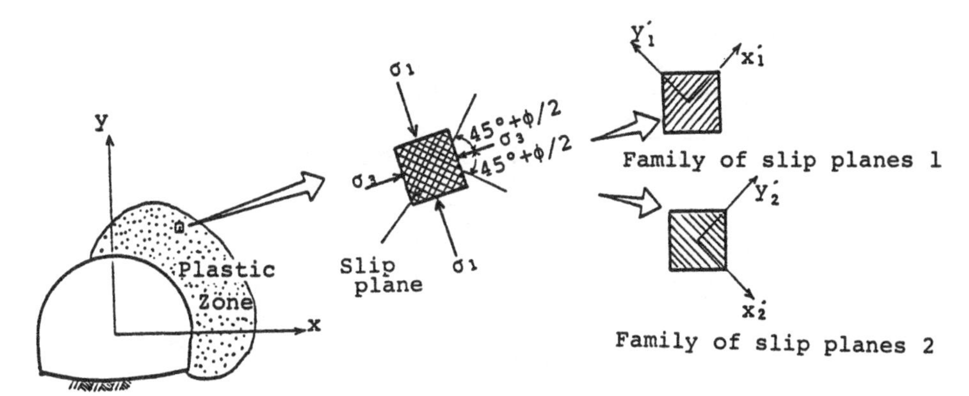

Fig. 2 Families of slip planes in plastic zone

where $\{\varepsilon_1\}$ and $\{\varepsilon_2\}$ are strains due to the families of two slip planes, respectively. Considering Eq. (3), Eq. (6) becomes,

$$\{\varepsilon\} = \frac{1}{2}\,[[D_1]^{-1} + [D_2]^{-1}]\{\sigma\} \quad (7)$$

This is the proposed stress-strain relationship for representing the plastic behavior of materials. Eq. (7) is expressed in the following common form.

$$\{\sigma\} = [D]\{\varepsilon\} \quad (8)$$

where

$$[D] = [\frac{1}{2}\{[D_1]^{-1} + [D_2]^{-1}\}]^{-1} \quad (9)$$

3 COMPUTER SIMULATIONS

In order to verify the validity of the constitutive equation for analyzing tunnelling problems, computer simulations have been conducted.

3.1 Spalling of joints

The tunnel under consideration here is shown in Fig. 3. A main discontinuity plane is located 4m above the crown of the tunnel. Joint element (Goodman et al 1968) is used to represent the mechanical behavior of the joints. The input data employed in the computer simulation is shown in Table 1.

The displacement distribution is obtained by the ordinary FE analysis. The displacements along the discontinuous plane are shown in Fig. 4. They clearly show the spalling of the joint due to tunnel excavation. The calculated displacements along the reference lines shown in Fig. 3 are considered "measured" displacements.

Back analysis for determining initial stress and material properties from the "measured" displacements is carried out to make the following error function minimum,

$$\delta = \sum_{i=1}^{N} (u_i^c - u_i^m)^2 \longrightarrow \min \quad (10)$$

where u_i^c and u_i^m are calculated and "measured" displacements, respectively. In the back analysis, two different constitutive equations are used, one is Eq. (3) taking into account the anisotropic parameter, and the other is the ordinary isotropic elastic material.

The back analysis results are shown in Fig. 5. It is clear that the displacements calculated by considering the anisotropic parameters and the "measured" displacements coincide, while the isotropic elastic constitutive equation of course can not simulate the spalling of joints.

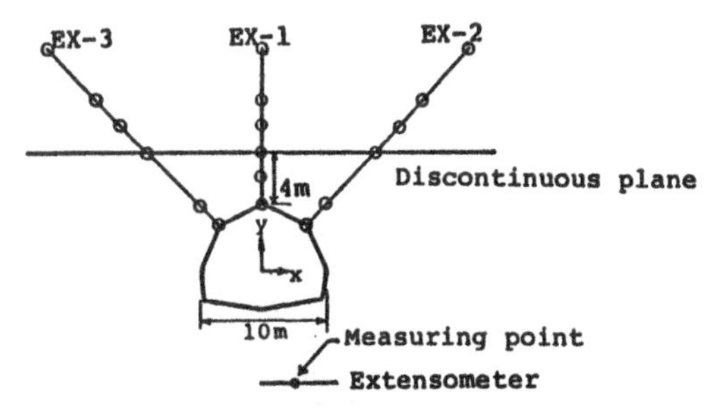

Fig. 3 Tunnel excavated near a discontinuous plane and location of measuring points

Table 1 Input data and back analysis results in case of spalling of joint

	FE ANALYSIS WITH JOINT	BACK ANALYSIS	
		Proposed method	Isotropic
σ_x kgf/cm²	-25.0	-25.7	-7.7
σ_y kgf/cm²	-50.0	-50.0	-50.0
τ_{xy} kgf/cm²	0.0	0.0	0.0
E kgf/cm²	10000.0	10546.3	6835.0
Poisson's Ratio ν	0.3	0.3(assumed)	
n	—	10.0	1.0
Data for joint			
Wall rock compressive strength (kgf/cm²)			-70.0
Ratio of tensile to compressive strength			0.1
Shear stiffness (kgf/cm³)			3850.0
Ratio of residual to peak shear strength			0.33
Maximum normal closure (cm)			10.0
Seating load (kgf/cm²)			-50.0
Friction angle of a smooth joint (degree)			30.0
Dilatancy angle (degree)			0.0

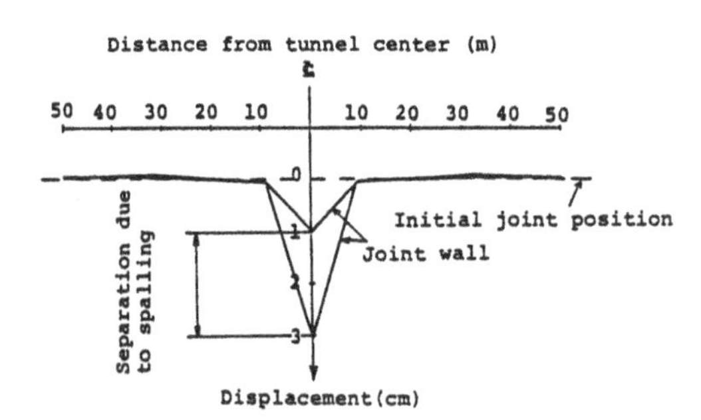

Fig. 4 Spalling of joint

3.2 Sliding along slip surface

The shallow tunnel shown in Fig. 6 is used as an example to demonstrate the validity of the proposed constitutive equation. The tunnel diameter is 8m, and the height of overburden is also 8m. The input data for material properties is given in Table 2. The slip surfaces are assumed as illustrated in Fig. 6, and the anisotropic parameter m is given for each slip zone.

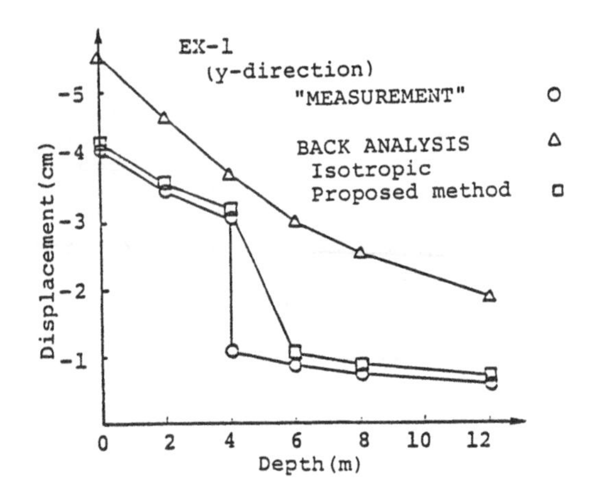

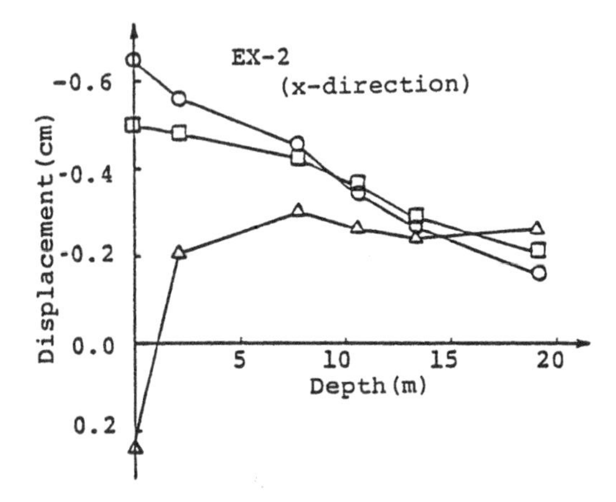

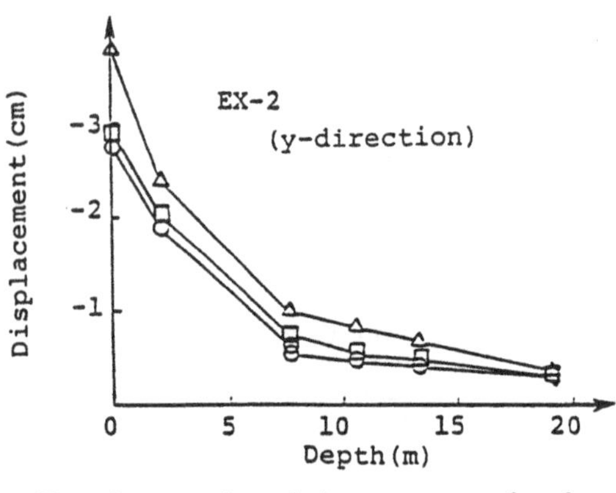

Fig. 5 comparison between measured and back-analyzed displacements

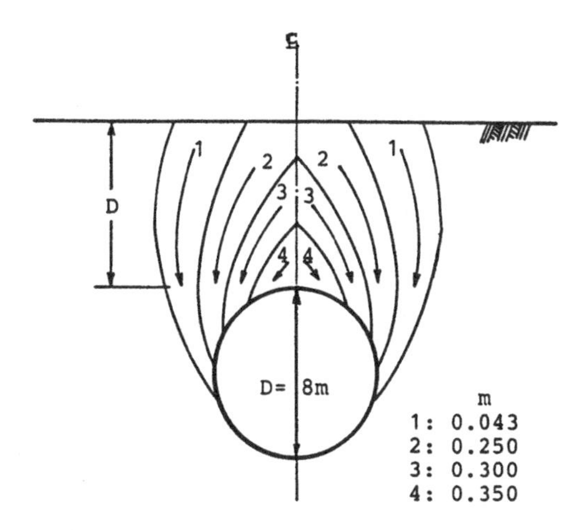

Fig. 6 Shallow tunnel excavated in sandy medium with slip surfaces

Table 2 Input data in case of sliding

Unit Weight(tf/m³)	2.00
Young's Modulus (tf/m²)	2,000
Poisson's Ratio	1/3
Coefficient of Earth Pressure at Rest K_0	0.5
m_1	0.043
m_2	0.250
m_3	0.300
m_4	0.350

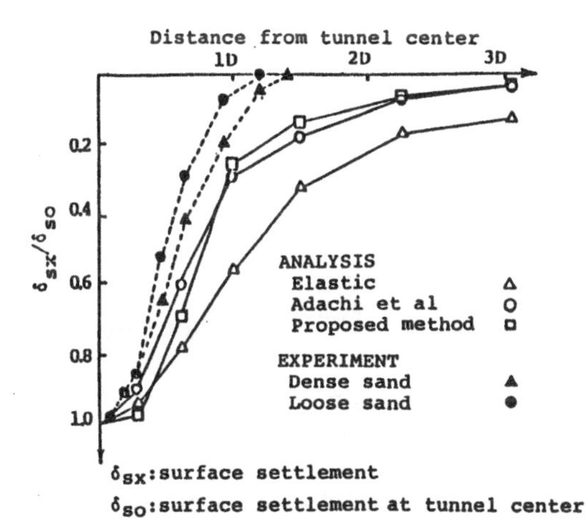

Fig. 7 surface settlement due to tunnel excavation

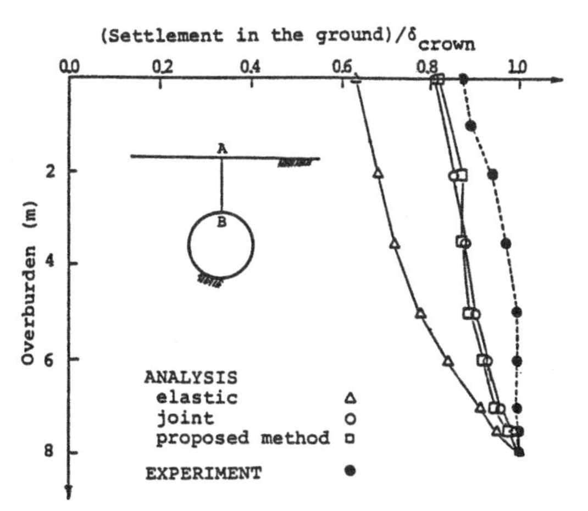

Fig. 8 Distribution of vertical displacement along line AB above tunnel crown

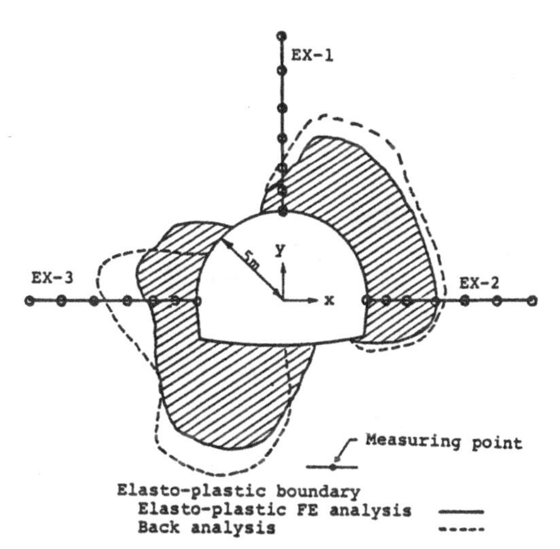

Fig. 9 Horseshoe-shaped tunnel excavated in an elasto-plastic medium

The FE analysis is carried out for given anisotropic parameters m, and the results are shown Figs. 7 and 8. Fig. 7 shows the surface settlement, and Fig. 8 is for the vertical displacement along the vertical reference line above the tunnel crown. In these figures, the behavior of isotropic elastic material is also shown for reference.

Adachi et al demonstrated that a shallow tunnel excavated in sandy material can not be analyzed by a continuum mechanics model, but instead, a discontinuous model of joint elements installed in between all the finite elements is recommended (Adachi et al 1986). The results obtained by considering the joint elements are also indicated in the figures.

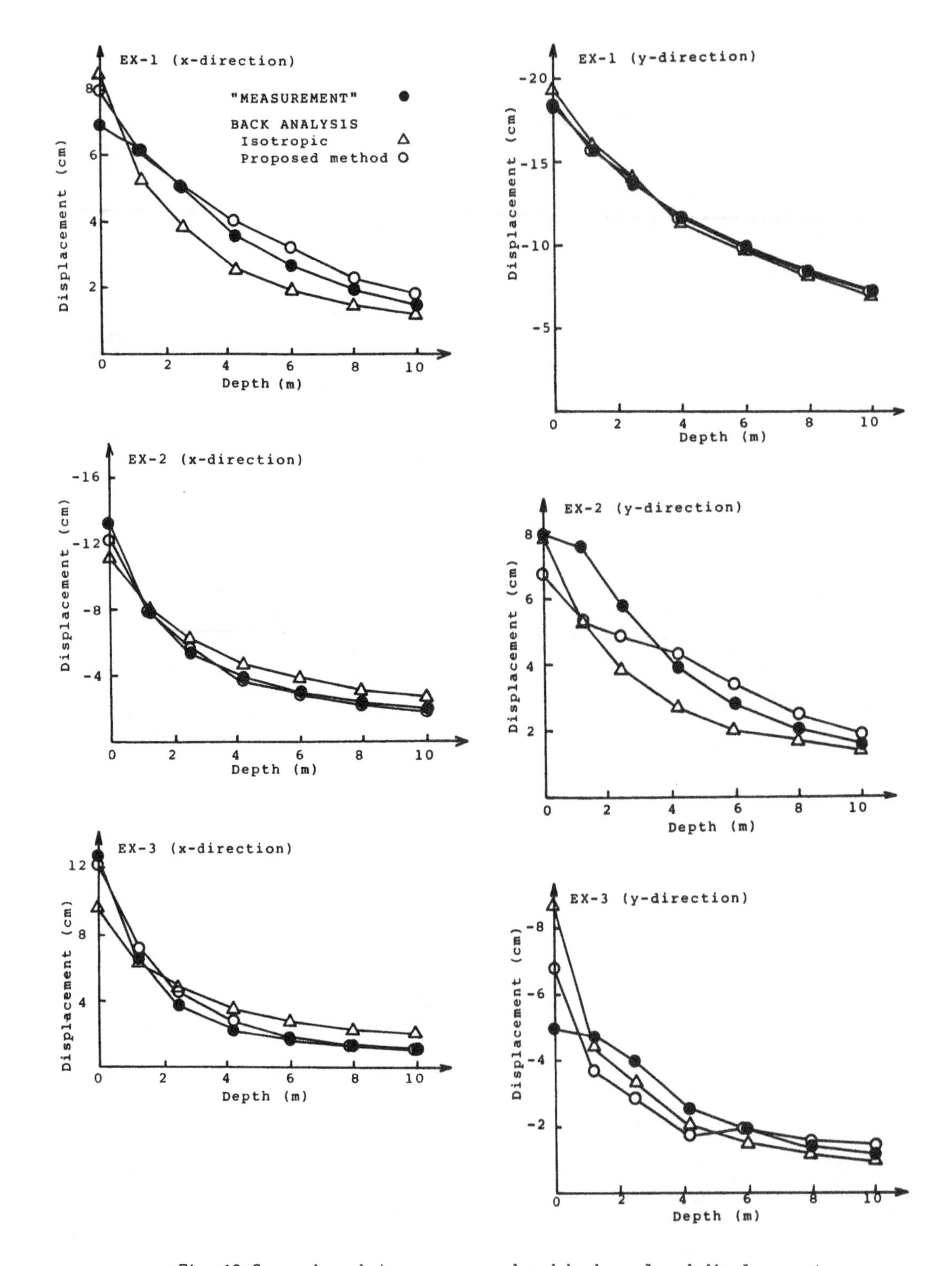

Fig. 10 Comparison between measured and back-analyzed displacements

Experimental results of displacement around a tunnel excavated in sandy material are plotted for comparison in Figs. 7 and 8. It is obvious that the isotropic elastic material behaves quite differently from the real behavior of sandy ground, while the proposed constitutive equation is acceptable for analysis of a shallow tunnel excavated in sandy materials. The joint model proposed by Adachi et al seems to be reasonable. However, the physical meaning of joint elements representing sandy materials is ambiguous.

3.3 Plastic flow

It is assumed that a horseshoe-shaped tunnel, 5m in radius, is bored in an elasto-plastic medium (See Fig. 9). Elasto-plastic FE analysis is first conducted by assuming the Drucker-Prager yielding function and Von Mises type of plastic potential function. The initial stress and material properties employed here are listed in Table 3. The calculated displacements along three reference lines (See Fig. 9) are taken as "measured" displacements, which are used as input for back analysis.

The plastic zone appearing around the tunnel is then back analyzed by the method previously proposed by one of the authors (Sakurai et al,1985), and is compared with the "real" plastic zone as shown in Fig. 9. The "real" plastic zone is one obtained by the ordinary elasto-plastic FE analysis.

Once the plastic zone is back analyzed, the proposed back analysis is conducted to determine the initial stress and material constants including anisotropic parameter m. It is noted in this analysis that the reduced parameter m is only valid in the plastic zone. The value $m = 1/2(1+\nu)$ is taken in the medium outside of the plastic zone.

The back analyzed displacements along the three reference lines are shown in Fig. 10 together with the "measured" displacements. In this figure the back analysis results obtained on the assumption of an isotropic elastic material are also shown for reference.

It is clear that the displacements back calculated by the proposed method and the "measured" displacements well coincide.

4 CONCLUSION

In this paper, an anisotropic constitutive equation has been proposed for use in the analysis of three different modes of deformation, i.e., (1) spalling of joints, (2)sliding, and (3) plastic flow.

Table 3 Input data and back analysis results in case of plastic flow

	ELASTO-PLASTIC FE ANALYSIS	BACK ANALYSIS	
		Proposed method	Isotropic
σ_x kgf/cm²	-25.0	-24.4	-28.4
σ_y kgf/cm²	-35.0	-35.0	-35.0
τ_{xy} kgf/cm²	8.66	11.26	10.02
E kgf/cm²	2000.0	1961.0	1257.4
Poisson's Ratio ν	0.3	0.3(assumed)	
Cohesive Strength C (kgf/cm²)	5.0	——	——
Internal Friction Angle ϕ (degree)	30.0	30.0	——
m	——	0.0962	$1/2(1+\nu)$ =0.3846
Critical Strain ε_0 %	——	0.866	——

The computer simulation together with some experimental results have demonstrated that the proposed constitutive equation is reasonably acceptable for analyzing the three different modes of deformation.

It is quite a surprise that such a simple constitutive equation based on continuum mechanics can satisfactorily represent the complex behavior of a discontinuous medium.

It is also demonstrated that the proposed constitutive equation can be used for back analysis of measured displacements to determine the initial stress and material constants.

REFERENCES

Adachi, T., T. Tamura, A. Yashima and H. Ueno 1986. Surface Subsidence above Shallow Sandy Ground Tunnel, Proceedings of JSCE, No.370/III-5:85-94.(in Japanese)

Goodman, R.E., R.L. Taylor and T.L. Brekke 1968. A Model for the Mechanics of Jointed Rock, J. Soil Mech. Found. Div., ASCE, vol. 94, No. SM3:637-659.

Sakurai, S. 1981. Direct Strain Evaluation Technique in Construction of Underground Opening, Proc. 22nd U.S. Symposium on Rock Mechanics, Rock Mechanics from Research to Application, MIT:298-302.

Sakurai, S. 1982. Monitoring of Caverns during Construction Period, Proc. ISRM Symposium, Rock Mechanics: Caverns and Pressure Shafts, Aachen:433-441.

Sakurai, S. and K. Takeuchi 1983. Back analysis of Measured Displacements of Tunnels, Rock Mechanics and Rock Engineering, vol.16:173-180.

Sakurai, S., N. Shimizu and K. matsumuro

1985. Evaluation of Plastic Zone around Underground Openings by Means of Displacement Measurements, Proc. 5th Int. Conf. Numer. Methods in Geomechanics, Nagoya, vol.1:111-118.
Sakurai, S., Niyom Deeswasmongkol and M. Shinji 1986. Back analysis for Determining Material Characteristics in Cut Slopes, presented at Int. Sympo. Engineering in Complex Rock Foundations, Beijin.

Computer and Physical Modelling in Geotechnical Engineering, Balasubramaniam et al. (eds)
© 1989 Balkema, Rotterdam. ISBN 90 6191 864 2

Special problems in the computer aided analysis of underground structures

Prabhat Kumar
Structural Engineering Research Centre, Roorkee, India

Bhawani Singh
Civil Engineering Department, University of Roorkee, Roorkee, India

ABSTRACT: The utility of computer aided analysis has been considered doubtful for several decades because of serious assumptions. Nevertheless computer aided analysis does give a feeling of the problem being analysed. Finite element method is a widely used tool in the computer aided analysis of underground structures. Although, many general purpose packages are readily available, most of them lack some of the special features and capabilities required for the analysis of underground structures. These additional features and some recent trends are outlined in this paper. The problems of reduction of a three-dimensional problem into a two-dimensional problem and analysis of unbounded domain can be handled by implementing quasiplane strain elements and infinite elements, respectively. A development of these special elements is given in some detail. The utility of infinite elements is demonstrated by solving problems of (a) deep underground openings (b) shallow underground opening and (c) parallel twin underground openings.

1 INTRODUCTION

The ground is in a state of prestress due to nonuniform solidification of its various layers at the time of its formation and the subsequent tectonic activities. In its virgin state, each rock block is restrained against movement by the adjacent blocks. In the underground excavation some rock mass is removed. Consequently, restraint of some of the rock block vanishes on at least one side. These rock blocks are then pushed by the adjacent blocks towards the no-restraint side. In the case of a highly jointed rockmass, some of the blocks may become unstable and may eventually fall down, thus, releasing constraint of more rock blocks. This process known as redistribution continues till a new equilibrium state is reached. This new state is characterized by concentration of compressive stresses in some locations and of tensile stresses in the other. The object of analysis of underground excavation is to determine these stress concentrations, the shape and size of the broken zone created by the excavation and the associated displacements. The input to this analysis are the opening shape, initial stress pattern, geology of the area and the mechanical strength characteristics of materials.

When rockmass is homogeneous and elastic, and opening shape is simple a closed form solution can be obtained by the theory of elasticity. More than often the first of these conditions is violated so that a numerical method such as the finite element method has to be employed in the analysis. In recent years, with the availability of large and fast digital electronic computes, the sophistication of the finite element method has also increased. It is now possible, at least in principle, to perform a non-linear time-dependent finite element analysis taking into account thermal effects and movement of pore water.

It must however be realised that the sophistication of finite element analysis alone is not sufficient to guarantee usefulness of the results of analysis, particularly so in the context of geo-technical problems. The material and geometrical modelling play at least an equally important role. The former of these is concerned with the adequate simulation of material behaviour while the latter deals with the modelling of geological features. It is believed that an adequate modelling followed by a somewhat simplified analysis may be of greater utility than a highly sophisticated analysis of a simplified numerical model.

Most of the readily available computer program packages of finite element analysis were developed from the perspective of a structural or mechanical engineer and, therefore, lack special features and analysis capabilities required in the solution of geotechnical problems. Some of these features and capabilities are described in the following sections. With the description of these special features and recent trends of analysis available, it should be possible to undertake development of the right kind of software for the analysis of geotechnical structures. It is not possible to describe in this paper recent hardware dependent developments and trends of computer programming such as parallel processing, array processing and symbolic manipulation. Due account to these must, however, be given before planning to develop new software.

2 ADDITIONAL NUMERICAL MODELLING FEATURES

The subsequently described features are usually present in most problems of geotechnical engineering. At times, it may be feasible to incorporate simplifications depending upon the conditions associated with a particular project. Still the software should be able to deal with these features as though all these are present.

2.1 Nonlinear Material Behaviour: (1-3)

For the analysis of stresses and displacements in the tunnel lining or in the surrounding rock or soil, the material properties must be considered as realistically as possible to provide safety and stability of the construction. The behaviour of rock and the lining is dominated in almost all cases by nonlinear material properties. The nonlinearity arise because of several factors some of which are given below:

- Time-dependent rheological deformations and stress changes in rock or soil
- Creep and shrinkage in the lining concrete
- Fissures, cracks and joints in rock, developing with the change of the state of stress
- Anisotropic material behaviour in connection th different strengths in different uirections
- Nonlinear characteristics of the stress-strain relationship
- Failure of rock due to local tension or unloading or separation of joints
- Limited deformation capability in compression and strain softening

It is obvious that a simplified analysis based on the assumption of homogeneous and isotropic material behaviour will be of little use. The problem is not only to device advance material models to accurately simulate the observed material behaviour but also to determine the numerical values of the associated parameters from a sample of matérial. Its implementation into an existing finite element analysis computer program may be done by the subsequently described strategies.

2.2 Special Solution Strategies: (4-7)

The solution strategies for the finite element analysis of geotechnical problems are complex and involved primarily because of a nonlinear constitutive relationship which is to be obeyed. To be specific, the solution strategy must be iterative in which the displacements, strains and stresses are progressively modified till these are uniquely related as per the requirements of the constitutive law. The reformulation and refactorization of the stiffness matrix at each iteration step may be prohibitively expensive. An alternative is to keep the same stiffness matrix in its factored form and workout a residual force vector caused by the violation of the constitutive law. This force vector is reapplied and diffused through the structure. This proceduere is to continue till the residue is acceptably small. The technique may, however, suffer from the disadvantage of slow convergence and at times the solution may diverge due to ill-conditioning of the system matrices. An alternative strategy is to use an incremantal analysis in which a small increment of load is applied at a time and the analysis is performed assuming the material behaviour to be linear between increments. Effectively, it corresponds to a piecewise linearization of the constitutive law. To make the strategy most effective, some iterations my also be performed between various load increments to precisely trace the load deflection curve. It is necessary to control the size of load increments, interval between stiffness matrix reformulations, the convergence tolerance and maximum number of cycles for equilibrium iteration.

Recently, various updating techniques based on the quasi-Newton variable metric methods have been proposed to avoid reformulation and refactorization of stiffness matrix. In such methods, a change in the solution due to a change in some part of the stiffness matrix can be computed directly. An availability of less expensive solution techniques and faster as well as bigger computers help in the solution of complex problems which could not have been solved a few years ago.

2.3 Initial Stress:

Most of the finite element analysis computer programs assume that the structure starts from a zero stress state. The stresses due to external loads are computed and stored in one vector, and later modified to account for a nonlinear behaviour of some of the elements. But as explained in the previous section, the geotechnical structures are in a state of prestress. In this situation, the stresses due to external loads have to be superimposed on the initial stress pattern to obtain effective stress pattern which determines the subsequent behaviour of various elements. The stress computation procedure in the presence of initial stresses become little more involved than it is without it. It is not guaranteed that such a procedure will be available in the existing finite element analysis packages. So an analyst has to find ways to deal with this situation before going ahead with the analysis.

2.4 Sequential Excavation and Construction: (8-14)

Most geotechnical structures are not constructed in one attempt. Various operations associated with a particular project may take place simultaneously but at different locations. The entire arrangement moves forward till the completion of project. This is very typical in tunnel construction. Research experience shows that there is a difference in the displacements and stresses when these are obtained by analysing the finished structure and by analysing it in actual construction steps, particularly so, when nonlinearities are involved. Such simulation of sequential construction and excavation is achieved by activating and deactivating a certain set of finite elements at appropriate stage in the calculations by the finite element method. It is quite obvious that the mesh to be employed in such an analysis must be drawn by taking the construction procedure into account.

2.5 Reduction of 3D Problem into 2D Problem: (15-17)

The actual problem of underground construction is three-dimensional in nature, the finite element analysis of which is expensive. If all the nonlinearities and sequential excavation procedures are also included in the analysis, its cost may be prohibitive. Methods have been devised to reduce the solution of actual 3D problem to either a series of 2D problems or to two different 2D analyses in two perpendicular planes. By doing an analysis in the longitudinal plane, the variation of longitudinal strain is determined in a ploynomial form which is then used in the analysis in transverse plane. This approach is further elaborated later on in this paper.

2.6 Analysis of Coupled Phenomenon: (18-20)

Most engineering analysis problems are coupled which are characterized by the dynamic interaction between fields so that response of the overall system has to be calculated concurrently. These problems have to be simplified, often drastically, in order to make them amenable to analysis. In this process the coupling is lost. The soil-structure-fluid interaction is typical example of a coupled phenomenon in which displacement and pressure are the variables in the solid and fluid domains, respectively. These variables are interrelated and influence numerical value of each other.

The coupled phenomenon is formally defined as follows (20) -

'Coupled formulations are those applicable to multiple domains and dependent variables which usually (but not always) describe different physical phenomenon and in which,

. Neither domain can be solved separately from the other
. Neither set of dependent variables can

be explicitely eliminated.'

In recent years, improvements chiefly in computing power, but also in understanding of algorithms and of physical processes have made it possible to look in more detail at these coupled effects. For this purpose, an altogether new brand of software is needed. For example, it is shown (20) that the resulting equations of coupled phenomenon can be expressed in standard form but the coefficient matrix becomes asymmetric so that the standard solution strategies require modification to handle the coupled effects.

2.7 Unbounded Domain of Analysis: (21-28)

In the case of most underground structures, the domain of analysis is the entire ground, the extent of which is large. It is not possible to include all of its in the finite element analysis. Initially, a truncation approach was employed in the analysis. This was based on the observation that the effect of an underground construction is not felt after a certain distance and that the error introduced by truncation is entirely due to the left-out portion of a decaying phenomenon. It can be controlled to be within a reasonable and acceptable limit, provided the truncation boundary is placed at a sufficiently large distance from the zone of disturbance. However, to determine a suitable location of the truncation boundary in most situations is an additional unknown which must be assigned a numerical value before an analysis can proceed.

An alternative lies in using a combination of finite and infinite elements to discreties the domain of analysis. The development of infinite elements and some applications are given later on in this paper.

2.8 Other Features:

In any finite element analysis, the preparation of input and interpretation of results are the most time consuming operations. These demand considerable specialization and skill of the analyst. It is possible to reduce the burden of the analyst by including the features like reading the node coordinates through a light pen. A preprocessor can also be written for automatic mesh generation and optimum node numbering for a minimum bandwidth of the resulting global stiffness matrix. A postprocessor for interactive plotting of output, analysis interruption and restsrt options and a host of other features will greatly expedite the analysis. These features, however, are common with other application of the finite element analysis.

3 DEVELOPMENT OF SPECIAL ELEMENTS

Some of the special tasks associated with the analysis of geotechnical structures, namely, reduction of a 3D analysis problem into a 2D analysis problem and that of modelling of unbounded domain of analysis can be accomplished by developing special elements and by implementing these in the element library of a finite element analysis package. Two such elements, for each of the abovementioned problems, are subsequently described.

3.1 Quasi Plane Strain Elements: (15,17)

With the help of these elements it is possible, in some cases, to reduce a 3D analysis problem into a 2D problem. It is particularly useful when all conditions of a plane strain formulation are not satisfied. Such is the case in a tunnel passing through anisotropic rock when a principal axis of material orthotropy does not coincide with the axis of tunnel, or alternatively, the direction of principal tectonic or initial stresses do not coincide with the axis of excavation.

The quasi plane strain analysis applies when all components of displacement exist but these are independent of the coordinate in the direction of tunnel axis. Then the analysis can be carried out in a two-dimensional plane but with fully three-dimensional displacement, strain and stress components.

Consider a completely three-dimensional situation. The standard equations are as follows. Let U be the displacement vector with its components as u,v and w in the three coordinate directions, x,y and z, respectively shown in Fig. 1. Then,

$$U = [u(x,y,z) \quad v(x,y,z) \quad w(x,y,z)] \tag{1}$$

$$\underline{\varepsilon} = L\,U \tag{2}$$

$$L^t = \begin{bmatrix} \frac{\partial}{\partial x} & 0 & 0 & \frac{\partial}{\partial y} & 0 & \frac{\partial}{\partial z} \\ 0 & \frac{\partial}{\partial y} & 0 & \frac{\partial}{\partial x} & \frac{\partial}{\partial z} & 0 \\ 0 & 0 & \frac{\partial}{\partial z} & 0 & \frac{\partial}{\partial y} & \frac{\partial}{\partial x} \end{bmatrix} \tag{3}$$

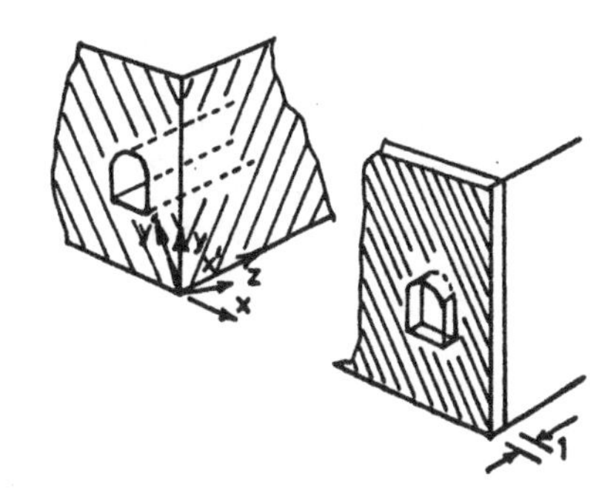

Fig. 1 Tunnel geometry

$$\underline{\varepsilon} = [\varepsilon_{xx}\ \varepsilon_{yy}\ \varepsilon_{zz}\ \varepsilon_{xy}\ \varepsilon_{yz}\ \varepsilon_{zx}] \quad (4)$$

$$\underline{\sigma} = [\sigma_{xx}\ \sigma_{yy}\ \sigma_{zz}\ \sigma_{xy}\ \sigma_{yz}\ \sigma_{zx}] \quad (5)$$

$$\underline{\sigma} = D\,\underline{\varepsilon} + \sigma_o \quad (6)$$

$$D = \begin{bmatrix} a_{11} & a_{12} & a_{13} & a_{14} & a_{15} & a_{16} \\ & a_{22} & a_{23} & a_{24} & a_{25} & a_{26} \\ & & a_{33} & a_{34} & a_{35} & a_{36} \\ & \text{sym.} & & a_{44} & a_{45} & a_{46} \\ & & & & a_{55} & a_{56} \\ & & & & & a_{66} \end{bmatrix} \quad (7)$$

For a displacement vector given in Eq. 1, the strain-displacement relations are given in Eq. 2. The L matrix is defined in Eq. 3. In this equation the superscript t denotes transpose. The strain and stress vectors are given in Eqs. 4 and 5, respectively. These vectors are related as in Eq. 6 where the D matrix is given in Eq. 7. In the D matrix, the coefficients about the main diagonal are symmetric. Various strain and stress components are coupled because of non-zero off-diagonal coefficients. In a plane strain situation the out-of-plane displacements w become zero, so that the above equations 1 to 7 simplify and are written as Eqs. 1p to 7p. The suffix p denotes a plane strain version of equation.

$$U = [u(x,y)\quad v(x,y)\quad 0] \quad (1p)$$

$$\underline{\varepsilon} = L\,U \quad (2p)$$

$$L^t = \begin{bmatrix} \frac{\partial}{\partial x} & 0 & \frac{\partial}{\partial y} \\ 0 & \frac{\partial}{\partial y} & \frac{\partial}{\partial x} \end{bmatrix} \quad (3p)$$

$$\underline{\varepsilon} = [\varepsilon_{xx}\quad \varepsilon_{yy}\quad \varepsilon_{xy}] \quad (4p)$$

$$\underline{\sigma} = [\sigma_{xx}\quad \sigma_{yy}\quad \sigma_{xy}] \quad (5p)$$

$$\underline{\varepsilon} = D\ \underline{\varepsilon} + \sigma_o \quad (6p)$$

$$D = \begin{bmatrix} a_{11} & a_{12} & a_{13} & a_{14} \\ & a_{22} & a_{23} & a_{24} \\ & \text{sym.} & a_{33} & a_{34} \\ & & & a_{44} \end{bmatrix} \quad (7p)$$

The quasi-plane strain situation lies in between these two extremes in which all components of displacement exist but these are independent of the coordinate in the direction of tunnel axis. Thus, the displacement vector is as given in Eq. 1q. Other relavent equations are as Eq. 2q to 6q. The suffix q denotes a quasi-plane strain version of an equation.

$$U = [u(x,y)\quad v(x,y)\quad w(x,y)] \quad (1q)$$

$$\underline{\varepsilon} = L\ U \quad (2q)$$

$$L^t = \begin{bmatrix} \frac{\partial}{\partial x} & 0 & 0 & \frac{\partial}{\partial y} & 0 & 0 \\ 0 & \frac{\partial}{\partial y} & 0 & \frac{\partial}{\partial x} & 0 & 0 \\ 0 & 0 & 0 & 0 & \frac{\partial}{\partial y} & \frac{\partial}{\partial x} \end{bmatrix} \quad (3q)$$

$$\underline{\varepsilon} = [\varepsilon_{xx}\quad \varepsilon_{yy}\quad 0\quad \varepsilon_{xy}\quad \varepsilon_{yz}\quad \varepsilon_{zx}] \quad (4q)$$

$$\underline{\sigma} = [\sigma_{xx}\quad \sigma_{yy}\quad \sigma_{zz}\ \sigma_{xy}\ \sigma_{yz}\quad \sigma_{zx}] \quad (5q)$$

$$\underline{\sigma} = D\ \underline{\varepsilon} + \sigma_o \quad (6q)$$

$$D = \begin{bmatrix} a_{11} & a_{12} & a_{13} & a_{14} & a_{15} & a_{16} \\ & a_{22} & a_{23} & a_{24} & a_{25} & a_{26} \\ & & a_{33} & a_{34} & a_{35} & a_{36} \\ & \text{sym.} & & a_{44} & a_{45} & a_{46} \\ & & & & a_{55} & a_{56} \\ & & & & & a_{66} \end{bmatrix} \quad (7q)$$

In the special case when the material is orthotropic (or isotropic) and a principal material axis coincides with the z-axis (Fig. 1), the quasi-plane strain problem uncouples into (a) the usual plane strain linear elastic problem and (b) a problem described by the out-of-plane displacements w(x,y) which determine σ_{yz} and σ_{zx} components of stress vector. The out-of-plane problem can be solved entirely separately from the plain strain problem and the results superposed only to compute principal stresses and strains. In a general case, the two problems must be solved at the same time. This approach leads to a more efficient solution than a general 3D case having three degrees of freedom per node. However, this approach fails when the problem is non-linear or when no axis of symmetry along z-axis exists.

$$U = [u(x,y) \quad f_z v(x,y) \quad 0] \qquad (3qp)$$

$$L^t = \begin{bmatrix} \frac{\partial}{\partial x} & 0 & 0 & \frac{\partial}{\partial y} & 0 & 0 \\ 0 & f_z\frac{\partial}{\partial y} & 0 & f_z\frac{\partial}{\partial x} & f'_z & 0 \\ 0 & 0 & 0 & 0 & 0 & 0 \end{bmatrix} \qquad (4qp)$$

$$\underline{\varepsilon} = [\varepsilon_{xx} \quad \varepsilon_{yy} \quad 0 \quad \varepsilon_{xy} \quad \varepsilon_{zy} \quad 0\,] \qquad (5qp)$$

$$\underline{\sigma} = [\sigma_{xx} \quad \sigma_{yy} \quad \sigma_{zz} \quad \sigma_{xy} \quad \sigma_{yz} \quad \sigma_{zx}] \qquad (6qp)$$

In an alternate formulation of quasi-plane problem (17), the displacement is assumed to be as given in Eq. 1qp. The strain-displacement relation is same as given in Eq. 2, however, the L matrix is now given by Eq. 3qp. The strain and stress vectors are given by Eqs. 4qp and 5qp, respectively. The function f_z in Eq. 1qp is determined by a separate finite element analysis in the y-z plane shown in Fig. 1. The rest of the analysis is as described in the preceeding paragraphs. It is seen that a quasi-plane strain formulation allows the variation in the z-direction to be taken into account without imposing the penalty of a full 3D analysis, provided certain conditions are met.

3.2 Infinite Elements:

The space surrounding an underground structure is usually large. It is practically impossible to take all of it into account in the finite element analysis. Consequently, the domain of analysis is truncated at a large but finite distance from the zone of disturbance as though nothing existed beyond that distance. Although this approach has been extremely useful, its working depends upon a judicious selection of the location of the truncation boundary. If it is located deep inside the far-field, the error in the solution is due to the left out portion of decay. It can be controlled to be within the limits of accuracy of any engineering analysis. If, on the other hand, it is located inside the near-field, absurd results of analysis may be expected. To determine an appropriate location of the truncation boundary for a problem may not be straightforward and may even necessitate numerical experimentation.

An alternative lies in the use of a combination of finite and infinite elements to represent the near-field and far-field, respectively. This approach has an added advantage in that the size of the resulting problem is relatively smaller for the same accuracy of the results of analysis. The infinite elements can be formulated in two different ways which are subsequently described.

In the conventional finite element formulation, two global to local transformations are employed. One is for the coordinates and the other for unknown function values. The coordinate transformation is used in computing Jacobian, which is then employed in computation of global derivatives of the shape functions. In the case of infinite element formulation, these two transformations are still employed, however, two different ways are possible each leading to a different formulation.

In one formulation, known as the displacement descent formulation, the coordinate transformation is based on the conventional shape functions eventhough some of the nodes of the infinite element exist at infinity (or at a large distance). The shape functions of the function value transformation are obtained by modifying the usual shape functions so that these now decay to infinity in a particular way as is achieved in Eq. 8. $M_i\,(\xi, \eta)$ are the original shape functions and $f_i(\xi, \eta)$ are known as decay functions. $N_i^I(\xi, \eta)$ are the shape functions for the infinite element. The subscript i correspond to the node number and (ξ, η) are the local coordinates.

$$N_i(\xi,\eta) = f_i(\xi,\eta)\ M_i(\xi,\eta) \qquad (8)$$

The decay function must be unity at its own node. In addition, N_i must tend to the far-field value at infinity. There is no restriction that the decay function take any special value at the other nodes but these must realistically describe behaviour of the problem. The following forms of decay functions given in Eqs. 9 to 11 are available (22).

Exponential decay in positive ξ - direction;

$$f_i(\xi,\eta) = \mathrm{Exp}\ (\frac{\xi_i - \xi}{L}) \qquad (9)$$

Exponential decay in both positive ξ- and η-direction;

$$f_i(\xi,\eta) = \mathrm{Exp}\ (\frac{\xi_i + \eta_i - \xi - \eta}{L}) \qquad (10)$$

Reciprocal decay;

$$f_i(\xi,\eta) = (\frac{\xi_i - \xi_o}{\xi - \xi_o})^n \qquad (11)$$

Physically the operation of Eq. 8 amounts to streching of a finite element to be an infinite element. A major disadvantage of this approach is that the so derived infinite elements require a special numerical integration scheme extended over a semi-infinite range in the evaluation of their stiffness and mass properties.

In the other formulation, the shape functions of the coordinate transformation are derived so that the infinite element in the physical plane is mapped into a convenient shape in the natural plane. Now the conventional shape functions are used in the function transformation. Physically this formulation amounts to compressing an infinite element in the physical plane to be a finite element of a regular shape in the natural plane. A major advantage of this approach lies in the evaluation of mass and stiffness properties which can be done by the conventional Causs-legendre numerical integration scheme.

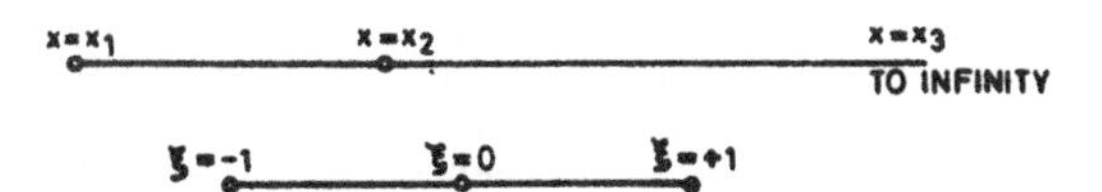

Fig. 2 One dimensional infinite element and its map

Consider the one-dimensional infinite element shown in Fig. 2. It is assumed that the function variable decays to zero at infinity.

$$N_1 = \frac{-2\xi}{1-\xi}\ ;\quad N_2 = \frac{1+\xi}{1-\xi} \qquad (12)$$

$$x = N_1x_1 + N_2x_2 + N_3x_3 \qquad (13)$$

$$\xi = 1 - \frac{2(x_2 - x_1)}{x - 2x_1 + x_2} \qquad (14)$$

$$\xi = 1 - \frac{2x_1}{x} \quad \text{or} \quad x = \frac{2x_1}{(1-\xi)} \qquad (15)$$

The shape functions are defined in Eq. 12. On substitution of these in the coordinate transformation Eq. 13 and solving for ξ, Eq. 14 is obatined. Let $x_2 = 2x_1$. This implies that the mid-side nodes are placed at twice the distance of node 1. Since the location of mid-side node is at the discretion of the analyst, it is equally convenient to choose it according to this requirement and it does not affect generality of the derivation. With this substitution, Eq. 14 reduces to Eq. 15. The mapping defined by this equation is described in Table 1. The Eq. 15 shows that the decay is of inverse type.

Table 1. Mapping by Eq. 15

x	ξ
x_1	- 1
x_2	0
infinity	+ 1

The infinite element shown in Fig. 2 can be extended to be a two-dimensional infinite element. Such an element along with the shape functions is shown in Fig. 3.

By following a procedure similar to that given in Eqs. 12-15, and using $x_4 = 2x_1$ and $x_5 = 2x_3$, the mapping produced by the shape functions of Fig. 3 can be shown to be as given by Eq. 16. The mapping is also described in Table 2. It can be seen that the infinite element in the physical plane reduces to a finite element of a regular shape in the natural or local plane. It can be integrated by Gauss-Legendre numerical integration scheme to evaluate

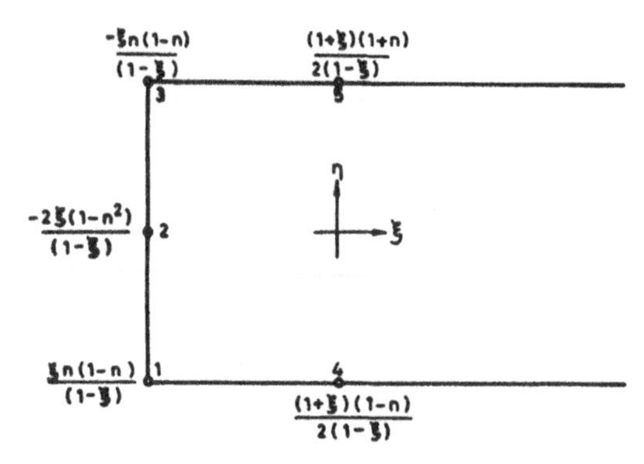

Fig. 3 Infinite element with (1/r) type decay

$$\xi = 1 - \frac{(x_1 + x_3) + \eta(-x_1 + x_3)}{x} \quad (16)$$

Table 2. Mapping by Eq. 16

ξ	η	x
+1	all η	Infinity
0	-1	$x = 2x_1 = x_4$
	+1	$x = 2x_3 = x_5$
-1	-1	$x = x_1$
	0	$x = \frac{x_1 + x_3}{2} = x_2$
	+1	$x = x_3$

its stiffness and mass properties. Further details of implementation and infinite elements with different types of decay are available elsewhere (28).

4 APPLICATION OF INFINITE ELEMENTS

In this paper some problems of practical interest are solved to demonstrate utility of the infinite elements. To be able to compare the results of analysis with the analytical results, a linear elastic isotropic and homogeneous medium is considered. Such a comparison is desirable to establish applicability of the finite/infinite element approach and also to develop guidelines for its most effective use in the solution of more involved problems.

4.1 Analysis of Deep Underground Openings: (29 - 30)

The deep openings of circular and elliptical shapes are analysed in a purely vertical initial stress field. The method, however, is applicable to any combination of vertical and horizontal stresses. In this study the modulus of elasticity of the medium is taken as 1000 kg/cm^2 and the Poisson ratio as 0.2. A 3x3 Gauss integration order is used for finite elements and a 2x2 integration is used for infinite elements. The finite/infinite element mesh configuration is shown in Fig. 4. The stress concentration factors obtained from this study are compared with their analytical values in Table 3. The agreement can be seen to be satisfactory.

4.2 Analysis of Shallow Underground Openings: (31-34)

A shallow tunnel is located close to the free surface and subsidence due to underground excavation may damage the structures on the ground. Also, any structure which is constructed after the tunnel construction will influence safety of the tunnel and its lining. The tunnels which are constructed to carry essential services like water supply, sewage, railway and road traffic are usually shallow. The forces considered in the analysis of such tunnels are the gravitational forces (unit weight = 2.79 gm/cc) and a horizontal initial stress field (parallel to free surface). According to one study (34), the magnitude of the horizontal initial stress field lies in the range (2.7 + 0.0018 z) Mpa and (40.5 + 0.0135 z) Mpa where z is the depth in meters below ground level. The analytical solution to the shallow tunnel problem are due to Mindlin (31,32). In this study the solution is obtained by the finite/infinite element method. The mesh configuration is shown in Fig. 5. The detail of the box portion are given in Fig. 6. The results of analysis for various mesh configurations are given in Table 4. The analysis shows that when the opening is placed very close to the free surface, tension does not develop anywhere along the free surface. Also, the finite/infinite element analysis can predict various stress concentration factors

Table 3. Stress concentration factors at selected locations for deep openings

Opening shape	Circular		Tall Elliptical		Long Elliptical	
Location	1	2	1	2	1	2
Analytical Values (30)	1.0	3.0	1.0	2.0	1.0	5.0
Numerical Values	1.0074	3.0173	0.8648	2.0486	1.011	4.7798

NOTE: Locations 1 and 2 are in the roof and side wall, respectively

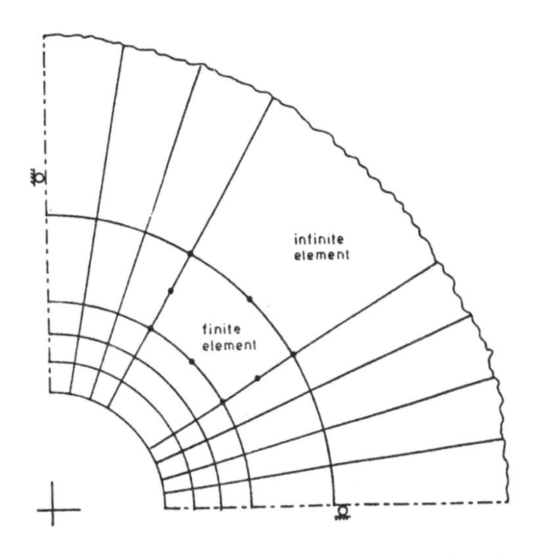

Fig. 4 A typical mesh for the analysis of deep circular openings

Fig. 5 Mesh for analysis of shallow tunnel openings

Table 4. Stress concentration factors for shallow circular opening with H/R = 1.1

Station	Theory	Experi-mental	Numerical computation with mesh type				
			A	B	C	D	E
m	0.3	0.38	1.1443	0.9173	0.5615	0.3271	0.3082
n	9.2	11.28	8.3974	8.4661	9.0143	9.1091	9.1269
o	-	-	0.8291	0.6996	0.8279	0.6982	0.7089
p	3.55	4.3	3.5668	3.5816	3.5683	3.5831	3.5838

NOTE: Stations are identified in Fig. 5

satisfactorily. In an study on shallow tunnels (33) it is shown that the suitability of the finite-infinite element subdivision of the analysis domain can be tested by simply moving the finite/infinite element interface.

4.3 Analysis of Deep and Adjacent Underground Openings: (35)

When needs grow to such an extent that the existing tunnel becomes inadequate, then, a new tunnel has to be excavated

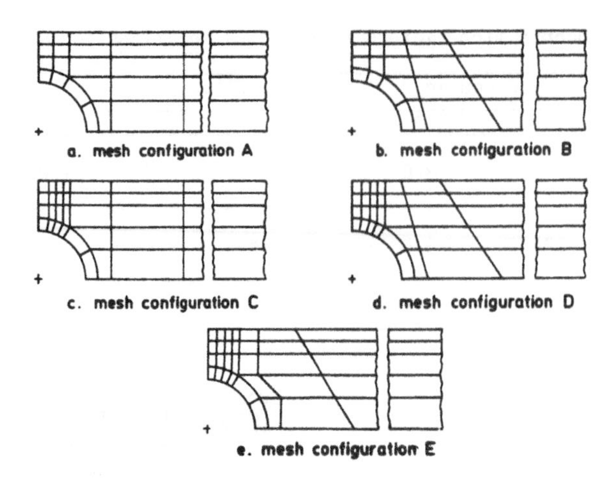

Fig. 6 Various mesh configurations used in Table 4

which may be conveniently located in the near vicinity of the existing tunnel. Even when multiple tunnel openings are planned, they are not constructed simultaneously. It is therefore important that the lining is designed to withstand interference from a new opening. A situation corresponding to the parallel excavation is shown in Fig. 7. It is analysed by the method of finite/infinite elements. The finite/infinite element mesh used in the study are shown in Fig. 8. This study is carried for the following values of the important non-dimensional parameters.

L/D_2 = 1.25 2.0 3.0

E_1/E_g = 100.0 10.0

D_1/t = 10.0 20.0

Other relevent constants are given the following values.

$D_1 = D_2 = 10.0$ m

$E_1 = 20.0\ E + 06\ KN/m^2$

$\nu_g = 0.3; \quad \nu_1 = 0.15$

$\sigma_h = \sigma_v = 10000\ KN/m^2$

Some of these symbols are identified in Fig. 7. E is the the modulus of Elasticity, ν = Poisson ratio, σ = initial stress intensity. The subscript g stands for ground, 1 for lining, h for horizontal direction and v for vertical direction. The results of analysis are shown in Fig. 9. The analysis reveals that the lining of the existing tunnel will be subjected to severe hoop tension and compression at crown and springing points, respectively. In this study, the lining is considered to be rigidly bonded to the existing tunnel opening. If interface elements are introduced at the lining-opening interface, it shall be possible to study the effect of a finite slip on the magnitude of tensile forces acting on the lining of existing tunnel.

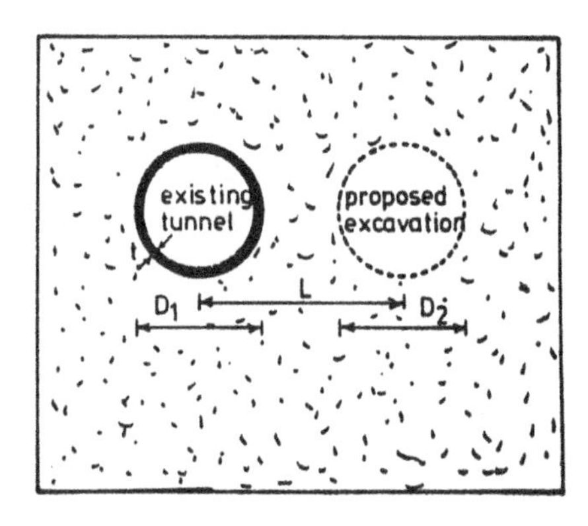

Fig. 7 Adjacent and parallel openings

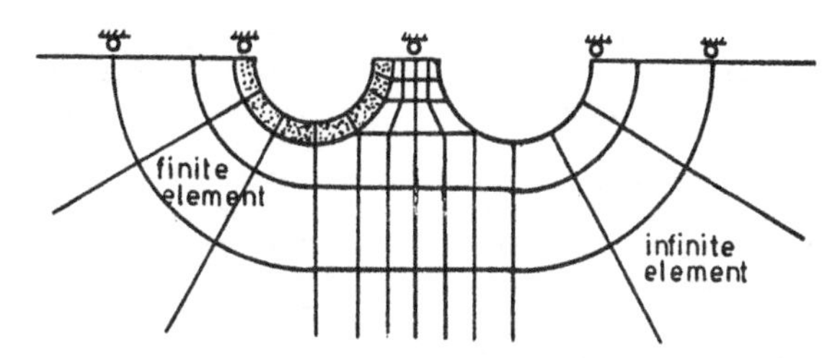

Fig. 8 Finite/infinite element mesh for adjacent and parallel openings

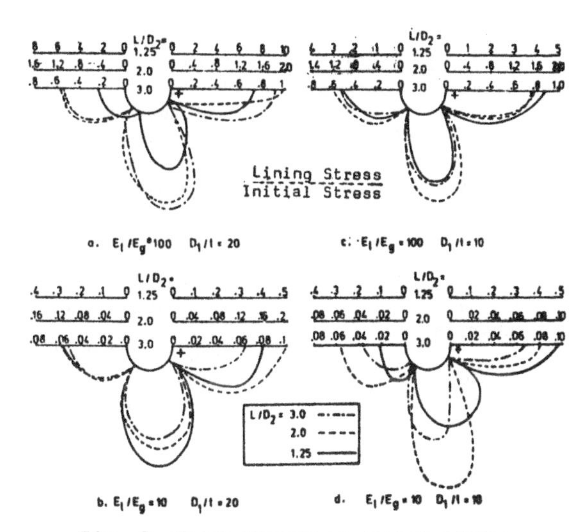

Fig. 9 Extra forces on tunnel lining

5 CONCLUDING REMARKS

The special features and capabilities required in the finite element analysis of underground structures are described in this paper. The development of quasi-plane strain elements and infinite elements is given. With the help of these a 3D problem can be reduced to a 2D problem and the unbounded analysis domain can be efficiently modelled. The latter aspect is demonstrated by solving three problems of practical interest.

REFERENCES

1. Singh,B. 1973. Continuum characterization of jointed rock masses. (Part I and II). International Journal of Rock Mechanics and Mining Sciences. Vol.10, pp. 311-349.
2. Christian, J.T. and Desai, C.S. 1977. Constitutive laws for geologic media, in numerical methods in geotechnical engineering. McGraw Hill Book Co., NY, pp. 65-115.
3. Zienkiewicz,O.C. and Pande,G.N. 1977. Time-dependent multilaminate models of rock - A numerical study of deformation and failure of rock masses. International Journal of Numerical Methods in Geomechanics. Vol.1, pp. 219-247.
4. Owen,D.R.J. and Hinton,E. 1980. Finite elements in plasticity: Theory and Practice, Pineridge Press Ltd., U.K.
5. Zienkiewicz,O.C. 1977. The finite element method. McGraw Hill Book Co., NY.
6. Goodman,R.E. 1975. Methods of Geological engineering in discontinuous rocks. West Publishing Co., St.Paul. Minn. USA.
7. Wilson,E.L. 1977. Finite elements for foundations, joints and fluids. In Finite Elements in Geomechanics. G. Gudehus (ed), John Wiley & Sons, London. pp. 319-350.
8. Duncan,J.M. and Clough,G.W. 1971. Finite element analysis of port allan lock. Journal of Soil Mechanics and Foundation Engineering, ASCE. Vol.97, No.SM4, pp. 1053-1068.

9. Ishihara,K. 1970. Relations between process of cutting and uniqueness of solutions. Soils and Foundations, Vol.10, pp. 50-65.
10. Christian,J.T. and Wong,T.H. 1973. Errors in simulating excavation in elastic media by finite elements, Soils and Foundations, Vol.13, pp. 1-10.
11. Chandrasekaran,V.S. and King,G.J.W. 1974. Simulation of excavation using finite elements, Journal of Soil Mechanics and Foundation Engineering, ASCE, Vol.100, pp. 1086-1089.
12. Ghaboussi,J.M. and Ranken,R.E. 1981. Finite element simulation of under ground construction. Proceedings of Symposium on Implementation of Computer Procedures and Strain-Strain Laws in Geotechnical Engineering, Chicago.
13. Dolezalova,M. 1979. The influence of construction work sequence on the stability of underground openings. 3rd International Conference on Numerical Methods in Geomechanics, Vol.II, p. 501.
14. Ghaboussi,J.M. and Pecknold,D.A. 1984. Incremental finite element analysis of geometrically altered structures, International Journal for Numerical Methods in Engng., Vol.20, pp. 2051-2064.
15. Zienkiewicz,O.C., Taylor,R.L. and Pande,G.N. 1978. Quasi-plane strain in the analysis of geological problems, Computer Methods in Tunnel Design, Institution of Civil Engineers (London).
16. Pack,S.C. and Mandel,J.A. 1983. 2D multiplane finite element technique for solving a class of 3D problems, International Journal for Numerical Methods in Engineering, Vol.19, pp. 113-124.
17. Dutta,A. 1985. Quasi 3D interactive analysis of strip footing on pile foundation. International Conference on Computer Aided Analysis and Design in Civil Engineering, Civil Engineering Department, University of Roorkee, Roorkee, India. pp. VI-65 to VI-71.
18. Gudehus,G. (ed). 1977. Finite elements in Geomechanics. John Wiley & Sons, London.
19. Hinton,E., Bettess,P. and Lewis,R.W. (eds.) 1981, Numerical methods for coupled problems, Proceedings of International Conference, University College of Swansea, Swansea, England.
20. Lewis,R.W., Bettess,P. and Hinton,E. (eds.) 1984. Numerical methods in coupled systems, John Wiley & Sons, London.
21. Bettess,P. 1977. Infinite elements. International Journal for Numerical Methods in Engineering, Vol.11, pp. 53-64.
22. Bettess,P. 1980. More on infinite elements. International Journal for Numericall Methods in Engineering, Vol.15, pp. 1613-1626.
23. Curnier,A. 1983. A static infinite element. International Journal for Numerical Methods in Engineering, Vol.19, pp. 1478-1488.
24. Zienkiewicz,O.C., Emson,C. and Bettess,P. 1983. A novel boundary infinite element. International Journal for Numerical Methods in Engineering, Vol.19, pp. 393-404.
25. Beer,G. and Meek,J.L. 1981. Infinite domain element. International Journal for Numerical Methods in Engineering, Vol.17, pp. 43-52.
26. Marques,J.M.M.C. and Owen,D.R.J. 1984. Infinite elements in materially non-linear problems. Computers and Structures, Vol.18, No.4, pp. 739-751.
27. Kumar,P. 1984. Novel infinite boundary element. International Journal for Numerical Methods in Engineering, Vol.20, pp. 1173-74.
28. Kumar,P. 1985. Static infinite element formulation. Journal of Structural Engineering, ASCE, Vol.111, pp. 2355-2373.
29. Kumar,P. 1986. Numerical modelling criterion for the analysis of underground openings using infinite elements. Applied Mathematical Modelling (London), Vol.10, pp. 357-366.
30. Overt,L. and Duvall,W.I. 1967. Rock mechanics and design of structures in rock. John Wiley & Sons, NY.
31. Mindlin,R.D. 1940. Stress distribution around tunnel. Transactions of American Society of Civil Engineers, Vol.105, pp. 1117-53.
32. Mindlin,R.D. 1948. Stress distribution around hole near the edge of a plate under tension. Proceedings of the Society for Experimental Stress Analysis, Vol.5, No.2, pp. 56-58.
33. Kumar,P. 1986. A contribution to an accurate analysis of shallow underground opening. Computers and Geotechnics (London), Vol.2, No.3, pp. 141-151.
34. Hoek,E. and Brown,P.T. 1980. Underground excavation in rock. Institution of Mining and Mettalurgy, London.
35. Kumar,P. and Sharma,S.P. 1985. Forces on the lining of an existing tunnel due to an adjacent excavation. IGC-85, Indian Geotechnical Conference, Vol.I, pp. 407-413.

Computer and Physical Modelling in Geotechnical Engineering, Balasubramaniam et al. (eds)
© 1989 Balkema, Rotterdam. ISBN 90 6191 864 2

The boundary element method for determining stresses around openings in non-homogeneous rock media

E.A.Eissa & A.Kazi
King Abdulaziz University, Jeddah, Saudi Arabia

ABSTRACT : A boundary element formulation is described to determine the stresses around underground openings in two and three layered media. In the former situation, consideration is given to the nearness of a thin stratum to an opening located in the host rock whilst in the latter instance the problem of an opening located in a central stratum confined between thin layers of rock is examined. It is shown that the distribution of stresses around underground openings in layered media is influenced by the thickness of the excavated layer in relation to the size of the opening and the contrast between the elastic properties of the excavated and surrounding layers.

The method has been applied to a situation encountered at the Abu Tartour phosphate mines in Egypt, where a bed of hard phosphate ore sandwiched between soft layers of shale is being mined. Recommendations are given for reducing excessive tensile stresses at both the roof and the floor of the adopted tunnel cross-sections by adjusting the heights of the excavations relative to the thickness of the phosphate layer.

1 INTRODUCTION

Boundary element methods have been successfully applied to many linear stress analysis problems, on the assumption that the domain is completely homogeneous (Bray, 1976; Brady and Bray, 1978-a; and Eissa, 1980). Conditions of non-homogeneous rock media (due to the presence of ore bodies surrounded by country rock) are of common occurrence in mining practice; such situations generally prevail in coal and phosphate mines, and frequently in bedded metallic deposits. Zienkiewicz et al. (1966) applied the finite element method of numerical analysis to problems in anisotropic media. The same method of numerical analysis was used for the determination of stresses around tunnels in two layered structures by Goodman (1966) and in three layered media by Agarwal and Boshkov (1969) and Barla (1972)

The boundary element method with its distinct advantages over the finite element and other methods of numerical analysis, as discussed by Brady and Bray (1978a), has been used in this paper to compute the stresses around the underground openings in layered rock. The type of problem examined in this study is illustrated in Figure 1, which shows a typical example of a mine opening located in a layer of thickness t which is subjected to a known state of uniform field stresses p_x and p_z. In order to avoid the effect of traction free surface on the stresses, the depth of the opening centre is considered to be greater than 12 a, where 2 a is equal to the height of the opening (Eissa, 1980). The rockmass is regarded as a composite structure where each layer is a homogeneous component with known values of E (Young's modulus) and ν (Poisson's ratio). It is assumed that the different layers remain in contact (bonded) during the induced deformation, thus the distribution of the stress and displacement components is considered to be continuous at the layer interface and no slip occurs between the layers.

2 BOUNDARY ELEMENT METHOD FOR NON-HOMOGENEOUS MEDIA

The solution of problems in the non-homogeneous media can be obtained in two simulations, by representing the homogeneous medium in which the excavation is located and the confining strata as follows:

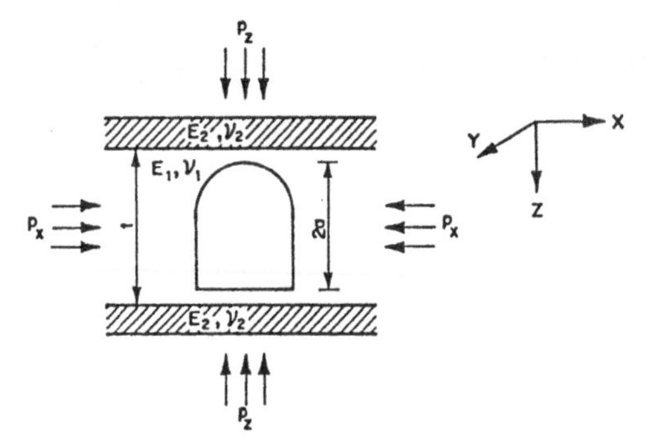

Fig.1 Statement of the problem

2.1 Excavated medium

The boundary element method of stress analysis in homogeneous media has been described by Bray (1976), Brady and Bray (1978a) and Eissa (1980). Figure 2a shows an opening of an arbitrary cross section to be excavated in a medium with known initial field stresses before excavation of p_x and p_z. The boundary element method consists of discretizing the excavation boundary into a number of discrete rectangular elements as shown in Figure 2b, where M, N are the local axes of one such element, the former is parallel to the boundary and the latter is normal to it.

The main requirement of the method is to find a distribution of fictitious forces applied to these elements where the process of excavating the material inside the proposed boundary will reduce the traction on each element to zero. The magnitudes and directions of these forces are chosen in such a way that the surface traction σ_n, τ_{mn} are equal to zero. These boundary conditions determine the stress and displacement components at all points, on and outside the boundary. The distribution of these forces can be approximated numerically by straight line elements with uniform distribution on each element (Figure 3). The stress components at any point p (x_i, z_i) due to uniformly distributed transverse (q_m) and normal (q_n) strip loads at element j can be obtained from expressions given in Appendix 1.

Referring to the local axes of each element M, N and hence for strip loads q_m, q_n, applied over all loaded elements j, the state of stress at the midpoint of element i is obtained by superposition of stress components associated with all the element loads (including element i itself), and the components of field stresses in the medium, as follows:

$$\sigma^i_m = p^i_m + \sum_{j=1}^{k} [q^j_m B^{i,j}_m + q^j_n B^{i,j}_n]$$

$$\sigma^i_n = p^i_n + \sum_{j=1}^{k} [q^j_m C^{i,j}_m + q^j_n C^{i,j}_n] \quad (1)$$

$$\tau^i_{mn} = p^i_{mn} + \sum_{j=1}^{k} [q^j_m D^{i,j}_m + q^j_n D^{i,j}_n]$$

where k represents the number of elements around the boundary of excavation,

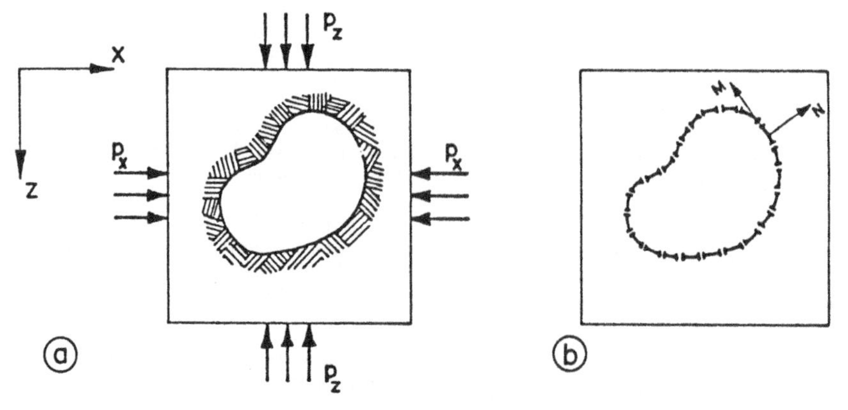

Fig.2 (a) Cross-section of an opening
(b) Discretized boundary showing local axes of boundary element

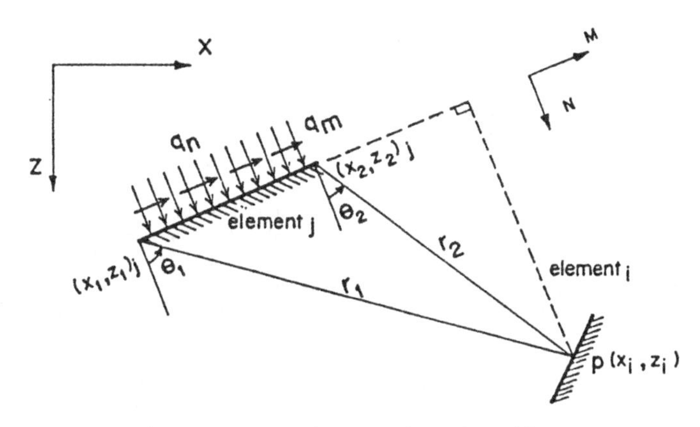

Fig.3 Geometric parameters determining the effect of strip loads on element j at the point $p(x_i,z_i)$

and p^i_m, p^i_n, p^i_{mn} are the field stress components at the midpoint of element i in local coordinates M, N. The coefficients B_m, B_n, C_n, ...etc. are calculated from the appropriate expressions given in Appendix 1 after transforming the stress components from the local coordinate axes of loading element j to the local axes of element i as shown in Figure 3.

The creation of an excavation is simulated by determining a set of strip loads q_m, q_n which will reduce σ_n, τ_{mn} to zero at the midpoint of element i, and they can be written as follows:

$$p^i_n + \sum_{j=1}^{k} (q^j_m C^{i,j}_m + q^j_n C^{i,j}_n) = 0$$

$$p^i_{mn} + \sum_{j=1}^{k} (q^j_m D^{i,j}_m + q^j_n D^{i,j}_n) = 0 \qquad (2)$$

Equations (2) constitute a set of 2k simultaneous equations (which can be solved for k values of q_m, q_n). These equations can be solved by iterative scheme. Upon the determination of these fictitious forces, the stress components at any point in the medium after excavation can be determined from equations (1).

2.2 Confining medium

The concept of a thin elastic seam (to be extracted) has been modelled by Crouch (1976), and Crouch and Starfield (1983) by considering an elastic element obeying a simple stress-strain relationship for compression and shear. They define the normal and shear stresses acting on the elastic seam element by the displacement discontinuity components in terms of normal closure and shear displacement. Their equations can be written in the equivalent form for the local coordinates M, N of the seam element as:

$$\sigma_n = \frac{E_s U_n}{t}, \quad \tau_{mn} = \frac{G_s U_m}{t} \qquad (3)$$

where

σ_n = normal stress

τ_{mn} = shear stress

E_s = modulus of elasticity of the seam

G_s = shear modulus of the seam

t = thickness of the seam

U_n = normal displacement (closure)

U_m = shear displacement

The method of quadrupole singularity described by Brady and Bray (1978-b), which takes account of the proximity of the adjacent boundaries in long narrow excavations is extended to the present solution of the boundary element method. Here the elements of thin confining layers are numerically modelled by uniformly distributed strip quadrupoles. Expressions for the component of stress and displacement due to these quadrupoles are given in Appendix 2.

Once again by discretizing the boundary of the problem (which includes the confining layer in this instance) into straight line elements, and determining the distribution of the fictitious quadrupoles which satisfy the boundary conditions, the problem can be written in a form similar to that expressed in equations (2) and by using the iterative routine, the fictitious singularities can be computed. In this condition the applied tractions are related to the induced normal closure and shear displacement as shown in equations (3) which in turn are also related to the applied fictitious forces and quadrupoles.

Thus for elastic confining layer (seam) elements, the following equations of the applied tractions have to be solved to satisfy the boundary conditions of reducing the boundary tractions to zero.

$$\sum_{j=1}^{l} \left[\left\{ C_m - \frac{E_s}{t} (Z_m^T - Z_m^B) \right\}^{i,j} Q_m^j + \left\{ C_n - \frac{E_s}{t} (Z_n^T - Z_n^B) \right\}^{i,j} Q_n^j \right] = 0$$

$$\sum_{j=1}^{l} \left[\left\{ D_m - \frac{G_s}{t} (X_m^T - X_m^B) \right\}^{i,j} Q_m^j + \left\{ D_n - \frac{G_s}{t} (X_n^T - X_n^B) \right\}^{i,j} Q_n^j \right] = 0 \qquad (4)$$

where

l denotes the total number of elements,

Q_n^j, Q_m^j correspond to the type of singularities,

Z_m^T, Z_m^B, ...etc. are coefficients relating to the top and bottom points of the confining layer elements respectively

Here elements 1 to K are of the fictitious force type of excavated medium, and those from K+1 to l are of the fictitious quadrupoles type of confining layer elements. After the determination of these fictitious loads, the induced stresses are calculated from equations identical to equations (1), taking into consideration the coefficients $B_m^{i,j}$, $B_n^{i,j}$, $C_m^{i,j}$... etc. which are determined from the appropriate expressions in Appendices 1 and 2 for the type of signularity involved. The total stress components are given by superimposing these induced stresses over the initial uniform field stresses in terms of the local axes of the various elements.

3 STRESS ANALYSIS

Although the boundary element method can be applied to a wide variety of simulations, but for the sake of simplicity a case of deep excavation with a circular cross-section has been considered for this example. A unit vertical stress field (p_z=1) is applied so that the numerical results for stress distribution can be normalized to this value.

3.1 Excavation in the proximity of a thin layer

Different configurations of a thin horizontal layer near a circular tunnel subjected to uniaxial field stress of unit intensity (p_z=1), as shown in Figure 4, have been investigated. Stress concentration σ_θ/p_z, (where σ_θ denotes the boundary stress) at the side wall, roof and floor of the tunnel (points S, R and F respectively) are shown in Table 1. These results suggest that in case 1, as the thin layer becomes soft, the excavated medium displaces greatly at Point S, thereby reducing the compressive stresses at this point, and slightly increasing the tensile stresses at point R (or F). In case 2, the compressive stresses at S do not change much with increasing softness of the thin layer, while a noticeable change occurs in the tensile stresses at R. However as thin layer approaches the roof as in case 3, there is a reduction in the tensile stresses at point R (with decreasing stiffness of the thin layer), while only minor changes in stresses are noticed at point F. Furthermore, there is an increase in the compressive stresses at the sidewall of the tunnel as the thin layer gets softer. Case 4, where the thin layer is farther away from the roof, is similar to case 3 except that in the former situation there are higher tensile stresses at the roof.

3.2 Excavation of a confined layer

Figure 1 illustrates the case where the mine opening is situated in a layer which is located between two confining layers. The middle layer may be considered to be softer or harder than the confining homogeneous strata. In both cases the stress

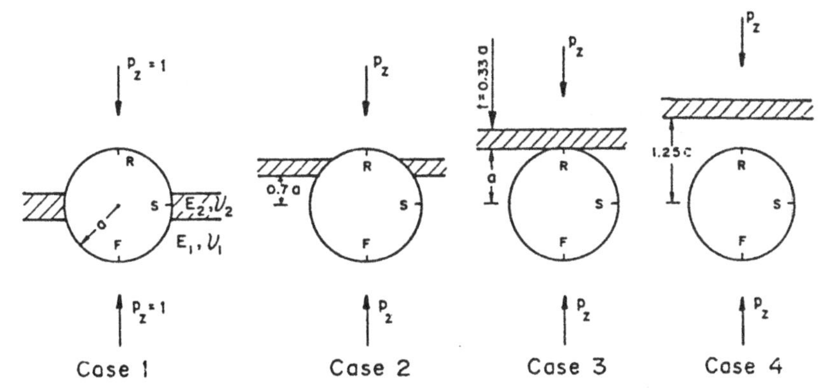

Fig.4 Thin horizontal soft layer near a circular tunnel under a unit vertical field stress. The different cases illustrate the position of this layer in relation to the roof of the tunnel

distribution around the excavation is governed by the thickness of the middle layer and the ratio of the elastic properties of this layer to that of the confining ones.

The results (Figures 5 and 6) are expressed in dimensionless quantities such as t/a (where t is the thickness of the middle layer and a is the radius of circular opening) and E_1/E_2 (where E_1 and E_2 are the Young's modulii of the middle layer and the confining homogeneous strata respectively). The values of t/a are taken to vary between 1 and 4 while those of E_1/E_2 are taken to range from 0.1 to 100. The ratios of $E_1/E_2 < 1$ indicate the presence of a soft middle layer, whereas those greater than 1 represent a relatively harder middle layer.

It may be seen from Figure 5 that for a given value of t/a, the softer the middle layer, the lower are the values of stress concentration at the sidewall of the opening. Furthermore the smaller the value

Table 1. Boundary stress concentration around a circular tunnel with thin horizontal soft seam as shown in Fig.4.

Case	E_1/E_2	ν_1	ν_2	Stress Concentration σ_θ/P_z		
				S	R	F
1	1	0.25	0.3	2.97	-0.98	-0.98
	2	0.25	0.3	2.65	-1.02	-1.02
	10	0.25	0.3	1.96	-1.15	-1.15
	100	0.25	0.3	1.38	-1.18	-1.18
2	1	0.25	0.3	3.08	-1.03	-1.01
	2	0.25	0.3	3.10	-1.07	-1.04
	10	0.25	0.3	3.15	-1.15	-1.07
	100	0.25	0.3	3.12	-1.42	-1.15
3	1	0.25	0.3	3.03	-1.01	-1.03
	2	0.25	0.3	3.12	-0.72	-1.05
	10	0.25	0.3	3.25	-0.43	-1.10
	100	0.25	0.3	3.90	-0.29	-1.16
4	1	0.25	0.3	3.02	-0.95	-1.02
	2	0.25	0.3	3.08	-0.97	-1.04
	10	0.25	0.3	3.29	-1.10	-1.08
	100	0.25	0.3	3.95	-1.65	-1.24

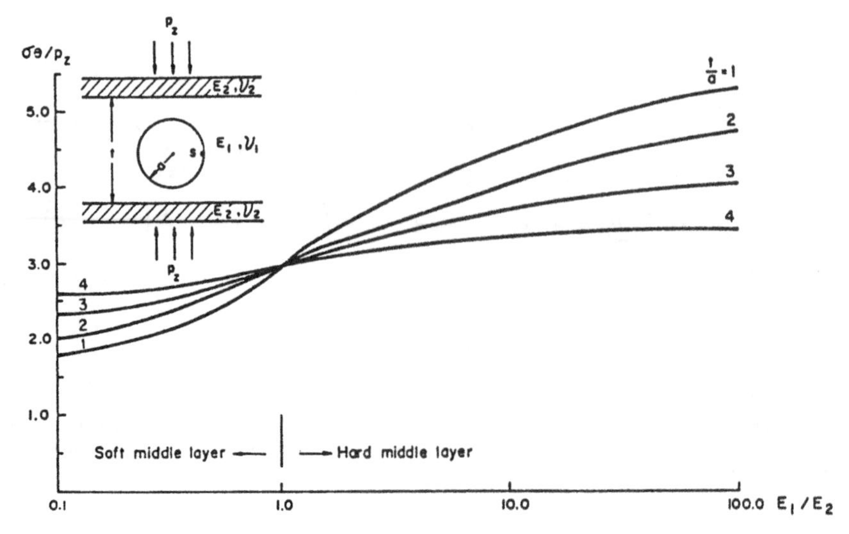

Fig.5 Stress concentration at the sidewall of a circular opening under a unit vertical field stress

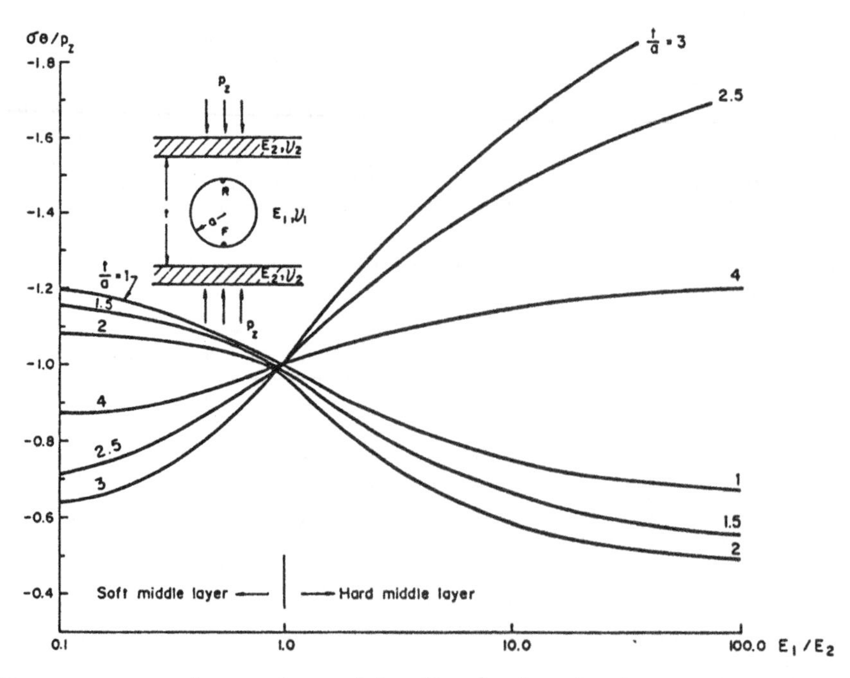

Fig.6 Stress concentration at the roof (or floor) of a circular opening under a unit vertical field stress

of t/a (when $E_1/E_2 < 1$) the lower the value of stress concentration, the opposite is true when the middle layer becomes harder than the confining layers ($E_1/E_2 > 1$). Figure 6 showing the magnitude of stress concentration at the roof (or floor) of an excavation, illustrates that for t/a varying between 1 and 2, the pattern of stress concentration is completely different from that when t/a ranges from 3 to 4. In the presence of the soft middle layer ($E_1/E_2 < 1$) with $t/a \leq 2$, the magnitude of tensile stresses is always greater than that corresponding to the homogeneous medium. For a given E_1/E_2, higher values of tensile stress are obtained with decreasing values of t/a. It can be seen that for $t/a > 2$, an increase in the thickness of the soft middle layer, results in a decrease in the values of the tensile stresses. However when t/a is greater than 4, the values of the tensile stresses approach that of the homogeneous medium irrespective of the values of E_1/E_2.

In a situation where the middle layer is harder than the confining strata ($E_1/E_2 > 1$), the values of the tensile stresses for $t/a \leq 2$ are much lower than those when $t/a > 2$. It is interesting to note that when t/a = 2.5, there is a significant change in the pattern of the stress at the roof of an opening. This may be due to the close proximity of the confining layers to the excavated medium. It is understood that in this situation, the influence of E_1/E_2 is far greater on stress concentration than in the instances where the confining layer is far away from the excavated medium.

4 PRACTICAL APPLICATION

The above method can be applied to a number of mining situations such as the extraction of phosphate and coal deposits. These deposits usually occur as hard inclusions between soft confining media. Sometimes nonmetalic mineral deposits have a similar occurrence or they may exist as soft layers surrounded by harder strata.

The usefulness of this method has been illustrated through its application to the mining of one of the biggest phosphate deposits known in Egypt. These deposits occur in the Abu-Tartour area which is located in the southwest of the western desert. Based on the geological investigation carried out by the Egyptian Geological Survey (Hermina, 1973), the stratigraphic succession encountered in the Abu-Tartour area is shown in Figure 7. The Phosphate formation lies between Nubian and Dakhla Formations and consists of three horizons. The upper and lower horizons containing the phosphate deposits are separated by gypsiferous shale. The upper horizon (0.7-7.3 m thick) is essentially made of phosphate clay deposit and is of no commercial value, on the other hand, the lower horizon which is sandwiched between gypsiferous shale and varigated clays (Figure 7) is productive. It occurs at a depth ranging from 60 to 280m, with an average thickness of 3.86m (Hermina and Wassif, 1975). The mining of this deposit has necessiated the excavation of gateroads, crosscuts, galleries and main adits. Due to the non-homogeneity of the rock media, it is obvious that the stress concentration around these excavations would be very much different than that expected if the rock media was completely homogeneous. This difference in the values of the stress concentration will depend upon the relative stiffness of the various layers and the thickness of the mined stratum with respect to the height of the excavation.

Figure 8 shows the two types of openings which currently exist at the Abu-Tartour Phosphate mine. Trapezoidal cross-sections are used for gateroads, crosscuts and outer ringways, while arched roof sections are used for radial trunk galleries. The modulus of elasticity of the phosphate bed is 45×10^3 Mpa, and that of the confining strata averages 16×10^3 Mpa.

Stress distributions around the boundary of the trapezoidal and arched sections at the sidewall, roof and floor are shown in Tables 2 and 3 for different values of K (horizontal stress ratio) and t/a. It must be mentioned that while adopting these ratios, the bottom width of the excavation has been kept fixed, equal to the bottom width of the existing excavation (Figure 8). According to the actual opening height and the average thickness of the phosphate bed, the value of t/a ranges between 2.4 (arched opening) and 2.9 (trapezoidal opening).

It may be seen from Tables 2 and 3 that the values of the stress concentration around the existing openings at these values of t/a are generally higher than those for the homogeneous media. The increased tensile stresses at both the roof and the floor of the these openings have added to the cost of support requirements. This problem can be resolved by increasing the opening height so that the values of $t/a \leq 2$. This will result in the reduction of the tensile stresses at the roof and floor regions. However, a slight increase in the compressive stresses will occur at the sidewalls but this will have only a minute effect on the overall stability of the excavation. Increasing the height will

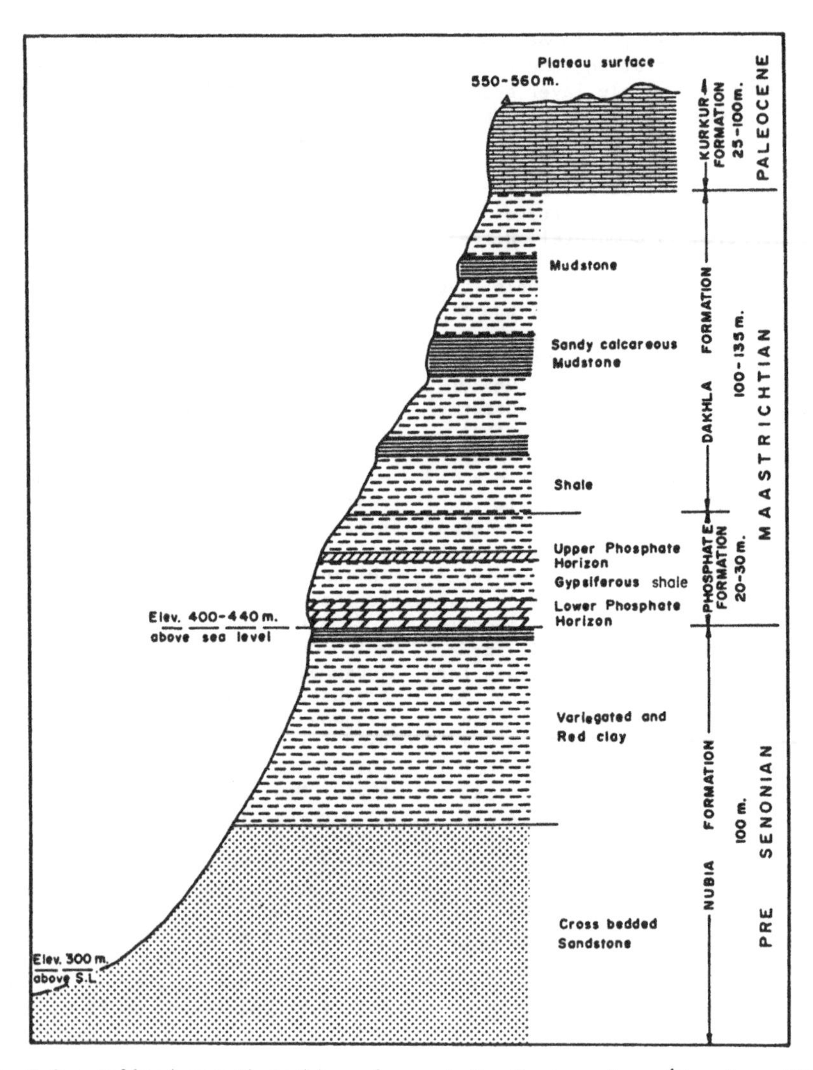

Fig.7 Generalized stratigraphic column at Abu Tartour Area (Hermina, 1973)

add to the cost of excavation but there will be a considerable saving in support requirements. Considerations of project economy and safety requirements should finally decide about the most practical alternative.

5 SUMMARY AND CONCLUSIONS

The boundary element method of analysis presented in this paper is of great help in the design of underground excavations in non-homogeneous media. Different hypothetical situations have been investigated including the one in which the excavation is mainly located in a layer which is either harder or softer than the confining strata. It has been shown that under uniaxial field stress conditions the creation of a circular opening in a hard middle layer causes an increase in the value of the stresses at the sidewall of the opening as compared to values which correspond to the homogeneous medium. However when t/a approaches 2, a significant reduction in the tensile stresses occurs at the roof and floor of the excavation. When $t/a > 2$ there is an overall increase in the tensile stresses, while the magnitude of the compressive stresses decreases. Both the tensile as well as the compressive stresses approach the values obtained for homogeneous mediums when $t/a > 4$. If the opening is located in a soft middle layer confined by hard layers, a relief of stress

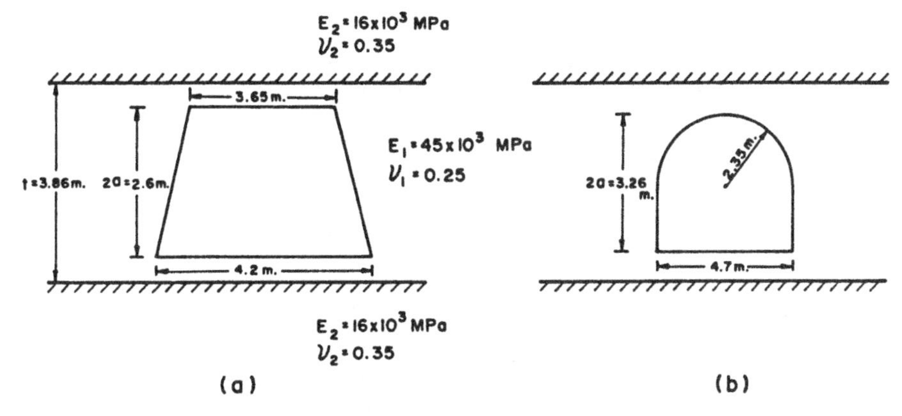

Fig.8 Geometry of mine openings at Abu Tartour Phosphate Mine
(a) Trapezoidal cross-section
(b) Arched roof cross-section

Table 2. Stress concentration around the trapezoidal cross section shown in Fig.8(a)

$\frac{t}{a}$	Stress Concentration σ_θ/P_z					
	K = 0.0			K = 0.333		
	R	S	F	R	S	F
1	-0.72	2.27	-0.83	-0.29	1.93	-0.56
2	-0.67	2.05	-0.71	-0.24	1.90	-0.52
2.5	-1.40	1.96	-1.62	-0.87	1.78	-1.35
3	-1.57	2.01	-1.70	-0.95	1.82	-1.58
Homogeneous Media	-0.83	1.87	-0.97	-0.31	1.58	-0.48

Table 3. Stress concentration around the arched roof opening shown in Fig.8(b)

$\frac{t}{a}$	Stress Concentration σ_θ/P_z					
	K = 0.0			K = 0.333		
	R	S	F	R	S	F
1	-0.75	3.03	-0.95	0.01	2.95	-0.53
2	-0.45	2.89	-0.87	2.02	2.47	-0.51
2.5	-1.35	2.82	-1.52	0.01	2.42	-1.12
3	-1.52	2.86	-1.72	0.10	2.43	-1.28
Homogeneous Media	-0.91	2.72	-0.98	0.01	2.39	-0.49

occurs at the sidewall of the opening, and this effect is enhanced by reducing the t/a ratio. It may however be emphasized that when t/a >2, the value of tensile stresses is reduced at both the roof and floor as compared to the homogeneous medium while the magnitude of the compressive stresses approaches a value corresponding to the homogeneous medium.

This method of analysis has been applied to the design of mine openings in the Abu-Tartour Phosphate mine, where the extraction of hard phosphate layers sandwiched between the soft strata has resulted in creating excessive tensile stress at the roof and floor of the openings, thereby adding to the cost of the necessary support. In order to minimize the support requirement, it is suggested that the height of the opening should be increased such that t/a approaches 2. However this will cause a slight increase in the compressive stresses at the sidewall especially for the trapezoidal section.

REFERENCES

Agarwal, R.K. & S.H. Boshkov 1969. Stresses and displacements around a circular tunnel in a three layer medium. Int. J. Rock Mech. Min. Sci. 6:519-540.

Barla, G. 1972. The distribution of Stress around a single underground opening in a layered medium under gravity loading Int. J. Rock Mech. Min. Sci. 9: 127-154.

Brady, B.H.G. & J.W. Bray 1978a. The boundary element method for determining stresses and displacements around long openings in triaxial stress field. Int. J. Rock Mech. Min. Sci. 15:21-28.

Brady, B.H.G. & J.W. Bray 1978b. The boundary element method for elastic analysis of tabular orebody extraction assuming complete plain strain. Int. J. Rock Mech. Min. Sci. 15: 29-37.

Bray, J.W. 1976. A programme for two-dimensional stress analysis using the boundary element method. Rock Mechanics progress report No.16, Imperial College of Science and Technology, London.

Crouch, S.L. 1976. Solution of plane elasticity problem by the displacement discontinuity method. Int. J. Num. Meth. Engng. 10: 301-343.

Crouch, S.L. & A.M. Starfield 1983. Boundary element methods in solid mechanics. London: George Allen and Unwin.

Eissa, E.A. 1980. Stress analysis of underground excavation in isotropic and stratified rock using the boundary element method. Ph.D. Thesis, University of London.

Goodman, R.E. 1966. On the distribution of stress around circular tunnels in non-homogeneous rocks. Proceedings of the first congress of the International Society of Rock Mechanics, Lisbon. 2: 249-255.

Hermina, M.H. 1973. Preliminary evaluation of Maghrabi-Liffiya Phosphates, western Egypt. Annals of Geological Survey of Egypt. 3: 39-74.

Hermina, M.H. & A. Wassef, 1975. Geology and exploration of the large phosphate deposit in Abu Tartour Plateau, the Libyan (western) desert, Egypt. Annals of the Geological Survey of Egypt. 5: 87-93.

Zienkiewicz, O.C., Y.K. Cheung & K.G. Stagg 1966. Stresses in Anisotropic media with particular reference to problems of rock mechanics. J. of Strain Analysis. 1: 172-182

Appendix 1 Stresses and displacements due to uniformly distributed transverse (q_x) and normal (q_z) infinite strip loads on element j

$$\sigma_x = \frac{1}{8\pi(1-\nu)}\left[\left\{(3-2\nu)\ \ln r^2 + \cos 2\theta\right\}q_x + \left\{4\nu\theta - \sin 2\theta\right\}q_z\right]_2^1$$

$$\sigma_z = \frac{1}{8\pi(1-\nu)}\left[-\left\{(1-2\nu)\ \ln r^2 + \cos 2\theta\right\}q_x + \left\{4(1-\nu)\theta + \sin 2\theta\right\}q_z\right]_2^1$$

$$\tau_{zx} = \frac{1}{8\pi(1-\nu)}\left[\left\{4(1-\nu)\ \theta - \sin 2\theta\right\}q_x + \left\{(1-2\nu)\ln r^2 - \cos 2\theta\right\}q_z\right]_2^1$$

$$U_x = \frac{1}{8\pi G(1-\nu)}\left[\left\{4(1-\nu)(X - Z_i\theta) - (3-4\nu)X\ln r\right\}q_x + (Z_i\ \ln r)\ q_z\right]_2^1$$

$$U_z = \frac{1}{8\pi G(1-\nu)}\left[(Z_i \ln r)q_x + \left\{(3-4\nu)(X - X\ \ln r) - 2(1-2\nu)\ Z_i\ \theta\right\}q_z\right]_2^1$$

where $r^2 = (X_i - X_j)^2 + Z_i^2$

$X1_j, X2_j$ are X coordinates of the ends of loaded element j

X_i, Z_i coordinates of the midpoint (P) of element of interest

G Shear modulus

ν Poisson's ratio

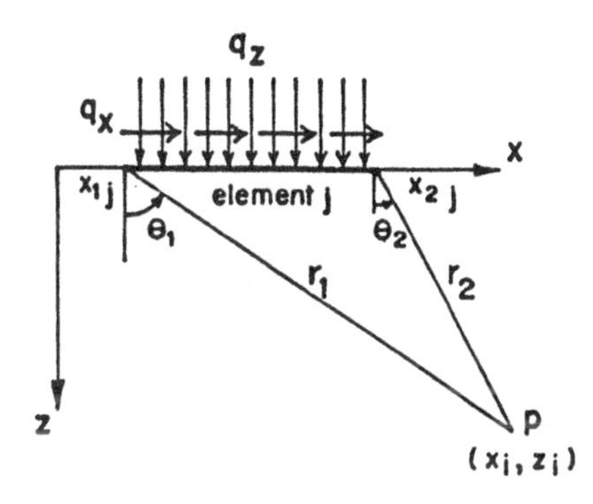

Appendix 2 Stresses and displacements due to uniformly distributed normal (a) and transverse (b) strip quadrupoles

$$\sigma_x = \frac{1}{4\pi(1-\nu)} \left[\left(\frac{6Z}{r^2} - \frac{4Z^3}{r^4}\right) S_x - \frac{(1-2\nu)}{(1-\nu)} \left(\frac{X}{r^2} - \frac{2XZ^2}{r^4}\right) S_z \right]_2^1$$

$$\sigma_z = \frac{-1}{4\pi(1-\nu)} \left[\left(\frac{2Z}{r^2} - \frac{4Z^3}{r^4}\right) S_x + \frac{(1-2\nu)}{(1-\nu)} \left(\frac{X}{r^2} + \frac{2XZ^2}{r^4}\right) S_z \right]_2^1$$

$$\tau_{zx} = \frac{-1}{4\pi(1-\nu)} \left[\left(\frac{2X}{r^2} - \frac{4XZ^2}{r^4}\right) S_x + \frac{(1-2\nu)}{(1-\nu)} \left(\frac{Z}{r^2} + \frac{2Z^3}{r^4}\right) S_z \right]_2^1$$

$$U_x = \frac{-1}{4\pi G(1-\nu)} \left[-\left\{2(1-\nu)\theta - \frac{XZ}{r^2}\right\} S_x + \frac{(1-2\nu)}{(1-\nu)} \left\{\ln r + \frac{Z^2}{r^2}\right\} S_z \right]_2^1$$

$$U_z = \frac{1}{4\pi G(1-\nu)} \left[-\left\{(1-2\nu)\ln r - \frac{Z^2}{r^2}\right\} S_x + \frac{(1-2\nu)}{(1-\nu)} \left\{2(1-\nu)\theta + \frac{XZ}{r^2}\right\} S_z \right]_2^1$$

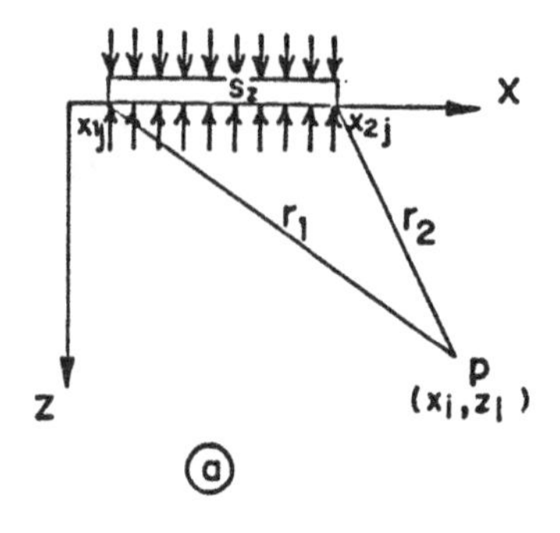

ⓐ

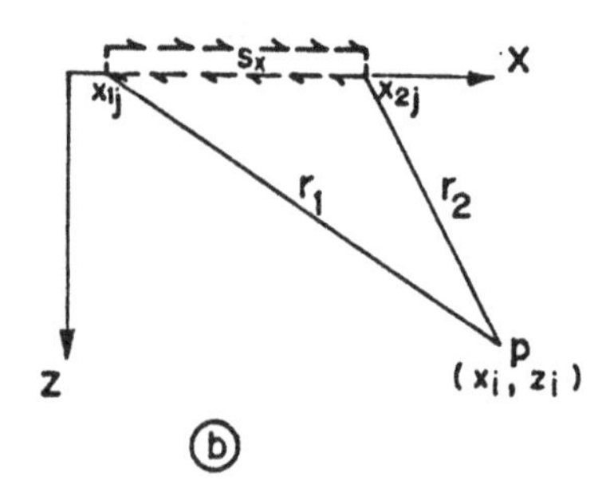

ⓑ

Computer and Physical Modelling in Geotechnical Engineering, Balasubramaniam et al. (eds)
© 1989 Balkema, Rotterdam. ISBN 90 6191 864 2

Interactive finite element mesh generation for three-dimensional tunnel analysis

Wolfgang G.Mertz & Gunter A.Swoboda
University of Innsbruck, Innsbruck, Austria

ABSTRACT: The paper describes a CAD program that was specially developed for the generation of three-dimensional finite element meshes as are used in tunnel construction. The program meets the specialized demands of tunnel construction works that otherwise make it difficult to implement the available CAD packages.

1 INTRODUCTION

In the last 15 years tunnel construction has shifted from purely empirical construction to a system supported by numerical models. The present generation of minicomputers, whose capacity ranges from 0.5 to 2.0 Mips, permits elastic and even plastic plane finite element calculations with an average of 5,000 unknowns in a time that is realistic for practical use, namely several hours CPU time. However, this type of analysis has the disadvantage of not being fully able to analyze the three-dimensional state of stress and displacement at the face. In the same way, the preliminary displacements which occur as a result of the high-grade three-dimensional excavation before the actual calculation cross section can be performed, can only be analyzed in a plane static system by introducing additional parameters [1]. These parameters, which describe the degree of the preliminary displacements, can be determined either by displacement measurements or by three-dimensional FEM calculations. Furthermore, three-dimensional tunnel calculations are also necessary for intersections (Fig.1), [2].

Initially, these meshes were drawn, all nodes, main and intersection nodes hand numbered and the coordinates then measured from the plan. This tedious work - the generation of a mesh normally took a week - was somewhat facilitated by generation programs. These programs assisted the designer in working out the geometry of the tunnel's lining, which was almost exclusively composed of basket arches. The irregular structures of the finite element mesh surrounding the arch, which stems on the one hand from the path taken by the ground's strata and on the other hand from the plastic zones appearing in the analysis. These zones have to show mesh refinement, and that made it impossible to perform automatic finite element mesh generation, so that the other elements adjoining the lining had to be added by hand.
Hardware and software advances in the field of computer graphics soon saw digitizer tablets come into use, with which the drawn meshes could be, usually together with the generation programs for the lining's geometry, read directly into the computer. This reduced the generation time to an average of one or two days. A disadvantage, however, was the fact that it was very difficult to change these meshes again.
The first CAD programs were developed about five years ago. These programs, written for mechanical engineering, enable the geometry of various mechanical components to be designed on the terminal screen, whereby the data is prepared and stored in such a way that it can usually be directly taken over by the NC machine. As a supplement to all these programs, modules are supplied that can automatically superimpose finite element meshes on the mechanical component. The original euphoria, brought on by the anticipation that these program systems could also be used to generate finite element meshes for tunnel construction has, unfortunately, been disappointed since the systems are especially intended for designing the geometry of mechanical engineering compo-

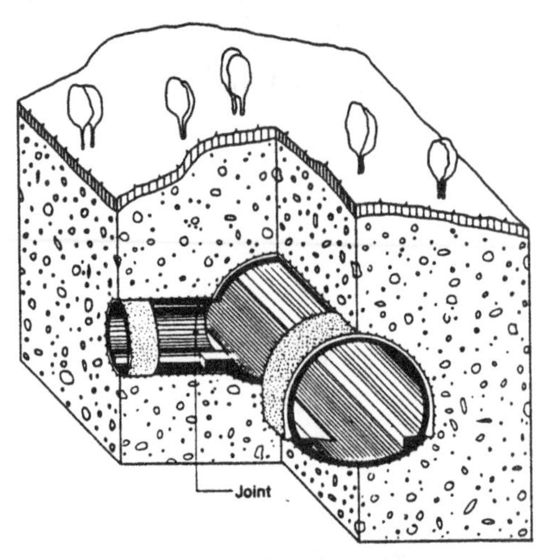

Fig.1: View of a tunnel intersection

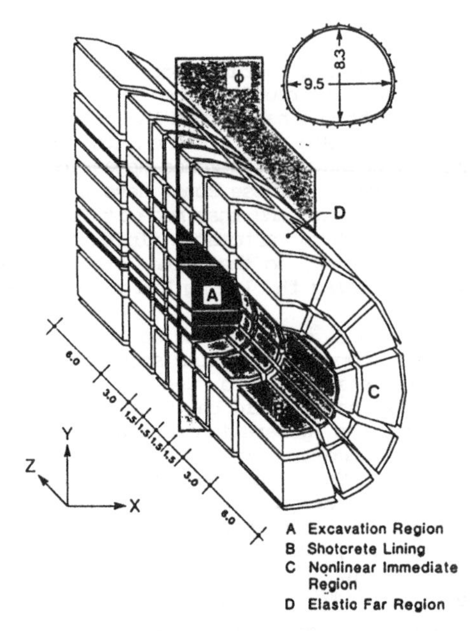

Fig.2: Three-dimensional model of a tunnel

nents and their incorporation into the generation of finite element meshes is impossible or unsatisfactory. A further disadvantage is posed by the fact that the component can usually only be provided with one "homogeneous" element type. In tunnel construction, however, it is necessary to have boundary elements on the surface or, in order to simulate the tunnel lining, to have shell elements in the structure.

Nevertheless, the generation of a finite element mesh, as for example in Fig.2, can only be performed with considerable difficulty in the absence of modern computer graphics. The problem entailed in a digitizer program is that a three-dimensional cross section can only be shown in an axonometric projection and thus the three-dimensional character of digitized coordinates is not entirely clear. The structure would therefore have to be digitized in several planes. However, the drafting work is not only complicated and prone to error, but also calls for complex post-processor programs that generate the spatial finite element mesh from the plane sections.

Thus, the problematic entailed here, whose geometry is simpler than that needed for mechanical engineering, made it necessary to develop a CAD program specially suited for finite element mesh generation of a tunnel's lining and for geological strata. In this way, it is not only possible to generate these meshes in a few hours, but they can also be modified on the terminal at any time or be composed or supplemented from various already stored mesh components.

2 HARDWARE AND SOFTWARE

As in all computer fields, computer graphics is characterized by a constant stream of new developments in the hardware sector. Nevertheless, software is harder to norm for graphic input and output equipment than for any other field. With certain reservations, the achievements brought for example by UNIX in the field of operating systems have not been met at all by the multitude of options and features offered by today's graphic terminals. The terminals differ from one another in their different types of resolution as well as their color, monochromatic, vector-storage tubes and raster screens. Moreover, a CAD work station can be recognized by whether its graphic input is performed by digitizer tablets, light-pen or by a mouse. Some work stations are equipped with two terminals, one for text and numerical input and one for graphic output. The main problems confronting the programmer are the different ESC sequences or control sequences that most terminals use for graphic in- and output.

The so-called GKS (Graphic Kernel System) makes an attempt at uniformity [3]. The graphic library's advantage is that programming can be performed completely independent of the equipment. It automatically translates the plot commands into the corresponding control commands. One disadvantage, however, is that it entails a certain overhead and, with it, time loss. So it is sometimes not possible to make optimum use of all the features of the different equipment. Moreover, GKS was unfortunately not implemented on all types of equipment, for example not on the authors' version of the VAXstation 100 by Digital Equipment that was used to develop the CAD program described below.

This terminal is a modern monochromatic graphic station with a high resolution (1088 x 864 Pixels) and 19" screen. The graphic input is performed using a mouse. The basic software permits graphics to be produced in two different modes; the menus can also be blended in and out [4]. This huge software package opens a wide range of graphic possibilities, whereby the menu technique permits text and numerical input as well as graphics on the same screen.

Despite the missing implementation of GKS, one programming goal was to make the major portion of the program as independent of the terminal as possible, so as to facilitate its use on other hardware later. For this purpose, a five-layer model was developed that is shown in the following diagram.

Routines	Layer	Datatransfer
Terminal-Independent Routines These routines take care of general program management as well as those tasks that are not directly involved in display on or input into the terminal.	Layer 1	Datatransfer by COMMON (Layers 1–3)
Terminal-Dependent Routines These routines are directly in charge of data display or input and have to be adapted for the existing terminal. For example, a display, pasteboard and window first have to be generated for a graphic output program on the VAXstation, while a Tektronix screen only has to have its graphic mode changed. The graphic commands naturally also differ.	Layer 2	
AST ROUTINES On many terminals, input is performed by means of so-called "Asynchronous System Traps", i.e. independent routines similar to Interrupt-Service Routines work asynchronous to the user program and, after they end, the main program is restarted.	Layer 3	
VAXstation User Library This layer contains routines that serve as user interfaces between the actual terminal routines. These are summaries of various calls of routines from Layer 5. This layer has to be rewritten for another work station.	Layer 4	Datatransfer by Parameter (Layers 4–5)
VAXstation System Library System routines and terminal driver.	Layer 5	

Tab.1: Program Layer Model

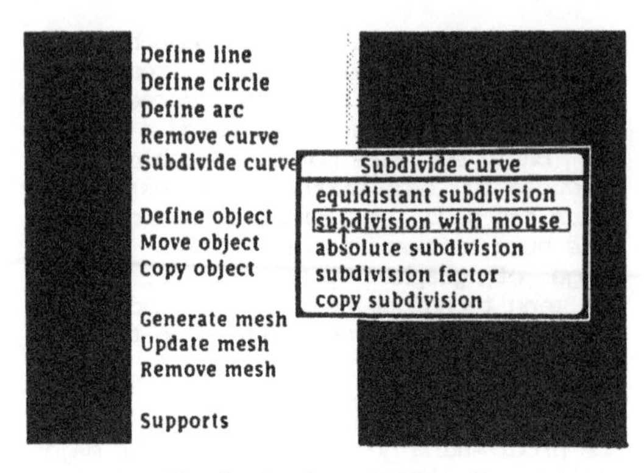

Fig.3: Regions and Menu

Layers 2 and 4 are the actual terminal-dependent parts of the program. While Layer 2 directly refers to the CAD program, Layer 4 contains general routines. These can be used to generate so-called regions, namely areas on the screen that are selected by the mouse. During selection, the program goes through the appropriate service routines. Menus are also generated. These are a number of regions that only appear on the screen after another input has been made. One of these regions is then chosen, causing the mouse to disappear from the screen (Fig.3). Other routines permit the input of texts and numbers as well as of mouse coordinates. The largest package however, constitutes the graphic output routines which make it possible to print lines, polygons, circles, arcs, arrows, markers and text. The parameter transfer of these routines was designed to be identical with that of other graphic terminals.

While Layer 4 can be summarized as a library and used by various programs, Layer 2 leaves several routines left over that have to be adapted to the particular work station in the user program. These are particularly routines that, for example, take over the generation of terminals on the screen. It is here that the various work stations differ the most.

3 INTERACTIVE OPERATIONS TO GENNERATE A FINITE ELEMENT MESH

As already mentioned in the Introduction, the task put to a CAD program that was written to draw up models of a finite element mesh of a tunnel's lining and the surrounding soil differs from that put to a CAD program for mechanical engineering components, because each program was designed to meet a different goal. While the purpose of the first program is the finite element mesh, in the second case the geometry is to be so defined that it can be directly taken over by an NC machine (numerical control) for production of the component.

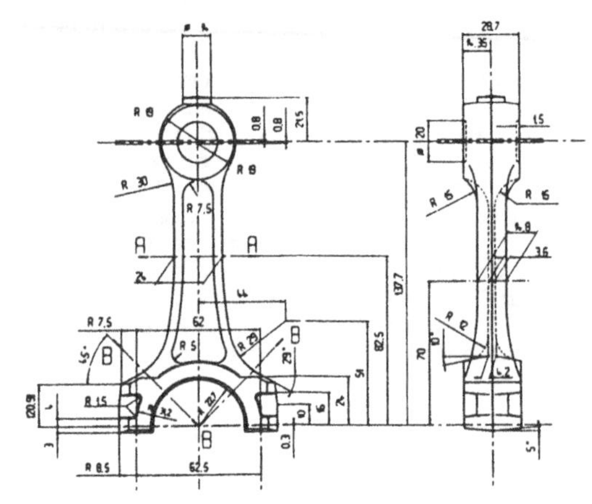

Fig.4: Typical component of a machine

Fig.4 shows a typical construction drawing by a CAD/CAM program. In contrast, the geometry of a geomechanical model is relatively simple. The tunnel lining usually consists of basket arches (Fig.5). The boundaries in the soil result from the surface structure (construction pits or buildings), from the makeup of the soil's strata (Fig.6) and also from the excavation sequence.

These few pieces of information are enough to work out the finite element mesh. The first task consists of reading the data in Figures 5 and 6 into the computer. The data that can be read from the drawings are coordinates, angles and radii. In most cases, the keyboard is used for coordinate input; radii and angles are generally derived from this information. All curves can now be generated, namely straight lines and arcs. Curves of a higher order (parabolas, splines, Bezier curves etc.) can almost always be disregarded, whereby in exceptional cases it suffices to approximate them using polygons.

Now all fixed points can be transferred from the drawing to the computer. These correspond to a plane cross section. The boundary lines are still missing. They are symmetry lines and the boundary to the infinite continuum, whereby its distance to the tunnel is dependent on the element library of the finite element program used. The program can describe the boundary by means of boundary elements or by supports. These lines are usually determined by the mouse with sufficient accuracy. From this cross section generated in one plane or, if for example an intersection is to be drawn up, from several such cross sections a spatial wire model must be created that is made up of one or more "logical cubes". Such a geometric figure can for example be seen in Fig.7. The edge polygons of this cube are now divided into parts, whereby opposing sides have to have the same number of intersection nodes. In the next step, these points will serve as the finite elements' corner nodes for automatic mesh generation.

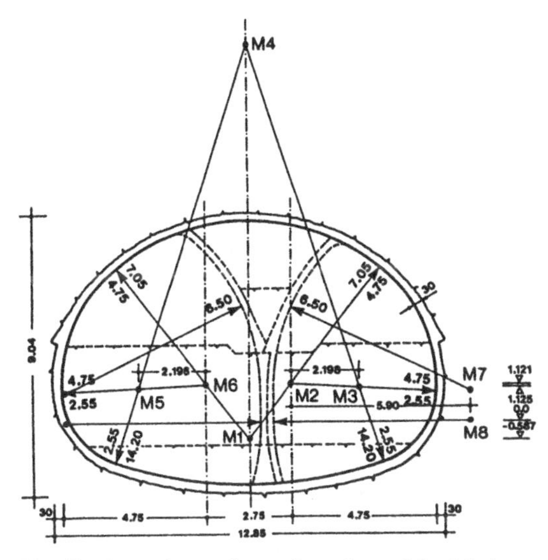

Fig.5: Geometry of a subwaytunnel's lining

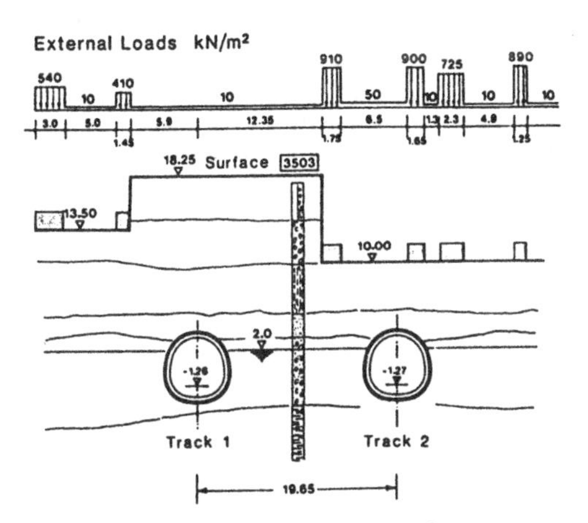

Fig.6: Geology and surface loads

4 AUTOMATIC MESH GENERATION

One of the most important parts of the program is the automatic mesh generation. It permits elements to be generated in parts of the structure, without individually reading each node or each element. The only boundary conditions are the polygons of the "logical squares" for surface and boundary elements or of the "logical cubes" for volume elements.

The procedure is as follows: The surface polygons are normalized, the dividing points calculated in the unit square or cube, and these then mapped back to the original coordinate system.

A simple way to do this is by using isoparametric curvilinear mapping of quadrilaterals, whereby the known formula in Equation (1) is used [5].

$$x = \sum_{i=1}^{8} N_i \cdot x_i$$

$$y = \sum_{i=1}^{8} N_i \cdot y_i \qquad (1)$$

$$z = \sum_{i=1}^{8} N_i \cdot z_i$$

However, the disadvantage of this method is that the shape function is only squared or at the most cubed. In this way, only low-order boundary polygons can be described. Thus, for curves of a higher

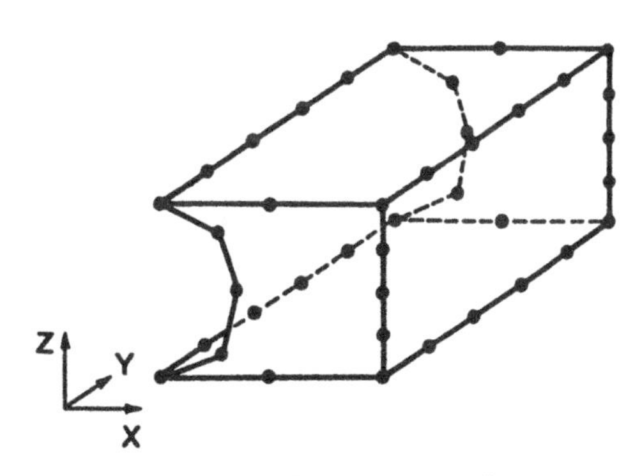

Fig.7: "Logical cube"

order the region has to be divided into several parts. Generally, this can only be done manually, which would also make it necessary to make further inputs in the algorithm.

A method that is better in this respect is illustrated in [6], [7] and [8] and was used by the authors as the basis for their CAD program.

The philosophy is as follows: The "logical cubes" shown in Fig.7 are bounded by six surfaces, such as the one in Fig.8. The criterium for dividing the internal nodal points is the division of the boundary polygons. Nodes between the individual polygon points are subject to linear interpolation.

First, these polygons are mapped from the X,Y,Z coordinate system to a normalized u,v coordinate system, whereby the minimal u,v coordinates are 0 and the maximimal are 1. The index in the direction of the u coordinates is i, and in v direction it is j. The number of nodes on polygons 1 and 2 is i-max, on polygons 3 and 4 it is j-max (Fig.8). This permits a mapping relation of the boundary polygons to be established between the two coordinate systems, which is defined by the distances between the polygon points (Equ.2).

Polygon 1

$$u_{i,1} = \frac{\sum_{n=2}^{i} \Delta l_{n,1}}{l_{ges,1}}$$

$$v_{i,1} = 0$$

Polygon 2

$$u_{i,j_{max}} = \frac{\sum_{n=2}^{i} \Delta l_{n,j_{max}}}{l_{ges,2}}$$

$$v_{i,j_{max}} = 1$$

Polygon 3 (2)

$$u_{1,j} = 0$$

$$v_{1,j} = \frac{\sum_{n=2}^{i} \Delta l_{1,n}}{l_{ges,3}}$$

Polygon 4

$$u_{i_{max},j} = 1$$

$$v_{i_{max},j} = \frac{\sum_{n=2}^{i} \Delta l_{i_{max},n}}{l_{ges,4}}$$

where

$$\Delta l_{n,1} = \left((X_{n,1} - X_{n-1,1})^2 + (Y_{n,1} - Y_{n-1,1})^2 + (Z_{n,1} - Z_{n-1,1})^2\right)^{\frac{1}{2}}$$

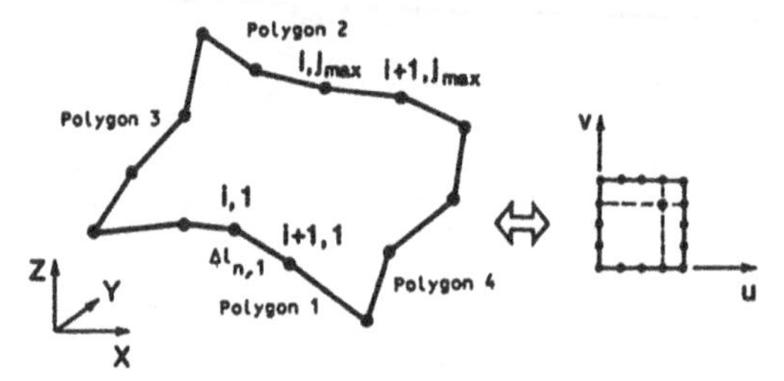

Fig.8: Mapping between space and unit square

In the next step, the missing u,v coordinates inside the unit square are obtained by linear interpolation (Fig.8).

To map the internal points, the linearly blended interpolation formula of Coons [6] is used. For the X coordinates it reads

$$\begin{aligned} X(u,v) = \; & (1-v)\cdot X(u,0) + v\cdot X(u,1) + \\ & (1-u)\cdot X(0,v) + u\cdot X(1,v) - \\ & (1-u)\cdot(1-v)\cdot X(0,0) - u\cdot(1-v)\cdot X(1,0) - \\ & (1-u)\cdot v\cdot X(0,1) - u\cdot v\cdot X(1,1) \end{aligned} \quad (3)$$

Similar formulas can be derived for the Y and Z coordinates. In (3), the coordinates X(0,0), X(0,1), X(1,0) and X(1,1) stand for the corner nodes of the "logical square". The coordinates X(u,0), X(u,1), X(0,v) and X(1,v) are the corresponding coordinates on the boundary polygons, which, as already mentioned, are laid linearly between the polygon points. Therefore, for example, for the coordinates X(u,1) it holds true that if

$$u_{i,j_{max}} \leq u \leq u_{i+1,j_{max}} \quad (4)$$

then

$$X(u,1) = \left(\frac{u - u_{i,j_{max}}}{u_{i+1,j_{max}} - u_{i,j_{max}}}\right)\cdot\left(X(u_{i+1,j_{max}},1) - X(u_{i,j_{max}},1)\right) + X(u_{i,j_{max}},1) \quad (5)$$

All the surface points have now been calculated. In the next step, the volumetric internal points will be generated in a similar manner, with the "logical cube" being mapped on a unit cube for this purpose (u,v,w coordinate system). The formula of Coons now reads

$$\begin{aligned} X(u,v,w) = \tfrac{1}{2}\cdot[\; & (1-v)\cdot X(u,0,w) + v\cdot X(u,1,v) + \\ & (1-u)\cdot X(0,v,w) + u\cdot X(1,v,w) + \\ & (1-w)\cdot X(u,v,0) + w\cdot X(u,v,1)\;] - \\ 2\cdot[\; & (1-u)\cdot(1-v)\cdot(1-w)\cdot X(0,0,0) + \\ & (1-u)\cdot(1-v)\cdot w\cdot X(0,0,1) + \\ & (1-u)\cdot v\cdot(1-w)\cdot X(0,1,0) + \\ & (1-u)\cdot v\cdot w\cdot X(0,1,1) + \\ & u\cdot(1-v)\cdot(1-w)\cdot X(1,0,0) + \\ & u\cdot(1-v)\cdot w\cdot X(1,0,1) + \\ & u\cdot v\cdot(1-w)\cdot X(1,1,0) + \\ & u\cdot v\cdot w\cdot X(1,1,1)\;] \end{aligned} \tag{6}$$

Here, X(u,0,w)... stands for the coordinates of the surface nodes and X(0,0,0)... for the coordinates of the corner nodes. The u,v,w coordinates in the volume are also generated linearly based on the division of the surface points.

5 INTENT OF THE PROGRAM

The basic idea was to write a CAD program suitable for creating three-dimensional finite element meshes. This FEMCAD (Finite Element Mesh Computer Aided Design) program was not only intended to fit into the FINAL [9] program system developed at the University of Innsbruck, but, if necessary, it could also be adapted to other finite element programs with minor changes. The same was true for the hardware. The coding took into account the fact that although the program was developed on a VAXstation 100 by Digital with the features of this graphic terminal being fully utilized, FEMCAD can also be implemented on other work stations.

Similar criteria predominated from the user's point of view, such as were given by Abel [10] as the demands made on an interactive graphic program:

- Completeness and flexibility of function - The program should contain all the functions necesary for drawing up a three-dimensional finite element mesh, or should indirectly provide for this by allowing several other functions to be combined.
- All inputs must be possible by mouse, from the keyboard or from files (for example digitized coordinates).
- All operations must give a quick graphic feedback. For longer calculations (for example mesh generation) an indicator must show the user that the operation is realy being performed.
- The display window must be adjustable and the drawing made more comprehensive by blending various parts in and out.
- The input menus must be easy to understand and supplementable by on-line documentation.
- Input must be minimized. This means, for example for mesh generation, that input has to be restricted to the definition of the eight corner nodes. Additional inputs should be necessary when this information is inadequate for mesh generation (two points can be connected in several ways, and the number of intersection nodes does not help the program know which connection is the right one).

6 EXAMPLE

The final example illustrates step by step the generation of the mesh used in [1] (Fig.2) to calculate the longitudinal displacement of a railway tunnel during excavation works conducted according to NATM. In order to simplify matters, various soil strata were not included.

Step 1 (Fig.9a): First, all the known center nodes and radii and thus the tunnel geometry have to be read.

Step 2 (Fig.9b): Next, the boundaries to the infinite continuum and to the symmetry lines are generated and the mesh size determined by subdividing the lines and arcs. In order to simplify input, only the corner nodes have to be defined for square or cube-shaped elements. Intersection nodes are calculated automatically by the program during mesh generation.

Step 3 (Fig.9c): The plane structure is duplicated along the tunnel's axis. In order to not have to copy every single curve, FEMCAD has an option with which curves can be combined to a so-called object and then moved or copied.

Step 4 (Fig.9d): In the last step before the actual mesh generation, the boundary lines along the tunnel's axis and their dividing points are determined. Here,

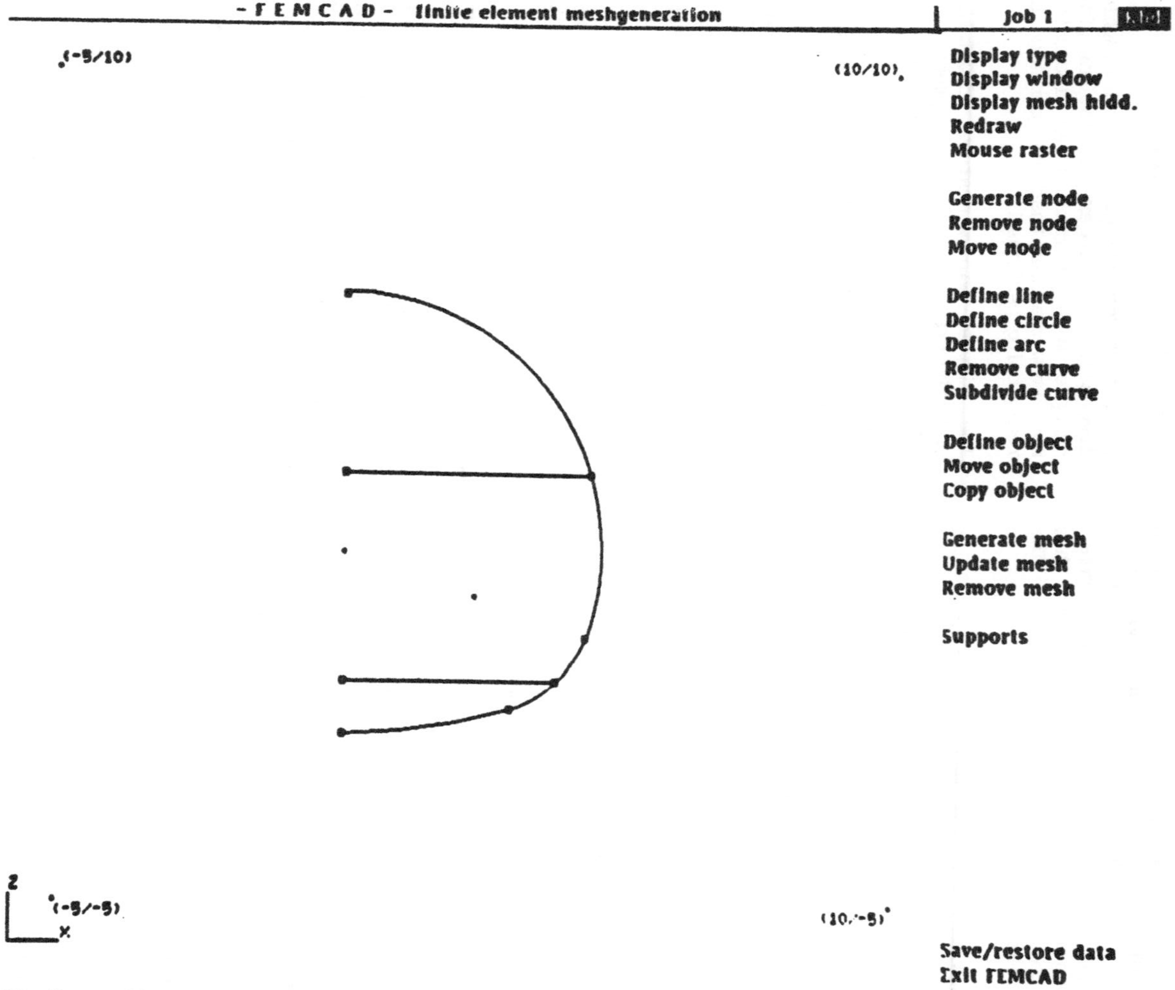

- F E M C A D - finite element meshgeneration

Job 1

(-5/10)

(10/10)

Display type
Display window
Display mesh hidd.
Redraw
Mouse raster

Generate node
Remove node
Move node

Define line
Define circle
Define arc
Remove curve
Subdivide curve

Define object
Move object
Copy object

Generate mesh
Update mesh
Remove mesh

Supports

z
(-5/-5)
x

(10/-5)

Save/restore data
Exit FEMCAD

Fig.9a: -Step 1- Geometry of the tunnel-lining.

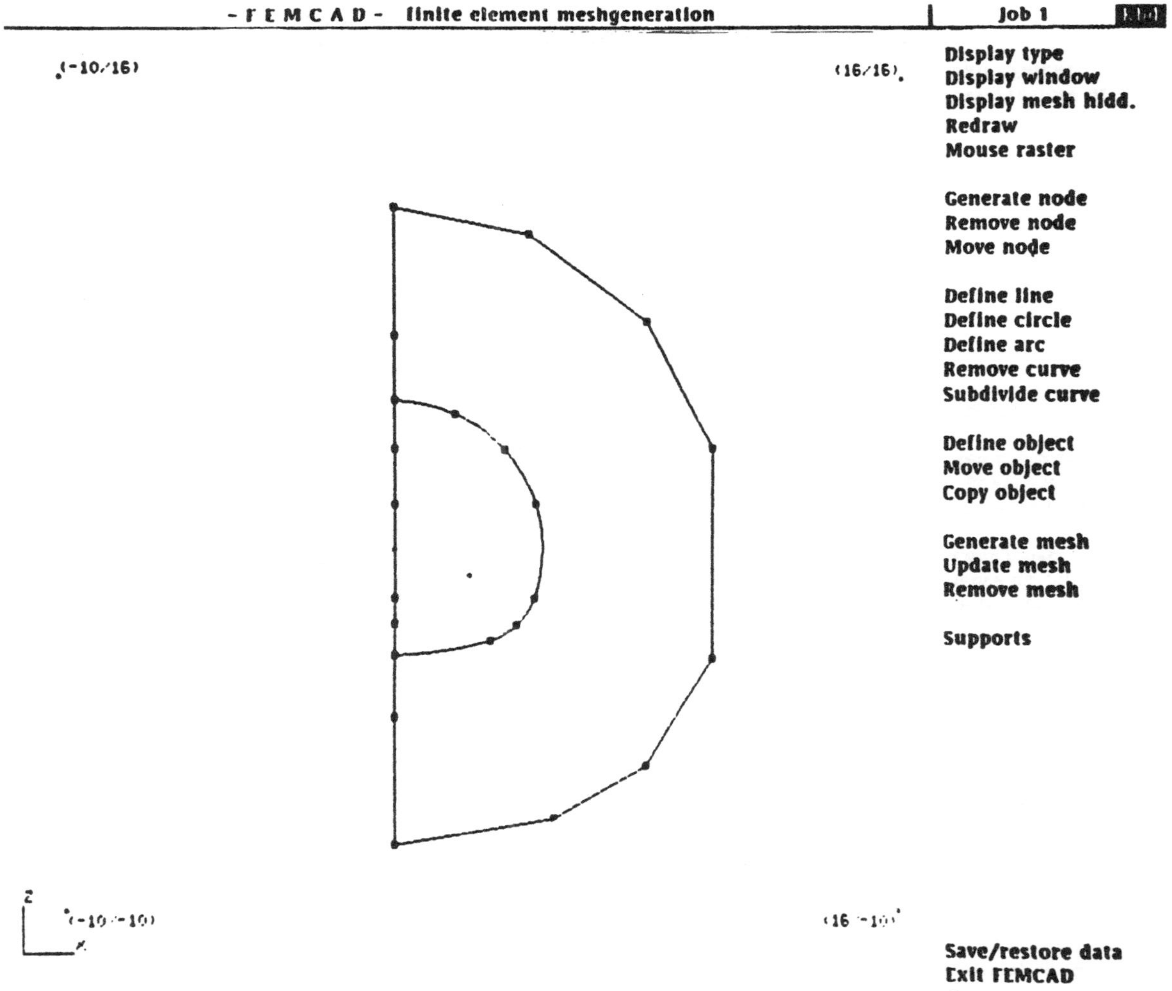

- F E M C A D - finite element meshgeneration | Job 1

(-10/16) (16/16)

Display type
Display window
Display mesh hidd.
Redraw
Mouse raster

Generate node
Remove node
Move node

Define line
Define circle
Define arc
Remove curve
Subdivide curve

Define object
Move object
Copy object

Generate mesh
Update mesh
Remove mesh

Supports

z x (-10/-10) (16/-10)

Save/restore data
Exit FEMCAD

Fig.9b: -Step 2- Boundaries of the infinite continuum with intersection nodes.

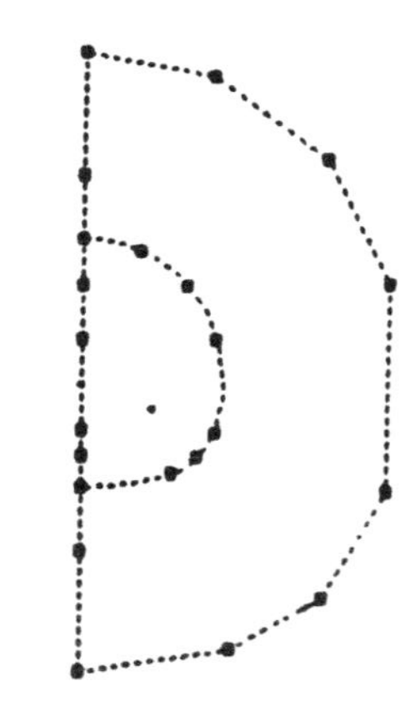

Fig.9c: -Step 3- Copy the plane structure as object (dashed lines) along the tunnel-axis.

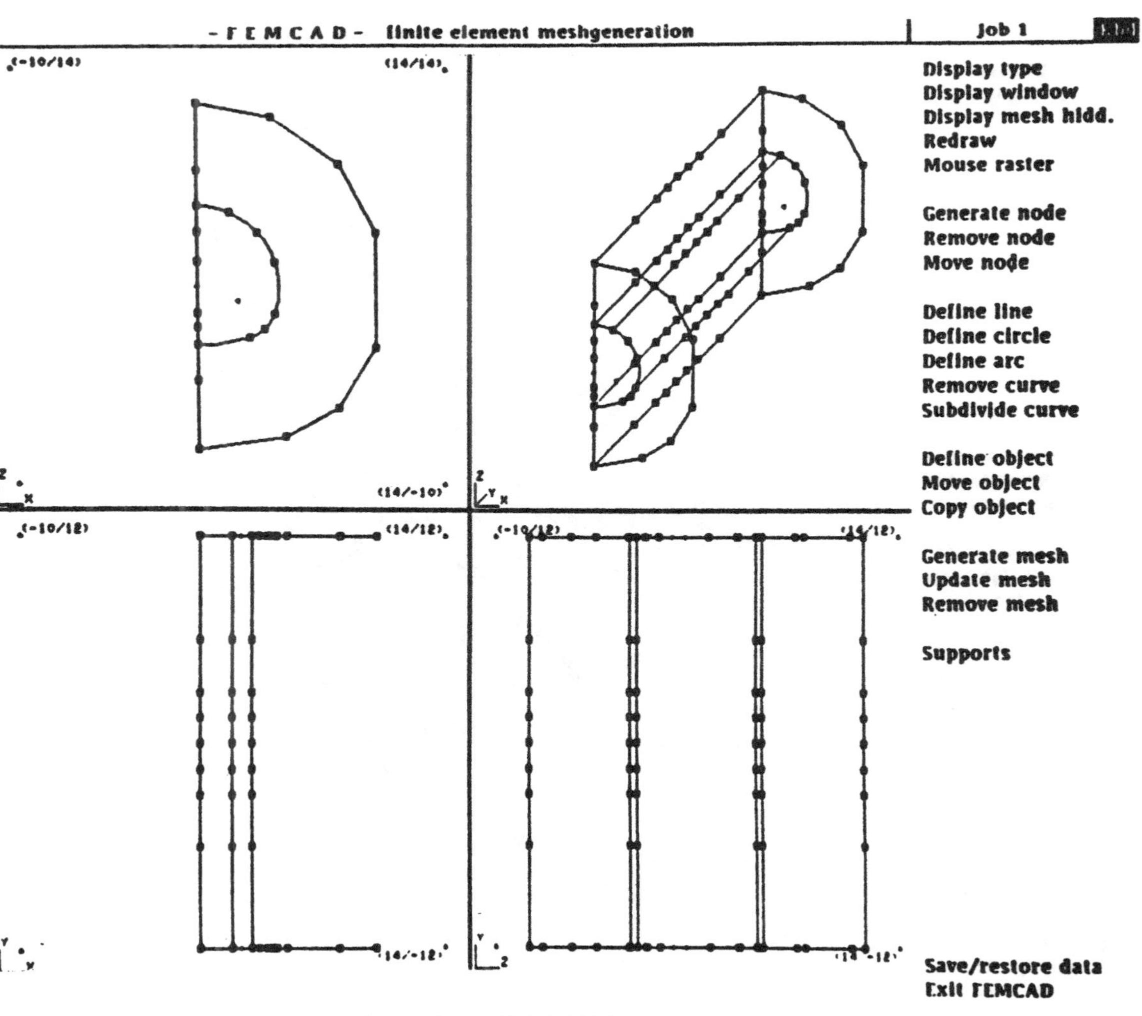

Fig.9d: -Step 4- Connecting lines along the tunnel.

Fig.9e: -Step 5- Automatic mesh generation.

too, FEMCAD can simplify matters: when a line's divisions has been determined, these can be copied onto any other line to be divided in the same way.

Step 5 (Fig.9e): Finally, the elements are generated. This is done by defining the "logical squares" for shell and boundary elements and the "logical cubes" for the volume elements. In a similar manner, "logical squares" also assist in generating the support conditions.

7 CONCLUSION

The rapid development of hardware and software not only makes it possible for us to constantly bring about new developments in the field of numerical analysis of stress and displacement in soil, but also permits adequate development of pre- and post-processors. The use of modern computer graphics enables us to generate finite element meshes within a few hours and to modify them just as quickly, instead of days, even weeks, that would otherwise be necessary. It is especially in the three-dimensional continuum that computer graphics has introduced a sensible solution to problems, since plane projections are hard put to show the complexity of spatial finite geometry while making it usable for the computer.

ACKNOWLEDGEMENT

The authors express their thanks to the "Österreichischen Fonds zur Förderung der wissenschaftlichen Forschung" for promoting this research project (no. P5723), making it possible to conduct essential fundamental investigations on the New Austrian Tunneling Method.

REFERENCES

[1] G.Swoboda, W.Mertz, G.Beer: Application of Coupled FEM-BEM Analysis for Threedimensional Tunnel Analysis, Proc. of the International BEM Conference, 1986, Beijing.

[2] H.Geisler, H.Wagner, O.Zieger, W.Mertz, G.Swoboda: Practical and Theoretical Aspects of the Three-dimensional Analysis of Finally Lined Intersections, Proc. of the 5th International Conference on Numerical Methods in Geomechanics, 1986, Nagoya (Japan).

[3] G.Enderle, K.Kansy, G.Pfaff: Computer Graphics Programming - GKS - The Graphics Standard, Springer Verlag, 1984

[4] Programming for the VAXstation Display system & VAXstation Native Graphics Procedures, Digital Equipment Corp., Malboro Massacusetts, U.S.A.

[5] O.C.Zienkiewicz, D.V.Phillips: An Automatic Mesh Generation Scheme for Plane and Curved Surfaces by Isoparametric Coordinates, International Journal for Numerical Methods in Engineering, Vol.3, p.519-528, 1971.

[6] S.A.Coons: Surfaces for Computer-Aided Design of Space Forms, Project MAC, MIT 1967, Available through Clearinghouse for Federal Scientific-Technical Information, Springfield, Va, USA.

[7] W.J.Gordon, C.A.Hall: Construction of Curvilinear Coordinate Systems and Applications to Mesh Generation, International Journal for Numerical Methods in Engineering, Vol.7, p.461-477, 1973.

[8] W.A.Cook: Body Oriented (Natural) Coordinates for Generating Three-dimensional Meshes, International Journal for Numerical Methods in Engineering, Vol.8, p.27-43, 1974.

[9] G.A.Swoboda: FINAL, Finite Element Analysis for Linear and Non-linear Structures, Version 5.7, University of Innsbruck.

[10] J.F.Abel: Interactive Computer Graphics for Applied Mechanics, Proceedings of the 9th U.S. National Congress of Applied Mechanics, 1982.

4. Computer controlled testing and investigation of soils

Computer and Physical Modelling in Geotechnical Engineering, Balasubramaniam et al. (eds)
© 1989 Balkema, Rotterdam. ISBN 90 6191 864 2

Porewater pressure monitoring during the screw plate testing in a soft clay

Gunther E.Bauer
Department of Civil Engineering, Carleton University, Ottawa, Canada

ABSTRACT: This paper presents the experimental results of screw plate test performed in a soft saturated clay deposit. The screw plate was outfitted with porewater pressure transducers in order to measure the generation and dissipation of porewater pressures during the test. Slow and rapid tests were performed in order to obtain drained and undrained soil parameters. The field results are given in graphical representations.

1 INTRODUCTION

The effective and economical design of structures requires a reasonably accurate estimation of soil parameters based upon field and laboratory investigations. Many times the structural requirements govern the criteria for design. In most instances, bearing pressure and settlement characteristics are the critical factors which control the soil behaviour. Several types of tests are available for the determination of soil parameters which can be performed in the laboratory or in situ. The purpose of these tests is primarily to evaluate the load-settlement response and the shear strength of the soil medium. These results are used to estimate the immediate and long term settlements, modulus of deformation and the bearing capacity of the soil. In situ tests involve little soil disturbance and avoid the difficulties associated with soil sampling to obtain undisturbed specimens for laboratory testing. The use of in situ testing devices therefore provide a useful and necessary function to verify the soil parameters obtained in laboratory tests. The accurate determination of the geotechnical properties of soil parameters from in situ testing techniques depends largely on the theoretical models available to interpret the test data of a particular test.

This paper presents a brief description of the screw plate testing technique, in particular the modified test and the interpretation and analysis of test data. Attention is focussed on the measurement and analysis of porewater pressure responses. The results presented could be used to estimate the in situ behaviour of prototype structures.

2 SCREW PLATE TEST

The screw plate test device originated in Norway where it was used to estimate settlement of an oil tank on sand (Kummeneje, 1956). Screw plate tests were also used to observe the settlement and the change in density of loose sand associated with soil densification due to blasting (Kummeneje et al. 1961). The device was later used to determine sand compressibilities and to correlate the test data to the plate bearing (Gould, 1968) and cone penetration tests.

Schmertmann (1970) suggested a method for screw plate testing. However, (Janbu et al. 1973) gave a detailed description of the theoretical principles involved and the interpretation procedures to determine in situ deformability and consolidation characteristics of silt and sand deposits.

Dahlberg (1975) used the test to ascertain settlement characteristics of preconsolidated natural sands. A general mathematical relationship between pressure, settlement and time was developed based on the screw plate tests carried out in soft clay and silt at different stress levels (Schwab et al. 1977). The theoretical assessment (Selvadurai, 1979), details of investigation and data for tests on soft clay (Selvadurai et al. 1979) and statistical analysis for the undrained

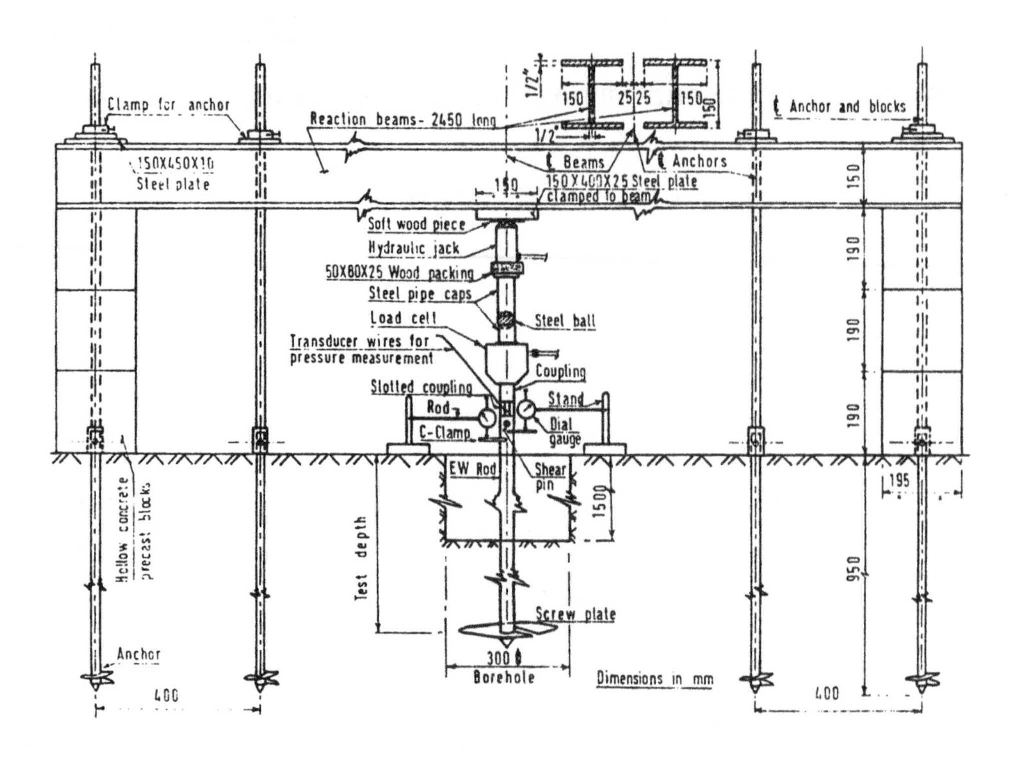

Figure 1: The Screw Plate Test Assembly

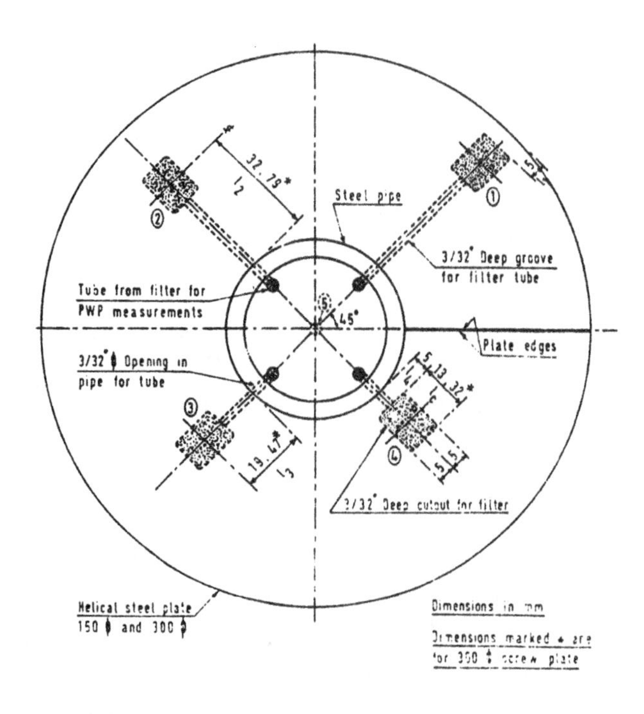

Figure 2: Screw Plate Plan View

modulus and the undrained shear strength values were subsequently presented (Nicholas, 1984; Selvadurait et al. 1981).

Recently, screw plate tests have been used to estimate undrained and drained moduli, undrained shear strength, coefficient of consolidation and settlement-time behaviour of stiff clays (Kay et al. 1982; Kay et al. 1980). When the screw plate test is performed in cohesive soils, the results of the test can be assessed by resorting to simplified theories based on linear elasticity and ideal plasticity. The results can also be interpreted by employing more sophisticated theories which incorporate the nonlinear load-displacement response of the screw plate test result. Regardless of which theory the investigator prefers, to this date there is no analytical method or methods available to describe fully the nonlinear material behaviour of saturated cohesive soils subjected to either slow or rapid load application by a circular plate. Probably the best theoretical analyses of the screw plate test results to date were given by (Selvadurai et al. 1979; Selvadurai et al. 1979; Nicholas, 1984) to estimate the undrained strength and deformation behaviour of soft cohesive soils. These studies take into consideration a number of important factors in their theoretical analysis, such as (1) the geometry and rigidity of the plate, (2) the bonding conditions between the plate and the soil and (3) the disturbance or remoulding of the soil above the installed plate. Details of the analysis are given in the three references cited above, but the results presented were in such a form that the undrained shear strength, S_u, and the undrained soil deformation modulus, t_u, could be estimated from the experimental load-deformation response.

The undrained shear strength is obtained from the observed ultimate load at failure, P_{ult}, using the following relationship:

$$S_u = \frac{4 P_{ult}}{\alpha \pi D^2} \qquad (1)$$

where D is the diameter of the screw plate and α is a factor which takes into account the uncertainties of interface and the flexibility conditions of the screw plate.

The undrained deformation modulus can be estimated from the linear portion of the load-displacement curve using the relationship:

$$E_u = \frac{\lambda 2P}{\pi D s} \qquad (2)$$

where P is any load within the linear load-displacement response,
s is the displacement corresponding to P, and
λ varies between 0.60 and 0.75 and takes into consideration the feasible conditions which may exist around the plate during the test.

The above equations are considered to be valid if undrained conditions prevail in the soil during a test. In soft cohesive soils under rapid loading conditions, this assumption might be true during the initial stage of the test, but it is very doubtful if undrained conditions prevail especially when the test is carried out over several hours.

In order to investigate the possible drainage or dissipation of porewater pressure during a test stage, the screw plate was outfitted with porous stones and pressure transducers in order to monitor the porewater response during a rapid test (i.e., undrained conditions). For the evaluation of the drained soil parameters, the load is applied very slowly such that the generation of excess porewater pressures is negligible.

A detailed description of the instrumentation of the plate, the calibration procedures were given by (Mitel, 1985). The same reference shows all the results obtained in the field testing program and also gives a detailed analysis and discussion of the various aspects of screw plate testing. In the following sections a brief description of the test set-up, the test procedure and some typical test results are presented. The analysis of data was accomplished with the aid of a desk computer.

3 SCREW PLATE SYSTEM

The screw plate system with porewater pressure monitoring capabilities consists of a helical plate, porous filters with pressure transducers and connecting drill rods. The plate itself consists of a single pitch helical auger. The pitch of the auger is such that it equals the downward movement per one revolution or turn of the plate. A plan view of the plate auger is shown in Figure 2 and an elevation is given in Figure 3. The plate has a detachable conical tip at its lower end. Figure 2 also indicates the location of the four porous stones for monitoring the porewater response at the plate/soil interface. Additional porous stones were located at the conical tip. The length and the diameter of the tip was

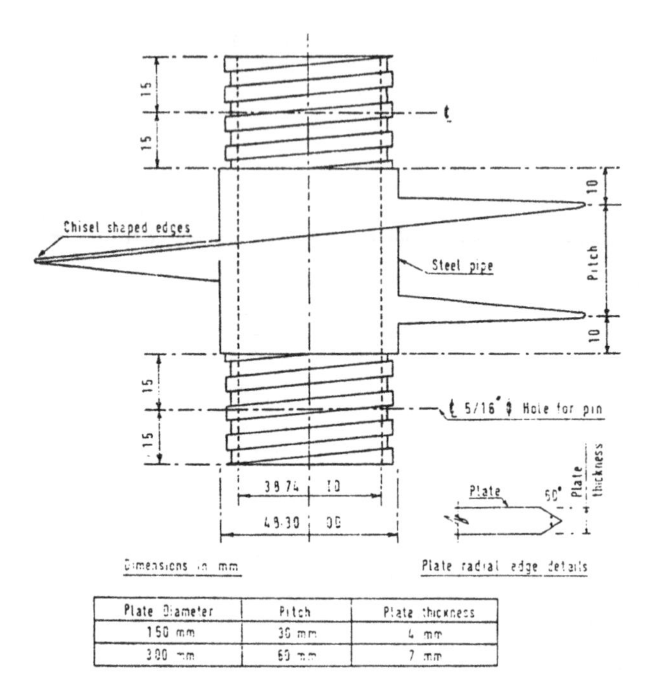

Plate Diameter	Pitch	Plate thickness
150 mm	30 mm	4 mm
300 mm	60 mm	7 mm

Figure 3: The Screw Plate - Elevation

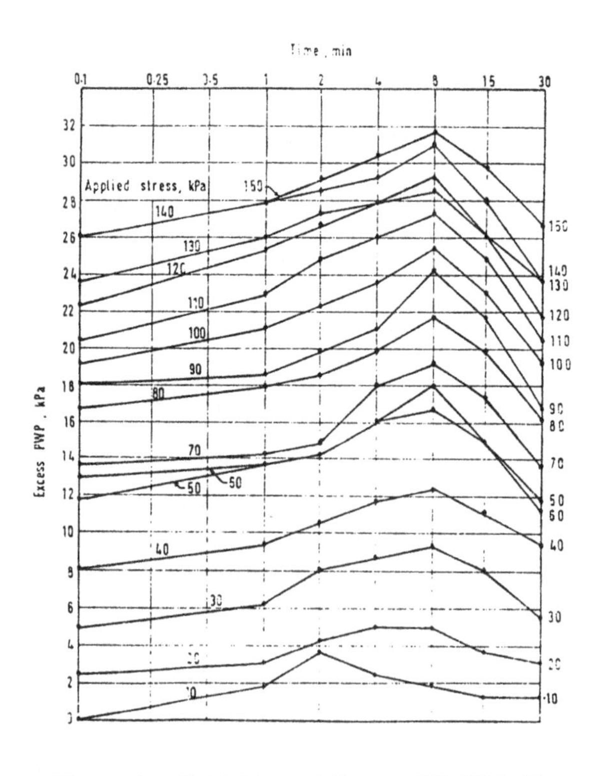

Figure 4: Variation of Excess PWP With Time - Test A4

TABLE 1
SUMMARY OF SCREW PLATE TESTS

Test	Type	Plate Diameter, mm	Test Depth, m	Remarks
A1-S	U	150	1.85	
A2-S	U	150	2.50	
A3-S	U	150	3.10	
A4-S	U	150	3.75	
B1-S	D	150	1.85	
B2-S	D	150	2.50	
B3-S	D	150	3.10	
B4-S	D	150	3.75	U-Undrained Test
C1-L	U	300	1.85	D-Drained Test
C2-L	U	300	2.50	
C3-L	U	300	3.10	
C4-L	U	300	3.75	S-Small Plate
D1-L	D	300	1.85	L-Large Plate
D2-L	D	300	2.50	
D3-L	D	300	3.10	
D4-L	D	300	3.75	
D5-L	D	300	5.25	
D6-L	U	300	6.75	
D7-L	U	300	8.25	
D8-L	U	300	9.85	

governed by the screw plate dimension. 150 mm and 3000 mm diameter plates were used in the field in order to study the size effect on soil parameters. The plate was connected to hollow drill rods. The drill rod couplings had shear pins which allowed the rods to be rotated clockwise and anticlockwise. This made retrieving of the auger plate possible after the completion of a test series.

The monitoring of the porewater pressure was accomplished with pressure transducers located inside the vertical shaft above the plate. The electrical leads were fed through the drill rods to the ground surface and were connected to a data acquisition system. The whole system was calibrated and checked out in the laboratory. In the field a trailer mounted drill rig was employed to penetrate the plate to the required depth. A hydraulic jacking system with anchored reaction beams were used to apply the load on top of the drill rods. The load was monitored by load transducer and the displacement was recorded with a LVDT or dial gauges (Figure 1).

4 FIELD TESTS

4.1 Description of site

The site for the field tests had been previously used for extensive field

TABLE 2

UNDRAINED MODULUS OF DEFORMATION, MPa

Test	E Min.	E Max.
A1-S	5.6	7.0
A2-S	9.5	11.8
A3-S	7.5	9.4
A4-S	12.3	15.3
C1-L	5.4	6.7
C2-L	7.5	9.4
C3-L	6.0	7.5
C4-L	10.6	13.2
D6-L	8.2	10.2
D7-L	10.6	13.2
D8-L	11.3	14.1

TABLE 3

EFFECTIVE MODULUS OF DEFORMATION, MPa

Determined From Excess PWP Considerations

Test	E_d Min.	E_d Max.
A1-S	4.8	7.2
A2-S	6.6	10.9
A3-S	5.6	8.5
A4-S	9.0	13.3
C1-L	3.4	5.1
C2-L	3.6	7.4
C3-L	4.0	6.4
C4-L	8.2	12.3
D6-L	5.7	9.0
D7-L	7.5	11.7
D8-L	6.6	10.2

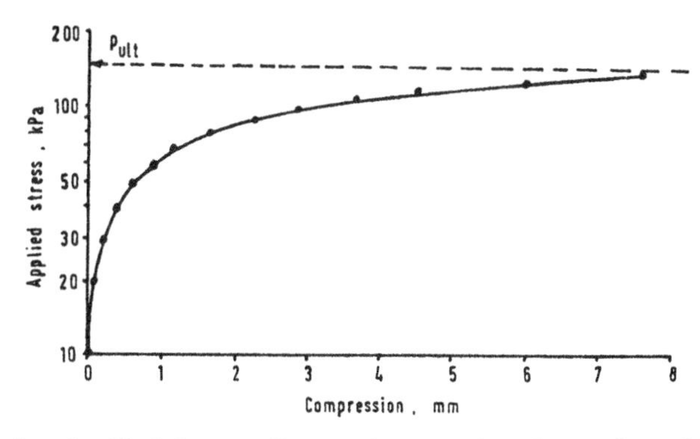

Figure 5: Applied Stress Versus Compression Curve for Ultimate Stress Determination

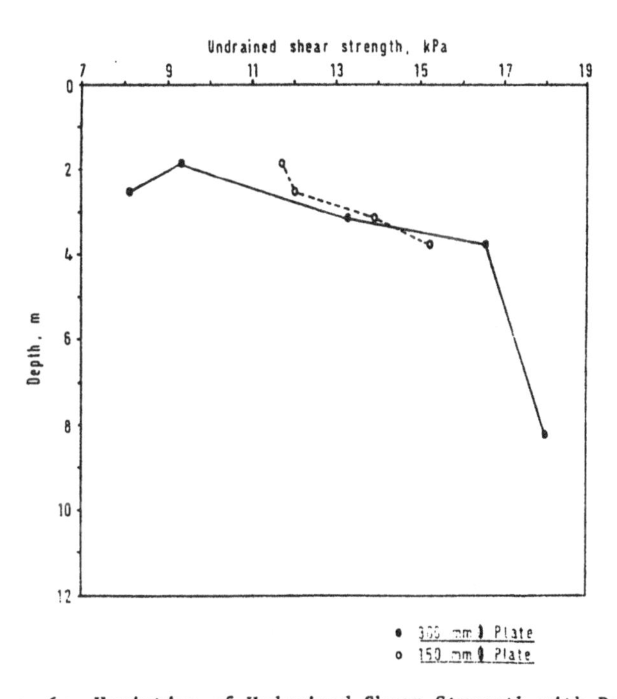

Figure 6: Variation of Undrained Shear Strength with Depth

investigations (Bauer et al. 1985; Demartinecourt et al. 1984), and is commonly known as the Gloucester Test Fill site at the south-east end of Ottawa, Canada. The soil in this area is a soft marine clay generally known as Leda clay or Champlain Sea clay. The clay is extremely soft and natural water content is in the order of 60 percent and is considerably higher than its liquid limit. The soil has negligible remoulded shear strength and it is termed as a quick clay. Initial void ratios are in the order of 2 to 3.

4.2 Test program

The field investigation program consisted of a total of twenty screw plate tests carried out in four boreholes to a maximum depth of 10 m. The groundwater table was at a depth of 1.4 m. Prior to a test the whole porewater pressure measuring system was saturated (i.e., draining and back-flushing of porous stones). The screw plate was rotated slowly into the soil with the drill rig, then the rig was removed and the loading mechanism (i.e., hydraulic jack bearing against a reaction

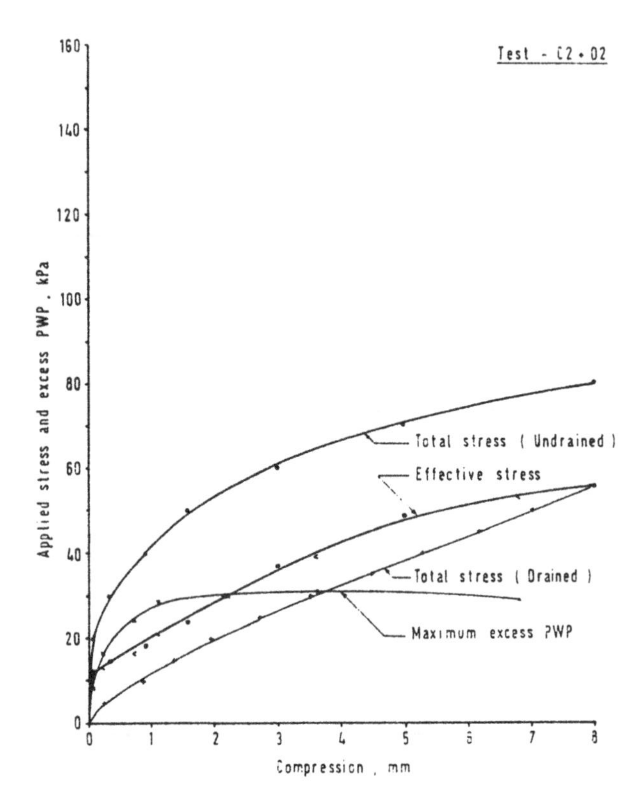

Figure 7: Applied Stress Versus Compression for Drained and Undrained Conditions

beam) was put into place. In the first three boreholes four tests each were performed at various depths. In borehole D, eight tests were carried out. Table 1 gives a summary of these tests indicating the type of test, plate size and test depth. Since the boreholes were spaced about 3 m apart from one another and the soil conditions are quite uniform in this area. Differences in test results can therefore be attributed to differences in test conditions (i.e., plate size and rate of loading).

5 TEST RESULTS

Porewater Pressures

The generation of excess porewater pressure and its dissipation with time were monitored for eleven plate tests. These measurements were recorded at depths of 1.85, 2.50, 3.10 and 3.75 m for the 150 mm diameter plate in borehole A and for the 300 mm diameter plate in borehole C. In borehole D, the water pressures were monitored for the 300 mm diameter plate at depths of 6.75, 8.25 and 9.85 m. A typical result of excess porewater pressure with time is given in Figure 4. The load was applied in increments of 10 kPa and was held constant for 30 minutes. The values of the maximum excess porewater pressure measured under the plate was about 40 percent of the applied stress. This did not change noticeably with depth or plate size. It was found that the porous stones located halfway between the edge and the stem of the plate yielded the highest pressure response. The drainage behaviour of the soil around the screw plate is influenced to a large degree by various factors such as soil disturbance, soil bonding wat the soil-plate interface and the rigidity of plate. A prerequisite, of course, is that the monitoring system is fully saturated. In all tests under so-called undrained conditions, some drainage did occur as the load increments were kept constant for 30 minutes.

5.1 Undrained shear strength

The undrained shear strength of the soil was evaluated from Equation (1). For the analysis of the test data, the failure load, P_{ult}, or the failure stress, P_u, was taken as the peak value of the applied stress-compression. Where this relationship did not assume a definite peak value, the asymptotic was chosen as shown in Figure 5. A typical result of undrained strength with depth is given in Figure 6.

5.2 Modulus of deformation

The undrained modulus of deformation was obtained from Equation (2). A summary of these values is provided in Table 2. Similar values were obtained for drained conditions, i.e., no excess porewater pressure was generated in the test, the values are given in Table 3. A comparison of applied versus compression for total stress and effective stress responses is given in Figure 7.

6 CONCLUSIONS

Continuous monitoring of load and pressure transducers was essential in testing a soft clay with a screw plate test device having water pressure monitoring capabilities. In this investigation five porewater pressure transducers, two L.V.D.T.'s and a load cell had to be continuously and simultaneously monitored in the laboratory. In the field two porewater pressure transducers, one at the tip and the other one at the base of the screw plate were monitored in the field together with the applied load and the deformation of the plate. From the data collected various total and effective soil parameters could be estimated. A study is presently underway to use the field results in order to obtain the consolidation behaviour of this clay.

ACKNOWLEDGEMENTS

This investigation was supported in part by a grant from the Natural Sciences and Engineering Research Council.

REFERENCES

Kummeneje, O., 1956. Foundation Behaviour of An Oil Tank in Drammen. Norwegian Geotechnical Institute (NGI), Publ.No.12, NOGPA.

Kummeneje, O. and Eide, O., 1961. Investigation of Loose Sand Deposits By Blasting, 5th International Conference on Soil Mechanics and Foundation Engineering, Vol.2, Paris, pp.491-497.

Gould, J.H., 1968. The Comparative Study of Screw Plate and Rigid Bearing Plate Tests, M.S. Thesis, Department of Civil Engineering, University of Florida, Gainesville, U.S.A.

Schmertmann, J.H., 1970. Suggested Method for Screw Plate Load Test, ASTM-STP 479, pp.81-85.

Janbu, N. and Senneset, K., 1973. Test Compressometer - Principles and Appli-Cations, 8th International Conference on Soil Mechanics and Foundation Engineering, Moscow, Vol.1.1, pp.191-198.

Dahlberg, R., 1975. Settlement Characteristics of Preconsolidated Natural Sands, Swedish Council for Building Research, Document D1.

Schwab, E.F. and Broms, B.B., 1977. Pressure-Settlement-Time Relationship By Screw Plate Tests In Situ, 9th International Conference on Soil Mechanics and Foudation Engineering, Tokyo, Vol.1, pp.281-288.

Selvadurai, A.P.S. and Nicholas, T.J., 1979. A Theoretical Assessment of the Screw Plate Test, International Conference on Numerical Methods in Geomechanics, Aachen, Vol.3, pp.1245-1252.

Selvadurai, A.P.S. Bauer, G.E. and Nicholas, T.J., 1979. Screw Plate Testing of a Soft Clay, 33rd Canadian Geotechnical Conference, Quebec City, pp.211-224.

Selvadurai, A.P.S. and Nicholas, T.J., 1981, Evaluation of Soft Clay Properties by the Screw Plate Tests, 9th International Conference on Soil Mechanics and Foundation Engineering, Stockholm, Vol. 2, pp.567-572.

Nicholas, T.J., 1984. Screw Plate Testing of Soft Clays, M.Eng. Thesis, Department of Civil Engineering, Carleton University, Ottawa, Canada.

Kay, J.H. and Avalle, D.L., 1982. Application of Screw Plate to Stiff Clays, ASCE Journal of Geotechnical Engineering, Vol.108, No.GT11, pp.145-154.

Kay, J.N. and Mitchell, P.W., 1980. A Downhole Plate Load Test for In Situ Properties of Stiff Clays, Australia-New Zealand Conference on Geomechanics, Wellington, Vol.1, pp.255-259.

Kay, J.N. and Parry, R.H.G., 1982. Screw Plate Tests In A Stiff Clay, Ground Engineering, Vol.15, No.6, pp.22-30.

Mital, S.K., 1985. A Screw Plate With Porewater Pressure Monitoring Capability, M.Eng. Thesis, Department of Civil Engineering, Carleton University, Ottawa, Canada.

Bauer, G.E. and Demartinecourt, J.P., 1985. The Application of the Borehole Shear Device (BSD) to a Sensitive Clay, Journal of Geotechnical Engineering, Southeast Asian Society, Vol.16, pp.167-189, June.

Demartinecourt, J.P. and Bauer, G.E., 1984.
The Modified Borehole Shear Device,
Geotechnical Testing Journal, Vol.6,
No.1, pp.24-29, March.

Computer and Physical Modelling in Geotechnical Engineering, Balasubramaniam et al. (eds)
© 1989 Balkema, Rotterdam. ISBN 90 6191 864 2

A weighted regression approach to model and estimate spatial variability of soil properties in one dimension

P.H.S.W.Kulatilake & P.Varatharajah
Department of Mining and Geological Engineering, University of Arizona, Tucson, Ariz., USA

ABSTRACT: A method is developed to estimate spatial variability of soil properties in one dimension. The applicability is limited to statistically homogeneous layers. The method is capable of incorporating both non-stationarity and stationarity portions of the spatial variation under one scheme. Under the non-stationarity portion, it is capable of incorporating the global mean trend as well as the non-constant variance around the mean trend. The estimation is accomplished through a general weighted regression approach. The method is applicable for both irregularly spaced data as well as regularly spaced data. An example is given to illustrate the use of the developed technique. With slight modifications, the method can be extended to estimate spatial variation in three dimensions.

1 INTRODUCTION

In deterministic analysis of geotechnical engineering problems, it is common to model the soil profile at a site in terms of homogeneous layers with constant soil properties. However, even within apparently homogeneous soil layers, engineering soil properties may show considerable variation from point to point. Risk and uncertainty analysis can incorporate the spatial variability of soil properties. Such analysis are important in improving the geotechnical performance prediction. Knowing of spatial variability of soil properties is also helpful in the design of soil exploration programs and in the evaluation of their effectiveness.

Uncertainties of soil properties arise from three major sources. The first source of uncertainty comes from the natural heterogeneity of the soil deposit. A second source of uncertainty can be attributed to the limited availability of data of the soil property in question. This leads to statistically uncertainty. This can be decreased at the expense of additional testing. A final source of uncertainty arises due to the difference between the measured values and the actual values. Sample disturbance, test imperfections and human errors are the main causes of this uncertainty. This third type of uncertainty is also possible when empirical correlations are used to estimate engineering properties from index properties.

This paper focuses primarily on the first type of uncertainty. The scope is limited to one dimension. A paper which deals with three dimensional spatial variability is under preparation. Random field theory in conjunction with regression analysis have been used to model spatial variability of soil properties (Alonso and Krizek 1975; Lumb 1975; Vanmarcke 1977; Tabba and Yong 1981; Baker 1984). All these models were developed for statistically homogeneous soil layers. In general soil property data may be characterized by the following equation:

DATA = TREND + SIGNAL + NOISE (1)

The TREND can be considered as the non-stationary component (Bendat and Piersol 1971) of the data and it provides the global systematic change of the spatial variation. The SIGNAL plus the NOISE represents the stationary stochastic portion of the data. The signal can be considered as the stationary correlated portion and provides the local systematic change of the spatial variation. Global and local trends are relative terms whose interpretation depends on site dimensions and availability of data with the spatial dimension. The noise can be considered as the stationary uncorrelated portion and is totally random.

Regression analysis using polynomial functions has been used to model the global trend (Alonso and Krizek 1975; Lumb 1975; Tabba and Yong 1981). Alonso and Krizek (1975), Lumb (1975), and Vanmarcke (1977) have modeled the stationary stochastic portion of soil properties in one dimension using regularly spaced data. Alonso and Krizek (1975) used two types of autocovariance functions as well as power spectra functions to model the stationary correlated portion. Lumb (1975) used first and second order autoregressive processes to model the same. Vanmarcke (1977) introduced the variance function from which the range of correlated values may be found.

Previous models suggest modeling trend and signal as two separate components. To do so, regression analysis has been used to model the trend. In performing regression analysis, it has been assumed that the variance around the mean trend is constant. However, some soil deposits show non-constant variance (Lumb 1974). Proper attention has not been paid to this issue in previous models. The procedure given in the paper addresses this issue and also show a technique to incorporate trend and signal under one scheme in estimating the spatial mean values and their variances.

2 SUGGESTED TECHNIQUE

Soil property data are considered to consist of the components given by equation (1). The non-stationary component may exist due to the presence of a global trend and/or a non-constant variance around the mean trend. Visual inspection of raw data (Vandaele 1983) and/or shape of the variogram for raw data (Journel and Huijbregt 1978) can be used to check the presence of non-stationarities. Under the most general condition, the spatial variability in one dimension may be modeled by a weighted polynomial regression model. The regression analysis procedure can be given as follows:

$$\underbrace{\begin{bmatrix} y_1 \\ y_2 \\ \vdots \\ y_i \\ \vdots \\ y_n \end{bmatrix}}_{[Y]} = \underbrace{\begin{bmatrix} 1 & x_1 & x_1^j & x_1^m \\ 1 & x_2 & x_2^j & x_2^m \\ \vdots & \vdots & \vdots & \vdots \\ 1 & x_i & x_i^j & x_i^m \\ \vdots & \vdots & \vdots & \vdots \\ 1 & x_n & x_n^j & x_n^m \end{bmatrix}}_{[X]} \underbrace{\begin{bmatrix} \beta_0 \\ \beta_1 \\ \vdots \\ \beta_j \\ \vdots \\ \beta_m \end{bmatrix}}_{[\beta]} + \underbrace{\begin{bmatrix} e_1 \\ e_2 \\ \vdots \\ e_i \\ \vdots \\ e_n \end{bmatrix}}_{[e]} \quad (2a)$$

$$E(e) = \begin{bmatrix} 0 \\ 0 \\ \vdots \\ 0 \end{bmatrix} \quad (2b)$$

$$V(e) = \begin{bmatrix} 1/W_1 & \rho_{12}/\sqrt{W_1W_2} & \cdots & \rho_{1n}/\sqrt{W_1W_n} \\ \rho_{21}/\sqrt{W_2W_1} & 1/W_2 & \cdots & \rho_{2n}/\sqrt{W_2W_n} \\ \vdots & \vdots & & \vdots \\ \rho_{n1}/\sqrt{W_nW_1} & \rho_{n2}/\sqrt{W_nW_2} & \cdots & 1/W_n \end{bmatrix} \quad (2c)$$

where y_i = property value at point i; $i = 1,2, \ldots n$

x_i = spatial variable value at point i

β_j = j th regression coefficient; $j = 1,2, \ldots m$

e_i = i th residual

E(e) = expected values of residual vector

V(e) = covariance matrix of the residuals

ρ_{ij} = correlation coefficient value between e_i and e_j due to signal portion (will be dealt later)

W_i, W_j = weighting factors for e_i and e_j due to non constant variance of the residuals (will be explained later)

It can be shown that it is possible to find a unique nonsingular symmetric matrix P (Draper and Smith 1981) such that

$$P'P = PP = P^2 = V = V(e) \quad (3)$$

where P' is the transpose of P.

Now by premultiplying Eqn (2a) by inverse of P, P^{-1}, the following new model can be obtained

$$Z = Q\beta + f \tag{4}$$

where $P^{-1}Y = Z$, $P^{-1}X = Q$ and $P^{-1}e = f$

It is possible to show that f satisfies the following two conditions (Draper and Smith, 1981).

$$E(f) = 0 \tag{5}$$

$$\text{and } V(f) = I\sigma_f^2 \tag{6}$$

where σ_f^2 is a constant. Basic least square theory can be applied to this new model to estimate β and other required parameters as given by the following equations (Draper and Smith 1981).

$$\hat{\beta} = b = (X' V^{-1} X)^{-1} X' V^{-1} Y \tag{7}$$

$$V(b) = (X' V^{-1} X)^{-1} \tag{8}$$

$$\hat{Y} = Xb \tag{9}$$

$$V(\hat{Y}) = XV(b)X \tag{10}$$

where symbol ^ is used to indicate the estimated values.

The estimated property values and the covariance matrix of the estimated property values are given by $\hat{Y}$ and $V(\hat{Y})$ respectively. In order to obtain these estimations, it is necessary to specify the elements of V(e). Procedure to obtain these elements is given in the next section.

2.1 Determination of V(e)

Before discussing the estimation of elements of V(e), it may be appropriate to explain the special cases of V(e). If there is no correlation between the residuals, (i.e. no signal portion) but there exist a non-constant variance of the residuals, then matrix V is given by equation (11). If correlation does exist between residuals and the variance of the residuals is constant, then the matrix V is given by equation (12). The two equations are as follows:

$$V(e) = \begin{bmatrix} 1/W_1 & 0 & 0 & . & . & . & 0 \\ 0 & 1/W_2 & 0 & . & . & . & 0 \\ 0 & 0 & 1/W_3 & . & . & . & 0 \\ . & . & . & . & . & . & . \\ . & . & . & . & . & . & . \\ 0 & 0 & 0 & . & . & . & 1/W_n \end{bmatrix} \tag{11}$$

and

$$V(e) = \sigma^2 \begin{bmatrix} 1 & \rho_{12} & \rho_{13} & \cdots & \rho_{1n} \\ \rho_{21} & 1 & \rho_{23} & \cdots & \rho_{2n} \\ . & . & . & & . \\ . & . & . & & . \\ . & . & . & & . \\ \rho_{n1} & \rho_{n2} & \rho_{n3} & \cdots & 1 \end{bmatrix} \tag{12}$$

where σ^2 is the constant variance of the residuals. If the variance is constant and there is no correlation, then matrix V is given by

$$V(e) = \sigma^2 I_n \tag{13}$$

where I_n is the identity matrix of order n. Regression with this condition of V(e) reduces it to ordinary unweighted regression.

To find the most suitable V(e), first it is necessary to perform the regression analysis under the assumption given by equation (13). Degrees one through a maximum degree may be tried in performing the regression analysis. Statisticians usually allow seven to ten data points for each regression coefficient estimated. This guideline may be used in deciding the maximum degree which should be tried. The multiple R square value and the residual mean square (RMS) value, can be used to obtain the best fit polynomial. Draper and Smith (1981) provide details in using these two parameters in detecting the best fit function. Such a regression analysis can be performed using a package program such as P5R of BMDP (1985). Output from this program can be used to obtain a plot between the magnitude of the residuals versus the spatial variable. This plot can be used to examine whether the variance of the residuals is constant or non-constant. Related to interpretation of this plot, the reader is referred to Draper and Smith (1981). If this plot shows a non-constant variance of the residuals, then the trend of the residuals should be estimated.

The trend of the residuals may be estimated by performing ordinary polynomial regression analysis between the magnitude of the residuals and the spatial variable. The resulting regression function can be considered to provide an estimate of the standard deviation for the residuals. Such a function can be expressed as

$$\hat{\sigma}_i = \left[X_{ij}\right]\left[\alpha_j\right] \tag{14}$$

where σ_i is the estimated standard deviation of the residual at point i and α_j are the estimated regression coefficients. Weight given to a data point should decrease as variance of the residual increases. Therefore, W_i values are estimated by

$$W_i = 1/\hat{\sigma}_i^2 \tag{15}$$

Existence of correlation is possible under two cases: (a) with constant variance residuals, (b) with non-constant variance residuals. To estimate the ρ_{ij} values, first normalized residuals are obtained. For case (a), the residuals are divided by the estimated constant standard deviation of the residuals (BMDP 1985) to obtain normalized residual values. For case (b), first it is necessary to perform another regression analysis between $\left[Y_i\right]$ and $\left[X_{ij}\right]$ with V(e) according to equation (11). Values of W_i required for this regression are obtained from equation (15). Each residual resulting from this regression analysis, when divided by $\hat{\sigma}_i$ obtained earlier, provides a normalized residual value. Estimation procedure for ρ_{ij} from normalized residuals depends on whether they are regularly spaced or irregularly spaced. For irregularly spaced data the method given in Agterberg (1970) may be used. For regularly spaced data ρ_{ij} can be estimated by

$$\hat{\rho}_{ij}(k) = \frac{\sum_{t=1}^{n-k} h(t)\ h(t+k)}{\sum_{t=1}^{n} \left[h(t)\right]^2} \tag{16}$$

where h denotes the normalized residuals and t and k are explained in Fig. 1. Values of ρ_{ij} fall between -1 and + 1. Values close to zero indicates insignificant correlation. This completes the formation of V(e).

3 APPLICATION

The purpose of the application is to illustrate the use of the developed estimation model. The soil property data considered in this section comes from testing performed in sand at Eglin Air Force Base, Florida (Schmertmann, 1969). Dutch cone penetration tests were performed in a sand deposit at the site. The general conclusion from the site investigation was that the entire site area can be considered as a statistically homogeneous, clean, coarse-grained sand deposit for the depth considered. Figure 2 shows the penetration results plotted with depth in the soil for one of the sounding logs. Point pressures are available at a regular interval of spacing of 0.4

$$\begin{array}{llllllll} x = & x_1 & x_2 & x_3 \ . \ . \ . & . & x_i \ . & . & . \ x_n \\ t = & 1 & 2 & 3 \ . \ . \ . & . & i \ . & . & . \ n \end{array}$$

k is a lag number and takes values 1, 2, ... (n-1)

Fig. 1 Definitions of t and k for regularly spaced data in one dimension

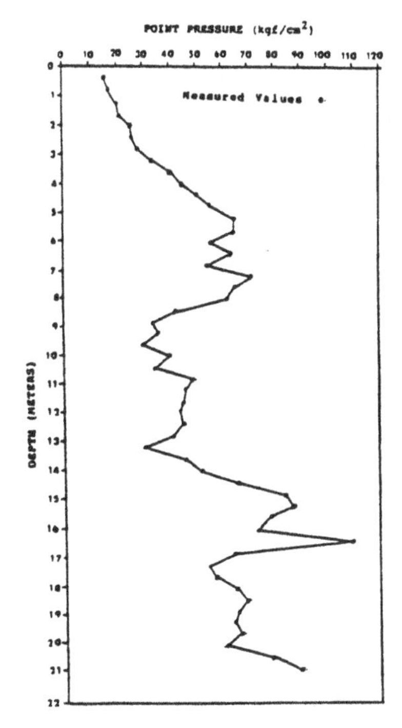

Fig. 2 Example of A typical dutch cone sounding log from Schmertmann (1969)

meters. These data were used to show an application of the developed model.

Figure 2 clearly shows that the data contain a global trend. Altogether this profile has fifty data points. Unweighted polynomial regression analyses were performed between point pressure and depth using degrees one through five for this data. This allowed a minimum of ten data points for each coefficient estimated. The results of the regression analysis are shown in Table 1. They indicate that the fourth degree polynomial represents the data best. This polynomial has the lowest RMS value and has a significant multiple R-square value. Figure 3 shows a graph of the values predicted by the fourth degree polynomial against the depth. The plot between the magnitude of the residuals versus the depth for this polynomial (Figure 4) shows that the variance of the residuals is approximately constant with depth. Therefore, estimation of W_i values is not required for this set of data. The RMS value of 112.84 found in the regression analysis (Table 1) provides an estimate for σ^2.

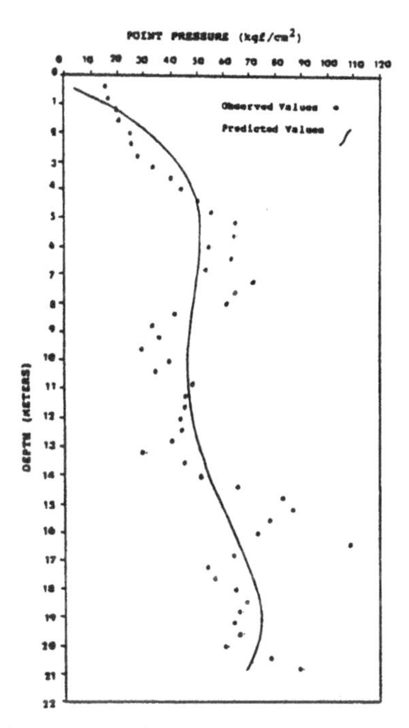

Fig. 3 Predicted values vs depth from fourth degree unweighted polynomial regression analysis.

Table 1. Results of the regression analyses between the point pressure and the depth

Degree of Polynomials	RMS	Multiple R-Square
1	151.287	.65433
2	131.037	.70684
3	119.183	.73903
4	112.841	.75829
5	113.844	.76156

As the next step the normalized residuals were obtained for this data by dividing each residual value by $\hat{\sigma}$. The obtained values are shown in Figure 5. This figure shows that the normalized residual values are free from non-stationarities. Since these normalized residual values are available at regularly spaced intervals, equation (16) was used to estimate autocorrelation coefficients. The obtained values are given in Table 2. These values were used to obtain ρ_{ij} values required for formation of V(e) according to equation (12). For example $\rho_{3,12}$ can be obtained from Table 2 by picking the autocorrelation coefficient value corresponding to k = 12 - 3 = 9.

Finally, the regression analysis was performed again between $\left[Y_i\right]$ and $\left[X_{ij}\right]$ with the V(e) formed according to equation (12). The final results along with the measured data are shown in Figure 6. The figure clearly

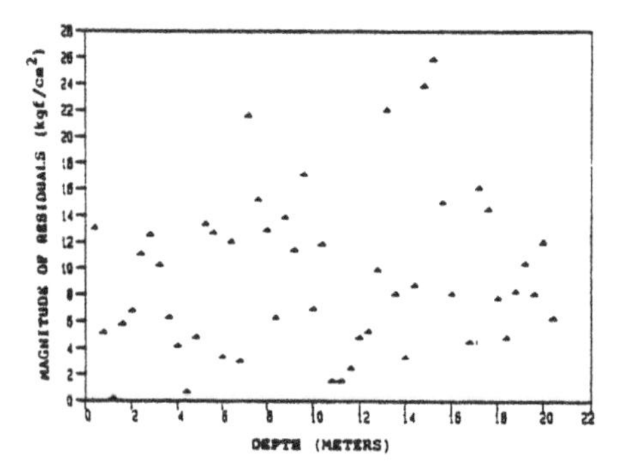

Fig. 4 Magnitude of the residuals .vs. depth from fourth degree unweighted polynomial regression analysis.

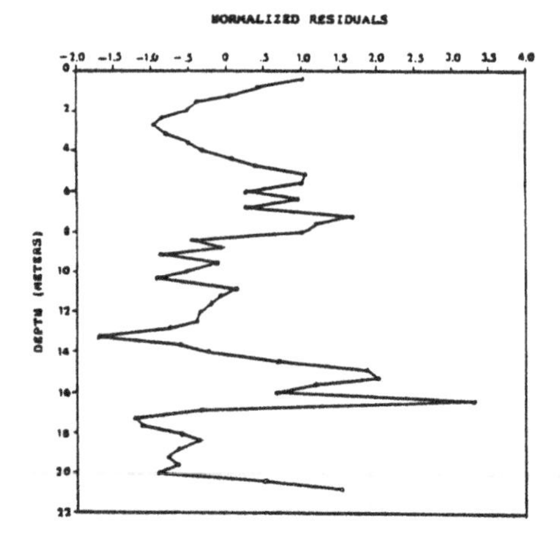

Fig. 5 Normalized residuals .vs. depth from fourth degree unweighted polynomial regression analysis.

Table 2. Autocorrelation coefficient estimates for the stationary portion of the data.

Lag Number (k)	Autocorrelation Coefficient	Lag Number (k)	Autocorrelation Coefficient
1	.5698	26	-.0419
2	.2426	27	-.0959
3	.1139	28	-.1706
4	-.0384	29	-.1957
5	-.1546	30	-.1751
6	-.3468	31	-.1480
7	-.3408	32	-.0994
8	-.2618	33	-.0434
9	-.2245	34	.0842
10	-.2125	35	.1492
11	-.2475	36	.0667
12	-.2466	37	.05
13	-.1857	38	.0764
14	-.0460	39	.0533
15	.1133	40	.0394
16	.0953	41	-.0169
17	.1476	42	-.0297
18	.1865	43	-.0279
19	.1797	44	-.0375
20	.2543	45	.0178
21	.1317	46	-.0258
22	.0882	47	-.0012
23	.0920	48	.0135
24	.0951	49	.0171
25	.0823		

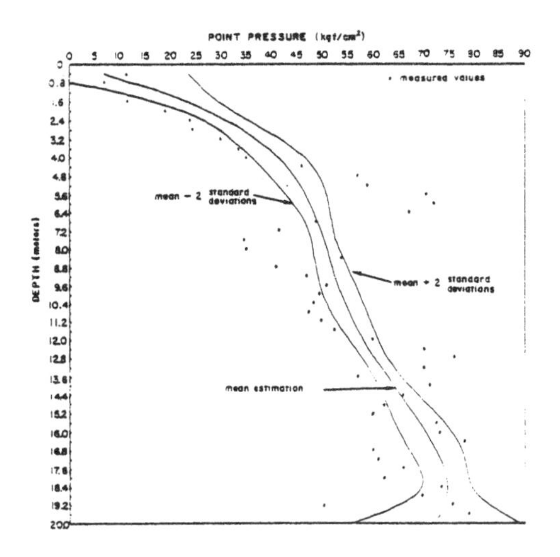

Fig. 6 Estimated values of the soil property variation with depth.

shows a low estimation variance indicating high confidence.

4 CONCLUSIONS

The technique introduced in this paper was found to be useful in estimating spatial variation of soil properties in one dimension. In applying this technique to a stratified soil deposit, first it is necessary to separate the deposit into statistically homogeneous layers. Then the technique can be applied to each layer separately. The method is capable of incorporating the global trend, non-constant variance around the mean trend and the autocorrelation structure under one scheme. With slight modifications, the method can be extended to estimate spatial variation in three dimensions.

REFERENCES

Agterberg,F.P., Autocorrelation functions in geology. D.F. Merriam, (ed.). Geostatistics, p. 113-141. New York: Plenum Press.

Alonso,E.E., and Krizek,R.J. 1975. Stochastic formulation of soil properties. Proc. Second Int. Conf. on Appl. of Stat. and Prob. in Soil and Structural Engr., 9-32.

Baker,R. 1984. Modeling soil variability as a random field. Mathematical Geology, 16: 435-448.

Bendat,J.S., and Piersol,A.G. 1971. Random data: Analysis and measurement procedures. New York: John Wiley Inc.

BMDP Statistical Software Manual. 1985. Berkeley: University of California Press.

Draper,N.R., and Smith, H.Jr. 1981. Applied regression analysis, 2nd ed. New York: John Wiley and Sons, Inc.

Journel,A.G. and Huijbregts,C. 1978. Mining Geostatistics. Academic Press.

Lumb,P. 1974. Applications of statistics in soil mechanics. I.K.Lee (ed.), Soil Mechanics: New horizons, p. 44-112. Newnes-Butterworth.

Lumb,P. 1975. Spatial Variability of soil properties. Proc. 2nd Int. Conf. on Appl. of Stat. and Prof. in Soil and Struct. Eng., Aachen, Germany, 397-421.

Schmertmann,J.H. 1969. Dutch friction-cone penetrometer exploration of research area t field 5, Eglin AFB, Florida. Contract Report S-69-4, U.S. Army Engineer Waterways Experiment Station, Vicksburg, Miss.

Tabba,M.M. and Yong,R.N. 1981. Mapping and predicting soil properties: Theory. J. Engrg. Mech. Div., ASCE, 107: 773-791.

Vandaele,W. 1983. Applied time series and Box-Jenkins models. Orlando: Academic Press, Inc.

Vanmarcke,E.H. 1977. Probabilistic modeling of soil profiles. J. of Geotech. Engrg. Div., ASCE, 103: 1227-1246.

Computer and Physical Modelling in Geotechnical Engineering, Balasubramaniam et al. (eds)
© 1989 Balkema, Rotterdam. ISBN 90 6191 864 2

Microcomputer automation of the hydrometer test for soil grading

V.U.Nguyen & F.J.Paoloni
Faculty of Engineering, University of Wollongong, Australia

ABSTRACT: A method of automating the hydrometer test for soil grading by the microcomputer is described. The method is based on the principle of counting electrical pulses resulting from collimated light beam transmitted across the moving stem of a hydrometer.
Of all laboratory soil tests, the hydrometer test is one of the most tedious and time-consuming, and whilst every application of automatic data-logging has been made to soil tests, little if any attention has been paid to the hydrometer test. This paper presents the principle involved in the automation of the hydrometer test by a microcomputer and some preliminary results of the work being carried out at the University of Wollongong.

1. DESCRIPTION OF HYDROMETER LOGGING

The hydrometer test is an extension of sieve analysis, used to obtain an estimate of the distribution of soil particle sizes from the No. 200 (0.075mm) sieve to around 0.001mm (1 micron). In the test, the diameter of a falling sphere at a depth L, at time t, is related to the specific weights of the sphere and of the fluid (normally water), and the viscosity of the fluid by Stokes' law:

$$D = \sqrt{\frac{18\eta V}{\gamma_s - \gamma_w}} \tag{1}$$

where η is the fluid viscosity in Pascal-second ($\eta_{H_2O} = 0.001\ Pa - s$), γ_s and γ_w are specific weights of soil sphere and water, respectively, and V is the falling velocity of the sphere over the depth L:

$$V = \frac{2\gamma_s - \gamma_w}{9\eta}\left(\frac{D}{2}\right)^2 = \frac{L}{t} \tag{2}$$

The proportion of soil grains having sizes smaller than D is still in suspension in the zone between the centre of the hydrometer bulb and the surface of water contained in the sedimentation cylinder, and it can be proved that the percentage of soil grains finer than D is proportional to the specific gravity of soil-water suspension at depth L.

Since the hydrometer displays the specific gravity of soil-water suspension at depth L, the percentage of soil grains finer than size D is thus computed from the hydrometer reading corrected for temperature, dispersing agent effect, and soil specific gravity. The more soil grains settle from suspension, the lesser the specific gravity of the suspension is, and the deeper the hydrometer sinks in the cylinder or the larger the distance L becomes. The principle of hydrometer automation is based on continual monitoring of the distance L with time. Figures 1 and 2 are photographs taken of the hydrometer automation system during the present development stage. The hydrometer automation system comprises mainly:

- A plastic cap holding an Aluminium collar with a 5mm hole giving about 1mm clearance for hydrometer movement. The cap is suspended on top of the cylindrical container (shown in Figure 2) whilst the collar inside the cap holds the light emitter.
- A DC-supply infrared light emitting diode

- A photo transistor receiver connected to a detecting circuit as shown in Figure 3.
- A hydrometer with the glass stem modified to contain a transparent graticule scale.

The infrared light is collimated to a 1.0 mm diameter beam by a steel cylindrical case fitted inside the hydrometer collar. It should be noted that virtually all current hydrometer designs have calibrated readings printed on a piece of white paper encased in the hydrometer stem. The modification thus involves mainly stem-cutting, removal of the paper scale, stem resealing, and re-marking with lines having width of size similar to that of the beam collimation, using black glass paint. Alternatively, the graticule scale can be made on a transparent tape then glued to the outside perimeter surface of the hydrometer stem. To avoid detachment of the scale tape upon contact with water, a coat of water-repellent varnish paint should be applied all around the stem before use. Figure 5 shows the hydrometer modified by using "scale tape" for the present test programme.

As the hydrometer moves downwards, light that reaches the photodetector, is alternately blocked and passed by the graticule scale on the stem. Electronics in the output stage conditions the photodetector response to TTL levels, i.e. 5 volts corresponding to no detected light and 0 volt to detected light. Communication to the microcomputer is by means of an analogue to digital (A/D) interface. Total downward movement of the hydrometer in steps equal to the width of the graticule scale is obtained by summing the total photodetector output transitions using a BASIC program in the microcomputer. Fluctuation in hydrometer readings due to upward and downward movements of the hydrometer can be easily filtered or adjusted in the BASIC program. The computer stores the progressive detector count and the time lapse from the beginning of the measurements. One other channel of the A/D board is interfaced with a thermocouple for temperature monitoring. The thermocouple is immersed in an adjacent water cylinder and its circuitry is shown diagrammatically in Figure 4.

2. PRELIMINARY RESULTS

Ideally, the width on the graticule scale should be equal to the width of the light beam, and improvements have recently been made to reduce the graticule scale width to 1.0mm being that of the collimation. Figures 6 and 7 give plots of total hydrometer movement (mm) and temperature (degrees Celsius) versus time during a simulation run in which the hydrometer is moved down the collar by hand. Figure 10 gives a listing of the fundamental BASIC code used in data-logging for hydrometer test.

A quick hydrometer test, based on A.S.1289-C6.3, following a conventional sieve analysis was also carried out just in time for this report, and the "full" grading curve is presented in Fig. 8. The hydrometer automation procedure used in grain size analysis is shown schematically in Figure 9.

3. SUGGESTIONS FOR FURTHER DEVELOPMENT

Continuing work of the present project consists mainly of:

(i) Effective designs of the graticule scale for hydrometer automation. Such designs may require custom-made etching and painting on the hydrometer stem to smaller widths, or gluing a photographic print strip of graticule scale to the hydrometer stem.

(ii) Improved design of electronic circuitry involving higher accuracy and gain, a smaller tip of light emitter, and a smaller collimation width.

ACKNOWLEDGEMENTS

The authors are indebted to Mr. A. Mowbray for his assistance with the electronic circuitry. The enthusiastic help from Maroun Yacoub and Doan Huu Thanh in finding ways to modify the hydrometer and in laboratory testing is acknowledged.

Figure 1. Circuitry of hydrometer movement counter & thermocouple, and aluminium collar for light emitter used in hydrometer automation

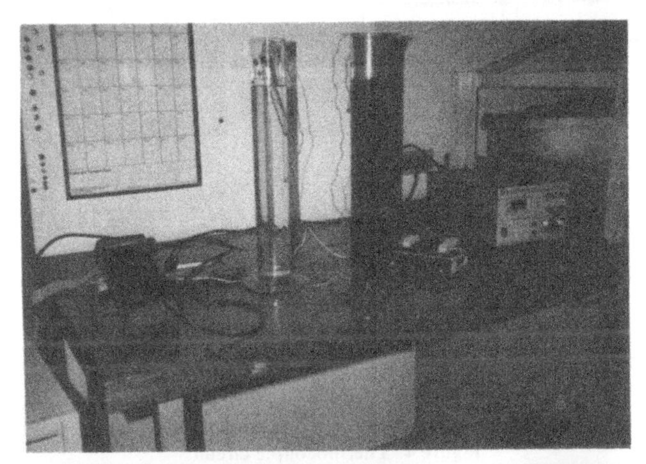

Figure 2. Test arrangement for hydrometer automation

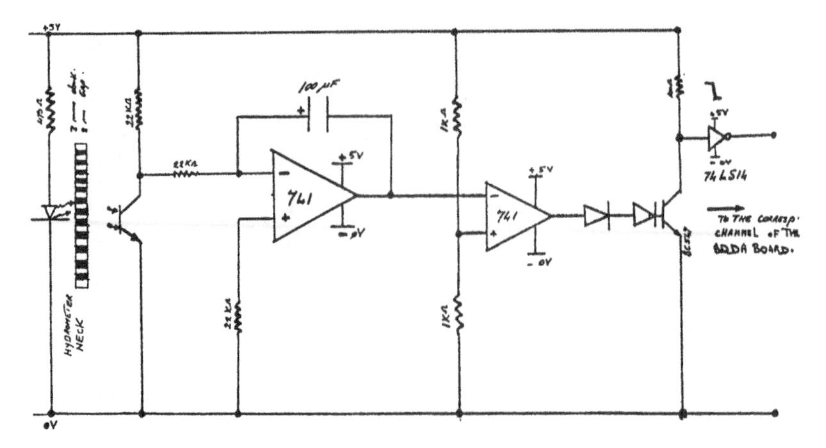

Figure 3. Electronic circuit for hydrometer automation

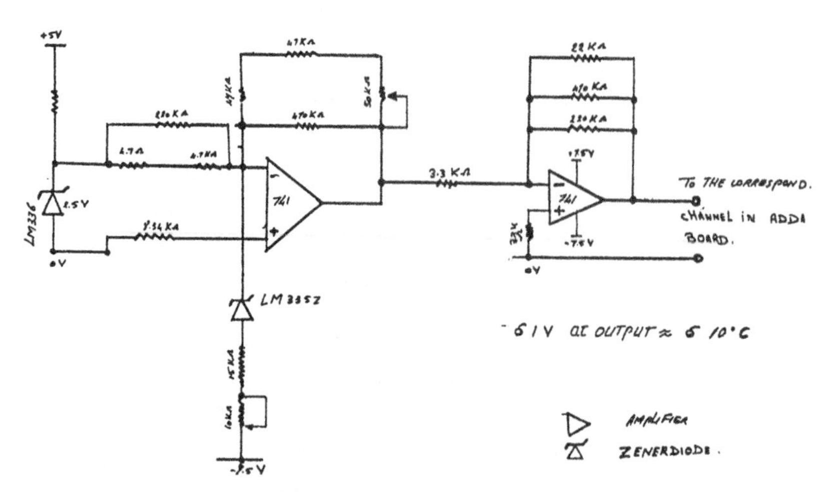

Figure 4. Thermocouple circuit

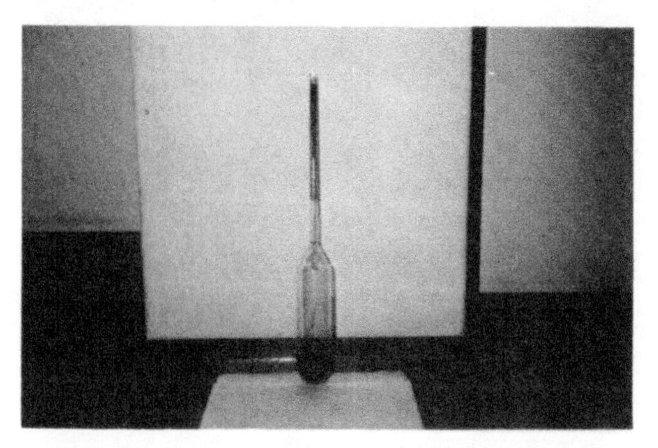

Figure 5. Hydrometer with stem modified for automation

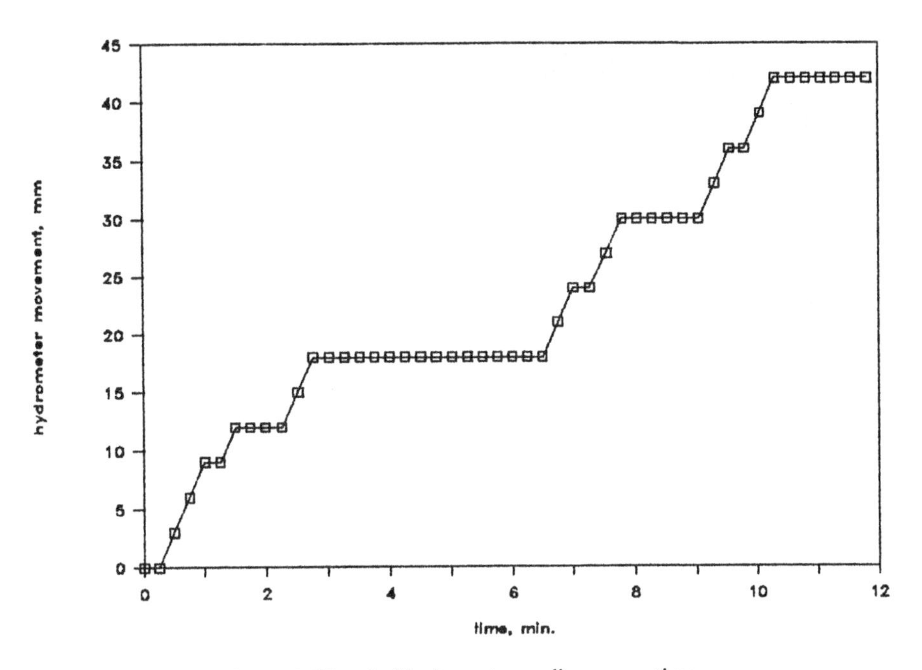

Figure 6. Electrical hydrometer reading versus time

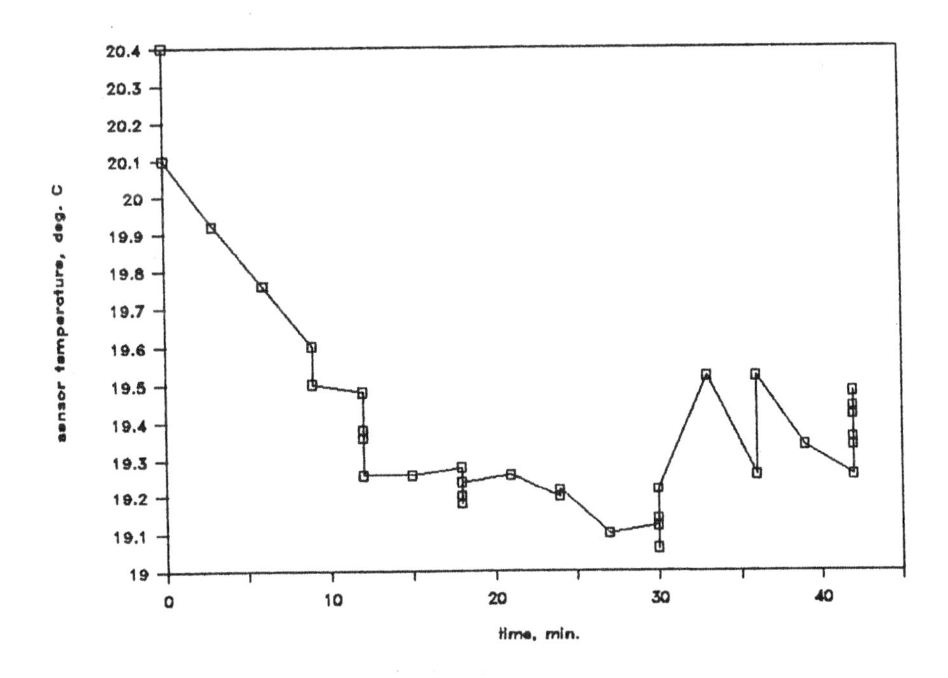

Figure 7. Thermocouple reading versus time

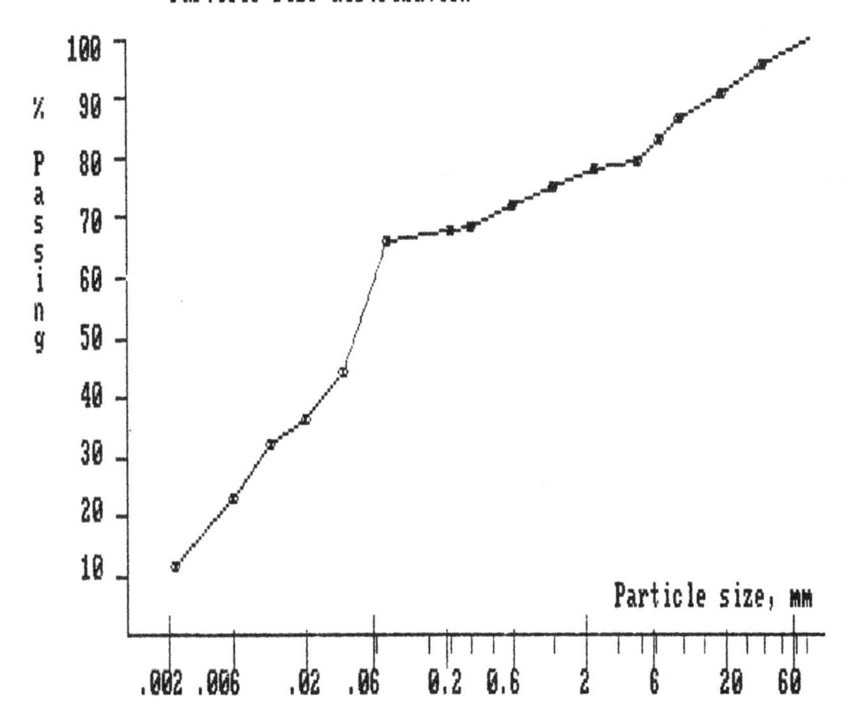

Figure 8. Full grading curve including automated hydrometer results

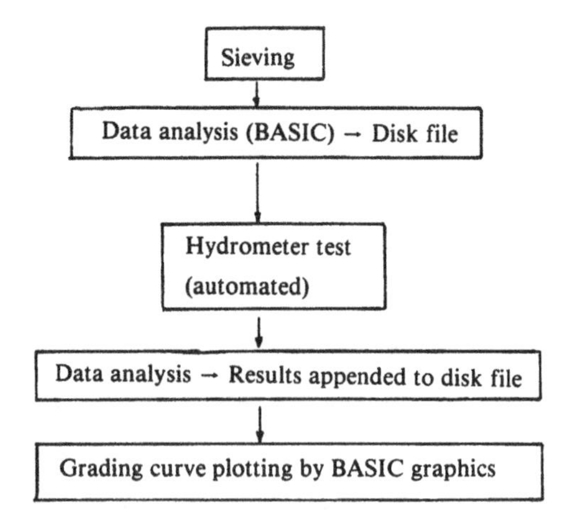

Figure 9. Microcomputer automation steps in soil grading

Figure 10. BASIC code for hydrometer data logging

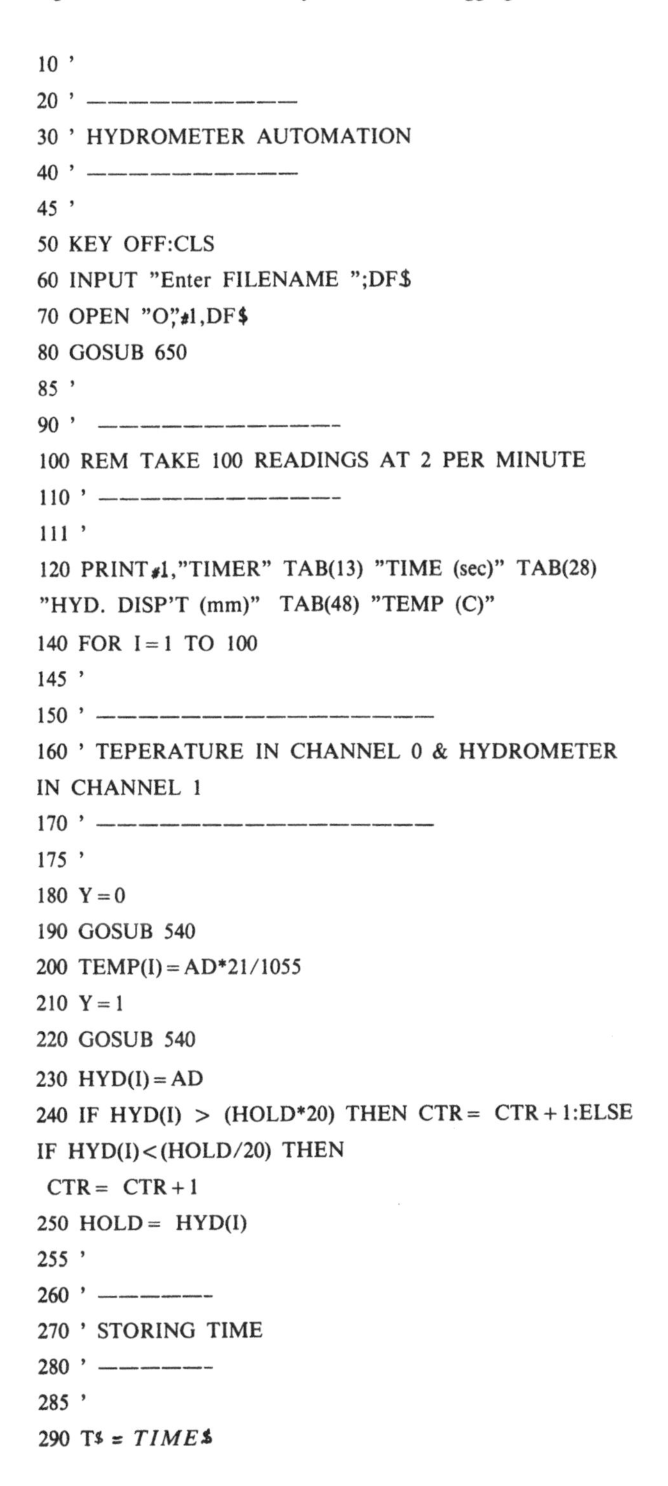

```
10 '
20 ' ------------
30 ' HYDROMETER AUTOMATION
40 ' ------------
45 '
50 KEY OFF:CLS
60 INPUT "Enter FILENAME ";DF$
70 OPEN "O",#1,DF$
80 GOSUB 650
85 '
90 '  --------------
100 REM TAKE 100 READINGS AT 2 PER MINUTE
110 ' ---------------
111 '
120 PRINT#1,"TIMER" TAB(13) "TIME (sec)" TAB(28)
"HYD. DISP'T (mm)"  TAB(48) "TEMP (C)"
140 FOR I=1 TO 100
145 '
150 ' ---------------------
160 ' TEPERATURE IN CHANNEL 0 & HYDROMETER
IN CHANNEL 1
170 ' ---------------------
175 '
180 Y=0
190 GOSUB 540
200 TEMP(I)=AD*21/1055
210 Y=1
220 GOSUB 540
230 HYD(I)=AD
240 IF HYD(I) > (HOLD*20) THEN CTR= CTR+1:ELSE
IF HYD(I)<(HOLD/20) THEN
 CTR= CTR+1
250 HOLD= HYD(I)
255 '
260 ' -------
270 ' STORING TIME
280 ' -------
285 '
290 T$ = TIME$
```

```
300 T(I)= VAL(T$)
310 H(I)= VAL(LEFT$(T$,2))
320 M(I)= VAL(MID$(T$,4,2))
330 S(I)= VAL(RIGHT$(T$,2))
335 '
340 ' ------
350 ' ELAPSE TIME
360 ' ------
370 DH(I)= H(I)-H(1)
380 DM(I)= ((M(I)+DH(I)*60))-M(1))
390 DS(I)= ((S(I)+DM(I)*60))-S(1))
400 ' ---------------------
410 ' PRINT AND STORE RESULTS (DATA) TO
SEQUENTIAL FILE 1
420 ' ---------------------
425 '
430 PRINT#1,T;TAB(15);DS(I);TAB(31);CTR;TAB(49);TEMP(I)
440 PRINT T;TAB(15);DS(I);TAB(31);CTR;TAB(49);TEMP(I)
450 ' ---------
460 ' TIMEOUT WAITING LOOP
470 ' ---------
475 '
480 FOR J=1 TO 29900
490 NEXT J:' this loop depends on speed of computer
500 NEXT I
510 CLOSE#1
520 END
526 '
530 ' -------
540 ' FETCH ROUTINE
550 ' -------
560 OUT BA,Y
570 A=INP(BA+3)
580 FOR I=1 TO 7:A=INP(BA+4):NEXT I
590 FOR I=1 TO 7:A=INP(BA+5):NEXT I
600 C=INP(BA+2)
610 HB=(C/16-INT(C/16))*16
620 LB=INP(BA+1)
630 AD=HB*256+LB
640 RETURN
```

```
645 '
650 ' ----------
660 ' INITIALIZE VARIABLES
670 ' ----------
677 '
680 DIM TEMP(1000), HYD(1000)
690 DIM DH(1000),DM(1000),DS(1000),T(1000),H(1000),M(1000),S(1000)
710 BA=632:REM Number 632 specifies the address
on the ADDA card
720 HOLD=0:' set initial value =0
730 CTR=-1:' control position is 1mm interval on hydrometer
740 RETURN
750 '
```

Computer and Physical Modelling in Geotechnical Engineering, Balasubramaniam et al. (eds)
© 1989 Balkema, Rotterdam. ISBN 90 6191 864 2

True triaxial dynamic soil test system aided by microcomputer

R.Kitamura, K.Jomoto, K.Nakamura & M.Hidaka
Department of Ocean Civil Engineering, Kagoshima University, Kagoshima, Japan

ABSTRACT: The true triaxial testing apparatus for dynamic soil test is explained. The shape of the specimen is cubical and three principal stresses can be applied six opposite faces of the specimen independently. The capacitance type's wave height meter is made use of to measure the change of water level in the burette from which the linear strain can be calculated. The microcomputer is introduced to acquire and to process the data of strain and excess pore water pressure automatically. The liquefaction test on Toyoura sand is carried out by using this testing apparatus and the results are shown.

1 INTRODUCTION

In the soil mechanics the mechanical characteristics of soil under the cyclic loading is one of the most important subjects to be solved. Many researches have been carried out since Niigata earthquake in 1964 which was occurred in Japan. Mróz et al. (1979) proposed two surface model based on the theory of elasto-plasticity and tried to apply it to the cyclic loading process. Towhata and Ishihara (1985) discussed the characteristics of sandy soil under the cyclic loading based on the triaxial torsional shear apparatus. Tatsuoka et al. (1986) discussed the method for the cyclic undrained triaxial and torsional test.

Ko and Scott (1967), Arthur and Menzies (1972), Yamada and Ishihara (1979) and Haruyama (1981) developed the box type's true triaxial soil testing apparatus which can generate three different pricipal stresses in the specimen. In our laboratory we are now trying to modify the box type's true triaxial testing apparatus which was produced by Haruyama (1981, 1985, 1987) to conduct the dynamic soil test under various stress conditions and automatically to acquire and to process the data by using the microcomputer.

In this paper a new measuring device of the deformation of the specimen, and a data acquisition and data processing system for the true triaxial dynamic soil testing apparatus are mainly explained. The dynamic soil test results are also shown which are drawn by the X-Y plotter.

2 TRUE TRIAXIAL DYNAMIC TESTING APPARATUS

2.1 Testing apparatus

Figure 1 shows the main part of the testing apparatus, which consists of cubical triaxial cell (7 in Fig.1), four volume change gauges (6 in Fig.1) and pressure transducer (8 in Fig.1). Figure 2 shows the cross sectional view of the bottom and side wall units of cubical triaxial cell whose internal dimension is 100 mm x 100 mm x 100 mm. The cubical specimen is placed in the cell formed by six wall assemblies whose opposite pairs are interconnected and led to three volume change gauges. Four vertical walls are first tightened by two horizontal clamp-bands and then the whole units are assembled by tightening the bottom and top clemp-plates. Each wall unit is made of a flexible rubber membrane, 0.8 mm in thickness, pressure chamber, an adjustable plate and an adjustable piston whith a nut. The adjustable plate lightly touches to the rubber membrane when the specimen is prepared. After the specimen is prepared, the adjustable plate is thrown back to fill the space between the membrane and the plate with water by the adjustable piston. Another volume change gauge is connected to two needles with porous stone,

2 mm in diameter, which are inserted at the opposite corner of the cubical cell and measure the volume change of specimen in the drained condition. Pressure transducer measures the excess pore water pressure in the undrained condition. An influence of membrane penetration on the volume change is neglected, considering the thickness of membrane, the stress level and the grading of material. The volume change, however, is corrected for the effect of expansion of all the tubing system under pressure.

2.2 Loading system

Three independent static pressure is supplied by the air-compressor, controlled by the air regulators which are connected with A1, A2 and A3 in Fig.1, and applied to six faces of the specimen through the rubber membrane containing water as shown in Figs.1 and 2. The dynamic pressure is applied by using the pressure generation system as shown in Fig.3. The function generator can generate three kinds of electric waves, i.e., the sine, triangle and square waves. The frequency can be varied between 0.0001 Hz and 100 kHz. The function generator has two channels and can vary the phase of two waves between 0° and 360°. The electric wave through the servo-controller is converted to the change of air pressure by the transducer. The change of air pressure is amplified by the ratio relay and is made more accurate by the servo-control system. Furthermore, the change of air pressure is converted to the change of water pressure at the air-fluid interface (3 in Fig.1) and applied to the specimen.

2.3 Measuring device of volume change

The volume change of water in the rubber membrane, which is occurred due to the deformation of specimen, is measured as the change of the water level in the volume change gauge. Then the principal strains in three orthogonal directions are calculated. In the static test the water levels in three directions are directly measured by the observer, but in the dynamic test the changes of water levels of three pressure gauges are so violent that the observer cannot measure the accurate levels at the same time. So, the principle of the capacitance type's wave height meter as shown in Fig.4 is introduced automatically to measure the water level in the volume change gauge. Figure 4 consists of the part of sensor (volume change gauge) and the part of data acquisition. The burette is set in the pressure chamber which is made of hollow cylindrical acrylic resin. The teflon wire, 0.3 mm in diameter, is strung in the center of the burette and the copper tape, 3 mm in width, is stuck on the wall of the burette. The pressure chamber is filled with paraffin oil and distilled water where the paraffin oil always occupies upper part because the density of paraffin is smaller than distilled water. Then the condenser is formed in the pressure chamber. The capacitance of the condenser changes as the water level changes, i.e., the change of the capacitance of the condenser can be linearly related to the change of the water level. Figure 5 shows the step voltage which is put in the condenser, and the response of the condenser. The frequency of the step voltage is 2 kHz. Figure 6 shows the way to make the step voltage, i.e., the circuit connected to the electric source is closed in the process 1 shown in Fig.5 and open in the process 2 shown in Fig.5. Figure 7 shows the relation between the water level in the burette and the output voltage. In Fig.7 the solid and dotted lines denote the relations in the cases that the water level changes statically and dynamically. It is found from Fig.7 that the linear relation exists in both cases of static and dynamic changes of the water level, and the gradient of lines are same. This means that this measuring device is valid in the dynamic soil test.

2.4 Data acquisition and data processing system

The change of the capacitance of the condenser, which is related to the change of the water level, is measured as the change of the voltage by the capacitance sensor shown in Figs.4 and 5. Furthermore, the analogue voltage is amplified and then then converted to the digital amount. Finally, the change of the capacitance of the condenser is stored in the microcomputer as the digital amount. Figure 4 shows this procedure of data acquisition. The present microcomputer has the user's memory of 256 kB and can continuously read eight kinds of digital data whose minimum sampling period is 0.1 second. The sampled data are stored in the floppy disc as the sequential file. Figure 8 is the flow chart which shows the procedure of data acquisition by the microcomputer. Figure 9 shows the detailed flow of the data sampling process shown in Fig.8. The method of moving average is used to minimize the influence of noise on sampling data when the data are sampled. The strain, the excess pore water pressure and effective stress are calculated based on the data stored in the floppy disc by the same microcomputer and stored in another floppy disc. Figure 10 is the flow chart which shows the graphic procedure by the microcomputer. The relation between stress and strain, strain and cyclic number, excess pore water pressure and cyclic number, and the effective stress path are drawn on the display and on the paper by the X-Y plotter in accordance with Fig.10.

3 DYNAMIC SOIL TEST

3.1 Material and test procedure

Test material is Toyoura sand whose specific gravity is 2.64, the maximum and the minimum void ratios are 0.938 and 0.582 respectively, and the coefficient of uniformity is 1.48. The saturated specimen is prepared by pouring Toyoura sand, which is boiled for several hours, in the cubical triaxial cell filled with distilled water. After the isotropic compression of 1 kgf/cm^2 is finished, the back pressure, 1.5 kgf/cm^2, is applied and the isotropic pressure is increased to 2.5 kgf/cm^2 at the same time, i.e., the effective isotropic pressure is 1 kgf/cm^2. Then the cyclic shear test under the undrained condition is carried out. Figure 11 shows the stress paths on Π-plane which are adopted in the test, where σ_z denotes the stress in the vertical direction, σ_x and σ_y denote the stresses in the horizontal direction. Three kinds of stress paths are adopted to investigate the influence of anisotropic initial fabric on the mechanical characteristics of liquefaction. The test with these stress paths are called Type-A, Type-B and Type-C respectively. The cyclic load of sine wave whose frequency is 0.1 Hz is applied under the constant total mean pricipal stress, 2.5 kgf/cm^2.

3.2 Results

Figures 12, 13 and 14 show examples of the relation between the strain and the cyclic number, where the compression is positive. The cyclic shear stress τ_{oct}=0.14 kgf/cm^2 is applied. Figures 15, 16 and 17 are the relations between the ratio of the excess pore water pressure to the initial mean effective principal stress, $\Delta u/\sigma'_o$, and the cyclic number. Figures 18, 19 and 20 are the effective stress paths.

4 CONCLUSIONS

In this paper the data acquisition and the data processing system aided by the micro-computer is described for the dynamic soil test which is carried out by using the cubical true triaxial testing apparatus. The capacitance type's measuring device is much useful to measure the water level in the burette which changes dynamically. The consideration should be done concerning the test results in near future.

The testing apparatus was manufactured by the grant-in-aid offered from the Ministry of Education. The authors wish to acknowledge the help of this goverment office.

REFERENCES

Arthur, J.R.F. and Menzies, B.K. 1972. Inherent anisotropy in sand. Geotechnique, Vol.22, No.1, pp.115-128.
Haruyama, M. 1981. Anisotropic deformation-strength characteristics of an assembly of spherical particles under three dimensional stresses. Soils and Foundations, Vol.21, No.4, pp.41-55.
Haruyama, M. 1985. Drained deformation-strength characteristics of loose Shirasu (volcanic sandy soil) under three dimensional stresses. Soils and Foundations, Vol.25, No.1, pp.65-76.
Haruyama, M. 1987. Effect of density on the drained deformation behavior of Shirasu (volcanic sandy soil) under three dimensional stresses. Soils and Foundations, Vol.27, No.1, pp.1-13.
Ishihara, K. and Yamada, Y. 1975. Sand liquefaction in hollow cylinder torsion under irregular excitation. Soils and Foundations, Vol.15, No.1, pp.45-59.
Ko, H-Y and Scott, R.F. 1967. A new soil testing apparatus, Geotechnique, Vol.17, No.1, pp.40-57.
Mróz, Z., Norris, V.A. and Zienkiewicz, O. C. 1979. Application of an anisotropic hardening model in the analysis of elasto-plastic deformation of soils. Geothechnique, Vol.29, No.1, pp.1-34.
Tatsuoka, F., Ochi, K., Fujii, S. and Okamoto, M. 1986. Cyclic undrained triaxial and torsional shear strength of sand for different sample preparation methods. Soils and Foundations, Vol.26, No.3, pp.23-41.

Towhata, I. and Ishihara, K. 1985. Shear work and pore water pressure in undrained shear. Soils and Foundations, Vol.25, No.3, pp.73-84.

Yamada, Y. and Ishihara, K. 1979. Anisotropic deformation characteristics of sand under three dimensional stress conditions. Soils and Foundations, Vol.19, No.2, pp.79-94.

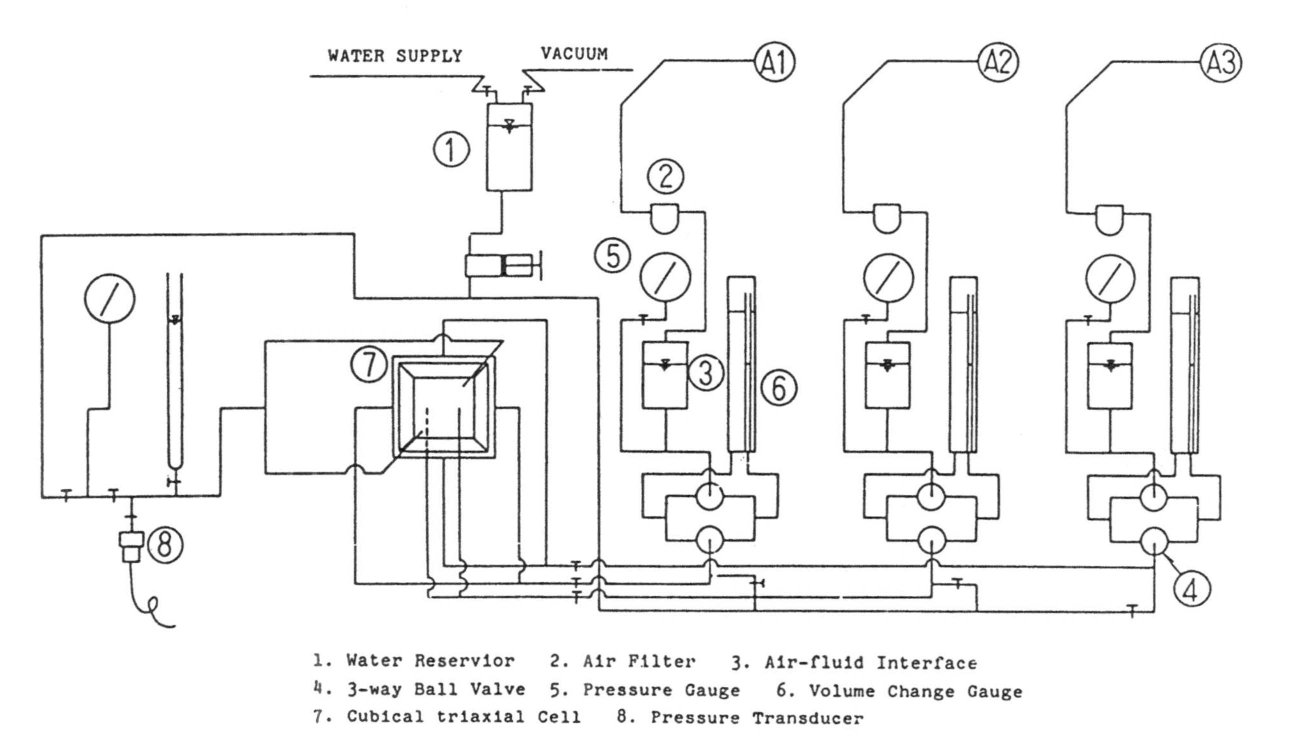

1. Water Reservior 2. Air Filter 3. Air-fluid Interface
4. 3-way Ball Valve 5. Pressure Gauge 6. Volume Change Gauge
7. Cubical triaxial Cell 8. Pressure Transducer

Fig.1 True triaxial testing apparatus

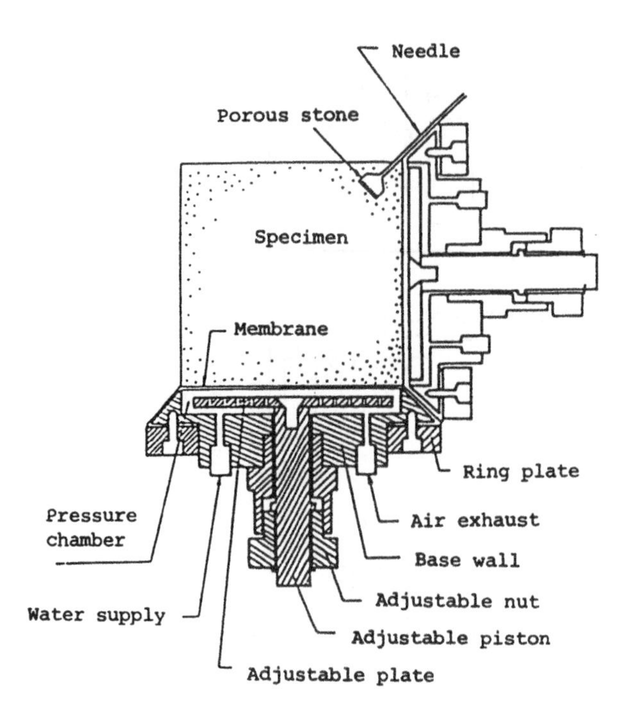

Fig.2 Cubical triaxial cell

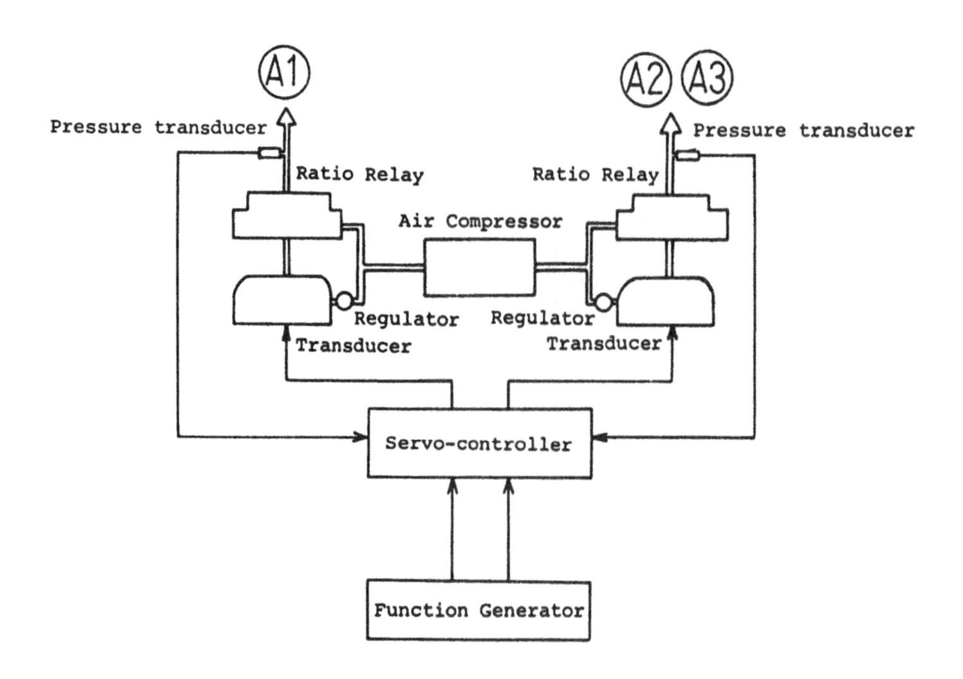

Fig.3 Pressure generation system for cyclic load

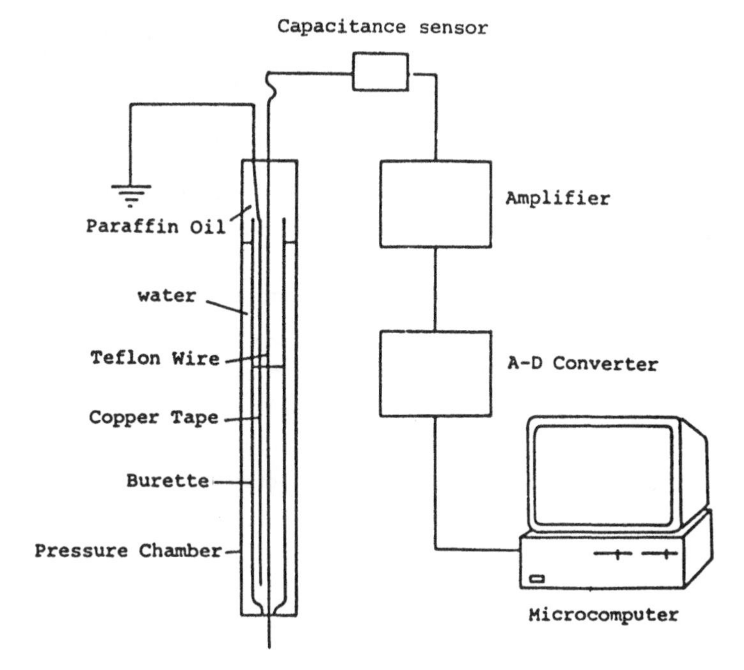

Fig.4 Volume change measuring device and data acquisition system

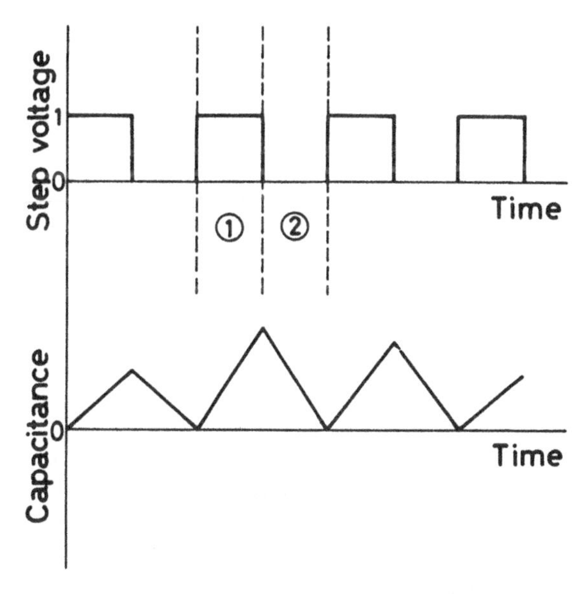

Fig.5 Relation between step voltage (input) and capacitance of condenser (output)

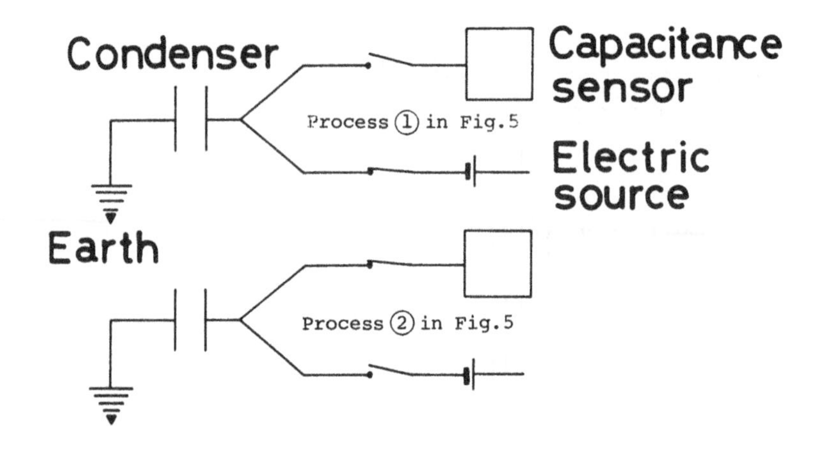

Fig.6 Circuit for generating step voltage

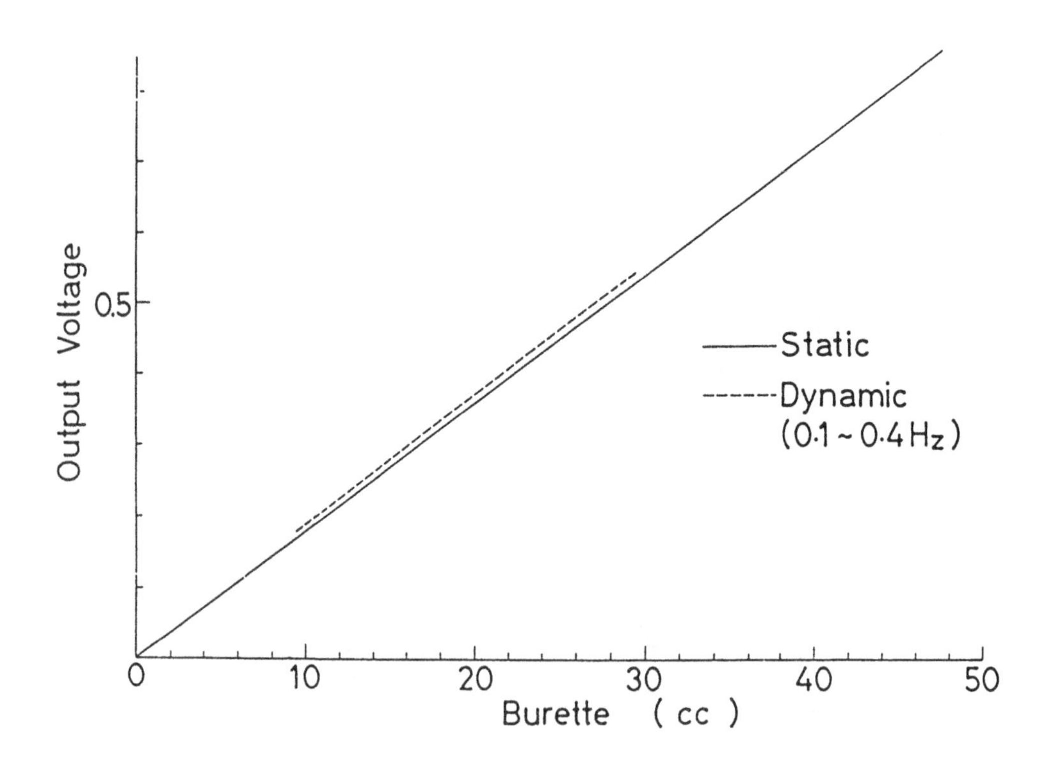

Fig.7 Relations between water level and output voltage

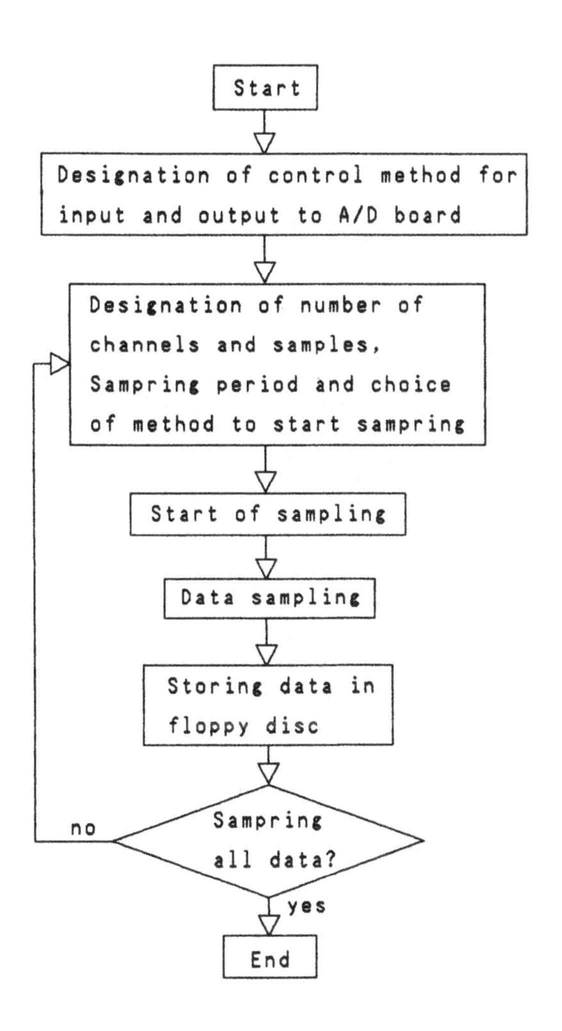

Fig.8 Procedure of data acquisition by microcomputer

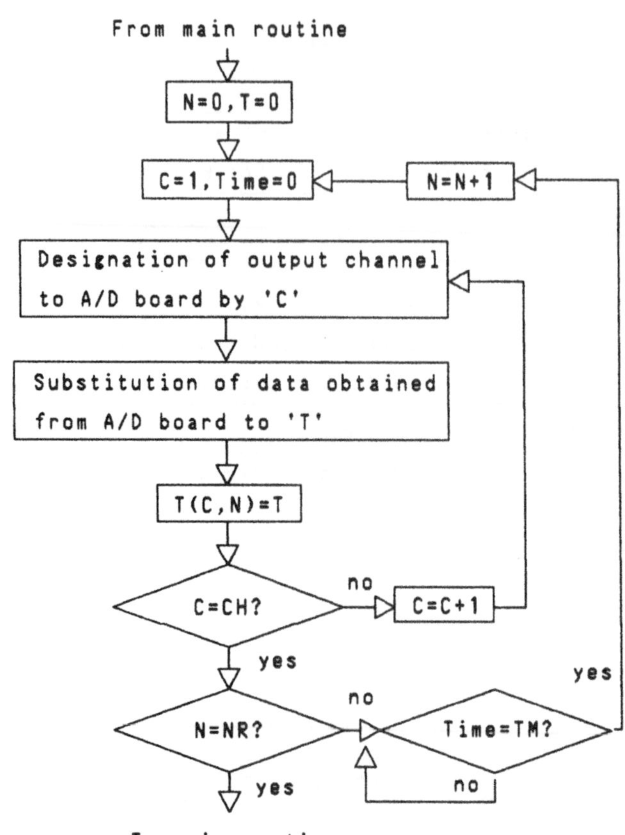

N; Dammy for sampling number, NR; Sampling number,
C; Dammy for channel number, CH; Channel number,
T; Digital sampling data, T(C,N); Stored data,
Time; Sampling time, TM; Designated sampling period.

Fig.9 Subroutine for data sampling

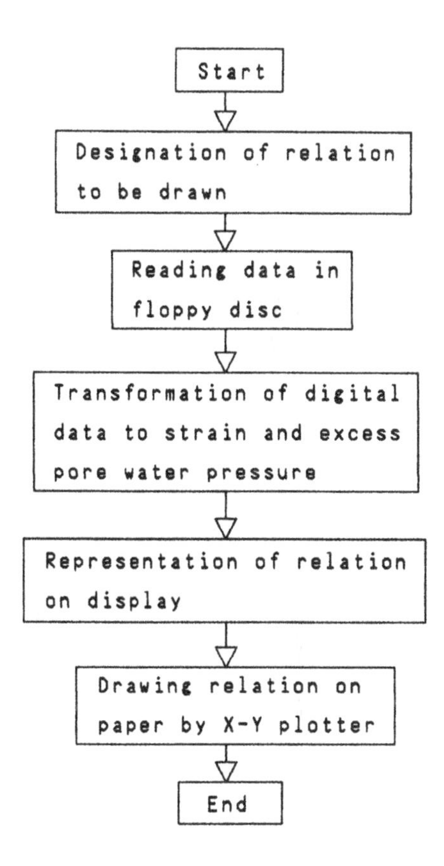

Fig.10 Graphic procedure of data by microcomputer

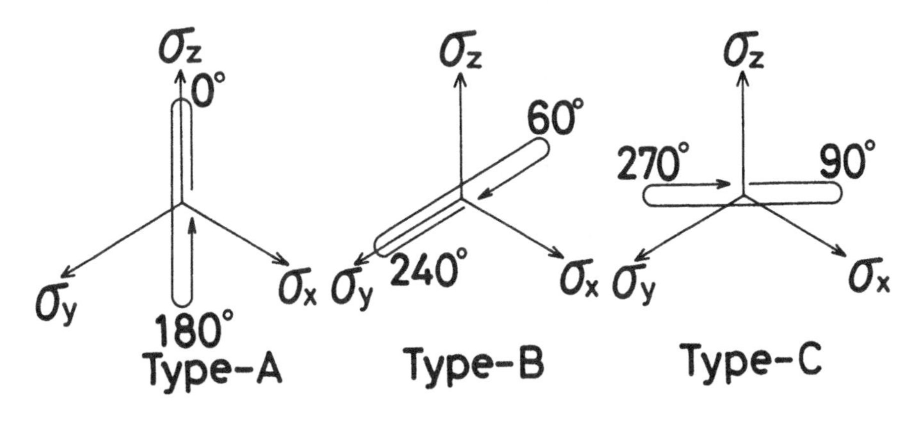

Fig.11 Stress path on Π-plane

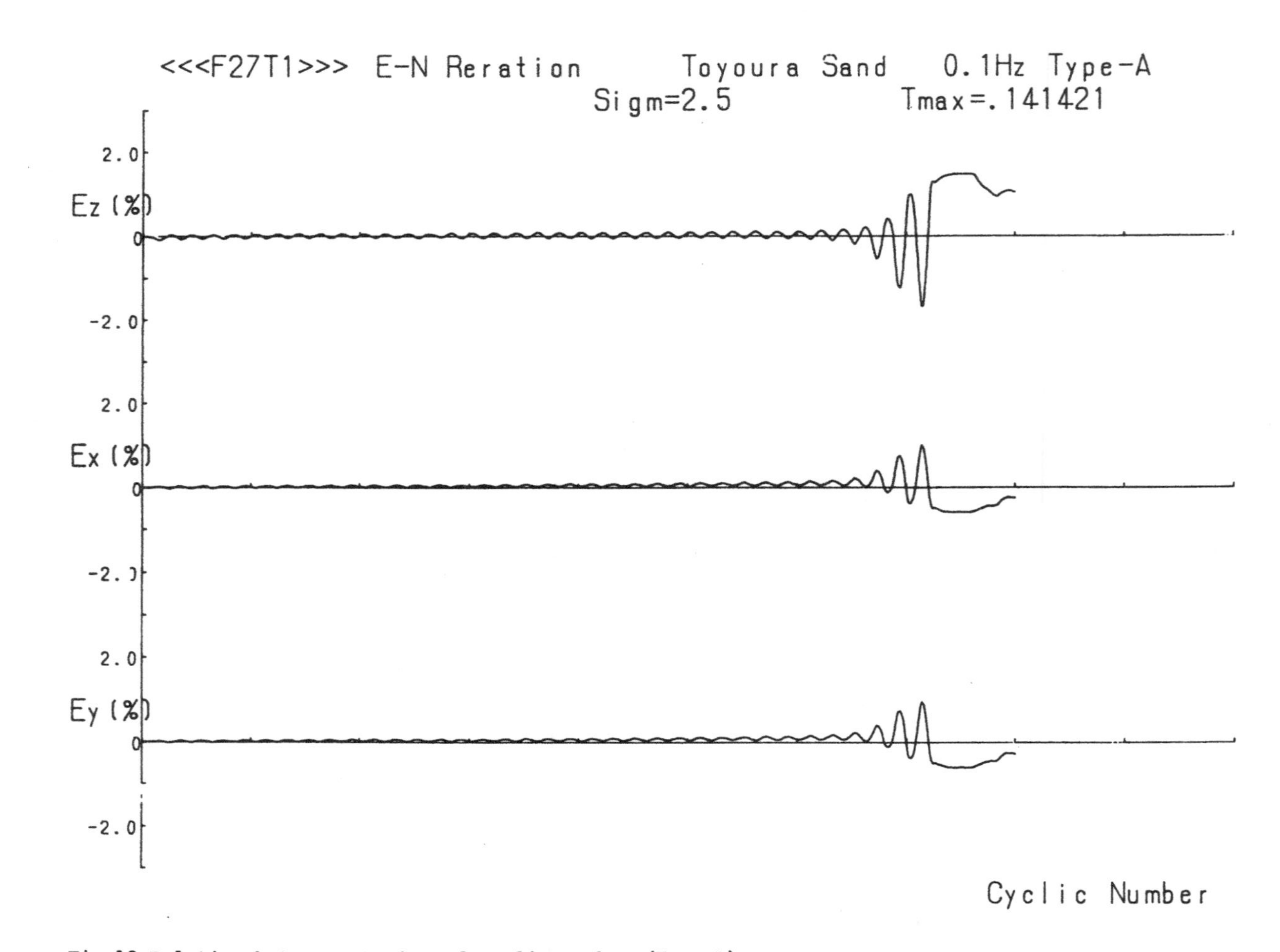

Fig.12 Relation between strain and cyclic number (Type-A)

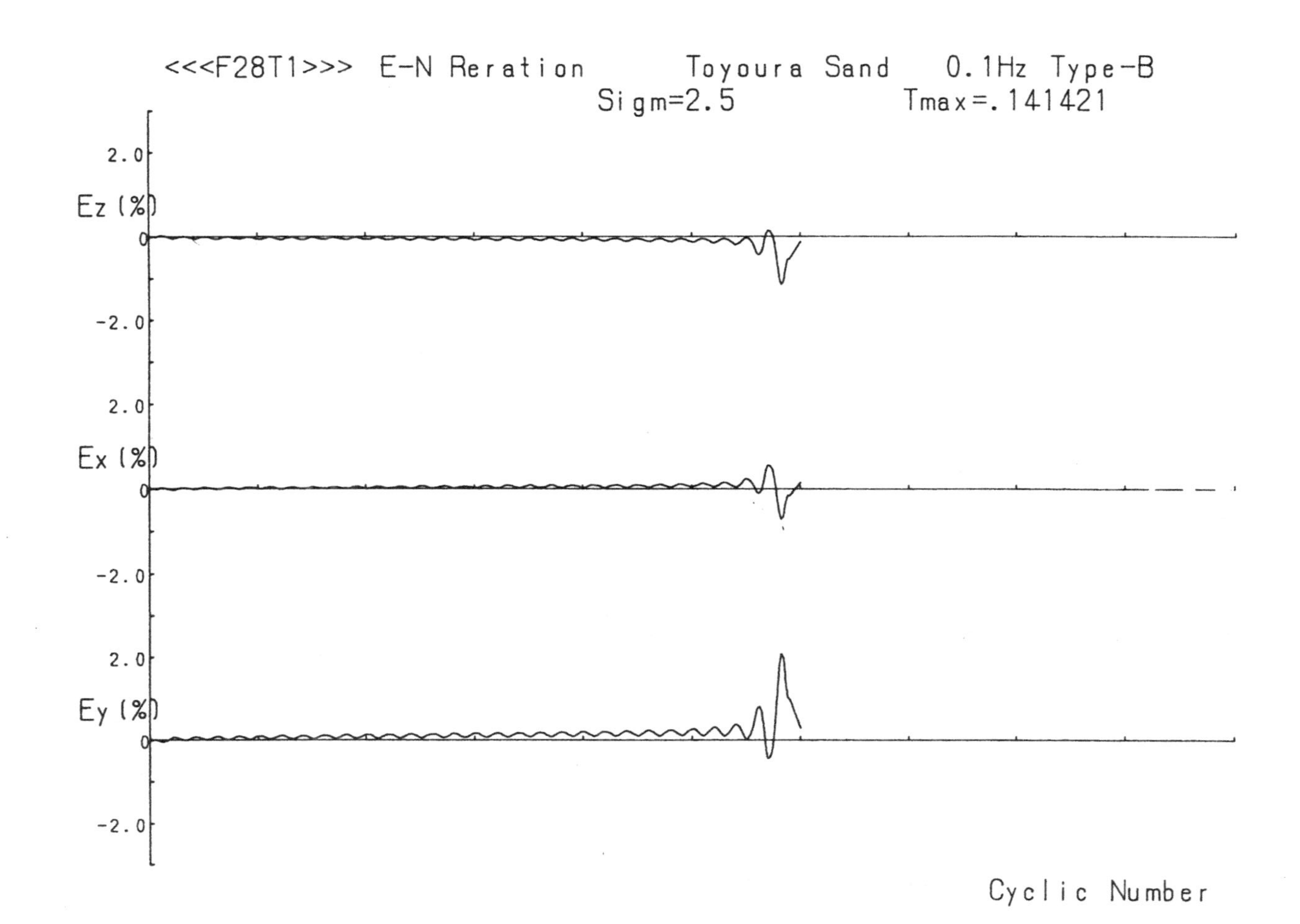

Fig.13 Relation between strain and cyclic number (Type-B)

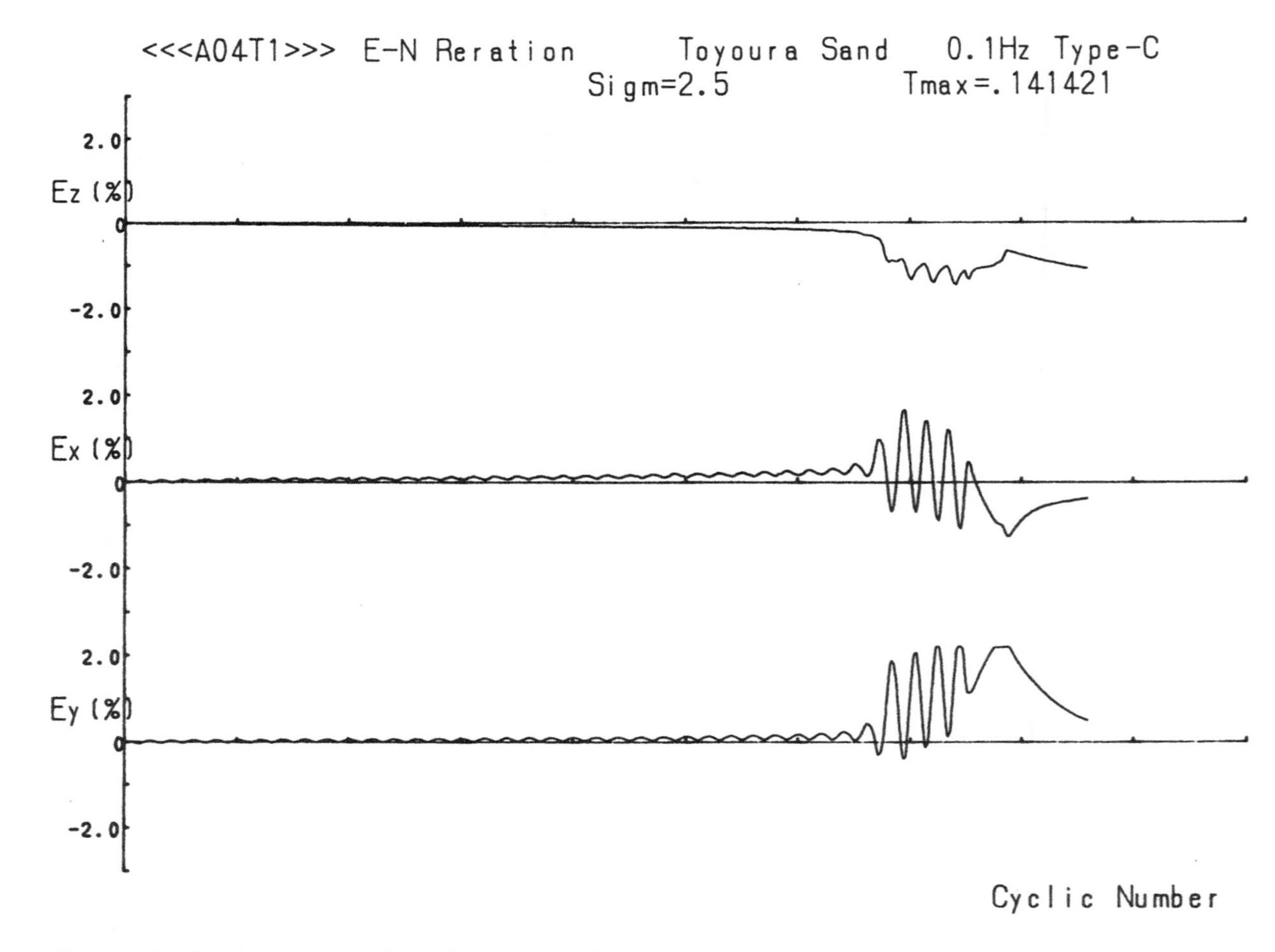

Fig.14 Relation between strain and cyclic number (Type-C)

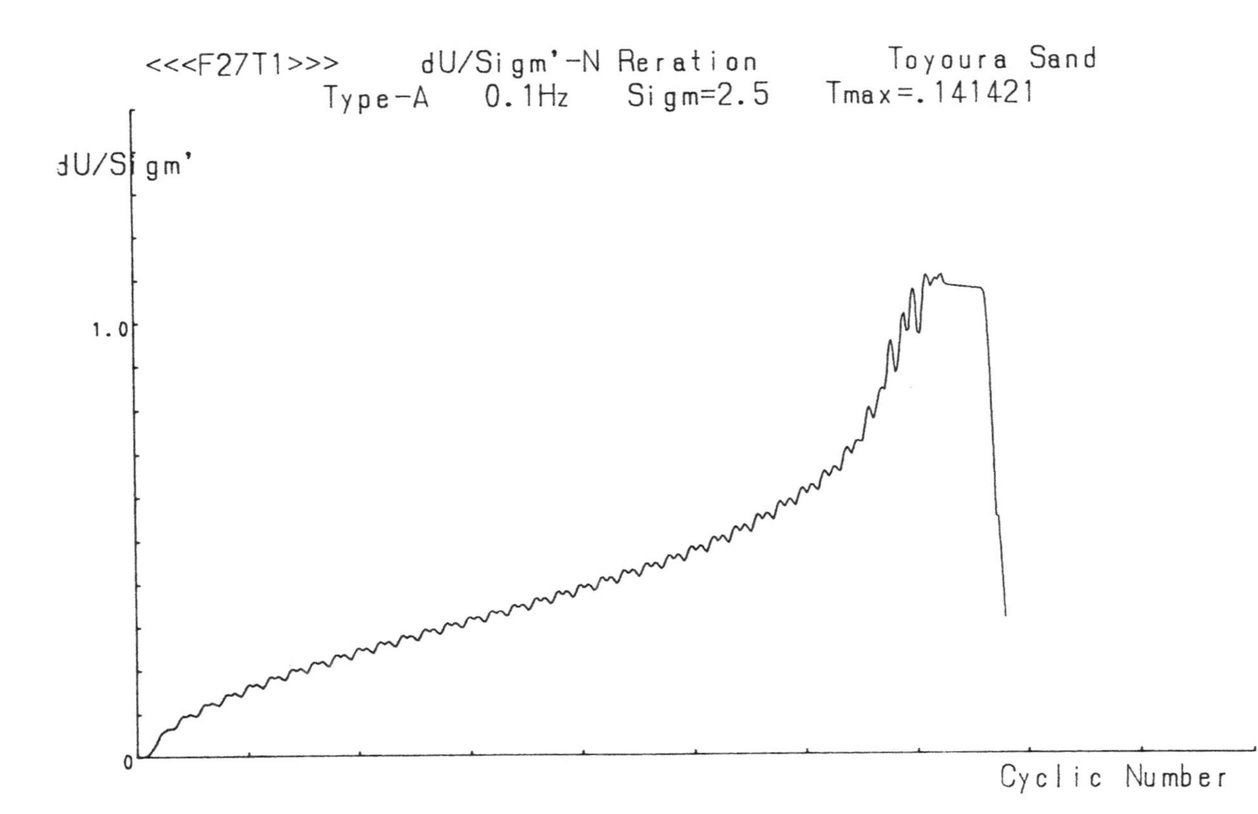

Fig.15 Relation between stress ratio, $\Delta u/\sigma_o'$, and cyclic number (Type-A)

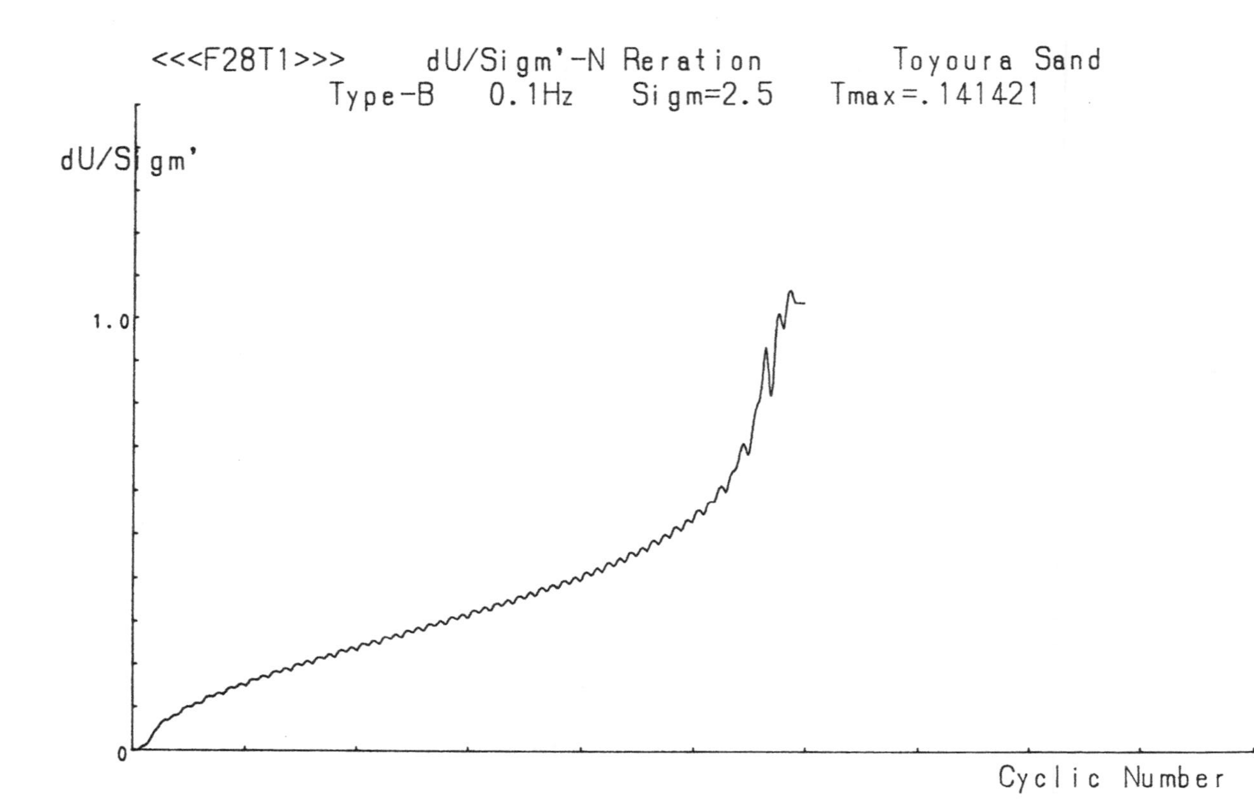

Fig.16 Relation between stress ratio, $\Delta u/\sigma_o'$, and cyclic number (Type-B)

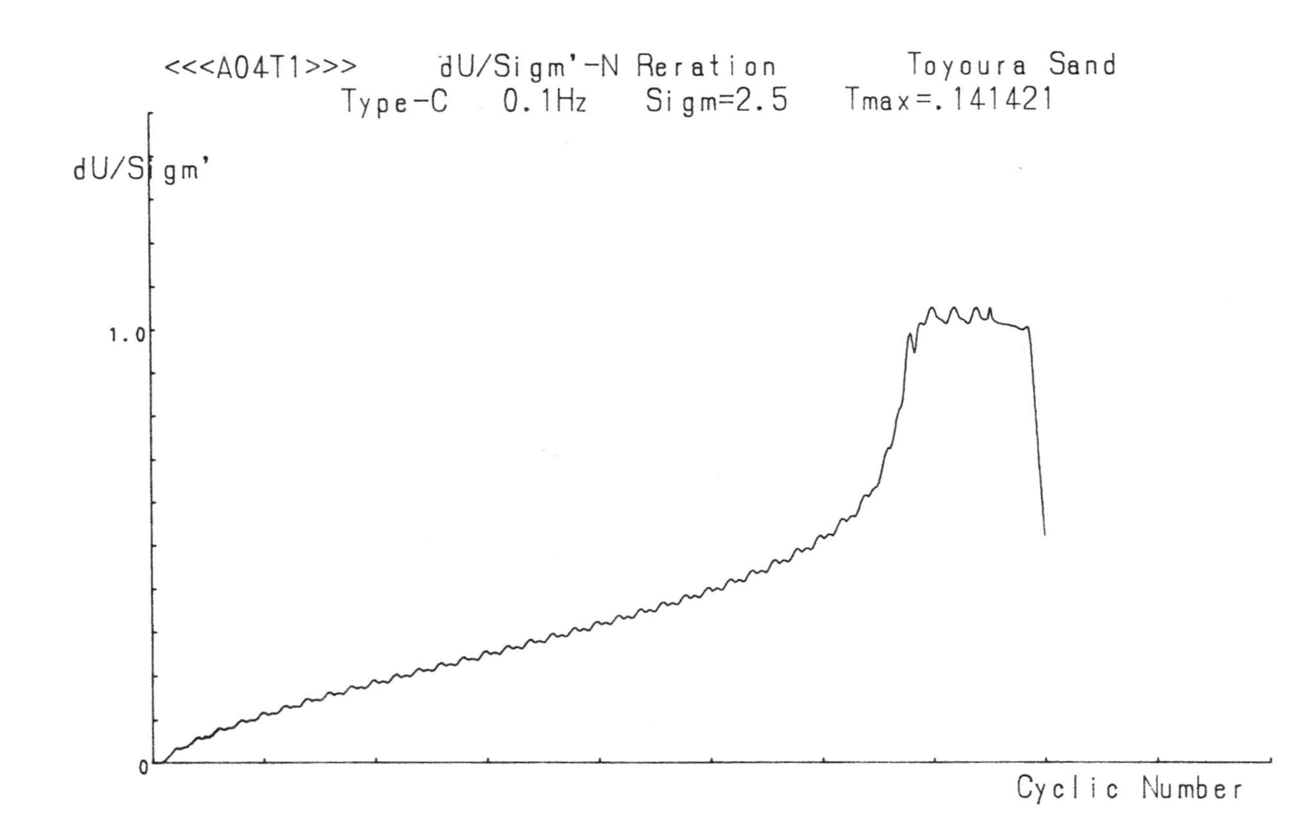

Fig.17 Relation between stress ratio, $\Delta u/\sigma_o'$, and cyclic number (Type-C)

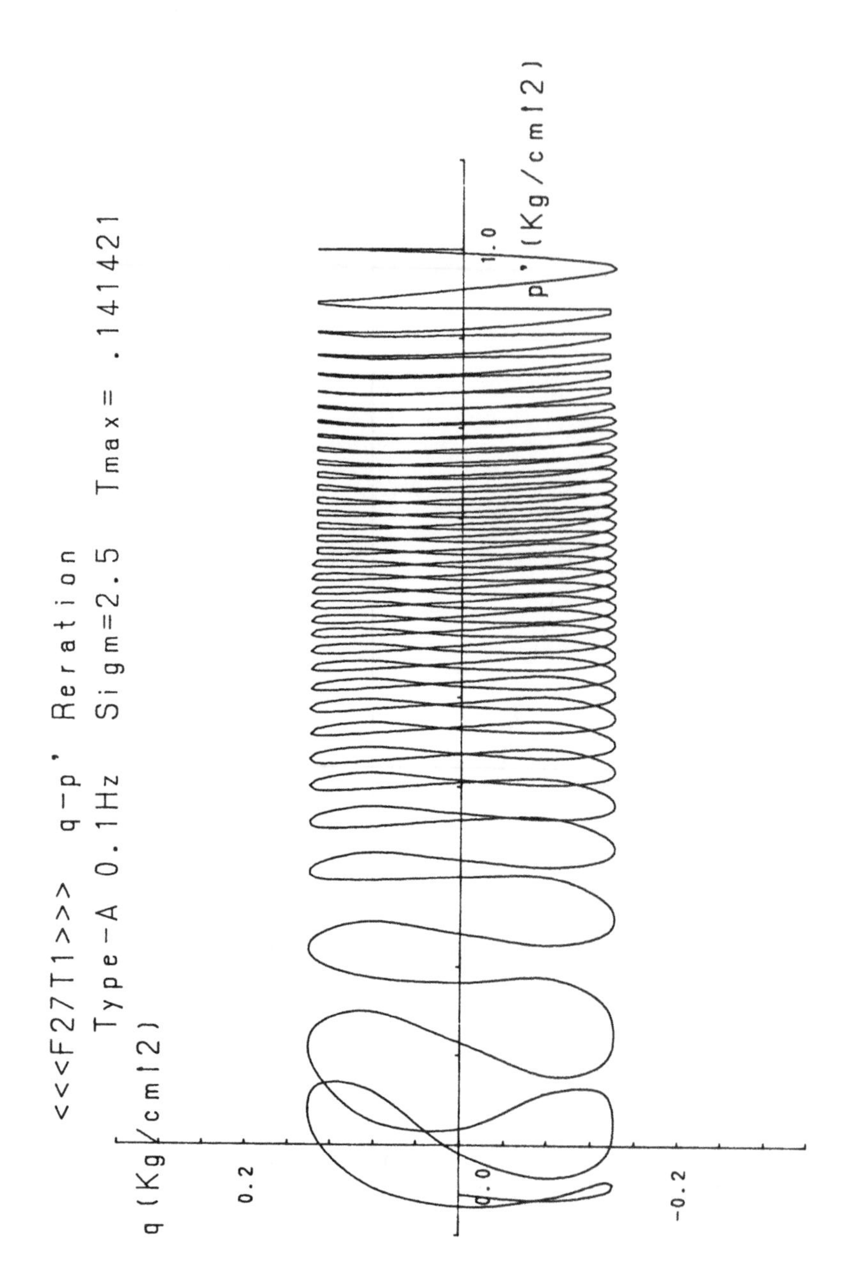

Fig.18 Effective stress path (Type-A)

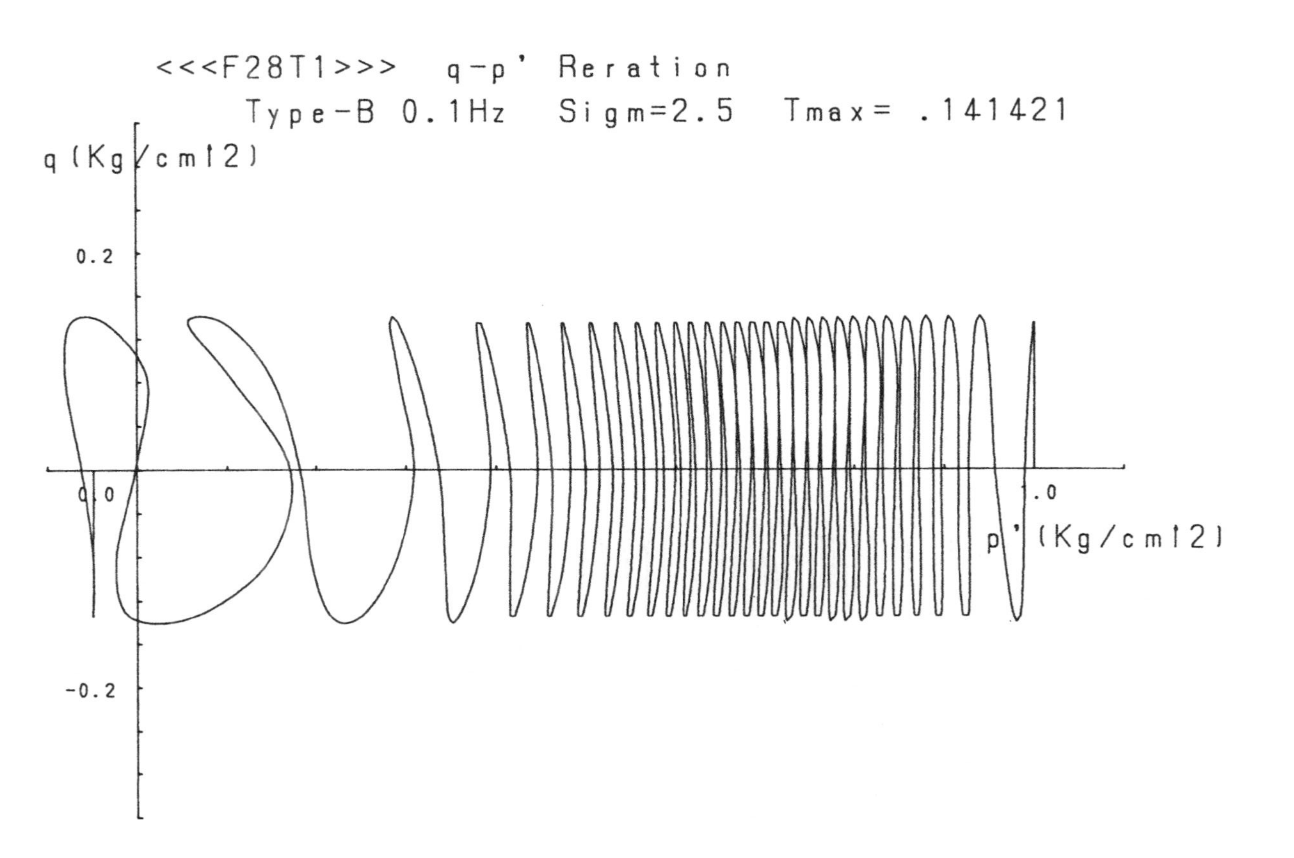

Fig.19 Effective stress path (Type-B)

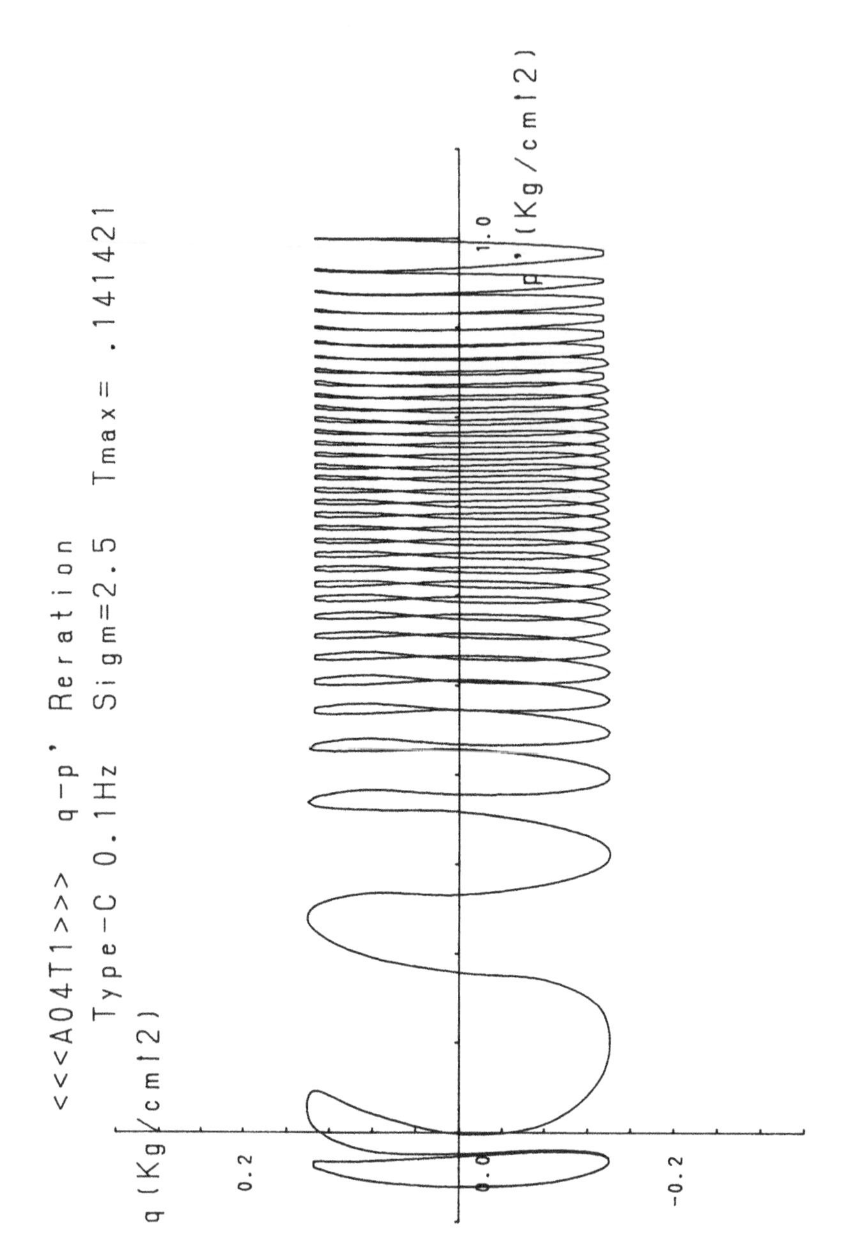

Fig.20 Effective stress path (Type-C)

Computer and Physical Modelling in Geotechnical Engineering, Balasubramaniam et al. (eds)
© 1989 Balkema, Rotterdam. ISBN 90 6191 864 2

An image analysis application for soil fabric* study

S.K.Bhatia, E.Nye & A.Soliman
Syracuse University, USA

ABSTRACT: It has been a known fact that many important properties of soil such as strength, compressibility and permeability are closely related to soil fabric. In this paper, a technique is presented which can be effectively used to measure soil fabric accurately.

1. INTRODUCTION

In the past decade, considerable progress has been made in geotechnical engineering toward understanding the behavior of granular material. Despite this progress, the topic of soil fabic* remains neglected. This is in spite of wide recognition of the fact that several properties of sands, such as strength (Oda, 1972a; Mitchell et. al., 1976), compressibility (Mahmood et al., 1976), and permeability (Bhatia and Nye, 1987), are significantly related to soil fabric.

Soil fabric may be defined as the basic framework or arrangement of individual constituents of an assemblage consisting of different components. A comprehensive description of the fabric of a soil would involve reconstruction of the complex, three-dimensional spatial arrangement of individual grains, which is extremely difficult to achieve. Perhaps the neglect of this topic is due in part to the complex nature of granular material.

In 1976 Oda discussed the two types of soil fabrics, i.e., homogeneous and heterogeneous. He suggested that sands having a heterogeneous fabric are composed of submasses with homogeneous fabric of different kinds and degree of particle configuration. Therefore, in order to describe a heterogeneous fabric, the three-dimensional distribution and orientation of homogeneous submasses must be known. He further noted that the description of the homogeneous fabric of sand should include at least two main features.

(1) orientation of individual particles, which is called orientation fabric; and

(2) mutual relationship of individual particles to other particles, which is called packing.

The orientation fabric of nonspherical particles can be represented by the inclination of the long and short axes of an individual particle with respect to a fixed reference axis (see figure 1a). The orientation of particles in sand cannot be uniquely defined, but the measurement of preferred orientation of long axes and short axes gives some indication of particle orientation.

To define packing, the mutual relation of one particle to another can be represented by the number of contacts and directions of normals (N_1,N_2,N_3 and N_4) which are normal to the tangents at the contacts, (C_1,C_2,C_3 and C_4) respectively (see figure 1b). The normals are defined by their inclination angles to the reference axis. Oda (1976) suggested that the number of contacts may be represented by the average coordination number, which might be closely related to void ratio or porosity. Some relationships between the coordination number and porosity

*Soil Fabric: The physical constitution of a soil material as expressed by the spatial arrangement of the solid particles and associated voids (Brewer, 1964).

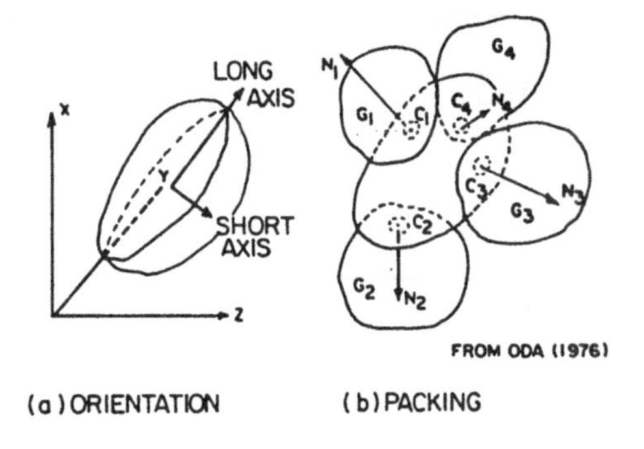

Figure 1 Fabric of Sand from Oda (1976)

has been obtained experimentally and theoretically (Oda, 1976). Therefore, for sands, it may be reasonable to use void ratio or porosity instead of coordination number to define the mutual relation of particles. In addition, it is a generally accepted procedure to express the directions of normals at contact points by the probability density function of the normals in three dimensions. Consequently, at least four fabric elements must be determined to define the fabric of cohesionless soil which consists of nonspherical particles:

1) the preferred direction of a fixed direction in an individual particle with respect to a reference axis;
2) the intensity of preferred orientation;
3) the probability density functions of the normal in three dimensions; and
4) void ratio or porosity to define the mutual relation of particles.

2. EXPERIMENTAL TECHNIQUES TO MEASURE ORIENTATION FABRIC AND PACKING

A wide variety of techniques have been used in the past to measure orientation fabric and packing for granular materials.

In 1963 Field studied the fabric of assemblies of rounded stones having different grain sizes, different gradings, and different void ratios. He poured molten wax into a sampler, and the wax was allowed to flow out from the container. A little was, however, was retained at the contacts among the rounded stones because of surface tension. The rounded stones firmly held in their position were taken out, and contacts were observed by naked eye to measure

(1) the average number of contacts per particle, and
(2) the inclination angle of the tangents at the contacts to the horizontal.

Field (1963) repeated this procedure for assemblies of various densities. A similar technique was also used by Lees (1970). He impregnated the soil sample with a pliable and sectile plastic. After the sample was cured, the soil particle network was removed. Measurements such as size and shape of voids were made. This method also had the advantage of identifying available flow channels. However, the technique used by Field (1963) and Lees (1970) is restricted to coarse grained material.

The micrometric method, the study of thin sections prepared by impregnation of resin in granular pores, is often used for the measurement of soil fabric. This method has been used by several investigators to measure orientation fabric and packing of sands. In 1970 Windisch and Soulie fixed the skeleton of sand with the help of chemical grout and proposed a thin section method to find relative density, homogeneity, isotropy, and grain size distribution. Oda (1976) used the thin section technique to measure both orientation fabric and packing for several different types of sands. Since 1976, several other investigators have used this technique successfully.

The thin sections can be studied in several ways, depending on the type and extent of information required and the nature and the size of the granular material. These techniques include electron microscopy, x-ray radiography (Bhatia and Nye, 1987), and optical microscopy (Shahinpoor and Shahrpass, 1982; Bhatia and Nye, 1987). The findings of some earlier studies reveal that for granular material, optical techniques provide maximum information. Lees (1970) and Windisch and Soulie (1970) used the optical technique to study pore structure of thin sections of granular material. They identified the void spaces of the thin section by color and measured the voids upon projection of the image.

Oda (1976) obtained photographs of thin sections by projecting the translucent sections on photographic paper. He measured the orientation fabric by measuring the orientation of the true longest and shortest axes of irregular-shaped grains with respect to a fixed direction. The measured apparent longest and shortest axes from the thin sections of granular material was presented in terms of axial ratio.

The technique proposed by Oda (1976) to quantify orientation fabric was employed by succeeding investigators to establish relationships between granular material fabric and several engineering properties (Mahmood et al., 1976; Mitchell et al., 1976). However, the relationship proposed between orientation fabric and the engineering properties of granular material was inadequate, because the complete description of a homogeneous fabric consists of two parts: orientation fabric and packing of pore space distribution. Orientation fabric alone is not sufficient to relate to engineering properties.
The thin section technique is not capable of providing a three dimensional picture of packing or all the pores or voids; however, void size distribution for a horizontal or vertical thin section has been obtained by Oda et al. (1972) for natural sand. The technique used by Oda et al. (1972) to measure void size distribution was based on identifying independent areas within the thin section by joining the center of gravities of all adjoining particles by straight lines. For each area, the ratios of the areas occupied by solid particles and voids was calculated. A frequency distribution of void ratio was calculated, (see figure 2a and 2b).

The concepts and techniques currently available to measure orientation fabric and packing using the micrometric (thin section) method are perhaps the best among available techniques. However, one of the major drawbacks of this technique is that excessive time is required to make measurements. In addition, human error may be significant in such tedious measurement. With the recent availability of image analysis systems most of the problems in measurement can be eliminated.

3. USE OF AN IMAGE ANALYSIS SYSTEM IN MEASURING SOIL FABRICS

Image analyzers have been commercially available since the early 1980's. A typical image analyzer consists of a video camera, video monitor, interface card, a personal computer (Apple IIe or IBM PC), and a dot matrix printer (see figure 3). The video camera can be attached to a microscope if higher magnification is required. The video camera scans the sample and acquires an electronic picture of the image with 254 X 192 picture point (pixel) resolution. The pixel density varies from system to system. Each pixel is assigned a grey-scale (brightness) value between 0 and 63. To select features of interest, a potentiometer is used to specify upper and lower brightness thresholds. Only those pixels whose brightness levels fall between the two thresholds will appear in the resulting "binary" image. Once acquired, the complete grey-scale image can be stored on disk for record-keeping accountability, and for reanalyzing images at a later date.

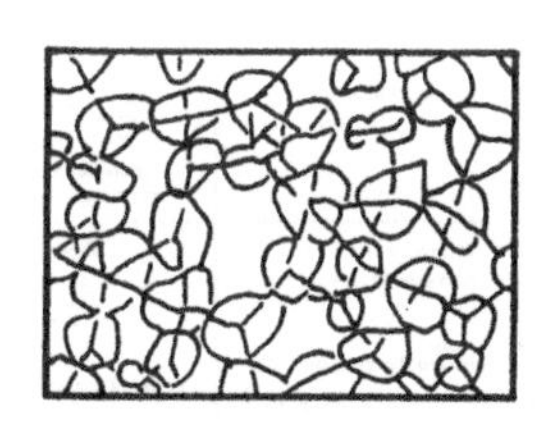

Figure 2a Determination of Pore Size Distribution

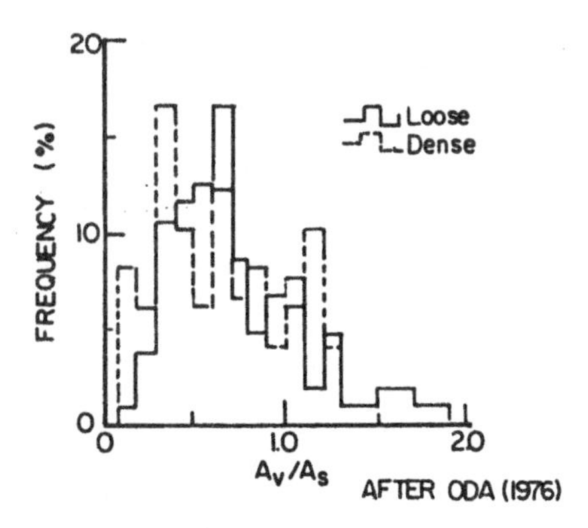

Figure 2b Distribution of Pore Size (sand $k^{1.19}_{0.84}$)

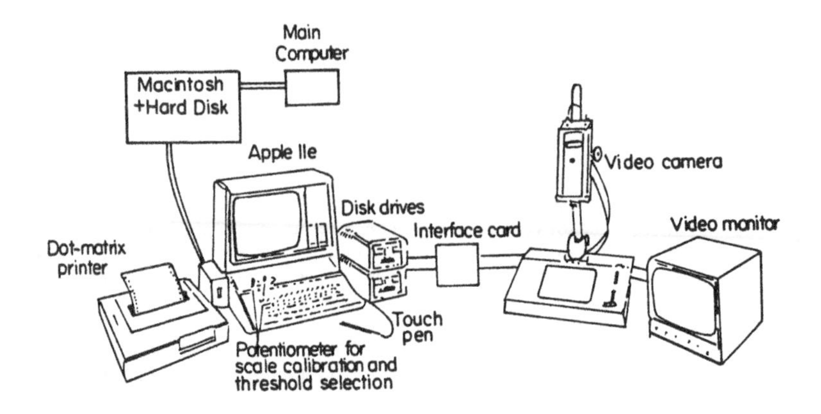

Figure 3. A Typical Image Analysis System

3.1 Features Measured by an Image Analyzer

Two kinds of measurement can be made, basic and derived, both of which are in the same format.

Basic measurements include area, number (full feature count and end feature counts), perimeter, and Feret diameters. These measurements are orientation dependent. In many image analysis systems, size distribution (grains or voids) are measured by chord sizing (Jongerius et al., 1972a, 1972b; Ismail, 1975). Void size distribution can be measured in more detail with a function computer. Voids are sized by area (actual number of picture points), perimeter, Feret diameter (see Figure 4), or projections, or a combination of these.

Many combinations of the derived measurements are possible. The following three listed are of a special significance for size measurements:

(a) Orientation. Vertical and horizontal Feret diameters of voids or particles are a good measure of their height and width and can be used to express their orientation. In addition, measurement of the longest and shortest axes of particles with respect to a fixed reference axis can also be used to express orientation.

(b) Irregularity. - Two separate aspects of the irregularity of voids can be considered. One is their digitate nature, as shown in figure 5, models A-D by Murphy et al. (1977). The other aspect is their serrate nature increasing from E to H (See Figure 5). Image analysis can be used to characterize both of these, either for individual voids or for the total voids in a field. This has been proved by Murphy et al. (1977) for cohesive soils.

(c) Form Separation (Pattern Recognition). Several parameters, each requiring a separate function computer, can be measured for each feature simultaneously. Features can be geometrically classified by combining two or more of their associated features.

In addition to the above mentioned measurements, a typical image analysis system has the capacity of calculating

Display	Measurement	Application
	Area (A)	Porosity: $\frac{A}{\text{Frame size}} \cdot 100$
	Number (N) (i) Full feature count (N_{FC}) (ii) End feature count (N_{EC})	Crenulation or serration $\frac{N_{EC} - N_{FC}}{N_{FC}} \cdot 100$
	Perimeter (Pe)	Orientation independent shape factors, e.g.: $\frac{A}{Pe^2}$ (for a circle $= \frac{1}{4\pi} = \cdot 0796$)
	Horizontal Feret's diameter (F_H) $= x$ Vertical projection (P_V) $= x + y + z$	Orientation $+\left(\frac{F_V}{F_H} - 1\right)$ or $-\left(\frac{F_H}{F_V} - 1\right)$ where numerator is the larger value
	Vertical Feret's diameter (F_V) $= x'$ Horizontal projection (P_H) $= x' + y' + z'$	Digitation $\frac{P_H + P_V}{F_H + F_V}$ (For regular convex objects =1)

From Murphy et. al., (1977)

Figure 4 Basic Measurement Made by the Quantimet 720 and some Derived Measurements Applicable to Voids.

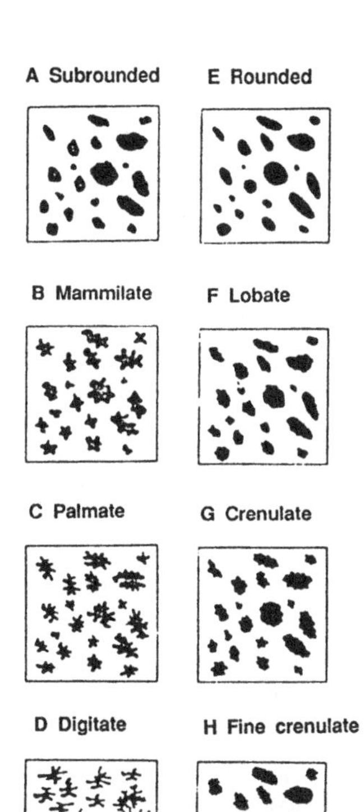

From Murphy et al., (1977)

Figure 5 Irregularity Models A-D shown increase in digitation, Models E-H increase in serration.

several other parameters, such as volume, surface area of the fitted ellipsoid or revolution, equivalent spherical diameter, mean intercept length, and number of features per unit volume.

At Syracuse University, research is in progress in the area of soil fabric measurements. The purpose of this research is to find better techniques of measuring soil fabric (orientation fabric and packing) using the thin section technique. In order to analyze the thin sections, an image analysis laboratory is being set up. In this paper, the preliminary results of this research are presented.

4. TYPE OF SAND USED

For this research, Ottawa C-190 has been selected. This is a clean, uniform sand with D50=0.7 mm, D10=0.6 mm and D60=0.75 mm. The specific gravity of this sand was measured as 2.65.

5 THIN SECTION PREPARATION TECHNIQUES

A permeameter was designed to better serve the goal of impregnating the sand sample with epoxy. The most important feature of the permeameter is the replacement of porous stones with fine screen to avoid impregnation of porous stones with the sample. There are a variety of epoxy types available in the market. The type used in this study was Castolite AC, produced by the Castolite Company, Woodstock, Illinois. This type is mentioned in geological literature on making rock thin sections and has been used successfully in research conducted at the Geology Department at Syracuse University. The epoxy comes in liquid form; how fast it hardens depends on the amount of catalyst added and heat applied.

To produce thin sections, epoxy is diluted to a viscosity capable of penetrating the pores of the rock sample. The first experimental procedure was based on that methodology. The idea was to find the permeability of the sample using an epoxy-acetone mixture at the viscosity of water. The sample could then be cured to a hard sample afterward.

Castolite has a viscosity of 600 centipoises, which would have to be highly diluted to achieve the viscosity of water of 1 centipoise. A 50% epoxy/50% acetone mixture, still having a viscosity higher than water, was used. To obtain a quality impregnated sample, the acetone had to evaporate from the sample, leaving the epoxy to harden. This posed a problem, because only some of the voids would contain epoxy after the acetone evaporated.

In attempting the technique, it was found that there was not enough surface area open to the atmosphere for the acetone to evaporate. Only the upper and lower surfaces of the sample were available for evaporation. The upper surface of the sample within the equipment was totally open to the atmosphere. Evaporation from the bottom of the sample was through the lower screen and perforated disc out the discharge hole in the bottom base. Samples containing the mixture and sitting at room temperature for a week were still moist. Since time was an important consideration, the samples were placed in the oven to quicken evaporation of the acetone and the hardening of the epoxy. Some times, exposing sample to higher temperature resulted in brittle samples.

Samples which were placed immediately in the oven had areas of hardened epoxy and other areas which remained wet. The hardened areas prevented the acetone from evaporating from the other areas. Because of the poor results in obtaining a fully impregnated sample hardened with epoxy. This procedure was rejected.

A method was described by Windisch and Soulie (1970), in which samples were first stabilized by a chemical grout named AM-9, was also used. According to Windisch and Soulie, this grout stablized the grains of the sample at their contact points. This would allow removal of the sample from confinement. Blocks of the sample containing areas of interest could then be cut. These blocks were then impregnated with epoxy. Some success was achieved using the procedure. After the grout gelled, the sample was removed from the glass tube and cut into slices. The grout within each slice was allowed to dry for 1-3 days, and the slice was then immersed in epoxy. Approximately 75% of the slices became impregnated.

The use of pure resin with black dye was the most promising method, because C-190 sand has voids large enough for pure resin to impregnate. The only precaution was that the sand be dry before it was impregnated. Wet samples impregnated with epoxy became brittle when hardened.

When samples were made with this procedure, the problem of a small amount of surface area open to the atmosphere was again encountered. If water in the sample was allowed to dry at room temperature, it would take from days to a couple of weeks to dry. The drying time was reduced by placing the equipment in an oven for about 8 hours at 110°C. The sample could then be impregnated with the epoxy resulting in fully impregnated sample. This procedure produced good solid samples and was chosen as the technique to be used.

The first step in this sectioning was to make an orientation mark along the side of the sample. This made it possible to put each thin section on a glass slide such that its position about an axis through the length of the sample was known. The sample was then marked, identifying where thin sections would be taken.

The sample was cut along the marks using a diamond tipped circular saw. Cylindrical blocks were produced. The height of each cylindrical block was measured and recorded. Each block was then labeled, identifying its position along the sample.

One side of each block was ground down to a smooth, flat surface by applying it to a rotating plate covered with grit and water. The blocks were then glued to a "frosted slide". (A frosted slide is one having a side ground down by rubbing it against a glass plate covered with grit and water.) The ground surface provided better adhesion between the sample and the slide. The orientation marks were all positioned in the same direction, and a label was placed identifying each block. The block were left to dry overnight.

After the glue dried, the samples could be cut closer to the slide. It was then ground down to the desired thickness.

From a 5-inch sample, about 15 horizontal thin sections could be taken; 8 from the 3-inch samples. Approximately 6-8 hours were needed to produce 15 thin sections over a 2-day period. All steps in the thin sectioning process were time dependent on the previous step. If the cut made to produce the individual blocks was irregular, more time was needed to grind down the surfaces. If the surface was not ground properly, more time was needed to glue the block to the slide without bubbles.

All thin sections were ground down to approximately -.100 mm thickness, although the uniformity in thickness across the section did vary. Other small problems were the inclusion of bubbles when the blocks were glued to the slide and the removal of some particle grains when the sample blocks were ground down, otherwise know as "particle pullout".

6 ANALYSIS OF THIN SECTIONS USING AN IMAGE ANALYZER

A model 2000 Image Analyzer was used to analyze the thin sections. This system consists of a high-resolution image and data monitors, an image processor for data extraction, and a computer for data reduction and evaluation.

The following information was derived from all thin sections analyzed: area of the pore space, area of the particles, pore size, and particle size distributions. For the distribution, each item was described by its area, perimeter, length, width, height, and breadth. The image analyzer used operated on the various degreees of greyness an object transmits to the camera. When a sample is placed underneath the camera, an image of the object is produced on a monitor. This image is made up of a matrix of small squares called pixels. Depending on the material, the pixels representing it in the image will hae a certain shade of grey. Each shade of gray is associated with a numerical grey level, with the darker shades having lower values.

By controlling the grey level, called image enhancement, the objects to be measured are determined. When a certain grey level is chosen, any pixels which have lower values are enhance, becoming bright on the monitor. It is these pixels which are used to separate the data. Variables measured are first counted in numbers of pixels and then converted to true scale by the computer.

Figure 6 .Thin Section Placed Under the Video Camera.

Figure 7' A Circular Window Chosen for Analysis.

6.1 Procedure

A. - The thin section was placed underneath the camera (see figure 6). At this time, the degree of magnification was chosen, and the sample image was brought into focus on the monitor. A scale calibration was made through the use of the cursor, or manually controlled cross hair. Two points of known distance were chosen with the cursor and typed into the computer.

B. - The type "window" used was chosen and the grey level set. The "window" was the area of the image on the monitor that would be analyzed by the equipment. A choice of the window's size and shape was available. The largest size window (see figure 7) was a 446 x 512 pixel rectangle. Smaller windows could be chosen in the shape of a rectangle or of a square of circle.

C. Image editing was performed. This step alters the image such that the data produced are more accurate, and it also permits the user to control the image measured.

Only those types of editing used in the analysis will be mentioned here:

1) Erosion/dilation: The erosion process took each enhanced image and removed a pixel from its boundary; the dilation process was the reverse of erosion and added a pixel to each enhanced image's boundary. The overall effect was removal of individual pixels unintentionally enhanced.

The following editing types were cursor controlled:

2) Add: A 3 X 3 pixel area of white enhancement was added to the image.

3) Erase: A 3 X 3 pixel area was removed from the image.

4) Cut: The operator could cut a one-pixel line through an enhanced area.

D. - Choice of either feature specific or field specific measurement was made. For feature specific, each individual enhanced area was measured

Figure 8 Different Components of a Typical Image Analysis System.

for its area, length, width, height, and breadth. The centroid position of the area could also be recorded. The field specific option determined the total area of the enhanced image.

E. Data was processed. For objects measured individually, the mean, standard deviation, and high and low value of the measured variable for the image analyzed were calculated. The image represented in pixels could also be printed. (Figure 8 shows the various equipment components).

Table I

Comparison of Void Ratios Found Manually to that derived Through Image Analysis for Section 5-1

	Void Ratio		Magnification used in Image Analyzer
	Manual	Image Analyzer	
1st square section	1.14	1.15	3x
2nd square section	0.88	1.08	3x
Total section	0.90	0.89	1x

6.2 Comparison of Manual Measurements to that of the Image Analyzer

The first thin sections to be analyzed were those from a sample containing 0% fines. Before using the image analyzer, the first thin section was photographed and enlarged such that the photo was of a diameter 13 times the original. Grains of the sample were then traced onto a sheet of paper, making it possible to measure the total grain and void area using a planimeter. Two separate square sections from within the traced sample were also measured for their respective grain and void areas. Comparison with void ratio values obtained through the image analyzer are shown in Table I

It is interesting to see that the results found at 1x magnification are relatively the same. Results obtained at 3x magnification, overall, were expected to be closer due to higher magnification. However, the accuracy was not as good as it was at 1x magnification. It can be stated that results from image analysis seem to compare well with manual measurements.

6.3 Typical Results

For the study undertaken at Syracuse University, around seventy thin sections were prepared from Ottawa C-190 sand. For some of the samples, a certain percentage of fine particles was added. All of these thin sections were analyzed by an image analyzer. A typical set of results for a section are given and discussed in this paper.

A photograph of thin section 5-1 is shown in figure 9. With the image analyzer, a photograph of a thin section or the thin section itself can be analyzed. As discussed in the previous section, an image analyzer operates on the various degrees of greyness an object transmits to the camera.

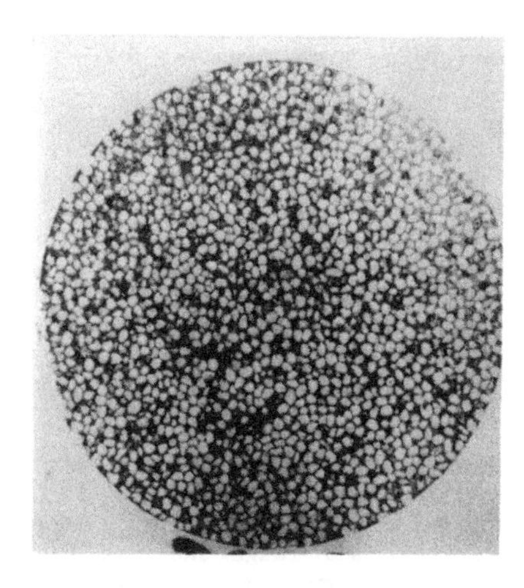

Figure 9 Photograph of Thin Section 5-1. (Particles are shown in white and voids in black or grey)

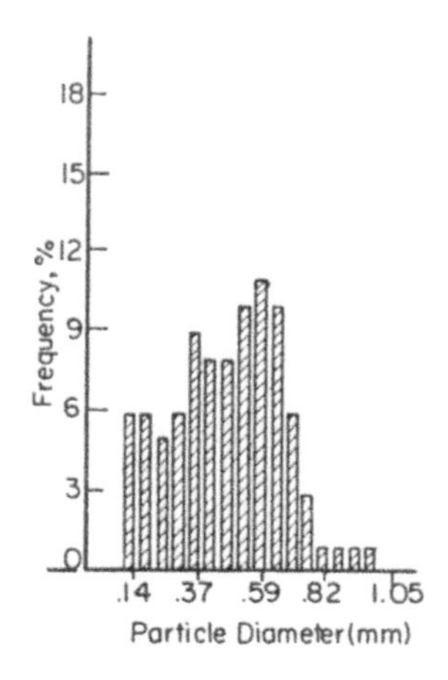

-Figure 11a Particle Diameter Histogram of Area 1, of Slide 5-1.

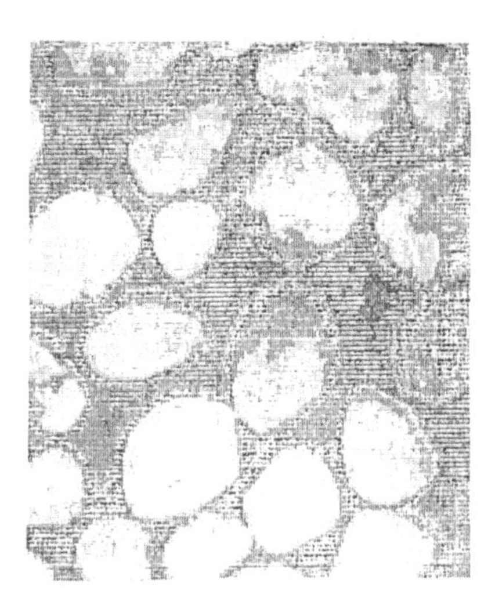

Figure 10 Typical Grey Image of a Small Part of Section 5-1.

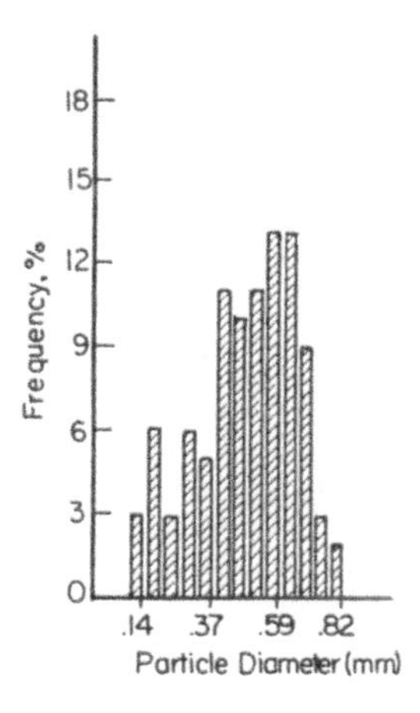

Figure 11b Particle Diameter Histogram using Data for all the Sections of Sample 5-1.

In figure 10, a grey image of a small area of section 5-1 is shown. For the thin sections analyzed, the grey level of grains is very different from the grey level of void space, these two features can be measured separately.

For the section 5-1, measurements were made of each grain and its areas, diameter, and parimeter. In figure 11a a histogram of frequency versus particle diameter is plotted for areas of section 5-1. In creating the histogram, the diameters were derived from circles having areas equivalent in area to that of a particle areas. The full range of particle diameters was then derived into 20 equal increments, and counts of diameters for each increment were made. Each histogram bar is plotted at the midpoint of each increment and represents the percentage of all particle diameters in that increment. Also, in figure 11b is presented a histogram of particle diameter for a sample. This histogram includes data from 12 different horizonal thin sections and five areas from each section. A comparison of figure 11a and 11b reveals that, for this sample, one area from one horizontal thin section can give a good representation of the particle areas in two dimensions for the whole sample.

For the measurement of fraction area or void ratio, section 5-1 was divided into four parts, and fraction areas and void ratio were calculated for all four areas. The variation of fraction area for different parts of section 5-1 is not significant. These kinds of measurements are extremely simple to make with an image analyzer.

Oda (1976) suggested a technique of measuring void ratio distribution for a given section. This measurement is important to project the variation of void ratio through a sample and also to define properties of voids such as their spatial arrangement, dimension, and shape.

The technique proposed by ODA (1976) was applied to five different areas of sample 5-1. A plot of frequency versus Av/As (Av = area of voids, As = area of solids) is given in figure 12. This plot reveals that the distribution area (of section 5-1) does not vary much from the average for all five areas. Nor did the distribution of the other four areas differ from the average.

The work of fabric measurement is in progress at Syracuse University. Based on our past two years' experience, we can say that for the measurement of soil fabric, an image analyzer will be essential since it facilitates measurement tremendously.

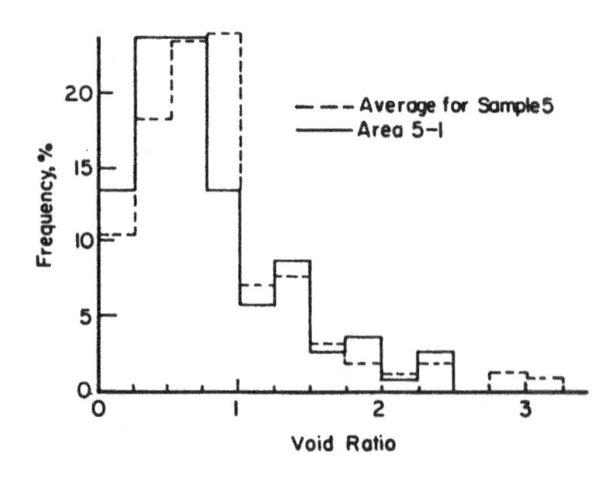

Figure 12 Frequency us Void Ratio using Oda's Method for Area 1 of slide 5-1.

7 CONCLUSIONS

Image analysis can be applied to the measurement and characterization of grains and voids in granular material, thus measuring the soil fabric. There are several advantages in the use of an image analyzer: (1) measured components (grains or voids) can be seen so that size, shape, and orientation can be assessed; (2) it is several times faster than manual measurements and provides additional measurement from which irregularity, shape, and orientation can be analyzed; and (3) it is more accurate than manual measurements. It would be incorrect to say that all image analysis systems are suitable and can be adopted for soil fabric work. However, additional research in this area will most of the difficulties.

ACKNOWLEDGMENTS

We thank Professor C. Tien, Professor of Cheical and Material Science at Syracuse University, for financial support. In additiion, we gratefully acknowledge the support provided by the Senate Research Committee of Syracuse University.

BIBLIOGRAPHY

Bhatia, S.K. & E. Nye 1987. Soil fabric and permeability. Paper under preparation

Brewer, R. 1964. Fabric and mineral analysis of soils. J. Wiley and Sons, Inc.

Field, W.G. 1963. Towards the statistical definition of a granular mass. Proc. 4th A. and N.Z. Conf. on Soil Mech., p. 143-148.

Ismail, S.N.A. 1985. Micromorphometric soil-porosity characterisation by means of electrical image analysis (Quantimet 720). Soil Survey Papers No. 9. Wageningen: Soil Survey Institute.

Jongerius, A., D. Schoonderbeek, & A. Jager 1927a. The application of the quantimet 720 in soil micromorphometry. The microscope 20, 243-54.

Jongerius, A., D. Schoonderbeek, D. Jager, A. Kowalinski 1972b. Electro-optical soil porosity investigation by means of quantimet B equipment. Geoderma 7, 177-98.

Lees, G., 1970. Studies of inter-particle void characteristics. W.J. Engineering Geology, 2, p. 287-299.

Mahmood, A., J.K. Mitchell & Ulf Limdbolm 1976. Effect of specimen preparation method on grain arrangement and compressibility in sand. Soil specimen preparation for laboratory testing, ASTM STP 599, p. 169-192.

Mitchell, J.D., J.M. Chatoian, & G.C. Carpenter 1976. The influence of sand fabric on liquefaction behavior. Report No. TE 76-1, University of California, Berkeley.

Murphy, C.P., P. Bullock, & R.H. Turner, 1977. The measurement and characeterization of voids in soil thin sections by image analysis, part 1, Principles and Techniques. J. Soil Sciences, 28, p. 498-504.

Oda, M., H. Kobayashi, Y. Yamazaki, & T. Onedera 1972. A new technique for determination of void ratio and its variation within deformed granular material. Report of Dept. of Found. Engrg., Faculty of Sci. and Engrg., Saitama University, vol 3, p. 39-50.

Oda, M. 1972a. Initial fabrics and their relations to mechanical properties of granular material. Soils and foundations, vol. 12., no. 1, p. 1-18.

Oda, M. 1972b. Deformation mechanism of sand in triaxial compression tests. Soils and foundations, vol. 12, no. 4, p 45-63.

Oda, M. 1976. Fabrics and their effects on the deformation behavior of sand. Report of Department of Foundation Engineering, Saitama University, Japan.

Shahinpoor, M. and A. Shahrpass 1982. Frequency distribution of voids in monolayer of randomly packed equal spheres. Bulk solids handling, vol. 2, 4, p. 825-838.

Windisch, S.J. & M. Soulie 1970. Technique for study of granular materials. Proc. of ASCE. vol 96, SM. 4, p. 1113-1126.

Computer and Physical Modelling in Geotechnical Engineering, Balasubramaniam et al. (eds)
© 1989 Balkema, Rotterdam. ISBN 90 6191 864 2

TestMast, an IBM-PC program to control and record computerised soil testings

Z.Indrawan
Christian University of Indonesia, Jakarta, Indonesia

ABSTRACT: This paper summarises a software called **TestMast**, an interactive program written in BASIC for the IBM-PC, to record and control four triaxial tests simultaneously, but independently from each other. However, it can also be easily modified to include other soil testings, such as consolidation test etc. The total number of tests that may be run at one time depends on the hardware's capacity, not on the software. Its present capability includes controlling eleven different types of triaxial test. Since it is written in modular structure, future addition to include other type of test may be added easily.

Introduction

The importance of a good software in a computerised soil testing system can not be overlooked. While a software is usually written to suit a particular system, there are some ideas employed in writing a particular software which could be of use to other researhers seeking to do similar thing.

This paper summarises a software called **TestMast**, written to run the computerised triaxial testing system explained in the accompanying paper (Pender and Indrawan, 1986). More explanations may be found elsewhere (Indrawan, 1986a & 1986b).

TestMast is an interactive software, written in BASIC for the IBM-PC to record and control several laboratory soil testings simultaneously, but independently from each other. The present version was written for the computerised triaxial system mentioned above. However, it can also be easily modified to include other soil testings, such as consolidation test etc.

At present it is capable of controlling and recording up to four triaxial tests simultaneously but it can be easily modified to handle more tests. The limitation is not in the software itself but more in the hardwares, such as the capacity of the A/D converter (maximum number of channels) and the maximum storage provided by the magnetic disks.

A total of eleven different types of triaxial test can be handled at the moment. Since it was written in modular structure, future addition may be added easily to the software as a BASIC subroutine.

Recently, a Turbo Pascal program has been incorporated, in a batch file, to record a dynamic stress-controlled triaxial tests.

TestMast's capability

TestMast was designed to distinguish various stages of each test. For example it can handle, simultaneously, test No. 1 which is at the saturation stage, test No. 2 which is at the consolidation stage and test No. 3 which is at the loading stage. The test operator may determine the reading/action interval of a particular test, independently from the others.

If, while doing the above duty another test, test No. 4, is to be started, the software can be terminated and re-started with the option to start a new test. Four other options are also available as will be explained later. After starting test No. 4, TestMast would add this test on to the list of tests being handled. Particulars of the new test are input interactively from the keyboard-screen. These details are kept in a disk file for future use.

If a test is controlled by TestMast, for example anisotropic consolidation test, TestMast does the necessary calculations to determine the number of pulses needed to bring the specimen to its next point and sends the pulses to the relevant stepper motor control circuit through the parallel port of the TecMar LabMaster (A/D converter).

TestMast is started in the usual way a BASIC program is run, i.e. load the BASIC interpreter, load TestMast and run it. The first response of the software would be to ask the test operator to press any key, which would lead to a display of the available options. The term 'start' refers to the condition when the first experiment is started; no other experiment is being conducted. Any other condition is called 'restart'. Therefore TestMast only needs to be started once, and it is active as long as at least one experiment is running.

When TestMast is running, it can be interrupted by pressing **Ctrl-Break.** The need to interrupt Testmast arises when it is necessary to ask the software to do something other than its normal routine duties, that is to take an action on a particular test at the appropriate time, for example at the end of the consolidation stage.

TestMast can be restarted using any of the available options except option #1 as obvious from the following list of the available options to follow:

Option **#1**: To start a new test, when no other test is active.
Option **#2**: To start another test.
Option **#3**: To resume duty.
Option **#4**: To do an 'Act-Now' and back to routine.
Option **#5**: To stop a test.

Options #1 and #2 are provided to start a new test. Option #1 is provided to save TestMast from searching for a non-existent master.log file as will be explained later. When one of these option is selected, TestMast would ask for the necessary details of the particular experiments. All these details are written to the disks (both A: and B:) to be accessed in the future, whenever that test is actioned.

The inputs are requested in an **interactive** manner, with clear messages as to what is requested.

Option #3 is used to restart TestMast, after any interrupt either by test operator (to modify a certain test) or by an accident such as power failure if the IBM-PC failed to re-boot, and bring the software to the state where it stops.

Option #4 is provided so that TestMast can be interrupted and restarted at any time to give the test operator the ultimate control on any of the experiment, such as changing the type of test from undrained to drained or changing the direction of a general stress path test etc.

It may also be used to restart TestMast if an immediate action on a particular test is needed, e.g. when a certain behaviour of the specimen is worth recording. If TestMast is restarted using this option, it will ask for the comment to be written at the end of that particular record. This is a useful facility for tracing back what has happened during an experiment.

Option #4 is also used to provide interrupts at the beginning and end of the standard stage of a triaxial test, such as saturation, consolidation and loading (shearing) stage. Therefore whenever option #4 is selected, TestMast would ask at what stage the experiment to be actioned is. The necessary procedures and calculations to be done at the end-of-saturation/start-of-consolidation and end-of-consolidation/ start-of-loading have been included in the software; the test operator would be guided interactively on what to do during these stages.

Option #5 is provided to stop an experiment. TestMast would erase the details of that particular experiment from the master.log file. The test.log file is kept on the disk. If that experiment is the only active test, the software would erase the master.log file.

How TestMast does the job

TestMast employs two types of **log file** to organise the tests.

The first log file is called Master.log; it contains the general information of each experiment, such as the next reading/action time, the filenames related to each experiments and the apparatus number. TestMast access this file after taking action to any of the experiment to update the reading times. TestMast also adds relevant information to the master.log file when a new experiment is started. Since master.log contains the reading times for each individual experiments as well as the next 'action' time, i.e. the time when TestMast would take an action on a particular experiment, changing the details of the master.log would dictate the time of action.

The second log file is called test.log where **test** represents the filename of any particular experiment. This log file contains the details of a particular test such as the type of test, initial cell pressure, back pressure, initial dimension of the specimen, initial values of measurement devices, etc. It also contains the reading interval, which determines how often TestMast is supposed to take action on that particular test. The contents of the test.log file can be changed at any time during the experiment to change various details of the experiment, such as changing the type of test from undrained to drained or instructing TestMast to take less frequent readings during the later part of an experiment.

Keeping the data

TestMast is the master program which organises various tests under its control. It reads raw data in [V] from LabMaster, converts them into engineering units, i.e. [kPa], [mm] etc. and write both the raw data and the engineering data to disk files; thus TestMast produces two types of output files for each test, the 'raw' file and the 'data' file. TestMast works in conjunction with another program, **Triaxial**, to analyse the data. Triaxial prints the result, e.g. stresses and strains to another output file, the 'result' file. Therefore, a total of three types of output file are produced for each test. These activities are performed every time TestMast takes the measurements of a test. The new raw data, data and results are appended to the existing files of the particular test being actioned.

Making use of the twin disk drives system, TestMast writes the data on both disks, A: and B:, to reduce the possibility of losing significant amount of data in the case of hardware failure. Thus a total of six files are accessed every time TestMast acts, thus for each test there are six disk files, three in each drive: not much chance to loose any precious test result.

Post-test Analysis

The software Triaxial may also be used to reanalyse any of the three types of data after the test is over, for example when the water content data become available. The final result may be printed or plotted or both.

Acknowledgement

The work reported in this paper was part of the author's PhD Research, under the supervision of Prof. M. J. Pender at The University of Auckland.

Scholarships were received from Okumenisches StudienWerk (OSW), Bochum and the Christian University of Indonesia, Jakarta.These are gratefully acknowledged.

References

Indrawan, Z. 1986a. Stress-strain and strength characteristics of an Auckland Soil. Phd Thesis. University of Auckland.

Indrawan, Z. 1986b. University of Auckland computerised triaxial testing system. User's Manual. University of Auckland, School of Engineering Research Report No.394.

Pender, M.J. and Z. Indrawan 1986. A computerised triaxial testing system. Symposium on Computer Aided Design and Monitoring in Geotechnical Engineering. Asian Institute of Technology, Bangkok.

Computer and Physical Modelling in Geotechnical Engineering, Balasubramaniam et al. (eds)
© 1989 Balkema, Rotterdam. ISBN 90 6191 864 2

A computerised triaxial testing system

M.J.Pender
University of Auckland, New Zealand

Z.Indrawan
Christian University of Indonesia, Jakarta, Indonesia

ABSTRACT: A versatile computerised triaxial testing system has been developed to perform various types of both stress-controlled and strain-controlled triaxial tests. The apparatus is described and the range of tests which can be performed is discussed. Some test results are presented to illustrate one of the capability of the system, i.e. to perform anisotropic consolidation. Measurement devices are mainly DC-LVDT based devices with the exception of the pressure transducer. Pressure supply is controlled using stepper-motor based devices. An IBM-PC is used to run the system; the interfacing device is TecMar LabMaster. The software developed for this system, **TestMast**, is described in the accompanying paper (Indrawan, 1986c).

1 INTRODUCTION

The versatility of triaxial apparatus is well recognised. Over the years there have been many efforts to improve both equipment and testing procedures. In recent years research has been particularly directed toward developing automated triaxial equipment (e.g. Mitchell, 1981; Coatsworth & Hobbs, 1984 and Atkinson, Evans & Scott, 1985). This paper summarises the development of a computerised triaxial testing system at the University of Auckland. Detailed explanations are reported elsewhere (Indrawan, 1986a and 1986b).

Underlying many of the reasons for computerising triaxial apparatus is the progress made in the understanding of soil behaviour and the growing awareness of the limitations of conventional triaxial equipment to simulate in situ soil behaviour. An example is the need to simulate Ko-consolidation prior to shearing the specimen. Routine triaxial test would, usually, consolidate a specimen under all-round pressure which produces an isotropic initial stress state.

Although it is well recognised that most naturally deposited soil would have undergone anisotropic consolidation during its history, the use of conventional triaxial equipment to perform anisotropic consolidation would be tedious and expensive exercise. Furthermore, manually performed anisotropic consolidation depends to a large extent on the magnitude of the stress increments and the intervals between increments. The advance of microcomputer technology has made it possible to eliminate some of the limitations of triaxial testing.

The automation process involves replacing various mechanical devices in the conventional triaxial apparatus with electromechanical devices, and employing microcomputer to take over most of the routine jobs from human operators.

Therefore, basically there are two categories of operation which need to be dealt with:

1.Automatic recording of various measurements during the test, e.g. axial load, pore water pressure, volume change and cell pressure.

2.Computer-controlled loading systems, which includes both axial load and lateral pressure, i.e. controlling cell pressure during the test and thus controlling the effective stress path.

2 THE TRIAXIAL SYSTEM

The schematic layout of the computerised triaxial system is illustrated in Figure 1. Fluid pressures are supplied from a compressed air pressure tank which is maintained at around 1000kPa by a

(a)

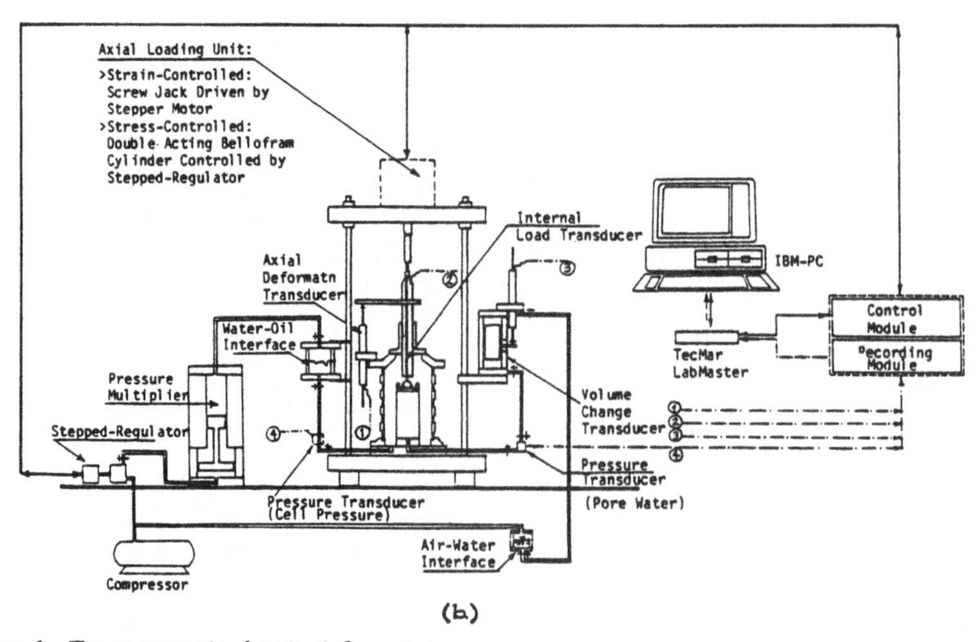

(b)

Figure 1. The computerised triaxial system.
a. View.
b. Schematic diagram.

compressor, with back up for failure provided by a second compressor which is connected to the emergency power supply.

The air pressure, controlled by constant bleed type air pressure regulator, is converted to water pressure in an air-water interface chamber. Small diameter high pressure plastic tube is used to connect the air-water interface pot to the triaxial cell. This arrangement reduces the possibility of air-diffusion through the membrane usually encountered in a long term triaxial test.

To further prevent osmosis, castor oil instead of de-aired water was used as cell fluid in this research. De-aired water was used on the back pressure line only.

Sometimes, cell pressures higher than

1000kPa are required. A pressure multiplier is used to multiply the pressure supply by a factor of about 2.2, to a maximum pressure of 2200kPa. Further explanations about this system will be given later in the paper.

As mentioned earlier, castor oil instead of de-aired water was used as the cell fluid in this research. An oil-water interface chamber was installed on the cell pressure line, using a Bellofram Pressure Diaphragm, to separate the water from the cell pressure supply and the castor oil in the triaxial cell.

The triaxial cell itself consists of three separable parts

1.The alluminum cell base, which houses all the necessary plumbings.

2.The cylindrical wall, consists of a reinfoced high capacity (3.5Mpa) perspex tube secured between two aluminum rings. This part is held down to the cell base by three stainless steel rods.

3.The cell top which guides the load ram. This part is secured to the top ring using Allen-screws.

Seals between these parts are provided by 'O' rings. An 'O' ring is also used to provide seal between the load ram and its guide (the cell top) since friction in this part is irrelevant as will be explained later. The soil specimen is mounted on a 25mm high pedestal. Although only 75mm diameter specimens were used in this research, the cell can be easily adopted for smaller diameter specimens by simply replacing the pedestal with one of suitable size.

Two axial loading systems were developed, namely stress-controlled and strain-controlled systems. Computer control of these systems was achieved by employing stepper motors. The stepper motors constitutes the mechanical to electrical interface for the system. A stepper motor rotates in a series of small angular increments, initiated by digital electronic pulses sent to its control circuit either from a computer or from a front panel switch.

For the strain-controlled system, the stepper motor drives a screw jack to move vertically and hence forces the specimen to deform axially at a rate determined by the rate of the digital pulses.

In the case of the stress-controlled system, a small stepper motor was used to rotate the air pressure regulator. This regulated air pressure was converted to axial force by means of a Bellofram Rolling Diaphragm Cylinder. Using the double acting version of this cylinder, the system is capable of applying compression as well as extension loading. This method of computer control of the air pressure regulator may also be used to control the cell pressure for a general stress path test.

Transducers used in this research are mainly DC-LVDT based devices with the exception of the fluid pressure transducer, which is a strain-gauge based device. A DC-LVDT is a direct-current version of LVDT, which is an electromechanical inductance device for measuring linear displacement. This device is employed as the measuring instruments in axial deformation, axial load and volume change transducers.

Control circuits for the stepper motors, signal conditioners for various transducers and power supplies for each of them are all contained in one single box. The A/D (Analog to Digital) Converter used was TecMar LabMaster which can handle up to 64 I/O single ended channels. An IBM-PC with dual disk drives, 192Kb memory was used to run this system.

A total of four triaxial cells were built, two stress-controlled and two strain-controlled systems. A computer program, **TestMast**, was written in BASIC to handle all four tests simultaneously. Each individual test could be started or stopped at any time, independently of any other. Actually, the total number of tests which can be handled by TestMast is only limited by the total number of I/O channels of the LabMaster. At present, up to 16 tests may be run simultaneously since one test requires 4 channels. Ten different types of test may be performed in this computerised triaxial system. The software TestMast is explained in the accompanying paper (Indrawan, 1986c).

3 MEASUREMENT DEVICES

As mentioned earlier, three out of four transducers used in this research were using DC-LVDT in one way or another. The DC-LVDT used was manufactured by Schevitz Engineering, New Jersey. The pressure transducer is a ready-to-use product. No development work was necessary. Only DC-LVDT based devices developed in this research are described.

3.1 Axial Deformation Transducer

Only minimum work is needed to use the DC-LVDT to measure linear displacement. It was necessary to provide extension rods at both ends of the DC-LVDT core and an internal non-magnetic compression spring to give the necessary force needed to

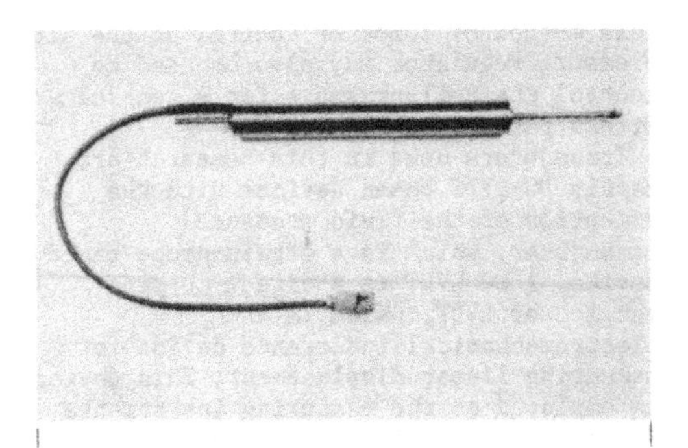

Figure 2. Axial Deformation Transducer.

press the rod hard against the reference surface.

This transducer (Figure 2) performed very satisfactorily over a long period of time. The type of DC-LVDT used has a 25mm linear range but its actual linear range was found to be well in excess of the specified value. The accuracy of this transducer is 0.017mm which gives 0.01% strain accuracy for a 150mm long specimen.

3.2 Load Transducer

The axial load acting on a specimen is measured using a DC-LVDT based load transducer which is located **inside** the loading ram (Figure 3). This internal load transducer eliminates the necessity to provide 'frictionless' bearing for the load ram and also reduces the amount of axial load's correction for the upthrust force, caused by the cell pressure, because of its small diameter.

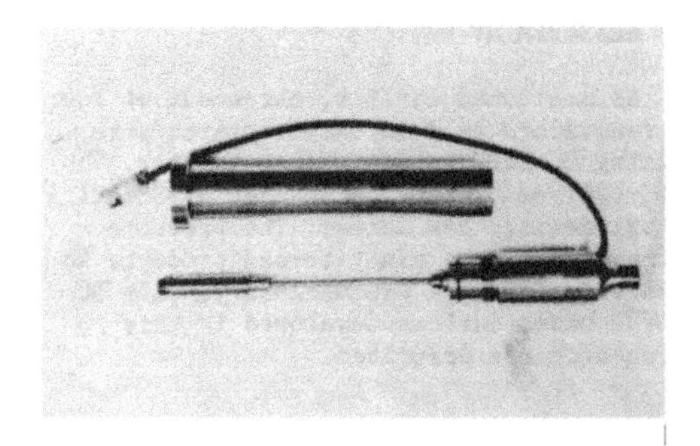

Figure 3. Load Transducer.

This load transducer was originally developed by Bacchus (1969) using a small AC-LVDT for dynamic triaxial tests. However, the AC-LVDT was found to be not suitable for long term experiments, so the DC version was chosen.

Two different capacities were designed. The high capacity transducer has 12.5kN capacity with 20N accuracy while the low capacity can withstand 1.8kN load with 4N accuracy. These give 6kPa and 1kPa accuracy of axial stress measurement for 75mm diameter specimen.

3.3 Volume Change Transducer

The conventional burette system was replaced with an electromechanical volume change device (Figure 4). It was developed following the ideas developed by researchers at Imperial College, London. Volume change of the specimen, which is represented by the movement of water in/out of the specimen, is converted into linear, vertical movement of a PVC piston.

Two Bellofram Rolling Pressure Diaphragms provide virtually frictionless support for this piston and also contain the fluids in the bottom and top chambers. The vertical movement of the piston is measured by an externally mounted displacement transducer mentioned earlier. An extra horizontal arm was provided on the other side of the cylinder for a dial gauge.

The transducer was designed to withstand high pressure without having a significant volume deformation. Calibrated at 700kPa and 1000kPa chamber pressures, it gives the same calibration factor.

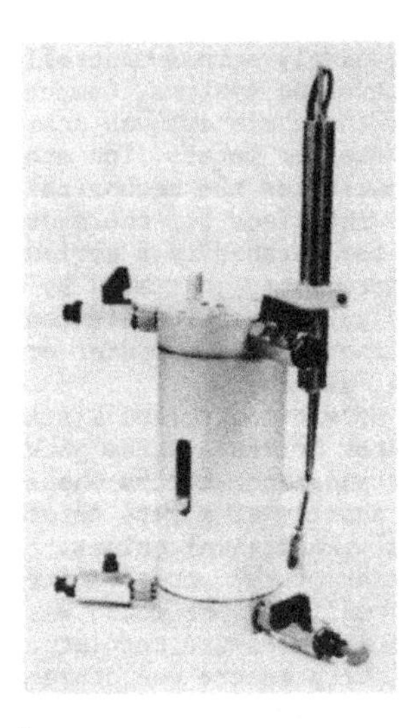

Figure 4. Volume Change Transducer.

(a)

(b)

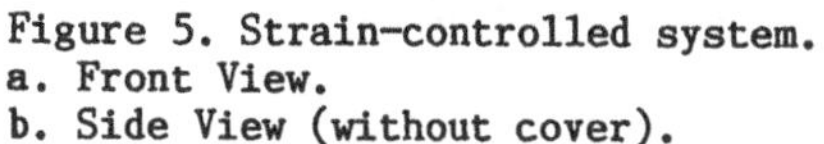

Figure 5. Strain-controlled system.
a. Front View.
b. Side View (without cover).

4 LOADING SYSTEMS

As mentioned earlier, the development of the automatic loading systems was based on the use of stepper motors. A high capacity Super Electric stepper motor, type M092-FC09 (max. torque = 2N.m) is used to turn a screw jack in the strain-controlled system, while only a relatively small capacity Phillips stepper motor, type 9904 112 33004, is needed to rotate the Bellofram air pressure regulator for the stress-controlled system.

4.1 Strain-Controlled System

The front and side views of the strain-controlled loading box are shown in Figure 5. The screw jack, worm gear actuator type, was manufactured by Templeton, Kenly & Co., Illinois, and has a capacity of 10kN. The maximum stroke is about 150mm. An harmonic drive with 110:1 ratio was installed between the stepper motor and the screw jack to improve both the load capacity and the resolution of this strain-controlled system. At present the strain-rate is controlled by a set of front panel switches. It is able to provide testing strain rate from 0.0001 to 0.01mm/min.

4.2 Stepped-Regulator

The compressed air supply is usually controlled using an air pressure regulator. Following the idea of Moore (1980) a **stepped-regulator**, i.e. an air pressure regulator driven by a stepper motor -which can therefore be controlled by computer- was developed. This stepped-regulator may be used to control the input pressures for the Bellofram Rolling Diaphragm Cylinder, to provide axial load for the stress-controlled system; and to automatically control the cell pressure for general stress path tests.

A view of the stepped-regulator is shown in Figure 6a. It consists of a small stepper motor which has a 7.5° step; a Bellofram pressure regulator; an harmonic drive (110:1 ratio), between the stepper motor and the regulator, to increase the resolution of the output pressure; and a small pressure gauge. The pressure gauge was added to give indications of the pressure during adjustments. It is not to be used for the actual pressure measurement.

For the type of air pressure regulator used in this research, the relationship between the rotation (revolution) and the output pressure is not linear. In Figure 6b, the typical relationship between the number of pulses and the output pressure is shown. For some regulators, the relationship at the lower pressure range does not quite fit the parabolic curve, which fits the rest of the data. If used to provide the axial load the working pressures are in this range since the Bellofram Cylinder has a very high cylinder to ram area ratio. For this purpose it is necessary to apply two

Figure 6a. Side view of stepped-regulator.

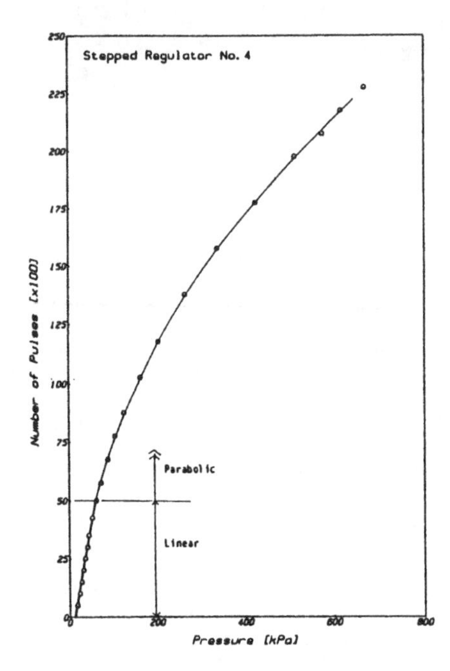

Figure 6b. Typical calibration curve of the stepped-regulator.

different curves, a straight line for low pressures and a parabolic curve for the high pressure range as illustrated in the above figure.

4.3 Stress-Controlled System

The stress-controlled system is shown in Figure 7. It consists of a stepped regulator and a Bellofram Rolling Diaphragm Cylinder. The type used is a double acting cylinder. By applying the right combination of pressures in its two chambers, both compression and extension

Figure 7. The stress-controlled system.

may be produced. The area ratio (cylinder area to ram area) is quite high. It produces an axial load of 14.03N for 1kPa of chamber pressure.

4.4 Controlling Cell Pressure

As mentioned earlier about 1000kPa of compressed air pressure is supplied by the compressor system. To provide higher cell pressures, a pressure multiplier is used on the cell pressure line. A stepped-regulator is installed to control the input pressure to the pressure multiplier. The pressure multiplier and stepped regulator combinations are shown in Figure 8. In this arrangement, the manual air pressure regulator on the front panel of the pressure multiplier is bypassed.

Figure 8. The pressure multiplier, controlled by the stepped regulator.

5 INTERFACING DEVICES

In Figure 9 the schematic diagram of interfacing between various devices to the IBM-PC microcomputer is illustrated. The power supply, amplifier and signal conditioner, and control circuit for stepper motors are all contained in one box.

This power supply and signal conditioner (PSSC) box has four modules, see Figure 10. One module contains power supply, amplifier and signal conditioner for pressure transducers. The output voltage from the pressure transducer is in the milli-Volt, mV range. Since the resolution of the A/D converter is 0.005V = 5mV, it is necessary to amplify this output to the Volt level. A 100x amplifier is used to bring this voltage to the order of 0.1V while the remaining 10x gain is done internally by the A/D converter.

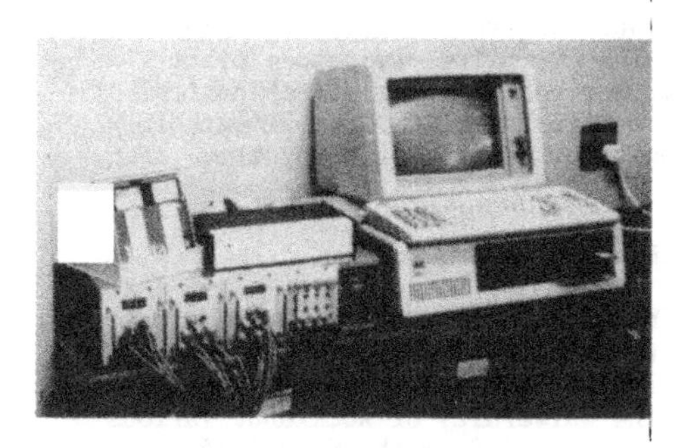

Figure 9a. A view of the IBM-PC, TecMar LabMaster and PSSC box.

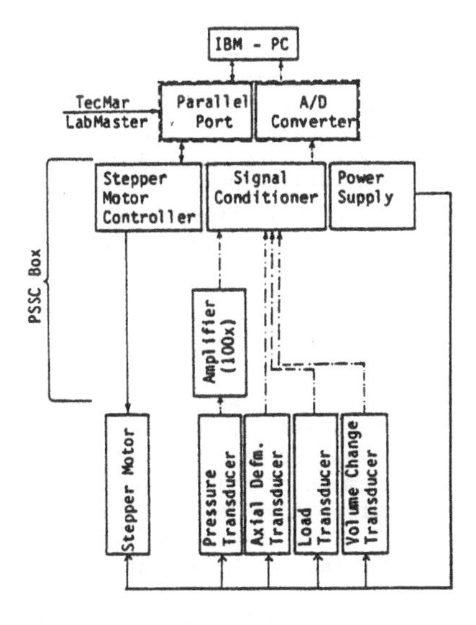

Figure 9b. Interfacing diagram.

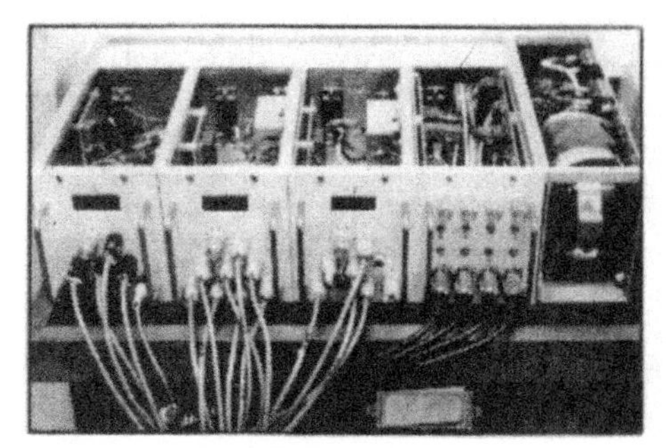

Figure 10. The PSSC box.

Two of the modules contain the power supply and signal conditioner for the DC-LVDT based transducers. The output voltage of these devices is in the order of 1 to 10V, so no amplifier is needed. In these two modules and also the pressure transducer module, low pass filters are installed to block the unwanted high frequency noises. Each of these three modules has 8 channels.

The fourth module contains the power supply and control circuit for the stepper motor. To prevent the stepper motor turning the pressure regulator out of range, limit switches are installed in the stepped-regulator box. The circuit to detect signals from these limit switches and send the signals back to the computer is also contained in this module, which has four channels.

Analog signals (output signals) from the measurement devices are converted to digital signals by TecMar LabMaster A/D converter, then sent to the IBM-PC. The A/D Converter has 16 single-ended I/O channels. An expansion board is needed to increase its capacity to 64 channels. To control the stepper motors, pulses from the IBM-PC are sent through the LabMaster's parallel port to the control circuit.

6 CAPABILITY OF THE TRIAXIAL SYSTEM

The use of computer to control the loading devices enables various types of tests to be performed in this equipment, in addition to the conventional undrained and drained tests.

The stress-controlled system is capable of performing any general stress path. The stress path is defined by its starting point, the slope in q/p' space and its maximum deviator stress, q max. In this way a soil specimen may be forced to follow a certain path up to a point, then, at that point, to turn and follow a

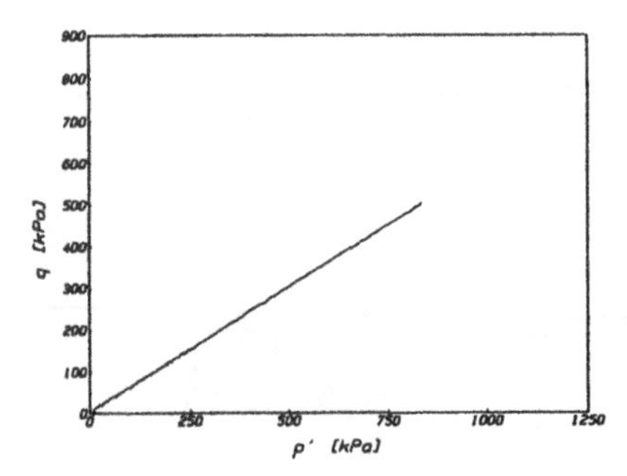

Figure 11. An illustration of the accuracy of anisotropic consolidation test achieved using stress-controlled equipment.

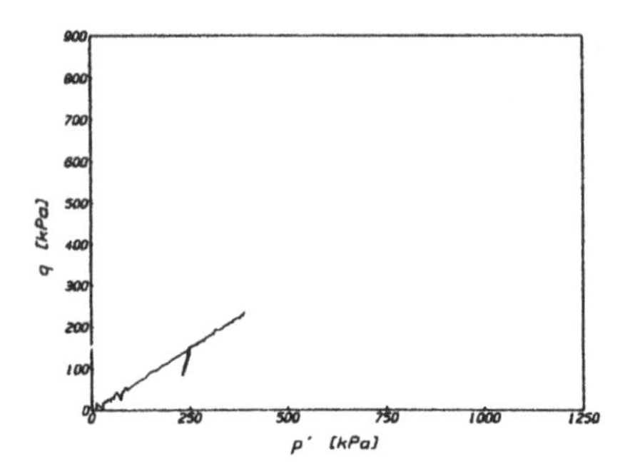

Figure 12. An illustration of the accuracy of anisotropic consolidation test achieved using strain-controlled equipment.

different path. This system may also be used to do an anisotropic consolidation test. Choosing an appropriate stress increment and reading interval, a very accurate q/p' constant stress path can be produced (Figure 11).

Another capability of the stress-controled system is to perform cyclic loading test. This is achieved by setting up a pair of values of deviator stress (minimum and maximum) to form the lower and upper limits. Any required loading rate may be performed by simply choosing a suitable interval between two loading increments.

The strain-controlled apparatus can also be used to do anisotropic consolidation test. This capability is needed when a specimen is going to be loaded to failure from an anisotropic consolidation state. For this purpose, the relationship between the deviator stress and axial deformation needs to be known. It may be obtained from an anisotropic consolidation test performed on the stress-controlled machine. Usually, this relationship can be piece-wised for the range of pressures of interest. Since the rate of strain is known (chosen), the rate of q against time can be calculated. Knowing the rate of q, the deviator stress at the next reading time may be estimated, hence the appropriate increase of cell pressure may also be estimated and the necessary number of pulses can be sent to the stepper motor.

Figure 12 shows an example of an anisotropic consolidation test performed on the strain-controlled apparatus. The result is comparable to the one produced by the stress-controlled system (see Figure 11), although at low pressures the stress-controlled system produced a better path.

The results shown in Figures 11 and 12 are directly plotted from the recorded data. No smoothing technique has been applied.

Connecting the stepper motor in the strain-controlled apparatus to the computer would enable direct control of the strain rate with improvement of results. This has not been done.

ACKNOWLEDGEMENT

The work reported in this paper was part of the PhD Research of the second author, under the supervision of the first author, at The University of Auckland. Various help received from the School of Engineering technician are gratefully acknowledged. Mr E Paul machined most of the mechanical parts while Mr G Carter built most of the electronic works.

Scholarships for the second author were received from Okumenisches StudienWerk (OSW), Bochum and the Christian University of Indonesia, Jakarta; while equipment grants were received from National Water and Soil Conservation Organisation (NWASCO) and the University Grants Committee, New Zealand. These are also gratefully acknowledged.

REFERENCES

Atkinson, J.H., J.S.Evans and C.R.Scott 1985. Developments in microcomputer controlled stress path testing equipment for measurement of soil parameters. Ground Engineering, Jan.85:1-22.

Bacchus, D.R. 1969. Cyclic deformation of a clay. PhD Thesis University of Auckland.

Coatsworth, A.M. and N.M.Hobbs 1984.

Computer-controlled triaxial soil testing equipment in a commercial laboratory. Ground Engineering, Oct.1984:19-23.
Indrawan, Z. 1986a. Stress-strain and strength characteristics of an Auckland soil. Phd Thesis University of Auckland.
Indrawan, Z. 1986b. University of Auckland Computerised Triaxial Testing System. User's Manual. University of Auckland, School of Engineering Research Report No. 394.
Indrawan, Z. 1986c. TestMast, an IBM-PC program to control and record computerised soil testings. Symposium on Computer Aided Design and Monitoring in Geotechnical Engineering, Asian Institute of Technology, Bangkok.
Mitchell, R.J. 1981. A New Control System for Soils Testing. In Laboratory Shear Strength of Soil, ASTM STP 740:180-190.
Moore, C.A. 1980. Modern Electronics for Geotechnical Engineers, Part 6, Process Control Applications. ASTM Geotechnical Testing Journal, Vol.3, No.4:159-162.

Computer and Physical Modelling in Geotechnical Engineering, Balasubramaniam et al. (eds)
© 1989 Balkema, Rotterdam. ISBN 90 6191 864 2

Seismic loading of foundations in a centrifuge

Pierre Morlier
Université de Bordeaux 1, France

Bernard Bourdin
CESTA (CEA), France

ABSTRACT : The conditions to obtain the solution of a dynamic problem with scaled-down models are given : they derive from the general equilibrium equations and from the rheological laws of the materials. The necessity of centrifugal testing for soil mechanics is shown. The experiments are then described specially when applied to seismic simulation.

A common practice in structural modelling (dams, bridges, ...) is that of substituting the prototype material with another, where deformability and strength properties are suitably choosen. Morever, it is possible to simulate the volume forces through experimental devices, for example with a system of tension roads. In geotechnical modelling, the complexity of the behavior of soils is such that it is lo longer possible to find equivalent materials.

The centrifugal system is therefore more and more used because it is able to set up geotechnical models employing the real soil and to create into them gravity forces in a consistent scale ; for this purpose the model is suspended at the end of a rotating arm, suitably loaded, in statics or dynamics, while measurements are performed.

I SIMILARITY FOR SOIL DYNAMICS

I.1 Similitarity conditions (after MANDEL - 1962)

We use the symbol (*) (scale of) which represents the ratio of one quantity in the model to the same quantity in the actual structure : for example L = 1/100 expresses a geometrical similarity of one hundreth. This geometrical similarity is the simplest but we can already notice that it will be difficult to represent some small details of the structure : how to simulate the rugosity of a given pile ?

I.1.1 Equilibrium

General equation of equilibrium are written

$$\sum_j \frac{\partial \sigma ij}{\partial xj} + \rho \left(gi - \frac{d^2 \xi i}{dt^2} \right) = 0 \qquad (1)$$

where ρ is the specific mass of the soil,
g is the gravity force per unit volume,
σij is the stress component,
ξi is the displacement.

When varying the scales, the first term is multiplied by $\sigma^* L^{*-1}$, the second by $\rho^* g^*$ and the third by $\rho^* \xi^* t^{*-2}$; équation (1) has to be verified with any unit, so

$$\sigma^* L^{*-1} = \rho^* g^* \qquad (2)$$
$$\text{and} \quad \xi^* = g^* t^{*2} \qquad (3)$$

ξ^* may be different from 1 (enlarged similarity)

Quasic-static problems

In static or quasi-static applications, inertial terms may be neglected :

$$\sigma^* L^{*-1} = \rho^* g^* \qquad (2)$$

In ordinary conditions, $g^* = 1$ and it is difficult to have large variations of ρ^* ($\rho^* = 1$), so $\sigma^* = L^*$.

Dynamic problems

The second term (ρg_i) of equation (1) may be neglected :

$$\sigma^* L^{*-1} = \rho^* \xi^* t^{*-2} \qquad (3)$$

when $\sigma^* = \rho^* = 1$ (same material) and $L^* = \xi^*$ (simple similarity),
$t^* = L^*$, hence $v^* = 1$.

I.1.2. Rheological relations

The second type of equations to be verified concerns the behavior of the material.

Elasticity

We can write

$$\frac{\Delta l}{l} = \frac{1}{E}\left[\sigma_1 - \nu\,\sigma_2 - \nu\,\sigma_3\right] ;$$

of course $E^* = \sigma^*$ and $\nu^* = 1$ if the same scales are taken for l and Δ l (this is necessary when studying large deformation problems).

Soil plasticity

The plasticity law for soils is

$$\tau = C + \sigma\, t_j\ \phi ;$$

of course $C^* = \sigma^*$ and $\emptyset^* = 1$.

Thus it is already difficult to design an equivalent material in elasticity or plasticity if σ^* is different from unit ; it becomes impossible to find such an equivalent materiel for an elasto-plastic problem.

The non-linear behavior of soils, the transition from brittleness to ductility for rocks, are other reasons to build the model with the actual material of the structure :

$$\sigma^* = E^* = C^* = 1 ,$$
$$\nu^* = 1$$
$$\emptyset^* = 1$$
$$\rho^* = 1$$

Equation (2) becomes

$$g^* = L^{*\,-1}.$$

So if we want to realize geometrically scaled-down model, it is necessary to increase gravity forces ; different methodes have been used, essentially :

- centrifugation of model (see II),
- hydraulic gradient method of ZELIKSON (1967).

It happens encountering materials the properties of which depend on their dimension (scale effect), in rock mechanics especially : it is then impossible to realize good scaled-down model and it becomes necessary to perform in situ tests.

I.1.3. Time representation

For dynamic problems, in simple similarity ($L^* = \xi^*$), using the actual material ($\sigma^* = \rho^* = 1$), we have already seen that

$$t^* = L^* \qquad (4)$$

If the problem involves water percolation, for example it includes a consolidation process, a new condition for time scale appears :

if we use the actual material with the actual fluid (water), the scale of consolidation coefficient (TERZAGHI) must be unit,

$$C_V^* = 1 \text{ with } C_V = \frac{k(1 + e)}{a_v\, \gamma_w} ;$$

as the duration of a consolidation process $t = \frac{h^2}{C_V} T$ where T is an adimensional term,

$$t^* = L^{*2} . \qquad (5)$$

It is clear that the conditions (4) and (5) are not compatible ; some authors have thought to modify either the permeability of the model soil (adding fine particles) or the viscosity of the liquid to try to restore this compatibility.

Another point of view concerns the liquefaction phenomenon, ofently involved in dynamic problems of soil mechanic ; experiments have shown that the influence of the frequency vanishes : for frequencies between 0.05 and 4 Hz, the cycle numbers before liquefaction is constant ; time does not affect liquefaction and we can follow the time scale of the laboratory : $t^* = 1$.

II. CENTRIFUGATION OF GEOTECHNICAL MODELS

II.I. Survey

As we have seen in the first paragraph, the necessity of scaled-down models and the importance of gravity forces in geotechnics have led to choose centrifugal testing : as early as 1930, studies were performed in USA and USSR, but it is the action of SCHOFIELD (CAMBRIDGE) in the sixties which decisively contributed to the rise of interest for this technique.

In the last ten years, the number of searchers and of countries concerned with the centrifugation of geotechnical models increases regularly and the results of their words are acknowledged by soil mechanicians. A Technical Committee on Centrifuge Testing has been founded inside the International

Society for Soil Mechanics and Foundation Engineering and organizes an International meeting, CENTRIFUGE 88 in France.

The table I (from CORTE 1986) gives a short description of large centrifuges devoted, partially (*) or totally, to geotechnical research ; since this publication centrifuges are planned at University of BOCHUM, LGM Delft, University of Boulder Colorado, ISMES, ...

Table I : Geotechnical Centrifuges on service in 1986

Organization and COUNTRY	radius (m)	g_{max} (g)	M_{max} (t)
Queen's University Ontario -CANADA-	3	300	0,1
Yangtze water conservancy & hydro power research institute -CHINA-	3	410	1
Nanning -CHINA-	2.5		
Engineering Academy -DENMARK-	2.3	80	1.25
Laboratoire des Ponts et Chaussées -FRANCE-	5.5	200	2
CESTA* (CEA)-FRANCE-	10	100	2
Simon Eng. Lab. Manchester -UK-	3.2	200	2
Cambridge University -UK-	4.8	160	1
Port and Harbour Research Institute -JAPAN-	3.8	115	2.75
M.I.I.T. Moscou -USSR-	2.5	320	0.18
A.Z.N.I.I.S.M. Baku -USSR-	5.5	500	1.5
Sanatro Laboratory* -USA-	8.0	240	1.5

g_{max} : maximum acceleration
M_{max} : maximum mass of the model

II.2. Description of the CESTA centrifuge

The CESTA Centrifuge was used for own experiments as well as for experiments of the Laboratory of Solid Mechanics (Palaiseau -FRANCE-) ; it was designed to simulate the behavior of large specimens both submitted to a static acceleration and different environmental conditions (temperature, vacuum, vibration, ...) ; it allows to transport large masses at the end of its arm (Table I and figure 1).

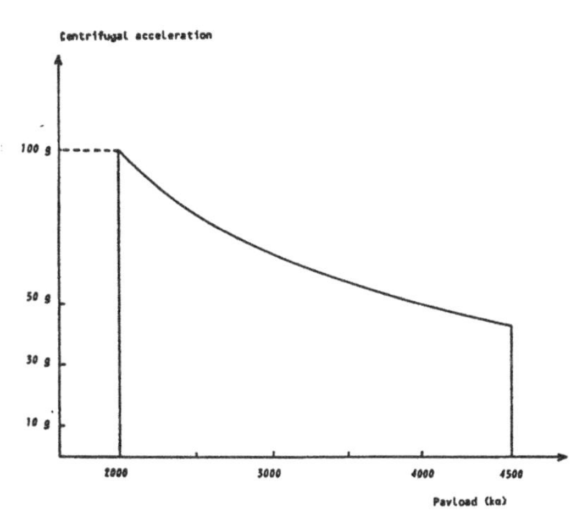

Figure 1 : Range of application of the CESTA Centrifuge

The centrifuge -figure 2- consists of a dissymetrical arm (mass = 40 tons) at the end of which is attached a rotating cradle; various measurements can be made inside the cradle (various gauges, thermocouples, accelerometers, ...) through more than hundred simple or coaxial wires and a slip ring system.

For soil mechanics studies, the cradle contains a soil mass the dimensions of wich are 1,30 m by 0,80 m by 0,40 m depth ; the walls of the cradleare covered of isomode so that the wall transfered parasitic vibrations be absorbed. The soil is a fine poorly graded sand from Fontainebleau, either saturated or wet ; the cell is filled by sprinkling the sand from two meters up to obtain an high density (1,6) ; the saturation is made by vertical uplift percolation.

The rotation of the cradle (figure 3 represents the planned one) permits an easy setting and disassembly of the model, allows the acceleration force to be constantly "vertical" for the model and the model

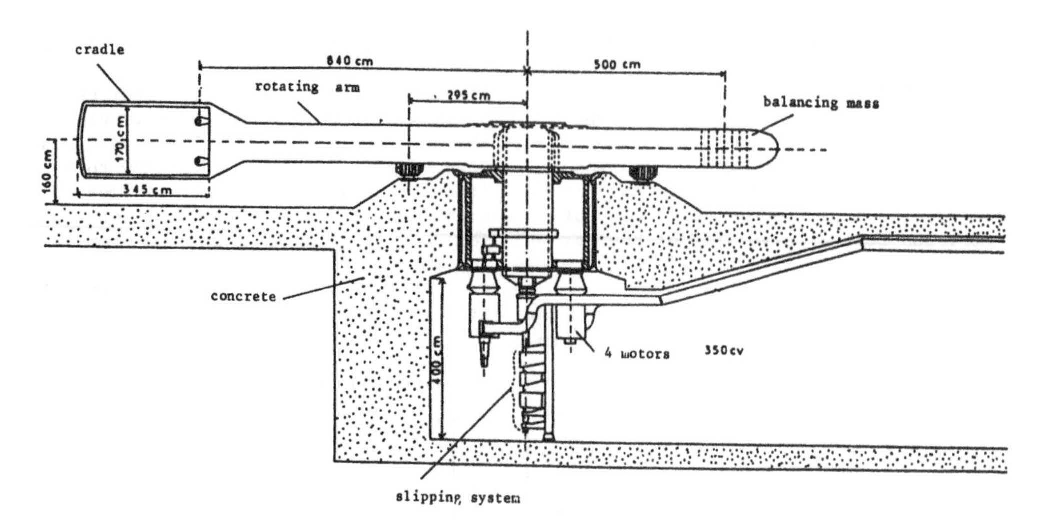

Figure 2 : C.E.S.T.A. Centriguge

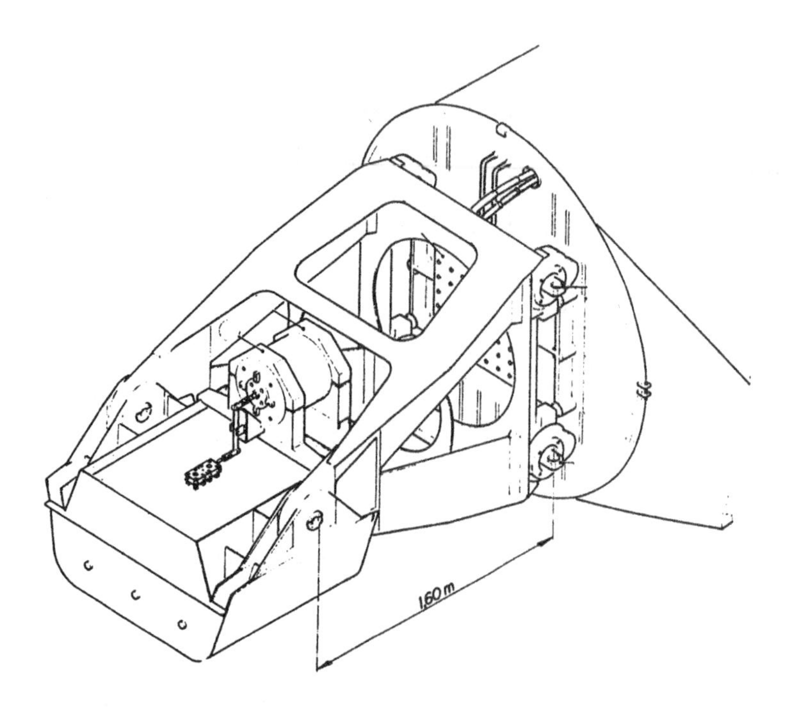

Figure 3 : Detailed view of the cradle : horizontal vibrations of a pile group

to be progressively loaded from 0 to 100 g.

After this description, it is possible to explain and compute the approximations of the centrifugal technique.

II.3. Approximations on the gravity forces

Radial and tangential components of the acceleration are

$$\gamma_r = \ddot{r} - r\,\dot{\theta}^2$$

$$\gamma_t = 2\dot{r}\,\dot{\theta} + r\,\ddot{\theta}$$

- As the arm of the centrifuge is stiff enough to ensure that the radius is constant, and as the experiments are made at a stabilized speed of rotation,

$$\gamma_r = r\,\dot{\theta}^2 \text{ and}$$

$$\gamma_t = 0$$

- The time necessary to reach the stabilized speed of rotation being about five minutes, the maximum of γt is negligible, something like 0,05 g.
- The aim of simulation is to create a "vertical" gravity field : the radius of the arm is large enough compared with the breadth of the cradle.
- The intensity of centrifuge forces γ_r depends on the distance to the axis of the machine but the depth of the model is small enough compared to the length of the arm to ensure a constant γ_r.
- When soil particles are in movement, the Coriolis forces can be estimated ; they are negligible.
- During the test, the model is also submitted to the gravity of the Laboratory but of course the last is not important at all.

This comments show that the dimensions of the cell have to be small with regard to the lenght of the arm in order to obtain a good "vertical" gravity fiel in the model.

III SEISMIC SIMULATION

The tests we have performed aimed to define the answer of a given structure, including soil-structure interaction, to a seism the intensity and duration of which are given ; it was thus necessary to design a system able to generate vibrations, equivavent to seismic wave trains, under a static acceleration of 100 g.

Conservation of the material gives the following similarity rules :

	$g^* = 100$,
lengths	$L^* = 1/100$,
displacements	$\xi^* = 1/100$,
time or durations	$t^* = 1/100$,
velocities	$\dot{\xi}^* = 1$,
frequencies	$f^* = 100$,
acceleration	$\ddot{\xi}^* = 100$.

III.1 Seismic generation

The seismic waves are generated in the soil by the transmission of a shock wave through a rubber diaphragm ; the shock wave is created by sequence of small charge blastings in a reverberation room where a mechanical filter eliminates frequencies greater than 100 Hz ; remember than for natural seisms the frequencies are between 0,5 and 15 Hz.

The sequence of firing may be programmed and allows either a simple blasting with a charge between 0,8 g and 10 g or a train of blastings with a maximum charge of 10 g; with a length scale $L^* = 100$, a charge of 5 g crates in the model a seismic energy equivalent to 5 tons in the prototype.

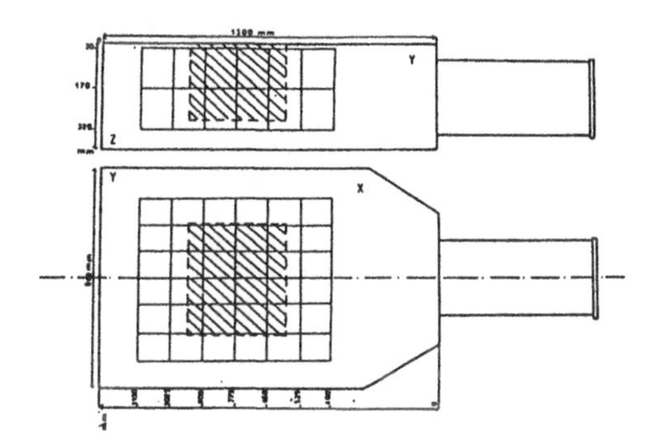

Figure 4 : cradle for direct seismic test showing - the network for accelerometers in the sand mass - the volume where the influ ence of lateral walls is not significant.

III.2 Type of waves

On the accelerogramms (figure 5) we can see two wave trains. For the first train, horizontal (along ox) and transversal (along oy) components increase with the depth whereas then vertical (along oz) component decreases ; the mean frequency is about 2000 Hz, the celerity is 830 ms^{-1} ; this train should be equivalent to primary waves.

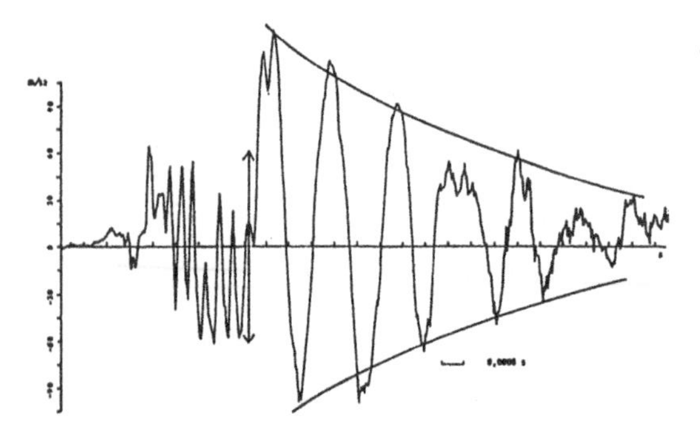

Figure 5 : Typical accelerogramm obtained in the centrifuge

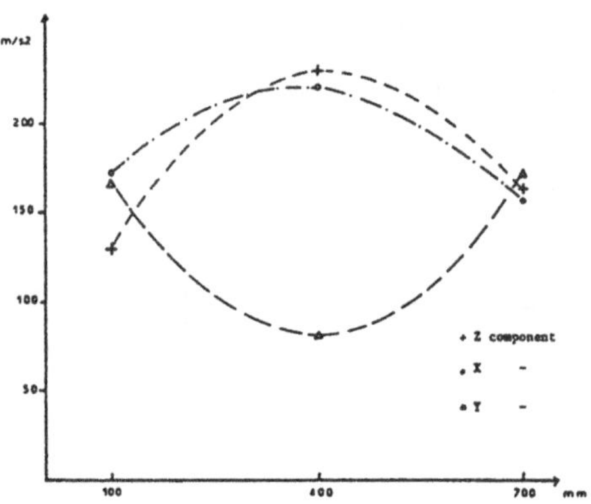

Figure 7 : Amplitude of the second wave train along OY.

For the second train, there is no great difference between the measured amplitude at the top and at the bottom of the cell ; the mean frequency is about 450 Hz and the celerity is 480 ms^{-1} ; this train should be equivalent to a Rayleigh wave especially because the phase difference between the horizontal and vertical components ; we can notice that the wavelength is more than twice greater than the depth of the soil.

RIVIERE (1983) studied the influence of the lateral walls ; figures 6 to 9 show the amplitude variation of the first or second train along Ox and Oz ; figure 10 gives the decay of the signal with the distance along Ox. In conclusion, RIVIERE defined a test volume inside the soil model, corresponding to a semi-infinite medium, shown on figure 4 : in this volume the influence of lateral walls in not significant.

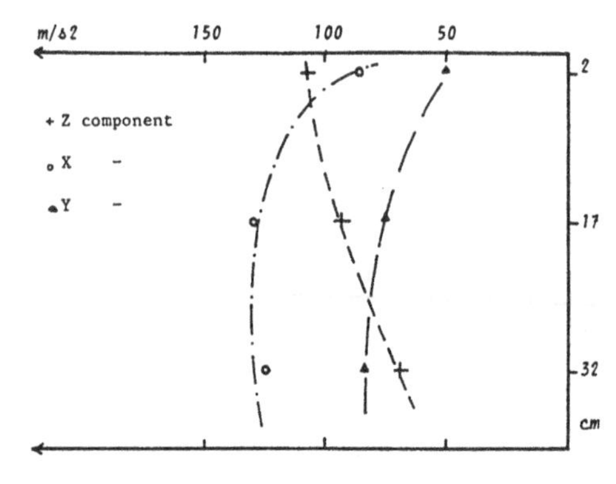

Figure 8 : Amplitude of the first wave train along OZ

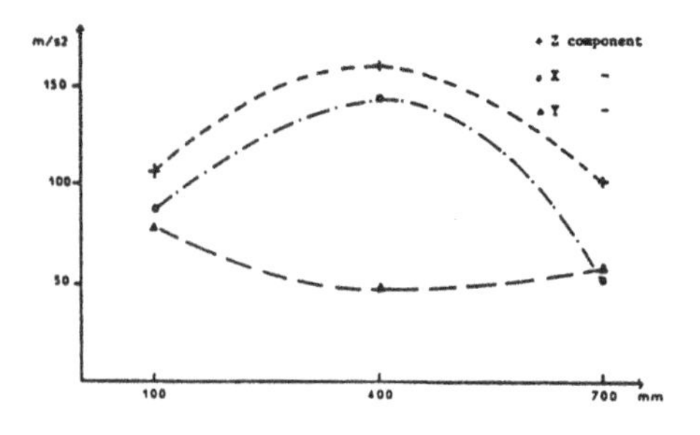

Figure 6 : Amplitude of the first wave train along OY.

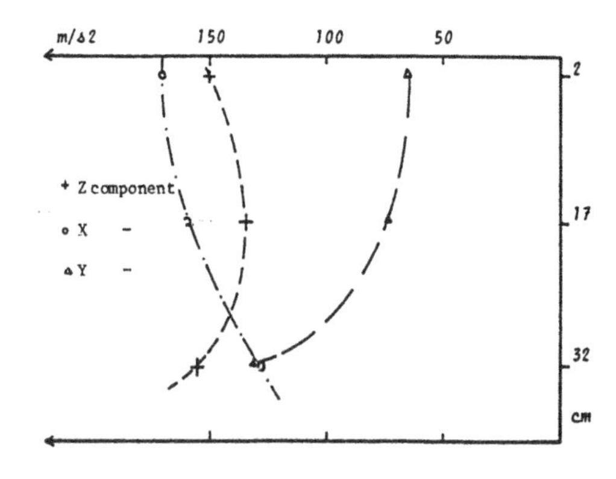

Figure 9 : Amplitude of the second wave train along OZ.

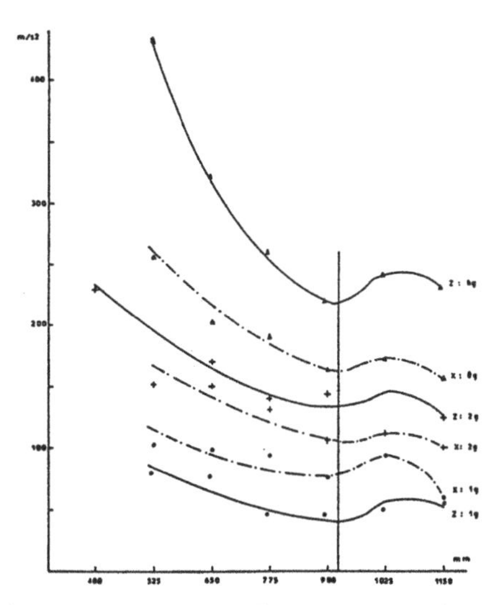

Figure 10 : Attenuation along OY with the different single blastings.

III.3 Comparison with true seisms

True seisms will be compared to vibrations generated in the test volume with three criteria : intensity, duration, frequency spectrum.

Intensity : the maximum amplitude of simulated accelerogramm depends on the explosive charge (figure 11).

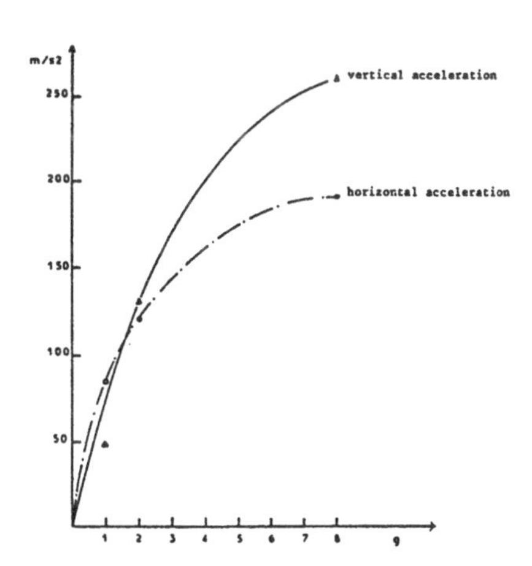

Figure 11 : Amplitude versus charge of single blastings.

Remember the definition of the modified MERCALLI intensity I_{MM} :

$$\log A_h = 0.014 + 0.3\ I_{MM}$$

where A_h, the horizontal maximum acceleration, is given in cms^{-2} ; the simulated seisms (explosive charge from 1 g to 8 g) are characterized by

$$6.5 < I_{MM} < 7.5$$

Seismic duration is used to measure the possible seismic damages, its definition may be the following (BOLT) : on the accelerogramm, we measure the cumulated duration of parts the amplitude of which is greater than 0.3 times the maximum amplitude ; for the SAN FRANCISCO seism (figure 12) the measured duration is about 35 seconds. In the simulated seisms, a single shot gives a duration about 1.8 s ; with train of blastings the duration reaches 4s.

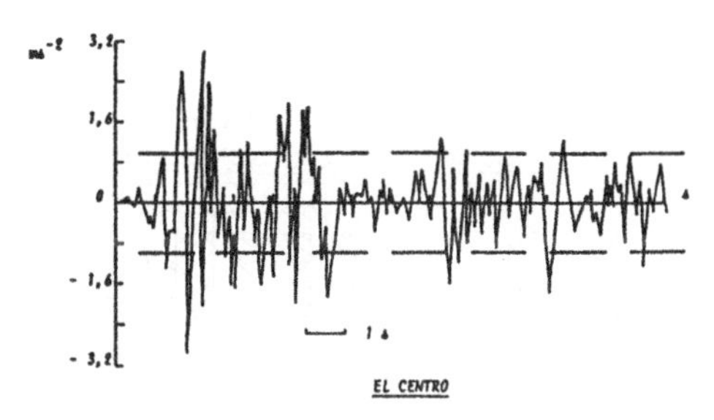

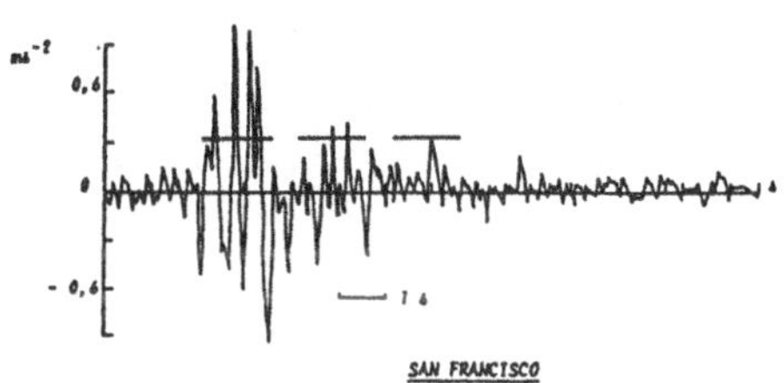

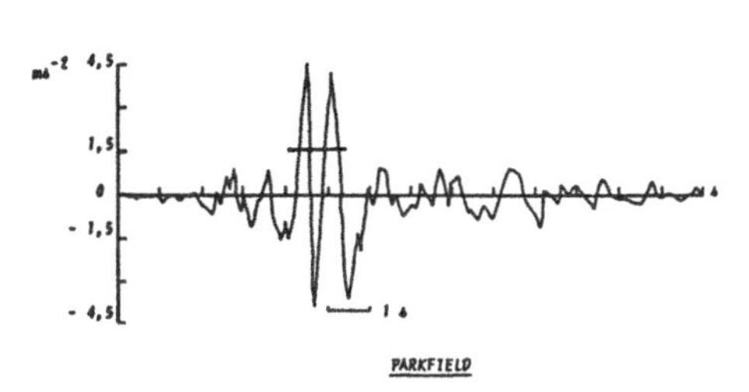

Figure 12 : True seisms

Response spectrum of the models can be compared to spectra of true seisms in the usual representation of figure 13. (RIVIERE 1983) or 14 (ZELIKSON & HABIB 1986) Thus the similary for seismic loadings is correctly achieved in the central volume of the cell.

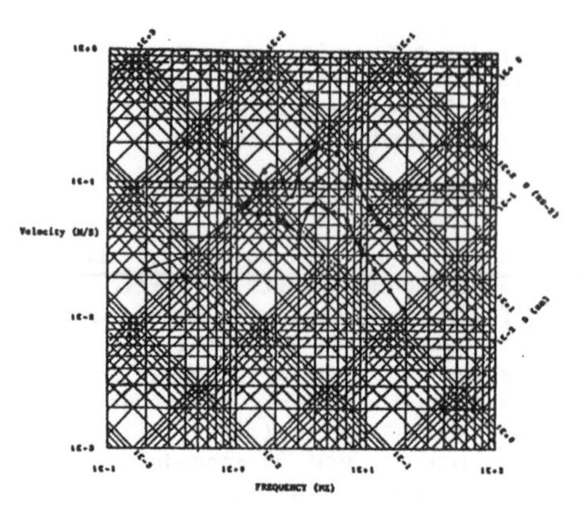

Figure 13 : Spectrum of San Francisco earthquake (0) compared with the one of a simulated seism (single bast 8 g).

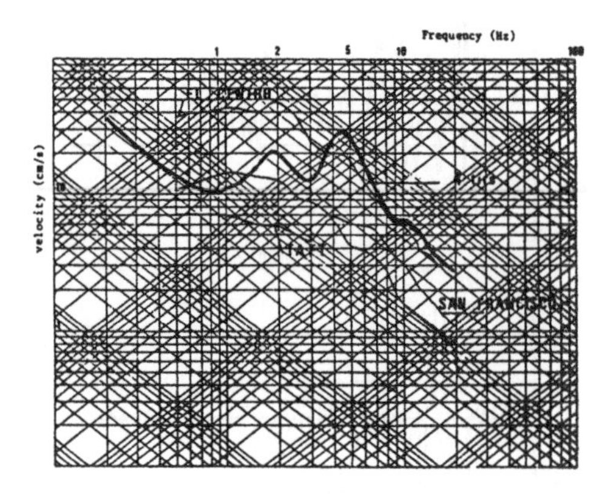

Figure 14 : (after HABIB and ZELIKSON) spectra of known earthquakes comparec to a four shot discharge in the centrifuge

Conclusion

This paper aimed to show that it is realistic to simulate seismic sollicitations on buildings and other construction works where it is necessary to take into account the gravity effects in the soil-structure system : centrifugal testing of scaled down models is a good method. For the time being (Bourdin 1987), a number of simulations have been performed for different companies or laboratories using either direct seismic simulation or vibrationg tests associated with modal analysis.

References

Mandel, J. 1962. Essais sur modèles réduits en mécanique des terrains. Rapport A.R.T.E.P. 7587.

Zelikson, A. 1967. Représentation de la pesanteur par gradient hydraulique dans les modèles réduits en géotechnique . Annales ITBTP 239.

Corte, J.F. & Garnier, J. 1986. Une centrifugeuse pour la recherche en géotechnique. Bull. Liaison Labo P. et Ch. 146.

Rivière, J. 1983. Interaction sol-structure lors d'un seisme ; simulation en centrifugeuse. Thèse Univ. Bordeaux.

Habib, P. & Zelikson, A. 1985. Modèles réduits en dynamique, in Génie Parasismique, Davidovici editor, Presses Ponts et Chaussées.

Bourdin, B. 1987. Vibrations horizontales de pieux en centrifugeuse. Thèse Univ. Bordeaux.

Computer and Physical Modelling in Geotechnical Engineering, Balasubramaniam et al. (eds)
© 1989 Balkema, Rotterdam. ISBN 90 6191 864 2

Anisotropic stress-strain behaviour of Bangkok sand and its application to finite element analysis

Ikuo Towhata, Tahir Mahmood Hayat & Kasem Jarupan
Asian Institute of Technology, Bangkok, Thailand

ABSTRACT: Triaxial compression tests were carried out in drained manners on laboratory reconstituted samples as well as undisturbed samples. Both types of samples were obtained with their axial directions either parallel or perpendicular to the plane of deposition, revealing anisotropic stress-strain behaviours. Comparison was made between two directions on Young modulus, Poisson ratio, and shear strength. The effects of soil anisotropy on numerical analyses were also examined.

1. INTRODUCTION

Since early 1960's two significant impacts have been hitting the geotechnical engineering discipline. One of them is the remarkable development in the finite element method (FEM) of deformation analysis with application to earth structures. The FE method has made it possible to calculate the stress, strain, and displacement of real earth structures without difficulty if appropriate data of soil property is available. The other impact is the improvement of the skill of laboratory soil testing, which has led to better understanding of the behaviour of soil under general stress states. It was felt, consequently, that an accurate and reliable prediction of the behaviour of any earth structures may be possible by constructing a stress-strain model of soil based on laboratory test data, and combining it with a sophisticated skill of numerical analysis.

In reality, the above expectation has not been achieved so far. This is mainly because elaborate and detailed laboratory testing on soil has indicated that behaviour of soil is more complicated than was expected before, and also that every stress-strain model ever proposed has limitations. One of the experimental findings which have been making the situation complicated is the anisotropy.

Two different types of anisotropy are there. One of them is the induced anisotropy, which means that a previous loading results in the increased stiffness in the same direction as the loading itself, while reduced stiffness in the directions orthogonal to it (Arthur et al., 1977). This kind of anisotropy is called induced anisotropy, and natural deposits of soil shows this anisotropy which is developed by the overburden pressures.

The other kind of anisotropy is called inherent, which is due to natural procedure of deposition. Oda (1972a) revealed that sand specimens prepared under vertical action of gravity have greater stiffness or Young modulus in the vertical direction than horizontal directions. Similar phenomenon was reported by Yamada and Ishihara (1979). Studies on anisotropic shear strength have revealed 2 to 3 degree difference in friction angles between vertical and horizontal directions (Arthur and Phillips, 1975; Ricceri and Soranzo, 1981).

In this text, a discussion is made of the experimental results which were recently derived by the authors on the anisotropy of Bangkok first-layer sand, followed by its application to deformation analysis of foundations.

2. SAMPLE PREPARATION AND TESTING PROCEDURES

The first layer Bangkok sand lies beneath the soft layer of Bangkok clay, and is very often employed as supporting layers of deep foundations due to its reasonable stiffness and shear strength. This sand is normally found to be cemented, although the extent of cementation varies from site to site. Both disturbed and undisturbed samples of this Bangkok sand were collected, for the purpose of laboratory testing, at a garbage dump pit which was excavated by the Bangkok Metropolitan Authority. The depth of the sampling site was 15.5m below the original ground level. This sand has Gs=2.64, maximum and minimum void ratios being 1.06 and 0.66, respectively, (Ng, 1986). The insitu dry density was found to be 1.43 g/cm^3 which gives the insitu void ratio of 0.846.

Since the sample preparation methods are influential on deformation of sand (Oda, 1972b), four different methods were employed for reconstituting samples in a rectangular prismatic mould which measured 10*10*16 cm^3 in sizes. The air-pluviation method had air-dry sand fall freely through a nozzle into the mould. The height of fall was adjusted so that the laboratory dry density may be identical with the insitu density. Air-dry sand was also compacted by hitting the sand surface (tamping), hitting four sides of mould (tapping), or pushing a rod into sand repeatedly (plunging). The compaction was followed by careful submergence and drainage of sand. The wet sand could now remain stable even when side plates of the mould were removed for the ease of sample collection.

The reconstituted samples were collected by the block-sampling technique which was conducted in both vertical and horizontal directions in the mould. The vertical sampling (Fig. 1) is essentially identical with Mori and Ishihara's (1979) method, in which a sample collecting tube is placed at the soil surface, and a soil column is trimmed out beneath the tube. When the diameter of the soil column is as small as that of the tube, it is possible to push down the tube with little force, causing no significant disturbance to the soil. The horizontal sampling was performed by pushing a sampling tube laterally toward the sand together with its wooden rest (Fig. 2). In this case also, sand was trimmed in advance to approximately same size as the tube. The collected samples were immediately frozen until the time of laboratory testing. Yoshimi et al. (1978) reported that the effects of freezing on the stress-strain behaviour of sand is negligible. The same technique of block sampling was employed for undisturbed sampling at the excavation pit. It should be noted that triaxial compression tests on vertically and horizontally collected samples corresponds to vertical and horizontal compression in the deposit, respectively.

3. TRIAXIAL TESTS ON LABORATORY-RECONSTITUTED SAMPLES

A series of triaxial compression tests were carried out on laboratory-reconstituted samples of Bangkok sand which were consolidated isotropically and sheared under drained conditions. Fig. 3 shows the test results derived from samples which were vertically collected and isotropically consolidated at 98 kPa. The direction of axial stress is orthogonal to the sedimentation plane in the mould. Of the four kinds of samples shown in the figure, the air-pluviated sample has the greatest stiffness, followed by tamping, tapping, and plunging. The pluviated sample also has the remarkable tendency to dilate. The secant Young modulus at 50% strength, E_{50}, was read from Fig. 3 and plotted against the void ratio at the end of consolidation (Fig. 4). Although the void ratio shows some scattering, it may be seen that the air-pluviation causes the highest value of modulus, tapping and tamping being similar, while plunging gives the least value. The modulus at the insitu void ratio of 0.846 were determined by interpolation in Fig. 4.

Fig. 5 indicates the deformation of reconstituted samples which were collected in the horizontal directions (Fig. 2). The axial direction of the samples agrees with the sedimentation plane in the mould. The measured E_{50} modulus were plotted against the initial void ratio in Fig. 6. The influence of different sample preparation methods on the value of the modulus is less substantial for horizontal samples than what was seen for vertical samples in Fig. 4. This observation suggests identical values of modulus for all the methods at void ratio of 0.846.

Fig. 7 illustrates the ratio of E_{50} moduli in vertical and horizontal directions (Figs. 4 and 6). The ratio for the pluviated sample is only 0.5, which means significant anisotropy, while plunging method gives approximately equal moduli in two directions. The other two methods lie between them. The Poisson ratio at 50% strength, which is a secant value, is seen in Fig. 8. Although the scattering is substantial, Poisson ratio takes on average values between 0.3 and 0.4, irrespective of sample preparation method.

Effects of confining pressure on the stress-strain behaviour of Bangkok sand was studied by consolidating air-pluviated samples under 98, 245, and 490 kPa isotropically. Fig. 9 indicates the stress-strain curves of vertically collected samples, in which the deviator stress was normalized by the confining pressure. It can be seen in the figure that the specimen tends to have greater deformations as the confining pressure increases, and that the specimen dilates more under lower confining pressures. The normalized deviator stress at failure is slightly less under higher pressure than under lower pressure, suggesting that the friction angle decreases when the confining pressure is raised.

Fig. 10 shows the deformation of air-pluviated specimens which were collected horizontally. The influence of confining pressure on deformation of sand seen in this figure is less significant than that in Fig. 9. However, the strain under 490 kPa is slightly greater than strains under lower confining pressures.

4. TRIAXIAL TESTS ON UNDISTURBED SAMPLES

Triaxial compression tests were also conducted on undisturbed samples which were collected by the block-sampling technique. The specimens were isotropically consolidated under various pressures. The test results from vertically-collected samples (Fig.11) indicates that, as the consolidation pressure is raised, the normalized deviator stress at failure decreases, the strain at failure increases, and the tendency to dilate is reduced. Horizontally collected samples showed similar trends as well (Fig.12).

5. STUDY ON YOUNG MODULUS AND STRENGTH UNDER DIFFERENT PRESSURES

It is not uncommon in practice to analyze in elastic manners the deformation of earth structures by using the secant Young modulus at 50% of strength, E_{50}, and a suitable value of Poisson ratio. With this in mind, a study is going to be made in this section on the Young modulus.

Fig.13 compares the E_{50} moduli derived from air-pluviated and undisturbed samples in both vertical and horizontal directions. For both kinds of samples, the vertical modulus is greater than the horizontal modulus. The modulus of undisturbed samples are apparently much greater than that of air-pluviated samples. This is probably because of the slight cementation in the undisturbed samples. It should be noted that, although the Young modulus of air-pluviated samples increases linearly with the consolidation pressure, modulus of undisturbed samples under higher consolidation pressures does not increase with pressure so much as under lower pressures. This is probably due to the breakage of cementation among sand particles caused by higher consolidation pressures and strains during consolidation (Saxena and Lastrico; 1978). Upto some pressure level, the cementation is intact, giving substantial stiffness to sand. However, when the consolidation pressure is high enough, particle movement or strain in the course of consolidation is large enough to break the cementation bonding among particles.

The ratio of vertical and horizontal Young moduli were plotted in Fig.14. The undisturbed samples has smaller ratio than pluviated samples, suggesting its greater extent of anisotropy. This is because the undisturbed samples involve both inherent and induced anisotropies.

Fig.15 compares the shear strength of air-pluviated and undisturbed samples. It is noteworthy that the difference between two types of samples is not so large as was seen for Young modulus. Particularly, the horizontal strengths are essentially identical. This may be because the cementation among sand particles was broken in the course of shear deformation, irrespective of consolidation pressure. The vertical strength is greater than the horizontal strength for both types of samples, although one exceptional data is seen at 500 kPa. The difference in friction angles between vertical and horizontal directions ranges roughly 2 to 4 degrees, which agrees with previously obtained values.

6. IMPROVEMENT OF LINEARLY ELASTIC FINITE ELEMENT ANALYSIS

The purpose of this section is to present an improvement measure of elastic finite element analysis so that anisotropic behaviours of soil may be taken into account. As was discussed before, one of the most attractive method of deformation analysis on earth structures is the nonlinear finite element analysis which makes use of perfectly reliable stress-strain models of soils. However, this approach of study is very difficult. All aspects of complicated soil behaviour have to be considered, and, at the same time, determination of soil properties has to be easy for practice, and computation time has to be reasonably short.

In view of this, finite element analyses being conducted at present on earth structures are split into two categories. The first category includes nonlinear or advanced analyses. Although the stress-strain model employed may not be perfect, errors occurring from this is regarded as negligible. Difficulty in data preparation and cost for computer run are also allowed. This type of analysis is normally performed when the project affords its cost.

The second category includes linearly elastic analyses. This kind of analysis is much easier and cheaper. Only two soil parameters are needed; Young modulus and Poisson ratio. However, the elastic analysis is of course less reliable from theoretical viewpoint than nonlinear analysis, and the difference in computed results between linear and nonlinear analyses becomes significant when stress level is close to failure, making nonlinear stress-strain behaviour substantial.

In spite of those deficiencies mentioned above, it is still possible to justify the use of linear elastic analysis. Earth structures are usually designed with a sufficient factor of safety; e.g. 2.0 to 3.0. When the deviator stress is only 1/3 to 1/2 of the strength, nonlinearity in stress-strain behaviour is not significant, and its linear approximation using E_{50} is good enough, as is suggested by those stress-strain curves in Figs. 3, 5, 9, 10, 11, and 12. Moreover, soil parameters required for nonlinear analysis are not always determined directly by measurement, whether insitu or in laboratory. They are sometimes evaluated through experiences in similar soils, and sometimes calculated with the aid of empirical correlations. Even if soil properties can be determined for some of finite elements by direct measurement, it is still impossible to determine them for all of them. Most elements accept soil properties which were guessed somehow from available informations. These procedures are apparently not so precise as intended in nonlinear programmings. Hence, it is possible that, although the computer program is very sophisticated, the reliability of results of analysis is deteriorated by soil data.

An improvement measure is going to be proposed for a linear elastic finite element analysis on earth structures. Firstly, Poisson ratio which is assumed in practice to lie between 0.3 and 0.4 is reasonable as is seen in Fig.8. Secondly, the anisotropic Young modulus can be taken into account by using Figs.13 or 14. When the vertical modulus is already known from vertically collected samples, the horizontal modulus is obtained by multiplying the modulus ratio (Fig.14) to the vertical value. When a horizontal modulus is somehow known, the vertical modulus can be derived in a similar manner. The orientation of the major principal stress in the ground can be calculated by elasticity theories without knowing the Young modulus (Poulos and Davis, 1974). According to the direction of major principal stress, different modulus values are assigned to each finite element.

7. EXAMPLE ANALYSIS

An example analysis was carried out on a strip loading placed at the surface of a level ground. The width of the strip is 2 m, and the intensity of surcharge is variable (Fig.16). Since the aim of the computation is the displacement of foundation soil caused by the surface load, different anisotropic modulus values have to be assigned to soil according to the direction, β , of the major principal stress induced by the surcharge. The β angle was calculated by an elasticity theory (Poulos and Davis, 1974).

$$\sigma_v = \frac{P}{\pi}\{\alpha + \sin\alpha\cos(\alpha + 2\delta)\}$$

$$\sigma_h = \frac{P}{\pi}\{\alpha - \sin\alpha\cos(\alpha + 2\delta)\}$$

$$\tau_{vh} = \frac{P}{\pi}\sin\alpha\sin(\alpha + 2\delta)$$

$$\beta = \tfrac{1}{2}\arctan\frac{2\tau_{vh}}{\sigma_v - \sigma_h}$$

in which P stands for the pressure of the strip loading, σ_v , σ_h , and τ_{vh} are vertical, horizontal and shear stress components, while and δ are angles illustrated in Fig.16.

Fig.16 illustrates the contour lines for β equal to 0, 30, 60, and 90 degrees, and domains among them are called A, B, and C. In the Domain A, the direction of loading of major principal stress, σ_1 , is more or less vertical, while the Domain C has approximately horizontal direction of loading. The Domain B is an intermediate one. It may be said that a vertical Young modulus is suitable for the Domain A, the horizontal modulus is appropriate for the Domain C, and an intermediate value such as an average of vertical and horizontal moduli are reasonable for the Domain B.

Fig.17 illustrates the finite element model of the ground. Only half of the ground is analyzed because of symmetry. In order to determine the Young modulus, the ground was divided into the upper layer between surface and 6 m deep, and the lower layer between 6 m and 15 m deep (Fig.18). By assuming the unit weight of soil, γ , equal to 1.6 t/m³ and Poisson ratio, ν , of 0.3, the minor principal stress before surcharge loading at the middle of each layer was calculated.

$$\sigma_3 = \frac{\nu}{1-\nu}\gamma z$$

in which z stands for the depth. The minor principal stress values at the middle of upper and lower layers, z = 3m and 10.5m, are substituted into empirical equations for vertical and horizontal moduli which are denoted by Ev and Eh, respectively.

$$Ev = 4400 \times \sigma_3^{0.40} \quad \text{(kPa)}$$

$$Eh = 900 \times \sigma_3^{0.58} \quad \text{(kPa)}$$

These equations were derived from data for undisturbed samples indicated in Fig.13. The Ev and Eh moduli thus derived for both upper and lower layers are denoted by Euv, Euh, Elv, and Elh. The subscripts, "u" and "l", stand for upper and lower layers, respectively. The calculated values of these moduli are listed below together with the average of vertical and horizontal moduli denoted by Eum and Elm.

In upper layers,

$$Euv = 1490 \ t/m^2$$
$$Euh = 525 \ t/m^2$$
$$\text{Hence,} \quad Eum = (Euv+Euh)/2 = 1008 \ t/m^2$$

In lower layers,

$$Elv = 2462 \ t/m^2$$
$$Elh = 1084 \ t/m^2$$
$$\text{Hence,} \quad Elm = (Elv+Elh)/2 = 1773 \ t/m^2$$

Finite element analyses were conducted on four cases in which different combinations of Young moduli shown above were assigned to the three domains in each of upper and lower layers. These combinations are shown in Table 1.

The case 1 is the most reasonable one in which the combination of modulus is consistent with the directions of major principal stress induced by the strip loading. The case 2 employs only vertical modulus in the analysis, corresponding to a situation in which soil modulus data is derived from triaxial compression tests on vertically-collected undisturbed samples. The study on case 3 was performed for a comparison with case 1 by using opposite combination of modulus. Although unrealistic apparently, vertical modulus is assigned to horizontally loaded elements and vice versa. This case is expected to indicate the effects of unsuitably determined modulus on calculated deformations of the ground. The case 4 has only horizontal modulus everywhere in the ground, representing a situation in which modulus is derived from horizontal loading of pressure as is carried out in pressuremeter testing.

8. DISCUSSION ON CALCULATED DEFORMATION OF THE GROUND

Fig.19 indicates the calculated settlement at the centre of the strip loading. As compared with the most reasonable case 1, the case 2 gives slightly less settlement, because the elements in domains B and C of case 2 are made stiffer with vertical modulus, preventing lateral flow of foundation elements more effectively than in the case 1. In contrast to more or less similar results from cases 1 and 2, however, cases 3 and 4 work out much greater settlement of the ground. This is because horizontal modulus are allocated to the domain A beneath the surcharge, making the foundation unreasonably soft. Therefore, it may be said that the determination of modulus values in the domain A is very important for a reasonable analysis, while modulus in domains B and C are less significant as suggested by the facts that similar results have been obtained from cases 1 and 2 as well as cases 3 and 4.

Fig.20 illustrates the lateral displacement calculated below the edge of the surcharge. The surcharge assumed here is 30 t/m^2. Similarly to Fig.19, cases 3 and 4 give greater displacements than cases 1 and 2. At the surface, soil moves toward the centre of the surcharge (negative value of displacements), because elements below the strip loading are dragging the surrounding elements when they move downwards. What is more important is that the outward movement below 1 m deep is greater for cases 3 and 4 than for cases 1 and 2. This is due to the fact that the smaller Young modulus of elements beneath the strip loading gives greater vertical strains in cases 3 and 4, resulting in turn in greater lateral expansion. This is true, whether the domain C in the upper layer has large modulus, Euv (case 3) or small modulus, Euh (case 4).

9. CONCLUSIONS

A series of triaxial drained tests were conducted on laboratory-reconstituted as well as undisturbed samples of the first-layer Bangkok sand. The conclusions drawn from them are as what follows.

1) Young modulus at shear stress of 50 % of failure was studied. Due to inherent and induced anisotropies in sand, the horizontal Young modulus is less than the vertical modulus.

2) Although the vertical Young modulus is influenced by sample preparation method very much, the horizontal modulus is reasonably independent of the preparation method.

3) The Poisson ratio takes values between 0.3 and 0.4.

4) The friction angle in a vertical loading is 2 to 4 degrees greater than in a horizontal loading.

5) A linearly elastic finite element analysis which considers the effect of anisotropy is possible, by allocating suitable modulus to elements, according to the orientations of major principal stress. This type of analysis is much easier than nonlinear analysis which employs anisotropic stress-strain model.

6) An example analysis conducted on a strip loading indicates that appropriate values of Young modulus are very necessary for those elements lying beneath the load, while modulus values for elements located aside are not very influential.

10. REFERENCES

Arthur,J.R.F., Chua,K.S. and Dunstan,T. 1977. Induced Anisotropy in a Sand, Geotechnique. 27. 1: 13-30.

Arthur,J.R.F. and Phillips,A.B. 1975. Homogeneous and Layered Sand in Triaxial Compression, Geotechnique. 25. 4: 799-815.

Mori,K. and Ishihara,K. 1979. Undisturbed Block Sampling of Niigata Sand, Proc. 6th ARCSMFE, 1: 39-42, Singapore.

Ng,F.H.A. 1986. personal communication.

Oda,M. 1972a. Initial Fabrics and Their Relations to Mechanical Properties of Granular Material, Soils and Foundations 12. 1: 17-36.

Oda,M. 1972b. The Mechanism of Fabric Changes during Compressional Deformation of Sand, Soils and Foundations 12. 2: 1-18.

Poulos,H.G. and Davis,E.H. 1974. Elastic Solutions for Soil and Rock Mechanics. p.36: Wiley.

Ricceri,G. and Soranzo,M. 1981. Anisotropic Behaviour of a Saturated Uniform Sand, Proc.10th ICSMFE. 1: 759-764, Stockholm.

Saxena,S.K. and Lastrico,R.M. 1978. Static Properties of Lightly Cemented Sand, Proc. ASCE. 104. GT12: 1449-1464.

Yamada,Y. and Ishihara,K. 1979. Anisotropic Deformation Characteristics of Sand under Three-Dimensional Stress Conditions, Soils and Foundations. 19. 2: 79-94.

Yoshimi,Y., Hatanaka,M. and Oh-oka,H. 1978. Undisturbed Sampling of Saturated Sands by Freezing, Soils and Foundations. 18. 3: 59-73.

Table 1. Young modulus assigned to elements in 4 case studies.

Layer	Upper layer			Lower layer			Remarks
Domain	A	B	C	A	B	C	
Case 1	Euv	Eum	Euh	Elv	Elm	Elh	Most reasonable combination
Case 2	Euv	Euv	Euv	Elv	Elv	Elv	Vertical modulus only
Case 3	Euh	Eum	Euv	Elh	Elm	Elv	Unrealistic combination
Case 4	Euh	Euh	Euh	Elh	Elh	Elh	Horizontal modulus only

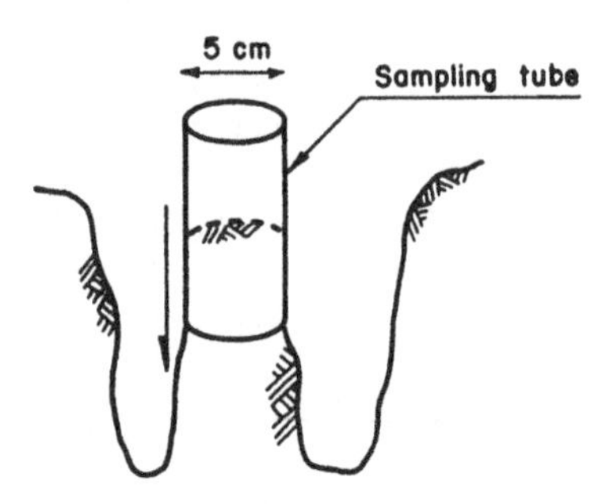

Fig.1 Block sampling being conducted in the vertical direction

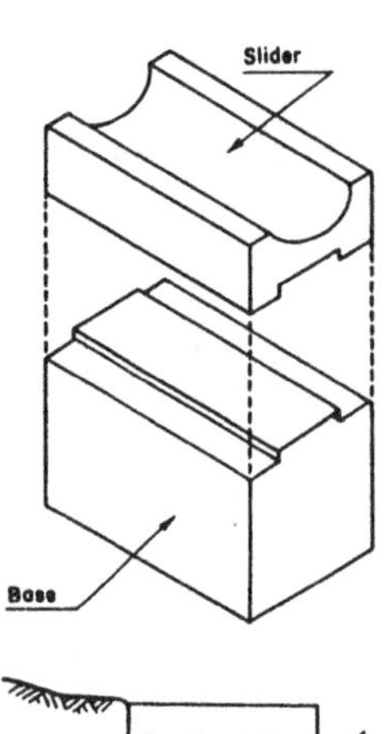

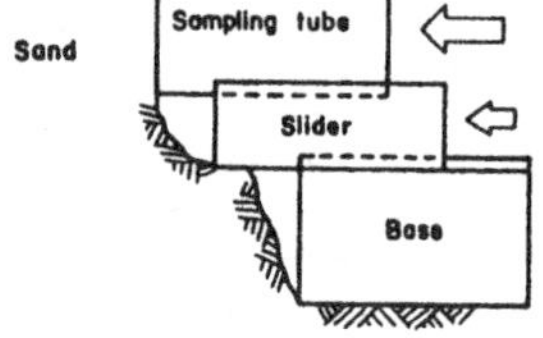

Fig.2 Wooden rest for horizontal sampling

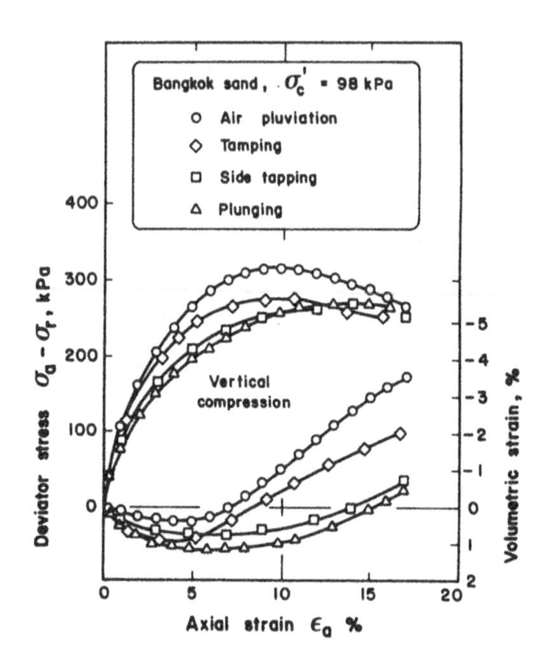

Fig.3 Deformation of reconstituted vertical samples

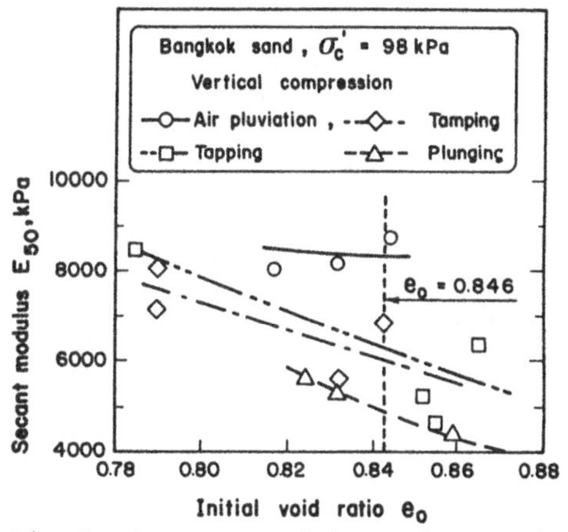

Fig.4 Secant modulus of vertical samples

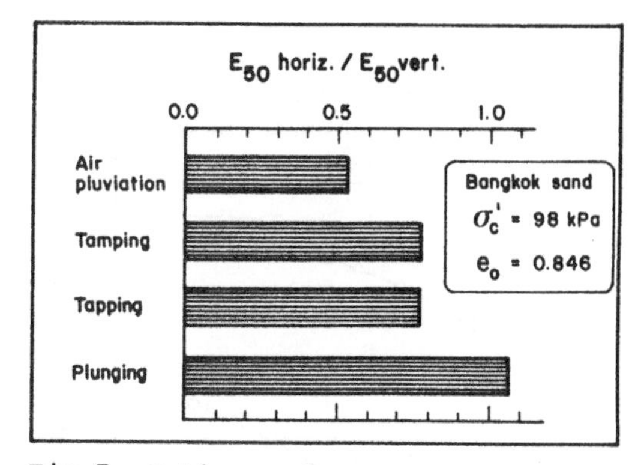

Fig.7 Ratio of secant moduli measured in vertical and horizontal directions

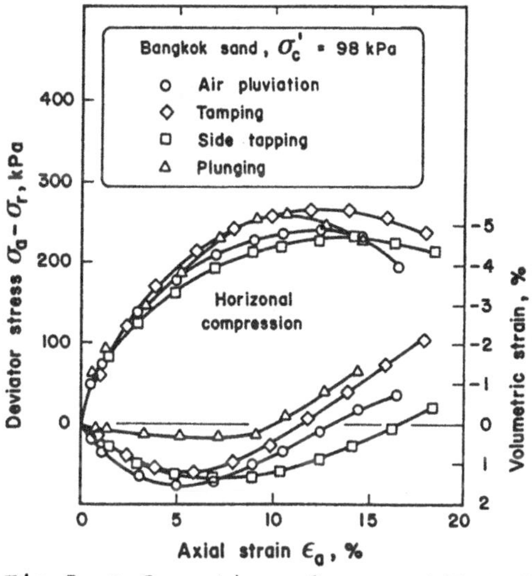

Fig.5 Deformation of reconstituted horizontal samples

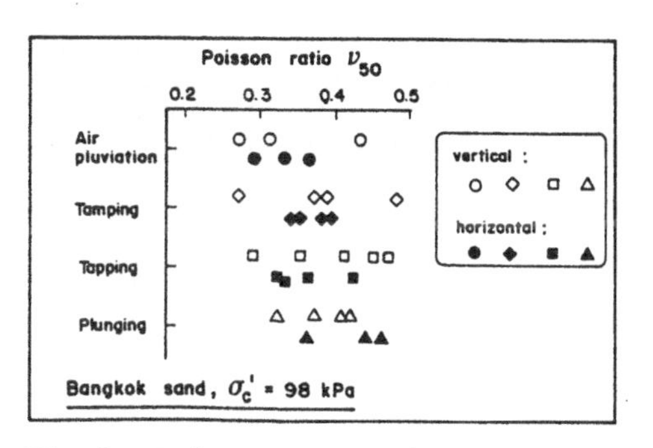

Fig.8 Poisson ratio at 50% strength

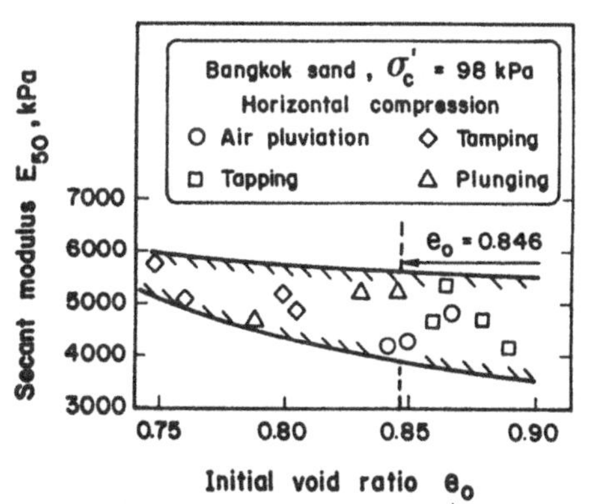

Fig.6 Secant modulus of horizontal samples

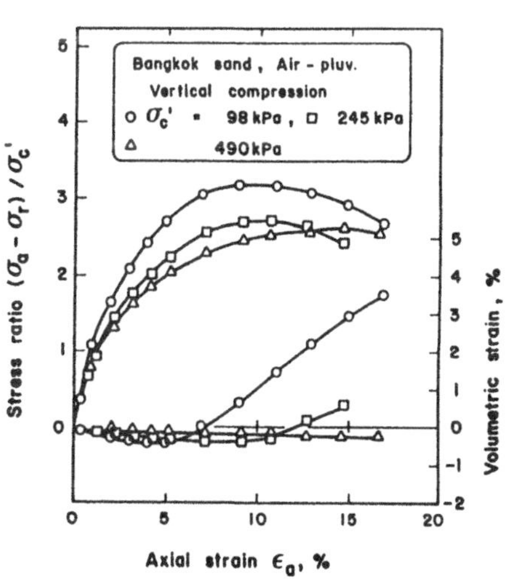

Fig.9 Behaviour of air-pluviated samples under different confining pressures in the course of vertical compression

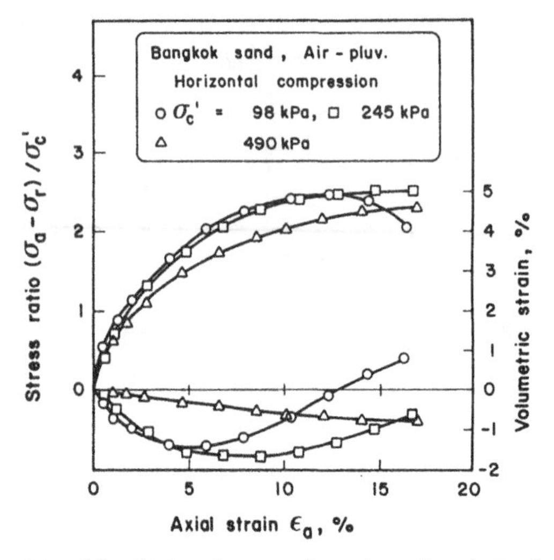

Fig.1Ø Behaviour of air-pluviated samples under different pressures in the course of horizontal compression

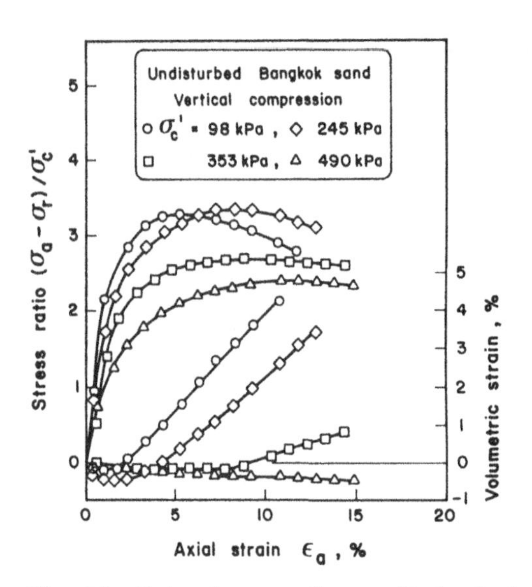

Fig.11 Behaviour of undisturbed vertical samples

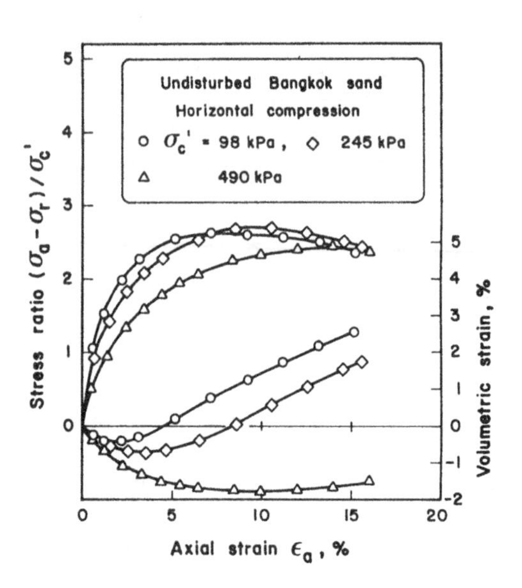

Fig.12 Behaviour of undisturbed horizontal samples

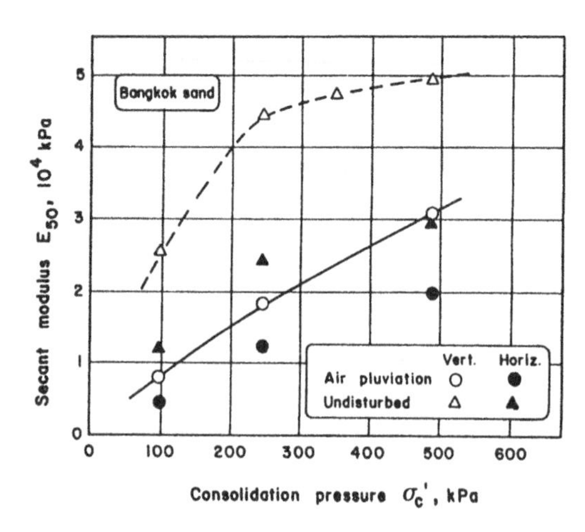

Fig.13 Effects of confining pressure on secant modulus

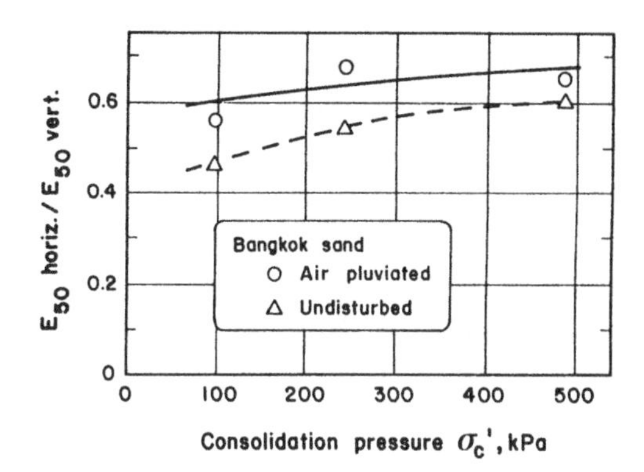

Fig.14 Effects of confining pressure on ratio of modulus between vertical and horizontal directions

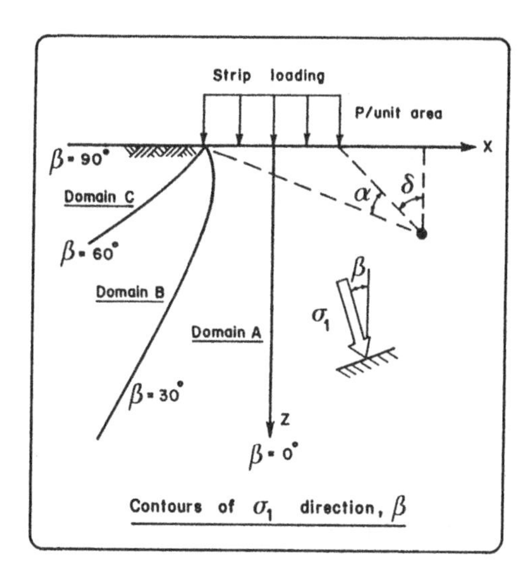

Fig.16 Direction of loading of major principal stress under infinite strip foundation

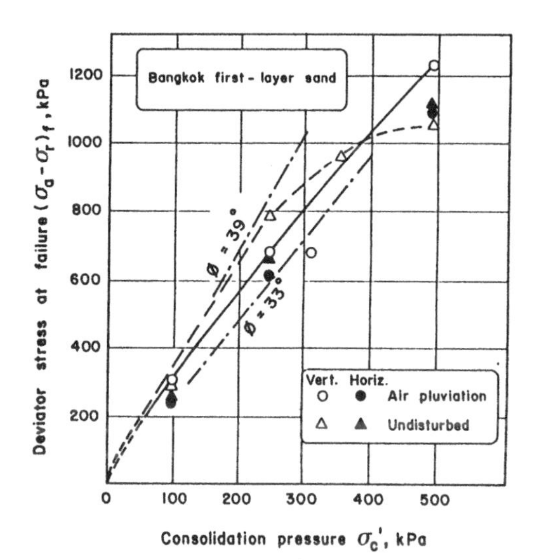

Fig.15 Effects of confining pressure on shear strength of samples

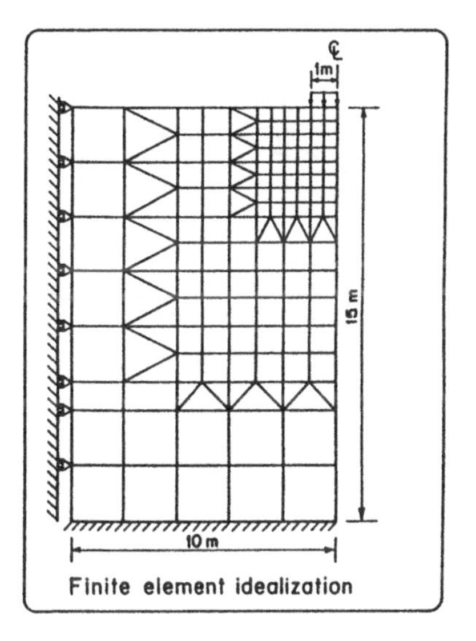

Fig.17 Finite element idealization of the ground

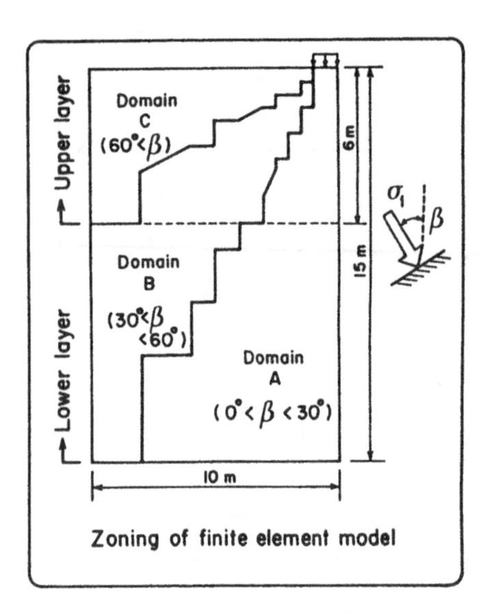

Fig.18 Zoning of the ground in accordance with the depth and direction of loading

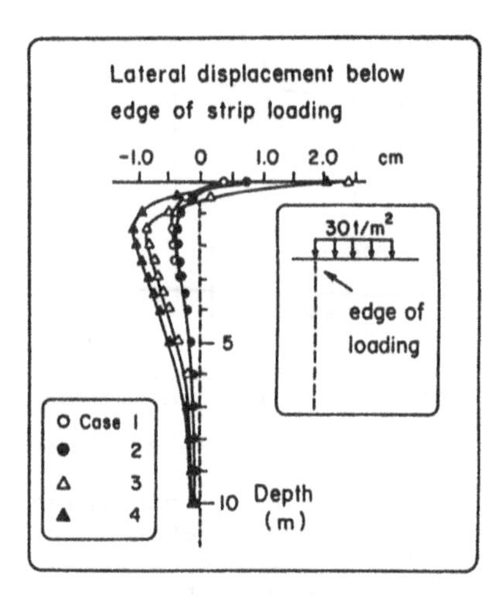

Fig.20 Calculated lateral movements beneath the edge of loading

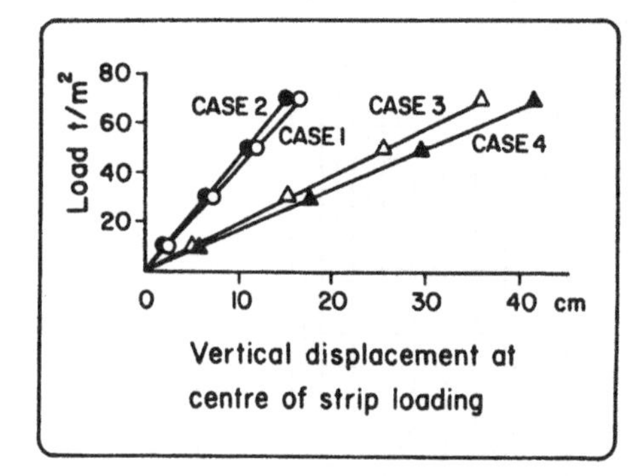

Fig.19 Relationship between surcharge and settlement at the centre of the load

5. Data acquisition and management in geotechnical engineering

Computer and Physical Modelling in Geotechnical Engineering, Balasubramaniam et al. (eds)
© 1989 Balkema, Rotterdam. ISBN 90 6191 864 2

Determination of significant material parameters for quality control of new techniques in geotechnical engineering using statistical methods

G.Maschwitz & G.Cunze
Institute of Soil Mechanics and Foundation Engineering, University Hannover, FR Germany

ABSTRACT: Applying new developed foundation techniques to real foundation problems one has to be sure that no damages will occur due to too high expectations. On the other hand an uneconomical overdimensioning caused by exaggerated safety philosophies should also be avoided. In such cases it is advisable to call in an independent expert, who will elaborate objective values of quality control, reliability and safety. For this purpose computer aided statistical methods should be used. How to succeed in this way this paper deals with a recently developed jet grouting method called Soilcrete as an example.

1 DEVELOPMENT OF A NEW FOUNDATION METHOD

The most important impulses for the development of the Soilcrete technique came from the research field of "Jet-Cutting". Applying the physical principles of these researches cutting and cleaning techniques were developed using high pressure jets of fuids. Therefore it also should be possible to cut and mix soils with jets consisting of e.g. a water-cementsuspension. The hydraulic feasibility of the suspension will then lead to a new hardening soil-cementmixture with a highly improved bearing capacity.

Starting from this more or less theoretical idea the technical realization follows in a second step using the experiences from other techniques, e.g. deep boring, high-pressure injections and grouting.

After a period of advanced in situ tests the new method of Soilcrete-columns then can be applied to a real construction problem. Before doing so, it is necessary to develop objective rules for controlling each single step of the works and of the complete product itself.

2 REALIZATION

2.1 Supporting or underpinning foundation blocks by soilcrete columns

Fig. 1 shows the arrangement of Soilcrete-columns which are to support a foundation block of about 3x3 m. The representative boring log shows, that the total length of each column will be up to 12 m with 2 m resting in bearing soil. The diameter of each column will be about 0,80 m.

2.2 Requirements to the bearing capacity and to the material strength

According to the chosen arrangement of the columns an allowable load of about 0,6 MN is required. This means, that the ultimate bearing capacity per column must exceed 1,2 MN at least.

Together with a calculated diameter of the columns of about 0,80 m this value requires an ultimate material strength of the Soilcrete-mixtures as follows:

$$\text{erf } q = \beta_R = \text{rd. } 1{,}2 \text{ MN/m}^2$$

2.3 Load-test results

The bearing capacity of the single columns was proved by load-tests whose results are shown in Fig. 2.

These load-tests were evaluated according to German standards. They show ultimate bearing capacities of about 1,4 MN till 1,6 MN. According to the planned allowable loads of 0,6 MN, this leads to sufficient factors of safety of more than 2 accom-

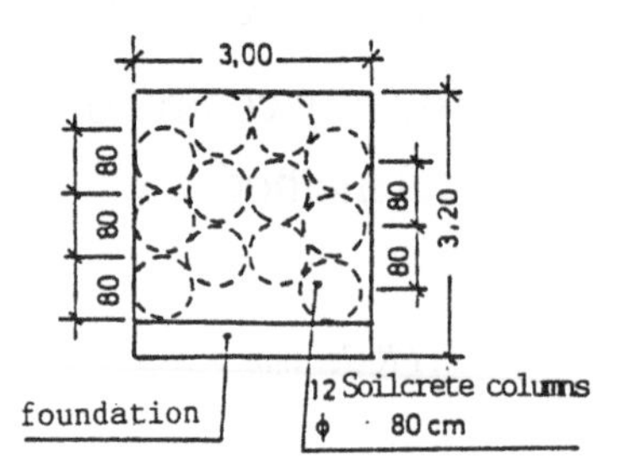

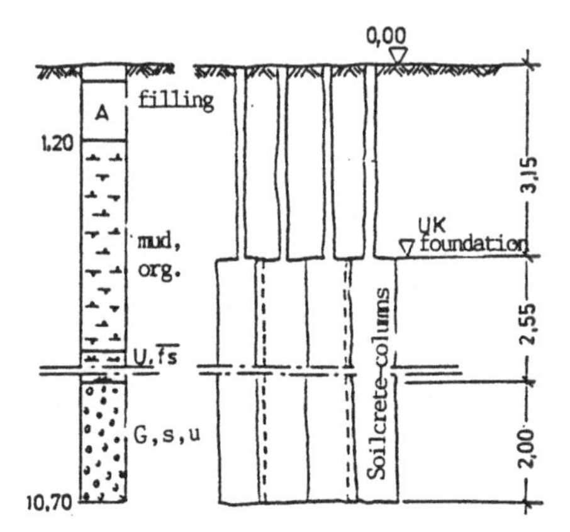

Fig. 1: Arrangement of the Soilcrete-columns supporting a foundation block

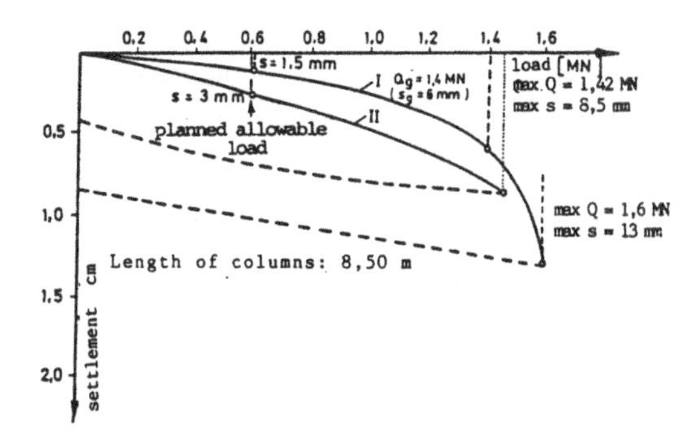

Fig. 2: Load-settlement-curves of two load-tests on Soilcrete-columns

panied by very small settlements (Fig. 2).

As a second result it can be assumed, that each other Soilcrete-column will perform respectively if they are carried out under the same conditions and show comparable properties such as suspension density, suspension volume, soil properties, column length, depth in bearing soil, column diameter etc.

3 MATERIAL STRENGTH

3.1 Testing of core samples and artificially produced soilcrete-cylinders

To test the material strength a few core samples were taken out of the hardened Soilcrete-columns directly. After that the samples were tested according to German standards for testing concrete (DIN 1048).

Additionally other parameters such as density, water content, void ratio etc. were determined.

As it is rather expensive to gain a sufficient number of core samples a lot of low-cost cylindrical samples were produced by filling the freshly mixed soil-suspension-material into suitable cylinders. After curing under similar conditions as in situ the 1-axial strength of these samples also were tested according the above mentioned standards. Thus a great number of test results also including values density, water content etc. were available.

3.2 Evaluation of the tests

The relatively wide scattering of the test results of core and "fresh" samples indicated a heterogene structure of the Soilcrete-material. This statement is a subjective impression even when the arithmetic mean value and the maximum and minimum values are considered.

In the concrete technology statistical and probabilistic methods are employed to get more general statements which additionally are resecured by means of probabilistic methods. Transfering these methods to the test results of the Soilcrete-samples it should be possible to get reliable factors of scale for the material strength of the new Soilcrete-material.

Assuming, that the results originate from one basic totality, the number of tests n, the arithmetic mean values x and the standard deviation s have to be determined. The indicating value of material strength then results from the following equation (1):

$$\beta_N = \beta_S \cdot t_{\alpha,n-1} \cdot s \qquad (1)$$

with:
- n = number of tests
- $\bar{x}$ = arithmetic mean value
- β_S = mean strength of the test series
- β_N = indicating value of material strength
- s = standard deviation
- $t_{\alpha,n-1}$ = t-value acc. to the t-distribution

3.3 Evaluation results

From the single results of the core samples an indicating value of material strength can be calculated according to equ. (1) with a probability of 95% as follows:

$$\beta_{N,1} = 4{,}65 - 1{,}73 \cdot 1{,}67 = 1{,}76 \text{ MN/m}^2$$

With the same probability the following indicating value of material strength of the "fresh" cylindrical Soilcrete-samples is calculated:

$$\beta_{N,2} = 4{,}44 - 1{,}75 \cdot 1{,}35 = 2{,}07 \text{ MN/m}^2$$

These results mean that with a probability of 95% only 5% of all values will be less than $\beta_{N,1}$ or $\beta_{N,2}$ respectively.

3.4 Additional tests

If quality control tests were possible using fresh made samples only, the test series could be quite comprehensive without rising costs. Therefore in a first step the difference of the relations of the variances of both random samples from a critical boundary value were examined using the Fisher-test:

$$F = \frac{s_1^2}{s_2^2} < F_{\alpha, n_1 - 1/n_2 - 1} \qquad (2)$$

with: s_1^2, s_2^2 = variances of the random samples 1 and 2

n_1-1, n_2-1 = degree of freedom of the random samples 1 and 2

F_{α}, n_1-1/n_2-1 = F-value from the F-distribution dependent on the degrees of freedom and on the chosen probability of error

According to equ. (2) the following values for the 1-axial strength of the core samples (random sample 1 with n_1 = 21 and s_1 = 1,67) and of the "fresh" samples (random sample 2 with n_2 = 15 and s_2 = 1,35) can be calculated:

$$F = 1{,}53 < 3{,}51 = F_{0{,}02;\ 20/14}$$

This result shows, that with a probability of 98% both random samples seem to belong to a joint basic totality with the same variance. If so, one must be sure that this is also valid for the two means respectively. Thus a joint density function is described by the variance and the mean. Therefore in a second step the scattering of the mean value of the basic totality was calculated according to equ. (3):

$$\mu = \bar{x} \pm t_{\alpha;\, n-1} \cdot \frac{s}{\sqrt{n}} \qquad (3)$$

with: μ = mean of the basic totality

$\bar{x}$ = mean of the random sample

s = standard deviation of the random sample

n = number of tests

$t_{\alpha, n-1}$ = t-value acc. to the t-distribution

According to this equation the mean of the core samples will range within the following interval:

$$3{,}59 \text{ MN/m}^2 \leq \mu \leq 5{,}71 \text{ MN/m}^2$$

Because the mean value of 1-axial strength of the "fresh" samples (random sample 2 with $\bar{x}_2$ = 4,44 MN/m²) also lies within this interval, it can be assumed that both random samples belong to a basic totality with the same mean value μ.

Both of these statistical control tests show that with a probability of 95% both random samples stem from the same basic totality with an according density function.

That means practically that it is sufficient to prove the material strength of the Soilcrete-columns by testing the artificially made "fresh" samples only.

3.5 Further parameters for quality control

For the further reliable quality control additional parameters can be determined by simple and unexpensive tests, as there are e.g. dry and wet density, void ratio and water content. These values can be determined from lump samples, taken directly of dug out columns.

With a probability of 99% equ. (3) shows, that the mean dry density γ_d of the core samples will lie within the following interval:

$$12{,}61\ kN/m^3 \leq \overline{\gamma}_d = \mu \leq 13{,}0\ kN/m^3$$

If the γ_d-values of lump samples also range within this interval, it can be assumed that these columns exhibit comparable material properties. Thus easily additional evident quality controls can be performed.

4 SUMMARY

Recently lots of new foundation techniques are developed. Normally no sufficient generally approved experiences are available for these methods. Therefore in most cases it is difficult to introduce such a new technique for real foundation problems without overdimensioning the foundation itself and without taking too high risk at the same time. This could be easier if there were objective values of quality control, reliability and safety.

This paper deals with the recently developed jet grouting method Soilcrete as an example. It shows how simple statistical methods can be employed to elaborate such objective values e.g. for the strength of the Soilcrete-material itself. Base of the evaluation are series of core samples, from boring holes, samples of "fresh" material which is filled in cylinders, cured and tested and mere lumps of material taken out of the Soilcrete columns. It is shown that it is sufficient to use the latter low-cost samples for quality and safety control.

There are more statistical methods which can be used for solving similar problems, e.g.:

- Analysis of variance
- Regression Analysis

These methods also can objectively be secured by statistical tests such as confidence intervals, errors of estimation and probabilities. Respective examples are given in the following list of publications.

REFERENCES

BEUTEL, P. 1980. SPSS 8, Eine Beschreibung der Programmversion 6, 7 und 8. G. Fischer Verlag, Stuttgart.

CLAUS, G., EBNER, H. 1974. Grundlagen der Statistik. Verlag Volk und Wissen, Berlin DDR.

GKN KELLER 1984. Soilcrete, Jet Grouting. Firmenprospekt und Referenzliste.

KEN, H.W. et al. 1977. Handbuch der Betonprüfung. Beton Verlag, Düsseldorf.

KRIZ, J. 1973. Statistik in den Sozialwissenschaften. Rowohlt-Verlag, Hamburg.

MASCHWITZ, G. 1983. Ein Beitrag zur Abschätzung des Tragverhaltens von unbewehrten pfahlartigen Tragelementen. Mitt. d. Inst. f. Grundbau, Bodenmechanik und Energiewasserbau (IGBE), Universität Hannover, H. 19.

NIE, H.H., Mull, C.H., JENKINS, J.G. 1970. Statistical Package for the Social Sciences. McGraw-Hill, New York, 2nd Ed.

RIZKALLAH, V., MASCHWITZ, G. 1979. Estimation of the Bearing Capacity of Large Bored Piles in Cohesive Soils Using Statistical Methods. Proc. 3rd ICASP, Sydney.

RIZKALLAH, V. 1981. Underpinning and Foundation near old Buildings. 10th ICSMFE, Stockholm, Vol. 3.

RIZKALLAH, V., MASCHWITZ, G. 1983. Entwicklungstendenzen bei der Herstellung von unbewehrten, pfahlartigen Gründungselementen. VII. Donau-Europäische Konf. ü. Bodenmechanik u. Grundbau, Kishinov, UdSSR.

RIZKALLAH, V. 1986. Unterfangungsmethoden, Einführung und Entwicklungen. Seminar "Unterfangung von Bauwerken", Haus der Technik, Essen.

SAMOL, H. 1986. Soilcrete-Verfahren. Seminar "Unterfangung von Bauwerken", Haus der Technik, Essen.

Computer and Physical Modelling in Geotechnical Engineering, Balasubramaniam et al. (eds)
© 1989 Balkema, Rotterdam. ISBN 90 6191 864 2

A note on the use of rigid-plastic analysis

H.B.Poorooshasb
Department of Civil Engineering, Concordia University, Montreal, Canada

T.Adachi
Department of Transportation Engineering, Kyoto University, Kyoto, Japan

N.Moroto
Department of Civil Engineering, Hachinohe Institute of Technology, Hachinohe, Japan

ABSTRACT: Using rigid plasticity the lack of existence of solutions for certain type of boundary value problems of the first kind demonstrated with the aid of an example: the expansion of a spherical cavity.

1 INTRODUCTION

The use of rigid plasticity in geotechnical analysis (i.e. an analytical procedure whereby the soil is assumed to behave as a rigid plastic material) is attractive in view of its relative simplicity and convenience. It has been used very extensively in solving eigenvalue problems but its application to the analysis of equilibrium or propagation problems has been rather limited. Where it has been applied the numerical evaluations have been found to be very simple and capable of being handled by a micro computer of limited memory capacity; see, for example, the papers by Poorooshasb and Yong (1983) and Poorooshasb et al (1985) (1985). Both papers were prepared with the aid of an Apple IIC computer with a total memory capacity of 48 kb.

The method is not without its own drawbacks however. For example, most of the current finite element programs use the nodal displacements as the unknown parameters of the problem. These programs can not be employed using a rigid plastic model for the soil behaviour: the matrix of the material properties is singular and it's inverse can not be evaluated. As a second example, and by the same token, mixed boundary value problems can not be approached. Another limitation is pointed out in the present paper: it is shown that a solution may not exist at all in the case of the boundary value problems of the first kind. The point is demonstrated by the aid of a simple example viz; the expansion of a spherical cavity in an infinite mass of sand subjected to an initial hydrostatic pressure P_0.

Although the example cited here is for the purpose of demonstration only, nevertheless, expansion of cavity solutions have contributed to the understanding and formulation of many geotechnical problems including the bearing capacity of deep foundations [Vesic (1975), Baligh (1976)], determination of soil properties [Gibson and Anderson (1961), Wroth and Hughes (1973)], cratering by explosives [Vesic (1965)], and the breakout resistance of anchors [Vesic (1971)]. It may be regarded as one of the fundamental problems in geotechnical analysis.

In the present paper the problem is formulated using the constitutive relation proposed by Poorooshasb et al (1966,67) and later modified by Poorooshasb (1971). It is subsequently shown that the formulation contains an initial stage in which a solution cannot be obtained, i.e a solution within the solution region does not exist!

2 FORMULATION OF THE PROBLEM

An infinite homogeneous and isotropic mass of cohesionless granular medium subjected to an initial hydrostatic pressure of P_0 is assumed. The origin contains a spherical cavity of radius r_c under an initial internal pressure of P_0. Since body forces are ignored in this study, then the whole system is in equilibrium and the solution region is r, the position vector of a typical element, such that $r_c \leq r \leq \infty$.

At a certain stage the cavity pressure is increased from its original value of P_0 to a higher value P, say, and it is plausible to assume that σ, the stress tensor associated with a typical element also varies such that σ_{rr}, $\sigma_{\theta\theta}$ and $\sigma_{\phi\phi}$ ($= \sigma_{\theta\theta}$) are the principal stresses and a function of r only. Thus the only equation of equilibrium to be considered is;

$$\frac{d\sigma_r}{d_r} = \frac{2(\sigma_\theta - \sigma_r)}{r} \tag{1}$$

where for convenience, σ_{rr} and $\sigma_{\theta\theta}$ have been replaced by σ_r and σ_θ, respectively. Kinematics of the problem requires that;

$$\frac{d\varepsilon_\theta}{d_r} = \frac{(\varepsilon_r - \varepsilon_\theta)}{r} \quad (2)$$

where $\varepsilon_r = -\partial u_r/\partial r$, $\varepsilon = -u_r/r$, the quantity u_r representing the radical component of the displacement vector and the minus sign used being in concordance with the current soil mechanics convention. Note that in this paper small strain theory is employed.

In the constitutive model proposed by Poorooshasb et al (1966, 67) and Poorooshasb (1971), one family of yield functions f are used in conjunction with a family of non-associated plastic potentials ψ in the forms;

$$f = \eta + m\, ln\, (1) \quad (3)$$

$$\psi = I \quad \bar{\psi}\, (\eta) \quad (4)$$

$$\eta = \frac{J^{\frac{1}{2}}}{I} \quad (5)$$

where $I = \sigma_{ii}/3$ and $J = \sigma'_{ij}\, \sigma'_{ij}$ are the first invariant of stress and the second invariant of stress deviation tensor. The stress deviation tensor is defined by the relation $\sigma'_{ij} = \sigma'_{ij} - I\delta_{ij}$, δ_{ij} being the Kronekers delta.

In terms of stress parameters σ_r, σ_θ, and σ_ϕ (= σ_θ), these invariants are $I = (\sigma_r + 2\sigma_\theta)/3$ and $J = 2/3\ (\sigma_r - \sigma_\theta)^2$ and equations (3) to (5) may be replaced by more suitable relations in the forms;

$$f = \eta + m\, ln\, \sigma_\theta \quad (3.a)$$

$$\psi = \sigma_\theta \quad \bar{\psi}\, (\eta) \quad (4.a)$$

$$\eta = \frac{\sigma_r}{\sigma_\theta} \quad (5.a)$$

It is remarked that m is a positive scalar constant which defines the curvature of the yield loci. It's value is a fraction of unity (≅ .4 το .6) [Poorooshasb (1971), Tatsuoka and Ishihara (1974)] and indeed in the original papers by Poorooshasb et al (1966, 67), it was assumed to have the value of zero. Fig. 1 shows the family of yield functions (for a value of m = 0.4) and the family of plastic potentials. The solid line marked k_0 is of special interest in soil mechanics and in the present study as it represents the so called at rest earth pressure condition.

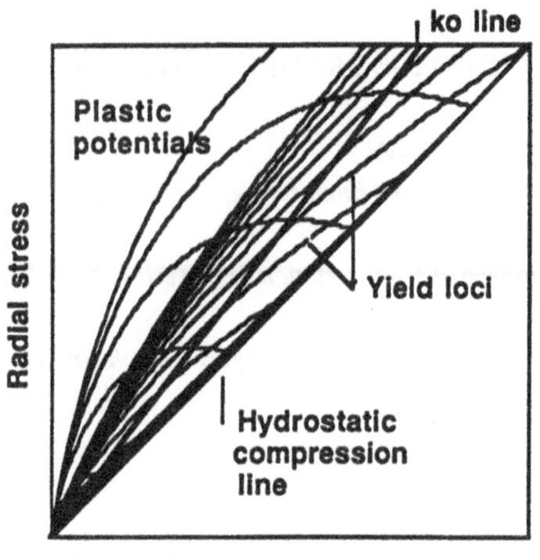

Fig. (1) Family of yield loci and plastic potentials (m = 0.4)

Note that along the k_0 line $\partial\psi/\partial\sigma_\theta = 0$. Assuming rigid plasticity the two strain increments $d\varepsilon_r$ and $d\varepsilon_\theta$ are obtained from the equations;

$$d\varepsilon_r = h\,.\,\frac{\partial\psi}{\partial\sigma_r}\left[(\frac{\partial f}{\partial\eta})\, d\eta + (\frac{\partial f}{\partial\sigma_\theta})\, d\sigma_\theta \right] \quad (6)$$

and

$$d\varepsilon_\theta = h\,.\,\frac{\partial\psi}{\partial\sigma_\theta}\left[(\frac{\partial f}{\partial\eta})\, d\eta + (\frac{\partial f}{\partial\sigma_\theta})\, d\sigma_\theta \right] \quad (7)$$

where h, the plastic modulus, $\partial\psi/\partial\sigma_r$ and $\partial\psi/\partial\sigma_\theta$ are functions of η only. The last two quantitues are given by the equations;

$$\frac{\partial\psi}{\partial\sigma_\theta} = \bar{\psi}\,(\eta) - \sigma_\theta\ \bar{\psi}'\,(\eta)(\frac{\partial\eta}{\partial\sigma_\theta}) = \psi\,(\eta) - \eta\ \psi'\,(\eta)$$

and

$$\frac{\partial\psi}{\partial\sigma_r} = \sigma_\theta\ \bar{\psi}'\,(\eta)(\frac{\partial\eta}{\partial\sigma_r}) = \bar{\psi}'(\eta)$$

Reverting to equation (1) and noting that, from equation (5,a), $\sigma_r = \eta\sigma_\theta$, results in;

$$\frac{\left(\frac{d\sigma_\theta}{dr}\right)}{\sigma_\theta} = 2\,(1-\eta)\,\frac{\left(\frac{d\eta}{dr}\right)}{r\eta}$$

With the aid of this equation and eq. (3.a), equations (6) and (7) may be rewritten as;

$$\frac{d\varepsilon_r}{dr} = H\,(\eta)\,R\,(r,\eta)\,\frac{\partial\psi}{\partial\sigma_r} \qquad (6.a)$$

$$\frac{d\varepsilon_\theta}{dr} = H\,(\eta)\,R\,(r,\eta)\,\frac{\partial\psi}{\partial\sigma_\theta} \qquad (6.b)$$

where for convenience $h(\eta)/\eta$ has been replaced by $H(\eta)$ and the auxiliary function $R(r, \eta)$ is equal to $[(\eta - m)\, d\eta/dr + 2m(1 - \eta)/r]$. Differentiating eq. (2) with respect to r leads to;

$$\frac{r d^2\varepsilon_\theta}{dr^2} = \frac{d\varepsilon_r}{dr} - 2\left(\frac{d\varepsilon_\theta}{dr}\right)$$

which in combination with equations (6.a) and (6.b) leads to;

$$r^2\frac{dx_1}{dr} = -\,2mH(\eta-1)\left(\frac{\partial\psi}{\partial\sigma_r} - \frac{\partial\psi}{\partial\sigma_\theta}\right) + r\left[2m\,(\eta-1)\,\frac{dx_2}{d\eta} + 2mH\frac{\partial\psi}{\partial\sigma_\theta} + (\eta - m)\,H\left(\frac{\partial\psi}{\partial\sigma_r} - \frac{\partial\psi}{\partial\sigma_\theta}\right)\right]\frac{d\eta}{dr}$$

where

$$x_1 = H(\eta - m)\left(\frac{\partial\psi}{\partial\sigma_\theta}\right)\frac{d\eta}{dr} \qquad \text{and} \qquad x_2 = H\frac{\partial\psi}{\partial\sigma_\theta}$$

Let,

$$\phi_1\,(\eta) = H(\eta - m)\,\frac{\partial\psi}{\partial\sigma_\theta} \qquad (8.a)$$

$$\phi_2\,(\eta) = 2m\,(\eta-1)\,\frac{dx_2}{d\eta} + 2m\,H\,\frac{\partial\psi}{\partial\sigma_\theta} + (\eta - m)\,H\left(\frac{\partial\psi}{\partial\sigma_r} - 2\,\frac{\partial\psi}{\partial\sigma_\theta}\right) \qquad (8.b)$$

$$\phi_3\,(\eta) = 2m\,H\,(\eta - 1)\left(\frac{\partial\psi}{\partial\sigma_r} - \frac{\partial\psi}{\partial\sigma_\theta}\right) \qquad (8.c)$$

Then the governing equation for the problem may be written as;

$$x_3 = \phi_1\,(\eta)\,\frac{d\eta}{dr}$$

$$r^2\frac{dx_3}{dr} = r\phi_2\,(\eta)\,\frac{d\eta}{dr} - \phi_3\,(\eta) \qquad (9)$$

Before proceeding further it is appropriate to make an observation regarding the function of $\phi_1(\eta)$ and $\phi_3(\eta)$. Since, as stated before, m is a fraction of unity and η is larger than unity, and since quantitues H, $\partial\psi/\partial\sigma_\theta$ and $(\partial\psi/\partial\sigma_r\ \partial\psi/\partial\sigma_\theta)$ are either positive definite or positive semi definite in the region bounded by the hydrostatic compression line for which $\sigma_r = \sigma_\theta$ and the k_0 line for which $\partial\psi/\partial\sigma_\theta = 0$, then it is concluded that both ϕ_1 and ϕ_3 are positive semi definite. This observation will be used in the arguments presented later.

Reverting to eq. (9) and noting that the problem is self similar, it is admissible to represent the product $r(d\eta/dr)$ by a unique function of η, say $F(\eta)$, i.e.;

$$\frac{d\eta}{dr} = \frac{F\,(\eta)}{r} \qquad (10)$$

Therefore eq. (9) reduces to;

$$r^2\left(F\,.\,\frac{\phi_1}{r}\right)_{,\,r} = \phi_2\,.\,F - \phi_3 \qquad (11)$$

where the subscript comma indicates differentiation with respect to r. Now, since ϕ_1, and F are functions of η only let the product $\phi_1 F$ be denoted by $\zeta(\eta)$. Then eq. (11) reduces to;

$$r\,\zeta'\,\frac{d\eta}{dr} - \zeta = \frac{\phi_2\,\zeta}{\phi_3} - \phi_3 \qquad (12)$$

which by the relation $r(d\eta/dr) = F = \zeta/\phi_1$ reduces to;

$$\zeta(\eta)\left[\zeta'(\eta) - (\phi_1 + \phi_2)\right] = -\phi_1\phi_3 \quad (13)$$

to be solved subject to the condition;

$$\zeta(1) = 0 \quad (14)$$

Once the function $\zeta(\eta)$ is determined the function $F(\eta) = \zeta(\eta)/\phi_1$ can be evaluated. Then the distribution of η within the solution region r is obtained from eq. (10);

$$r = r_c \exp\left(-\int_{\eta}^{\eta_c} \frac{d\eta}{F(\eta)}\right) \quad (15)$$

Where η_c is the value of σ_r/σ_θ at $r = r_c$, the radius corresponding to the boundary of the cavity.

3 DISCUSSION

Let N define a range of η values given by;

$$1 \le \eta < k^{-1}$$

where as stated before k_0 is the so called coefficient of at rest earth pressure. Note that the range does not include the $\eta = k_0^{-1}$. Let M define a range of η given by;

$$1 \le \eta < k^{-1}$$

where $k^{-1} < k_0^{-1}$. Obviously, M is a subdomain of N. Introduction of these ranges facilitates the discussions that are to follow.

Equation (13) is a special form of Abel's equation of the second kind with a solution in the form;

$$\zeta(\eta) = u(\eta) + v(\eta) \quad (16)$$

where

$$v(\eta) = \int_1^{\eta}\left[\phi_1(v) + \phi_2(v)\right] dv \quad (17)$$

is the solution to equation $\zeta' - \phi_1 - \phi_2 = 0$. Substituting for ϕ_1 and ϕ_2 from (8.a) and (8.b) and carrying out the required integration results in;

$$v(\eta) = 2m\left[(\eta - 1) H \frac{\partial\psi}{\partial\sigma_\theta}\right] + \int_1^{\eta} (v - m) H(v) \left[\frac{\partial\psi}{\partial\sigma_r} - \frac{\partial\psi}{\partial\sigma_\theta}\right] dv \quad (18)$$

Since for $\eta \in N$ both $\partial\psi/\partial\sigma_\theta$ and $[\partial\psi/\partial\sigma_r - \partial\psi/\partial\sigma_\theta]$ are positive semi definite then $v(\eta)$, eq. (18), is positive semi definite. It is also of class at least C^1 and has the property that;

$$v(1) = v'(1) = 0$$

Substituting for $\zeta(\eta)$ from eq. (16) in eq. (18) leads to;

$$\left[u(\eta) + v(\eta)\right] u'(\eta) = -\phi_3 \quad (19)$$

But ϕ_3 is positive definite for all $\eta > 1$. Therefore from eq. (19) either of the following two conditions must be satisfied;

condition (i): $\left.\begin{array}{l}\text{if } u'(\eta) > 0 \\ \text{then } u(\eta) + v(\eta) < 0\end{array}\right\} \eta \in N$

condition (ii): $\left.\begin{array}{l}\text{if } u'(\eta) < 0 \\ \text{then } u(\eta) + v(\eta) > 0\end{array}\right\} \eta \in N$

Since $v(\eta)$ is positive then $u(\eta)$ must be negative if condition (i) is to be adopted. In this case the following relations must hold;

$$\left.\begin{array}{r} u'(\eta) > 0 \\ u(\eta) < 0 \\ u(1) = u'(1) = 0 \end{array}\right\} \quad (20)$$

No "function" of class C^1 can be found that satisfies the set of relations (20). Condition (i) is, therefore, discarded as incompatible. It now remains to evaluate condition (ii). Analogous to the set of expressions (20) one may write;

$$\left.\begin{array}{r} u'(\eta) < 0 \\ u(\eta) < 0 \\ u(1) = u'(1) = 0 \end{array}\right\} \quad (21)$$

subject to the condition that abs (u) < v. The conditions demanded by eq. (21) are indeed compatible and it is concluded. therefore, that at least in the range M (a subdomain of N, see Fig. 2),

$$\zeta(\eta) > 0 \quad \eta \in M \tag{22}$$

Before closing this section it may be noted that if the coefficient m in the expression for the yield function, eq. (3), is assumed to be zero [the original formulation proposed by Poorooshasb et al (1966, 67)] then $u(\eta) = 0$ and $\zeta(\eta) = v(\eta)$ is positive semi definite in $\eta \in N$ i.e.;

$$\zeta(\eta) > 0 \quad \eta \in N$$

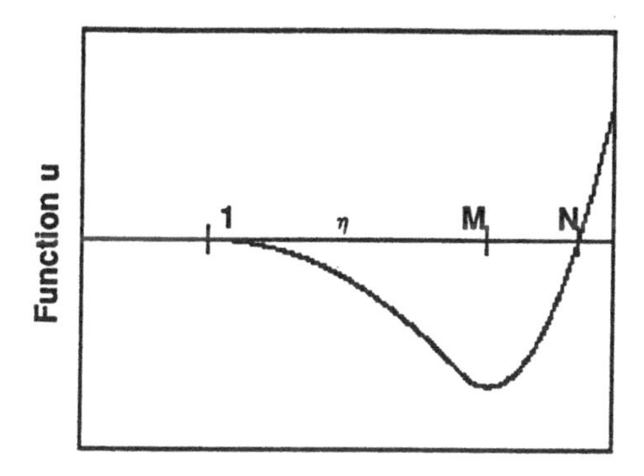

Fig. (2) Variation of u with η

4 CONCLUSION

The function $\phi_1(\eta)$ is positive semi definite in the range $\eta \in m$. Therefore;

$$\int_{\eta}^{\eta_c} \left[\frac{\phi_1(\eta)}{\zeta(\eta)}\right] d\eta > 0 \tag{23}$$

using expression (22). If this last result is used in conjunction with eq. (15), it yields;

$$r < r_c$$

That is, a solution for r within the solution region which is bound by r_c and $r \to \infty$ is unattainable! This result is of value from a computational point of view. For much valuable time can be wasted trying to obtain a solution which does not exist in the first place.

ACKNOWLEDGEMENTS

The financial support from the Natural Science and Engineering Council of Canada and the Japan Society for Promotion of Science, is gratefully acknowledged.

REFERENCES

Baligh, Mohsen, M., "Cavity Expansions in Sands with Curved Envelopes" J. Soil Mech. Found.Div., ASCE, Vol. 102, No. Gtii, Proc. Paper 12536, Nov. 1976.

Gibson, R.E. and Anderson, W.F., "In Situ Measurements of Soil Properties with the Pressuremeter", Civil Engineering and Public Works Review, Vol. 56, 1961.

Poorooshasb, H.B., "Deformation of Sand in Triaxial Compression", Proc. Fourth Asian Regional Conference on Soil Mech. and Found. Eng., Bangkok, Thailand, Volume 1, 1971.

Poorooshasb, H.B., Holubec, I. and Sherbourne, A.N., "Yielding and Flow of Sand in Triaxial Compression", Canadian Geotechnical Journal, Part I, Vol.3, No. 4, 1966, Parts II and III, Vol. 4, No. 4, 1967.

Poorooshasb H.B. and Yong, R.N., "Plastic-Rigid Analysis of Vertical Piles in Normally Consolidated Clays", Soils and Foundations, Japanese Society of Soil Mech. Found. Eng. Vol. 23, No. 2, 1983.

Poorooshasb, H.B., Pietruszczak and Ashtakala, B., "An Extension to Pasternak Foundation Concept", Soils and Foundations, J. Japanese Society for Soil Mech. Found. Eng. Vol. 25, No. 3, 1985.

Tatsuoka, F. and Ishihara, K., "Yielding of Sand in Triaxial Compression", Soils and Foundations, Vol. 14, No. 2, 1974.

Vesic, A.S., "Cratering by Explosives as an Earth Pressure Problem", Sixth International Conference on Soil Mech. Found. Eng., Montreal, Canada, 1965.

Vesic, A.S., "Breakout Resistance of Objects Embedded in Ocean Bottom", J. SoilMech. Found. Div., ASCE, Vol. 97, No. SM9, Proc. Paper 8372, 1971.

Vesic, A.S., "Expansion of Cavities in Infinite Soil Mass", J. Soil Mech. Found. Div., ASCE, Vol. 98, No. SM3, Paper 8790, 1972.

Vesic, A.S., "Principles of Pile Design", Lecture Series on Deep Foundations, Cambridge, Mass., USA, 1975.

Computer and Physical Modelling in Geotechnical Engineering, Balasubramaniam et al. (eds)
© 1989 Balkema, Rotterdam. ISBN 90 6191 864 2

Liquefaction characteristics of tailings

H.B.Poorooshasb
Department of Civil Engineering, Concordia University, Montreal, Canada

K.Ishihara
Department of Engineering, Tokyo University, Tokyo, Japan

ABSTRACT: The development of the excess pore water pressure in the body of a deposit of tailing caused by an earthquake type excitation is the subject of the study presented in this paper. An element of the deposit is assumed to behave as an elasto-plastic material following a modified version of the constitutive model proposed by Poorooshasb and Pietruszczak (1986). The equation governing the mode of propagation of the shearing stresses and the excess pore water pressures is derived in detail. This equation is written in the finite difference code for a special case and subsequently solved with the aid of a *microcomputer* of limited memory capacity. The results of the analysis are shown graphically and clearly demonstrate the effect of the "amplitude" of the excitation force on the nature of the propagated shearing forces, stress paths followed by the various elements and the location of the liquified zone, should such zones develop during the period of ground shaking.

1 INTRODUCTION

Tailing deposits as a rule consist of fine grained cohesionless particles packed at a very loose state. In the event of an earthquake this open structure would tend to collapse leading to the development of high pore water pressure caused by the inability of water to escape the pore space. If the developed excess pore water pressure exceeds the total overburden pressure, then the material liquifies and this may lead to the instability of the whole system with heavy financial and environmental losses and possibly human life losses. In the present paper a method is presented which is applicable to the analysis of such a system. The method is relatively simple to apply and for the case presented in the paper use was made of a microcomputer of useful memory capacity of less than 128 Kbyte. The paper is written in the following sequential sections. First the material properties, as used in here, are outlined. Then the governing equation of the problem is derived. Next, this equation is broken into its finite difference form and certain comments regarding the solution of the resulting equation provided. Finally, a set of graphs showing the distortion of the shape of shear waves as a function of time and position (height above the base at which position excitation is applied) and the stress paths followed by a number of elements at certain locations are presented. These graphs also indicate the stage at which a layer liquifies. It is seen that the depth of liquefication can not be determined a priori being dependent on a number of factors including the form of the excitation imposed on the system,

2 MATERIAL PROPERTIES

Since the height of the deposit, as a rule, is very small compared to its width, then at points sufficiently away from the retaining dams (starting dam and secondary dams constructed to retain the tailing) the problem may be treated as one dimensional. Thus, representing the y axis along the height of the deposit and the x axis parallel to the direction of excitation the three components of the stress tensor which play paramount roles in this study may be represented by σ_x, σ_y and σ_{xy}, and the corresponding strain increments by $d\varepsilon_x$, $d\varepsilon_y$ and $d\varepsilon_{xy}$. The constitutive model employed in this paper assumes a set of relations between the strain increment tensor and the stress increment tensor in the form;

$$d\varepsilon_x = h\left(\frac{\partial\psi}{\partial\sigma_x}\right)\left[\left(\frac{\partial\eta}{\partial\sigma_x}\right)d\sigma_x + \left(\frac{\partial\eta}{\partial\sigma_y}\right)d\sigma_y + 2\left(\frac{\partial\eta}{\partial\sigma_{xy}}\right)d\sigma_{xy}\right] + $$
$$+ C_1 d\sigma_x - C_2 d\sigma_y$$

$$d\varepsilon_y = h\left(\frac{\partial\psi}{\partial\sigma_y}\right)\left[\left(\frac{\partial\eta}{\partial\sigma_x}\right)d\sigma_x + \left(\frac{\partial\eta}{\partial\sigma_y}\right)d\sigma_y + 2\left(\frac{\partial\eta}{\partial\sigma_{xy}}\right)d\sigma_{xy}\right] + $$
$$+ C_1 d\sigma_y - C_2 d\sigma_x$$

$$d\varepsilon_{xy} = h\left(\frac{\partial\psi}{\partial\sigma_x}\right)\left[\left(\frac{\partial\eta}{\partial\sigma_x}\right)d\sigma_x + \left(\frac{\partial\eta}{\partial\sigma_y}\right)d\sigma_y + 2\left(\frac{\partial\eta}{\partial\sigma_{xy}}\right)d\sigma_{xy}\right] +$$

$$+ C_3\, d\sigma_{xy} \qquad (1)$$

where the scalar h is the "plastic modulus", the function ψ is the plastic potential, and the function η is the yield function. The coefficients C_1, C_2 and C_3 are the elastic moduli and their inclusion in the set of expressions (1) is essential: otherwise certain inversion processes involved in the analysis could not be performed. Denoting by σ and τ the first and second invariants of the stress tensor, i.e. $\sigma = (\sigma_x + \sigma_y)/2$ and $\tau = [(\sigma_x - \sigma_y)^2/4 + \sigma_{xy}{}^2]^{1/2}$, the plastic potential and the yield functions can be expressed by equations:

$$\psi = \sigma \,.\, \phi(\eta) \qquad (2)$$

$$\eta = \frac{\tau}{\sigma} \qquad (3)$$

Strictly speaking, it may be noted that the set of equations (1) is, only correct if it is assumed that the intermediate principal stress is equal to the mean of the other two principal stresses, Actually, this condition is obtained quite quickly as the loading proceeds and the savings in computational times are enormous. Furthermore, the difficulties associated with the correct assessment of the various parameters (such as the elastic moduli and the exact form of η and ψ at the lower stress levels) are so many that the errors caused by making this assumption are likely to be of second order of importance and will not, materially, change the arguments. Under these circumstances, the elastic moduli C_1, C_2 and C_3 are equal to $(1 - \nu^2)/E$, $\nu(1 + \nu)/E$ and $(1 + \nu)/E$ respectively, where ν is Poisson's ratio and E the Young's modulus, the latter being assumed to be linear function of the mean stress σ.

Before closing this section, it is appropriate to make certain remarks with respect to the two functions h and ψ. During virgin loading the function h is calculated from the equation;

$$h(\eta) = \frac{b}{(\eta_f - \eta)^2}$$

an expression derived from the so-called hyperbolic law. During stress reversals, however, h is replaced by h^r where h^r is related to a datum stress point σ_d and a conjugate stress point σ_c, both of which are located on the "bounding surface", Dafalias (1979), by the relationship:

$$h^r = h^b\left[\frac{(\delta_d - \eta)}{(\delta_d + \delta_c)}\right]^n$$

where δ_d and δ_c are associated with σ_d and σ_c as shown in Fig. (1). The modulus h^d is the plastic modulus corresponding to the stress point σ^c and the exponent n is of the order of 3 to 4.

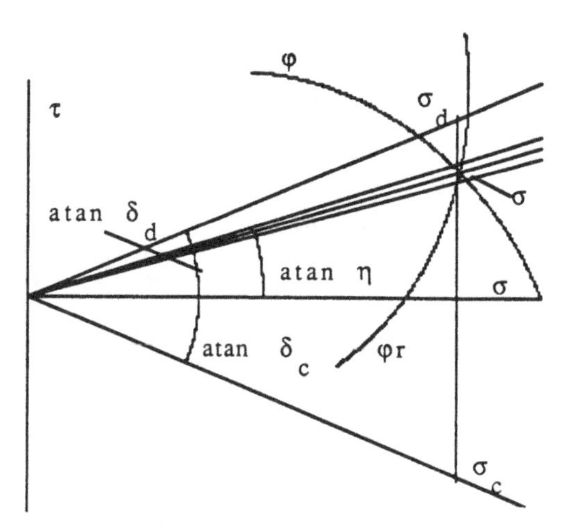

Fig.1 Datum and conjugate stress points plastic potential and reflected curves

In Fig. (1) the curve denoted by the symbol ψ^r and represents the so-called reflective plastic potential. When plotted graphically, as indeed is done so in Fig. (1), it traces a curve which is the mirror image of the plastic potential curve ψ, passing through the current stress point σ. Its gradient at the current stress point is used to obtain the components of the plastic strain increment tensor during the unloading processes. For a general formulation of this constitutive model, the interested reader may consult a paper by Poorooshasb and Pietruszaczk (1986).

Having given a brief account of this constitutive relation, it is now convenient to proceed with the formulation of the problem.

3 FORMULATION OF THE PROBLEM

Using the argument of similarity in conjunction with the compatibility condition, it may be shown that;

$$d\varepsilon_x = 0$$

Thus, from the first set of Eqs,(1) the stress increment components $d\sigma_x$ is evaluated to be;

$$d\sigma_x = \left[\frac{(h\psi,_x\eta,_y - C_2)}{(h\psi,_x\eta,_x + C_1)}\right] d\sigma_y - \left[\frac{2h\psi,_x\eta,_{xy}}{(h\psi,_x\eta,_x + C_1)}\right] d\sigma_{xy}$$

where for convenience $\partial\psi/\partial\sigma_x$, for example, has been replaced by the notoation $\psi,_x$. Next, note that $\partial\sigma_{xy}/\partial_x$must be equal to zero. This follows from the argument of similarity of the elements at a constant elevation y. Thus, if the incremental change of the pore water pressure is denoted by du and considering that the body forces in a vertical direction are effectively constant, then equations of equilibrium demand that $d\sigma_y = -du$. Substituting for $d\sigma_y$ in the last equation and denoting the expressions preceeding $d\sigma_y$ and $d\sigma_{xy}$ by a_1 and a_2 respectively, this equation may be rewriten in the form;

$$d\sigma_x = a_1\, du - a_2\, d\sigma_{xy} \tag{4}$$

In a fairly similar way the third equation of the set (1) may be used to yield;

$$d\varepsilon_{xy} = a_3\, du + a_4 d\sigma_{xy} \tag{5}$$

where

$$a_3 = h\psi,_{xy}(\eta,_x a_1 - \eta,_y)$$

and

$$a_4 = h\psi,_{xy}(2\eta,_{xy} - a_2\eta,_x) + C_3$$

Next, consider the incremental equations of equilibrium in the horizontal direction, i.e. along the x axis;

$$\frac{\partial\sigma_{xy}}{\partial y} = -(\frac{\rho}{g})(\frac{\partial^2 U}{\partial t^2}) \tag{6}$$

where U_x is the component of the displacement along the x axis, r is the total unit mass (liquid plus solid particles) and ρ is the acceleration due to gravity. Note that in Eq. (6) the current geotechnical sign convention is observed and that it is tacitly assumed that the solid and the liquid phases move with the same acceleration. This assumption is made quite often even if the mixture theory is used to establish the governing equations of motion, see, for example, Ishihara and Towhata (1980). Differentiating the two sides of Eq,(6) with respect to y and substituting for $\partial(\partial U_x/\partial y)/\partial t = -2\partial\varepsilon_{xy}/\partial t$ from Eq, (5) results in;

$$\frac{\partial^2\sigma_{xy}}{\partial y^2} = (\frac{2\rho}{g}) \cdot \frac{\partial(\frac{a_3\partial u}{\partial t} + \frac{a_4\partial\sigma_{xy}}{\partial t})}{\partial t} \tag{I}$$

A second equation in terms of σ_{xy} and u may be obtained through the second equation of set (1) by noting that;

$$\frac{\partial\varepsilon_y}{\partial t} = \frac{-K\partial^2 u}{\partial y^2} \tag{7}$$

where K is the coefficient of permability of the deposit. Thus, substituting for $d\sigma_x$ from Eq. (4) in the second equation of the set (1) and rearranging terms yields;

$$\frac{\partial\varepsilon_y}{\partial t} = \frac{a_5\partial u}{\partial t} + \frac{a_6\partial\sigma_{xy}}{\partial t} \tag{8}$$

where

$$a_5 = h\psi,_y(A_1\eta,_x - \eta,_y) - C_1 - a_1 C_2$$

and

$$a_6 = h\psi,_y(2\eta,_{xy} - a_2\eta,_x) + a_1 C_2$$

Equating the right hand side of the two equations (7) and (8) provides a required relation viz;

$$\frac{-K\partial^2 u}{\partial y^2} = \frac{a_5\partial u}{\partial t} + \frac{a_6\partial\sigma_{xy}}{\partial t} \tag{II}$$

The two equations (I) and (II) govern the mode of development of the pore water pressure and the shearing stresses within the mass of the deposit. They are coupled and while their simultaneous solution is fairly straight forward, they can not be handled by a microcomputer: the relatively large number of iterations involved makes the time required for completion of computations for even small duration excitations impractical. A great deal of simplicity is introduced by noting that in general the left hand side of Eq. (II) is quite small and may be considered to be equal to zero. This follows from the fact that tailing deposits are composed of extremely fine particles with a very low coefficient of permeability and that near the surface of the deposit where the gradients $\partial u/\partial y$ and $\partial^2 u/\partial y^2$ are likely to be values of large magnitude usually a "baked crust" is present which does not follow the above constitutive relations and hence may not be considered to belong to the elasto-plastic solution domain. Stated otherwise, it is proposed that for short durations involved in this type of study (of the order of seconds or even fraction of seconds) it is quite rational to assume that the process of loading is under the so-called undrained conditions. Thus, eliminating the term $\partial u/\partial t$ between Eqs. (I) and (II) and replacing the symbol σ_{xy} by a single notation τ the governing equation of the problem is derived in the form;

$$\frac{\partial^2 \tau}{\partial y^2} = \left(\frac{2\rho}{g}\right) \frac{\partial \left(\frac{\alpha \partial \tau}{\partial t}\right)}{\partial t} \tag{9}$$

where

$$\alpha = \frac{a_4 - a_3 a_6}{a_5} \tag{9.a}$$

Equation (9) must be solved subject to the following boundary conditions:

(i) At the elevation $y = 0$, the value of τ (t) is specified. The variation of acceleration with respect to time is given and the function τ (t) is evaluated using expression (6) in conjunction with the value of τ evaluated in the vicinity of $y = 0^+$.

(ii) At the elevation marking the boundary between the "baked crust" and the body of the tailing deposit, no discontinuity in any of the stress components is permitted.

(iii) At the surface of the deposit (the upper surface of the "baked crust" the stress components τ and σ are equal to zero.

The initial conditions are simply that at $t = 0$, the value of τ is equal to zero for all values of y.

Finally, the deposit is said to have become unstable if at any elevation y the value of the normal stress component σ_y reduces to a level smaller than 2% of its original value (the insitu vertical stress component prior to the dynamic loading). At this stage the layer in question is said to have liquefied.

In the following section, the numerical scheme used in evaluating Eq. (9) is outlined.

4 THE NUMERICAL PROCEDURE

Figure (2) shows a typical finite difference node (i,j). Note that while the time axis t is divided into equal segments, the spacial y axis is not. This is due to the fact that the Δy increments are determined on the basis of the speed of travel of the initial waves which do so at decreasing speeds. Therefore, for equal Δt increments, the Δy increments assume smaller values as the y ordinate increases.

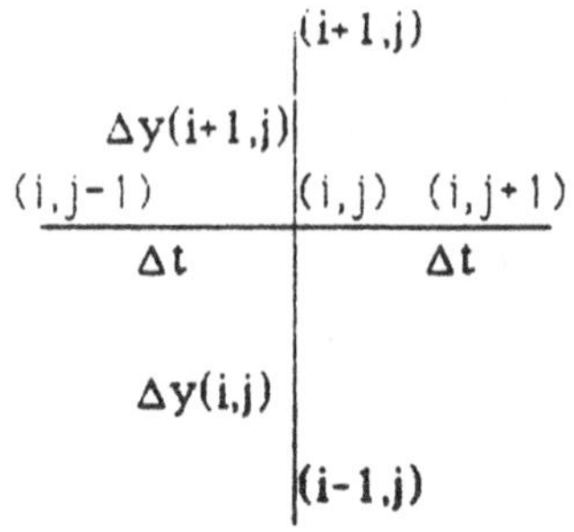

(i+1,j)
Δy(i+1,j)
(i,j-1) (i,j) (i,j+1)
Δt Δt
Δy(i,j)
(i-1,j)

Fig. 2 Nodal point (i,j)

Referring to Fig. (2) the finite difference form for the node (i,j) may be written as:

$$\kappa\left[\Delta y_i (\tau_{i+1,j} - \tau_{i,j}) - y_{i+1} (\tau_{i,j} - \tau_{i-1,j})\right] +$$
$$+ (\alpha_{i,j} + \alpha_{i,j-1}) (\tau_{i,j} - \tau_{i,j-1}) =$$
$$= (\sigma_{i,j} + \sigma_{i,j+1}) (\tau_{i,j+1} - \tau_{i,j}) \tag{10}$$

where

$$\kappa = \frac{2g\Delta t^2}{\rho}\left[\Delta y_i \Delta y_{i+1} (\Delta y_{i+1} + \Delta y_i)\right] \tag{10.a}$$

Assuming that all the calculations prior and up to the nodal point (i, j) have been completed, it may be noted that the left hand side of Eq. (10) is a known quantity. Of the right hand side of the equation, the quantities $\tau_{i,j+1}$ and $\alpha_{i,j+1}$ are unknown. The coefficient $\alpha_{i,j+1}$ is determined by a process of iteration. That is, its value of $\tau_{i,j+1}$ is now used to evaluate the new α, and the process is repeated until convergence is achieved.

5 RESULTS AND DISCUSSIONS

In the examples shown in Fig. (3) to (5), the deposit is assumed to have a height of 10 meters with a crust of 1.5 meters. The base of the deposit is assumed to be subjected to an excitation of the form;

$$\tau(0, t) = a_0 t \,.\, \mathrm{Exp}(-\beta t) \,.\, \sin(mt)$$

in which a_0 is a measure of the size of the excitation, β is the damping factor and m is the frequency coefficient. In the present study the values of β and m are assumed to remain constant and equal to 1 and 15, respectively, in all the three cases studied. The value of a_0 is taken to equal 0.1 for the case shown in Fig. (3) and 0.3 and 0.5 for the situations shown in Fig. (4) and (5), respectively. The fact that the excitation agent is expressed in terms of shearing stress at the base, rather than the acceleration, is intentional since now a direct comparison of results, investigating the influence of the size of the input excitation, is possible.

The plots on the left hand side of the figure show that small oscillations continue even at stages where the excitation force has died out. Although some decay in the amplitude of these waves is observed, the rate of decay is much slower than would be noticed in nature. This is due to the fact that for larger time values, the positive excess pore water pressure would tend to dissipate with an increase in the stiffness of the material, a factor that the present analysis does not take into account. They would, of course, be accounted for had a coupled set of Eqs. (I) and (II) been solved. As mentioned before, however, the operation would be beyond the practical capacity of a microcomputer.

Fig. (4) shows the response of the deposit to an excitation of medium size, $a_0 = 0.3$. The program stops after a period of about 1.1 seconds. This indciates that a layer has liquefied and indeed a reference to the graph on the right hand side of the figure indicates this point (see the stress path associated with the 5th layer from the top of the layer). The fact that the stress path has not gone through the origin of the stress space, is that the program plots the stress path belonging to the

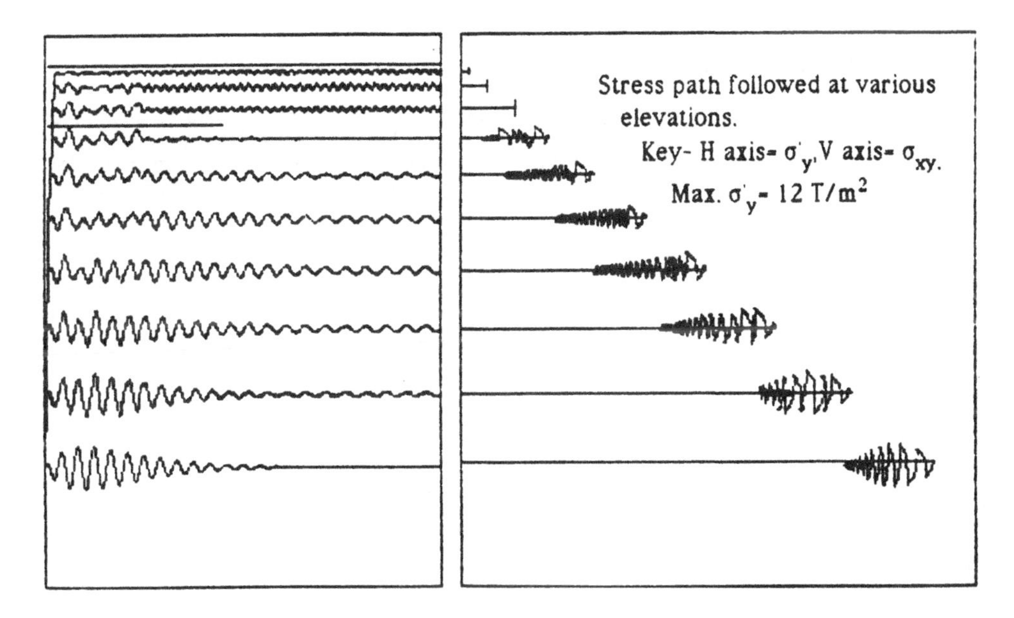

Fig. 3 Response to small size excitations - Diagram on the LHS is shear stress time response. Width of this diagram = 10 sec.

In Fig. (3), the response of the system to a small excitation is shown. As may be seen, although some distortion of the shape of the input wave is observed, the system remains stable. This fact may be deduced from the graphs plotted at the right hand side of the figure, which show the stress paths followed by the elements located at various heights.

elements at selected positions, and these positions may or may not include the liquefied element. In Fig.(4), for example, the liquefied layer is at a slightly higher elevation than the position of the 5th layer.

Finally, in Fig. (5), a large size excitation is imposed on the system. Again, the system became

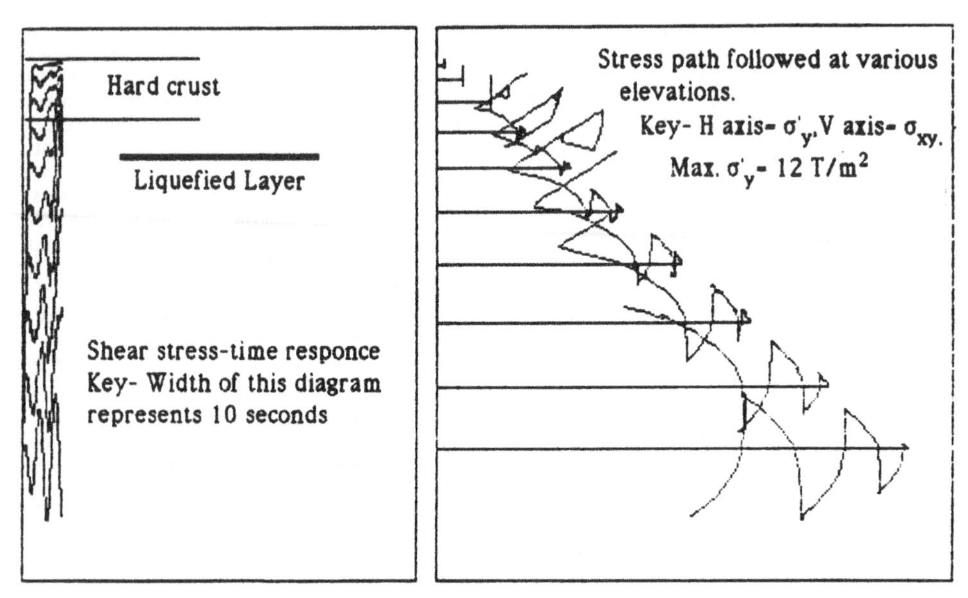

Fig. 4 System response to a medium size excitation

unstable as is evident from the stress path followed by the 2nd layer from the bottom. In this case, the stress path does converge towards the origin, and indeed, the indicator on the left hand side of the figure shows that the elements in the 2nd layer have liquefied.

6 CONCLUDING REMARKS

Microcomputers may be used to analyze certain dynamic loading elasto-plastic problems, provided some simplified assumptions are in order. One such case is the tailing deposits, since by virtue of their

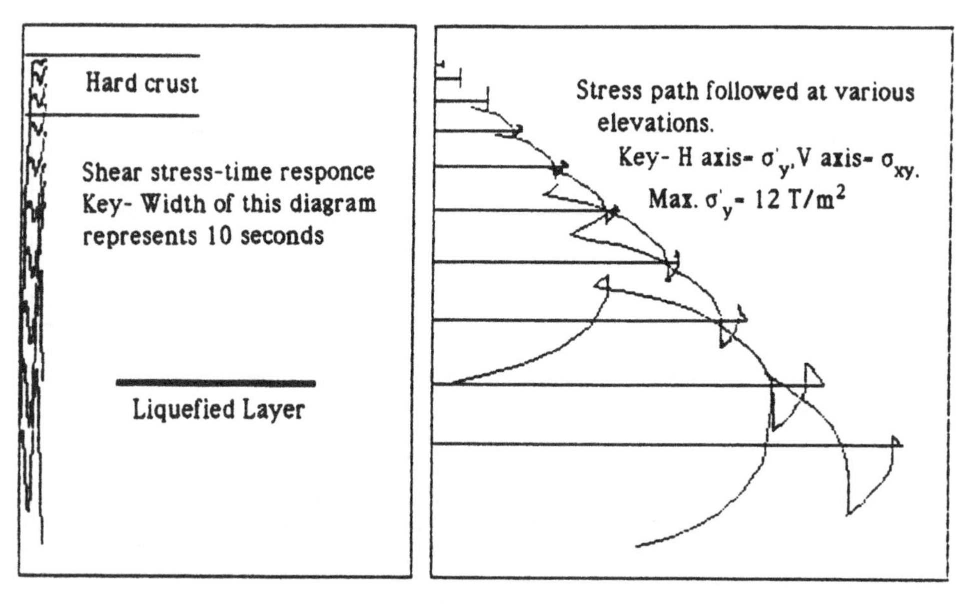

Fig. 5 System response to a large size excitation

geometry, they may be treated as one dimensional problems, and also since they may be assumed to undergo loading under undrained conditions in view of their low permeability.

Large size excitations may cause instability of the deposit by inducing liquefication in certain layers. If the system as a whole (i.e. including the starter dam and the secondary dams) is not capable of resisting the forces exerted at elevations above the liquefied layer, it may fail with catastrophic results. The position of the liquefied layer, however, appears to be as much a function of the nature of excitation, as is dependent on the physical properties of the material of the tailing. Once a layer has liquefied, it would, at least theoretically, reject further excitations from the datum but would undergo vibrations in the pocket above the liquefied layer. These vibrations may lead to the development of further liquefied layers.

ACKNOWLEDGEMENT

The financial support from the Natural Science and Engineering Council of Canada, and the Japan Society for Promotion of Science, is gratefully acknowledged.

REFERENCES

Dafalias, Y.F. (1979), A model for soil behavior under monotonic and cyclic loading conditions. Trans. 5th Int. Conf. on Struct. Mech. in Reactor Tech., No. K1/8.

Ishihara, K., and Towhata, I. (1980), One-Dimensional soil response analysis during earthquakes. J. Fac. of Eng. Tokyo University, Vol. XXXV, No. 4.

Computer and Physical Modelling in Geotechnical Engineering, Balasubramaniam et al. (eds)
© 1989 Balkema, Rotterdam. ISBN 90 6191 864 2

Identification of outliers in geotechnical data

C.Cherubini, V.Cotecchia, C.I.Giasi & G.Todaro
Institute of Applied Geology and Geotechnics, University of Bari, Italy

ABSTRACT: This paper provides an analysis of the methodologies allowing for the identification of outliers in univariated and multivariated samples. A brief example is also presented in which, starting from pairs of data of effective cohesion and friction angle an attempt is made to identify those data that differ unexpectedly from most of the others.

1 INTRODUCTION

In the field of geotechnics, and in particular in studies of vast areas which are becoming more and more important in relation to particular types of intervention or use of territory, there is a growing need to have available the right sort of numerical and statistical techniques capable of processing large quantities of data in order to obtain the necessary information.

One of the problems that crops up in the "filtering" phase of the data is that of outliers, i.e. of data that "differ, unexpectedly, from most of the others" (Anscomb 1968). Sometime the "outlier" may be considered as being a manifestation of the natural variability of the soil, while on other occasions it may lead one to doubt the hypotheses on which the model for interpreting a given soil was based. If we think, for example, of a large-scale geotechnical study with any given specific aim (territorial planning, seismic microzoning, etc.) in which a considerable number of data must be analysed deriving from various geognostic field trips, thus carried out at different times, by different operators and sometimes using different techniques, it is easy, in such a context, to find outliers which may radically alter the results of the analysis. The situation may be further complicated by the fact that these anomalies may not necessarily be of a striking nature and may thus be hard to identify by whoever is processing or interpreting the data.

Hence the need for the availability of statistical methodologies capable of reducing the influence of outliers in the evaluation of the parameters that one is concerned with, or else of identifying such data which can then be left to the technician whose task is to discover the reasons for their presence (Rethati 1983).

In the case of geotechnical data deriving from in situ studies or from the lab_ ratory, the presence of outliers may be attributed to the following causes (Cheru bini and Ottaviani 1985):

1) an error in sampling through the attribution of a given sample to a for mation that it does not in fact belong to; this problem is particularly common in transition zones, though it may sometimes represent a genuinely "singular case" within the ambit of the formation itself;
2) error in the attribution of the quality coefficient of the sampling;
3) gross differences due to poor handling of data by the operator, to calculation errors or imperfections in the equipment used.

Obviously, the problem of outliers does not come to an end with the definition of the anomalies of the suspect data, but

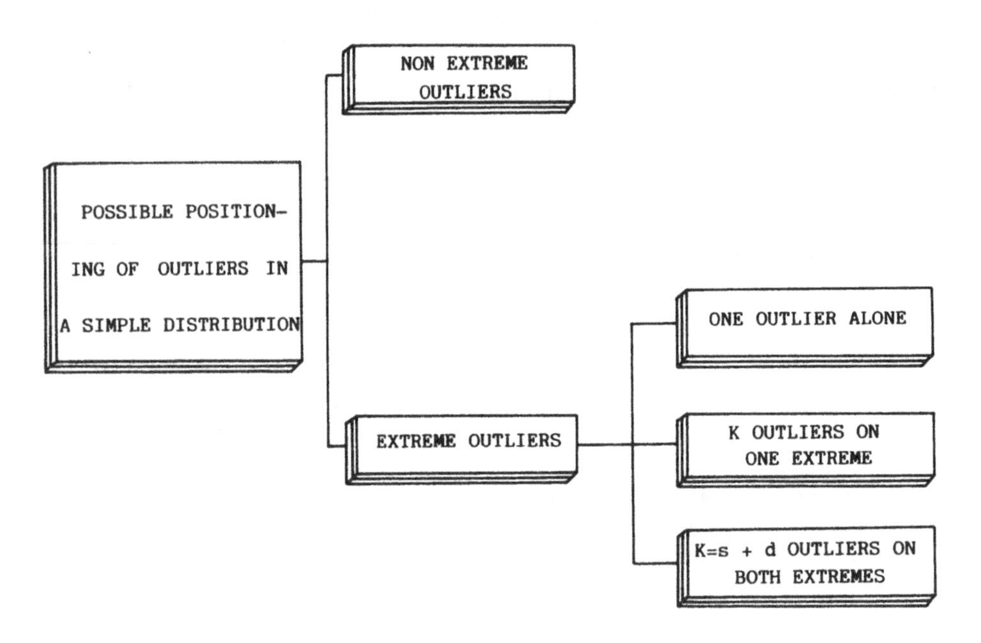

Fig. 1. Possible positioning of outliers in a set of data

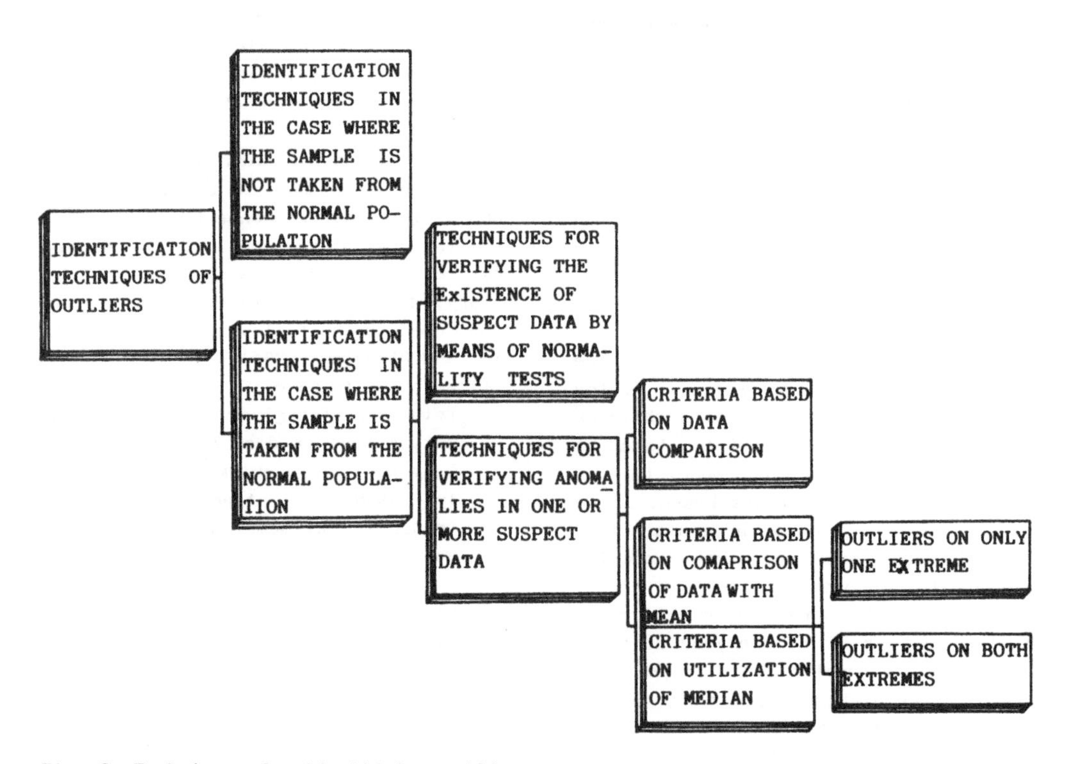

Fig. 2. Techniques for identifying outliers

there is also the question of how to treat these anomalous data in subsequent processing, and in particular of how to use them in order to obtain sampling estimates of the population parameters.

In fact, if one concludes that there is an error in the data, no major difficulties arise in so far as the model chosen to represent the data as a whole still remains valid, and the influence of the error may be annulled or else reduced by eliminating the relevant data or by modifying the data through techniques of varying sophistication. If, on the other hand, the conclusion is that it is the model that is at fault, either one has to modify the model itself by choosing one that is suited to representing all of the data, or else by verifying an elimination (e.g. in the case of a sample taken from a transition zone) or else one decides that one has come across a rare value.

The second solution, i.e. that of an inadequate model, is hardly acceptable because the choice of model is the last result in a series of past experiments or of studies previously carried out (for example, we know that the percentage of water of a "homogeneous" soil normally varies by 15-20% around its mean value, and thus, if one comes across a greater variability in a given site, the causes of this anomaly must be found).

Lastly, it is worth noting that in general, in geotechnical studies, the acquisition of data is carried out through the identification of several physical or mechanical characteristics on each observational unit thus often making it necessary to use multivariated analysis techniques capable of simultaneously considering variations in different properties by searching for outliers or by identifying single geotechnical units.

2 Operative choices in the presence of outliers

In Fig. 1 there is an example of a possible position for outliers in an ordered sample which may or may not be of an extreme nature. In any case, we are only interested in identifying extreme outliers in so far as it has been shown that they have a higher possibility of occurrence.

The statistical methods available to the researcher are either techniques for identifying outliers (Fig. 2) or techniques capable of reducing the influence of anomalous data in calculating the parameters of the population that are deemed relevant.

One of the first contribution to the problems of outliers was that of Chauvenet (1863) who identified a criterion for the rejection of doubtful observations in which those data which present certain standard differences that are greater than a given amount are eliminated.

Subsequently it was realized that the elimination of outliers from any type of processing is only one of many possible decisions, and scholars began to speak of identifying outliers rather than of rejecting them, in order to decide whether to accept them, reject them or modify them (Gumbel 1960).

At the moment, there are numerous techniques available for identifying outliers proposed in statistical literature, in particular those which hypothesize that the data to be processed are a sample obtained from a normal population. In fact, the hypothesis of normal distribution is commonly accepted in geotechnics and has been convalidated by various authors (Magnan and Baghery 1982).

The methods for identifying outliers can be distinguished as follows (Barnett and Lewis 1984):

1) Anomaly verification techniques of one or more suspect data;
2) techniques for verifying the existance of outliers through normality tests.

Anomaly verification techniques of one or more suspect data are based on a comparison of data; with their arithmetical mean (Dixon 1950, Grubbs 1950), with the median (Ottaviani and Soccorsi 1981, Ottaviani 1983) or with some other term of comparison (Ottaviani 1980). Some of these tests are specific for outliers at only one extremity, while others can be applied independently of the position assumed by the anomalous data. With this aim in mind, statistics have been introduced in order to measure the diversity of the outliers from the various terms of comparison proposed; these statistics are either distances or ratios between sums of power of distance. With these statistics adhoc tests have been devised in order to identify one or more outliers: the zero hypothesis of such tests is that all the data have been extracted from the same population, i.e. they are homogeneous.

The application of such tests creates a few problems: first of all, the value of K, i.e. of the number of outliers must be known. In actual fact, the value of K is unknown and its determination is uncertain; an error in the presumed value of K may lead either to an underestimate or to an overestimate of the number of outliers.

The normality tests for verifying the existence of outliers, unlike those already seen which presuppose that the researcher should concentrate his attention on just a few data and decide whether or not they are anomalous, poses a more general kind of problem, i.e. of whether outliers exist, in other words if the elements in the sample all come from the same normal population or if it has been contaminated by normal populations with a different average or with diverse variance. For this purpose normality tests are utilized, and in the literature the ones reported as being the most powerful are the Fergusson criteria (1961).

The latter utilize the sample coefficient of Skewness and the sample index of Kurtosis respectively when there is the suspicion of contamination with a population having a different average and when there is the suspicion of contamination with a population having a diverse variance.

3 Robust estimators

The problem of outliers does not come to an end with the identification but it naturally follows into the estimation phase of the parameters of the population.

The problem of the search for outliers thus logically leads into the search for "robust estimators".

The use of robust techniques is particularly important in that they can be applied even when the distribution function of the population from which the sample has been taken is unknown.

Most of the estimators used are of the "L type", i.e. they are obtained from a linear combination of the ordered sample.

They are many estimators of this sort reported in the literature. The ones most frequently used for calculating the location parameter are:

- the "trimmed estimator", obtained eliminating on both extremities of the ordered sample of a certain number of data (Afifi and Azen 1979);
- the "Winsorized estimator". All of the observations that are considered as excessively external are shifted to a pre-arranged position near the central observations; in this way, the external observations are not completely discarded (Afifi and Azen 1979);
- the "B.E.S.", which is an average weight of the first quartile, the median and the third quartile (Wonnacott and Wonnacott 1977).

Another class of estimators is the "M" type, including Hampel's Piecewise Linear M Estimator, which may be considered as a pondered arithmetical average of the data to which are attributed decreasing weights as they get further away from the estimate of the average.

4 Procedures in multivariated analyses

The current trend in the analysis of geotechnical data is that of applying multivariated analysis procedures in that such methods make it possible to operate on variations in different properties simultaneously.

Initial, multivariated data were analysed considering each variable separately in that the procedures for multivariated analysis turn out to be more complex. In fact, a computer must be used for such analyses.

Also in this case most of the procedures assume that the data belong to a multinormal population.

Some of the techniques for analysing outliers derive directly from the corresponding techniques for univariated analysis while others have had to be worked out expressly for such analyses.

For example, the χ^2 test is utilized in the search for outliers when the mean vector (μ) and the covariance matrix (Σ) are known (Afifi and Azen 1979).

In most cases the (μ) and (Σ) parameters are not known and so it is more appropriate to adopt a procedure that makes use of the sample of distance of Mahalanobis (Afifi and Azen 1979).

In this procedure all of the samples are tested singly, evaluating each distance from the rest and comparing a particular calculated statistic of distribution F with the value of the percentile, to the preordained level of significance, of distribution F.

Sample is thus eliminated to which corresponds the greater value of F in that it is identified as being anomalous, and the procedure is repeated on those remaining until it is found that none of the data presents any anomalies.

5 Application

Some treated tests have been applied to 23 samples of clay coming from the location of "Lioni" (Chiocchini and Cherubini 1986). Only the cohesion and the effective friction angle have been considered.

The data are reported in Table 1. In Table 2 the values of some statistical parameters calculated on these data are reported.

In Table 3, the results relating to the application of the tests for identifying outliers are reported. It can be noted that in a univariated analysis, the only test that identifies outlies is Chauvenet's criterion which identifies 31.5 as being anomalous.

In Table 4 it can be noted that all the robust estimators of the mean both in the case of C' and of $\emptyset'$, more or less coincide with the mean value.

The application of the test for identifying outliers by using Mahalanobis's distance identifies as anomalous samples, at a level of significance of 5%, samples 13 and 20, and from a data analysis it can be noted that for some samples the ratio between the two measurements differs a lot from other samples. In fact, samples 13 and 20 in particular are those that present at the same time the highest values of $\emptyset$ and the lowest values of c.

Table 1. Set of geotechnical data to be examined

Sample	c'	Ø'
1	0.23	21
2	0.18	23.5
3	0.23	25.9
4	0.17	25
5	0.20	21.5
6	0.12	20.3
7	0.17	21.5
8	0.24	26
9	0.18	23.4
10	0.17	26.5
11	0.18	25.5
12	0.13	27
13	0.06	29.7
14	0.23	23
15	0.24	23.3
16	0.12	21.2
17	0.27	25.9
18	0.19	20.3
19	0.17	25
20	0.05	31.5
21	0.27	26.4
22	0.26	23.3
23	0.31	27.3

Table 2. Sample statistics

Statistic	c'	Ø'
Mean	0.19	24.55
Min. Value	0.05	20.3
Max. Value	0.31	31.5
Median	0.18	25
Skewness	0.42	0.47
Kurtosis	3.24	2.93
Stand.Dev.	0.06	2.96
Coef.of Var.	0.34	0.12

Table 3. Identification of outliers

Method		c'	Ø'
Chauvenet			31.5
Dixon	5%		
	10%		
Tiet-Moore			
Fergusson	b1		
	b2		
Ottaviani			

Table 4. Robust estimators

Robust Estimators	c'	Ø'
B.E.S.	0.19	24.47
Trimming	0.19	24.38
Winsorizing	0.19	24.54
HPL M-estim	0.19	24.33

References

Afifi, A.A. & S.P. Azen 1979. Statistical analysis. A computer oriented approach. Academic Press.

Anscomb, F.J. 1969. Special problems of outliers in statistical analysis. International Encyclopedia of Social Sciences, vol. 15.

Barnett, V. & T. Lewis 1984. Outliers in statistical data. Wiley, N.Y.

Chauvenet, W. 1863. A manual of spherical and practical astronomy. Vol. 2°, J.B. Lippincott. Philadelphia.

Cherubini, C. & M.G. Ottaviani 1985. I dati anomali in geotecnica. Gruppo Nazionale di coordinamento per gli studi di Ingegneria Geotecnica. Roma.

Chiocchini, U. & C. Cherubini 1986. Seismic microzoning of the Lioni village destroyed by the November 23, 1980 earthquake (Irpinia, Campano-Lucano Apennine). Proc. of the Int. Symposium on Engineering Geology Problems in Seismic Areas. Bari.

Dixon, W.J. 1950. Analysis of extreme values. The Annals of Mathematical Statistics. Vol. 21.

Ferguson, T.S. 1961. Rules for rejection of outliers. International Statistical Revue. Vol. 3.

Grubbs, F.E. 1950. Sample criteria for testing outlying observations. The Annals of Mathematical Statistics. Vol. 20.

Gumbel, E.J. 1960. Discussion on the papers of messers Anscombe and Daniel. Technometrics. Vol. 2.

Magnan, J.P. & S. Baghery 1982. Statistiques et probabilites en mecanique des sols: etat des connaissances. Laboratoire Central des Ponts et Chausses, Paris, Rapport de Recherche LPC, n. 109.

Ottaviani, M.G. & R. Soccorsi 1981. Sull'utilizzazione della mediana per la individuazione di un dato anomalo estremo. Metron, vol. XXXIX.

Ottaviani, M.G. 1983. Sull'utilizzazione della mediana per l'individuazione contemporanea di più dati anomali estremi. Statistica, Anno LXIII, n. 1.

Ottaviani, M.G. 1980. Sull'individuazione dei dati anomali delle distribuzioni statistiche semplici. Quaderni di Statistica e Ricerca Sociale "C.Gini", Roma.

Réthati, L. 1983. Asymmetry in the distribution of soil properties and its elimination. Fourth Int. Conf. on Appl. of Stat. and Prob. in Soil and Struct. Eng. Firenze. Vol. 2.

Wonnacott, T.H. & R.J. Wonnacott 1977. Introductory Statistics. Wiley, N.Y.

6. Computer aided solutions for some special problems in engineering

Computer and Physical Modelling in Geotechnical Engineering, Balasubramaniam et al. (eds)
© 1989 Balkema, Rotterdam. ISBN 90 6191 864 2

Test control and data acquisition of reinforced foundations

G.E.Bauer, A.D.Abd-El Halim & A.P.S.Selvadurai
Department of Civil Engineering, Carleton University, Ottawa, Ontario, Canada

ABSTRACT: The concept of soil reinforcement is known for some time in road construction and reinforcing elements such as wood, steel strips, geotextiles and membranes have been used in the past. In recent years the application of geogrids to strengthen the base material has become cost-effective mainly due to the increase in costs of granular aggregates and the fact that grids will interlock with the granular material much better than pliable membranes or steel strips. Geogrids also have the advantage that they do not deteriorate or corrode under adverse climatic conditions.

This paper reports on the test program which was carried out at Carleton University to investigate the deformation behaviour of reinforced and unreinforced road bases under cyclic and static load conditions. In order to monitor the load-deformation response and the stresses within the reinforcing elements, an extensive data acquisition system was used. This paper describes the computer aided methodologies that were employed to acquire, analyse and present the data pertaining to this investigation.

1 INTRODUCTION

A basic requirement for reinforcement used in permanent reinforced road bases is the ability to resist sustained load without causing intolerable large deformations or tensile failure of the reinforcement elements. The design life of roads varies from municipality to municipality and from country to country, but an average design life of 60 years is accepted by many organizations. Also, there is no generally accepted criterion regarding the deformation of roads under sustained cyclic loading, but as a general rule, 25 mm (1 in.) has been accepted before resurfacing of the asphalt layer is needed. At present the vast majority of reinforced soil structures use galvanized mild steel strips and the service life of these structures is controlled by the rate of corrosion of the reinforcing elements. The problem of corrosion can become quite severe, especially in countries where salting of roads during the winter months has become a normal practice. It was, therefore, quite a natural step to turn to the use of plastic grids or non-corrodible materials as soil reinforcement. The proposition of employing plastic as soil reinforcement has been the topic of several research studies and conferences over the last decade. So far very little is known about the behaviour of plastic geogrids in road bases under sustained cyclic loading.

The geotechnical group at Carleton University, Canada, has been involved over the last few years investigating the behaviour and properties of geogrids within soil media. The main objectives of the study reported in this paper were to verify or postulate a soil/reinforcement interaction mechanism, to provide experimental data for analyzing the load-deformation response of reinforced soil bases and to provide a basis for future research and field trials.

The results from the experimental and theoretical modelling will be fully documented in present and future publications. This paper on hand will outline the basic features of the experimental program and will highlight the manner in which computer-aided data acquisition and processing procedures can be used to enhance the experimental research program.

2 THE EXPERIMENTAL PROGRAM

The experimental research program was carried out in a test bin assembly. The bin

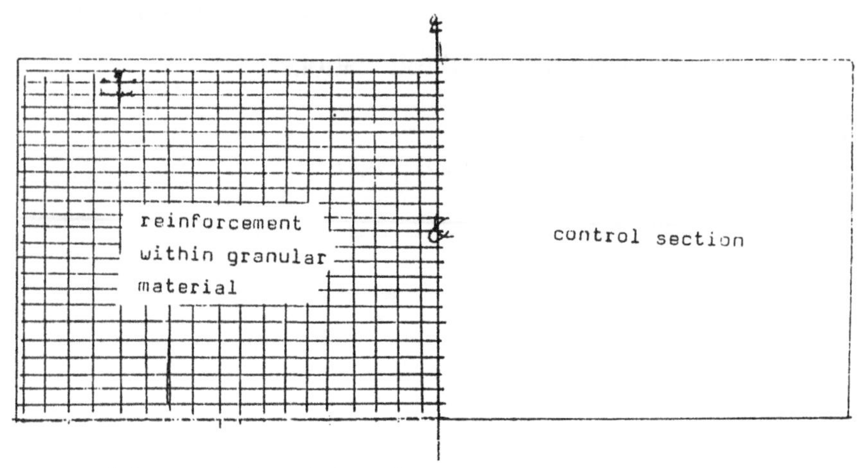

Figure 1(a): Plan View of Test Bin

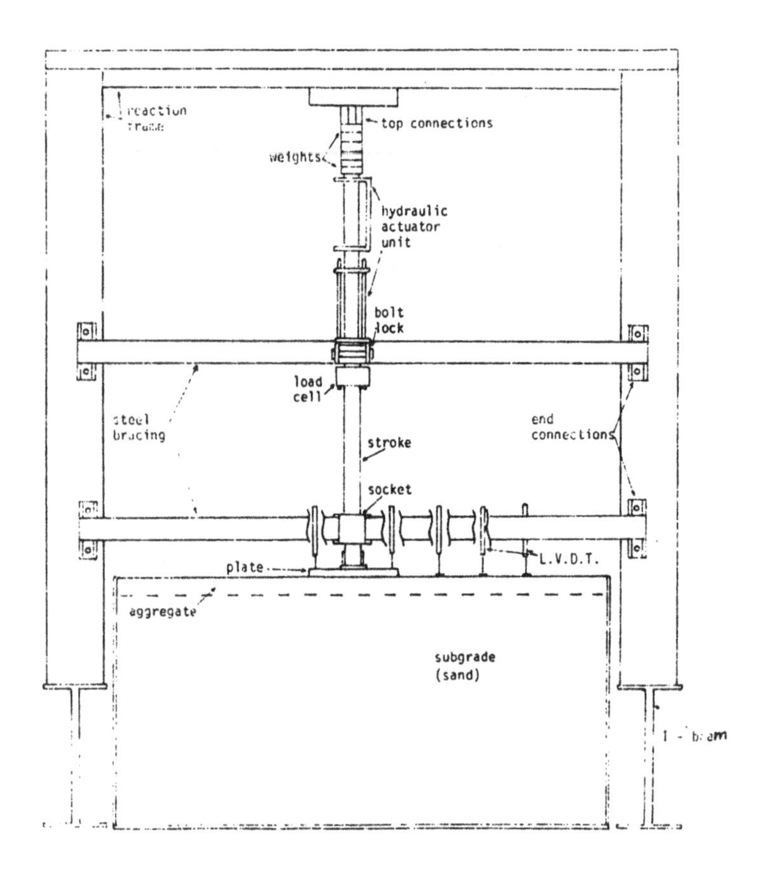

Figure 1(b): Section of Test Assembly

was 1820 mm by 4600 mm in plan and had a height of about 1000 mm. Figure 1 shows a plan view and a section through the test assembly. One half of the bin was always used as a control section, which meant that all tests were duplicated in the unreinforced control section for direct comparison of results. The test program was to achieve (1) an explanation of the concepts and mechanisms of base reinforcement, (2) give an evaluation of the soil and material parameters, (3) present the major results showing the effects of reinforcement on the performance of road base materials with respect to deformations, strains, deflections and other responses, and (4) compare the reinforced section to the unreinforced control section in order to estimate the increase in service life of the road base.

The principal components of the experimental facility are briefly outlined in the ensuing sections.

2.1 The test box

The test box or bin was made of laminated plywood and was reinforced with angle braces in order to minimize any lateral deflection under load conditions. The inside of the box was lined with galvanized steel for moisture retention and for minimizing side wall friction. The plan of the box and a section are given in Figure 1. The bin was 4600 mm long and 1880 mm wide.

2.2 Sub-base material

The test bin was filled with a fairly uniform medium size sand to a height of 75 mm. The sand was compacted in 150 mm lifts to its maximum dry density and optimum moisture content (i.e., γ_d=1.95g/cm^3 and W=10%).

2.3 Granular base

The granular base material consisted of a well graded crushed limestone material with maximum particles of 13 mm (0.5 in.). Only 12% of the material passed the No.50 sieve size and 2% passed the No.200 sieve. The maximum dry density was 2.35g/cm^3 with a corresponding optimum moisture content of 5%. The thickness of this crushed stone material was varied from 75 mm to 3000 mm according to the test series and placement of the reinforcement grid.

2.4 Geogrid

The reinforcing mesh used in this investigation was a high density, high strength polymer (TENSAR AR1). The mesh openings were 60 by 50 mm. The grid has strands which are about 1 mm thick and 10 m wide at the narrow part and about 10 mm wide and and 2 mm thick at the nodal points. This geogrid was chosen as the most suitable reinforcement for the aggregates on hand. This selection was based on a previous investigation (Bauer and Mowafy, 1986). In summary, the TENSAR AR1 has high tensile strength, a high modulus of elasticity and resistant to chemical and biological attacks.

2.5 Hydraulic loading system

The loading system consisted of an MTS function generator, a servo-hydraulic controller and an actuator assembly. The hydraulic actuator assembly was equipped with a load cell and an internal LVDT (linear voltage displacement transducer). A 305 mm diameter (12 in.) rigid steel plate was attached to the lower end of the actuator stroke. The plate was 25 mm thick and in order to prevent any tilting or lateral movement of the actuator, the plate and piston was held in place. The maximum load applied was 440 kN (9000 lbs) and the maximum frequency was in the order of 3 Hz.

2.6 Load cell and L.V.D.T.

The applied load was monitored by a load cell at the end of the actuator and the stroke displacement or plate penetration was recorded with the built-in L.V.D.T. Both transducers were connected simultaneously to the MTS function generator and to the data acquisition computer. The load-displacement could therefore be read at both display panels. The load and corresponding displacement were recorded at certain intervals during the dynamic load application and a complete incremental load-displacement record was obtained during a static load test.

2.7 The data acquisition/computer system

The array of displacement transducers monitoring the surface displacements, the strain gauges mounted on the geogrid, the load cell and the L.V.D.T. monitoring the stroke displacement, were connected to a HP3457A data acquisition subsystem. This system had 240 channels and can be enhanced with additional cards if need arises. The

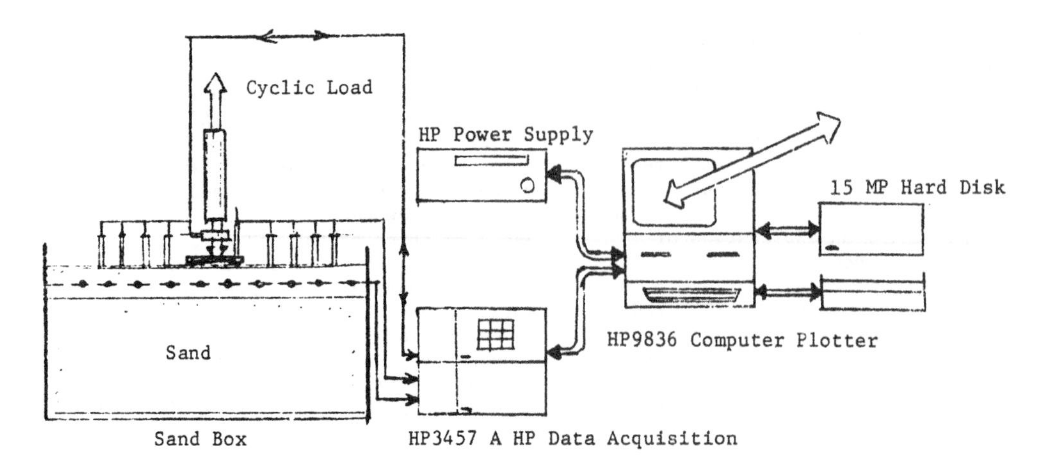

Figure 2: Flow Chart of Data Acquisition

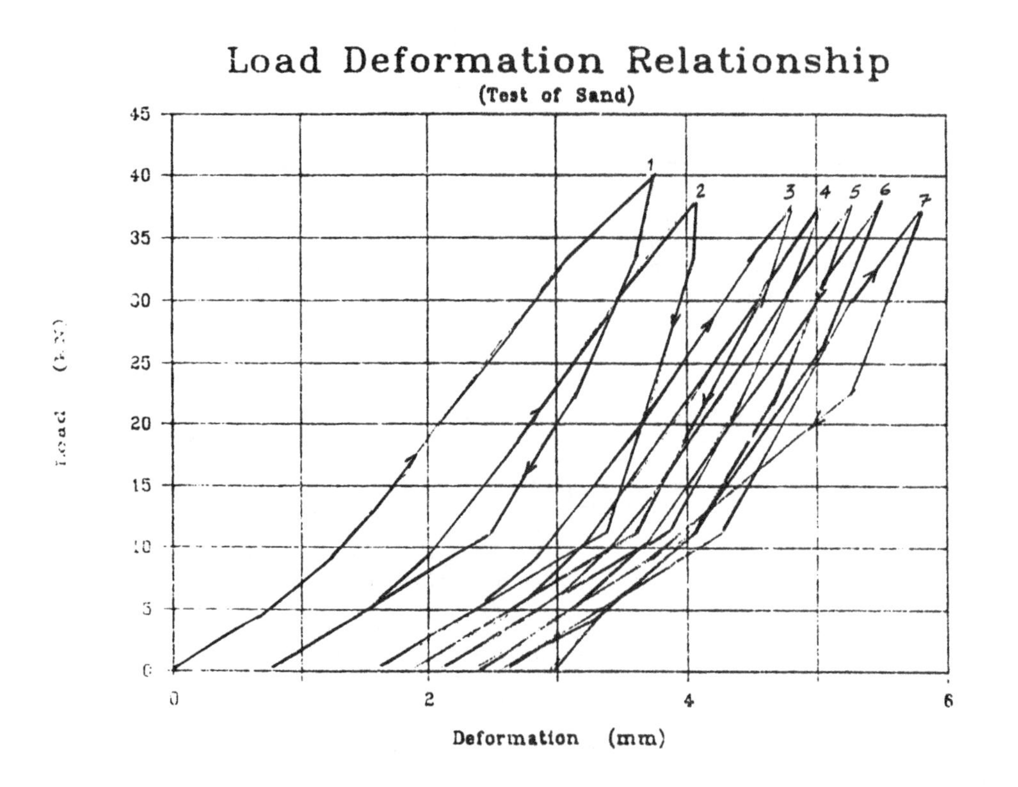

Figure 3: Load-Deformation Behaviour of Sand

HP3457A subsystem in turn was connected to a HP9836 computer, which monitored the test and stored the obtained data. Data storage was accomplished with a HP9145 15 Mbyte Winchester hard disk and the results were plotted on a HP82905B dot matrix printer. The data acquisition/computer system was housed on and inside a desk cabinet located next to the test facility. During test set-up and placement of the soil within the sand box, the data acquisition system was removed in order to prevent dust from getting into the system. The data was stored on the hard disk and in order for the extensive data to be processed on the University's mainframe, a Honeywell CP6 computer system, the data from the hard disk was transferred to a 8 mm (1/4 in.) streaming tape by using a stream tape drive connected to the HP9836 computer. This transfer took about six hours. The streaming tape drive was then connected to an HP900 computer, model 500, which was connected to a 13 mm (1/2 in.) track magnetic tar tape drive. This process, streaming tape to tar tape, took approximately five hours. To make the stored data on the tar tape accessible to the CP6 computer, the data had to be transferred to the mainframe. This was usually done at night time during the off-peak hours of the main computer.

3 SOFTWARE DEVELOPMENT AND DATA PROCESSING

The data acquisition and data processing systems had to satisfy the basic requirements dictated by the experimental research program. These requirements were as follows:

(1) to control and to monitor the load application of the actuator from zero to 40 kN (0 to 9000 lbs.) during the static and cyclic phases of the test,
(2) to monitor the strains of the eighteen resistance strain gauges attached to the reinforcing polymer grid,
(3) to monitor the nine displacement transducers during a test loop,
(4) to control and to monitor the number of load cycles with time,
(5) to provide a visual image on the computer's CRT of the load-displacement response, and the number of cycles of load applications, and elapsed time,
(6) to store all the data from the various transducers and strain gauges on the hard disk, and finally,
(7) to continue the process of recording and storage of data in case of a short-term power failure.

A schematic flow chart of the data acquisition/control system is shown in Figure 2.

(2) The LVDT's and the strain gauge readings were recorded for every static load increment. Both load values and stroke displacements were recorded for every dynamic and static load change together with elapsed time. Several computer programs were written to sort the data and place them into two files. The file with the static load data was called "STEP" and the file for the dynamic loading phase was called "CYCLE". The data specified for the two files was given in the following format:

(a) Static Test
STEP1, STEP2, STEP3 and STEP4 as Time (7 integer digits), load (4 interger digits), stroke displacement (4 real digits), 9-LVDT readings (4 real digits each), 18-strain gauge readings (5 real digits each), 18-strain gauge readings (5 real digits each), and

(b) Dynamic Test
CYCLE1, CYCLE2, CYCLE3 and CYCLE4 as Time (7 integer digits), load (4 integer digits), stroke displacement (4 real digits) 9-LVDT readings (4 real digits each), 18-strain gauge readings (5 real digits each).

As aforementioned the data from the hard disk was transferred to the CP6 Honeywell computer via the HP9000 computer system. Several FORTRAN computer programs were used to retrieve and process the data stored on the mainframe computer. The program to process the data of the static load tests had to fulfill the following requirements:

(1) read the data line by line of the static load test (STEP),
(2) take an initial value (from the first line value),
(3) subtract the initial value from subsequent values,
(4) type out the values with corresponding titles,
(5) to allow manual selection of certain data points for curve plotting or tabulation.

The program to process the dynamic loading data had to satisfy similar basic requirements as in the static test, except that an additional requirement was to select data points corresponding to a pre-set cycle number.

The experimental testing program was stopped by the computer, when either of the two pre-determined failure criteria were reached. Failure was defined when either the permanent deformation under the steel

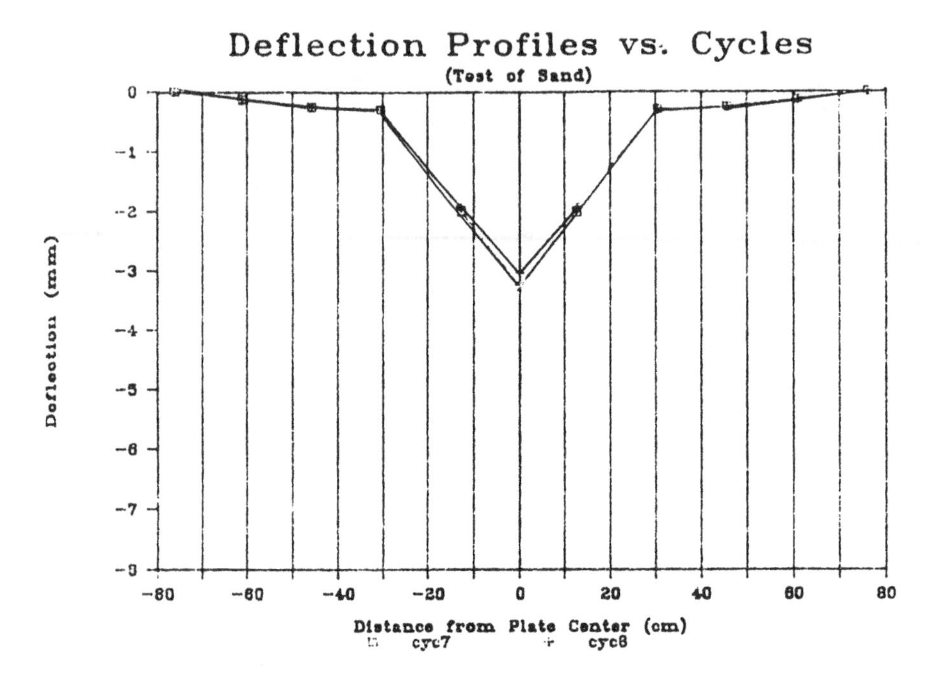

Figure 4: Deflection of Sand Surface (Sub-Base)

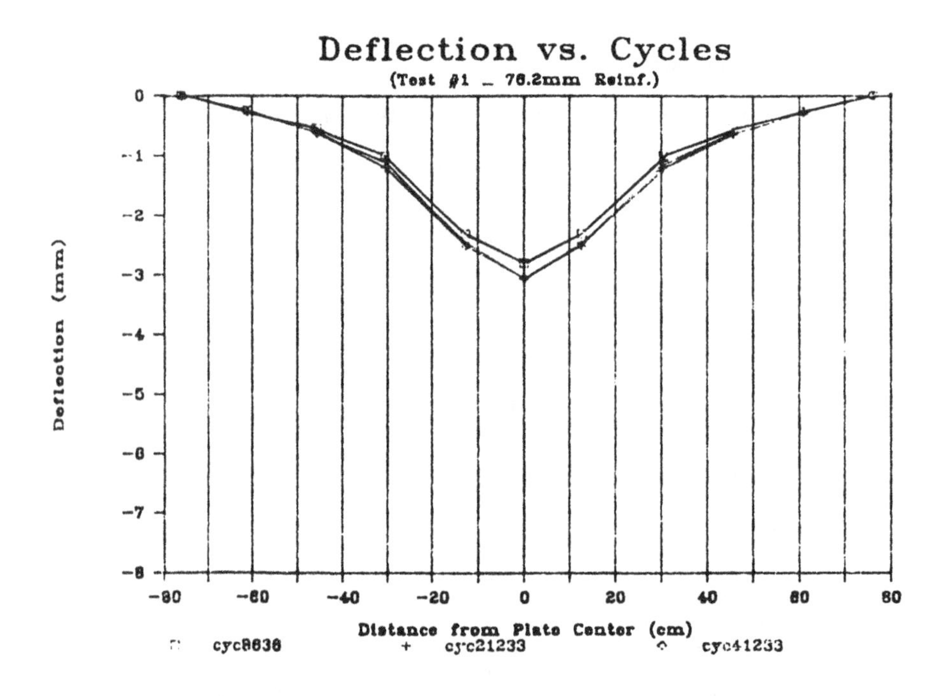

Figure 5: Deflection With Reinforcement

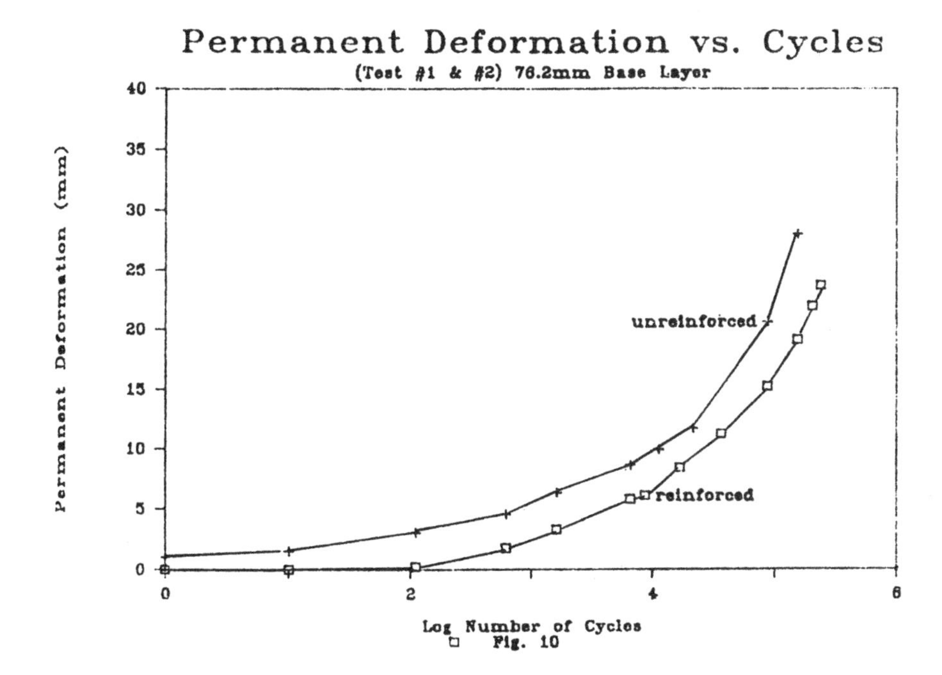

Figure 6: Deformation of Reinforced and Unreinforced Sections

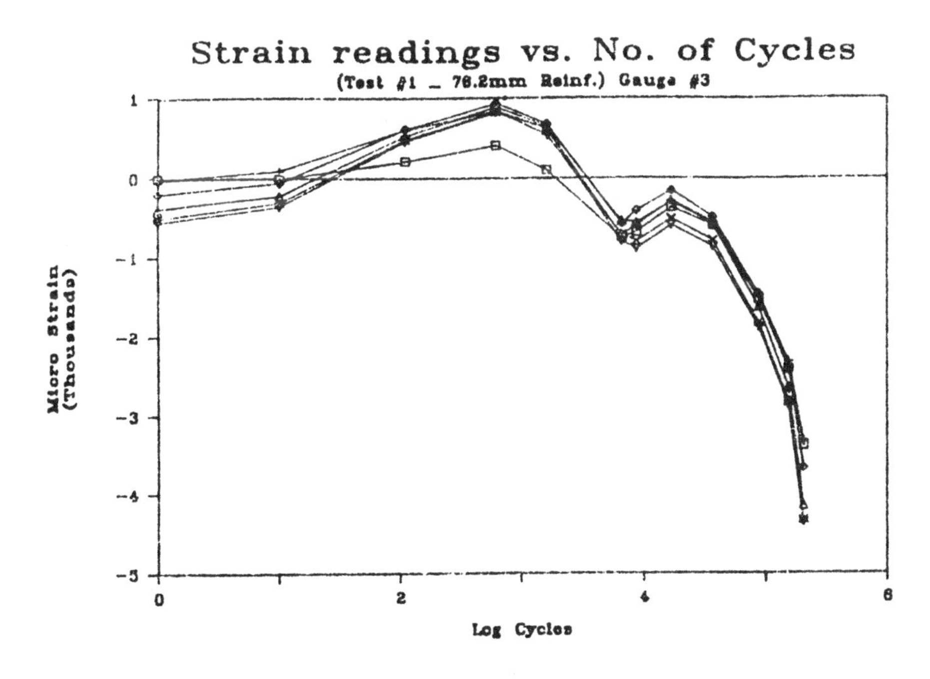

Figure 7: Mobilization of Strain of Geogrid (TENSAR AR1)

plate had reached 25 mm, or when the number of cycles had come to 1 million cycles.

4 RESULTS

The recording and storage of data was quite extensive. In order to select the pertinent information, the preliminary data were screened and the majority of the test results were presented in graphical form. A typical load-deflection response of the steel plate on the unreinforced sand subgrade is given in Figure 3 for eight cycles. The permanent deflection profile of the surface of the sand is shown in Figure 4 after various numbers of load cycles. A similar deformation profile is given in Figure 5 for the case where the geogrid was placed at the interface between the subgrade and a 76 mm (3 in.) base layer. A comparison of permanent deformation and number of load cycles is given in Figure 6 for an unreinforced and a reinforced section where the granular base material was 76 mm (3 in.) thick. Similar graphs were obtained for 152 mm (6 in.), 228 mm (9 in.) and 305 mm (12 in.) thick bases and different reinforcement placements. These graphical presentation gave a quick visual reference, which combination of thickness of base material and location (depth) of reinforcement yielded the best load-displacement behaviour. These data then could be re-analyzed on a more detailed basis.

A typical response of the strains mobilized in the geogrids is given for one strain gauge in Figure 7. One would, if desired, plot strain contours from the results of the eighteen strain gauges. None of the strains monitored, even after one million load applications were quite low. Once the data were stored and sorted, graphical representations and relationships for most parameters are possible. These results then serve as input parameters for several theoretical models which have been postulated. The comparison of experimental to theoretical results will be the object of future publications.

5 CONCLUSIONS

This paper outlines the methodologies which have been used in the computer-aided experimental investigation dealing with the load-deformation responses of reinforced and unreinforced road bases. The experimental testing facility attempts to give a realistic simulation of road bases under traffic loading. Since these studies are time-dependent and involve an extensive monitoring program with continuous scanning and collecting of data over several months, it is essential to resort to a data acquisition and computer control system.

It is most expedient that before such a long-term experimentation is undertaken, detailed plans to design and to develop an efficient system for data acquisition and processing, including the necessary software. Computer-aided data acquisition and processing perform invaluable and essential tasks in long-term experimentation where data collection is continuous and extensive.

ACKNOWLEDGEMENTS

The study described in this paper was financially supported by a group research grant from the Natural Sciences and Engineering Research Council of Canada awarded to the authors and a group of researchers at the University of Waterloo under the direction of R. Haas. The authors wish to acknowledge the assistance of S. Conley, research engineer, who was responsible for developing the data acquisition (control system) and Messrs. Alkhatib and Chan, graduate students, for carrying out the experimental work and the analysis of data.

REFERENCES

Bauer, G.E. and Mowafy, Y.M., 1986. Behaviour of Reinforced Earth Walls Under Self-Weight and External Loading. 2nd International Conference on Numerical Models in Geomechanics, Ghent, Belgium, March.

Computer and Physical Modelling in Geotechnical Engineering, Balasubramaniam et al. (eds)
© 1989 Balkema, Rotterdam. ISBN 90 6191 864 2

An analytical investigation of the behaviour of a soil-steel structure

Mohammad Rezaul Karim
Department of Civil Engineering, BUET, Dhaka, Bangladesh

ABSTRACT: Corrugated metal culverts are quite flexible and they derive a considerable portion of their load carrying capacity through interaction with the surrounding backfill. Hence the name soil-steel structure. A computer program has been developed for a two-dimensional non-linear plane strain analysis of soil-steel structures using finite element method. The non-linear properties of soil are obtained from triaxial test and represented by a set of graph of octahedral shear stress and strain. These are mathematically formulated in terms of cubic spline functions and followed by an incremental analysis technique. Geometric non-linearity of the structural member is incorporated using stability functions.

A series of analyses are then performed to study the effect of such important parameters as depth of cover above crown, shape of culvert, stiffness of culvert material and poisson's ratio of the surrounding soil. From the analyses the following conclusions could be drawn. For corrugated metal culverts, there is an optimum depth of culvert over the crown at which stresses are minimum. Circular shape gives better performance than elliptical shape from the structural point of view. With increasing stiffness, load carried by the culvert increases and poisson's ratio of soil has insignificant effect on the behaviours of soil-culvert system.

INTRODUCTION

Corrugated metal culverts can be economical replacements for short-span bridges. In recent years long-span metal culvert structures are being used increasingly in railroads and highways. Traditionally, culvert design has been largely empirical, but with the increase in demand for long-span structures, the need for rational analytical procedures, to model their operational mechanism realistically, has grown. In this paper an attempt has been made to study the fundamental behaviour of corrugated metal culverts by an interactive analytical scheme.

As far as the behaviour under load is concerned, buried culvert and its surrounding soil form an integral system. A study of the interaction between the two leads to a better understanding of the behaviour of the system. A one-step analysis of the culvert and soil as an integral component of a single physical system makes it possible to account for the interaction that may take place between the components. In this study, the finite element method is used to represent the soil and the culvert by discrete elements having appropriate characteristics. The non-linear behaviour of the soil is incorporated in the analysis by means of an incremental solution technique.

In this paper results of analyses involving appropriate parameters relevant to the design of a soil-steel structure are presented. Moreover, possible reason for observed behaviour are also discussed.

FINITE ELEMENT IDEALIZATION

Two basic types of element namely straight beam-column element (to represent culvert) and isoparametric quadrilateral element (to represent the soil) are employed to represent soil-steel structures. Finite element mesh used in the analysis to idealize the system is shown in Fig. 1. Lateral boundary were placed at a horizontal distance of approximately six conduit radii from the centre line of the soil-culvert system. The bottom boundary were placed at three to four conduit radii vertically below the springline to simulate an infinite depth of homogeneous soil mass. Lateral boundary and line of symmetry were restrained against horizontal movement and were free to displace vertically. The bottom boundary were assumed to be fixed.

The details of the finite element analysis are presented in a thesis by the author (Karim, 1985).

MATERIAL PROPERTIES

For all the analyses a culvert section shown in Fig. 2 have been used. The only variation were in the thickness of the plate which was assumed to be made of mild steel. The culvert material were assumed to be linearly elastic, having modulus of Elasticity, $E = 29x10^6$ psi. However, the modulus of elasticity of the material of culvert was suitably adjusted to conform to plane strain behaviour. Soil properties were obtained from the octahedral stress-strain curve shown in Fig. 3 (Duncan et al., 1970). Unless otherwise stated, in all the analyses the Poisson's ratio of soil was taken as 0.45.

LOADING PATTERN USED IN THE ANALYSES

The values of equivalent continuous loading employed in the analysis described in this paper were taken as similar to those employed by Duncan (1979), which are reproduced in Table 1. These were determined by calculating the intensity of the continuous load that produces the same peak vertical stress on the crown of the culvert as does the HS-20 vehicle.

Table 1. Equivalent lane load for HS-20 vehicle.

Cover depth in feet	Live load in kip/ft
3	3.6
7	2.4
10	2.0
20	1.3
40	0.75
70	0.48

RESULTS AND DISCUSSIONS

Distribution of stresses

Study of the results of the analyses show that the distribution of the pattern of stresses in the culvert does not vary significantly with the variation of soil and/ or structural parameter. The variation of structural stress around the culvert is shown in Fig. 4. From the figure it is evident that the only significant stress is the axial stress, bending and shearing stresses being much smaller and, therefore, may be considered negligible. Hence, most of the discussions hereafter concerns axial stress only.

The distribution of horizontal normal stress on a vertical plane above crown is shown in Fig. 5. for an elliptical culvert 20 ft by 30 ft in size at a cover depth of 40 ft. The distribution of vertical normal stress on a horizontal plane through springline for the same culvert is shown in Fig. 6. It has been found that horizontal stresses in the vicinity of the culvert is reduced from those which would have generated if there were no culvert. This may be due to the movement of soil shown schematically in Fig. 7. The interior prism of soil moves downwards relative to exterior prism thus transferring part of the load to the exterior prism through shear. It has been found that in most of the analyses specially with low poisson's ratio, tensile stresses in soil occur below the culvert at the area which is shown marked with crosses (x) in Fig. 7. This might be due to the deformation of the culvert. The deformed shape of the culvert is shown in Fig. 8.

Effect of shape of culvert

To study the effect of the shape of culvert, the ratio of the major axis 'a' to the minor axis 'b' (Fig. 9 (inset)), for an elliptical culvert having constant cross-sectional area, were varied. The depth of cover was kept constant at 20 ft. Results of the analyses, presented in Fig. 8, shows that at 'a' by 'b' ratio equal to unity, i.e. for a circular x-section, stress distribution in the culvert is more favourable (minimum), than that at 'a' by 'b' ratios other than unity. For culverts whose 'a' by 'b' ratio is other than unity, stress distribution is found to be more favourable (smaller) when the major axis is in the vertical direction than when it is in the horizontal direction. This may be due to the fact that when the major axis lies in the horizontal direction, area subjected to direct vertical loading effect is more than when the culvert is placed with its major axis in the vertical direction.

Effect of depth of cover above crown

To study the effect of depth of cover over crown on the performance of buried culvert, depth of cover over a 20 ft by 30 ft elliptical culvert was varied from 5 ft to 60 ft. The results of the analyses are shown in Fig. 10. It has been found that for the size and shape of the culvert studied, there is an optimum depth of cover at which stresses are less than those for the same culvert with other depths of cover.

Stress in structural members are due to the effect of both live and dead loads. As the depth of cover increases, effect of dead load increases due to increase in soil pressure. On the otherhand, effect of live load decreases as the depth of cover over the crown increases. This is due to the fact that live loads are dispersed over a larger area when depth of cover increases. Combination of this divergent state of stress conditions due to dead load and live load is considered to be responsible for yielding an optimum depth at which stresses are minimum.

Effect of structural stiffness

To ascertain the effect of relative stiffness of soil and structure, stiffness of the culvert material has been varied keepng that of soil constant, for an elliptical shape of culvert (20 ft x 30 ft) at two cover depths 5 ft and 30 ft. Result of the analyses are presented in Figs. 11 and 12. It has been found that with the increasing stiffness of the culvert, axial load in the culvert increases although stresses decreases (due to reduction in area). This means that a stiffer culvert will carry a greater amount of load than a relatively flexible culvert when embedded in the same soil.

Effect of Poisson's ratio of soil

As stated earlier, all the analyses, described so far, were performed for various culverts buried in a soil having constant elastic parameters. In each analyses Poisson's ratio of soil was kept constant at 0.45. As part of the study, it was decided to observe the sensitivity of the interaction to changes in Poisson's ratio of the soil and examine the necessity of incorporating Poisson's ratio as a variable in the analyses. Accordingly, Poisson's ratio of the soil was varied for a particular size (20 ft x 30 ft elliptical) of culvert at a particular depth of cover (30 ft) above the crown. The result of the analyses are shown in Figs. 13 and 14. From Figs. 13 and 14 it is evident that the governing stress for the design of culvert (axial stress at springline) does not vary significantly with the variation of Poisson's ratio, although other stresses in soil and structure vary significantly. Due to a three fold variation of Poisson's ratio (from 0.15 to 0.45), variation of axial stress at springline was only 14.55%.

Variation of Poisson's ratio of soil changes the stiffness of the soil. As the stiffness of the culvert material remain unchanged, the relative stiffness of soil and culvert changes with Poisson's ratio of soil. This causes redistribution of stress in culvert and the soil.

CONCLUSIONS

The following conclusions may be drawn from the study reported in this paper.

1. Shear and flexural stresses in a flexible culvert are negligible in comparison to axial stress. Governing axial stress being that at springline.
2. The shape and orientation of the culvert affect the qualitative distribution of stresses vis a vis the magnitude of stresses in the culvert. Circular shape gives better performance (measured by smaller stresses) than elliptical shape.
3. For the same surface live load, there is an optimum depth of cover over the crown for which the stresses are minimum.
4. With increasing stiffness of culvert material with respect to surrounding soil mass, load carried by the culvert increases.
5. Variation of Poisson's ratio of soils within the practical range does not affect significantly the result of the analyses for soil-steel structure. Therefore, any reasonable estimate of Poisson's ratio will give satisfactory results.

REFERENCES

Duncan, J.M. & C.Y. Chang 1970. "Non-linear analysis of stress and strain in soils". Journal of the Soil Mechanics & Foundation Division, ASCE Vol.96, SM5.

Duncan, J.M. 1979. "Behaviour and Design of Long-span Metal Culverts". Journal of the Geotechnical Engineering Division, ASCE. Vol. 105 No. GT3.

Karim, M.R. 1985. "An Investigation of the Behaviour of Soil-steel Structures". Unpublished M.Sc. Engg. Thesis, Dept. of Civil Engineering, BUET, Dhaka.

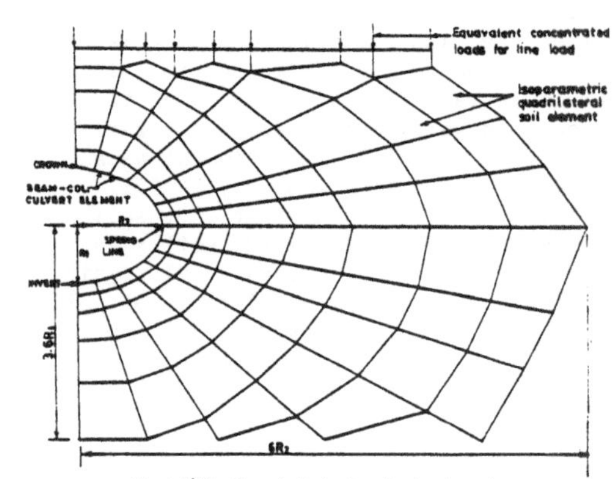

Fig. 1 Finite Element Mesh Used in the Analysis.

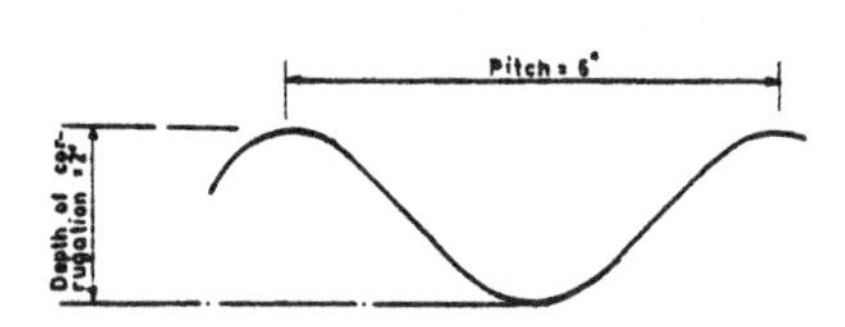

Fig. 2 Culvert Section Used in the Analysis.

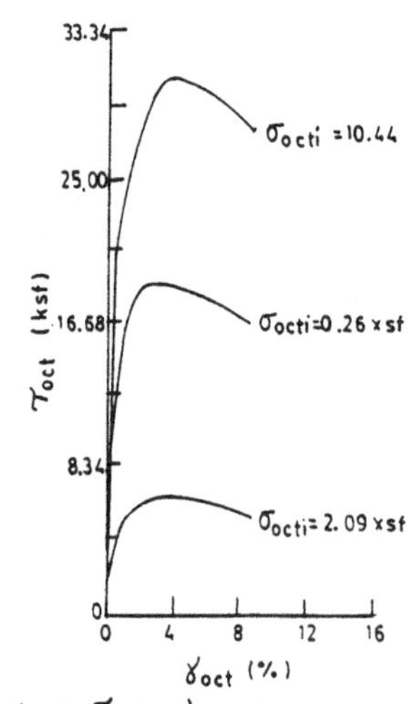

Fig. 3 $\tau_{oct} - \gamma_{oct}$ Curves used in the analysis.

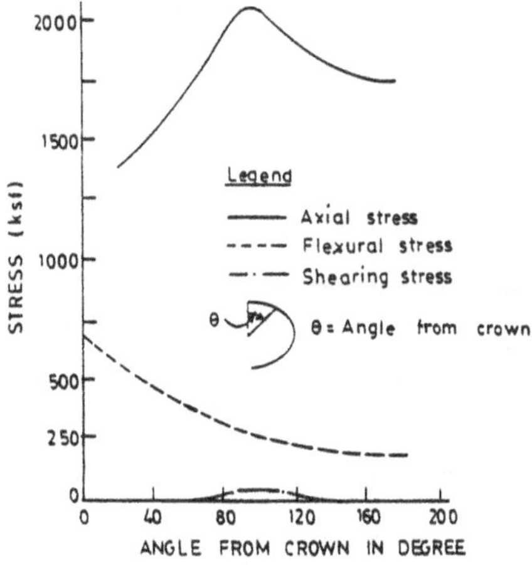

Fig. 4 Variation of stresses around the culvert.

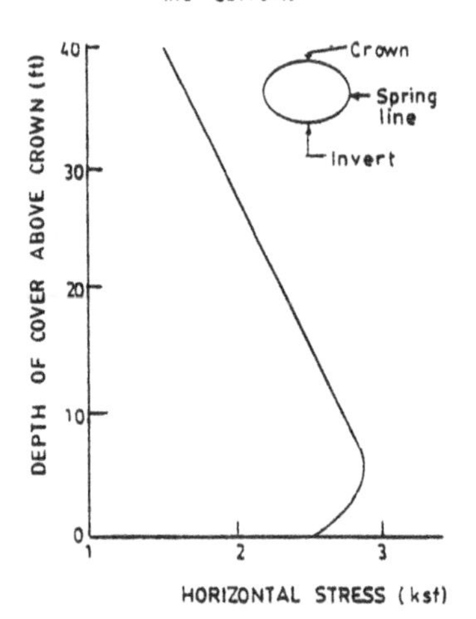

Fig. 5 Horizontal normal stress on a vertical plane through crown.

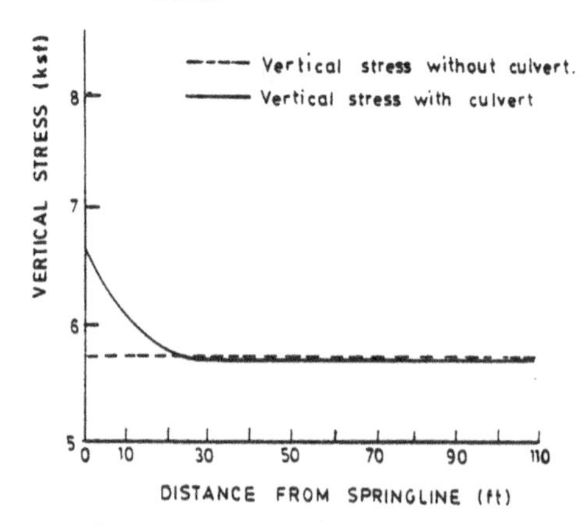

Fig. 6 Vertical stress on a horizontal plane through spring line.

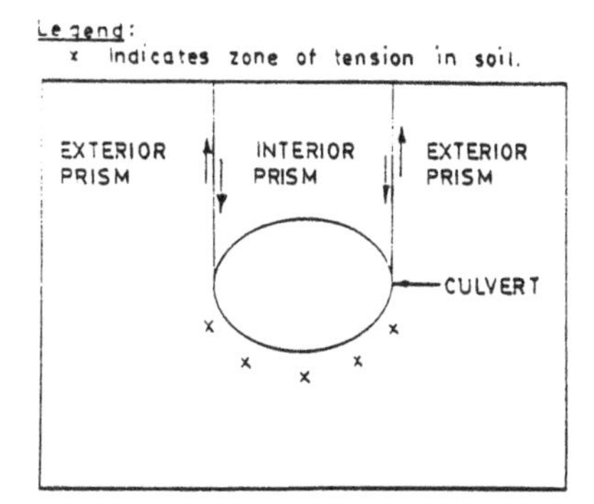

Fig. 7 Movement of soil due to applied load.

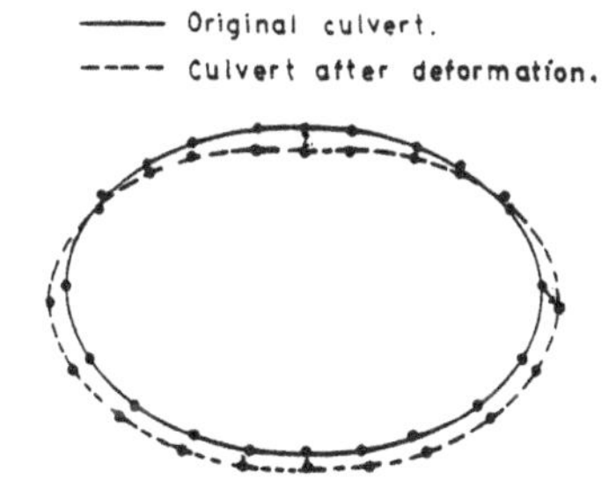

Fig. 8 Deformation of culvert.

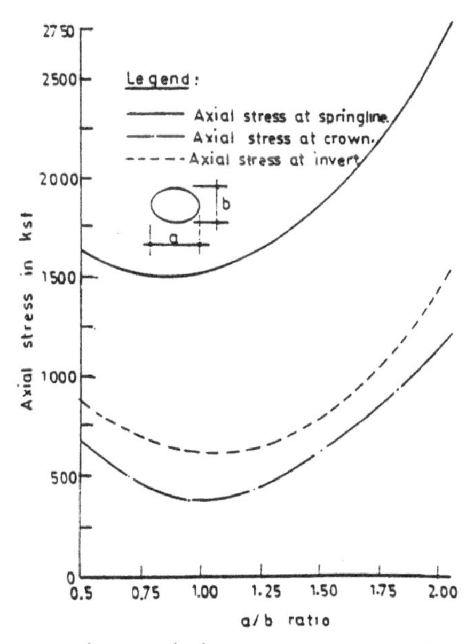

Fig. 9 Variation of culvert stress with shape of culvert.

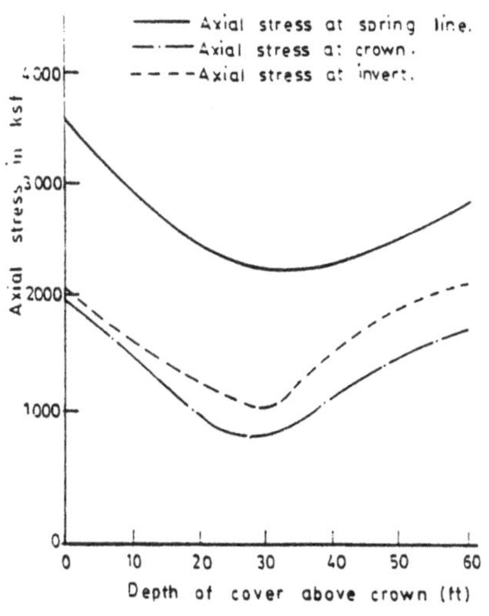

Fig. 10 Variation of culvert stress with depth of cover above crown.

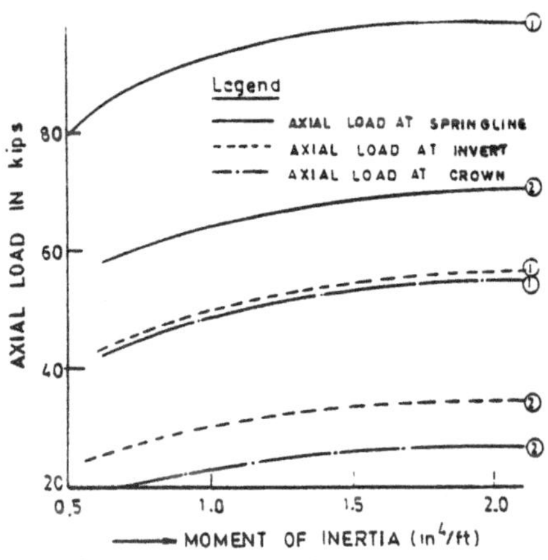

Fig. 11 Variation of axial load with stiffness

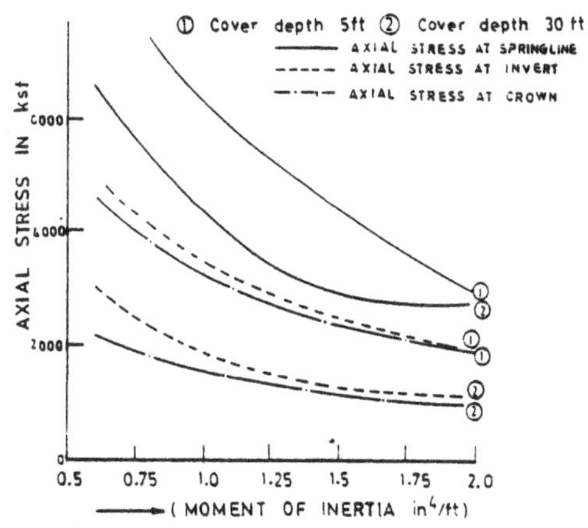

Fig. 12 Variation of axial stress with stiffness.
① Cover depth 5 ft ② Cover depth 30 ft

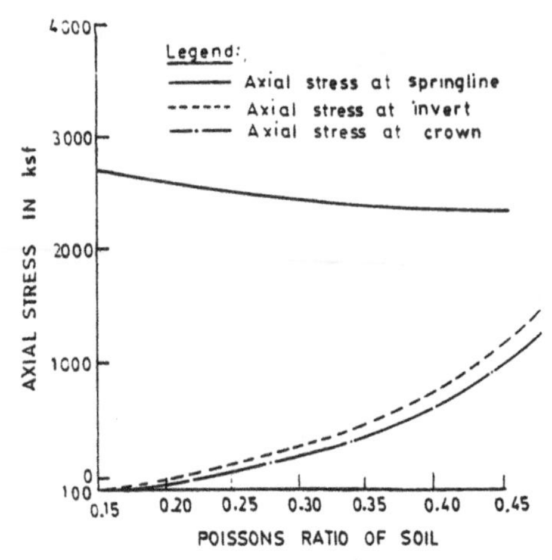

Fig. 13 Variation of axial stress with poissons ratio.

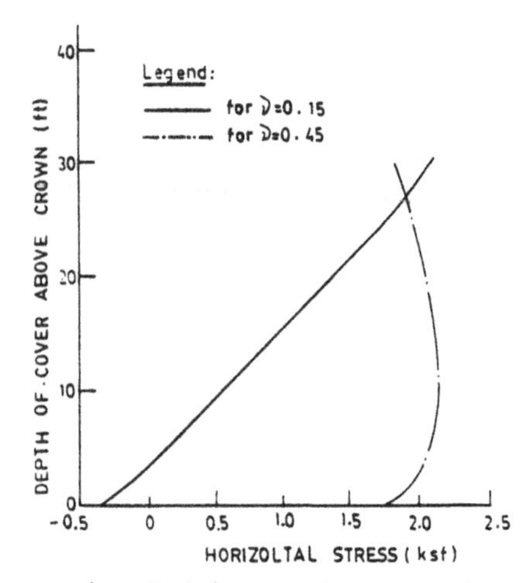

Fig. 14 Variation of Horizontal stress in a Vertical plane through crown.

Computer and Physical Modelling in Geotechnical Engineering, Balasubramaniam et al. (eds)
© 1989 Balkema, Rotterdam. ISBN 90 6191 864 2

Geotechnical modelling – An engineering approach to problems of mechanical interaction of construction work and natural geological environment

Ján P.Jesenák
Slovak Technical University, Bratislava, Czechoslovakia

ABSTRACT: On the basis of theoretical considerations, general rules are given for transforming the engineering-geological model of the geological environment into the geotechnical model of an engineering problem within this environment. Categories of geotechnical models are defined and illustrated with examples. The engineering character of geotechnical work is emphasized.

1 INTRODUCTION

Protection of the natural environment has become one of the most serious challenges of Nature to Man: success or failure in solving this problem may decide about the survival of future human generations in large areas of the planet. Inconsiderate exploitation of natural resources has already caused much incorrigible damage in different parts of the Earth: to restrict further impacts and, wherever possible, to restore and to improve original conditions should be a undisputed imperative of any human activity.

However, the development of human society cannot be stopped nor reversed, and so the manifold interaction between the natural environment and human life will continue. Much work has been done in the field of environmental control. The serious problems which arise are discussed and solved on governmental and even at an international level. These trends must continue and increase, and environmental reasoning must become an integral part of general human thinking.

Construction work is one of the branches of Man's activities with most serious impact on the natural environment. The mechanical interaction of construction work with the geological environment is a matter of geotechnics: the environmental aspects of construction have always been an integral part of well-designed geotechnical work. The last few decades have presented old problems in new dimensions: the accelerating development of human society all over the world has caused an immense demand for construction work, requiring more effective methods of site assessment, planning, design and construction.

There are two scientific disciplines concerned with problems affecting the interaction of construction work and the natural geological environment, namely engineering geology and geotechnical engineering. Engineering geology is a branch of geological science: it presents models of the geoenvoronment defined in different ways and on different taxonomical levels. These models serve as basic information for regional planning, construction, environmental control and partly for mining, too. Geotechnics is an engineering science with an interdisciplinary character: it solves concretely defined engineering problems in concretely defined geological conditions, and these can be modelled by engineering geology. Mutual understanding between these two sciences is often far from ideal: they have different ways of deliberation, a different terminology and, most important of all, a different level of accuracy.

This may cause problems when transforming the engineering-geological model of the geological environment into the geotechnical model of an engineering task within this environment.

Improvements in the social efficiency of the whole complex of geological-geotechnical works for construction purposes require a realistic modelling both in the engineering-geological and in the geotechnical phase of the work. Considering geological and geotechnical work to be one complex problem, the objective of further development will be both technological and economical optimization, starting with geological assessment and ending with completed construction.

This paper is a short generalization of experience gained in situations requiring close cooperation between engineers and engineering geologists in the course of the last 25 years in Czechoslovakia.

2 A GENERAL SCHEME FOR GEOTECHNICAL SOLUTIONS

A well-known but often not fully respected peculiarity of geotechnical engineering is that it works not only with man-made materials like steel or concrete, and uses not only structures with more or less clear statical effects, but is for the most part concerned with soils and rocks, and the results of natural geological processes. The geological structure of the site as well as the considerably varying properties of individual subsoil strata, and also the scatter of their characteristics have to be stated in each individual case by expensive and time-consuming assessment, field tests, laboratory exploration and testing. All these only give information on individual points of the subsoil (samples, field tests), or in vertical, eventually inclined or horizontal, straight lines (sounding, boring). To define the engineering-geological model (EG model) of the site, extrapolation on the level of a professional estimate is necessary. Data on the geological environment therefore cannot provide exact information in relation to the construction work, and this represents a limiting factor for any soil-structure interaction problem.

There are two principal decisive factors for any geotechnical solution:

1. a correct estimation of the physical substance of the interaction, and
2. a realistic EG model of the site, i.e. an engineering-oriented abstraction of the natural geoenvironment.

On the basis of these considerations, a general scheme for the solution of geotechnical problems had been compiled (Fig.1). The scheme is universal. In individual cases some marginal steps may not be of use. The interdisciplinary side of the solution is emphasized, and assumes that geotechnicians are familiar with the engineering aspects.

Simultaneous evaluation of the physical substance of interaction, the claims of the designed construction work, and the EG model of the site allows the geotechnical

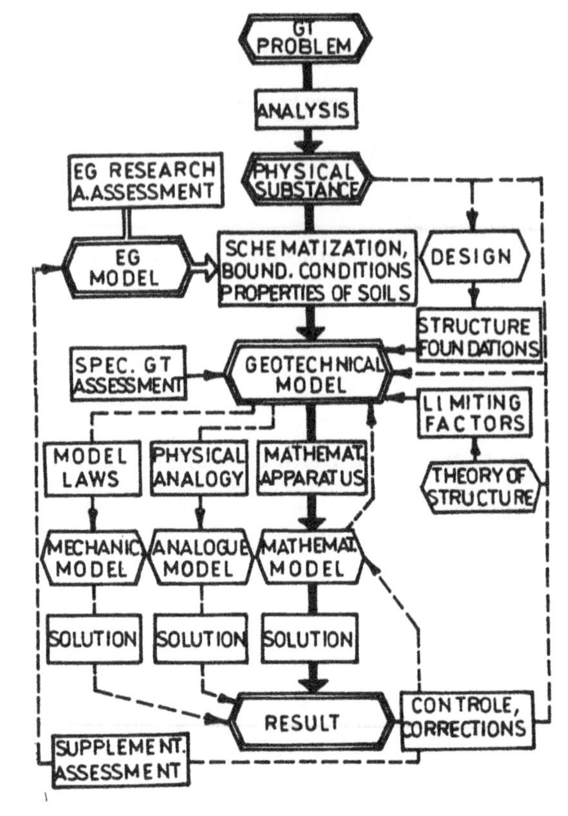

Fig.1 A general scheme for the solution of geotechnical problems

model of the problem to be defined. The term "geotechnical model" (GT model) was introduced as a basic model for the solution of a given geotechnical problem, defined merely in engineering terms. The GT model represents a further step of abstraction of the natural geoenvironment, evaluated with respect to a certain, often very narrowly defined engineering problem.

The main way of finding a solution is the mathematical one: an adequate mathematical apparatus makes it possible to define the mathematical model (M model) of the solution. Note that between the EG model and the GT model there is only a one-way dependence, whereas a backfeed exists between the GT and the M model: the GT model must be defined already with respect to a certain appropriate method of mathematical solution, though in special cases experimental model-solutions may prove advantageous.

As an important conclusion we can state that realistic geotechnical prognoses depend primarily on a correct estimate of the physical substance of the interaction and on realistic input data. The level of the mathematical solution itself plays a secondary role: lack of information cannot be compensated for by sophisticated mathematics. The benefits of more perfect mathematical solutions may be obtained only if adequate input data are available.

3 GEOTECHNICAL MODELLING

3.1 Theoretical considerations

Both EG and GT models interpret the natural geological conditions for purposes of planning and construction. This interpretation is not easy: it must overcome the interdisciplinary gap between the classical geological disciplines as currently practised with the methods of the natural sciences - and, civil engineering, which solves its problems with technical, and often technological, methods and procedures. Engineering geology belongs to the group of geological sciences: EG models are, in principle, geological models, which interpret natural conditions in terms understandable for technicians, and if possible in a numerical manner. Geotechnics belongs unambiguously to the group of engineering sciences. GT models must therefore be defined merely in engineering terms,in spite of the fact that they reflect not only a technical problem, but also the natural geoenvironment. There are some peculiarities concerning GT models: they are always defined from the standpoint of an individual, often very narrowly specified geotechnical problem, and they must be defined in a manner fully corresponding to a certain supposed mathematical or experimental method of solution.

So the EG model is the basis of the solution, and the M model is a medium for it. The GT model, however, is an integral part of the engineering solution, and has a decisive influence on producing realistic results.

Both EG and GT models are abstracted from the same natural geological environment: they represent two steps of the assessment and adjustment of this environment for engineering purposes. Primary information about natural conditions provides the EG model. The GT model (or, more correctly, the GT models since they relate to different purposes and solution methods) must strictly respect them. In this sense the GT model plays a secondary role. Sometimes it is possible to define the GT model without a previously compiled EG model, e.g. if solving a detailed problem on the basis of a special geotechnical assessment. Of course, this does not alter the secondary, derived character of the GT model as a special purpose engineering interpretation of the natural geoenvironment. It is to be emphasized that interdisciplinary difficulties on the boundaries of the geological and engineering approaches can be economically mastered only by strict conservation of the correct sequence of EG and GT modelling in the course of the solution.

3.2 Schematization of the real conditions

A model is an abstraction of the reality. It conserves the mean properties of the reality, which are assumed to be decisive from a given point of view, and neglects the secondary characteristics in order to make the model open for mathematical or experimental treatment. The mean way of abstraction is schematization. Simplifications refer to

1. the geological structure of the zone of interaction,
2. the hydrogeological conditions of the site,
3. the physical properties and mechanical behaviour of the soils and rocks,
4. the boundary conditions of the problem,
5. the interaction between construction work and geoenvironment.

The way and the size of schematizations depend on several conditions:

1. the relative spatial density of EG and GT information within the zone of interest,
2. the quality of this information,
3. the expected accuracy of the solution,
4. the optimum scale of the model under given conditions,
5. the kind of problem,
6. the level of the available mathematical or experimental apparatus.

A detailed analysis of these circumstances, though very important in practice, will be omitted here.

3.3 Transformation of the EG model into the GT model

The EG model of the zone of interest is from the interdisciplinary standpoint the first important basis for the GT model. In accordance with the nature of the problem and the requirements of the chosen way of solution (mathematical, experimental and combined methods are possible) it is necessary to transform the EG model, i.e.

to adapt it for a particular engineering problem and to define it quantitatively in engineering terms. This transformation of the EG model of the geological environment into the GT model of a special engineering problem within this environment is the most important part of the solution, at least from the geotechnical point of view: it determines whether the GT model will be sufficiently representative, and will be able to provide realistic and accurate results at the required level of precision.

In individual cases, the way of transformation will be different. Only technical principles may be generalized. The following steps should be adhered to:

1. specification of the physical substance of the interaction,
2. definition of the zone of interaction,
3. estimation of the field of interaction intensity,
4. schematization of the geological structure within the zone of interest,
5. determination of the properties of soils and rocks,
6. definition of the boundary conditions.

Some explanations may prove useful:

1. A realistic estimate of the physical substance of interaction and its adequate schematization play a decisive role and therefore require great attention, especially in cases exceeding the level of conventional solutions. For example, the stability of the slope of a tailings impoundment may be described as sliding along slip surfaces, and this may be correct from a mathematical point of view; but this description may lead to confusion, if a breakdown by liquefaction or destruction by seepage and suffosion should occur. If, however, the physical substance of the interaction has been correctly defined, even severe simplifications and simple mathematical solutions may lead to realistic results.

2. The EG model describes, as a rule, a substantially larger zone than that directly participating in the mechanical interaction. The zone of interest of the GT model includes the area affected directly by the interaction and its surroundings (over not too wide an area) and is stated as a professional estimate on the basis of engineering experience, with respect to the requirements of the chosen method of solution.

3. The intensity of mechanical interaction may or may not be uniform within the zone of interaction. The field of interaction intensity has a great influence on the admissible measure of schematization: areas with low intensity of interaction tolerate much more schematization without substantial influence on the precision of the results.

4. The model of the geological structure must be geometrically defined with simplifications according to the interaction intensity and with respect to the facilities of the method of solution.

5. At the time of an EG assessment, the method of solution is often still unknown, and sometimes even the GT problems have still not been defined clearly. In such cases, the properties of soils and rocks have to be determined later by a special geotechnical assessment of the site.

6. Simple calculations are less susceptible to correct boundary conditions than complicated mathematical methods. Great attention has to be paid to making an adequate choice of boundary conditions when using the method of finite elements (FEM). A lack of experience should be compensated for by checking the influence of the boundaries with parametric studies.

All the factors mentioned above are closely connected with the schematization of both natural conditions and the interaction, and may have considerable influence on producing realistic results. The fundamental basis for a successful transformation is, however, a well defined realistic EG model of the natural geological environment.

The EG and GT assessment of sites are, as a rule, carried out in several stages. In the course of the assessment and the design, the EG, the GT, as well as the M model become gradually more and more realistic. The requirements and the conditions of schematizations remain valid, and always rank successively on higher qualitative level.

3.4 Categories of GT models

Geotechnical problems occur at different stages of planning, design and construction, under difficult conditions, even in the post-construction period. If planning and design have not been supported by representative geotechnical information, the consequences appear later, during the construction period, or after the job has been finished. Additional changes in the distribution of structures in housing areas affect the ideas of the designer and the environment of future inhabitants, and they may also cause trouble in the planned flow of technology in industrial plants. Unforeseen changes of foundation methods and additional site stabilization always lead to a considerable increase of costs and a loss of time.

It is therefore very important to have geotechnical information available for the earliest stages of country and urban planning and design, when deciding on the selection of the site, the distribution of individual objects in new-built housing districts or industrial zones, or the alignment of roads, railways, channels and levees. The EG basis for this stage are EG maps of typolocigal zoning. Lower taxonomical units of EG typological zoning-subrayons and districts - are already directly connected with construction works and they represent an appropriate level of EG models, which are qualitatively homogeneous (Matula 1976). Transforming them, we get "GT type models". If the latter are represented graphically in plane, maps of geotechnical typological zoning will arise. This kind of GT modelling will be advantageous - first of all on large sites with relatively simple geological structure, when (a) the detailed distribution of buildings is still unknown, but the types of objects, the loads and their transfer into the subsoil are given; or(b) the types, loads and the distribution of objects are known, and the kind and intensity of the interaction and its technological and economical consequences are to be estimated.

The second fundamental case arises at the interaction of a single object with the subsoil. The zone of interaction is relatively small and spatially determined, and the distribution of interaction intensity is (a) known - when a conventional GT problem in concretely defined geological conditions should be solved; or (b) still unknown, when a new unusual case of interaction occurs. For such cases, the term "special GT model" has been introduced.

The third case originates when the relative influence of variable input data on the result of the solution should be stated. This case leads to "parametric GT models", which solve concretly defined geotechnical problems in altering idealized conditions by iteration. The results of parametric GT models supply essential information when simplifications have to be introduced.

Each of the three categories - GT type models, GT special models, and GT parametric models - occur in practice in different modifications and at different scales. Each group has, however, its specific properties, which distinguish it clearly from the others. In the following part of the discussion attention will be paid to these categories, illustrating them by several examples.

4 GEOTECHNICAL TYPE MODELS

4.1 The principle hypothesis

The basis of GT type modelling is the hypothesis that locally variable parameters of the subgrade (i.e. kind, thickness and sequence of soil strata, their properties, the ground water level, etc.) may be replaced for construction purposes on limited areas by unified type values of these parameters in order to obtain the required level of accuracy on a model unit of the subsoil, which may be considered, from the standpoint of a certain geotechnical problem (e.g. the foundation of structures), to be qualitatively homogeneous. The practical consequence of homogeneity is the possibility of solving a certain geotechnical problem throughout the whole area in the same way (e.g. equal type and dimensions of foundations for equal structures).

The validity of this hypothesis has been proved in practice. The advantages of the method both for designers and contractors are evident.

4.2 Geotechnical typology

GT typological systems may be compiled for different purposes and for different scales. They must fulfil the following requirements:

1. universality: they must include all types of soils and subgrade structures, which may occur in practice, at least on a given site,
2. typification of Quaternary and pre-Quaternary soils and rocks, and typification of the geological structure of the subgrade within the zone of interest;
3. unification of the geotechnical properties of each soil type, and unification of the subgrade structure within each type, unification of conditions for a certain geotechnical task;
4. compatibility with obligatory standards and regulations, and compatibility with an appropriate system of engineering geological typology and zoning;
5. adaptability for different scales, for changes of classification and typification criteria, and for various numbers of soil types;
6. system access: logical (when stating suitable types of soils and subgrade structures) , and formal (when marking types with brief and instructive code numbers to be used in maps and sections).

The compiling of the typological system is one of the most responsible parts of GT typological modelling, and has a decisive influence on the practical serviceability of the resulting GT type models and the GT typological zoning of the site.

4.3 Small-scale GT type models

In Czechoslovakia a number of GT typological systems was developed for different scales and for different purposes.

Logical connections between small-scale EG type models and the appropriate GT type models are shown in Fig.2.

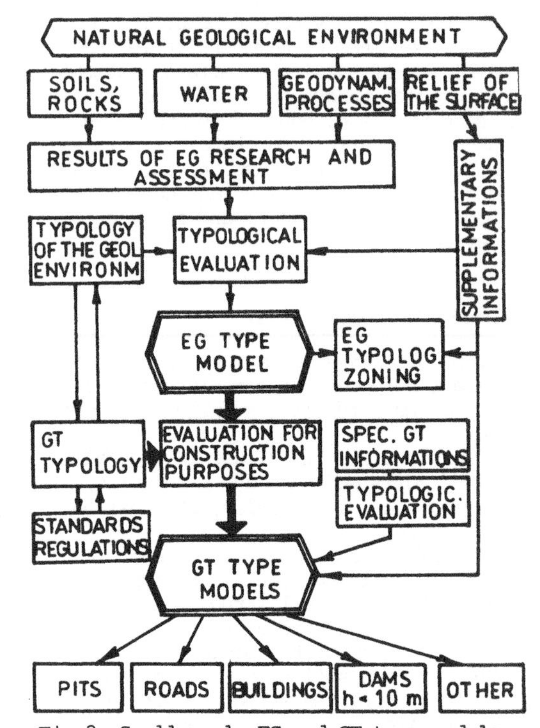

Fig.2 Small scale EG and GT type models

For systematical engineering geological mapping for the purpose of regional and landuse planning and environmental control the scale 1:25 000 is used. It is very useful to complete these by giving basic geotechnical information. The typological system used here, a result of cooperation of engineering geologists and geotechnicians, was developed and first used before 1970, and was later improved and entirely coordinated with the system of EG typological zoning. Only the main principles of the system can be given here:

Eight types of Quaternary soils and eight types of pre-Quaternary bedrock are distinguished, the first marked with small letters, the latter with capitals (Table 1 and 2).

For Quaternary soils, the thickness is distinguished in three steps, marked with arabic numbers, written behind the symbol of the soil type. Similarly, the depth of the bedrock below the surface is given

Table 1 Quaternary soil types

No	Type	Code
1	Gravel, sandy gravel	g
2	Sand	p
3	Cohesive soils, gravel with cohesive fill, soft	m
4	Ditto, stiff or hard	t
5	Alternating thin cohesive and loose strata	k
6	Loess, loess loam	s
7	Boulders, cobbles (sandy fill)	b
8	Soils not suitable for foundations (organic, very soft layers a.s.)	o

Thickness	Code	Type thickness
1 to 2 m	1	1.75 m
2 to 5 m	2	3.50 m
> 5 m	-	7.0 m (max)+
		14.0 m (max)+

+Note: max. 7 m if bedrock lies within 5-10 m, max. 14 m if it lies within 10-20 m below surface

Table 2 Bedrock types

No	Type	Code
1	Hard rock, compact	S"
2	Ditto, fissured, wheathered	S
3	Weak rock	B
4	Residual soils	Z
5	Gravel, sandy gravel	G
6	Sand	P
7	Cohesive soils, hard	T
8	Alternating cohesive and loose strata	K

Depth b.s.	Code index
< 5 m	1
5 to 10 m	2
10 to 20 m	3

in three steps and marked with arabic numeral indices. For preliminary calculations (e.g. of the settlement of appartment houses in new living areas), the type thickness of the Quaternary soil strata is estimated so as to produce results which will be on the safe side. Type values of the basic properties of soils are derived from Czechoslovak State Standards for soil classification and foundation of structures (not given here).

The introduction of type codes proved advantageous: they are easy to memorize and are instructive. For example, a section with 1.9 m stiff silty loam and 6.6 m sandy gravel resting upon Tertiary claystone gets a type code

$$t1gB_2$$

This code number marks simultaneously all sections with the same soil types, with the upper stratum being 1-2 m thick, the second > 5 m thick, and with the bedrock surface lying between 5 and 10 m below the surface of the terrain. A comparison of possible variations of the real section with the unified type section is shown in Fig.3: at a scale of 1:25 000 the differences are not really substantial.

If qualitatively homogeneous type models are presented in plane, maps of typological zoning arise. At a scale of 1:25 000

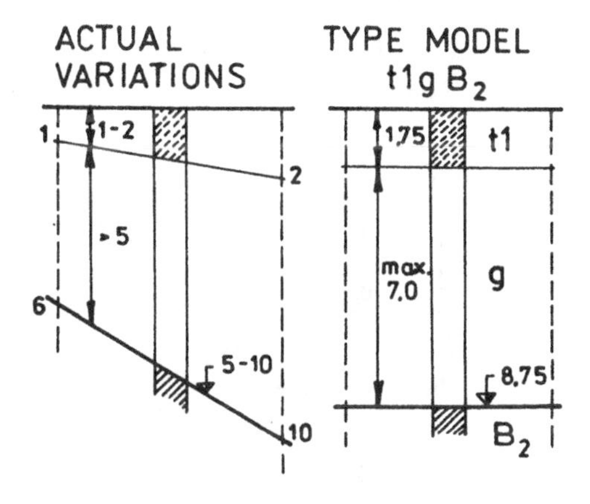

Fig.3 Comparison of possible variations of the actual section with the type section $t1gB_2$

the maps of EG and GT typological zoning are identical. Geotechnical data are here given for each distinguished type model of the subgrade (i.e. for EG subrayons and districts) only as a complement. On the basis of calculations using the type values of soil parameters, semi-quantitative scales were compiled in order to allow a preliminary estimate of conditions for different geotechnical works. For instance, the effects of the subgrade on column footings of skeletal structures were classified on the basis of the allowable settlement. The appropriate foundation work was defined as follows:

Degree	Definition
Low	individual footings in minimum depth, contact area <5 m^2,
Medium	contact area >5 m^2, or foundation in greater depth wanted,
High	pile foundation necessary,
Very high	foundation on large diameter piles or raft foundation are necessary.

Similarly, scales were compiled for structures with load-bearing walls (strip footings), for pits and trenches (recommended slope inclinations and expected water inflow are given), for roads (the suitability of the subgrade as a base for pavement constructions, as a material for the construction of embankments, as subgrade for embankments, and the workability of soils were estimated) and for low dams and levees (the suitability of soils for constructing the impervious, or the pervious parts of dams, and the properties of the subgrade were estimated).

On occasion, also problems involving other principles are to be solved. During a longlasting flood on the Danube river underpiping caused two breaks of the left-bank levee. About 1000 km^2 of most fertile agricultural land, some 40 villages and two towns were inundated for several months, and the losses were very high. Afterwards, a geotechnical basis for reconstruction and strengthening of the round 150 km long Czechoslovak section of the left-bank alignment of defense was required.

The unusual task had been solved on the basis of archives; data (about the geological structure of the subsoil, hydrology, history of dam construction from its beginnings in the 13th century, and observations of seepage and piping effects during floods), and investigation of the terrain. Six factors were considered to be important:

1. the cross section of the dam (single or double dam, without or with short or long diaphragm walls against seepage),
2. the height of the dam (up to 7 m),
3. the material and properties of the dam body (old parts only slightly compacted, locally pervious sandy layers),
4. the permeability of the upper soil stratum (sandy or silty),
5. the seepage and piping phenomena registered,
6. the situation of dam alignment related to the river (broad or narrow forefield, material pits, convex or concave side).

Each of these factors was expressed with several degrees, marked by increasing numeral codes. Sections with constant degrees of all six factors were considered to be qualitatively homogeneous, i.e. guaranteeing equal safety against damage by means of longlasting high floods on the river. The codes of the six factors gave the code number of the section: it is instructive and easy to memorize. For mutual comparison of different sections a complex characteristic was introduced. This code we called the "group index": this is a number obtained as the simple sum of six individual codes forming the code number of the section. The group index value served as a criterion for declaring that during long-lasting floods the section would be perfectly safe, not fully satisfactory, or in a critical state. Reconstruction was started on sections falling into the latter category.

The results of the research were graphically presented in a longitudinal section 1:50 000/500. The method of this GT typological zoning had already been described in detail (Jesenák,1974).

Similar cases may be successfully solved only if a sufficient amount of information is available, and if its coverage of the whole of the zone of interest is approximately uniform. A thorough knowledge of the natural geological environment also plays a primary role here.

4.4 Large-scale GT type models

The advantages of large-scale GT type models (1:5 000 up to 1:500 and more) are evident mostly on large sites with a relatively simple geological structure, and with repeated types of buildings (e.g. appartment houses); but they are important also if different objects are designed. Zoning of the alignment of roads, railways and levees is also beneficial.

The connections of large-scale GT type modelling with the results of EG research and assessment are shown in Fig.4. Presentation of GT type models in plane leads to GT typological zoning, which at this scale may or may not be directly connected with an appropriate EG typological system.

Each GT type model contains an engineering abstraction of real geological conditions. For a certain type of geotechnical work, however, a different geological structure of the subsoil may represent equivalent conditions. For example, for flat foundations or even for pile foundations of buildings it is often indifferent, whereas the firm subsoil below the compressible upper strata is gravel, hard clay or rock. If, as a further abstraction, in such cases homogeneous units are defined

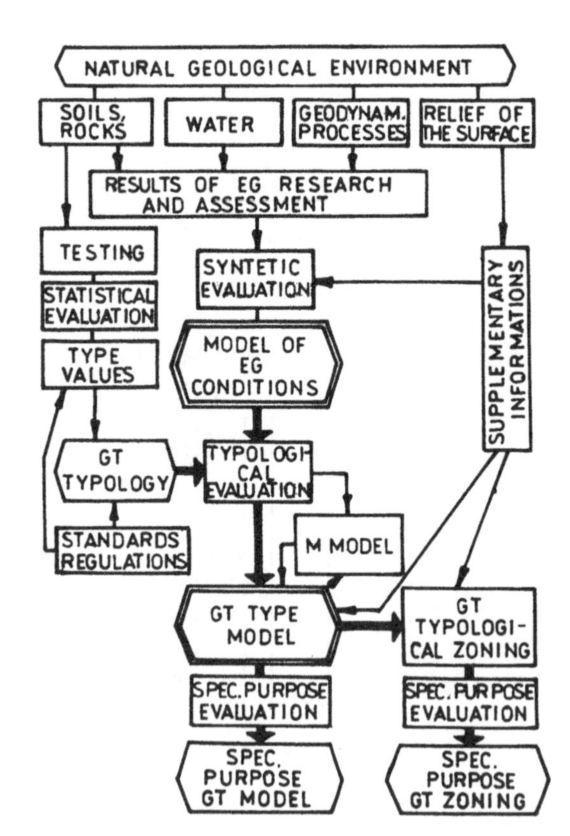

Fig.4 Large scale models of EG conditions and GT type models

on the basis of a merely geotechnical criterion (e.g. a certain settlement, or the necessary length of piles), single-purpose special GT typological modelling and zoning of the site will result. Geotechnical criteria of this sort represent, from a special geotechnical point of view, the "complex characteristics" of the subsoil. Only a complex characteristic allows a qualitative comparison of various GT type models (compare with the "group index" in the Danube-levee example).

Fig.5 demonstrates schematically the procedure of prognosing the probable amount of different types of foundations on the large site of a new housing area. The upper scheme shows the relative series of appartment houses according to the project, and the bottom scheme shows the series of available types of foundations. Typological evaluation of the geological structure of the site renders a series of GT type models (A to N, middle right), which give, after evaluation through a suitable complex characteristic, a series of six types of conditions for different types of

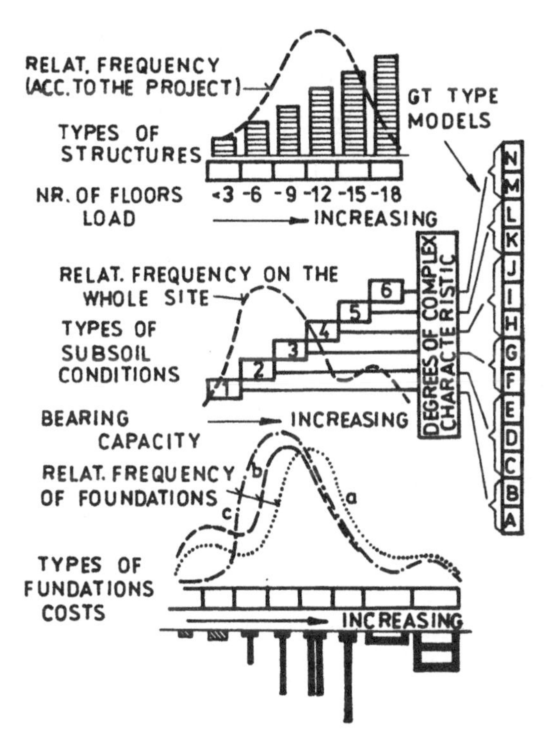

Fig.5 Prognosing the amount of different foundation types on the basis of GT typological zoning

foundations, with a certain relative frequency according to real conditions on the site (middle scheme). According to the originally planned distribution of buildings on the site, for each category of loading there is an optimum type of foundation: this gives line "a" for the frequency of foundations on the site (bottom scheme). Should a greater number of heavy structures be constructed in poor subsoil conditions, situation corrections will be unavoidable(b). Technological and economical efficiency demands some restrictions in the number of different foundation methods and determines the final distribution of various types of foundations on the site (c).

As another example, the typological evaluation of the subsoil of a new housing district, is shown in Figures 6 to 9 (originally at the scale of 1:2 000). Fig.6 presents the results of the GT typological zoning, in principle according to Tables 1 and 2. Only the code $\bar{t}$ was introduced for gravel with cohesive fill in order to demonstrate that the fill material is the same, like the overlying clay (both of Quaternary age). The upper zone of the Tertiary bedrock (claystone and clay shales) goes to depth of about 3 m and is uniformly wheathered. (This is not marked in maps). The marginal parts of the site are locally threatened by stabilized and active slope sliding. Fifteen different types of subsoil occure on the site (see Fig.6) : the GT typological zoning shows a mosaic of type models of the subsoil which are irregular in shape, but qualitatively homogeneous. Figures 7, 8 and 9 are parallel maps of single-prupose GT zoning: for 2-floor sceletal structures (schools, store houses, services, etc.), for 6-floor and 12-floor appartement houses. As a "complex characteristic" for flat foundations (block or strip footings), different values of settlement (up to the maximum allowable value according to Czechoslovak State Standards specifications = 6 cm), and for pile foundations the necessary length of 38 cm diameter VUIS-type vibro--bored cast-in-place concrete piles are used.

Figures 6 to 9 show clearly, that the map of GT typological zoning is more complicated, like the maps of special purpose GT zoning. At the same time the first has the advantage of reflecting the real geological structure of the subsoil, and may therefore serve for different GT purposes. The latter are very simple and very advantageous for designers and contractors, but they can only serve the purpose they were compiled for.

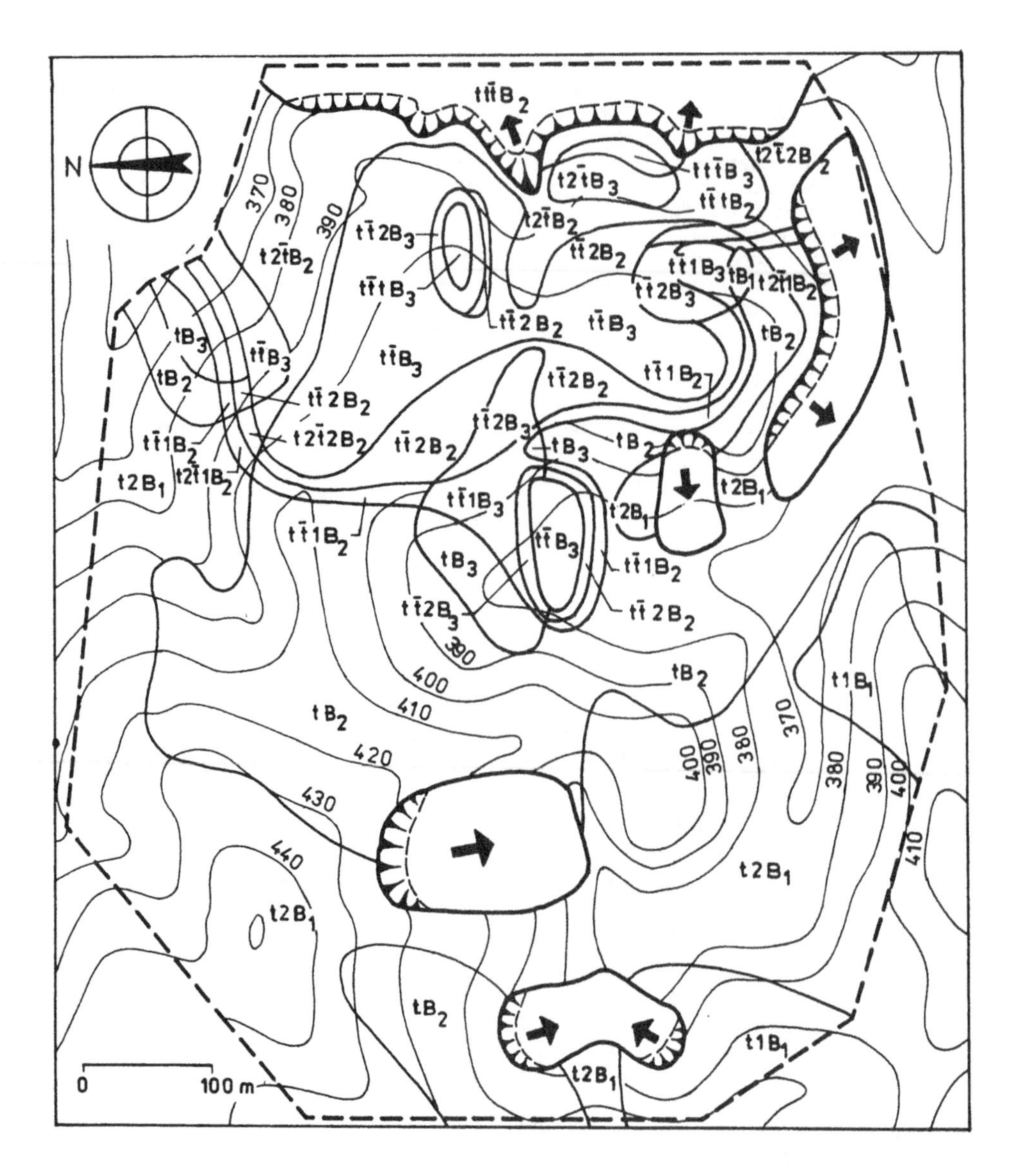

Fig.6 Geotechnical typological zoning map of a large site (originally at scale 1:2000)

4.5 Diagrams for rapid computations

The evaluation of GT typological zoning maps requires a lot of geotechnical calculation. For example, for constructing the special purpose maps in Figures 7 to 9 the settlement of all types of designed structures in all types of the subsoil had to be computed. Where the resulting settlement exceeded the maximum allowable value, it was calculated again for pile foundations for different length of piles.

In order to make such calculations more rapid and to simplify them so that they may be done by technicians with lower professional qualifications, a series of diagrams were compiled. To facilitate settlement calculations, first the results for the elastic halfspace and those obtained according to one-dimensional methods were compared for average conditions. Since the differences were smaller than errors caused by oscillations of soil parameters, using the simple and more adjustable one-dimensional method diagrams for a settlement influence factor "K"

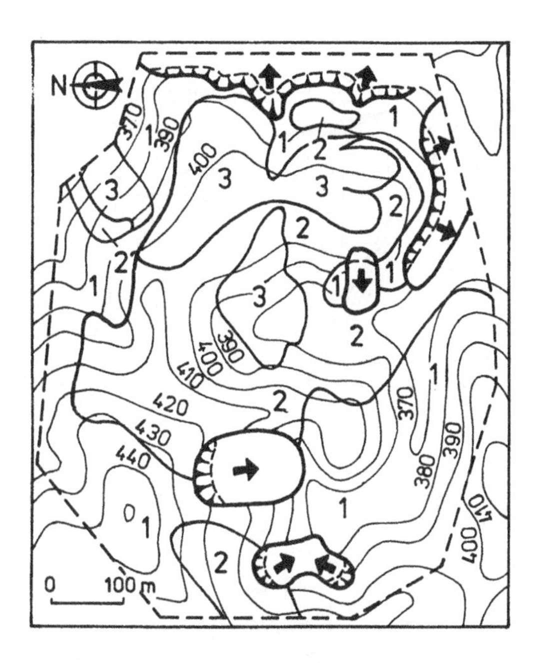

Fig.7 GT zoning for 2-floor skeletal structures: 1-settlement < 3.5 cm, 2-settlement 3.5-5.0 cm, 3-settlement > 5 cm

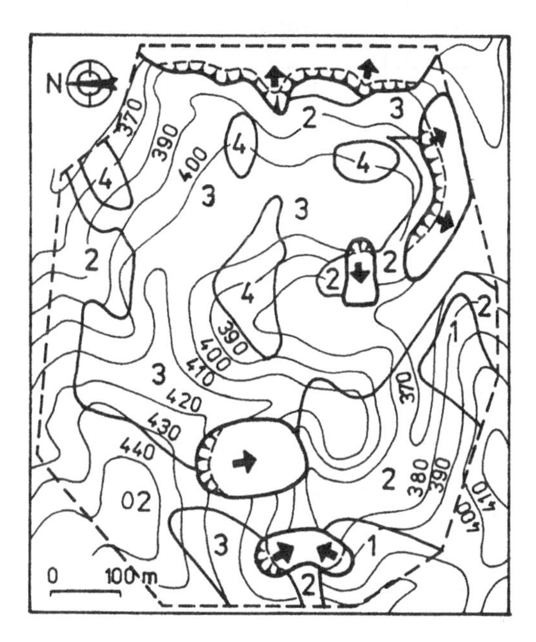

Fig.8 GT zoning for 6-floor appartment house: 1-strip, s < 3.5 cm, 2-strip, s= 3.5-5.0 cm, 3-strip, s > 5 cm, 4-piles Ø 38 cm, l=6 m

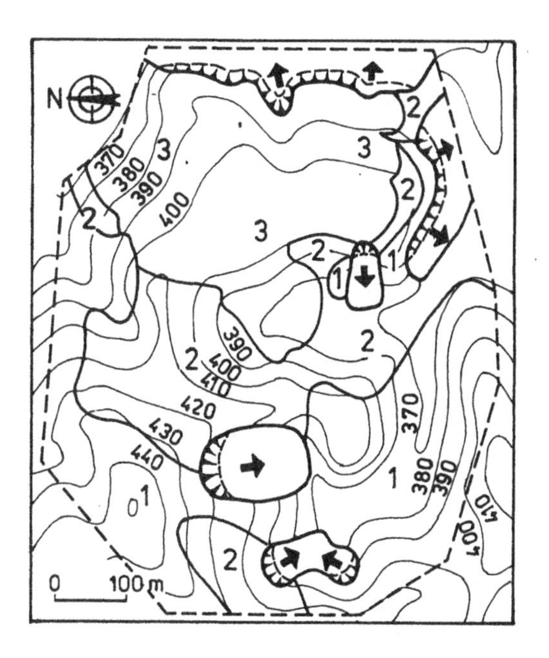

Fig.9 GT zoning for 12-floor appartment house: piles Ø 38 cm, 1-l=8 m, 2-l=10 m, 3-l=12 m

were worked out. The following cases were considered: the corner of a uniformly loaded rectangle, the characteristic point (Fig.10a), the uniform and triagnular strip loads, and the uniform load on a circular area. A lot of similar diagrams may be found in the literature, mostly compiled for homogeneous elastic subsoil and therefore hardly applicable for layered geologic structures. The influence of neighbouring loads was also computed and presented in diagrams for different axial distance x/B of interfering footings (Fig.10b). Similarly, diagrams were compiled for the settlement of road and highway embankments with cross sections according to Czechoslovak State Standards specifications(Jesenák,1980), for arbitrary dam sections (by superposition), for consolidation, for water inflow in pits and trenches, for the stability of their slopes, etc.

When using diagrams at suitable scales, the results are fully equivalent to the results of computations, the probability of errors is much lower, and the saving of time,is for simple cases about 90 %, for more complex ones about 75-80 % (Jesenák,1978, 1979). The graphical presentation of generalized results of computations proved a powerful tool for mastering the large numbers of calculations connected with the geotechnical evaluation of subsoil type models.

5 GEOTECHNICAL SPECIAL MODELS

5.1 Definition and specification

The conventional engineering approach to soil-structure interaction problems is by means of GT special models.

The measure of schematizations of the geological structure, hydrogeological conditions, material properties, boundary conditions, and of the kind of interaction is here restricted to the extent which is unavoidable from the standpoint of the solution method.

The GT special model is then an engineering abstraction of an interaction problem, defined quantitatively with respect to the physical substance of the interaction and to the actual structure and properties of the zone of interaction, in accordance with the requirements of the chosen method of solution and of the expected level of accuracy. The main attributes of GT special models distinguishing them clearly from other categories of GT models are

1. they give a true description of real conditions throughout the whole model,
2. they have a unique character, eliminating repeated application.

GT special models are always defined at large scales. They are inevitable when detailed solutions of interaction problems of large and pretentious structures with the subsoil are required (bridges, tall buildings, silos, nuclear, thermal or hydroelectric power plants, heavy and big structures), or when the behaviour of large dams, tailings dumps or impoundments, open pit mines, etc., has to be prognosed. GT special models are used also for geotechnical solutions at smaller dimensions: retaining structures, sloped, braced and sealed pits, cuts, embankments, detailed research on flat or deep foundations, anchoring, etc. Also, problems of seepage and drainage under natural or artficial conditions are, as a rule, solved on the basis of GT special models.

From the above considerations it follows that the GT model will be "special" in two cases:

1. if there is a unique geotechnical problem to be solved, where typification is not possible or absurd,
2. if on the required level of accuracy, the detailed knowledge of the structure, the behaviour of the geological environment and of their interaction are necessary (the problem itself may be a conventional one).

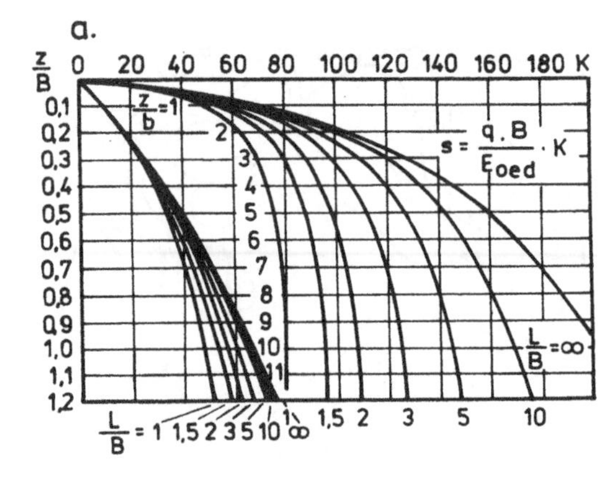

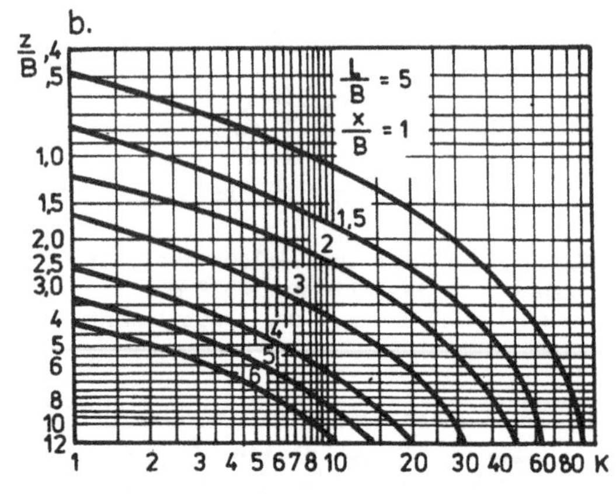

Fig.10 Settlement influence factor K:
a - for the characteristic point,
b - for the influence of neighbouring foundations

Both cases need a detailed model of EG conditions (Fig.11). Also, special geotechnical assessment and research are often necessary, and in individual cases new geomechanical solutions are required: their results have to be evaluated simultaneously with the results of engineering geology.

5.2 Solution methods

The method of solution may be simple and only approximate when a conventional problem under special geological conditions is to be solved, or when a new simple solution for an unusual case of interaction has been found. Correct mathematical solutions can be derived, as a rule, only for

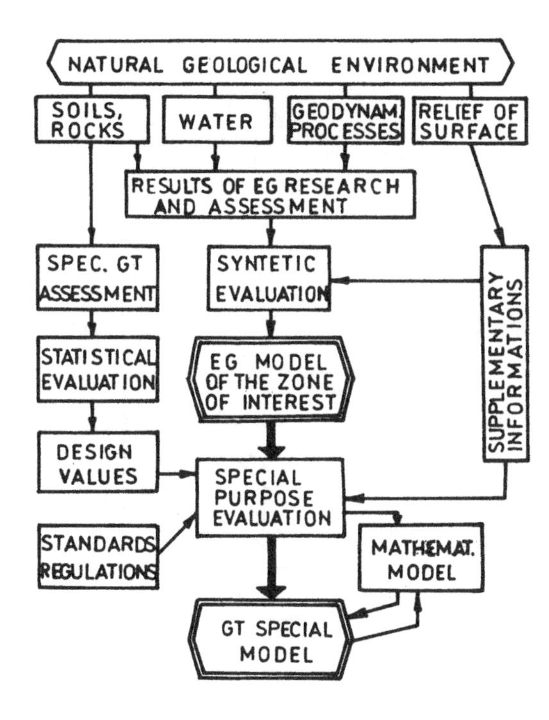

Fig.11 Connections between EG results and GT special model

a simple structure of the zone of interaction, and for simple boundary conditions. In spite of this, correct solutions of geomechanics have a fundamental importance: they may serve as reference methods for testing the results of approximate or numerical procedures. Nevertheless, in practice the possibility of their application is limited. The great progress in the qualitative and quantitative analysis of geotechnical problems during the last decades has brought numerical methods adapted for computers, and most importantly the method of finite elements. It applies well not only in the case of complicated geological structure and boundary conditions, but also allows the use of highly sophisticated constitutive laws. As a consequence of its universality, the FEM seems at present overestimated: it is often used in cases where no equivalent input data are available, and the results remain inadequate for the pretentious mathematical solution. It must be emphasized, however, that from the standpoint of geotechnical engineering it is indifferent whether a mathematical solution is approximate or correct, or whether it was obtained with an analytical or numerical procedure.

The important criterion is the extent to which the results agree with the actual interaction under natural conditions. Mathematics cannot inherently be "good" or "bad", but only adequate or inadequate. Besides, providing a physically correct estimate of the interaction, the reliablility of the results is given by the representativeness of the information. A lower level of input data may be partly compensated for only if decisions are made on the basis of results obtained by parametric studies and engineering experience.

It is clear that the amount and the quality of information concerning the natural geoenvironment also represents a limiting factor in the case of GT special modelling.

5.3 Examples of GT special models

GT special models solving conventional problems under concrete geological conditions only contribute quantitatively to geotechnical knowledge and experience. Qualitative progress in geotechnics is put forward only by new solutions. Because there are still neither exact nor approximate criteria for deciding whether in any concrete case more detailed site assessment is required or whether the covering of uncertainties by reasonable overdimensioning of the foundations and/or the structure would be more efficient, and seeing that there are still no proved methods of optimization for the whole complex of geological-geotechnical works for construction purposes, simple approximate but physically realistic solutions will conserve their importance for practical geotechnical engineering even in future. Simple solutions allow rapid checking of the influence of various factors in order to select the best solution, or, if necessary, to decide which alternative should be solved in detail in a more complex way. Some examples - only the results may be given here - may illustrate the foregoing considerations.

1. For the foundation of HT and EHT electric transmission towers on two parallel prismatic walls, the ultimate oblique pulling load of a stiff, vertical body embedded in the soil had to be determined.

The problem was solved on the basis of results of small-scale laboratory model tests field tests, and mathematical modelling by FEM for a non-linear, elastic medium with joints between foundation and soil, for angles of inclination of the pulling force within the entire range from the vertical to the horizontal: a GT special model solved by pretentious combined methods. On the basis of the results, a simple but well suiting approximate method was developed, now incorporated into the appropriate Czechoslovak State Standards. The method has already been published and will therefore not be analysed here (Jesenák e.a.,1981, 1986).

Generalizing, we can conclude that after sufficient experience has been gained, the GT special model may turn into a GT type model, ready for application under different conditions.

2. In another case, the ultimate bearing capacity of saturated clays below the slope of a high dump was to be checked in order to prescribe the allowable slope inclination. The task was solved by assuming deep, vertical cracks through the slope, and failure of the base in undrained conditions. If the strength of the subsoil c_u is constant, and the slope is considered to be a triangular strip load, the ultimate bearing capacity is given by the term

$$q_u = 7.36\ c_u$$

and the ultimate slope inclination β_u:

$\beta_u = \arc\tan(3.23\ c_u/\gamma d)$,

where γ is the unit weight of dump material and d the thickness of the clay stratum.

The investigation was extended also for undrained strength increasing linearly with depth, and also for trapezoidal loads (Jesenák 1984, 1985). For rapid calculation, tables, diagrams and approximate formulae were given. Some results are shown in Fig.12.

The simple method derived on the basis of laborious calculation is suitable for checking the possibility of base failure below rapidly increased dumps and embankments or below excentrically loaded strip footings. After failure occurs, the residual strength of the subsoil can be determined.

The problem - originally a special GT model - was here again transformed into an engineering type solution.

3. The final settlement of the power station of a newly built hydroelectric plant had to be determined. The dimensions of the powerhouse were 242.0 x 78.5 m, and the foundation depth was 32.5 m below the surface. Since the subsoil was about 300 m thick with a very pervious complex of sandy gravels and a single horizontal clay stratum 89 m below the terrain, the powerhouse was built in a sloped pit sealed with vertical diaphragm walls and a pan-like grouted bottom (at present the greatest sealed pit of the world). On the basis of loading tests in boreholes, car-

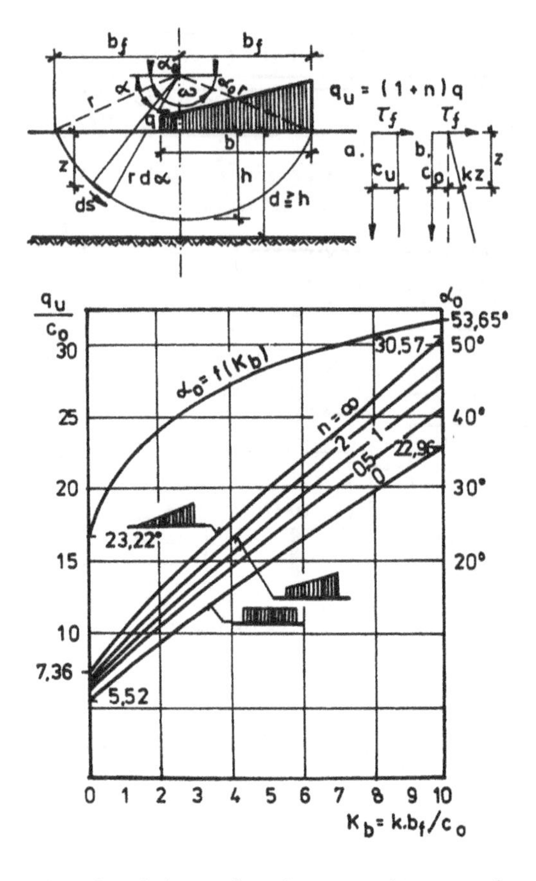

Fig.12 Ultimate bearing capacity q_u and angle α_o in depencence on the parameter K_b=k.b_f/c_o

ried out to a maximum depth of 40 m, the deformation modulus of the gravel complex was supposed to follow a parabolic law:

$$E_{def} = 27.47.z^{0,4}\quad [MPa]$$

Oedometric tests gave an average value of the deformation modulus for the clay stratum at about 20 MPa. Young's moduli were in general 2.5-times higher.

Settlements were calculated for different stages of unloading and reloading, parallelly with the conventional one-dimensional method (using diagrams) and with FEM (for a continuously non-homogeneous, linearly elastic medium - sandy gravel - with a single different layer - clay). The supposition of linear elasticity was justified by the fact that unloading by excavation exceeds reloading by the structure. The differences between the results of both methods are - surprising-

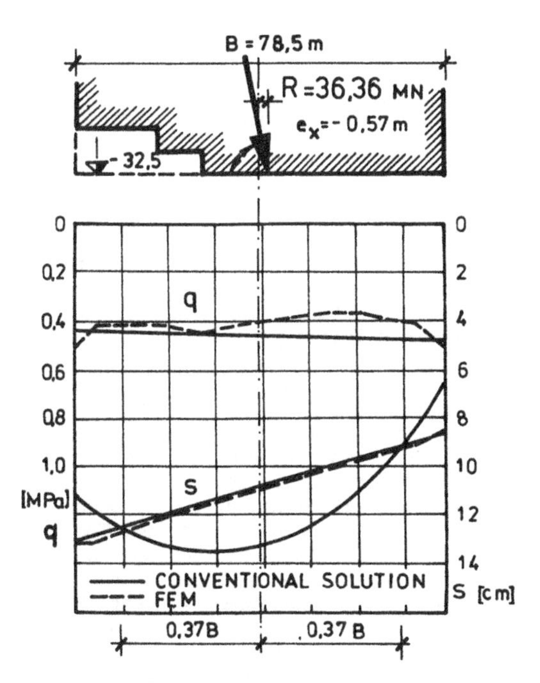

Fig.13 Final settlement of a hydroelectric power plant: comparison of conventional solution and FEM result

ly - quite negligible (see Fig.13 - contact pressure q and vertical displacement s for normal working conditions of the plant; the one-dimensional settlement is given for complety stiff and for an elastic structure. Dipping is caused by the weight of water in the sealed upper reservoir).

Of course, these results cannot be generalized. But they call attention to the fact that there are still no criteria for deciding where the relatively cumbersome and expensive FEM is really efficient, and where simple approximate methods render also quite satisfactory results, even when using "non-scientific" settlement charts.

4. The investigation of seepage through the slope of a 90 m high, valley-type tailings impoundment showed a hyperbolic decrease of the average value of permeability toward the pond.

Using Dupuit´s simplifications, formulae were derived for two-dimensional seepage through continuously non-homogeneous media, with permeabilities changing according to different laws. For general cases, with permeabilities changing both in horizontal and in vertical direction, an approximate engineering solution respecting the shape of cross-sections of the valley was derived (Jesenák 1985).

This example confirms that in individual cases even difficult problems may be solved with simple approximate methods.

6 GEOTECHNICAL PARAMETRIC MODELS

GT parametric models do not serve for the direct solution of geotechnical problems in geological conditions connected with a real site. They investigate problems in idealized conditions, which could theoretically occur anywhere. They are a tool for realizing the influence of different variable parameters on the result of an individual method of solution. Another application is to compare the results when using different solutions for a specified geotechnical problem. Such mathematical experiments may be used to investigate for example the influence of varying boundary conditions, geological structure, the properties of individual strata, and extreme combinations of acting factors; or they may test the reliability of approximate solutions or the boundaries of their application, etc.

GT parametric models are time- and work-consuming, but they render directly generalized results, thus contributing to theoretical and applied knowledge in geotechnical engineering. As an illustration, two simple examples should suffice:

1. The influence of the foundation depth on the distribution of stresses below flat foundations is often neglected, or respected according to theoretical solutions supposing undisturbed halfspace above the loaded area (Kézdi 1952, Škopek 1961). The competent Czechoslovak State Standards recommend an approximate method derived originally by Jelínek (1951). A parametric study was run for homogeneous subsoil according to these methods, controled by FEM (Fig.14). The results show clearly that only the approximate method renders realistic results: the physical substance of the "theoretically correct" solutions is not realistic.

2. The influence of a thin, much stiffer or much weaker layer on the settlement of the subsoil was checked by FEM for a linearly elastic medium. The width of the uniformly loaded strip was B, the thickness of the stiffer (weaker) layer B/4, its relative depth below surface varied from z/B = 0 to z/B = 2, and the thickness of the compressible stratum was 4B.

In Fig.15 the integral curves of vertical displacements are given, expressed as a percentage of the entire settlement of a homogeneous layer (dashed line). A detailed analysis of the results showed that

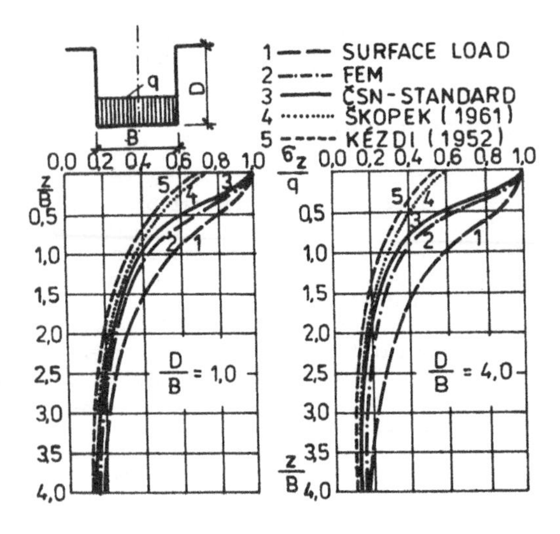

Fig.14 Influence of foundation depth on vertical normal stress σ_z

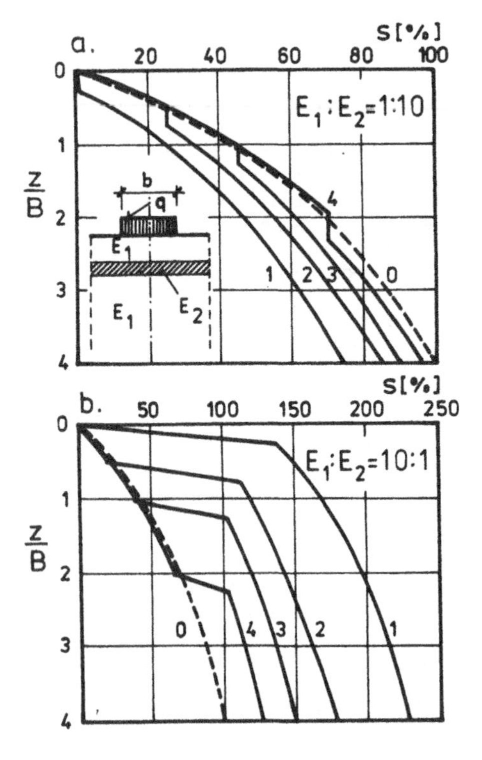

Fig.15 Influence of a stiff (a) or weak (b) layer on vertical displacements (related to total settlement of homogeneous medium)

a stiff layer has only a little influence on settlements, but it alters the distribution of stresses substantially. A much weaker layer has the opposite effect.

7 CONCLUSIONS

Environmental aspects of geotechnical engineering include also the problems of the mechanical interaction between construction work and the natural geological environment. The interdisciplinary character of these problems is the reason why there is still often a gap in mutual understanding between geologists and engineers. To overcome interdisciplinary difficulties, engineering geologists should define more accurate and more realistic models of the geological environment, at scales adequate to the given geotechnical problem. The efforts of geotechnical engineering should tend to a better understanding of the peculiarities of the geoenvironment. Realistic estimates of the physical substance of soil-structure interaction, well-suited transformation of engineering geological models of the geological environment into geotechnical models of concretely defined engineering problems within this environment, representative parameters of the mechanical behaviour of geomaterials, as well as fully adequate mathematical methods are fundamental requirements for successful solutions.

Geotechnical modelling, as developed in Czechoslovakia during the last decades, is an engineering approach to problems of mechanical (and hydromechanical) interaction. GT type modelling at small and at large scales, as well as GT special modelling of detailed problems, are closely connected with appropriate engineering geological models of the geoenvironment. GT parametric models serve first of all for better understanding of interaction problems, for conventional as well as for particular solutions. With more engineering geology, more geotechnical experience, and with adequate mathematics serious problems have been solved: geotechnical information is included in EG maps already at a scale of 1:25000, EG and GT typological zoning of large sites is gradually beeing introduced in practice, problems of construction on unstable slopes have repeatedly been solved successfully, sliding areas stabilized, seepage in granular tailings impoundments controlled, and geotechnical problems concerned with high dams, bridges, roads, channels, nuclear, thermal and hydroelectric power plants have been mastered in difficult conditions in a quite satisfactory manner.

The close cooperation of geologists and engineers leads to a better mutual understanding, and good understanding is the basis of successful cooperation. The strategic object - the optimization of the whole complex of geological - geotechnical work for construction purposes - is however still waiting to be solved.

ACKNOLEDGEMENT

The author is indebted to pay distinguished thanks to M. Masarovičova M.SC. Ph.D. for energical support in preparing the paper.

REFERENCES

Hulman,R., Jesenák J. 1971, Katalogisierung der Baugrundverhältnisse und statistische Voraussagen im Grundbau, Proc.4th CSMFE Budapest, 629-638

Jelinek, R.1951, Der Einfluss der Gründungstiefe und begrenzter Schichtmächtigkeit auf die Druckausbreitung im Baugrund, Bautechnik, 125

Kézdi,Á. 1952, Einige Probleme der Spannungsverteilung im Boden, Acta technica Hung., Tom. II, Fasc. 2-4

Jesenák,J. 1974, Geotechnische Rayonisierung von Hochwasserdämmen, Proc.4th Danube-European CSMFE, Bled, 23-27

Jesenák,J., Hulman,R. 1979, Geotechnical models of the subsoil and their evaluation, Proc. "Engineering geological investigation of rock environment and geodynamical phenomena", ed. M.Matula, VEDA Bratislava, 279-291 (in Slovak, English summary)

Jesenák,J. 1980, Setzungen des Baugrundes unter Strassendämmen, Proc.6th Danube-European CSMFE, Varna, 2/14, 139-148

Jesenák,J., Masarovičová,M. 1980, Spannungen und Setzungen bei inhomogänem Baugrund, Proc.6th Danube-European CSMFE, Varna, 2/15, 149-158

Jesenák,J., Kuzma,J., Masarovičová,M. 1981, Prismatic foundation subjected to oblique pull, Proc.10th ICSMFE, Stockholm, 5/29, 145-150

Jesenák,J. 1983, Contribution to the hydraulics of a continuously nonhomogeneous medium, Vodohosp. čas. 31, No.3, VEDA Bratislava, 626-642, (in Slovak, English summary)

Jesenák,J. 1984, Grundbruch unter einem Böschungsfuss (φ_u=0-Analyse), Proc.6th CSMFE, Budapest, 99-104

Jesenák,J. 1985, Load bearing capacity of ideal cohesive subsoil below a strip load, Staveb. čas. 33, No.4, VEDA Bratislava, 275-291 (in Slovak, English summary)

Jesenák,J. 1985, Contribution to the hydraulics of tailings impoundments, Proc. 11th ICSMFE, San Francisco, 1189-1192

Jesenák,J., Masarovičová,M., Bojsa,M. 1986, Foundation of HT and EHT towers on prismatic walls and piers, CIGRE 1986, Group 22-Overhead lines, Prefer. subject 22

Matula,M. 1976, Principles and types of engineering geological zoning, Mem. Soc. Geol. It., 14, 327-336

Matula,M., Hrašna,M. 1979, Engineering geological mapping and typological zoning, Proc. "Engineering geological investigation of rock environment and geodynamical phenomena", ed. M.Matula, VEDA Bratislava, 261-277 (in Slovak, English summary)

Škopek,J. 1961, The influence of foundation depth on stress distribution, Proc. 5th ICSMFE, Paris, Vol.I., 3A/42, 815-818

Computer and Physical Modelling in Geotechnical Engineering, Balasubramaniam et al. (eds)
© 1989 Balkema, Rotterdam. ISBN 90 6191 864 2

CAD hardware and graphics software systems – A state of the art

Erik L.J.Bohez
Production Engineering, Asian Institute of Technology, Thailand

The center of a CAD system is the engineering workstation. The engineering workstation consists of a graphic display device, a display controller and the interaction devices (keyboard, mouse, tablet, dials, ...).

According to a recent estimate (ref.. 2), 7330 CAD workstations were shipped in Western Europe in 1982, with an annual 25.1 percent projected growth to 1988, leading to 28,100 units sold in that year. (For graphic terminals alone, the corresponding figures are 16,360 for 1982, 58,400 for 1988.) For those familiar with the usual installation and acclimatization problems of new computer systems in industrial organizations, this indicates that the systems bought satisfied a real need and were able to do so in an efficient, congenial and cost-effective way. This has been very much helped by the proliferation of relatively low-cost systems (costing less than $100,000 each). In 1981 about 580 of these were installed in the USA, by 1986 their annual sales are expected to rise to 10,600, accounting for 20 per cent of the total CAD/CAM market revenues.

HARDWARE

1. DISPLAYS

There are three main types of display technology based on the cathode ray tube (CRT). The picture is drawn on a CRT screen by controlling the deflection voltage so that the lines, text and other graphic elements which make up the picture are traced by the beam. The electron beam excites the phosphor, which glows for a short period. To maintain a steady, flicker-free image the picture must be redrawn usually 30 or 50 times/s. The picture is described by instructions and data stored in an area of memory called a display file. These are executed by a display controller which drives the display deflection system via special vector and character generators and digital to analog converters. Early systems used part of the host computers CPU memory (central processing unit memory) for the display file, but the imposed heavy load on the computers input/output system led to the adoption of a refresh memory in the display (called refresh buffer). The need to refresh the picture places a limit on the number of vectors that can be shown without flicker. Early displays could draw only a few hundred lines; now some displays can refresh many tens of thousand lines, although this requires expensive circuiting. Local computing power is also needed to process interrupts (from interaction devices) and to store and manipulate the display file. These factors have meant that refresh displays have always remained expensive and are unlikely to gain wide acceptance. At the end of the 1960's the emergence of the direct view storage tube (DVST) had a huge impact on computer graphics. The big advantage of the storage tube terminal was that it usually had a simple serial interface and required no local computing power to refresh the picture. A new type of CRT, was used in which the picture was stored as a charge on a grid located behind the screen's surface.

The storage tube had two particular disadvantages compared with the refresh display. First it had no selective erasure. The only way to rub out a part of the picture was to clear the screen and redraw the picture minus the deleted part - and second, it was slow when used via RS232 interface. A consequence of the popularity of the storage tube display was that many of the programming techniques which had evolved for refresh display were cast aside as they simply would not work on a storage tube. Although a display file in the terminal is not needed to refresh the picture, it is better to have one to avoid re-transmitting the entire picture from the host computer and redraw times of less than a second are possible.

During the 1970's raster display technology began to emerge as the dominant type. Nearly all recent developments in the display field have been in this area. On a raster display the picture is drawn by scanning from the top of the screen to the bottom in a manner similar to a domestic television. The picture is made up of an array of dots, called picture elements or pixels. Each pixel has a value which can be set by a number of bits. One bit per pixel allows a black and white display. More than one bit per pixel allows color. The value of each pixel is stored in a pixel memory (frame buffer) in the display.

The reducing cost of this memory and the use of standard TV technology have resulted in the predominance of raster-scan displays.

A typical colour raster display is shown in Fig. 1. The pixel memory is arranged conceptually as a series of planes, one for each bit in the pixel value. The memory is read in scanline order by a display controller, whose output drives a television monitor via DAC (Digital to Analog Converters). The resolution of the display (number of pixels) determines the quality of the picture. The jaggedness of lines at angles other than multiples of 45 degrees is a well-known feature of raster devices (this phenomena is also called staircasing). The aliasing problem diminishes as the resolution increases, at a higher cost for storage.

Flat-panel displays are appearing in the market, the resolution is still low but high resolution displays are announced.

One of the disadvantages of the CRT is that, in general, the depth of the tube is the same as the screen size. Although CRTs will be around for a long time, the four following technologies are being used to produce a flat-panel display. Because they are based on solid-state, rather than electron beams, they consume less power and are more rugged than CRTs. The slim

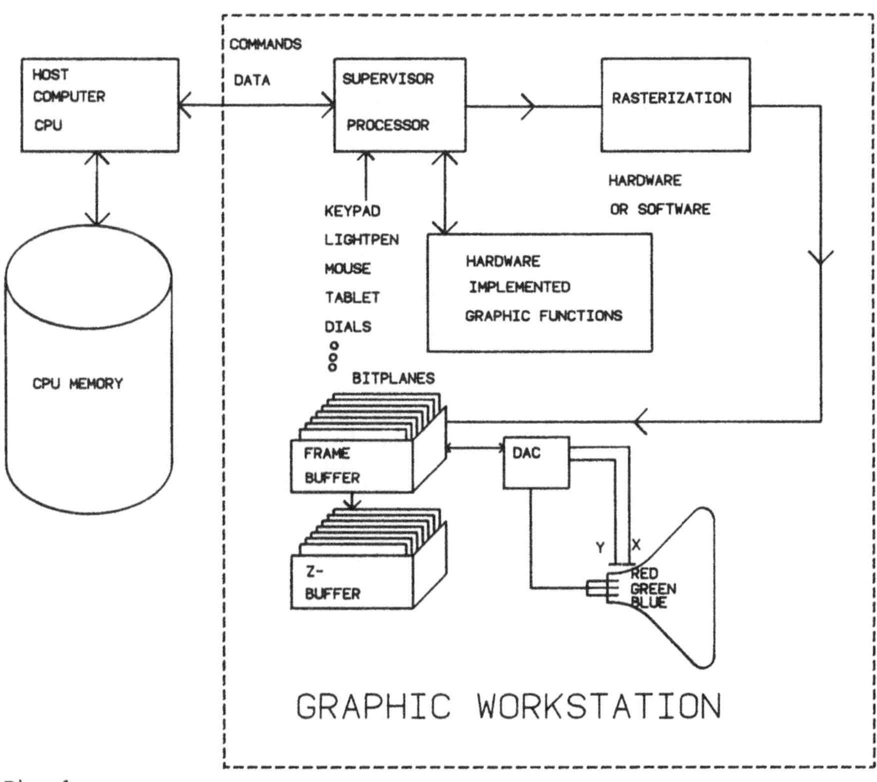

Fig. 1

profile and light weight make them ideal for portable computers.

Plasma display panel (PDP) - Plasma panels represent the most advanced of the flat-panel technology. They consist of two glass plates with finely spaced electrodes. Sealed in between the glass panels is neon gas and trace elements of argon or xenon. When an addressed pixel is subjected to an electric field, the gas mixture glows with its characteristic orange colour at that point. IBM has a three-inch thick screen. Resolutions of 512 by 512 are available with 1024 by 1024 in development. In the planning stage it has a 2 square meter screen with 2000 (45x45) pixels in each square centimeter. A big advantage is that the principle has an inherent memory effect. Between the two addressing grids a permanent sustaining voltage is applied, to light a pixel momentarily a voltage higher than the sustaining voltage is applied on the two addressing wires of the pixel, so that it is sufficient to start the electric arc. To switch of a pixel a voltage lower than the sustaining voltage is applied between the wires to address the pixel. So here selective erasure is possible. A disadvantage is that the price is still very high (5,000 US$, 512 x 512 pixels).

Electroluminescent displays (EL) - Instead of a gas layer between the glass panels, a metallic layer of sulfide and manganese is sandwiched in the panels. Vertical wires are imbedded in one panel and horizontal wires in the other. When a current is sent through the appropriate wire, an electric field is established at a particular point and causes a dot in the metallic layer to glow. Resolutions of 320 by 240 are available on 12-inch by 12-inch panels. The colour produced is orange or yellow-orange.

Liquid crystal displays (LCDs) - Liquid crystal material is deposited between two sets of polarizers and electrodes. Normally, light shines through the polarizers, but when a current is passed between the electrodes, the resulting field aligns the crystals, thus preventing the light from shining through. LCDs require very little power, but have limited viewing angles. Present capability is about 250,000 pixels.

Electrophoretic displays (EPD) - As with PDP and EL panels, the EPD has two plates with a material sandwiched in between. In this case, electrically charged particles of pigment are suspended in a fluid of contrasting colour. When current is applied, the electrical field pulls the charged particles to the front of the screen, creating a coloured dot. The chief advantage of this method is that the screen has an inherent memory. Once a particle is "turned on", no further power is required to maintain it. Once an image is displayed, it can remain for days. The image is erased by reversing the electrical field.

Although flat panels require less power than CRTs, one of the major problems is the large number of connections required to address all the pixels, in many cases, thousands of wires. Another problem is the light output of the display.

2. WORKSTATION INTELLIGENCE

The tendency to implement graphics functions into the workstations hardware instead of in the software gives highly intelligent workstations which can operate long periods without host computer support. Although they are still at a relatively early stage of development, CAD systems and production facilities for VLSI-chips (Very Large Scale Integration) should allow us to design and build display systems with almost every graphics software function implemented in the hardware. An example of a step in this direction is Clark's Geometry Engine, a chip for performing 3D geometric transformations (Ref. 4).

Optical disc-based systems which can store 1000 megabytes of user data per disc side will surely influence future CAD systems.

3. PLOTTERS

An important part in the CAD/CAM hardware is the ways to obtain hardcopy output for the generated drawings. There are four basic alternatives available:

* pen plotters
* electrostatic plotters
* laser printers
* photographic devices.

I will briefly review each category and describe certain device features to assist in selecting the optimum solution. In reaching a decision on a plotter following criteria need to be considered:

* speed of producing hardcopy output
* accuracy and/or resolution in hardcopy output
* cost-per-copy considerations

* end use of the output
* anticipated volume
* CPU-use

Pen plotters have lower initial purchase cost but they produce very accurate drawings with-high resolution. Potential disadvantages are that they are quite slow and need a relatively high level of maintenance.

Electrostatic plotters operate with technology similar to that found in common office copiers and can use either paper or reproducible drafting media. The plot is created point by point and "written" continuously onto the plotting media as the media moves in one direction out of the plotter. This approach is in contrast to a pen plotter, where the plot is created by moving the pen back and forth over the paper many times in much the same way a very organized, disciplined artist might create an image.

These plotters also require a different type of software interface between the CAD file and the plotter. This interface must take information that is essentially vector in nature and "rasterize" it; that is, convert it to a point-by-point representation of the drawing, much the way a TV picture is created. Once rasterized, the CAD created information is in a format suitable for point-by-point plotting on the electrostatic printer.

Electrostatic plotters offer the following advantages:

* They produce drawings very quickly
* They can sustain a high volume of output
* They require little support during operation.

The following are some potential disadvantages to consider:

* Additional computer resources may be required because of the need to rasterize the information to be plotted.
* Output is of somewhat lower resolution and accuracy than pen plotters, attributable to
- The number of spots per inch at which the plotter can operate, which directly effects image "sharpness" or resolution.
_ The feed of paper through the device, which has an impact on the drawing accuracy, especially along the length of the plot.

Laser printers are similar to electrostatic plotters, except that the image is "written" using laser technology. These plotters also require software that, at a minimum, rasterizes the information to be plotted. Depending on the size of the output media used by the printer, the software may need to execute other actions -- reducing the drawing size or segmenting it into pieces, for example -- before an image can be plotted.
Laser printers offer certain advantages:

* They are very fast, even faster than electrostatic plotters, and are particularly useful for producing
- Fast proofs
- Quick, multiple copies of drawings
* They offer the potential of easily combining both graphic and text information, using composition software and other support to create illustrated publications on a single pass through the printer.

Of course, there are potential disadvantages:

* The media size may be limited, and quite small by drafting standards, requiring reduction or conversion to multiple sheets to portray a single large drawing.
* As with electrostatic plotters, additional computer overhead is required to process the information into a format usable by the plotter.

There are two basic alternatives for producing photographically imaged output of CAD drawings. One alternative is commonly referred to as COM, or Computer Output on Microfilm. The image is created directly from computer information and transferred to film, without the intermediate step of creating hardcopy before photographing it. Once processed, the microfilm image is placed in a special reader to magnify it for reading or examination or, if properly equipped, to produce a quick print of the image.

The following are advantages of microfilm:

* The quality of microfilm images is very high.
* Output can be created very quickly, though a modest amount of time is required for processing before the image is usable.
* The output is very small, making it easy and relatively inexpensive to distribute and store.
* The images are easily reproduced.

The following are disadvantages of microfilm:

* Special reading devices are required to magnify and make legible the information contained on microfilm.
* Multiple microfilm images or frames may be required if the drawing being outputed is long or otherwise oversized.

Of course, microfilm images can also be created by photographing actual hardcopy rather than via the COM approach. The same advantages and disadvantages apply, though the actual process of creating the microfilm image will be more difficult because of the necessity of manually photographing the image and manually performing certain activities that are automatic in a COM environment, for example, photographing multiple frames for large drawings.

The second alternative uses similar equipment but produces a larger photographic print directly from the image created by the computer information, rather than a microfilm image. These prints are typically large enough to be read with the naked eye and are used in technical publications that must incorporate CAD produced drawings.

These photographic prints, sometimes referred to as RC prints (after the type of photographic paper frequently used), can then be used in the publishing process. Negatives are created from these prints and inserted among the text as specified in the document layout. Printing plates are then made, and documents including both text and CAD produced drawings can then be printed.

The plotting alternatives previously discussed are largely aimed at producing plots of reasonably high quality, suited for most uses. There often is a need for making fast plots for quick review, etc. Alternatives that lend themselves to such needs include:

* Buffer plots on small electrostatic plotters
* Soft copy (for example on very cheap flat panel displays)

The rapid growth of colour raster graphics has created the problem of how to obtain colour hard copy. One solution is to use a modified dot-matrix printer with coloured ribbons. Another is to attach a camera system to the colour monitor, with output on Polaroid prints, or 35 mm transparencies. This latter solution is useful for business graphics, but is not very practical for CAD. Yet a third option is to use a pen plotter and draw filled coloured areas with cross-hatching. Two more recent developments hold greater promise for CAD: the ink-jet plotter, and the colour electrostatic plotter.

The ink-jet plotter was first developed in Sweden in the early 1970's. It works by scanning a sheet of paper in a raster fashion and squirting small drops of coloured ink at it in order to build up an image. There are two basic designs. The first is like a modified conventional printer, with ink jets in place of the print head. Continuous roll or fan-fold paper can be used with this arrangement. The second uses a rotating drum to which a sheet of paper is fixed.

Two ink-jet technologies have been developed. One employs a continuous stream of electrically charged drops of ink and a deflection system to direct unwanted drops away from the paper. Unwanted ink is recirculated. The second uses a drop-on-demand method, where ink drops are only 'fired' at the paper when required. The latter method is being generally adopted as more reliable and less prone to clogging.

Plots are made using the subtractive primary colours yellow, cyan and magenta. Mixing combinations of these give a total of eight colours (including the white of the paper). In practice, most plotters also have additional ink jets for black ink, because the black obtained by mixing all three primaries is usually a rather dirty colour. About 64 additional hues and shades can be obtained with dithering techniques but although these work well for coloured areas, they are not effective for line drawings.

Typical resolutions for ink-jet plotters are between 80 and 160 dots/linear inch on A4 or A3 paper, or up to 22 x 34 inches for the Applicon plotter. Suppliers include Applicon, Tektronix, ACT, Siemens, Sharp and a number of other Japanese companies.

The colour electrostatic plotter is a development of the already well-established monochrome versions, such as those marketed by Versatec and Benson. These provide a resolution of 200 dots/linear inch, and this will soon rise to 400. At present, Versatec seem to have beaten their competitors to the market with a colour plotter, although it is expensive. The device operates by making multiple passes over the paper to print each primary colour. Registration marks are made on the edge of the paper to allow accurate registration of subsequent passes. Additional colours are obtained by dithering; the Versatec has around 400 hues/shades and can produce plots 40 inch wide.

4. CABLING CONSIDERATIONS

Terminals and plotters need to be connected to the host computer and there are a variety of alternatives available:

* Coaxial cable
* Fiber optics
* Microwave
* Cable (CATV) or closed circuit TV (CCTV)
* Common carrier (telephone lines)

Several criteria need to be considered before making a selection:

* What is the location of the terminals and/or plotters with respect to the computer?
* How fast must you be able to pass data back and forth for acceptable operations?
* What is the cost involved?
* What communications interfaces are required?
* What is the nature and extent of the negotiations that may be required with communications vendors (e.g., common carriers)?
* Will the alternative provide a level of security that meets your requirements?
* What sort of maintenance is involved?
* Which of the alternatives is already in use in your organization, if any?
* What sort of special staffing is required (for example, a licensed radio operator for microwave)?

Networking is a dominant trend in linking both design workstations in the office and manufacturing cells in the factory. The struggle to develop an application of the ISO Open Systems Interconnect (OSI) protocol has been arduous and frustrating. Two standards are now emerging - Ethernet and Manufacturing Automation Protocol (MAP) - both of which are compatible at the higher levels of OSI and plans are inhand for a connection device (to be known as a router) at the lower levels. MAP is rapidly gaining momentum under the direction of General Motors (GM), and a large number of companies (including IBM) have committed themselves to it. GM (which expects to be completely MAP-integrated by 1988/90) claims to prefer a broad band token bus network because of its industrial quality and the fact that the token bus time frame is deterministic, but it must also have been influenced by the existence of over 60 GM broad band installations.

Many companies have been installing broad band cable networks in anticipation of the completion and adoption of the MAP protocol and because such cable is well shielded from electromagnetic interference and allows more than one network on different channels (e.g. voice, data, video). Special MAP devices on a board will be available from companies such as Intel by 1986 and these will cover the bottom four layers of the OSI seven-layer model. These VLSI devices should reduce the cost of MAP token bus from the present $3000 per connection to nearer the cost of Ethernet connections (presently about $1000).

5. COMPUTER

The heart of a workstation is the computer which controls the station. The first computers to run CAD/CAM programs were the mainframe of the early 1960's. Today machines running CAD/CAM software span an even broader range, from low-cost personal computers that make the technology affordable to even the smallest of firms, to the most powerful super-computers. Computers still can be broadly classified according to physical size, purchase prize, processing speed, and accessible memory. The four main types of computers generally used for CAD/CAM include microcomputers, minicomputers, mainframes, and supercomputers. Selecting one type over the others is almost always a trade-off between cost off the machine and the computational load it can handle.

The least expensive CAD/CAM systems are based on high-end micro-computers generally called personal computers. These are typically 16-bit machines with at least 256 Kbytes of random access memory (RAM). Most PC-based CAD/CAM systems use the IBM PC (either the XT or more powerful AT version) or one of several other machines from Apple, Digital Equipment Corp., Texas Instruments, Hewlett-Packard, Compaq, AT&T, NEC, Victor, Columbia, Eagle, Tandy, and Corona.

CAD/CAM systems based on PCs have significant advantages compared to larger systems. The computer itself costs only a few thousand dollars, so the price of a complete system is generally less than $10,000 compared with $50,000 to $500,000 for a larger minicomputer-based system. PC systems usually have simpler functions, making them easier to use and faster to learn than larger ones with more "bells and whistles." Response time in a PC-based system is more consistent because computing resources are rarely shared and, thus, not degraded by sharing the processor with other users.

A growing number of PC-based systems are being used for such computation-intensive applications as solid modelling, finite-element analysis, and circuit simulation.

As the power of the PC grows, prices for 32-bit workstations are dropping so sharply that distinctions between PCs and workstations are blurring. Workstations such as Appollo's DN3000, DEC's VAX station II/RC, Sun Micro Systems' 3/50 M and IBM's RT PC all fall within the $10,000 to $20,000 range. If IBM introduces an 80386-based addition to its PC family, the price/performance level should be comparable to today's low-end 32-bit workstations.

At the moment, the IBM PC AT is the undisputed leader at the low end of the CAE/CAD market. The AT's major constraint is the DOS memory limitation of 640 K bytes. The PC is also limited in its graphics capability. Most vendors now support the IBM Enhanced Graphics Adapter, which provides a 16 color, 640 x 350 pixel resolution on a 13 in monitor. That's much more limited than the displays typically provided by 32-bit workstations, which are of the order of 1024 x 800 pixels. Another serious concern for many users is the inability of DOS to provide multitasking, preventing users from running a background task future versions of DOS are expected to remove both deficiencies. Recently workstation competition has been intensified with a new contender from IBM the introduction of a 32-bit Unix-based workstation (January 1986) the IBM PC RT, based on Reduced-Instruction-Set Computer (RISC) concept, the IBM PC RT is a powerful multi-user workstation. It features a proprietary RISC micro-processor that runs upto 2 MIPS, 1 to 4 Mbytes of system memory, on optionally 80286 coprocessor that runs IBM PC software and a 40-bit wide virtual memory management unit (MMU) chip that allows addressing upto 1 Tbyte (1 million Mbyte). Manufacturers of mainframe computers traditionally meet the demand for greater performance by increasing the complexity of their machines. But increased complexity makes the systems harder to build and harder to use. Longer design times for the developper are matched by heavier training and maintenance requirements for the user. All of this adds up to higher performance at higher cost. By providing enhanced performance with a reduced instruction set computer (RISC) architecture, vendors can provide the same or superior performance RISC architecture features a simple regular instruction set which allows a combination of instructions to be executed faster than the equivalent complex instruction. This lowers system purchase cost but can also reduce training and maintenance requirements.

Because the IBM RT uses RISC concepts it is difficult to compare statistics such as Million of Instructions per Second (MIPS) with the Complex Instruction Set Computer (CISC) MIPS. IBM quotes an execution rate of 1.6 to 2.1 Mips, but for computational Intensive tasks, the RT will run approximately twice as fast as IBM PC AT. Detractors point out inadequacies detected in current RT products. These stem from the small display size and general lack of a high-speed network environment: PC Net is available but it provides only one-fifth of the data transferrate of Ethernet, commonly used in the workstation world. Designed to meet high-performance multi-user requirements, the RT PC system, combined with the IBM 5080 workstation from a workstation with stand alone ability. By supplying application packages that are compatible with those running on IBM mainframes a station like this cost around US$40,000. Also in January Digital Equipment Corporation (DEC) introduced its VAX station II/GPX which adds high-performance graphics to its successful Micro VAX II Line. SUN micro systems introduced a low-cost version of its sun-3 Unix based workstations. Appollo Computer announced a low-end engineering workstation the DN 3000 Personal Workstation. Fig. 2 gives an overview of some of the features of these workstations.

Most CAD/CAM systems continue to rely on minicomputers as the processing machine. These are generally 32-bit computers that access up to 16-Mbytes of main memory. Considerable third-party software is available, spanning the rang of CAD/CAM

SOME WORKSTATIONS AT GLANCE					
WORKSTATION TYPE	IBM PC RT	APOLLO 3000	SUN 3/52 M	VAXSTATION II/GPX	IBM PC AT
PROCESSOR	PROPRIETARY RISC 32 BIT PROCESSOR	MOTOROLA 68020	MOTOROLA 68020	MICROVAX II	INTEL 80286
COPROCESSORS	FLOATING POINT IBM PC AT IBM 5085 GRAPHIC PROCESSOR	MOTOROLA 68881	MOTOROLA 68881	VLSI GRAPHIC GPX COPROCESSOR	INTEL 80287
OPERATING SYSTEMS	AIX (UNIX SYSTEM V) PC DOS	AEGIS UNIX SYSTEM V.4.2	UNIX BERKELY 4.2 BSD PC DOS	ULTRIX (UNIX 4.2 BERKELY) VMS	XENIX PC DOS
PERFORMANCE	1.6 TO 2.1 MIPS	1.2 TO 1.5 MIPS	1.5 MIPS	1.5 MIPS	0.3 MIPS
SYSTEM MEMORY	1 TO 4 MBYTES	2 TO 4 MBYTES	4 MBYTES	3 TO 9 MBYTES	2 TO 16 MBYTES
VIRTUAL ADDRESSING	1 TBYTE	64 MBYTES	256 MBYTES PER PROCESS	32 BIT ADRESS	16 MBYTES / TASK
MAX. NO. OF USERS	8 VIA ASCII TERM	1	1	MUTMULTI USER	1
DISK STORAGE	40 TO 70 MBYTES FIXED 1.2 MBYTE FLOPPY	86 MBYTE WINCHESTER 1.2 MBYTE FLOPPY	71 MBYTE DISK 45 MBYTE 1/4 IN TAPE	71 MBYTE DISK TAPE DRIVE	20 MBYTES DISK 1.2 MBYTES FLOPPY
GRAFICS RESOLUTION	14 IN 720X512 COLOR 15 IN 1024X768 MONO	15 IN 1024X800 COLOR 19 IN 1280X1024 MONO	19 IN 1152X900 MONO	19 IN 1024X864 COLOR	11 IN 640X200 MONO 11 IN 320X200 COLOR
NETWORKING	PC NET	APOLLO DOMAIN,ETHERNET TCP/IP,X.25,SNA	ETHERNET , TCP/IP ARP, SUNLINK TO SNA	DECNET,SNA GATEWAY ETHERNET,TCP/IP,X.25	PC NET
PRICE RANGE	US DOL 11,000 - 20,000	US DOL 10,000 20,000	US DOL 15,000	US DOL 35,000	US DOL 5,000 7,000

Fig. 2

programs including geometric modelling, drafting, finite-element analysis, kinematics, and NC programming as well as electronic circuit design and analysis.

Vendors in the market include Digital Equipment Corp., Data General, Harris, Control Data, Prime, Gould, Hewlett-Packard, Apollo, Perkin Elmer, Celerity, Ridge, Masscamp, Adra, and Elxsi. Most vendors offer a family of minicomputers ranging from small desktop, deskside, or underdesk units dedicated to single workstations, to so-called superminicomputers with speed and memory comparable to that of some mainframes.

Some of the fastest minis operate at speeds of 3 to 5 million instructions per second (MIPS) using emitter-coupled logic (ECL) circuit. This range of processor permits firms to start with relatively low-cost, entry-level machines priced below $15,000 and move up to more powerful equipment priced from $125,000 to $500,000 while keeping the same software.

Minicomputers are generally considered to have the best price/performance ratio of any computer class and have some other compelling advantages for most CAD/CAM applications. While mainframes are often located in corporate data-processing facilities, mini-computers are generally placed within the engineering department so the CAD/CAM system does not depend on outside computing resources. One significant feature of most minicomputers is the virtual memory operating system, which is essential to run extremely large programs characteristic of task such as finite element analysis.

Mainframes are used in applications requiring substantial data processing and large memory capacities. They typically support numerous peripheral devices such as printers, plotters, terminals, tape drives, and disk drives. Most mainframes have 32 or even 64-bit word lengths and huge memory capacities capable of handling data-intensive tasks such as financial operations as well as CAD/CAM applications.

Mainframes are generally the largest of computers, typically requiring room-size facilities that are air-conditioned to maintain the proper ambient temperature and humidity. Some units are even cooled with a circulating refrigerant to carry heat away from the processing circuits.

Mainframes cost from $700,000 to $2 million, with processing speeds in the neighbourhood of 4 to 6 MIPS. They are often offered as a family of products with a wide range of prices and processing power.

Rather than being used for interactive operation, mainframes are generally used to provide processing capabilities after a problem has been set up with the aid of a distributed minicomputer or even a microcomputer system. A finite-element model might be created on one of these smaller systems and then sent to the mainframe for analysis, after which output data is routed back to the smaller system for a display of results. In this manner, mainframes are used to link together distributed minicomputer-based systems throughout the company, providing access to powerful processing capabilities as well as a shared database for several users.

Supercomputers are the fastest processors, using the most recent advances in electronic circuits, processing techniques, and memory organization to attain extremely high processing rates. In general, supercomputers can sustain an average rate of 20 megaflops (million floating-point operations per second) over a range of problems and can hit over 400 megaflops in short bursts for well-structured problems.

Supercomputers use 64-bit word lengths and generally cost from $10 to $15 million. Processors in this class include the Control Data Cyber series and the gray machines. In performing a finite-element analysis problem, for example, supercomputers complete the task three to ten times faster than a mainframe, with computation speeds increased 100 times processing capabilities. Such speed makes supercomputers up to 500 times less expensive than mainframes, provided they are kept busy.

As a result, complex problems in CAD/CAM that were not even considered several years ago because of prohibitive processing costs can now be handled economically. Furthermore, many problems can be solved so rapidly that several iterations can be performed to increase accuracy or to evaluate different alternatives. Typical applications include high-end modelling and simulation tasks such as solid modelling, kinematics, finite-element analysis, model analysis, and fluid-flow simulations.

Supercomputers attain such high speed because of data pipelining, high speed circuits, and large internal memories. In pipelining, data elements are streamed through the processor in blocks instead of being handled one at a time as in conventional computers. This is done with a so-called vector-processing approach where ordered groups of data elements (called arrays or vectors) are arranged in blocks of sequential memory locations. A few instructions then initiate the same operation on all elements of the vector, which is much faster than issuing a separate command for each individual data element.

In supercomputers, clock-cycle time (the fixed interval of time that controls the movement of data) is increased through the use of special high-speed silicon chips that have a relatively short one-nanosecond gate delay (the time required to turn on and off). In addition, the high density of these large-scale ICs lowers processing and memory costs.

Since the mid- 1970's it has become apparent that the potential exists for constructing an optical computer, a computing device in which signals are transmitted by beams of laser light rather than electric currents. There is a powerful incentive for developing such an optical computer it might operate 1,000 times faster than an electronic computer. In an electronic computer virtually all the switches are transistors, and the fastest transistors now in use cannot be made to change states in less than about a nano second. An optical device analogous to the transistor could switch in about a pico second. Such an optical transistor could be employed to build the computers in the future.

SOFTWARE

Computer Graphics is a concept that has many roots in the hardware. However, software is the photosynthesis that gives it life. Ivan Sutherland is widely regarded as the person who spawned the industry, whose "SKETCHPAD" Ph.D. thesis at MIT was the seminal data-structure work that laid the software-theoretical foundation for computer graphics. He is also co-founder of a leading company in CAD/CAM : EVANS and SUTHERLAND COMPUTER CORP.

The PDP-9 (1966) and its successor, the PDP-11, laid the groundwork for DEC's reputation in engineering graphics support computers. The PDP-11 opened the door to CAD for DEC.

Applicon introduced its first system using the PDP-11 in the early 1970s and was followed a few years later by Intergraph and McAuto. All these companies have upgraded to the VAX series and were joined by Autotrol and Calma as DEC-based turnkey suppliers.

Lockheed Corporation, along with General Motors and McDonnel Douglas, were early pioneers in IBM's CAD effort. Today, Lockheed is one of the largest users of CAD workstations and has many turnkey IBM systems, some with as many as 100 workstations on a single mainframe.

Lockheed's subsidiary, CADAM, Inc., is a major third party software vendor, and CADAM software is sold under license by IBM. The 370, 30XX, and 4300 series round out the past 20 years of progress for IBM in CAD/CAM computer support.

Apollo, another manufacturer of networked microprocessor-based systems, has made a significant impact on CAD. Started in 1979, Apollo supplies workstation hardware to Autotrol and Calma, as well as several of the new Computer-Aided Engineering turnkey electrical vendors. Data General, Univac, CDC, Prime, and Hewlett Packard are also shown as major factors in CAD. These companies represent a substantial portion of the industry, but are by no means a complete list.

There are three types of workstation hardware including refresh storage tube, and colour raster crts. IBM had been the major proponent of refresh technology until the introduction of its 5080 colour raster product in 1983, and very recently its Microcad stations 5550 and 5560. Stroke systems offer significant speed-of-response advantages for fast picture update, and are well-matched to the high speed I/O channel and CPU performance of the IBM and Univac mainframes.

It is very important to understand the software configuration of a CAD system (Fig. 3) to understand the different levels of standardisation taking place.

CAD software is concerned mostly with the development of techniques and algorithms. As mentioned previously, the most famous contribution occurred in 1963 with Ivan

Sutherland's "sketchpad." Equally important was GM's DAC-1 project, developed in 1962. This project was important because it demonstrated the usefulness of computer-based techniques in major design and manufacturing environments.

Another early contribution was the work of Chasen and his colleagues at the Lockheed Georgia Company. Inspired by "Sketchpad" and DAC-1, their two principal applications were the production of tapes for running NC machine tools and the solution of structural analysis problems. Interactive graphics techniques were used by both designers and machine tool process engineers. They also showed, by using a Univac 418 computer, that a large computer was not required for effective computer graphics.

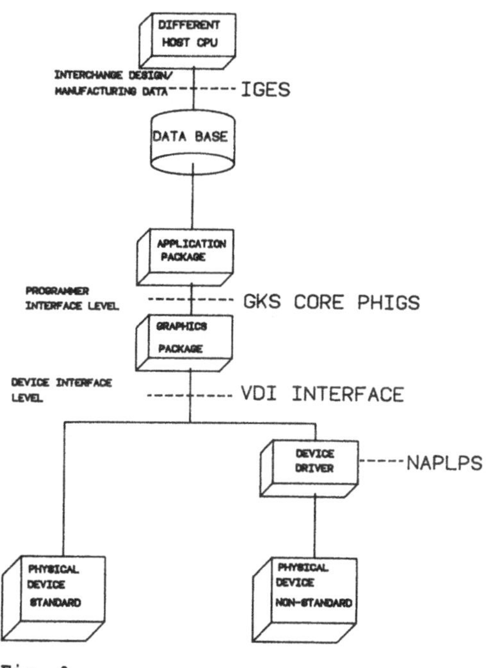

Fig. 3

Also involved in the early development of computer graphics was the Boeing Commercial Airplane Co. They simulated cockpit views by using a computer to generate the frames in an animated film sequence.

Of course, these early attempts did not produce portable, multipurpose software. The programmer still had to produce both the graphics software and the application software. Applications programmers needed to develop new or better algorithms for solutions to their particular problems. Surface modeling algorithms were developed in response to the need for automobile designers. An early example of this was the development in 1964 of a surface-modeling system that used four-sided, curved surface elements, which are now known as "Coon's patches."

The modelling process in computer graphics is growing through its third phase. The first phase began in the mid-1960s as a few straight lines on display. These lines could represent just about anything from the circuit path on a printed circuit board to the object lines of a drawing. As the two-dimensional elements of line and curves grew into three-dimensional wire-frame models, the mid-1970s saw the evolution of the second phase of computer-aided modelling. The second phase in the modelling process evolved as the design continued to grow in complexity, representing more and more information about the precise shape, size, and surface contour of the parts required.

Solid modelling represents the third step in the evolution of computer-aided modelling. All of the edges, surfaces, and holes of an object are knitted together to form a cohesive whole. The computer can determine the inside of the object from the outside. Perhaps more importantly, it can automatically trace across the object and readily find all intersecting surfaces and edges.

There are three distinct methods of representing solid objects used
in today's solid-modelling system (Ref. 7)

a. Construction Solid Geometry (CSG). the essence of this approach uses well-defined three-dimensional objects as building blocks. Various sizes of blocks and cylinders are added and subtracted from each other to form the desired part.

The advantage of this process is the fast description of the shape of a part and the ease of modification. Unfortunately, modelling the full range of part shapes is difficult if not impossible. Cast and forged parts with tapered sides and rounded corners are good examples of parts extremely difficult to model with the CSG approach.

b. Boundary Representation (B-Rep). In this process, every vertex, edge and face is explicity defined. The connectivity

(Topology) showing the relationship between each of these elements provides the glue to turn the list of elements into a geometrically solid object.
This process is inherently more flexible but requires rigorous algorithms to guarantee the construction of a valid object. However, the easy access to individual surfaces for sampling or display is an advantage of B-Rep solid modellers. For manufacturing applications, B-Rep is preferred to CSG method because a B-rep database is many times larger and better suited to the variety of analyses that must be performed on a design for manufacturing.

c. Spatial Subdivision. It divides object space into three-dimensional volume elements called "voxels" (comparable to "pixels" for two-dimensional elements. If this approach is used directly by subdividing object space into small (resolution) elements, large amount of storage is required to achieve adequate resolution for many engineering applications.

Systems using spatial subdivision reduce this memory requirement by using encoding scheme which recognise that the regions which separate material containing space from empty space are relatively large and continuous. While CSG and B-Rep models are used by almost all of the commercial modellers today, only one commercial solid modelling system, Phoenix Data Systems INSIGHT, uses a spatial subdivision form of representation.

The benefits derived from solid modelling are: It enables the visualization and communicating of complex geometry, assures design accuracy, and creates complex, unambiguous information needed for a common database. In short, solid modelling will allow better, faster and more accurate design and ultimately, design can be integrated with manufacturing.

In the United States, DATAQUEST, a market research firm, believes that there are four underlying reasons why solid modelling has not caught on among users.

a. Solid modelling is treated as an application.
b. Current modellers lack uniformity with existing CAD/CAM systems.
c. Today's solid modellers need improvement in response time and interactivity.
d. Solid modelling software costs too much.

Figure 4 gives some information about solid Modellers.

On the CAM side, the first computer operations were primarily aimed at production control. In 1955, Pennsylvania Railroad leased an IBM 705 to deal with the millions of pieces of railroad paperwork. Also in 1955, Sylvania set up the world's first company data processing center, leasing 12,000 miles of Western Union Telegraph lines to connect more than 60 plants and laboratories in 11 states with a Univac 1 in Camillus, NY.

By 1957, the oil industry had become a major computer use and not only for accounting and management. Tide water's Delaware "Refinery of the Future" used an IBM 650 process punched-tape output from digital data loggers, producing 24-hour summaries of the enormously complex refinery operations in a few minutes.

In 1959 Texaco was using a TRW-RW-300 computer to control a polymerization unit at its Port Arthur refinery. This was the first example of direct digital process control (DDC).

In the CAD area, software standards (Fig. 3) are directed at the following:

GKS (Graphical Kernel System) - Most hardware technology in use today have been available since 1965. However, during the early years of graphics development, the cost of equipment was high and the quantity of users limited. With the increasing usage of low-cost raster displays, the situation has reversed itself, and the need for software standards is quite obvious. As a result, GKS has gained acceptance as a worldwide standard for computer graphics. It was originally developed by the Deutches Institute for Normung, the official standards-making body of West Germany. In essence, GKS is a standard for a graphics programming interface, a standard graphics subroutine package.

GKS divides the output of graphics into two distinct parts. The first produces output on a virtual device space. The second allows individual workstations to interpret this virtual space in a way specified by the applications program.

GKS was designed to allow portability of graphics systems between installations. The concept of a workstation is fundamental to the realization of GKS as a device independent yet practical standard.

Solid MODELLER	CV's SOLIDE-SIGN	DASSAULT's CATIA	MAGI's SYNTHA-VISION	MATRA's EUCLID	SDRC's GEOMOD	SHAPE DATA's ROMULUS
Blended Surfaces	Not Attempted	Yes (1)	Not Attempted	Yes	Yes	Yes
FE Mesh Gen'n & Analysis	Yes	Yes	Not Attempted	Yes	Yes	Not Attempted
Sheet Metal Nesting	Yes	Not Attempted	Not Attempted	Yes	Not Attempted	Not Attempted
NC Data Genn'n	Yes (2)	Yes	Not Attempted	Yes	Not Attempted	Yes
Modelling of Assemblies	Yes (3)	Yes (4)	Yes (4)	Yes (3)	Yes (4)	Not Attempted
CPU Time for Hidden Line Display	Unknown	Unknown	15 min.	1.3 min.	0.7 min.	3.3 min.

NOTES: (1) The CATIA surface modelling operators were used to produce the blends.

(2) CV was the only vendor to attempt and successfully perfrom threaded shaft tool turret interference check.

(3) Interference checking was visually, not through an analysis program.

(4) Interference checking was done with an analysis program.

Fig. 4

Various levels of capability are defined for workstations within GKS. A maximally configured OUTPUT/INPUT workstation has the following properties:

a. A physical display surface
b. A mapping onto the display surface
c. Support for a variety of graphic primitives and associated attributes
d. Provision for picture segment storage and manipulation
e. Incorporation of the GKS logical input device model
f. Additional capabilities accessed by the generalized drawing primitive and escape functions

There are three different strategies for converting software to the GKS standard. The first is to make GKS look and act similar to an existing package. The next is to convert the application to GKS. And, of course, the third is combining the above two methods.

CORE - Proposed by SIGGRAPH "ACM/SIGGRAPH CORE Graphics System Recommendation" - is a standard that was proposed in competition with GKS. While GKS has been more universally accepted, there are still some groups still promoting the CORE concept.

While both standards have similarities, there are enough differences to make note of them:

a. GKS has wide acceptance and is more of a worldwide standard.
b. The VDM and VDI standards support the needs of GKS, but will not fully support CORE.
c. FORTRAN binding (list of routine names, called sequences, and data types) found in GKS is not part of the CORE standard.
d. GKS is two-dimensional and CORE is three-dimensional. However, an effort to add 3-D extensions to GKS is taking place.

e. The concept of workstations as presented in GKS represents an advance over CORE. And in GKS the interface between device-independent and device-specific portions of the system is identified.

IGES (Initial Graphics Exchange Specification) - is another standard addressing CAD/CAM compatibility. Specifically, IGES is for the interchange of design and manufacturing data.

PHIGS (Programmer's Hierarchical Interface for Graphics) - is a proposed computer graphics standard that serves as functional specification of the control and data interchange between an application program and its graphics support system. It is designed to satisfy the needs of applications whose functional requirements cannot be satisfied by GKS. Integral to the PHIGS proposal are the following features:

a. High interactivity
b. Hierarchical structuring of graphics data
c. Real-time modification of graphic data
d. Support for geometric articulation
e. Adaptability to distributed user environments
f. 3-D as well as 2-D graphical data

Application areas for the PHIGS standard are CAE, control, simulation, and scientific.

VDI (Virtual Device Interface) - The two primary levels of graphics standards are the programmer and the device interface levels. GKS provides the standard interface between the application program and graphic utilities. VDI standardizes the interface between the graphic utilities and device drivers. VDI makes all input/output devices identical virtual graphics devices by defining a standard input/output protocol. With VDI one has object-code portability. This allows a multitude of computers (or at least a family of computers) to run the same graphics program.

The VDI standard includes the following basic concepts:

a. VDI enhances graphics standardization by supporting drawing and input graphics primitives with individual attribute control.
b. VDI supports raster devices with fill and cell (pixel) array primitive.
c. VDI controls the attributes associated with each output primitive.
d. VDI provides the interface through which the application program controls physical devices by redirecting graphics I/O at any time.
e. To aid the programmer, the VDI also provides an "inquire" operation that allows the application program to determine the current operating state, primitive attributes, viewing operations, and transformations, as well as device capabilities.

VDM (Virtual Device Metafile) - A metafile is a representation of a picture in a device-independent form. VDM is a way to transport the definition of a picture over space and/or time. The VDM consists of the required set of VDI functions and a fixed selection of the nonrequired VDI functions.

NAPLPS (North American Presentation Level Protocol Syntax) - is well on its way to becoming a North American standard. It was developed by the Canadian government with AT&T and other computer communications companies as the basis for electronic journalism, home banking, and tels-shopping applications commonly referred to as teletex and videotext. However, it is totally unconcerned with the physical details of communication, addressing itself instead to mapping particular streams of code to unique graphics without any underlying assumptions about the hardware involved.

The NAPLPS is a means for encoding and freely intermixing text and graphics in an ASCII format. NAPLPS is a code for defining, storing, and transmitting computer graphics information.

There are five major features to NAPLPS:

a. Compactness - The NAPLPS code requires only about 10% of the space used by other graphics codes for defining the same graphics.
b. Resolution - NAPLPS graphics are resolution independent. A graphics application created on a low resolution display will have high resolution and clarity when shown on a high-resolution display.
c. Standardization - All of the codes used to describe a NAPLPS-coded picture are ASCII signals.
d. Integration - NAPLPS is designed to be integrated into the larger communications network through videotape, telephone, television signal, and computer

data base. NAPLPS also integrates graphics and text.

e. Colour - Included is a colour look-up table that allows the colour of any part of a graphic to be altered instantaneously. NAPLPS includes several colour modes for establishing foreground and background colour for both text and graphics.

In future Expert Systems and Artificial Intelligence will play an important role in CAD/CAM. The first applications seems to appear on the market. A question which is often asked by a company who acquires a CAD/CAM system is what do we do with our previous designs? Can we digitize our previous designs and use them with the CAD/CAM system? This is the field of the automatic digitizing scanner. In this technique drawings are digitized automatically through vector scanning or raster scanning. Vector scanning was the first technique to emerge. Employing a technique based on manual digitizing, vector scanning uses "electronic following". Simplicity is crucial with line following because the system requires operator guidance when a line intersects one or more other lines. Dashed and dotted lines present problems as each dash is treated as a separate line. Likewise, the system is unsuitable for symbols and alphanumerics, since they have intersecting lines. Presently, raster scanning offers the most promise in the automatic intelligent digitizing field. The problems are reduced to pattern recognition and storing the result in the right data format. Here the use of the IGES - standard would be very appropriate.

Another important field is the automatic functional dimensioning. Advanced CAD systems allow to model complete assemblies in one model (or set of hierarchical models) and each function of the individual parts can be specified by the designer. Through the use of an expert system automatic functional dimensioning can be achieved. The designer only works with nominal dimensions when he creates his objects. The expert system takes care of putting out dimensioned drawing with the right tolerancing. The above hypothetical expert system would have two very important improvements in the design and manufacturing process:

a. Design times would be reduced by 50% or more

b. The designer has the tendency to put his tolerances on the safe side, so that often the manufacturing people cannot achieve the strict tolerance and one ore more reviews are to be passed before a compromise tolerance is achieved. The automatic system would not need to do so.

When choosing a workstation there are five important questions you should ask a prospective supplier:

QUESTION 1 What standards does the workstation conform to?
Workstations are complex pieces of electronics, and there are several industry standards relevant to their use and operation.

Probably the most important of these is the UNIX operating system. Unfortunately the original UNIX written by Berkeley University in the USA has been embellished in different areas by different organizations. The most common (and most advanced) of these is Berkeley 4.2 bsd with system V extensions. The Tektronix UTek operating system is an enhanced version of this operating system.

UNIX was originally designed for software development and it is the logical choice for workstation environments for a number of reasons:

* Interactivity
- A response is initiated as soon as a command is issued.
* Multitasking
- Workstation users can initiate new processing tasks without waiting for the current tasks to be completed.
* Multiusers
- More than one user can be supported simultaneously which promotes information sharing between workstations within a UNIX-based network.

Many users will 'graduate' from a personal computer to a workstation. It is therefore also important that the chosen workstation supports the two most popular operating systems: MS-DOS and CP/M-86*.

Another important standard to consider is that of Graphics software algorithms. There are two major graphic standards in use today. In Europe the preferred standard the Graphical Kernel System (GKS) whilst in the USA It is the proposed SIGGRAPH 'Core' standard. Some workstations conform to both standards, giving the user maximum flexibility.

As indicated previously, at some stage the user will wish to network stations together. The token ring system used by some manufacturers has two major drawbacks.

* The network has to be formed into a complete circle in order to pass the 'token' between all workstations
* There is no one standard used by several manufacturers.

On the other hand the IEEE 802 ETHERNET network (which is a collision detect system) does not need to be formed into a complete loop. It is also supported by several major manufacturers including Tektronix and Digital Equipment.

While the vast majority of engineering programs are currently written in Fortran, there are several other languages which should be supported by an engineering workstation. Pascal is a structured language for efficient software design; CRT is a fast efficient language preferred by system programmers and Basic is easy to learn and use, as well as having a large program base. All these languages should be supported by the chosen system.

QUESTION 2 Is the workstation compatible with existing hardware?

The vast majority of new users of workstations have already got a terminal connected to a mini, mainframe or timesharing bureau. Quite understandably, they would much prefer to be able to integrate their existing hardware, not only terminals, but minicomputers and printers/plotters as well, into any new workstation they might purchase.

QUESTION 3 What application software is available?

2D, 3D, Finite Element Modelling, Mechanical & Electrical Engineering ...

QUESTION 4 What packages in project management, spreadsheeting, document and manual preparation, preparation of meeting presentation material etc ... is available on the workstation? (PC-DOS)

A designer spends 60% of his time on this.

QUESTION 5 What does the future have in store?

Read/write optical storage, flat screen displays, artificial intelligence, software functions implemented in the hardware.

Ivan E. Sutherland, once remarked about his favourite subject : "I think of a computer display as a window on Alice's Wonderland in which a programmer can depict either objects that obey well-known natural law or purely imaginary objects that follow laws he has written into his program". Through that same window we can see already the "factory of the future" completely unmanned and controlled by computers, humans left with a lot of time, to use even more creatively than before, or ...

REFERENCES

1. "Computer Graphics and Displays" by R.J. Hubbold CAD, volume 16, May 1984.

2. "Graphics rise mushrooms as users watch their data" by M. Park, Computer Weekly (September 15, 1983).

3. "Computer Software for Graphics" by Andries van Dam, Scientific American, September 1984.

4. "Computers for CAD/CAM : the list gets longer" by John K. Krouse, Machine Design, September 26, 1985.

5. "The CAD/CAM Primer" by Daniel J. Bowman, edited by Welborn Associates.

6. "The CADAM Managers Guide to System Implementation" CADAM Inc.

7. "Solid Modelling : A state-of-the-Art Report" by Robert H. Johnson.

Computer and Physical Modelling in Geotechnical Engineering, Balasubramaniam et al. (eds)
© 1989 Balkema, Rotterdam. ISBN 90 6191 864 2

Formulation and implementation of bounding surface model

W.F.Chen & W.O.McCarron
School of Civil Engineering, Purdue University, USA

ABSTRACT: Recently, a class of elastic-plastic material models known as Bounding Surface models have been introduced. These models include subsurface plastic behavior in the material response. Since soils often exhibit such behavior, it is expected this class of model may provide an improved representation of soil behavior. This paper presents a particular Bounding Surface model and computer code necessary for implementation.

1. INTRODUCTION

While the application of the classical theories of plasticity to soils have enjoyed some success, the assumption of a distinct zone of elastic behavior is contrary to many observations of soil response. Dafalias and Popov (1975), and Kreig (1975) have introduced the Bounding Surface formulation to provide some mechanism to model the occurrence of plastic behavior within the yield surface (Chen and Baladi, 1985).

The Bounding Surface model may be viewed as an extension of the sublayer models or the multi-surface models of Mroz (1967), Iwan (1967) and Prevost (1981). However, in the Bounding Surface formulation the elastic-plastic response within the bounding surface is continuous. In the following, a Bounding Surface model falling within the Critical State classification is introduced.

2. MATHEMATICAL FORMULATION

The formulation of the constitutive relationship for the Bounding Surface model closely follows that for the classical plasticity models (Chen and McCarron, 1986), requiring the definition of the elastic material response, bounding (loading) surface, flow rule, and hardening rule. In addition it is necessary to define the direction and magnitude of plastic deformations occurring within the bounding surface. This is accomplished with the aid of a projection rule and plastic modulus.

The purpose of the projection rule is to associate a state of stress within the bounding surface with an image point in the surface. Once this association is made, the direction of the plastic strain increment is assumed to be the same as that for the image point on the bounding surface. For instance, if an associated flow rule is in use, the plastic strain increment is in the direction of the normal to the bounding surface at the image point.

The plastic modulus controls the magnitude of the plastic strains within the bounding surface. The plastic modulus is dependent on the distance between the state of stress and it's image point. For large distances from the bounding surface, the plastic modulus is large, resulting in essentially an elastic response. As the state of stress approaches the bounding surface, the plastic modulus approaches the value defined by the bounding surface and the classical plasticity formulation. When the state of stress is on the bounding surface, the formulation reduces to a classical plasticity formulation.

Plastic deformations within the bounding surface are assumed to occur when the loading function f is positive. The loading function is defined as

$$f = \frac{1}{K_p} \frac{\partial F}{\partial \bar{\sigma}_{ij}} d\sigma_{ij} \qquad (1)$$

where K_p is the plastic modulus, F is the expression for the bounding surface, and $\bar{\sigma}_{ij}$ is the stress at the image point on the bounding surface. The plastic modulus K_p is retained in Eq. (1) to allow for unstable behavior when both $\frac{\partial F}{\partial \bar{\sigma}_{ij}} d\sigma_{ij}$ and K_p are negative. This treatment of softening is a simplification which may require further investigation.

The plastic constitutive relation is derived as (Dafalias and Herrmann, 1982).

$$C^p_{ijkl} = \frac{C^e_{ijtu} \frac{\partial F}{\partial \bar{\sigma}_{rs}} \frac{\partial G}{\partial \bar{\sigma}_{tu}} C^e_{rskl}}{K_p + \frac{\partial F}{\partial \bar{\sigma}_{mn}} C^e_{mnpq} \frac{\partial G}{\partial \bar{\sigma}_{pq}}} \quad (2)$$

where G is the plastic potential function, and the plastic modulus K_p has the form

$$K_p = \bar{K}_p(\bar{\sigma}_{ij}, e) + H(\sigma_{ij}) \quad (3)$$

where $\bar{K}_p$ is the contribution to the plastic modulus due to the image stress, H is a correction term dependent on the current state of stress (distance from bounding surface) and the void ratio e, is a hardening parameter.

Once the forms of the bounding surface, potential surface, and plastic modulus K_p are selected, the formulation of the constitutive relation (Eq. (2)) is the same as for the classical plasticity models. Further, the simplification of the constitutive relation to matrix form (e.g. Chen, 1982) also takes the same form.

Figure 1 presents the form of the Bounding Surface model to be employed here. The hardening surface is described by an ellipse of the form

$$F_1 = I_1^2 + \left(\frac{\bar{R}-1}{N}\right)^2 J_2 - \frac{2I_1 I_o}{\bar{R}} + \frac{(2-\bar{R})}{\bar{R}} I_o^2 = 0 \quad (4)$$

where I_o is the intersection of the ellipse with the hydrostatic axis and $\bar{R}$ is a geometric parameter related to the aspect ratio of the ellipse. The second surface, controlling the strength and deformation characteristics of overconsolidated soils, is described by

$$F_2 = \frac{J_2}{N^2} + \frac{2J_2^{1/2} I_o}{N}\left(\frac{1}{\bar{R}} + \frac{A}{N}\right) + \frac{2AI_o^2}{NR} - I_1^2 + \frac{2I_1 I_o}{\bar{R}} = 0 \quad (5)$$

where N is the slope of the Critical State line in the $J_2^{1/2} - I_1$ plane, and A is a material parameter controlling the surface shape.

$\bar{R}$ does not have the same definition as the parameter R commonly used in relation to the Cap model (McCarron and Chen, 1985). However, under some circumstances relations between the hardening surface parameters for the Cap and Bounding Surface models exist. For example, if the hardening surface of the Cap model intersects a Drucker-Prager type surface with an apex at the origin of the stress space, then the following relations exist:

$$R = \frac{\bar{R}-1}{N}; \; x = I_o; \; \ell = \frac{I_o}{\bar{R}} = L \quad (6)$$

The projection rule used in the present formulation is known as the radial projection rule. This rule determines the image point on the bounding surface by passing a line through the origin of the stress space and the current state of stress. The image point on the bounding surface is the point of intersection of the surface and the line. Thus, the image point is on the hardening surface if the state of stress is below the projection of the critical state line, and on the limiting surface if the state of stress is above the critical state line.

For the two surfaces presented above, and the radial projection rule, Dafalias and Herrmann (1982) have presented a method for determining the image point given the state of stress. The image stress is determined as

$$\bar{I}_1 = \gamma(\theta) I_o \quad (7)$$

where

$$\gamma(\theta) = \frac{1 + (\bar{R}-1)[1 + \bar{R}(\bar{R}-2)x^2]^{1/2}}{\bar{R}[1 + x^2 + \bar{R}(\bar{R}-2)x^2]}, \quad \theta < N$$

$$= \frac{x - xy - [(x-y-1)^2 + (x^2-1)y^2]^{1/2}}{\bar{R}(x^2-1)} \quad \theta > N \quad (8)$$

where

$$y = \frac{\bar{R}A}{N}; \; x = \frac{\theta}{N}; \; \theta = -\frac{J_2^{1/2}}{I_1} \quad (9)$$

and then

$$J_2^{1/2} = \theta \bar{I}_1 \quad (10)$$

The $\bar{K}_p$ contribution to the plastic modulus is evaluated as

$$\bar{K}_p = - \frac{\partial F}{\partial \varepsilon^p_{ij}} \frac{\partial F}{\partial \bar{\sigma}_{ij}} \qquad (11)$$

Since the hardening parameter is assumed to be only a function of the volumetric plastic strain, Eq. (11) becomes

$$\bar{K}_p = - \frac{\partial F}{\partial \varepsilon^p_{kk}} \delta_{ij} \frac{\partial F}{\partial \bar{\sigma}_{ij}} = - 3 \frac{\partial F}{\partial I_o} \frac{\partial I_o}{\partial \varepsilon^p_{kk}} \frac{\partial F}{\partial \bar{I}_1} \qquad (12)$$

For the Critical State formulation (Ko and Sture, 1981)

$$\frac{\partial I_o}{\partial \varepsilon^p_{kk}} = \frac{1}{I_o} \frac{\kappa - \lambda}{1 + e_o} \qquad (13)$$

where

$$\kappa = \frac{C_c}{2.303}; \; \lambda = \frac{C_r}{2.303} \qquad (14)$$

and C_c and C_r are the soil compression and rebound indices, respectively.

For the specific surfaces presented here,

$$\bar{K}_p = -12 \frac{I_o^3}{\bar{R}} \frac{(1 + e_o)}{\lambda - \kappa} (\gamma + \bar{R} - 2)(\gamma - \frac{1}{\bar{R}}), \; \theta \leq N \qquad (15)$$

$$= -12 \frac{I_o^3}{\bar{R}} \frac{(1 + e_o)}{\lambda - \kappa} (\gamma - \frac{1}{\bar{R}}) [\gamma(1 - x - xy) + \frac{2A}{N}] \quad \theta > N$$

The correction term is assumed to have the form

$$H = hp_a [1 + |\frac{N}{\theta}|^m] [9(\frac{\partial F}{\partial \bar{I}_1})^2 + \frac{1}{3} (\frac{\partial F}{\partial \sqrt{\bar{J}_2}})^2] \frac{\delta}{\delta_o - \delta} \qquad (16)$$

where h and m are material parameters, p_a is one atmosphere of pressure in the appropriate units, δ is the distance between the current state of stress and the image point, and δ_o is a reference distance.

3. COMPUTER IMPLEMENTATION

The bounding surface model described in the previous sections has been coded in FORTRAN 77. This computer code is presented in the Appendix along with a Guide to Data Input, flowcharts for the program, and descriptions of the responsibilities of the various subroutines.

4. APPLICATIONS TO BOSTON BLUE CLAY

The application of the bounding surface model described here as well as the application and computer implementation of a cap model presented elsewhere (Chen and McCarron, 1986) to Boston Blue Clay are given in the papers by McCarron and Chen (1987).

REFERENCES

Chen, W.F., 1982. Plasticity in Reinforced Concrete, McGraw-Hill, New York.

Chen, W.F. and McCarron, W.O., 1986. Modeling of Soils and Rocks Based on Concepts of Plasticity, Proceedings of the International Sympsoium on Recent Developments in Laboratory and Field Tests and Analysis of Geotechnical Problems, Bangkok, 6-9 December, Balasubramaniam, A.S., Chandra, S. and Bergado, D.I. (Editors), A. A. Balkema, Rotterdam, pp. 467-510.

Chen, W.F. and Baladi, G.Y., 1985, Soil Plasticity: Theory and Implementation, Elsevier, Amsterdam.

Dafalias, Y.F., and Popov, E.P., 1975. A Model for Nonlinearly Hardening Materials for Complex Loading, Acta Mechanics, Vol. 21, pp. 173-192.

Dafalias, Y.F., and Herrmann, L.R., 1982. Bounding Surface Formulations of Soil Plasticity, Soil Mechanics - Transient and Cyclic Loads, Pande, G.N., and Zienkiewicz, O.C. (Eds.), John-Wiley, New York.

Krieg, R.D., 1975. A Practical Two-Surface plasticity Theory. Journal of Applied Mechanics, Vol. 42, pp. 641-646.

McCarron, W.O., and Chen, W.F., 1985. Documentation for the Cap Model Routines, CE-STR-86-5, Purdue University, School of Civil Engineering, West Lafayette, IN.

McCarron, W.O., and Chen, W.F., 1987. Application of a Bounding Surface Model to Boston Blue Clay, Computers and Structures, Vol. 21, To appear.

McCarron, W.O., and Chen, W.F., 1987. A Cap Model Applied to Boston Blue Clay, Canadian Geotechnical Journal, Vol. 24, No. 4, November.

Mroz, Z., 1967. On the Description of Anisotropic Hardening, Journal of Mechanics and Physics of Solids, Vol. 15, pp. 163-175.

Iwan, W.D., 1967. On a Class of Models for the Yielding Behavior of Composite Systems, Journal of Applied Mechanics, Vol. 34, pp. 612-617.

Prevost, J.H., 1981. Constitutive Theory for Soil, Limit Equilibrium, Plasticity and Generalized Stress-Strain in Geotechnical Engineering, Yong, R.N., and Ko, H-Y., (Eds.), ASCE, pp. 745-814.

Appendix A
INPUT GUIDE AND FLOWCHARTS

This Appendix contains the guide for data input for bounding surface program presented in Appendix B. Also presented here are flowcharts of the program. The program will also save information concerning the stress and strain response for plotting or other purposes. See McCarron and Chen (1985) for compatible plot routines.

The program requires the definition of the array PROP. This array contains 25 variables defining the elastic-plastic response, material history, Initial state of stress, etc. Actually only 22 parameters are required since the array is `oversized' to allow for future modification of the program and input.

The input parameters are defined as:

```
INDNL             = 0 for linear analysis (not applicable here)
                    1 for material nonlinearity only
                    2 not available
                    3 not available
                    4 for updated lagrangian

ITYP2D            = 0 for axisymmetric conditions
                    1 for plane strain conditions
                    2 for plane stress conditions
                    (not available here)

PROP(1) = XK1     = bulk modulus parameter
    (2) = XK2     = bulk modulus parameter
```

$$K = XK1*2.303 \left(\frac{1+e_o}{Cr}\right) \frac{XI}{3} + XK2$$

```
    (3) = XG1     = bulk modulus parameter
    (4) = XG2     = shear modulus parameter

                    G = XG1 * K + XG2

    (5) = CC      = compression index
    (6) = CR      = rebound index
    (7) = XIOL    = limiting position past which cap may not
                    contract (i.e. XIO <= XIOL)
    (8) =           open
    (9) = XN      = slope of critical state line in I - SJ2 spc
   (10) = R       = cap parameter
   (11) = A       = limit surface parameter
   (12) = S       = elastic zone parameter
   (13) = H       = hardening parameter coefficient
   (14) = XPA     = atmospheric pressure coefficient
   (15) = XM      = hardening parameter coefficient
   (16) = XIOI    = initial cap position
                  > 0 to compute n.c. position in IEP214
                  < 0 to specify initial position for o.c. soil
   (17) = XI1     = minimum value for computing bulk mod and
                    plastic strains
   (18) = EO      = initial void ratio
   (19) = A1      = parameter to compute in-situ stress
   (20) = A2      = ditto
   (21) = A3      = ditto
                          SIGZ = A1*z + A2
                          SIGX = SIGZ*A3
                          SIGY = SIGZ*A4
```

```
(22) = IDR    = parameter for drained or undrained analysis
              > to compute both tension & comp pp
              < to compute only compr pp
(23) =          open
(24) =          open
(25) = TCUT   = maximum tension principal stress
```

Figures A.1 through A.3 present flowcharts outlining the operation of the bounding surface model program listed in Appendix B. These routines are capable of producing path, or a strain path for a given stress path.

Figure A.1 shows the overall operation of the program. Figure A.2 shows the responsibilities of the key subroutines. Figure A.3 shows the responsibilities of the material model subroutines.

The subroutines required for use in solving boundary value problems with finite element codes are:

EP2D14	MD2D14	MAXMIN
IEP214	EPL214	
ED2D14	DEVPRT	

BSM DOCUMENTATION

```
Record 1:  Title Card

Record 2:  ITYP2D, INDNL

           ITYP2D = 1 For Axisymmetric Case
                    2 For Plane Strain Case
                    3 For Plane Stress Case

           INDNL  = 1 For Material Nonlinearity
                  = 4 For Material and Geometric Nonlinearity

Record 3:  PROP(I) I=1,NCON

           PROP(I) = Bounding Surface Model Parameters

Record 4:  SOLTYP

           SOLTYP = `stress' For Stress Path Problem
                  = `strain' For Strain Path Problem
```

FOR STRAIN PATH

Record 5: NTIMES, (DSTRAIN(I), I=1,5)

NTIMES = The Number of Times to Apply Dstrain(i)
DSTRAIN = The Incremental Strain Vector
= {eyy, ezz, 2eyz, exx, w}

(Any Number of Record Type 5 May Be Entered)

FOR STRESS PATH

Record 5: RTOL, ITMAX, IREF

RTOL = Solution Tolerance For Convergence
= ||unbal stress||/||total stress||
ITMAX = The Maximum Number of Iterations Allowed
IREF = The Maximum Number of Reformulations of the Constitutive Relation

Record 6: KSWTCH, NTIMES, (DSIG(I), I=1,4)

KSWTCH = 0 For Computing Appropriate Constitutive Relation
= 1 For Computing Elastic Constitutive Relation
NTIMES = Number of Times to Apply DSIG(I)
= (syy, szz, syz, sxx)

(Any Number of Record Type 6 May be Entered)

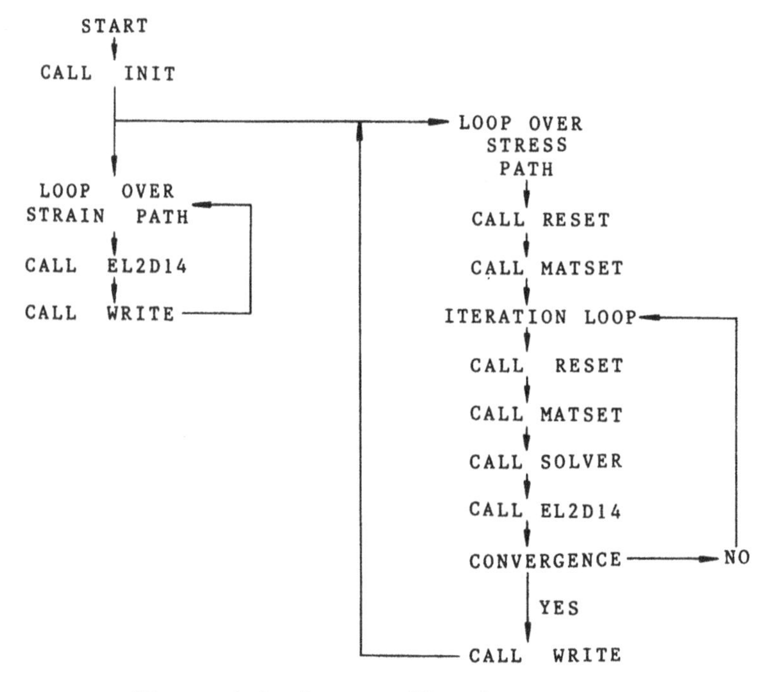

Figure A.1 Program Flowchart.

SUROUTINE	OPERATION
MAIN	Driver for program
INIT	Initialize parameters, read and echo print material data
EL2D14	Access to material model
RESET	Stores current material state on disk
MATSET	Sets coefficients in equilibrium equations and load vector (incremental stress)
SOLVER	Forward/Backsubstitution of equilibrium equations
WRITE	Print material response to output Write material response to disk file for plotting program

Figure A.2 Responsibility of Principal Subroutines.

SUBROUTINE	OPERATION
EL2D14	Access to material model
IEP214	Initialize material history
EP2D14	Driver for material model
EPL214	Compute material response for given strain increment
MD2D14	Form constitutive relation
MAXMIN	Determine principal stresses
DEVPRT	Determine stress invarients and deviatoric Components

Figure A.3 Responsibility of material model subroutines

Appendix B
PROGRAM LISTING

```
C
      COMMON A(60)
      COMMON /VAR   / NG,KPRI,MODEX,KSTEP,ITE,ITEMAX,IREF,IEQREF,INOCMD
      COMMON /EL/ IND,ICOUNT,NPAR(20),NUMEG,NEGL,NEGNL,IMASS,IDAMP,ISTAT
     1           ,NDOF,KLIN,IEIG,IMASSN,IDAMPN
      COMMON /DIMEL / N101,N102,N103,N104,N105,N106,N107,N108,N109,N110,
     1               N111,N112,N113,N114,N120,N121,N122,N123,N124,N125
      COMMON /MATMOD/ STRESS(4),STRAIN(5),D(4,4),IPT,NEL
      COMMON /BOUNSF/ ISR,IST,A1,A2,B1,B2,G,C2,BM
      COMMON /UNLOAD/ TRV(100),NTURNS,KSWTCH
      COMMON /DPA/ NDPA
      COMMON /GAUSS / XG(4,4),WGT(4,4),EVAL2(9,2),EVAL3(27,3),E1,E2,E3
```

```
      COMMON /EM/ S(136),XM(16),B(4,16),NOD(8),NODM(8),RE(16),
     1            EDIS(16),AEM(204),XX(16),NOD5M(4)
C
      DIMENSION DSIG(4),SS(4,4),DSTRN(5), SIGTOT(4),RHS(4)
      DIMENSION IA(1)
C
C
      CHARACTER *6 SOLTYP
C
      EQUIVALENCE (A(1),IA(1)),(INDNL,NPAR(3)),(ITYP2D,NPAR(5)),
     1            (NCON,NPAR(17))
C
      OPEN (8,STATUS = 'SCRATCH',FORM='UNFORMATED')
      OPEN (9,FILE='PATH',STATUS='UNKNOWN',FORM='UNFORMATED')
      DATA MTOT,STRAIN,SIGTOT / 60,9*0./
      DATA D,RHS,NCVG /20*0.,0/
C
C       ITYP2D = 0 FOR AXISYMETRIC PROBLEM
C              = 1 FOR PLANE STRAIN
C              = 2 FOR PLANE STRESS
C
C       INDNL  = 1 FOR MATERIAL NONLINEARITY ONLY
C              = 2 NOT AVAILABLE
C              = 3 NOT AVAILABLE
C              = 4 FOR U.L.(J)
C
C       NCON   = THE NUMBER OF CONSTANTS NECESSARY TO DESCRIBE THE
C                MODEL
C
C       STRESS = TOTAL STRESS = ( SY, SZ, SYZ, SX )
C
C       KSWTCH = 0 TO CALCULATE THE APPROPRIATE STIFFNESS
C              = 1 TO CALCULATE THE ELASTIC STIFFNESS
C
C
C     INITIALIZE THE PROGRAM DATA AND READ NFAP AND MATERIAL DATA
C
      CALL INIT (MTOT)
      PRINT*, 'INITIALIZATION COMPLETE '
C
C     READ THE PROBLEM TYPE
C
      IND = 4
      READ(5,100) SOLTYP
      IF(SOLTYP .EQ. 6HSTRESS) GO TO 500
      IF(SOLTYP .NE. 6HSTRAIN) GO TO 890
C
C     THE STRAIN INCREMENT HAS BEEN SPECIFIED
C
      NSTEP = 0
   10 PRINT*, 'STRAIN PATH SPECIFIED'
   11 READ*, NTIMES,DSTRN
      IF(NTIMES .EQ. 0) GO TO 900
      WRITE(6,650) NTIMES,DSTRN
      WRITE(6,640)
      CALL WRITE(NSTEP)
C
      DO 30 I= 1,NTIMES
      NSTEP  = NSTEP + 1
C
      IF(INDNL .EQ. 4) GO TO 20
C
      DO 15 J   = 1,5
   15 STRAIN(J) = STRAIN(J) + DSTRN(J)
```

```
      GO TO 25
C
   20 DO 21 J    = 1,5
   21 STRAIN(J) = DSTRN(J)
C
   25 CALL EL2D14
C
   30 CALL WRITE(NSTEP)
      GO TO 11
C
C     THE STRESS PATH IS SPECIFIED
C
  500 PRINT *, 'THE STRESS PATH IS SPECIFIED'
      READ  *, RTOL,ITMAX,IREF
      PRINT *, 'RTOL    = ',RTOL
      PRINT *, 'ITMAX   = ',ITMAX
      PRINT *, 'IREF    = ',IREF
C
      RTOL    = RTOL * RTOL
      NSIZE   = 4
      MBAND   = 4
      ICOUNT  = 4
      NSTEP   = 0
      IF(ITYP2D .GT. 0) NSIZE = 3
      IF(ITYP2D .GT. 0) MBAND = 3
C
      DO 501 I  = 1,4
  501 SIGTOT(I) = STRESS(I)
C
      WRITE(6,640)
      CALL WRITE(NSTEP)
C
  502 READ *, KSWTCH, NTIMES,DSIG
      IF(NTIMES .EQ. 0) GO TO 900
C
      PRINT *, 'KSWTCH = ',KSWTCH
      PRINT *, 'NTIMES = ',NTIMES
      PRINT *, 'DSIG   = ',DSIG
C
C
C     CALL MATERIAL MODEL TO OBTAIN INITIAL STIFFNESS
C
      ICOUNT   = 4
      CALL EL2D14
C
C
      DO 600 N = 1,NTIMES
      IT       = 0
      NREF     = IREF
      NSTEP    = NSTEP + 1
      KK       = 1
C
C     WRITE THE INITIAL CONDITIONS TO TAPE 8
C
      CALL RESET(MTOT,1)
C
C     SET THE TERMS IN THE BANDED STIFFNESS AND RHS
C
      CALL MATSET(DSIG,RHS,SS,D,SIGTOT,RT,NSIZE,1)
C
      IF(INDNL .NE. 4) GO TO 505
      DO 503 I  = 1,5
  503 STRAIN(I) = 0.
C
```

```
  505 IT          = IT + 1
      IF(IT .GT. ITMAX) GO TO 880
C
C     RESET THE MATERIAL PARAMETERS
C
      CALL RESET(MTOT,2)
C
C     SOLVE THE SYSTEM
C
      CALL SOLVER (RHS,SS,NSIZE,MBAND,KK)
C
      KK = 2
C
C     SET THE STRAIN FOR MATERIAL NONLINEARITY
C
      DO 510 I    = 1,4
  510 STRAIN (I) = STRAIN(I) + RHS(I)
C
C     CALL THE MATERIAL MODEL TO EVALUATE THE STRESS INCREMENT
C
      CALL EL2D14
C
C
C
C     EVALUATE THE UNBALANCED STRESSES
C
      DO 525 I = 1,NSIZE
  525 RHS(I)    = SIGTOT(I) - STRESS(I)
C
C     EVALUATE THE UNBALANCED NORM
C
      RI = 0.
      DO 530 I = 1,NSIZE
 530  RI = RI +  RHS(I) * RHS(I)
C
      TOL     = RTOL * RT
      IF(RI .LE. TOL) GO TO 540
      IF(IT .LT. ITMAX) GO TO 505
C
      NREF = NREF - 1
      IF(NREF .LT. 0) GO TO 880
C
C     RECALCULATE STIFFNESS AND TRY ITERATION AGAIN
C
      IT      = 0
      CALL MATSET(DSIG,RHS,SS,D,SIGTOT,RT,NSIZE,2)
      KK      = 1
      GO TO 505
C
C
  540 CALL WRITE (NSTEP)
C
  600 CONTINUE
C
      GO TO 502
C
  880 PP  = 0.
      II  = N110 + 4
      XIO = A(N110 + 8)
      EO  = A(N110 + 9)
      FAC = A(22)
      IPEL=IA(N110 + 11)
      IF(NSTEP .EQ. 0) GO TO 885
      IF(ITYP2D .EQ. 2) GO TO 885
```

```
      EKK = A(II) + A(II+1) + A(II+3)
      IF(EKK .GT. 1. E-9 .AND. FAC .LE. 1. E-9) GO TO 885
      PP        = EKK * BM * ABS(FAC)
C
  885 WRITE(6,120) RI,RT,IPEL,SIGTOT,STRESS,PP,XIO,EO
      GO TO 900
  890 WRITE(6,121) SOLTYP
  900 CONTINUE
  100 FORMAT (A6)
  120 FORMAT (28H*** ITERATION LIMIT EXCEEDED,/,
     +         28H    INCREMENTAL NORM        =,E12.4,/,
     +         28H    TOTAL NORM              =,E12.4,/,
     +         28H    IPEL                    =,I12,/,
     +         28H    SIGTOT (APPLIED)        =,4E12.4,/,
     +         28H    STRESS (TOTAL)          =,4E12.4,/,
     +         28H    PORE PRESSURE           =,E12.4,/,
     +         28H    XIO                     =,E12.4,/,
     +         28H    EO                      =,E12.4)
  121 FORMAT (43H***ERROR*** INVALID SOLUTION TYPE; SOLTYP = ,A6)
  610 FORMAT(4(5X,E13.6))
  640 FORMAT(//,7H  NSTEP,3X,3HEYY,9X,3HEZZ,9X,3HEYZ,9X,3HEXX,10X,2HPP,
     1     9X,3HSYY,9X,3HSZZ,9X,3HSYZ,9X,3HSXX,9X,3HXIO,6X,4HIEPL,//)
  650 FORMAT(//,8H  NTIMES,2X,4HDEYY,8X,4HDEZZ,8X,4HDEYZ,8X,4HDEXX,10X,
     1               2HDW,/,I5,5E12.4)
      STOP
      END
C
C
C
      SUBROUTINE INIT (MTOT)
C
C        SUBROUTINE TO INITIALIZE PROGRAM PARAMETERS
C
      COMMON A(1)
      COMMON /ELSTP/  TIME
      COMMON /VAR   / NG,KPRI,MODEX,KSTEP,ITE,ITEMAX,IREF,IEQREF,INOCMD
      COMMON /EL/ IND,ICOUNT,NPAR(20),NUMEG,NEGL,NEGNL,IMASS,IDAMP,ISTAT
     1           ,NDOF,KLIN,IEIG,IMASSN,IDAMPN
      COMMON /DIMEL / N101,N102,N103,N104,N105,N106,N107,N108,N109,N110,
     1                N111,N112,N113,N114,N120,N121,N122,N123,N124,N125
      COMMON /MATMOD/ STRESS(4),STRAIN(5),D(4,4),IPT,NEL
      COMMON /BOUNSF/ ISR,IST,A1,A2,B1,B2,G,C2,BM
      COMMON /UNLOAD/ TRV(100),NTURNS,KSWTCH
      COMMON /DPA/ NDPA
      COMMON /GAUSS / XG(4,4),WGT(4,4),EVAL2(9,2),EVAL3(27,3),E1,E2,E3
      COMMON /EM/ S(136),XM(16),B(4,16),NOD(8),NODM(8),RE(16),
     1           EDIS(16),AEM(204),XX(16),NOD5M(4)
      DIMENSION IA(1)
      DIMENSION TITLE(80)
      CHARACTER *1 TITLE
      EQUIVALENCE (A(1),IA(1)),(INDNL,NPAR(3)),(ITYP2D,NPAR(5)),
     1           (NCON,NPAR(17)), (IDW,NPAR(18))
      DATA XX    /1.,0.,-1.,0.,-1.,-1.,1.,-1., 8*0./
      DATA XG /              0.0,                0.0,               0.0,
     ^                       0.0,
     ^          -0.5773502691896,    0.5773502691896,               0.0,
     ^                       0.0,
     ^          -0.7745966692415,                0.0,   0.7745966692415,
     ^                       0.0,
     ^          -0.8611363115941,   -0.3399810435849,   0.3399810435849,
     ^           0.8611363115941/
      DATA WGT /             2.0,                0.0,               0.0,
     ^                       0.0,
     ^                       1.0,                1.0,               0.0,
```

```
     ^                  0.0,
     ^          0.5555555555556,    0.8888888888889,    0.5555555555556,
     ^                  0.0,
     ^          0.3478548451375,    0.6521451548625,    0.6521451548625,
     ^          0.3478548451375/
C
C        SET SOME PARAMETERS TO MAKE THE SUBROUTINE COMPATIBLE WITH
C        NFAP
C
      N107 = 1
      N109 = 1
      IND  = 0
      NPAR(10) = 1
      NDPA = 1
      IPT  = 1
      NEL  = 1
      MODEX= 1
      KPRI = 1
      KSWTCH = 0
      TIME   = 0.
      ICOUNT = 3
      NCON   = 25
      IDW    = 12
      N110   = N109 + NCON
C
C       READ AND PRINT TITLE
C
      READ(*,500) TITLE
      WRITE(6,500) TITLE
C
C       READ SOME ADDITIONAL CONTROL PARAMETERS FOR NFAP
C
      READ*, NPAR(5),NPAR(3)
C
C       INPUT ECHO
C
      PRINT*, 'ICOUNT = ',ICOUNT
      PRINT*, 'NCON   = ',NPAR(17)
      PRINT*, 'IDW    = ',NPAR(18)
      PRINT*, 'ITYP2D = ',NPAR(5)
      PRINT*, 'INDNL  = ',NPAR(3)
      PRINT*, 'KPRI   = ',KPRI
C
C       IF(NCON .GT. 24) PRINT*,'CHECK MEMORY ALLOCATION; NCON .NE. 20'
C       IF(NCON .GT. 24) STOP
C
C       READ THE MATERIAL PROPERTIES FOR THE PROBLEM
C
      READ*, (A(I),I=1,NCON)
      PRINT*, 'PROP(1)  = AK1   = ', A(1)
      PRINT*, 'PROP(2)  = AK2   = ', A(2)
      PRINT*, 'PROP(3)  = AG1   = ', A(3)
      PRINT*, 'PROP(4)  = AG2   = ', A(4)
      PRINT*, 'PROP(5)  = CC    = ', A(5)
      PRINT*, 'PROP(6)  = CR    = ', A(6)
      PRINT*, 'PROP(7)  = XIOL  = ', A(7)
      PRINT*, 'PROP(8)  = OPEN  = ', A(8)
      PRINT*, 'PROP(9)  = AN    = ', A(9)
      PRINT*, 'PROP(10) = R     = ', A(10)
      PRINT*, 'PROP(11) = A     = ', A(11)
      PRINT*, 'PROP(12) = S      = ', A(12)
      PRINT*, 'PROP(13) = H    = ', A(13)
      PRINT*, 'PROP(14) = XPA  = ', A(14)
      PRINT*, 'PROP(15) = XM    = ', A(15)
```

```
      PRINT*, 'PROP(16) = XIOI   = ', A(16)
      PRINT*, 'PROP(17) = XIL    = ', A(17)
      PRINT*, 'PROP(18) = EO     = ', A(18)
      PRINT*, 'PROP(19) = A1     = ', A(19)
      PRINT*, 'PROP(20) = A2     = ', A(20)
      PRINT*, 'PROP(21) = A3     = ', A(21)
      PRINT*, 'PROP(22) = IDR    = ', A(22)
      PRINT*, 'PROP(23) = OPEN   = ', A(23)
      PRINT*, 'PROP(24) = OPEN   = ', A(24)
      PRINT*, 'PROP(25) = TCUT   = ', A(25)
C
      PRINT*, ' '
      PRINT*, 'INITIALIZING MATERIAL PARAMETERS'
      PRINT*, ' '
C
C     CALL MATERIAL MODEL TO INITIALIZE VARIABLES
C
      CALL EL2D14
       WRITE(6,111) (A(I) ,I=1,MTOT)
 111   FORMAT(10E13.5)
C
      PRINT* , ' '
      PRINT*, 'PROP = '
      WRITE(6,620) (A(I),I=1,NCON)
      II = NCON + 1
      III= II   + 3
      PRINT*, 'SIG = '
      WRITE(6,620) (A(I),I=II,III)
      II = III + 1
      III= II  + 3
      PRINT*, 'EPS ='
      WRITE(6,620) (A(I),I=II,III)
      I   = III + 1
      II  = I   + 1
      III = II + 1
      IIII= III + 1
      PRINT*, 'XIO = ',A(I), 'EO = ',A(II), 'FAC= ',A(III),
     1        'IPEL =',IA(IIII)
      PRINT*, ' '
      PRINT*, 'BEGINNING SOLUTION'
      PRINT*, ' '
      PRINT*, 'INITIAL STRESSES'
      WRITE(6,660)STRESS
      RETURN
  500 FORMAT(80A1)
  620 FORMAT(5(5X,E13.6))
  660 FORMAT(/,11X,3HSYY,9X,3HSZZ,9X,3HSYZ,9X,3HSXX,/,5X,4E12.4,////)
      END
C
C
C
      SUBROUTINE MATSET (DSIG,RHS,S,D,SIGTOT,RT,NSIZE,KKK)
C
C        SUBROUTINE TO SET THE COEFFICIENT MATRIX ,RHS AND SIGTOT
C
      COMMON /MATMOD/ STRESS(4),STRAIN(5),DD(4,4),IPT,NEL
      DIMENSION D(4,4),DSIG(1),RHS(1),S(4,4),SIGTOT(1)
C
      DO 10 I = 1,4
      DO 10 J = 1,4
   10 S(I,J)  = 0.
C
      DO 20 I = 1,4
      DO 20 J = I,NSIZE
```

```
      K = J - I + 1
   20 S(I,K)  = D(I,J)
C
      IF(KKK .EQ. 2) RETURN
      RT = 0.
      DO 30 I  = 1,NSIZE
      SIGTOT(I) = SIGTOT(I) + DSIG(I)
   30 RT        = RT + SIGTOT(I)*SIGTOT(I)
C
      DO 40 I = 1,NSIZE
   40 RHS(I)  = SIGTOT(I) - STRESS(I)
C
      RETURN
      END
C
C
C
      SUBROUTINE WRITE(NSTEP)
C
      COMMON A(1)
      COMMON /BOUNSF/ ISR,IST,A1,A2,B1,B2,G,C2,BM
      COMMON /EL/ IND,ICOUNT,NPAR(20),NUMEG,NEGL,NEGNL,IMASS,IDAMP,ISTAT
     1           ,NDOF,KLIN,IEIG,IMASSN,IDAMPN
      COMMON /DIMEL / N101,N102,N103,N104,N105,N106,N107,N108,N109,N110,
     1              N111,N112,N113,N114,N120,N121,N122,N123,N124,N125
C
      DIMENSION CV(2)
      DIMENSION IA(1)
      CHARACTER *2 CV
      EQUIVALENCE (A(1),IA(1)),(INDNL,NPAR(3)),(ITYP2D,NPAR(5)),
     1            (NCON,NPAR(17))
      DATA CV /2H  ,2HNC/
      NCVG = 0
      XIO = A(N110 + 8)
      EO  = A(N110 + 9)
      FAC = A(22)
      IPEL=IA(N110 + 11)
      I   = N110
      II  = N110 + 4
C
C     COMPUTE PORE PRESSURE
C
      PP  = 0.
      IF(NSTEP .EQ. 0) GO TO 30
      IF(ITYP2D .EQ. 2) GO TO 30
      EKK = A(II) + A(II+1) + A(II+3)
      IF(EKK .GT. 1. E-9 .AND. FAC .LE. 1. E-9) GO TO 30
C
      PP        = EKK * BM * ABS(FAC)
C
   30 WRITE(6,6)NSTEP,(A(J),J=II,II+3),PP,(A(J),J=I,I+3),XIO,CV(NCVG+1),
     1          IPEL
      WRITE (9) (A(J),J=II,II+3),PP,(A(J),J=I,I+3),XIO
      RETURN
  6   FORMAT(I5,5E12.4,5E12.4,1X,A2,I2)
      END
      SUBROUTINE SOLVER (R,S,NSIZE,MBAND,KK)
C
C           THE COEFFICIENT MATRIX MUST BE SYMMETRIC
C           AND STORED IN BANDED FORMAT IN THE MATRIX 'S'
C           THE TERMS S(I,1) CORRESPOND TO THE DIAGONAL TERMS
C
C           NSIZE = THE NUMBER OF ROWS (COLUMNS)
C           MBAND = THE BAND WIDTH
```

```
C              KK      = 1 FOR FORWARD REDUCTION AND BACKSUBSTITUTION
C                      = 2 FOR BACKSUBSTITUTION ONLY
C
      DIMENSION S(4,4),R(4)
C
       GO TO (700,800), KK
C
C          FORWARD REDUCTION OF MATRIX (GAUSS ELIMINATION)
C
  700 DO 790 N = 1,NSIZE
      DO 780 L = 2,MBAND
C
      IF (S(N,L) .EQ. 0.) GO TO 780
C
      I          = N + L - 1
      C          = S(N,L)/S(N,1)
      J          = 0
C
      DO 750 K = L,MBAND
      J          = J + 1
  750 S(I,J)     = S(I,J) - C*S(N,K)
      S(N,L)     = C
C
  780 CONTINUE
  790 CONTINUE
C
C       FORWARD REDUCTION OF CONSTANTS (GAUSS ELEMINATION)
C
  800 DO 830 N = 1,NSIZE
      DO 820 L = 2,MBAND
C
      IF (S(N,L) .EQ. 0.) GO TO 820
C
      I          = N + L - 1
      R(I)       = R(I) - S(N,L) * R(N)
  820 CONTINUE
C
  830 R(N)       = R(N)/S(N,1)
C
C       SOLVE FOR UNKNOWNS BY BACK SUBSTITUTION
C
      DO 860 M = 2,NSIZE
      N          = NSIZE + 1 - M
C
      DO 850 L = 2,MBAND
C
      IF (S(N,L) .EQ. 0.) GO TO 850
C
      K          = N + L - 1
      R(N)       = R(N) - S(N,L) * R(K)
C
  850 CONTINUE
  860 CONTINUE
      RETURN
      END
C
C
      SUBROUTINE RESET (MTOT,K)
C
C         SUBROUTINE TO READ AND WRITE CURRENT MODEL CONFIGURATION
C
      COMMON A(1)
C
      GO TO (1,2), K
```

```
C
    1 REWIND 8
      WRITE(8) (A(I),I=1,MTOT)
      RETURN
C
    2 REWIND 8
      READ (8) (A(I),I=1,MTOT)
      RETURN
      END
C
C
C
      SUBROUTINE MAXMIN (STRESS,P1,P2,AG)
      DIMENSION STRESS(1)
C
C        AG = MAXIMUM PRINCIPAL STRESS DIRECTION MEASURED FROM + X-AXIS
C        P1 = MAXIMUM PRINCIPAL STRESS
C        P2 = MINIMUM PRINCIPAL STRESS
C
      CC=(STRESS(1)+STRESS(2))*0.5
      BB=(STRESS(1)-STRESS(2))*0.5
      T=STRESS(3)
      CR= SQRT(BB**2+ T**2)
      P1=CC+CR
      P2=CC-CR
      IF(ABS(T) .LE. 1.E-4) GO TO 20
      X=BB/T
      Y= SQRT(X*X + 1.)
      YY = -X + ABS(Y)*ABS(T)/T
      AG = ATAN ( YY)
      RETURN
   20 CONTINUE
      AG=0.
      SS = 1.0
      IF(BB .LT. 0.) AG = 2.* ATAN(SS)
      RETURN
      END
      SUBROUTINE FUNCT2(R,S,H,P,NOD5,XJ,DET,XX,NEL)
C . . . . . . . . . . . . . . . . . . . . . . . . . . . . . . . . . . .
C .
C .   P R O G R A M
C .
C .      TO FIND INTERPOLATION FUNCTIONS ( H )
C .      AND DERIVATIVES ( P ) CORRESPONDING TO THE NODAL POINTS
C .      OF A 4- TO 8-NODE ISOPARAMETRIC QUADRILATERAL
C .
C .      TO FIND JACOBIAN ( XJ ) AND ITS DETERMINANT ( DET )
C
C .
C .                     NODE NUMBERING CONVENTION
C
C .
C .             2                   5                   1
C .
C .               0 . . . . . . . . 0 . . . . . . . . 0
C .               .                                   .
C .               .                                   .
C .               .                 S                 .
C .               .                 .                 .
C .               .                 .                 .
C .             6 0                 . . . R           0 8
C .               .                                   .
C .               .                                   .
C .               .                                   .
```

```
C .                    .                           .
C .                    .                           .
C .                    0 . . . . . . . . 0 . . . . . . . . 0
C .
C .                  3                 7                  4
C .
C .
C . . . . . . . . . . . . . . . . . . . . . . . . . . . . . . . . . . . .
C
C
      COMMON /MATMOD/ STRESS(4),STRAIN(5),D(4,4),IPT,NELE
      COMMON /TODIM/ BET,THIC,DE,IEL,NND5,LSTM
      DIMENSION H(1),P(2,1),NOD5(1),IPERM(4),XJ(2,2),XX(2,1)
      DATA IPERM/2,3,4,1/
      DATA IEL,LSTM /4,0/
C
      RP = 1.0 + R
      SP = 1.0 + S
      RM = 1.0 - R
      SM = 1.0 - S
      R2 = 1.0 - R*R
      S2 = 1.0 - S*S
C
C
C     INTERPOLATION FUNCTIONS AND THEIR DERIVATIVES
C
C     4-NODE ELEMENT
C
      H(1) = 0.25* RP* SP
      H(2) = 0.25* RM* SP
      H(3) = 0.25* RM* SM
      H(4) = 0.25* RP* SM
      P(1,1)=0.25*SP
      P(1,2)=-P(1,1)
      P(1,3)=-0.25*SM
      P(1,4)=-P(1,3)
      P(2,1)=0.25*RP
      P(2,2)=0.25*RM
      P(2,3)=-P(2,2)
      P(2,4)=-P(2,1)
C
      IF (IEL.EQ.4) GO TO 50
C
C     ADD DEGREES OF FREEDOM IN EXCESS OF 4
C
      I=0
    2 I=I + 1
      IF (I.GT.NND5) GO TO 40
      NN=NOD5(I) - 4
      GO TO (5,6,7,8), NN
C
    5 H(5) = 0.50* R2* SP
      P(1,5)=-R*SP
      P(2,5)=0.50*R2
      GO TO 2
    6 H(6) = 0.50* RM* S2
      P(1,6)=-0.50*S2
      P(2,6)=-RM*S
      GO TO 2
    7 H(7) = 0.50* R2* SM
      P(1,7)=-R*SM
      P(2,7)=-0.50*R2
      GO TO 2
    8 H(8) = 0.50* RP* S2
```

```
      P(1,8)=0.50*S2
      P(2,8)=-RP*S
      GO TO 2
C
C     CORRECT FUNCTIONS AND DERIVATIVES IF 5 OR MORE NODES ARE
C     USED TO DESCRIBE THE ELEMENT
C
   40 IH=0
   41 IH=IH + 1
      IF (IH.GT.NND5) GO TO 50
      IN=NOD5(IH)
      I1=IN - 4
      I2=IPERM(I1)
      H(I1)=H(I1) - 0.5*H(IN)
      H(I2)=H(I2) - 0.5*H(IN)
      H(IH + 4)=H(IN)
      DO 45 J=1,2
      P(J,I1)=P(J,I1) - 0.5*P(J,IN)
      P(J,I2)=P(J,I2) - 0.5*P(J,IN)
   45 P(J,IH + 4)=P(J,IN)
      GO TO 41
  50  IF(LSTM.EQ.0) GO TO 60
C
C     CORRECTIONS FOR LINEAR STRAIN TRIANGLE
C     NODE 1 = 2 = 5 ,IEL = 8
C
      DH   = R2*S2/8.
      DHDR = -2.*R*S2/8.
      DHDS = -2.*S*R2/8.
C
C     SHAPE FUNCTIONS
      H(3) = H(3) + DH
      H(4) = H(4) + DH
      H(7) = H(7) - 2.*DH
C
C     DERIVATIVES
      P(1,3) = P(1,3) + DHDR
      P(2,3) = P(2,3) + DHDS
      P(1,4) = P(1,4) + DHDR
      P(2,4) = P(2,4) + DHDS
      P(1,7) = P(1,7) - 2.*DHDR
      P(2,7) = P(2,7) - 2.*DHDS
C
C     EVALUATE THE JACOBIAN MATRIX AT POINT (R,S)
C
  60  DO 100 I=1,2
      DO 100 J=1,2
      DUM = 0.0
      DO 90 K=1,IEL
   90 DUM = DUM + P(I,K)* XX(J,K)
  100 XJ(I,J) = DUM
C
C     COMPUTE THE DETERMINANT OF THE JACOBIAN MATRIX AT POINT (R,S)
C
      DET = XJ(1,1)* XJ(2,2) - XJ(2,1)* XJ(1,2)
      IF(DET .GT. 1.0D-08) GO TO 110
      WRITE(6,2000) NEL,IPT,DET
  110 CONTINUE
C
      RETURN
C
C
 2000 FORMAT(/52H ***WARNING***  NEGATIVE OR ZERO JACOBIAN IN ELEMENT,
     &     I4,19H, INTEGRATION POINT,I3,13H, DETERMINANT,E15.5)
```

```
C
      END
      SUBROUTINE DEVPRT(SIG,S1,S2,SS,S4,SM,SBAR)
      DIMENSION SIG(1)
C
      SM = (SIG(1) + SIG(2) + SIG(4))/3.
      S1 =  SIG(1) - SM
      S2 =  SIG(2) - SM
      SS =  SIG(3)
      S4 =  SIG(4) - SM
      SBAR= .5*(S1*S1+S2*S2+S4*S4) + SS*SS
      IF(SBAR.LE.0.) RETURN
      SBAR = SBAR**0.5
      RETURN
      END
      SUBROUTINE EL2D14
      REAL A
C                              M O D E L    =    1 4
C
C        E L A S T O P L A S T I C   B O U N D I N G   S U R F A C E
C                               M O D E L
C
C
      COMMON /EL/ IND,ICOUNT,NPAR(20),NUMEG,NEGL,NEGNL,IMASS,IDAMP,ISTAT
     1           ,NDOF,KLIN,IEIG,IMASSN,IDAMPN
      COMMON /DIMEL / N101,N102,N103,N104,N105,N106,N107,N108,N109,N110,
     1                N111,N112,N113,N114,N120,N121,N122,N123,N124,N125
      COMMON /MATMOD/ STRESS(4),STRAIN(5),D(4,4),IPT,NEL
      COMMON /DPA/ NDPA
      COMMON /VAR   / NG,KPRI,MODEX,KSTEP,ITE,ITEMAX,IREF,IEQREF,INOCMD
      COMMON A(1)
      DIMENSION IA(1)
C
      EQUIVALENCE (NPAR(10),NINT),(NPAR(3),INDNL)
      EQUIVALENCE (A(1),IA(1))
C
C     FOR ADDRESSES N101,N102,N103,...  SEE SUBROUTINE TODMFE
C
      IDW  = NPAR(18)*NDPA
      IDW1 = NPAR(18)
      NCON = NPAR(17)
      NPT  = NINT*NINT
      MATP = IA(N107 + NEL - 1)
C ///////////////// SHOULD REMOVE FOR NFAP USE /////////
      MATP = 1
C \\\\\\\\\\\\\\\\\\\\\\\\\\\\\\\\\\\\\\\\\\\\\\\\\\\\\\\\\\\\\\\\\\\\\\\\\\\\\\\\\\\\\\\\\\\\\\\\\\\\\\\\\\\\\\\\\\\\\\\
      NM   = N109 + (MATP-1) * NCON*NDPA
      IF (IND.NE.0) GO TO 100
      IF(NPAR(17) .EQ. 25) GO TO 20
      WRITE(6,101) NPAR(17)
      STOP
C
C
C     I N I T I A L I Z E  W A   W O R K I N G   A R R A Y
C
   20 NN=N110 + (NEL - 1)*NPT*IDW
C
      CALL IEP214 (A(NN),A(NN),A(NM),NINT,IDW,IDW1)
C
      RETURN
C
C
C     F I N D   S T R E S S - S T R A I N   L A W   A N D   S T R E S S
C
C
```

```
  100 NS=N110 + ( (NEL-1)*NPT + (IPT-1) ) * IDW
      CALL EP2D14 (A(NM),A(NS),A(NS+4*NDPA),A(NS+8*NDPA),A(NS+9*NDPA),
     1             A(NS+10*NDPA),A(NS+11*NDPA))
C
      RETURN
  101 FORMAT(4X,16HERROR - NCON IS ,I5,12HSHOULD BE 25 )
      END
      SUBROUTINE IEP214 (WA,IWA,PROP,NINT,IDW,IDW1)
      COMMON /EL/ IND,ICOUNT,NPAR(20),NUMEG,NEGL,NEGNL,IMASS,IDAMP,ISTAT
     1          ,NDOF,KLIN,IEIG,IMASSN,IDAMPN
      COMMON /VAR   / NG,KPRI,MODEX,KSTEP,ITE,ITEMAX,IREF,IEQREF,INOCMD
      COMMON /DPA/ NDPA
      COMMON /EM/ S(136),XM(16),B(4,16),NOD(8),NODM(8),RE(16),
     1            EDIS(16),AEM(204),XX(16),NOD5M(4)
      COMMON /MATMOD/ STRESS(4),STRAIN(5),D(4,4),IPT,NEL
      COMMON /GAUSS / XG(4,4),WGT(4,4),EVAL2(9,2),EVAL3(27,3),E1,E2,E3
      COMMON /TODIM/ BET,THIC,DE,IEL,NND5,LSTM
      DIMENSION WA(IDW1,1),IWA(IDW,1),PROP(1)
      DIMENSION H(8),P(2,8),XJ(2,2)
      EQUIVALENCE (NPAR(3),INDNL)
      DATA AC/ 0.0 /
C
C     LOADING SURFACE FUNCTIONS
C
      F1(XI,XIO) =  ((2.*XI*XIO/R - (2. - R)*XIO*XIO/R -
     1               XI*XI)*AM*AM/(R - 1.)**2)**.5
C
      F2(XI,XIO) = -((XI - XIO/R)**2 * AM*AM + (AA*XIO)**2)**.5 -
     1               XIO*(AA + AM/R)
C-----------------------------------------------------------------------
C     FOR DESCRIPTION OF THE DATA IN THE ARRAY PROP(I)
C     SEE SUBROUTINE EPL214
C
C     SET THE TERMS IN THE WORKING ARRAY (MATERIAL HISTORY) TO
C     REFLECT THE CURRENT (INITIAL) CONDITIONS
C
C    WA( 1,J) = SY      WA( 2,J) = SZ     WA( 3,J) = SYZ    WA( 4,J) = SX
C    WA( 5,J) = EY      WA( 6,J) = EZ     WA( 7,J) = EYZ    WA( 8,J) = EX
C    WA( 9,J) = XIO     WA(10,J) = EO     WA(11,J) = P FAC  WA(12,J)=IPEL
C-----------------------------------------------------------------------
      IF(INDNL .NE. 2 .AND. INDNL .NE. 3) GO TO 10
      WRITE(6,1000) INDNL
      IF(MODEX .NE. 0) STOP
C
   10 KJ = 11*NDPA+1
      J  = 0
      AM = PROP(9)
      R  = PROP(10)
      AA = PROP(11)
      RFI= (R -1.)/AM
      EO = PROP(18)
      CAP= PROP(16)
C
C     LOOP OVER THE GUASS POINTS
C
      DO 40 LX = 1,NINT
            E1 = XG(LX,NINT)
      DO 40 LY = 1,NINT
            J  = J + 1
            E2 = XG(LY,NINT)
C
C     COMPUTE THE DEPTH OF THE CURRENT INTEGRATION POINT
C
      CALL FUNCT2 (E1,E2,H,P,NOD5M,XJ,DET,XX,NEL)
```

```
            Z   = 0.0
        DO 30 I   = 1,IEL
   30       Z   = Z + XX(2*I)*H(I)
C
C       SET INITIAL STRESSES TO IN-SITU VALUES
C       THE Z-DIRECTION IS ASSUMMED VERTICAL
C
        WA(2,J)   = PROP(19)*Z + PROP(20)
        WA(1,J)   = PROP(21)*WA(2,J)
        WA(4,J)   = PROP(21)*WA(2,J)
        WA(3,J)   = 0.
C
C       SET THE INITIAL STRAINS TO ZERO
C
        WA( 5,J) = 0.
        WA( 6,J) = 0.
        WA( 7,J) = 0.
        WA( 8,J) = 0.
C
C       SET INITIAL STRESS STATE TO  -ELASTIC-
        IWA(KJ,J) = 1
C
C       SET THE PORE PRESSURE COEFFICIENT
C
        WA(11,J)   = PROP(22)
C
C       INPUT CHECK FOR INITIAL STRESSES
C
        STRESS(1) = WA(1,J)
        STRESS(2) = WA(2,J)
        STRESS(3) = 0.
        STRESS(4) = WA(4,J)
C
        CALL DEVPRT(STRESS,SRR,SZZ,SRZ,STT,SM,SJ2)
        CALL MAXMIN(STRESS,S1,S2,T)
        SJ1          = 3.*SM
C=====================================================================
C       SET INITIAL VOID RATIO AND CAP POSITION
C
        IF(CAP .GT. 0. ) GO TO 33
C--------------------------------------------------------------------
C       OVER CONSOLIDATED SOIL
C       SET VOID RATIO
        WA(9,J)   = CAP
        WA(10,J) = EO
        XIO         = CAP
        GO TO 35
C----------------------------------------------------------------------
C       COMPUTE CAP POSITION FOR NORMALLY CONSOLIDATED SOIL
C
   33 A             = 1. - (AM*RFI)**2
      BB            = -2.* (SJ1 - AM*AC*RFI**2)
      C             = SJ1**2 + (RFI*SJ2)**2 - (RFI*AC)**2
      WA( 9,J)   = -C/BB
      IF( WA( 9,J) .GT. 0.) WA( 9,J)   = 0.
C
      IF(ABS(A) .LE. 1.E-9) GO TO 34
C
      EL1           = (-BB + (BB*BB - 4.*A*C)**0.5)/(2.*A)
      EL2           = (-BB - (BB*BB - 4.*A*C)**0.5)/(2.*A)
      IF(EL1.LE. 0. .AND. EL1 .GE. SJ1) WA(9,J) = EL1
      IF(EL2.LE. 0. .AND. EL2 .GE. SJ1) WA(9,J) = EL2
   34 WA( 9,J)   = WA( 9,J) * R
      XIO           = WA( 9,J)
```

```
C
C     SET VOID RATIO
C
      WA(10,J) = EO
C=====================================================================
C     CHECK INITIAL STRESSES AGAINST YIELD CONDITIONS
C
   35 XIO = XIO*1.001
      IPEL = 1
      IF(S1  .GT. PROP(25)) IPEL = 0
      IF(SJ1 .GT. XIO/R) GO TO 36
      IF(SJ1 .LE. XIO) IPEL = 3
      IF(SJ1 .LE. XIO) GO TO 37
      IF(SJ2 .GT. F1(SJ1,XIO)) IPEL = 3
      GO TO 37
 36   IF(SJ2 .GT. F2(SJ1,XIO)) IPEL = 2
 37   IF(IPEL .NE. 1) WRITE(6,100) IPEL,NEL,J,STRESS,EO,XIO
C     FFF1 = F1(SJ1,XIO)
C     FFF2 = F2(SJ1,XIO)
C     PRINT*, 'SJ1=',SJ1,'XIO=',XIO,'FFF1=',FFF1,'FFF2=',FFF2
      IF(IPEL .NE. 1 .AND. MODEX .NE. 0) STOP
C
  40  CONTINUE
C
      RETURN
  100 FORMAT(  18H ERROR *** IPEL = ,I2,11H IN ELEMENT,I4, 7H AT IPT ,
     1       I2, 9H STRESS= ,4E11.4, 7H  EO = ,E11.4, 5H XIO= ,E11.4,
     2       /,40H            DURING INPUT CHECK IN IEP214    )
 1000 FORMAT(1X,17HERROR -- INDNL =  ,I5,11H IN EL2D14  )
C
      END
      SUBROUTINE EP2D14 (PROP,SIG,EPS,XIO,EI,FAC,IPEL)
      COMMON /EL/ IND,ICOUNT,NPAR(20),NUMEG,NEGL,NEGNL,IMASS,IDAMP,ISTAT
     1           ,NDOF,KLIN,IEIG,IMASSN,IDAMPN
      COMMON /MATMOD/ STRESS(4),STRAIN(5),C(4,4),IPT,NEL
      COMMON /BOUNSF/ ISR,IST,A1,A2,B1,B2,G,C2,BM
      COMMON /UNLOAD/ TRV(100),NTURNS,KSWTCH
      COMMON /ELSTP/  TIME
      COMMON /VAR   / NG,KPRI,MODEX,KSTEP,ITE,ITEMAX,IREF,IEQREF,INOCMD
      COMMON /IGRAV / BF,BOUY,XFAC,IGRV,H(8),GTOT,GGRS
      COMMON /CONST /DT,DTA,JNK(21),IOPE
      DIMENSION        PROP(1),SIG(1),EPS(1)
      DIMENSION        TAU(4),DELEPS(5),STATE(7)
      EQUIVALENCE      (ITYP2D,NPAR(5)),(INDNL,NPAR(3)),(W,STRAIN(5))
      CHARACTER *7 STATE
      DATA NGLAST /1000/
      DATA STATE /7HTENSION,7HELASTIC,7HLIMIT  ,7HCAP    ,7HCRITCAL  ,
     1             7HP-CAP  , 7HP-LIMIT /
C
C     THE FOLLOWING PROVIDE THE VALUE OF SQRT (J2) GIVEN SJ1
C
      F1(XI,XIO) =  ((2.*XI*XIO/R - (2. - R)*XIO*XIO/R -
     1                XI*XI)*XN*XN/(R - 1.)**2)**.5
C
      F2(XI,XIO) = -((XI - XIO/R)**2 * XN*XN + (A*XIO)**2)**.5 -
     1               XIO*(A + XN/R)
C
C     SJ1  = THE FIRST STRESS INVARIENT
C     SJ2  = SQRT OF SECOND DEVIATORIC STRESS INVARIENT
C
C     ISR IS THE NUMBER OF STRAIN COMPONENTS
C     IST IS THE NUMBER OF STRESS COMPONENTS
C
      IF (IPT.NE.1) GO TO 110
```

```
      IST = 4
      IF(ITYP2D.EQ.2) IST = 3
      ISR = 3
      IF(ITYP2D.EQ.0) ISR = 4
C
C     SET SOME MODEL PARAMETERS
C
      A  = PROP(11)
      R  = PROP(10)
      XN = PROP( 9)
C=================================================================
C     CALCULATE INCREMENTAL STRAINS
C
  110 IF(INDNL .EQ. 4) GO TO 125
      DELEPS(4) = 0.0
      DO 120 I  = 1,ISR
  120 DELEPS(I) = STRAIN(I) - EPS(I)
      DELEPS(5) = STRAIN(5)
      GO TO 127
C
C     STRAIN(5) IS THE ROTATIONAL TERM FOR LARGE DISPL ANALYSIS
C
C     FOR INDNL = 4, STRAIN CONTAINS THE INCREMENTAL VALUES
C
  125 DELEPS(4) = 0.
      DO 126 I  = 1,ISR
  126 DELEPS(I) = STRAIN(I)
      DELEPS(5) = STRAIN(5)
C
  127 DO 130 I  = 1,4
      STRESS(I) = SIG(I)
  130 TAU(I)    = SIG(I)
C=======================================================================
C
C     C O M P U T E   E L A S T I C  -  P L A S T I C   M A T E R I A L
C                        R E S P O N S E
C
C     IPEL = 0 = TENSION REGION;  IPEL = 1 = ELASTIC REGION
C     IPEL = 2 = LIMIT SURFACE ;  IPEL = 3 = HARDENING SURFACE
C     IPEL = 4 = CRITICAL STATE;  IPEL = 5 = SUBSURFACE YIELD
C
      CALL EPL214 (PROP,TAU,DELEPS,XIO,EI,IPEL,DFDI,DFDJ,DK)
C
C.......................................................................
C
C       U P D A T E   S T R E S S  /  S T R A I N   V A L U E S
C
      DO 250 I  = 1,IST
      STRESS(I) = TAU(I)
  250 SIG(I)    = TAU(I)
C
      IF(INDNL.GE.4) GO TO 270
      DO 260 I  = 1,ISR
  260 EPS(I)    = STRAIN(I)
      IF(ITYP2D.EQ.2) EPS(4) = STRAIN(4)
      GO TO 290
C
  270 DO 280 I = 1,ISR
  280 EPS(I)   = EPS(I) + STRAIN(I)
      IF(ITYP2D .EQ. 2) EPS(4) = STRAIN(4)
C......................................................................
C     COMPUTE PORE PRESSURE AND ADD TO `STRESS' TO FORM TOTAL STRESSES
C     `TAU' AND `SIG' WILL STILL REPRESENT THE EFFECTIVE STRESSES
C
```

```
  290 IF(ITYP2D .EQ. 2) GO TO 300
      EKK = EPS(1) + EPS(2) + EPS(4)
      PP  = 0.
      IF(EKK .GT. 1. E-9 .AND. FAC .LE. 1. E-9) GO TO 300
C
      PP         = EKK * BM * ABS(FAC)
      STRESS(1) = STRESS(1) + PP
      STRESS(2) = STRESS(2) + PP
      STRESS(4) = STRESS(4) + PP
C
  300 IF(KPRI.EQ.0) GO TO 700
      IF(ICOUNT.EQ.3) RETURN
C=================================================================
C
C        F O R M   T H E   M A T E R I A L   L A W
C
      KK = IPEL
      IF(KSWTCH .EQ. 1) KK = 1
C.................................................................
C
C        ELASTIC-PLASTIC MATERIAL
C
      CALL MD2D14 (TAU,DELEPS,C,DFDI,DFDJ,DK,KK)
C
C     MODIFICATIONS FOR UNDRAINED ANALYSIS
C     ADD A LARGE TERM TO THE STIFFNESS - REPRESENTS INCOMPRESSIBILITY
C
      IF(ITYP2D .EQ. 2) RETURN
C
      BMM    = ABS(FAC)*BM
      IF(EKK .GT. 1. E-9 .AND. FAC .LE. 1. E-9) BMM = 0.
      C(1,1) = C(1,1) + BMM
      C(1,2) = C(1,2) + BMM
      C(2,1) = C(2,1) + BMM
      C(2,2) = C(2,2) + BMM
C
      IF(ITYP2D .NE. 0) RETURN
      C(4,1) = C(4,1) + BMM
      C(4,2) = C(4,2) + BMM
      C(1,4) = C(1,4) + BMM
      C(2,4) = C(2,4) + BMM
      C(4,4) = C(4,4) + BMM
      RETURN
C==============================================================
C
C     P R I N T I N G   O F   S T R E S S E S
C
  700 CONTINUE
      CALL DEVPRT(TAU,DY,DZ,DS,DX,DM,SJ2 )
      SJ1 = 3.*DM
C
      FT1 = -1.
      FT2 = -1.
      IF(SJ1 .LE. XIO) FT1 = XIO - SJ1
      IF(SJ1 .LE. XIO/R .AND. SJ1 .GT. XIO) FT1 = SJ2 - F1(SJ1,XIO)
      IF(SJ1 .GT. XIO/R) FT2 = SJ2 -F2(SJ1,XIO)
      XX  = 3.*DM - XIO/R
      XX  = ABS(XX)
      IF(IPEL .EQ. 2 .AND. XX .LE. .001*ABS(XIO/R)) IPEL = 4
      IF(IPEL .EQ. 3 .AND. XX .LE. .001*ABS(XIO/R)) IPEL = 4
C
      IF (NG.NE.NGLAST) GO TO 802
      IF (NEL.GT.NELAST) GO TO 806
      IF (IPT-1) 810,808,810
C
```

```
  802 NGLAST = NG
  808 WRITE (6,2003)
C
  806 NELAST=NEL
      WRITE (6,2004) NEL
C
  810 CALL MAXMIN (TAU,SX,SY,SM)
C
      II = IPEL+1
      WRITE (6,2007) IPT,STATE(II),TAU(4),(TAU(I),I=1,3),
     1               SX,SY,FT1,FT2,XIO,SM
      IF(ABS(PROP(22)) .GT. 1.E-9) WRITE(6,2008)PP
      RETURN
C
 2003 FORMAT ( 90H ELEMENT  STRESS
     1                                        8X,14HYIELD FUNCTION /
     2 54H NUM/IPT   STATE   STRESS-XX    STRESS-YY    STRESS-ZZ    ,
     3 47H     STRESS-YZ   MAX STRESS  MIN STRESS      CAP ,5X,8H LIMIT
     4    ,6X,2HXL ,4X,5HANGLE /)
 2007 FORMAT (5X,I2,2X, A7 ,4E13.5,2E12.4,3E11.4,F6.2)
 2008 FORMAT (9X,7HPORE P=,E13.5)
 2004 FORMAT (I4)
C
      END
      SUBROUTINE MD2D14(TAU,DEPS,DP,A,B,DFDK,KK)
C
C=====================================================================
C
C        THIS ROUTINE DEVELOPES THE ELASTIC-PLASTIC STRESS-STRAIN
C        RELATION FOR AN ELASTIC-PLASTIC MATERIAL OF THE I1 - J2 TYPE
C
C        THE PROCEDURE AND TERMINOLOGY IS THAT USED BY MIZUNO +
C        CHEN IN THE REPORT CE-STR-82-15, PURDUE UNIVERSITY
C
C     A          = DF/DI1
C     B          = DF/DJ2
C     AL         = 3*LAM + 2G = ELASTIC PARAMETER
C     A1         = (K + 4G/3)
C     B2         = LAME'S CONST = (K - 2G/3)
C     G          = SHEAR MODULUS
C     DFDK       = PLASTIC HARDENING EFFECT
C     KK        <= 1 TO FORM ELASTIC STIFFNESS
C
C     ITYP2D     = 0 FOR AXISYMMETRIC PROBLEMS
C                = 1 FOR PLANE STRAIN PROBLEMS
C                = 2 FOR PLANE STRESS PROBLEMS
C
C=====================================================================
      COMMON /BOUNSF/ ISR,IST,A1,A2,B1,B2,G,C2,BM
      COMMON /MATMOD/  STRESS(4),STRAIN(5),C(4,4),IPT,NEL
      COMMON /EL/ IND,ICOUNT,NPAR(20),NUMEG,NEGL,NEGNL,IMASS,IDAMP,ISTAT
     1            ,NDOF,KLIN,IEIG,IMASSN,IDAMPN
      DIMENSION TAU(1),DEPS(1),DP(1)
      DIMENSION H(4)
C
      EQUIVALENCE (ITYP2D,NPAR(5))
C
C.....................................................................
C     IF KK <= TO 1 COMPUTE ONLY THE ELASTIC STIFFNESS
C
      HH     = 1.
      IF(KK .LE. 1) GO TO 85
C.....................................................................
C
```

```
      CALL DEVPRT(TAU,SY,SZ,SS,SX,SM,SJ2)
      AL     = 3.*B2 + 2.*G
      HH     = 3.*A*A*AL + 4.*B*B*G*SJ2*SJ2 + DFDK
      H(1)   = A*AL + 2.*G*B*SY
      H(2)   = A*AL + 2.*G*B*SZ
      H(4)   = A*AL + 2.*G*B*SX
      H(3)   =        2.*G*B*SS
C
C     COMPUTE THROUGH THICKNESS STRAIN FOR PLANE STRESS CASE
      IF(ITYP2D.NE.2) GO TO 80
      DP(13) = B2 - H(1)*H(4)/HH
      DP(14) = B2 - H(2)*H(4)/HH
      DP(15) =    - H(3)*H(4)/HH
      DP(16) = A2 - H(4)*H(4)/HH
      DEPS(4)= -(DP(13)*DEPS(1)+ DP(14)*DEPS(2)+ DP(15)*DEPS(3))/DP(16)
C
C     D-LAMBDA TERM
C
   80 X1 = 3.*BM * A * (DEPS(1) + DEPS(2) + DEPS(4))
     1   + 2.*B*G*(SY*DEPS(1)+ SZ*DEPS(2)+ 2.*SS*DEPS(3)+ SX*DEPS(4))
      X1 = X1/HH
C     PRINT*,'DLAM =',X1
      GO TO 100
C
   85 DO 90 I=1,4
   90 H(I)   = 0.
C.....................................................................
C     FIRST COLUMN OF PLASTIC RELATION
  100 CONTINUE
C     PRINT*,'KK=',KK,'A2=',A2,'B2=',B2,'G=',G,'HH=',HH
C     PRINT*,'TAU=',(TAU(II),II=1,4)
C     PRINT*,'SY=',SY,'SZ=',SZ,'SX=',SX,'SS=',SS
C     PRINT*,'H=',H
      DP( 1) = A2 - H(1)*H(1)/HH
      DP( 2) = B2 - H(1)*H(2)/HH
      DP( 3) =    - H(1)*H(3)/HH
      DP( 4) = B2 - H(1)*H(4)/HH
C
C     SECOND COLUMN
      DP( 5) = DP( 2)
      DP( 6) = A2 - H(2)*H(2)/HH
      DP( 7) =    - H(2)*H(3)/HH
      DP( 8) = B2 - H(2)*H(4)/HH
C
C     THIRD COLUMN
      DP( 9) = DP( 3)
      DP(10) = DP( 7)
      DP(11) =  G - H(3)*H(3)/HH
      DP(12) =    - H(3)*H(4)/HH
C
C.....................................................................
C     FORTH COLUMN
      DP(13) = DP( 4)
      DP(14) = DP( 8)
      DP(15) = DP(12)
      DP(16) = A2 - H(4)*H(4)/HH
C
      IF(ITYP2D.LT.2) RETURN
C.....................................................................
C     PLANE STRESS / MODIFY DP MATRIX
C
      DO 120 I=1,3
      AA = C(I,4)/C(4,4)
      DO 120 J=I,3
```

```
      C(I,J)  =  C(I,J) -C(4,J)*AA
  120 C(J,I)  =  C(I,J)
      STRAIN(4)=STRAIN(4) + DEPS(4)
C
      RETURN
      END
      SUBROUTINE EPL214 (PROP,TAU,DEPS,XIO,EO,MTYPE,DFDI,DFDJ,DK)
C . . . . . . . . . . . . . . . . . . . . . . . . . . . . . . . . . . .
C .                                                                    .
C .      IST      NUMBER OF STRESS COMPONENTS                          .
C .      ISR      NUMBER OF STRAIN COMPONENTS                          .
C .      TAU      STRESSES AT THE END OF THE PREVIOUS UPDATE           .
C .               TAU(1) = YY COMPONENT                                .
C .               TAU(2) = ZZ COMPONENT                                .
C .               TAU(3) = YZ COMPONENT                                .
C .               TAU(4) = XX COMPONENT                                .
C .                                                                    .
C .      EPS      STRAINS  AT THE END OF THE PREVIOUS UPDATE           .
C .               EPS(1) = YY COMPONENT                                .
C .               EPS(2) = ZZ COMPONENT                                .
C .               EPS(3) = YZ COMPONENT (ENGINEERING SHEAR STRAIN )    .
C .               EPS(4) = XX COMPONENT                                .
C .                                                                    .
C .      RATIO    CORRECTION FACTOR USED TO RETRUN STRESS STATE        .
C .               TO FAILURE SURFACE                                   .
C .      DEPS     INCREMENT IN STRAINS                                 .
C .                                                                    .
C .     IPEL      = 0, MATERIAL IN TENSION REGION                      .
C .      OR         1, MATERIAL IS ELASTIC                             .
C .     MTYPE       2, MATERIAL ON LIMIT SURFACE                       .
C .                 3, MATERIAL ON HARDENING SURFACE                   .
C .                 4, CRITICAL STATE                                  .
C .                 5, MATERIAL PROJECTION ON LIMIT SURFACE            .
C .                 6, MATERIAL PROJECTION ON HARDENING SURFACE        .
C .                                                                    .
C .    INDNL      = 0 FOR LINEAR ANALYSIS (NOT APPLICABLE HERE)        .
C .                 1 FOR MATERIAL NONLINEARITY ONLY                   .
C .                 2 NOT AVAILABLE                                    .
C .                 3 NOT AVAILABLE                                    .
C .                 4 FOR UPDATED LAGRANGIAN                           .
C .                                                                    .
C .    ITYP2D     = 0 FOR AXISYMMETRIC CONDITIONS                      .
C .                 1 FOR PLANE STRAIN CONDITIONS                      .
C .                 2 FOR PLANE STRESS CONDITIONS (NOT AVAILABLE HERE) .
C .                                                                    .
C .     PROP (1) = XK1   = BULK MODULUS PARAMETER                      .
C .          (2) = XK2   = BULK MODULUS PARAMETER                      .
C .          (3) = XG1   = BULK MODULUS PARAMETER                      .
C .          (4) = XG2   = SHEAR MODULUS PARAMETER                     .
C .          (5) = CC    = COMPRESSION INDEX                           .
C .          (6) = CR    = REBOUND INDEX                               .
C .          (7) = XIOL  = LIMITING POSITION PAST WHICH CAP MAY NOT    .
C .                        CONTRACT (I.E. XIO <= XIOL)                 .
C .          (8) =         OPEN                                        .
C .          (9) = XN    = SLOPE OF CRITICAL STATE LINE IN I - SJ2 SPC .
C .         (10) = R     = CAP PARAMETER                               .
C .         (11) = A     = LIMIT SURFACE PARAMETER                     .
C .         (12) = S     = ELASTIC ZONE PARAMETER                      .
C .         (13) = H     = HARDENING PARAMTER                          .
C .         (14) = XPA   = ATMOSPHERIC PRESSURE                        .
C .         (15) = XM    = HARDENING PARAMETER                         .
C .         (16) = XIOI  = INITIAL CAP POSTION                         .
C .                      > 0 TO COMPUTE N.C. POSTION IN IEP214         .
C .                      < 0 TO SPECIFIY INITIAL POSITION FOR O.C. SOIL.
```

```
C .         (17) = XIL  = MINIMUM VALUE FOR COMPUTING BULK MOD AND      .
C .                         PLASTIC STRAINS                              .
C .         (18) = EO   = INITIAL VOID RATIO                             .
C .         (19) = A1   = PARAMTER TO COMPUTE IN-SITU STRESS             .
C .         (20) = A2   = DITTO                                          .
C .         (21) = A3   = DITTO                                          .
C .                                   SIGZ = A1*Z + A2                   .
C .                                   SIGX = SIGZ*A3                     .
C .                                   SIGY = SIGZ*A4                     .
C .                                                                      .
C .         (22) = IDR  = PARAMTER FOR DRAINED OR UNDRAINED ANALYSIS     .
C .                     > 0 TO COMPUTE BOTH TENSION & COMP PP            .
C .                     < 0 TO COMPUTE ONLY COMPR PP                     .
C .         (23) =        OPEN                                           .
C .         (24) =        OPEN                                           .
C .         (25) = TCUT = MAXIMUM TENSION PRINCIPAL STRESS               .
C .                                                                      .
C . . . . . . . . . . . . . . . . . . . . . . . . . . . . . . . . . . . .
C
C  INSERT COMMON
      COMMON /BOUNSF/  ISR,IST,A1,A2,B1,B2,G,C2,BM
      COMMON /MATMOD/  STRESS(4),STRAIN(5),C(4,4),IPT,NEL
      COMMON /EL     / IND,ICOUNT,NPAR(20),NUMEG,NEGL,NEGNL,IMASS,IDAMP,
     1                 ISTAT,NDOF,KLIN,IEIG,IMASSN,IDAMPN
      DIMENSION        DTAU(4),PROP(1), TAU(1), DEPS(1)
      EQUIVALENCE      (ITYP2D,NPAR(5)),(INDNL,NPAR(3))
C
C=================================================================
C
C     ELASTIC MODULI FUNCTIONS
C
      BULK(XI ,EO) = XK1*2.30259*(1. + EO)*ABS(XI)/(3.*CR) + XK2
      SHRM(XI ,EO) = XG1*BM + XG2
C................................................................
C
C     FUNCTIONS FOR LOADING SURFACES
C
C     F1 = CAP FUNCTION ; F2 = BACK SURFACE FUNCTION
C     THESE FUNCTIONS RETURN THE VALUE OF SJ2 GIVEN XI AND XIO
C
      F1(XI,XIO) =  ((2.*XI*XIO/R - (2. - R)*XIO*XIO/R -
     1               XI*XI)*XN*XN/(R - 1.)**2)**.5
C
      F2(XI,XIO) = -((XI - XIO/R)**2 * XN*XN + (A*XIO)**2)**.5 -
     1               XIO*(A + XN/R)
C                 ..............................
C     FF1 = CAP FUNCTION ; F2 = BACK SURFACE FUNCTION
C     THESE FUNCTION RETURN THE VALUES OF FAILURE FUNCTION
C     NOT THE VALUE OF SJ2
C
      FF1(XI,XIO,SJ2) = XI*XI +((R - 1.)*SJ2/XN)**2 - 2.*XI*XIO/R +
     1                  (2. - R)*XIO*XIO/R
C
      FF2(XI,XIO,SJ2) = XI*XI - (SJ2/XN)**2 - 2.*XI*XIO/R -
     1             2.*XIO*XIO*A/(R*XN) - 2.*(1./R + A/XN)*SJ2*XIO/XN
C................................................................
C
C     FUNCTIONS FOR LOADING SURFACE DERIVATIVES
C
C     THE ARGUMENTS XIO AND XI FOLLOW NORMAL SIGN CONVENTION
C     TENSION = +VE
C
C     THE ARGUMENT GT WILL BE +VE IF XIO IF -VE (USUALLY)
C
C     NOTE    SJ2 = -GT*XIO      = -GAMMA * THETA * XIO
```

```
C           XIBAR = -GAMM * XIO = THE IMAGE POINT
C
C      NOTE DFDJ IS THE DERIVATIVE WRT J2 NOT SQRT J2
       DF1DI(XIO,GAM)      =  2.*XIO*(GAM - 1./R)
       DF1DJ(XIO,GAM)      =  ((R - 1.)/XN)**2
C
       DF2DI(XIO,GAM)      =  2.*XIO*(GAM - 1./R)
       DF2DJ(XIO,GT,SJ2)   = -XIO*(-GT + XN/R +A)/(XN*XN*SJ2)
C
C      DF1DIO(XIO,XI)      =  (4.*XIO -2.*R*XIO -2.*XI)/R
C      DF2DIO(XIO,XI,SJ2) = -(4.*XIO*A/XN + 2.*XI)/R -
C     1                       2.*SJ2*(XN/R + A)/(XN*XN)
C.......................................................
C
C      FUNCTIONS FOR HARDENING MODULI
C
C      THE ARGUMENTS FOLLOW THE FOLLOWING CONVENTIONS
C
C      XIO    = +VE FOR TENSION
C      XI     = +VE FOR TENSION
C      GAM    = POSITIVE IF XI AND XIO ARE OF SAME SIGN
C      THETA = ABS ( SJ2/XI ) IS +VE
C      X      = THETA / XN      IS +VE
C
       HF1(XIO,GAM,EO)    = -12.*XIO**3*(1. + EO)*(2. - R - GAM)*
      1                      (GAM - 1./R)*2.30259/((XW*CR - CC)*R)
C
       HF2(XIO,GAM,X,EO) =  12.*XIO**3*(1. + EO)*(GAM - 1./R)*
      1                      (GAM*(1 - X - X*Y) + 2.*A/XN)*
      2                      2.30259/((XW*CR - CC) *R)
C
C      CAN'T USE HC IF SJ2 = THETA = 0 --- ZERO DIVIDE
C      THIS IMPLIES NO SUBSURFACE YIELD FOR HYDROSTATIC LOADING
C
       HC(THETA,SJ2,DFDI,DFDJ,D,DO) =
      1                     H* XPA *(1.+ (XN/THETA)**XM) *
      2                     (9.*DFDI**2 + (DFDJ*SJ2)**2 *4./3.)*
      3                     D/(DO - D)
C.............................................................
C
C      THE REFERENCE DISTANCE PARAMETER
C
C      DD WILL BE -VE IF XIO IS +VE (WATCH OUT)
C
       DD(XIO,THETA) = -XIO * ((1.+THETA**2)/(1.+27.*THETA**2))**0.5
C.......................................................
C
C     SCALING FUNCTIONS TO DETERMINE IMAGE POINTS
C
C      X = THETA /XN       Y = R*A/XN
C
C     THE SCALING FUNCTIONS ARE SET UP TO RECIEVE +VE ARGUMENTS
C     APPARENTLY BECAUSE DAFALIAS ASSUMES +VE = COMPRESSION
C
       GAMM1(X) = (1. + (R - 1.)*(1 + R*(R -2.)*X*X)**0.5)/
      1            (R*(1 + X*X + R*(R - 2.)*X*X))
C
C      CAN'T USE GAMM2 IF X = 1 -- ZERO DIVIDE
C
       GAMM2(X) = (X -1. + X*Y -((X -Y -1.)**2 + (X*X -1.)*Y*Y)**0.5)/
      1            (R*(X*X -1.))
C=======================================================
C
C      SET MATERIAL PARAMETERS
```

```
C
      EI   = EO
      A    = PROP(11)
      CC   = PROP(5)
      CR   = PROP(6)
      H    = PROP(13)
      R    = PROP(10)
      S    = PROP(12)
      XPA  = PROP(14)
      XG1  = PROP(3)
      XG2  = PROP(4)
      XIL  = PROP(17)
      XK1  = PROP(1)
      XK2  = PROP(2)
      XN   = PROP(9)
      XM   = PROP(15)
      XIOL= PROP(7)
      Y    = R*A/XN
C.............................................................
C
C     CALCULATE ELASTIC PARAMETERS
C
C     PRINT*,'---ENTERING EPL214--- DEPS=',(DEPS(I),I=1,4)
      CALL DEVPRT(TAU,SYY,SZZ,SYZ,SXX,SM,SJ2)
      XI   = 3.*SM
      XI1  = AMIN1(XI + XIL, 0.) - XIL
      XW   = XIO/(AMIN1(XIO + XIL, 0.) - XIL)
C     PRINT*,'XW =',XW,'XIL =',XIL,'XI1 =',XI1,'XI=',XI,'XI1=',XI1
      IF(XW .LT. 0.) WRITE(6,2000) XW,XIO,XIL
      IF(XW .LT. 0.) XW = 1.
C
      BM   = BULK(XI1,EO)
      G    = SHRM(XI1,EO)
C
      A2   = BM + 4.*G/3.
      B2   = BM - 2.*G/3.
      C2   = G
      PV   = (3.*BM - 2.*G)/(2.*(3.*BM + G))
      IF(PV .LT. 0.) WRITE(6,1010)PV
      IF(ITYP2D.EQ.2) GO TO 10
C
C     PLANE STRAIN - AXISYMMETRIC
      A1   = A2
      B1   = B2
      GO TO 15
C
C     PLANE STRESS
  10  YM   = 9.*G*BM/(3.*BM+G)
      A1   = YM/(1. -PV*PV)
      B1   = A1*PV
  15  D2   = PV/(PV - 1.0)
C=============================================================
C
C     TRIAL STRESSES
C
C     PRINT*, 'BM = ',BM, 'G= ',G
      IF(ITYP2D.EQ.2) DEPS(4) = D2*(DEPS(1)+DEPS(2))
      DE   = DEPS(1) + DEPS(2) + DEPS(4)
C
C     W    = ROTATIONAL TERM FOR LARGE DISPLACEMENT ANALYSIS
C
      W    = DEPS(5)
      IF(INDNL .LE. 3) W = 0.
      SYY = 2.*W*TAU(3)
```

```
      SYZ = W*(TAU(1) - TAU(2))
C
      DTAU(4) = 0.
      DTAU(1) = A2*DEPS(1) + B2*(DEPS(2) + DEPS(4)) + SYY
      DTAU(2) = A2*DEPS(2) + B2*(DEPS(1) + DEPS(4)) - SYY
      DTAU(3) = C2*DEPS(3)                          - SYZ
      IF(ITYP2D .EQ. 2) GO TO 20
      DTAU(4) = A2*DEPS(4) + B2*(DEPS(1) + DEPS(2))
C
   20 DO 25 I = 1,IST
   25 TAU(I)  = STRESS(I) + DTAU(I)
C     PRINT*, 'DEPS=', (DEPS(II),II=1,5)
C     PRINT*, 'DTAU=', (DTAU(II),II=1,4)
C     PRINT*, 'TAU=', (TAU(II),II=1,4)
C
C     DETERMINE THE DEVIATORIC COMPONENTS AND INVARIENTS
C     AT RESULTANT STRESS STATE
C
      CALL DEVPRT(TAU,SYY,SZZ,SYZ,SXX,SM,SJ2)
      XI     = 3.*SM
      EO     = EO + DE *(1. + EO)
      MTYPE  = 1
C--------------------------------------------------------------
C
C     CHECK TRIAL STRESSES FOR ELASTIC RESPONSE
C
C     COMPUTE IMAGE POINT
C
      THETA = 1000.
      IF(ABS(XI) .GT. 1.E-5) THETA = ABS(SJ2/XI)
      X     = THETA/XN
      IF(THETA .GT. XN) GAM =  GAMM2(X)
      IF(THETA .LE. XN) GAM =  GAMM1(X)
C
C     D IS THE DISTANCE BETWEEN THE STRESS AND IMAGE POINTS
C     IF D < 0 THEN STRESS STATE IS BEYOND THE BOUNDING SURFACE
C
      D = (1. + THETA*THETA)**0.5 * ( XI - GAM*XI0)
C     PRINT*,'D=',D
C     PRINT*, 'XI=',XI, 'XI0=',XI0,'GAM=',GAM
      IF(D .LE. 0.) GO TO 165
      IF(S .LE. 1.) GO TO  35
C
      XIBAR  = GAM * XI0
      IF(THETA .GT. XN) SJBAR = F2(XIBAR,XI0)
      IF(THETA .LE. XN) SJBAR = F1(XIBAR,XI0)
C
C     RR IS THE DISTANCE BETWEEN THE IMAGE POINT AND CENTER
C     OF PROJECTION
C
      RR = (XIBAR*XIBAR + SJBAR*SJBAR)
      IF(RR .GT. 0.) RR = RR**0.5
C     PRINT*,'GAM',GAM,'THETA',THETA,'RR',RR,'XBAR',XBAR,'SBAR',SJBAR
      IF(D  .GE. RR/S) RETURN
C--------------------------------------------------------------
C     CHECK TO SEE IF INITIAL STRESS STATE IS OUTSIDE
C     THE ELASTIC ZONE AND RESPONSE IS UNLOADING
C
   35 CALL DEVPRT(STRESS,SYY,SZZ,SYZ,SXX,SM,SJ2)
      XI     = 3.*SM
      THETA = 1000.
      IF(ABS(XI) .GT. 1.E-5) THETA = ABS(SJ2/XI)
      X      = THETA/XN
      IF(THETA .GT. XN .AND. XI .GT. XI0/R) GO TO 50
```

```
C..............................................................
C     MATERIAL IS IN THE HARDENING REGION
C
      GAM   = GAMM1(X)
C
C     CHECK FOR UNLOADING  (XL < 0)
C
      DFDI  = DF1DI(XI0,GAM)
      DFDJ  = DF1DJ(XI0,GAM)
      D0    = DD(XI0,THETA)
C
      CALL DEVPRT(DTAU,DYY,DZZ,DYZ,DXX,DM,DJ2)
      DJ2   = SYY*DYY + SZZ*DZZ + 2.*SYZ*DYZ + SXX*DXX
      XL    = DFDI*3.*DM + DFDJ*DJ2
      IF(XL .LE. 0.) RETURN
      GO TO 70
C..............................................................
C     MATERIAL IS IN THE LIMITING  REGION
C
   50 GAM   = GAMM2(X)
      GT    = GAM*THETA
C
C     CHECK FOR UNLOADING  (XL < 0)
C
      DFDI  = DF2DI(XI0,GAM)
      DFDJ  = DF2DJ(XI0,GT,SJ2)
      SJBAR = -GT*XI0
      D0    = DD(XI0,THETA)
      DK    = HF2(XI0,GAM,X,E0) + HC(THETA,SJBAR,DFDI,DFDJ,D,D0)
C
      CALL DEVPRT(DTAU,DYY,DZZ,DYZ,DXX,DM,DJ2)
      DJ2   = SYY*DYY + SZZ*DZZ + 2.*SYZ*DYZ + SXX*DXX
      XL    = DFDI*3.*DM + DFDJ*DJ2
      IF(XL .GT. 0.) GO TO 70
      XL    = XL/DK
C     PRINT*,'XL =',XL, 'DK=',DK,'GT=',GT,'SJBAR=',SJBAR,'D=',D
C     PRINT*,'DM,DJ2',DM,DJ2,'DFDI,DFDJ',DFDI,DFDJ,'THETA=',THETA
      IF(XL .LE. 0.) RETURN
C=============================================================
C
C     MATERIAL RESPONSE IS LOADING BUT WILL REMAIN
C     WITHIN THE BOUDING SURFACE
C
C     ARBITRARILY DIVIDE STEP INTO FOUR INCREMENTS
C
   70 M = 4
      GO TO 179
C----------------------------------------------------------------
C
C     MATERIAL RESPONSE MAY INTERSECT THE BOUNDING SURFACE
C
C     ARBITRARILY DIVIDE STEP INTO 10 INCREMENTS
C
  165 M     = 10
C----------------------------------------------------------------
C     RETURN TO INITIAL STATE
C
  179 E0        = EI
      DO 180 I = 1,IST
  180 TAU(I)    = STRESS(I)
C
C     SET INCREMENTAL STRAINS
C
      DO 190 I = 1,5
```

```
  190 DEPS(I)   = DEPS(I)/M
      W         = DEPS(5)
C.............................................................
C     BEGIN LOOP OVER SUBINCREMENTS
C
      DO 300 I = 1,M
C     PRINT*,'---------- TOP OF SUBINCREMENT-------------'
C
C     CALCULATE ELASTIC PARAMETERS
C
      CALL DEVPRT(TAU,SYY,SZZ,SYZ,SXX,SM,SJ2)
      XI    = 3.*SM
      XI1   = AMIN1(XI + XIL, 0.) - XIL
      XW    = XI0/(AMIN1(XI0 + XIL, 0.) - XIL)
      IF(XW .LT. 0.) WRITE(6,2000) XW,XI0,XIL
      IF(XW .LT. 0.) XW  = 1.
      BM    = BULK(XI1,E0)
      G     = SHRM(XI1,E0)
C
      A2    = BM + 4.*G/3.
      B2    = BM - 2.*G/3.
      C2    = G
      PV    = (3.*BM - 2.*G)/(2.*(3.*BM + G))
C
      IF(ITYP2D.EQ.2) GO TO 191
C     PLANE STRAIN - AXISYMMETRIC
      A1    = A2
      B1    = B2
      GO TO 195
C
C     PLANE STRESS
 191  YM    = 9.*G*BM/(3.*BM+G)
      A1    = YM/(1. -PV*PV)
      B1    = A1*PV
 195  D2    = PV/(PV - 1.0)
C
      THETA = 1000.
      IF(ABS(XI) .GT. 1.E-5) THETA = ABS(SJ2/XI)
      X      = THETA/XN
C..............................................................
C     CHECK TO SEE IF STRESS EXCEEDS ELASTIC REGION
C
      IF(THETA .GT. XN) GAM =  GAMM2(X)
      IF(THETA .LE. XN) GAM =  GAMM1(X)
C
C     D IS THE DISTANCE BETWEEN THE STRESS AND IMAGE POINTS
C     IF D < 0 THEN STRESS STATE IS BEYOND THE BOUNDING SURFACE
C     IF S <= 1 THEN THERE IS NO ELASTIC ZONE
C
      D = (1. + THETA*THETA)**0.5 * ( XI - GAM*XI0)
      IF(D .LE. 0.) GO TO 196
      IF(S .LE. 1.) GO TO 196
C
      XIBAR  = GAM * XI0
      IF(THETA .GT. XN) SJBAR = F2(XIBAR,XI0)
      IF(THETA .LE. XN) SJBAR = F1(XIBAR,XI0)
C
C     RR IS THE DISTANCE BETWEEN THE IMAGE POINT AND CENTER
C     OF PROJECTION
C
      RR = (XIBAR*XIBAR + SJBAR*SJBAR)
      IF(RR .GT. 0.) RR = RR**0.5
C
C     IF ( D > RR/S ) THEN THE RESPONSE IS ELASTIC
```

```
      IF(D  .GE. RR/S) MTYPE = 1
      IF(D  .GE. RR/S) GO TO 220
C
  196 IF(THETA .GT. XN .AND. XI .GT. XI0/R) GO TO 200
C.............................................................
C     MATERIAL IS IN THE HARDENING REGION
C
      MTYPE = 6
      GT    = GAM*THETA
      DFDI  = DF1DI(XI0,GAM)
      DFDJ  = DF1DJ(XI0,GAM)
      SJBAR = -GT*XI0
      D0    = DD(XI0,THETA)
      D     = (1. + THETA*THETA)**0.5 * ( XI - GAM*XI0)
      IF(D .LT. 0.) D = 0.
      DK    = HF1(XI0,GAM,E0)
      IF(THETA .GT. 0.) DK = DK + HC(THETA,SJBAR,DFDI,DFDJ,D,D0)
      GO TO 220
C.............................................................
C     MATERIAL IS IN THE LIMITING REGION
C
  200 MTYPE = 5
C     PRINT*,'--- ENTERING LIMIT REGION-- MTYPE = ',MTYPE
      GT    = GAM*THETA
      DFDI  = DF2DI(XI0,GAM)
      DFDJ  = DF2DJ(XI0,GT,SJ2)
      SJBAR = -GT*XI0
      D0    = DD(XI0,THETA)
      D     = (1. + THETA*THETA)**0.5 * ( XI - GAM*XI0)
      IF(D .LT. 0.) D = 0.
      DK    = HF2(XI0,GAM,X,E0) + HC(THETA,SJBAR,DFDI,DFDJ,D,D0)
C..............................................................
C     CALL CONSTITUTIVE RELATION
C
  220 CALL MD2D14(TAU,DEPS,C,DFDI,DFDJ,DK,MTYPE)
C     PRINT*,'AFTER CALL TO MD2 MTYPE=',MTYPE
C
C     SPIN TERMS FOR JAUMAN STRESS RATE
      SYY       = 2.*W*TAU(3)
      SYZ       = W*(TAU(1) - TAU(2))
C     PRINT*,'XI0=',XI0,'XI=',XI
C     PRINT*,'TAU=',(TAU(II),II=1,4)
C
C     COMPUTE NEW STRESSES
      DO 225 K  = 1,IST
      STRESS(K) = TAU(K)
      DO 225 J  = 1,ISR
  225 TAU(K)    = TAU(K) + C(K,J)*DEPS(J)
C     PRINT*,'BM=',BM,'G=',G
C     PRINT*,'TAU=',(TAU(II),II=1,4)
C
      TAU(1)    = TAU(1) + SYY
      TAU(2)    = TAU(2) - SYY
      TAU(3)    = TAU(3) - SYZ
C
      CALL DEVPRT(TAU,SYY,SZZ,SYZ,SXX,SM,SJ2)
      XII = XI
      XI  = 3.*SM
      DXI = XI - XII
C.............................................................
C     UPDATE VOID RATIO
C
      DE   = DEPS(1) + DEPS(2) + DEPS(4)
      EI   = E0
```

```
      EO    = EO + DE * (1. + EO)
C..........................................................
C     UPDATE HARDENING PARAMETER
C
      DEE   = DXI/(3.*BM)
      DEP   = (DE - DEE)*(1. + EI)
      DXIO = 2.30259*DEP*XIO/(XW*CR - CC)
      XIO   = XIO + DXIO
      IF(XIO .GT. XIOL) XIO = XIOL
C     PRINT*,'UP FOR SSY DEE= ',DEE,'DEP=',DEP,'DXIO=',DXIO,'XIO=',XIO
C..........................................................
C     CHECK TO SEE IF STRESS EXCEEDS BOUNDING SURFACE
C
      IF(XI .LE. XIO/R) FF = FF1(XI,XIO,SJ2)
      IF(XI .GT. XIO/R) FF = FF2(XI,XIO,SJ2)
      IF(FF .LE. 0.   ) GO TO 300
C----------------------------------------------------------
C     PROVIDE FINAL CORRECTION TO SURFACE IF NO CONVERGENCE
C
      THETA  = 1000.
      IF(ABS(XI) .GT. 1.E-5) THETA  = ABS(SJ2/XI)
      X      = THETA/XN
      MTYPE  = 3
      IF(THETA .GT. XN .AND. XI .GT. XIO/R) MTYPE = 2
      IF(XI .LE. XIO) GO TO 270
C..........................................................
C     SCALING FOR STRESS STATE OVER BOUNDING SURFACE
C
      IF(THETA .GT. XN) SJBAR = F2(XI,XIO)
      IF(THETA .LE. XN) SJBAR = F1(XI,XIO)
      RATIO  = SJBAR/SJ2
      SYY    = SYY*RATIO
      SXX    = SXX*RATIO
      TAU(3) = SYZ*RATIO
      TAU(1) = SYY + SM
      TAU(4) = SXX + SM
      TAU(2) = -SYY - SXX + SM
C     PRINT*, 'RATIO=',RATIO
      GO TO 300
C..........................................................
C     SCALING FOR STRESS STATE PAST CAP
C
  270 IF(THETA .GT. XN) GAM =  GAMM2(X)
      IF(THETA .LE. XN) GAM =  GAMM1(X)
      XIBAR  = GAM * XIO
      IF(THETA .GT. XN) SJBAR = F2(XIBAR,XIO)
      IF(THETA .LE. XN) SJBAR = F1(XIBAR,XIO)
      RR = XI*XI + SJ2*SJ2
      D  = XIBAR*XIBAR + SJBAR*SJBAR
      IF(RR .GT. 0.) RR = RR**0.5
      IF(D  .GT. 0.) D  = D **0.5
      D  = D/RR
C
      DO 275 K = 1,IST
  275 TAU(K)   = TAU(K)*D
C     PRINT*,'D/RR=',D
C
  300 CONTINUE
C     PRINT*, 'FF=',FF, 'SJ2 = ',SJ2 , 'RATIO= ',RATIO
C     RATIO = 55555.
C==========================================================
C     CALCULATE PARAMETERS NECESSARY TO CALL MD2D14
C
      CALL DEVPRT(TAU,SYY,SZZ,SYZ,SXX,SM,SJ2)
```

```
      XI    = 3.*SM
      THETA = 1000.
      IF(ABS(XI) .GT. 1.E-5) THETA = ABS(SJ2/XI)
      X     = THETA/XN
      IF( THETA .GT. XN) GO TO 320
C.....................................................................
C     MATERIAL IS IN THE HARDENING REGION
C
      GAM   = GAMM1(X)
      GT    = GAM*THETA
      DFDI  = DF1DI(XI0,GAM)
      DFDJ  = DF1DJ(XI0,GAM)
      SJBAR = -GT*XI0
      D0    = DD(XI0,THETA)
      D     = (1. + THETA*THETA)**0.5 * ( XI - GAM*XI0)
      CALL DEVPRT(TAU,SYY,SZZ,SYZ,SXX,SM,SJ2)
C     FF= SJ2 - F1(3.*SM,XI0)
C     PRINT*,'D= ',D,'GAM=',GAM,'FF=',FF,'XI0=',XI0,'R=',R
      IF(D .LE. 0.) D = 0.
      DK    = HF1(XI0,GAM,E0)
C     PRINT*,'DK =',DK
      IF(THETA .GT. 0.) DK = DK + HC(THETA,SJBAR,DFDI,DFDJ,D,D0)
      IF(ABS(D) .LE. .00001) MTYPE = 3
C     PRINT*, 'MTYPE=',MTYPE,'DJDI=',DFDI,'DFDJ=',DFDJ,'DK=',DK
      RETURN
C.....................................................................
C     MATERIAL IS IN THE LIMITING REGION
C
  320 GAM   = GAMM2(X)
      GT    = GAM*THETA
      DFDI  = DF2DI(XI0,GAM)
      DFDJ  = DF2DJ(XI0,GT,SJ2)
      SJBAR = -GT*XI0
      D0    = DD(XI0,THETA)
      D     = (1. + THETA*THETA)**0.5 * ( XI - GAM*XI0)
      IF(D .LE. 0.) D = 0.
      DK    = HF2(XI0,GAM,X,E0) + HC(THETA,SJBAR,DFDI,DFDJ,D,D0)
C/////////////////////////////////////////////////////////////////////
C ---------- YOU MAY REMOVE IF YOUR EQUA SOLVER MAY HANDLE SOFTENING
      IF(DK .LE. 0.) DK = 0.
C\\\\\\\\\\\\\\\\\\\\\\\\\\\\\\\\\\
      IF(ABS(D) .LE. .00001) MTYPE = 2
      RETURN
C
C
 1010 FORMAT(1X,'ERROR POISSON RATIO LESS THAN ZERO PV = ',E12.4)
 2000 FORMAT(1X, 40HERROR XW < 0. IN EPL214, CHANGING TO 1.
     1     ,/,1X,  5HXW = ,E12.4,6HXI0 = ,E12.4,6HXIL = ,E12.4)
  END
```

Computer and Physical Modelling in Geotechnical Engineering, Balasubramaniam et al. (eds)
© 1989 Balkema, Rotterdam. ISBN 90 6191 864 2

Construction control of embankment on the wall type improved ground by deep mixing method

N.Noriyasu
Chuden Engineering Consultants Co., Inc., Hiroshima, Japan

S.Hayashi
Department of Civil Engineering (SUIKO), Kyushu University, Fukuoka, Japan

ABSTRACT: Deep Mixing Method that forms a unified underground mass has been recently adopted as stabilization of marine soft ground in Japan. Stabilizing mechanism of the improved ground by Deep Mixing Method differs from the ordinary ground. Therefore construction control system is necessary to be established from a new viewpoint. In this paper authors present construction control system that have been developed with the intention of stability control of revetment construction on the reclamation from the sea.

1 INTRODUCTION

Deep Mixing Method which stabilizes the soft ground accumulated thickly in the sea bottom has been recently developed in Japan, and attained to the practical use stage. Deep Mixing Method, which we call "D.M.M." from now on, makes the firmly improved ground by means of compulsory mixing the soft ground in-site with chemical stabilizer such as cement, lime etc.
A principle of D.M.M. is utterly different from other improvement methods such as replacement method, accelerated consolidation method etc, and is to cement artificially soil skeleton. D.M.M. saves natural resources for soil improvement because of changing the existing old soft ground itself into the firm one. D.M.M. has a lot of advantages as follows : (1)It is able to shorten a long term construction, because the effects of improvement appear rapidly on the strength of the mixed ground. (2)It has little influence on environment such as noise, vibration around the construction works. (3)And it improves deformation properties such as compression and horizontal displacement of the partially improved ground as well as the entirely improved ground.
Especially in harbor works, as the foundation supports very heavy upper structures and horizontal external force acting on the ground is very large, the three improved types, block type , wall type and grid type as shown in Fig. 1-1, which are formed a kind of the underground structures by partially duplicating high strengthened soil units, have been frequently adopted as foundation of revetments.
D.M.M. is a comparatively new method as above mentioned, so till now the cases of monitoring introduced into the D.M.M. improved ground have been applied for researches and experiments with the intention of verifing a design method. However, there have been a few of observational procedure cases which is enough to verify a current design method, and the design method currently proposed dose not necessarily explain the stability of the D.M.M. improved ground in all various situations as mentioned in section 3. Hereafter, D.M.M. is substantially attaining to the practical use stage, still more it is necessary to apply the construction control system to the D.M.M. improved ground, from the results of which we are able to judge continuation or modification or interruption of the construction.
In this paper we present an application case of construction control system utilizing computer system to the D.M.M. improved ground under shore protection on the reclamation from the sea.

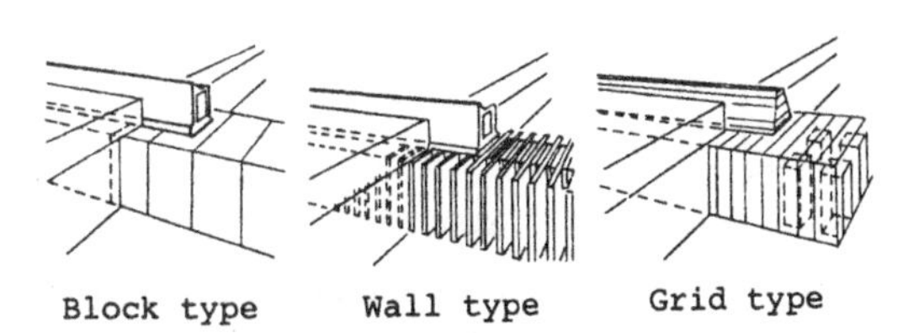

Fig. 1-1 Improved types of D.M.M.

2 AN OUTLINE OF D.M.M. CONSTRUCTION

An aspect of reclamation is shown in Fig. 2-1. In the A and the C type of revetments which are close to residential areas, D.M.M. was selected because of it's low noise. As shown in Fig.2-2, the improvement type is economical wall type that has stiffener to keep the unimproved part between the improved parts in the shape of wall from slip. A soil boring log in this site is shown in Fig. 2-3. Since it is possible to improve the ground with the N-value less than 12 by D.M.M., there remains unimproved silty understratum from 2 to 4 meters in thickness under gravel stratum having N-value more than 12. Therefore, it is impossible for the D.M.M. improved ground to be supported with bearing stratum, hence the silty understratum was very much concerned how to influence the stability of the D.M.M. improved ground.

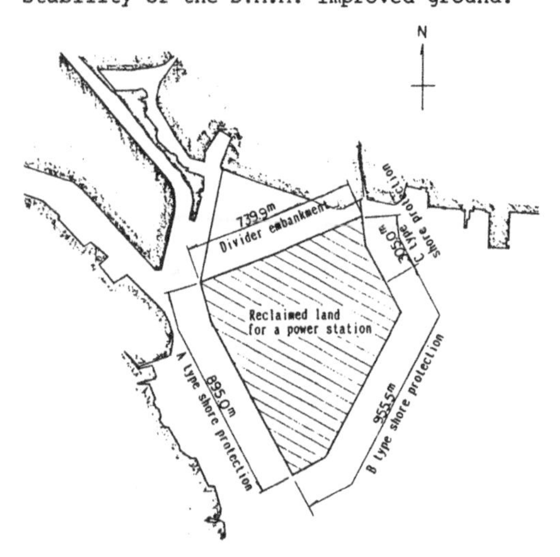

Fig. 2-1 An aspect of reclaimed land

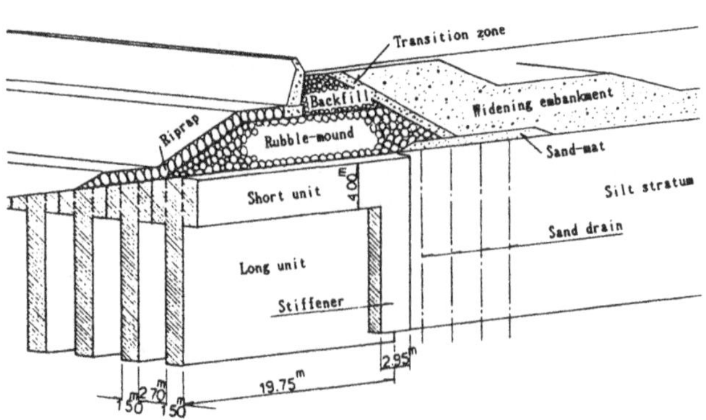

Fig. 2-2 A perspective view of D.M.M. shore protection

3 SIGNIFICANCE ON INTRODUCTION OF CONSTRUCTION CONTROL SYSTEM

We consider that it is necessary to introduce the construction control system to the D.M.M. improved ground. And significance on introduction of it is as follows.

(1) In current design method, flow chart of which is assumed to be a unified rigid ground is assumed to be a unified rigid body, and firstly stability of a rigid body under external forces is investigated, that is called external stability, and secondly stress generated in the stabilized ground itself is investigated to be less than allowable stress, that is called internal stability. However, current design method does not necessarily assure the stability of the D.M.M. improved ground in all situations. For examples in current design method external forces to investigate external stability as well as internal stability are estimated in ultimate plastic equilibrium. However in the design method of caisson foundation which is treated underground structure in the same case as the D.M.M. improved ground, subgrade reaction is estimated in consideration of rotation of the structure. In the D.M.M. which dose not be supported with bearing stratum as the case in this site, it may be sufficiently possible for rotation of D.M.M. to generate. Then because of rotation of it, external forces may be supposed to be different from those in current design method.

(2)Because strength and deformation properties of the silty ground around the D.M.M. improved ground change according to

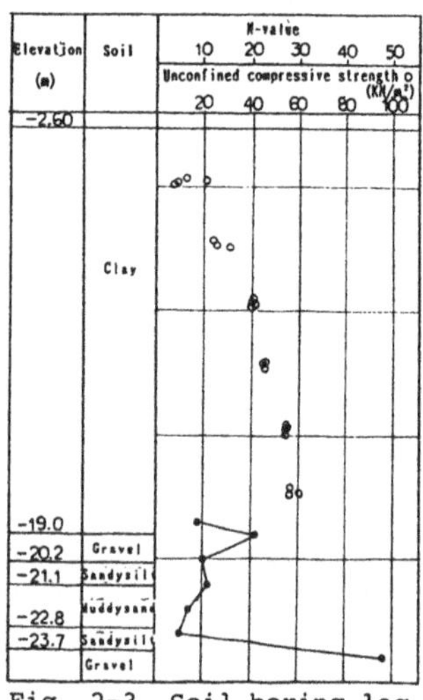

Fig. 2-3 Soil boring log in this site

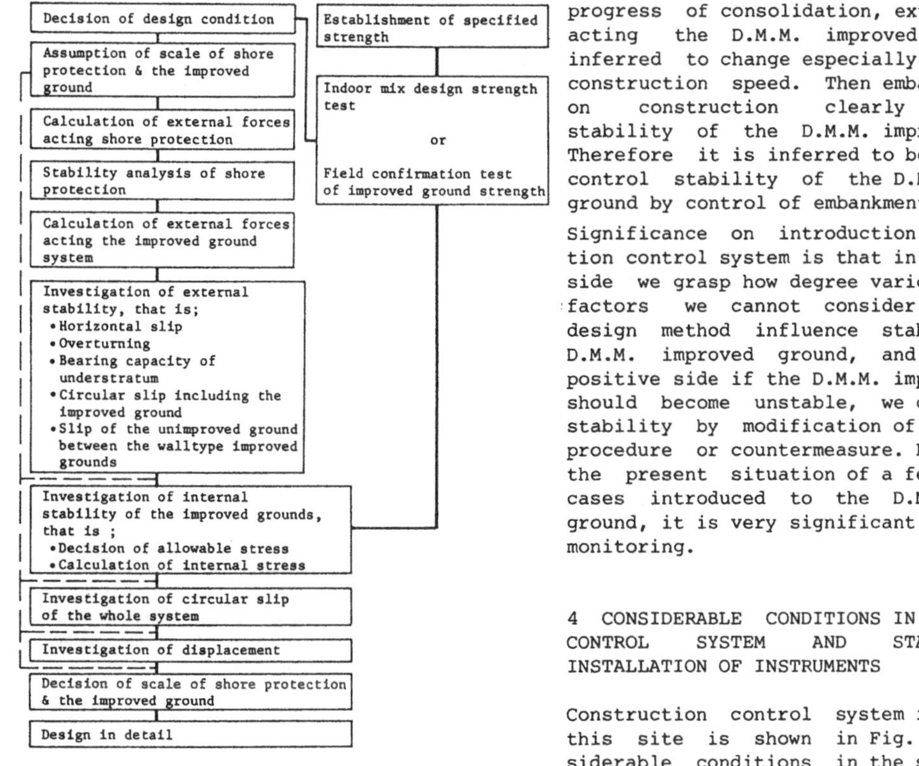

Fig. 3-1 Flow Chart of current design method

progress of consolidation, external forces acting the D.M.M. improved ground are inferred to change especially according to construction speed. Then embankment speed on construction clearly influences stability of the D.M.M. improved ground. Therefore it is inferred to be possible to control stability of the D.M.M. improved ground by control of embankment speed.

Significance on introduction of construction control system is that in the negative side we grasp how degree various uncertain factors we cannot consider in current design method influence stability of the D.M.M. improved ground, and that in the positive side if the D.M.M. improved ground should become unstable, we could control stability by modification of construction procedure or countermeasure. Especially in the present situation of a few monitoring cases introduced to the D.M.M. improved ground, it is very significant to introduce monitoring.

4 CONSIDERABLE CONDITIONS IN CONSTRUCTION CONTROL SYSTEM AND STANDPOINT ON INSTALLATION OF INSTRUMENTS

Construction control system introduced in this site is shown in Fig. 4-1 and considerable conditions in the system are as follows.

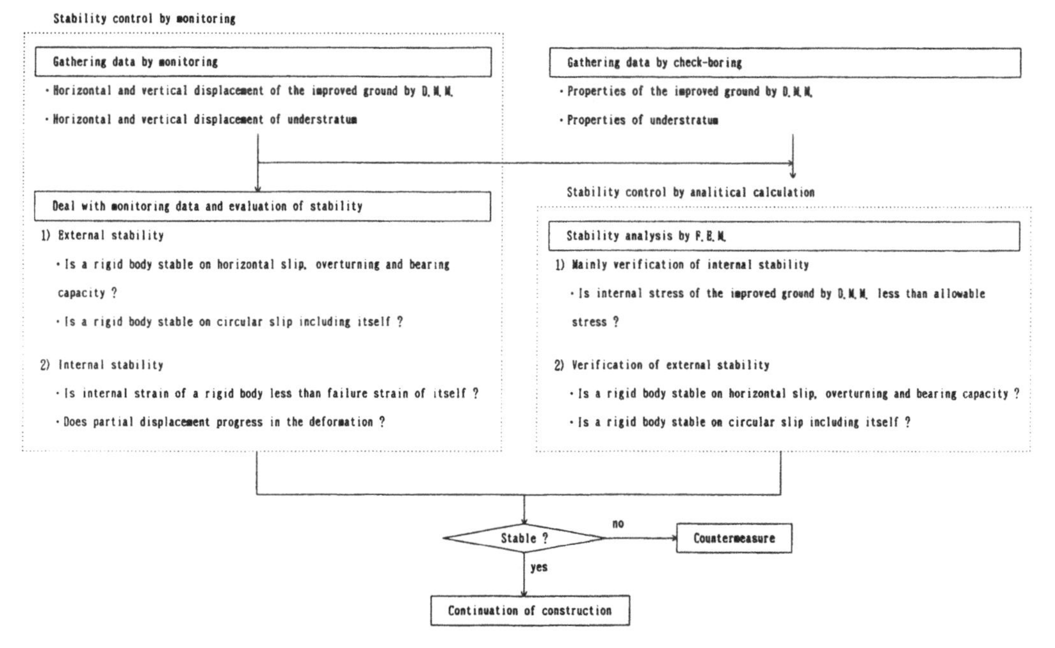

Fig. 4-1 Construction control system

(1)In construction control system, all possible failure modes in the D.M.M. improved ground, that is, all failure modes not only on external stability but also on internal stability, is necessary to be grasped.

(2)Stabilizing mechanism of the D.M.M. improved ground differs from it of the ordinary ground. Therefore, in order to grasp each of failure modes in the D.M.M. improved ground, instruments are to be suitably installed in it and appropriate control methods are necessary to be established. (The control method is described in detail in section 5.)

(3)In this site we attached greater importance to a practical use of stability control rather than verification of design method. In this point of view it is necessary to save labour in monitoring and to cut down to expenses. Therefore precise instruments were selected.

A view of installed instruments is shown in Fig. 4-2. Instruments were selected and arranged on the basis of fundamental viewpoints mentioned above. Failure modes of the D.M.M. improved ground is roughly divided into one on external stability and the other on internal stability. From now our way of thinking on construction control system and viewpoints of instruments arrangement are concretely explained on respective stability above mentioned.

4.1 Control of External Stability

When the D.M.M. improved ground is assumed as a rigid unified body, the stability on horizontal slip, overturning, bearing capacity, and circular arc slip including a rigid body are called external stability as a generic name.

In monitoring of the ground on construction of earth structures, displacement as an object of monitoring is mainly observed as usual. This is caused by following reasons : displacement as an object of monitoring is more precise than the others such as pressure, strain, and so on, and various kinds of construction control methods are proposed on displacement information. Similarly when we control external stability of the D.M.M. improved ground, it is obvious that failure of the D.M.M. improved ground has close relation to generated displacement as well as it's history before failure. And because the instrument for displacement is superior to the others in precision and endurance, it is effective to observe displacement. Therefore we decided to select displacement as an object of monitoring.

On control of external stability, viewpoints of observing displacement are as follows.

(1)Because the stabilized ground contribute to external stability as a unified

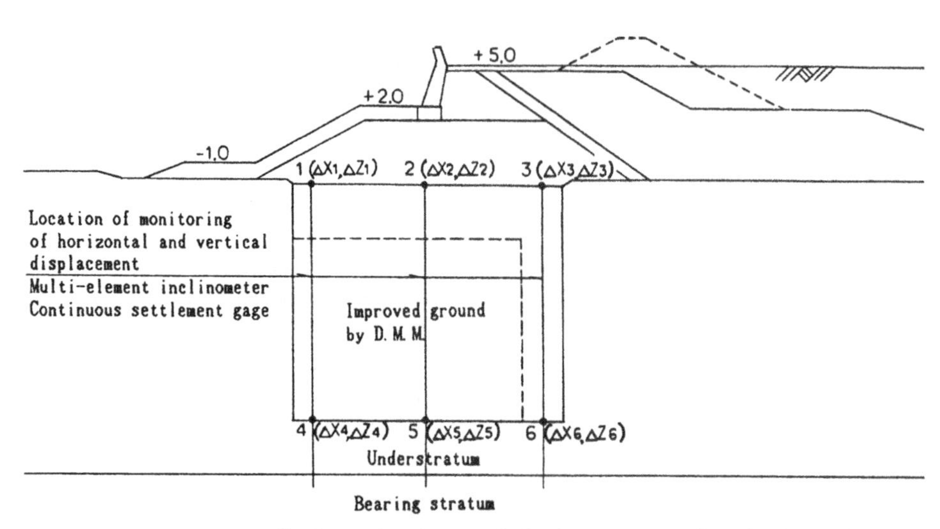

Fig. 4-2 A view of installed instruments

mass, it is necessary to grasp distribution of displacement at some locations in the stabilized ground itself and to grasp behaviour of a mass. Therefore horizontal and vertical displacement at plural locations are to be observed. As shown in Fig. 4-2, we observed horizontal and vertical displacement on three lines of the D.M.M. stabilized ground itself. On observation of the vertical displacement, we observed those at both top and bottom points of the D.M.M. stabilized ground. In this case of installed instruments, points at which we were able to grasp both horizontal and vertical displacements simultaneously were six. It is considered that as observing points increase, we are able to grasp more precisely behaviour of the D.M.M. stabilized ground as a mass. However we consider that on sufficient control of external stability, it is necessary to observe displacement at least on three lines of the stabilized ground.

(2)Deformation of the D.M.M. improved ground itself is little, and deformation of the D.M.M. improved ground mostly depends upon it of understratum. Therefore on judging failure modes of the D.M.M. improved ground, especially deformation of understratum offers useful information. That is, because the deformation pattern of understratum differs owing to failure mode of the D.M.M. improved ground as shown in Fig. 4-3, it is indispensable to know deformation of understratum as well as the whole ground. However we have to pay attention to that in this case of installed instruments as shown in Fig. 4-2, we cannot know discontinuous displacement between the stabilized ground and understratum.

(3)In a wall improved type, interaction between wall improved part and unimproved parts sandwiched by improved parts remains in unresolved problem. Therefore in design interaction between those is considered to be disadvantageous one in a viewpoint of stability. However in a viewpoint of monitoring it is effective to observe deformation of wall improved part that mainly contributes external stability. Therefore we observed deformation of the improved part. Because an improved pattern in this site has stiffener, it is not necessary to investigate slip of unimproved part. Therefore we did not observe deformation of the unimproved part.

4.2 Control of Internal Stability

If internal stability of the D.M.M. stabilized ground itself should not be secured, external stability of a rigid body as above mentioned could not be kept up. Therefore in order to secure internal stability, various ways not only in design but also in construction are considered. In current design method we decide allowable stress of the D.M.M. stabilized ground that is drastically reduced average real stress and use it on design in the same way as design of concrete structure, which differs from ordinary design of earth structure. On the other hand in soil improvement work quality as well as quantity of stabilizer and degree of mixing are severely controlled and in order to improve the ground as a unified structure, not only construction technology but also construction control method on partial duplication of units

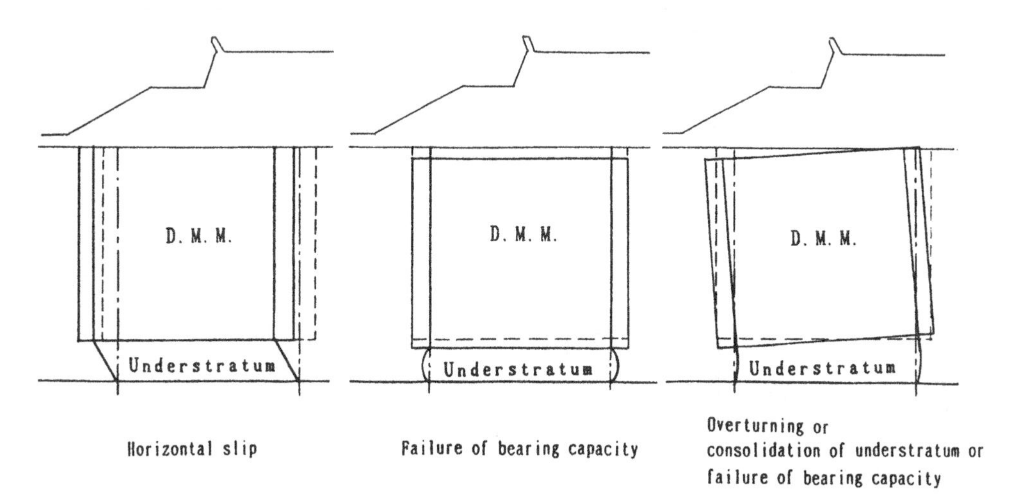

Fig. 4-3 Elemental deformation pattern of the improved ground

which consist of stabilized columns had been developed with scrupulous care. The speciality in control of internal stability is that internal stability considerably depends on construction technology as well as construction control on soil improvement work, which are necessary conditions to control of internal stability before introduction of monitoring. On monitoring to control internal stability, there is no doubt but that the quality of the stabilized ground is precondition.

As the results of D.M.M. constructions till now, it was reported that the stresses in the stabilized ground itself were unevenly distributed. There are some unsolved points in failure pattern of the stabilized ground itself or in behaviour of it after failure. These some unsolved points are able to control by means of construction control. Viewpoints of construction control of internal stability are as follows.

(1)In control of internal stability, it is necessary to grasp strain distribution on some segments in stabilized ground itself in the same way as control of external stability. We are able to roughly grasp some strains between monitoring points as shown in Fig. 4-2. And from deformation at plural locations in the stabilized ground itself, we can judge that the whole stabilized ground behaviour as a unified mass. Especially under situations of unevenly distributed strength of the stabilized ground, control of internal stability with strain gauges installed at some points in the stabilized ground is not practical because of inferior precision of strain gage, expensive efforts and heavy cost. Therefore we exclude monitoring with strain gauges.

(2)On construction control of internal stability, F.E.M. analysis is very effective. After soil improvement, as quality control of the stabilized ground check-boring are usually conducted. Then we can grasp the properties of the improved ground as well as the information as for displacement in monitoring. With using these real information gained after soil improvement, we can verify internal stability by F.E.M. analysis, which differs from it in design. This verification supplements strain distribution on some segments which is roughly grasped in the instruments arrangement as shown in Fig. 4-2.

As mentioned above we can control internal stability by means of not only monitoring but also F.E.M. analysis.

5 TECHNICAL METHODS OF CONSTRUCTION CONTROL AND IT'S PROCEDURE

Stabilizing mechanism of the D.M.M. improved ground differs from it of the ordinary improved ground by vertical drain method or sand compaction pile method. Therefore in technical methods of construction control we have to consider this mechanism. Because notable feature of the D.M.M. improved ground is less deformation, the D.M.M. improved ground is inferred to fail in less degree of deformation. Therefore technical methods of construction control which are able to steadily predict failure of the D.M.M. improved ground are very important.

5.1 Technical Methods of External Stability Control

The stabilized ground by D.M.M. are able to be regarded as a unified rigid body. Generally deformation of a rigid body, as long as a condition of unification is satisfied, consists of displacement, rotation and distortion as shown in Fig. 5-1. From the points aimed at displacement and rotation of those, we can control external

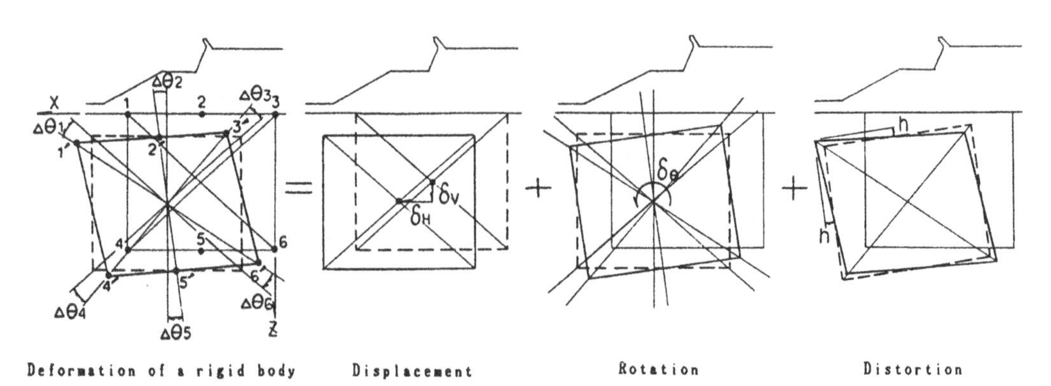

Fig. 5-1 Deformation of a rigid body

stability of a rigid body. When we know displacement and rotation at each point of a rigid body $(\Delta Xi, \Delta Yi, \Delta Zi) i=1-6$ as shown in Figs. 4-2 and 5-1, we can estimate displacement and rotation of the center of gravity of it, δh, δv and $\delta\theta$ by following Eqs. (5-1), (5-2) and (5-3);

horizontal displacement :

$$\delta h = \frac{\Delta X1+\alpha\Delta X2+\Delta X3+\Delta X4+\alpha\Delta X5+\Delta X6}{4+2\alpha} \tag{5-1}$$

vertical displacement :

$$\delta v = \frac{\Delta Z1+\alpha\Delta Z2+\Delta Z3+\Delta Z4+\alpha\Delta Z5+\Delta Z6}{4+2\alpha} \tag{5-2}$$

rotation :

$$\delta\theta = \frac{\Delta\theta 1+\alpha\Delta\theta 2+\Delta\theta 3+\Delta\theta 4+\alpha\Delta\theta 5+\Delta\theta 6}{4+2\alpha} \tag{5-3}$$

Where δh, δv and $\delta\theta$ are respectively horizontal displacement, vertical displacement and rotation at the center of gravity of a rigid body, $(\Delta Xi, \Delta Zi, \Delta\theta i) i=1-6$ are respectively horizontal displacement, vertical displacement and rotation at each point of a rigid body, and α is weight of the rigid body. Here, displacement at the center of gravity of a rigid body is regarded as to be integrated displacement of each particle all over whole section area and to be divided by whole section area. Because instrument at the center is considered to represent twice area as large as instrument at the side, we considered $\alpha=2.0$ as weight of displacement at the center.

Some of diagrams on control of external stability are shown in Fig. 5-2, 5-3, and 5-4. These diagrams were immediately drawn with X-Y plotter after dealing with monitoring data by a personal computer.

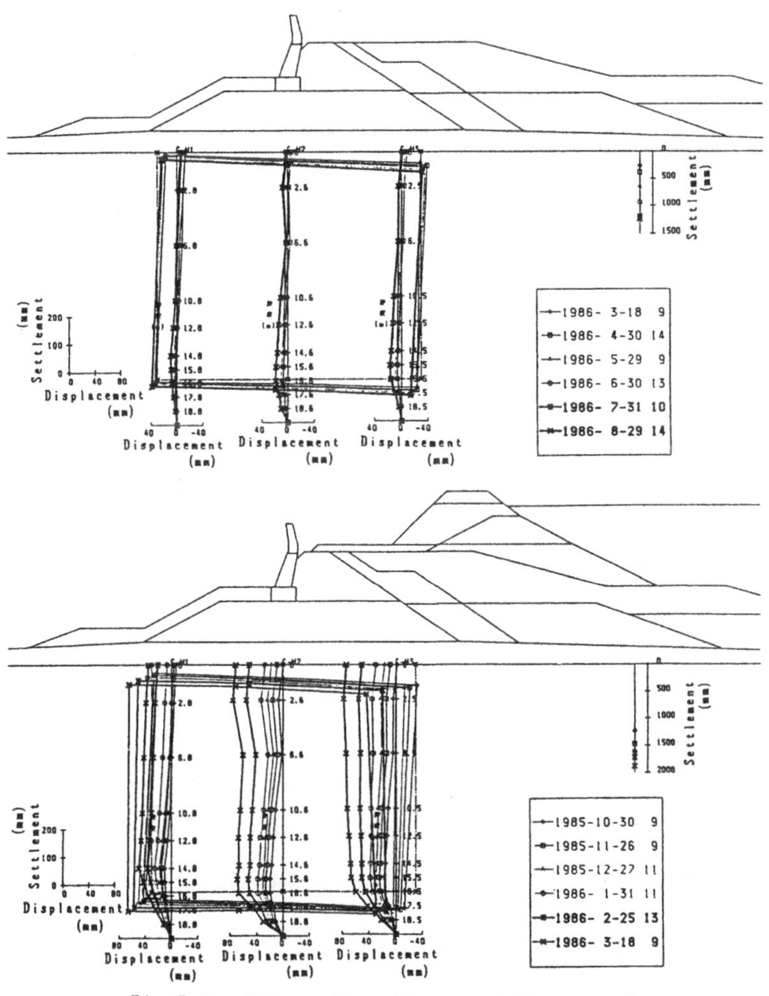

Fig. 5-2 Deformation diagram of the ground

From a deformation diagram as shown in Fig. 5-2, we can roughly grasp deformation of the D.M.M. improved ground as well as the whole ground, caused by construction. It is obvious that deformation of understratum 3 meters in thickness is comparatively large, and that deformation as well as stability of the D.M.M. improved ground greatly depend on the understratum. A deformation diagram offers fundamental and general information for judgmemt of stability.

Diagrams shown in Fig. 5-3 describe height of embankment, some of horizontal deformation, δh, δv, $\delta \theta$ and $\delta h/\delta v$, $\delta \theta/\delta v$ to elapsed time. The horizontal front and rear deformations were measured by multi element inclinometers and the one point broken line shows a bottom of the stabilized ground in Fig. 5-3. On the control of external stability meaning and points of this diagram are as follows.

(1)Considering stabilizing mechanism of the D.M.M. improved ground, it is reasonable for us to aim at δh, δv, and $\delta \theta$. For δh, δv and $\delta \theta$ are closely related with each of failure mode of the D.M.M. ground. That is, horizontal slip is related with δh, overturning is related with $\delta \theta$, bearing capacity is related with δv, and $\delta \theta$ and circular arc slip including the stabilized ground is related with δh and $\delta \theta$. When we aim at δh, δv and $\delta \theta$ which represent behaviour of the whole D.M.M. improved ground and regard as index for construction control, we can effectively control each of failure mode of the D.M.M. improved ground.

(2)When there is consolidation stratum under the D.M.M. stabilized ground as in this case, δh, δv, $\delta \theta$ complexly behave with relation to construction. Owing to consolidation or shear of understratum, δh, δv, $\delta \theta$ correlatively behave one another. Therefore it is useful on judgement of failure mode to aim at the ratio of $\delta h/\delta v$ and $\delta \theta/\delta v$.

For example, increase of the ratio $\delta h/\delta v$ means probability of horizontal slip, while stop of increase means stability of horizontal slip. On the other hand reduction of the ratio $\delta h/\delta v$ means proceeding of settlement owing to consolidation or shear of understratum. Here, a phenomenon of reduction of the ratio $\delta h/\delta v$ owing to consolidation of understratum means to become stable, while the similar phenomenon owing to shear of understratum means probability of failure of bearing capacity. Moreover because moment force ordinarily acts the D.M.M. improved ground, δv and $\delta \theta$ correlatively behave each other. That is, because owing to moment force reacting force acting understratum show uneven distribution, consolidation or shear of understratum generate rotation $\delta \theta$ of a rigid body. A phenomenon of increase of the ratio $\delta \theta/\delta v$ obviously means probability of overturning or failure of bearing capacity at the edge of the stabilized ground. A phenomenon of that $\delta \theta$ progresses and that the ratio $\delta \theta/\delta v$ is constant are considered to be caused by consolidation of understratum. This phenomenon in spite of progress of $\delta \theta$ means to become stable. As above mentioned, aiming at the ratio $\delta h/\delta v$ and $\delta \theta/\delta v$ we can judge each of failure mode and control external stability.

(3)On judgement of failure mode, deformation of understratum as well as whole ground as shown in Figs. 5-2 and 5-3 offer useful information. In this site deformation of the improved ground suggest failure mode of horizontal slip.

Diagrams as shown in Fig. 5-4 describe height of embankment, displacement, δh, δv, $\delta \theta$, $\delta h/\delta v$, $\delta \theta/\delta v$, displacement rate, $\dot{\delta h}$, $\dot{\delta v}$ and rotation rate $\delta \theta$ to elapsed time. In this control diagram we aim at displacement rate and rotation rate at the center of gravity of the D.M.M. improved ground. That is, it is obvious that these rates are closely related with external stability of the D.M.M. improved ground. For example, the case that these rates increase as time elapses or the case that these rates continue to progress at the level of more than some value means instability, while the case that these rates stop to continue or reduce means stability.

5.2 Procedure of External Stability Control

Procedure of external stability control with using technical methods as above mentioned is as follows.

(1)Procedure in the period from Sept. 1985 to the first half of Mar. 1986. In this period rubble mound and back fill on the D.M.M. improved ground and widening of embankment behind it was constructed. In the period of construction of rubble mound and back fill - from Sep. 1985 to the first half of Oct. 1985 - δh did not almost progress. Owing to unevenly distributed load, $\delta \theta$ progressed to the rearward. However δv progressed too and the ratio $\delta \theta/\delta v$ was almost constant after each of the construction. Therefore we judged that progress of $\delta \theta$ to the rearward was not caused by failure of bearing capacity but was caused by consolidation of understratum. Because each of these rates $\dot{\delta h}$, $\dot{\delta v}$ and $\dot{\delta \theta}$ were small in this period, we judged the D.M.M. improved ground stable.

In the period of construction of widening of embankment, - from the later half of Oct. 1985 to the first half of Mar. 1986 -, immediately after each of construction δh progressed and since then there was a tendency that δh stopped to progress. These phenomenon was obvious since the ratio $\delta h/\delta v$ increased immediately and since then showed almost constant value. Therefore we judged the D.M.M. improved ground stable on horizontal slip mode. While $\delta\theta$ progressed immediately after the construction and since then $\delta\theta$ progressed to rearward. However the ratio $\delta\theta/\delta v$ showed almost constant value, therefore we judged that this phenomenon was caused by consolidation of understratum. In this period each of these rates δh, $\delta\theta$ has tendency to decrease, then we judged the D.M.M. improved ground to be stable.

(2)Procedure in the period from the later half of Mar. 1986 to Aug. 1986. Since the half of Mar.1985, dredged soil was started putting into the pond behind the shore protection, and after raising of banking step by step, dredging soil was being put into the pond. In this period the behaviour of the D.M.M. improved ground differs from it up to the first half of Mar. 1986. That is, δh as well as $\delta\theta$ began to progress to the front side. As raising of banking and dredging progressed, consequently δh, δv and $\delta\theta$ were gradually getting larger. And because in this period the ratio $\delta h/\delta v$ and $\delta\theta/\delta v$ was gradually getting larger, we had to watch on the stable of the D.M.M. improved ground on each of failure mode. As shown by the process of rates, $\dot{\delta}h$, $\dot{\delta}v$, $\dot{\delta}\theta$ in Fig. 5-4, on the construction of banking +5.0 meters in hight, these rates were comparatively small, and especially have no tendency to increase. Therefore the construction was progressed as we continued to watch on the stability by monitoring. After construction of banking +7.0 meters in height and putting dredging soil into the place behind the monitoring point, the rate $\dot{\delta}h$ uncontinuously increased up to the range from 0.8 mm/day to 1.0 mm/day at the peak, and since then $\dot{\delta}h$ reduced immediately or gradually. Therefore we judged the D.M.M. improved ground stable as to the mode of horizontal slip. And because $\dot{\delta}\theta$ showed the same tendency as $\dot{\delta}h$, we judged the improved ground stable as to the modes of overturning as well as bearing capacity. On the construction of banking +8,7 meters in height and putting dredging soil into the pond, the rate $\dot{\delta}h$ continued at the comparatively high range from 1.0 mm/day to 1.3 mm/day. Therefore dredging was stopped for a while, since then the rate $\dot{\delta}h$ had tendency to reduce. As above mentioned it is very effective method on the control of external stability for us to aim these rates.

5.3 Technical Method and Procedure of Internal Stability Control

As mentioned in section 4 the speciality in control of internal stability is that internal stability considerably depends on construction technology as well as construction control on soil improvement work. Significance on construction control is that - if the stabilized ground should fail and the whole improved ground could not secure stability - we would deal with modification of construction procedure, for example construction speed or design section. We controlled internal stability by following methods. Diagrams as shown in Fig. 5-5 describe each process of internal strains between monitoring points in the D.M.M. improved ground. The situation that some of these internal strains as shown in Fig. 5-5 has tendency to increase as time elapsed, or become large more than failure strains shows the stabilized ground unstable. Fig. 5-6 shows a distribution map of failure strains in unconfined compression test of the samples gained by checkboring. As shown in Fig. 5-6 failure strains were distributed unevenly and average failure strain is 0.66 % and standard deviation is 0.29 %. It was reported that because confined pressure has little influence on the strength in UU conditions, it was not almost problem that shear strength in UU conditions was estimated as the half of strength in unconfined condition.

While it is not necessarily obvious to what degree confined pressure has influence on failure strain in UU condition. Judging from the results under various confined pressure as shown in Fig. 5-7, confined pressure has little influence on failure strain. So we may judge that failure strains of the D.M.M. stabilized soil under confined pressure is almost equal to those of unconfined test. It is obvious as shown in Fig. 5-5 that though some of internal strains of the D.M.M. stabilized ground has tendency to increase, all of those is less than average failure strain.

Essentially it is necessary to direct our attention to partial strains of the stabilized soil rather than internal strains between monitoring points. For generally partial strains are considered to be larger than the internal strains. Instruments installed in this site are not necessarily

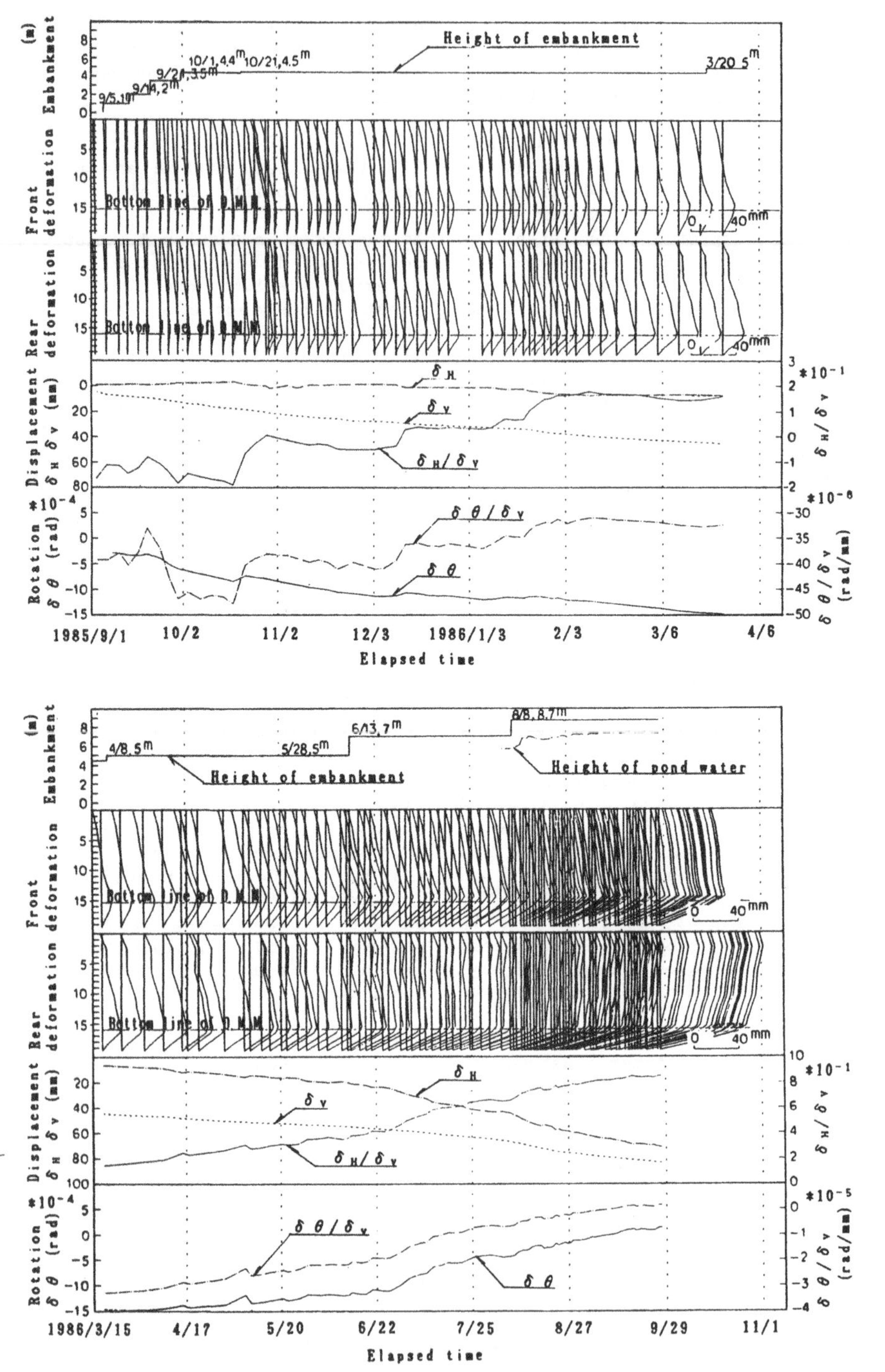

Fig. 5-3 Embankment, deformation, δ_H, δ_V, δ_θ, δ_H / δ_V, δ_θ / δ_V ~t diagram

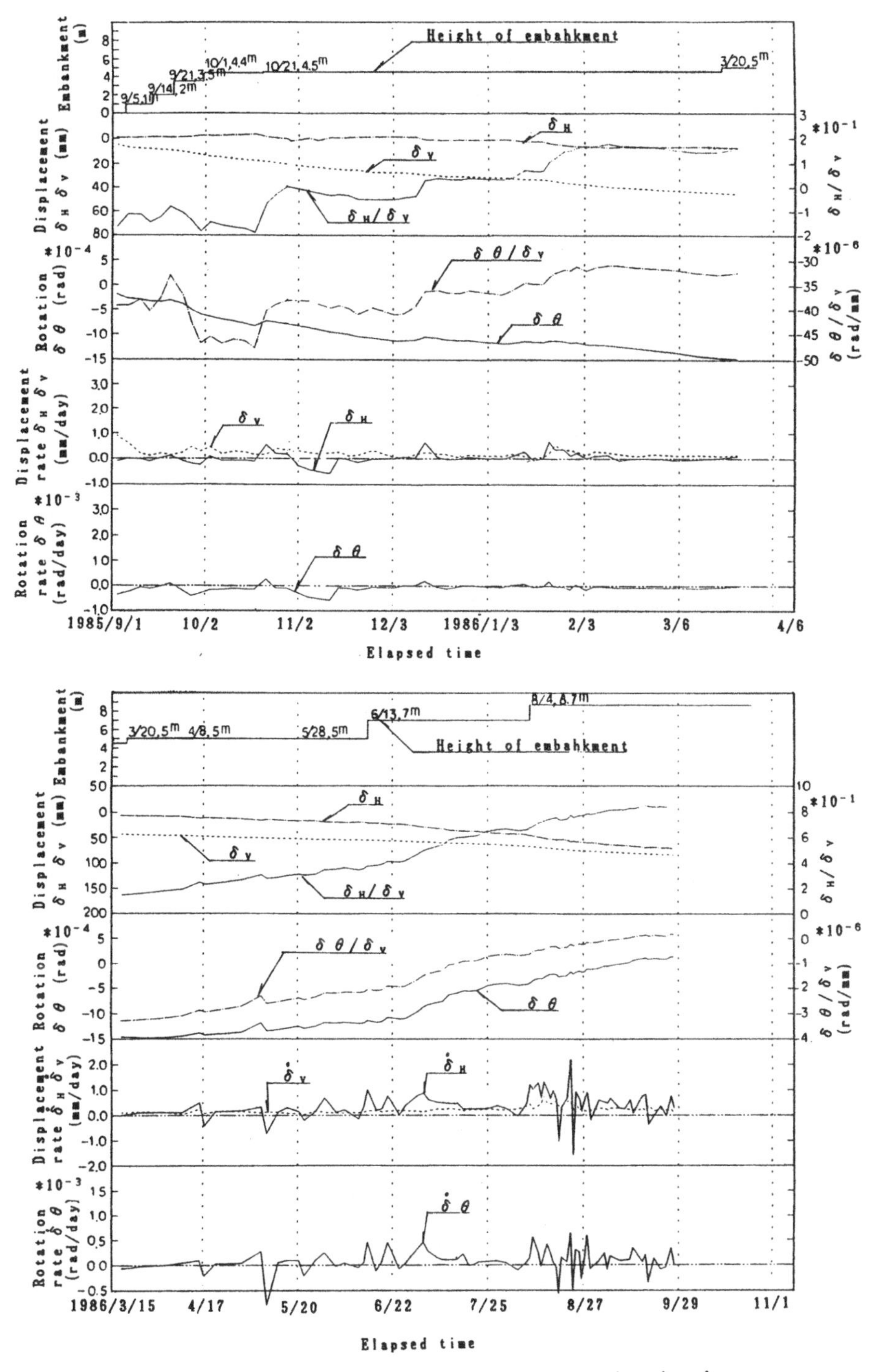

Fig. 5-4 Embankment δ_H, δ_V, δ_θ, δ_H / δ_V, δ_θ / δ_V, $\dot{\delta}_H$, $\dot{\delta}_V$, $\dot{\delta}_\theta$ ~t diagram

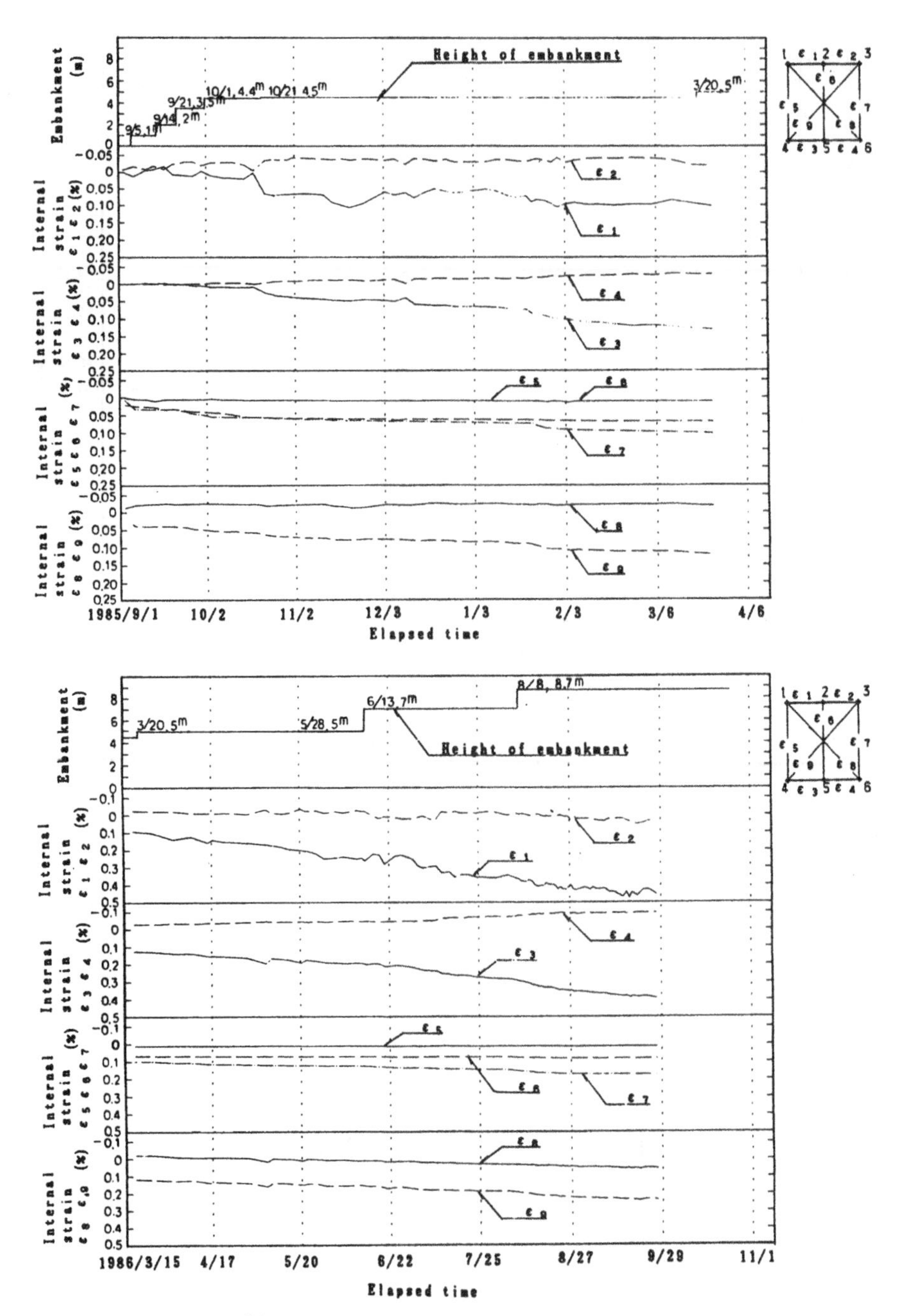

Fig. 5-5 Embankment, ε_i ~t diagram

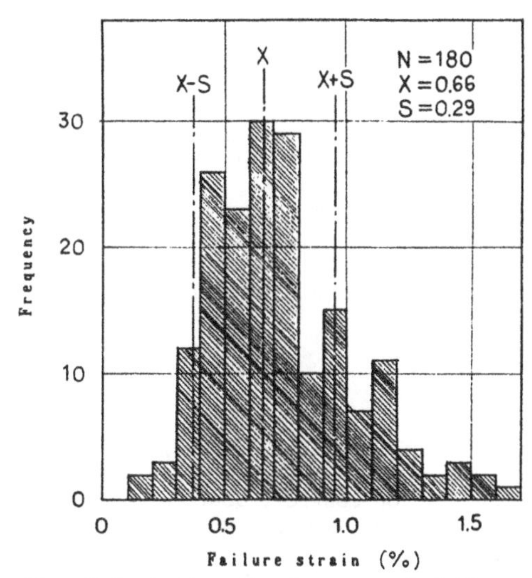

Fig. 5-6 A distribution map of failure strains at unconfined test

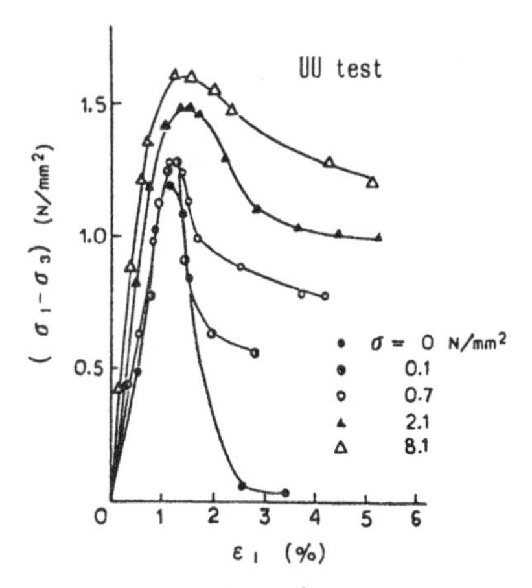

Fig. 5-7 $(\sigma_1-\sigma_3)\sim\varepsilon_1$ relation at UU test

sufficient to grasp a strain distribution of the D.M.M. stabilized soil itself, but offers important information to grasp how the stabilized ground behave as a unified mass.

And otherwise we verified the internal stresses of the stabilized ground itself by F.E.M. analysis. Considering monitoring procedure as well as the results of F.E.M. analysis, we controlled internal stability.

6 CONCLUSIONS

When from now on D.M.M. is substantially attaining the practical use stage, we consider that it is necessary to apply construction control system to the D.M.M. improved ground. Since stabilizing mechanism of the D.M.M. improved ground differs from it of the ordinary ground, construction control system is necessary to be established from a new viewpoint. Our fundamental viewpoints of construction control system introduced in this site and results of applying it are as follows.

[Significance on Introduction of Construction Control System]

(1)In current design method external design force acting the D.M.M. improved ground are inferred to differ owing to consolidation of surrounding silty soil as well as rotation of the D.M.M. stabilized ground itself. Therefore it is necessary to grasp the influence caused by various uncertain factors that are not considered in design.

(2)It is obvious that construction speed has influence on stability of the D.M.M. improved ground. So from the results of monitoring if the D.M.M. improved ground should be judged unstable, we could control the stability of the improved ground by means of monitoring of firstly modification of construction speed as well as secondary countermeasure.

[Considerable Conditions of the System]

(3)Construction control system is necessary to be grasped all possible failure modes on external stability as well as internal it.

(4)Especially in this site we attached greater importance to a practical use of stability control, and selected precise and confidential instruments.

[Our Way of Thinking of Construction Control System and Standpoint on Installation of Instruments]

(5)On control of external stability, generated displacement as well as it's history before failure has close relation to failure of the improved ground, and instruments for displacement is superior to the others in precision and endurance. Therefore we selected displacement as an object of monitoring. Because the stabilized ground contributes to external stability as an unified mass, it is necessary to grasp displacement distribution of stabilized ground itself. Therefore we observed horizontal and vertical displacement at plural locations.

(6)In wall improved type, interaction between improved wall part and unimproved part sandwiched by improved parts remains in unsolved problem. In this site it is effective to observe displacement of improved wall part that mainly contributes external stability. So we observed those of improved wall part.

(7)On control of internal stability, it is necessary to grasp strain distribution on some segments in the stabilized ground itself. We are able to roughly grasp some strains between the same points as those of external stability control. Moreover we can verify internal stresses of the stabilized ground itself by means of F.E.M. stability analysis with using the real properties of the stabilized ground gained after soil improvement as well as information as for displacement in monitoring. Synthesizing these points we control internal stability.

[Technical Methods of External Stability Control and It's Procedure]

(8)Deformation of a rigid body consists of displacements, rotation and distortion, as long as a condition of a unification is satisfied. From the points aimed at displacement and rotation of those, we can control external stability. We estimate horizontal, vertical displacement and rotation at the center of gravity of a rigid body : δh, δv, $\delta\theta$. δh, δv, and $\delta\theta$ have close relation to failure modes. Therefore it is reasonable for us to aim at δh, δv, and $\delta\theta$.

(9)Owing to consolidation or shear of understratum, δh, δv,and $\delta\theta$ behave correlatively one another. Therefore it is useful on judging failure modes to aim at the ratios of $\delta h/\delta v$ and $\delta\theta/\delta v$.

(10)Horizontal,vertical displacement rates $\dot{\delta h}$, $\dot{\delta v}$ and rotation rate $\dot{\delta\theta}$ are very useful index to control external stability. That is , the case that these rates increase as time elapsed, or the case that these rates continue to progress at the level of more than some value, means instability, while the case that these rates stop to increase or reduce,means stability.

(11)On external stability control, using technical methods as above mentioned in items of from (8) to (10), we used control diagrams drawn with X-Y plotter after dealing monitoring data with a personal computer immediately. Consenquently it became obvious that these technical methods were very effective to judge failure mode and to control external stability.

(12)We controlled internal stability by means of a diagram showing the process of each internal strain between monitoring points in the D.M.M. stabilized ground. And otherwise we verified the internal stresses of the stabilized ground itself in detail.

ACKNOWLEDGMENTS

This construction control system was applied to a reclamation work from the sea for the Yanai thermal power plant of the Chugoku Electric Power Co., Inc.. The authors gratefully acknowledge the helpful support and cooperation provided by Mr. Shibata, a head of the construction office and Mr. Shintani, a chief of the construction section. And the authors express the gratitude to Mr. Tahara, a director of the Chuden Engineering Consultants Co., Inc. for his various advices and supports and to Mr. Ueda and Mr. Okahara, engineers of ditto for their assistances of making soft-programs.

REFERENCES

CDM Research Committee (1985), Robotizing of Deep Mixing Method and Automatic Construction Control, Jour. of Foundations Engineering, Vol.13, No.2, pp.22-29, (in Japanese).

Kimura, H., Tahara, M. and Noriyasu, N. (1983), Construction Control of Embankment on the Improved Ground by D.M.M., Proc. of 21th Annual Meeting of JSSMFE, pp.1083-1086, (in Japanese).

Matsuo, M. and Kawamura, K. (1977), Diagram for Construction Control of Embankment on Soft Ground, Soils and Foundations, Vol.17, No.3, pp.37-52.

Noto, S., Kuchida, N. and Terasi, M. (1983), Consideration of Practices and Problems of Deep Mixing Method - Practical Cases of Deep Mixing Method - , Jour. of JSSMFE, Vol.31, No.7, pp.73-80.

Shugano, T. (1980), Monitoring Test in-Site of the Improved Soft Muddy Ground in the Seabed by Deep Mixing Method Utilizing Cement-Stabilizer, Proc. of Symposium on Testing Method of the Stabilized Soils, JSSMFE, pp.53-60, (in Japanese).

Shibata, K. (1986), A Summary of the Reclamation Work for the Yanai Thermal Power Plant, Jour. of Electric Power Civil Engineering, No.200, pp.89-94, (in Japanese).

Terashi, M., Tanaka, H., Matsumoto, T., Niidome, Y. and Honma, S. (1980), Fundamental Properties of Lime and Cement Treated Soils (2nd Report), Report of the Port and Harbour Research Institute, Vol.19, No.1, pp.33-57.

Terashi, M., Hushetani, H., and Noto, S. (1983), Consideration of Practices and Problems on the Deep Ground Improvement, Practices and Problems of Deep Mixing Method - An Outline of Deep Mixing Method - , Jour. of JSSMFE, Vol.31, No.6, pp.57-64, (in Japanese).

Terashi, M. (1984), State-of-the-Art Report on Deep Mixing Method, Proc. of Symp. on Strength and Deformation of the Composite ground, JSSMFE, pp.1-12, (in Japanese)

Terashi, M. and Kitazume, M. (1984), Design External Forces for D.M.M. of Wall Type, Proc. of Symp. on Strength and Deformation of the Composite Ground, JSSMFE, pp.75-83, (in Japanese).

Tomita, I., Shegawa, M., Katayama, S. and Katoh, K. (1984), Construction Research, Construction Experiment of Deep Mixing Method, Civil Engineering Construction, Vol.25, No.5, pp.11-19, (in Japanese).

Yano, K. and Hanazono, H. (1984), Monitoring of Soil - Monitoring in-Site and Construction Control - Practical Cases of Construction Control by Monitoring in-Site, Jour. of JSSMFE, Vol.32, No.7, pp.69-76, (in Japanese).

Computer and Physical Modelling in Geotechnical Engineering, Balasubramaniam et al. (eds)
© 1989 Balkema, Rotterdam. ISBN 90 6191 864 2

Localisation theory and its application to the fracturing process in geomechanics

T.Kawamoto & Y.Ichikawa
Nagoya University, Nagoya, Japan

T.Ito
Toyota Technical College, Toyota, Japan

ABSTRACT: There are many difficulties in the analysis of global failure in the evaluation of the bearing capacity of foundations and the stability of slopes by finite element method. However, the localisation theory has been recently started to be adopted to bridge the local failure to the global failure in geotechnical engineering structures. The purpose of this study is to simulate the global failure in the load bearing capacity and slope stability problems by an elasto-plastic analysis employing this theory. In addition, the adaptive mesh method has been used to increase the accuracy of solutions by the finite element method.

1. INTRODUCTION

Finite element method is capable of solving almost all problems in continuous medium and is applied to various problems in the field of geotechnical engineering. In addition, numerical techniques and constitutive laws for materials are developed by exerting mutual effects on each other. However, there are still problems remaining regarding the calculation of the ultimate resistance of the geotechnical engineering structures such as the load bearing capacity problems of foundations and the stability of slopes. As the analysis by the finite element method is originally based on the continuous medium approach, the analysis of the global failure at the ultimate state such as in the case of the load bearing capacity of foundations and slopes is a problem so far. Nevertheless, the following methods are considered in relation with the above problems:

1)Method for the prediction of the limiting value with the use of non-linear constitutive law in the finite element analysis.

2)Method employing the lower and upper value theorems in the finite element analysis.

3)Sliding surface method and others coupled with the finite element method.

Method 1 is for obtaining the limiting value by carrying out the calculation up-to a point that the calculation becomes impossible together with the use of a non-linear constitutive analysis in the finite element analysis. This method largely depends upon the constitutive equation used and finite element mesh. There are some problems on the difference between calculated results due to the accurracy of calculations.

As for method 2, plastic finite element methods have been proposed, based upon the upper and lower value theorem which regards the body as perfectly plastic body and it is assumed that the load at the ultimate state is path-independent (Tamura & Kobayashi 1984, Lysmer 1970).

In addition to these, there are other methods attempting to predict the global failure by using specific type of finite elements. Among these, there is the cracked triangular element method consisting of a elasto-plastic joint element to simulate the sliding or separation planes resulting from stress state (Obara et al. 1982).

Strain localisation theory has been mainly applied to metals particularly to buckling problems and necking problems in the metal forming (Needleman & Rice 1978). The localisation theory has a long past and the condition of the localisation was proposed by Hill (1962) in relation with the stability and uniqueness of the solution for elasto-plastic bodies. Since then, there is number of theoretical and numerical studies (Rudnicki & Rice 1975, Rice 1976). However, the physical meaning of the localisation condition is not clarified. In addition, there is a need for the evaluation of stresses and deformations at a highly reliable accuracy when the occurence and the propagation of the local failure are studied by the localisation theory. Therefore, the finite element optimisation technique is tried to be employed in order to reduce the error arising from the discretisation procedure and to increase the accuracy of the analysis (Kikuchi, 1985, Diaz et al. 1983).

In this study, based on the localisation theory of Rice (1976), the technique for the application of the localisation condition to elasto-plastic finite element analysis with the consideration of the dilatancy of rock-like materials is described and the applicability of the localisation condition to the bearing capacity of rock foundations and rock slope stability is considered and discussed.

2. LOCALISATION THEORY

2.1 Localisation condition

The localisation of deformation is herein regarded as an instability that can be predicted in terms of the pre-localisation consititutive equation of the material. The material is assumed to be isotropic and homogeneous. The stress and strain increments are assumed to be continuous throughout the material. And the condition under which the deformation concentrates into a shear band is sought.

Let us consider an infinitely thin shear band V* with the normal ν in a body occupying the space region V as shown in Fig. 2.1. The deformation is uniform and the incremental field quantities are permitted to take on different values throughout this shear band from those outside this band.

Discontinuity quantities across the band must satisfy the geometrical compatibility condition. Let $[d\varepsilon]$ denote the discontinuous strain increment across the shear band and be expressed by

$$[d\varepsilon] = (d\varepsilon)_b - (d\varepsilon)_0 \qquad (2.1)$$

where $(\ )_b$ denotes the field quantities within the band, $(\ )_0$ denotes respective quantities outside the band.

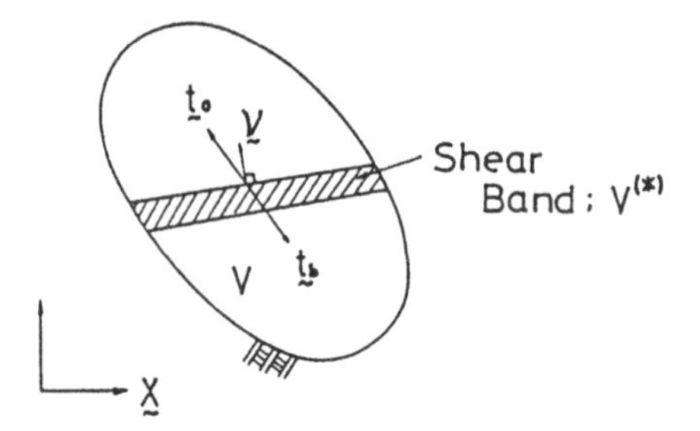

Fig.2.1 Illustration of shear band

Geometrical compatibility condition is written as

$$[d\varepsilon] = \boldsymbol{b} \otimes \boldsymbol{\nu} \qquad (2.2)$$

where b is an arbitrary vector.

From the equilibrium condition across the band, we obtain the following

$$d\boldsymbol{t}_o + d\boldsymbol{t}_b = \mathbf{o} \qquad (2.3)$$

where

$$d\boldsymbol{t}_o = d\boldsymbol{\sigma}_o^T \boldsymbol{\nu} \qquad (2.4)$$

$$d\boldsymbol{t}_b = -d\boldsymbol{\sigma}_b^T \boldsymbol{\nu} \qquad (2.5)$$

In the region V and V*, the constitutive relations are written in the following forms

$$d\boldsymbol{\sigma}_o = \boldsymbol{D}_o^{ep} d\boldsymbol{\varepsilon}_o \qquad (2.6)$$

$$d\boldsymbol{\sigma}_b = \boldsymbol{D}_b^{ep} d\boldsymbol{\varepsilon}_b \qquad (2.7)$$

Combining Eqs. (2.1)-(2.7) yields

$$(\boldsymbol{\nu}^T \boldsymbol{D}_b^{ep} \boldsymbol{\nu})\boldsymbol{b} = \{(\boldsymbol{D}_o^{ep} - \boldsymbol{D}_b^{ep})d\boldsymbol{\varepsilon}_o\}\boldsymbol{\nu} \qquad (2.8)$$

As it is previously assumed that the material is homogeneous at the initial state, we have

$$\boldsymbol{D}_o^{ep} = \boldsymbol{D}_b^{ep} = \boldsymbol{D}^{ep} \qquad (2.9)$$

and Eq. (2.8) is rewritten as

$$(\boldsymbol{\nu}^T \boldsymbol{D}_b^{ep} \boldsymbol{\nu})\boldsymbol{b} = \mathbf{o} \qquad (2.10)$$

If the vector b has a value except zero, the discontinuous strain increment across the shear band $[d\varepsilon]$ must exist. The localisation condition is then written as

$$det(\boldsymbol{\nu}^T \boldsymbol{D}_b^{ep} \boldsymbol{\nu}) = 0 \qquad (2.11)$$

2.2 Stability of elasto-plastic deformation process

Let us consider the relationship between the stability of the elasto-plastic deformation process and the localisation condition by Eq.(2.7). The elasto-plastic deformation process is assumed to be stable at any point in the body as long as the following condition is satisfied:

$$d\sigma \cdot d\varepsilon > 0 \tag{2.12}$$

The total strain increment $d\varepsilon$ can be decomposed into the elastic strain increment $d\varepsilon^e$ and the plastic strain increment $d\varepsilon^p$, thus we have

$$d\sigma \cdot d\varepsilon = d\sigma \cdot d\varepsilon^e + d\sigma \cdot d\varepsilon^p \tag{2.13}$$

The elasticity tensor D is positive definite, then the elastic part of the Eq.(2.12) is written for non-zero $d\varepsilon$ as

$$d\sigma \cdot d\varepsilon^e = d\varepsilon^e \cdot (\boldsymbol{D} d\varepsilon^e) > 0 \tag{2.14}$$

Therefore the deformation process is stable so that the following condition is satisfied:

$$d\sigma \cdot d\varepsilon^p \geq 0 \tag{2.15}$$

This is so-called the Drucker's stability condition, and if the flow rule is applied, the plastic potential must be equal to the yield function and be convex.

Let us consider the condition (2.12) for some elasto-plastic materials. In the case of hardening type materials, the associated flow rule is given by

$$d\varepsilon^p = \lambda \frac{\partial F}{\partial \sigma} \tag{2.16}$$

where $F(\sigma;\kappa)$ is yield function and κ is the hardening parameter. The plastic work done can be written using Eq. (2.16) as

$$d\sigma \cdot d\varepsilon^p = \lambda \frac{\partial F}{\partial \sigma} \cdot d\sigma \geq 0 \tag{2.17}$$

As long as this relationship is satisfied, the deformation process is said to be stable. Inserting Eq.(2.17) into Eq.(2.13) gives

$$d\sigma \cdot d\varepsilon = d\sigma \cdot d\varepsilon^e + \lambda \frac{\partial F}{\partial \sigma} \cdot d\sigma \geq 0 \tag{2.18}$$

The equality is valid only when $d\sigma$ is zero.

On the other hand, the plastic potential is not associated with the yield function in the case of the non-associative flow rule. Let F be of the Drucker-Prager's type yield function given as

$$F = \alpha\bar{\sigma} + S + K(\varepsilon^p) = 0 \tag{2.19}$$

in which α material constant, s is the deviatoric stress tensor, $\bar{\sigma}$ is the mean stress tensor, and let G be the same type plastic potential. Then, we have

$$\frac{\partial F}{\partial \sigma} = \boldsymbol{m} + \alpha \boldsymbol{n}$$
$$\frac{\partial G}{\partial \sigma} = \boldsymbol{m} + \beta \boldsymbol{n} \tag{2.20}$$

where $m = s / |s|$, $n = \bar{\sigma} / |\bar{\sigma}|$ and β, called dilatancy function is the ratio of the volumetric strain to the deviatoric strain (Ichikawa et al. 1985). In this case, we have the plastic work done as

$$d\sigma \cdot d\varepsilon^p = \lambda(\beta - \alpha) d\sigma_n + \lambda \frac{\partial F}{\partial \sigma} \cdot d\sigma \tag{2.21}$$

where $d\sigma_n = d\sigma \cdot \boldsymbol{n}$ Therefore, elasto-plastic work done is written as

$$d\sigma \cdot d\varepsilon = d\sigma \cdot d\varepsilon^e + \lambda(\beta - \alpha) d\sigma_n + \lambda \frac{\partial F}{\partial \sigma} \cdot d\sigma \tag{2.22}$$

The value of the dilatancy function β is generally smaller than the parameter α. the second term of the right hand side of Eq.(2.22) has the negative value, and $d\sigma \cdot d\varepsilon$ can be zero even though $d\sigma$ is not zero.

It must be noted that, in the case of non-associated flow rule using the Drucker-Prager's yield function and the plastic potential, there is a dissipated work done corresponding to $\lambda(\beta - \alpha) d\sigma_n$, and $d\sigma \cdot d\varepsilon$ can be zero even though the loading process is on.

Let us now consider the relation between the condition $d\sigma \cdot d\varepsilon = 0$ and the localisation condition. The constitutive equation is written as

$$d\sigma = \boldsymbol{D}^{ep} d\varepsilon \tag{2.23}$$

and

$$d\sigma \cdot d\varepsilon = d\varepsilon \cdot (\boldsymbol{D}^{ep} d\varepsilon) \tag{2.24}$$

If $d\varepsilon$ is arbitrary, the instability condition $d\sigma \cdot d\varepsilon = 0$ is valid whenever the following condition for $\boldsymbol{D}^{ep}$ is satisfied :

$$det \boldsymbol{D}^{ep} = 0 \tag{2.25}$$

Let us consider the discontinuity of strain increment $[d\varepsilon]$ across the shear band. The work for the strain increment $[d\varepsilon]$ is written as

$$[d\varepsilon]\cdot[d\sigma] = [d\varepsilon]\cdot D^{ep}[d\varepsilon] \quad (2.26)$$

and using Eq.(2.2) yields

$$[d\varepsilon]\cdot[d\sigma] = b\cdot\{(\nu^T D^{ep}\nu)b\} \quad (2.27)$$

Eq.(2.27) is zero as long as the following condition is satisfied

$$det(\nu^T D^{ep}\nu) = 0 \quad (2.28)$$

This condition is equivalent to the localisation condition (2.11). Therefore, it is understood that the instability condition (2.25) is the sufficient condition for the localisation.

3. ADAPTIVE MESH METHOD

Errors involved in the finite element analyses are considered to be mainly caused by the discretisation of space, numerical integration and the convergency of the non-linear analysis, and others. Of these, the errors due to discretisation can be reduced by using the adaptive mesh method. In this study, this method is employed in the finite element analysis to obtain more accurate solutions.

3.1 Error estimation

It is necessary to estimate the discretisation error for using the adaptive mesh method. Herein we will describe the procedure how to estimate the error in the case of the analysis for a linear elastic material.

The virtual work equation for the virtual displacement δu can be written in the following

$$a(u,\delta u) = f(\delta u) \quad (3.1)$$

where

$$a(u,\delta u) = \int_V (D\varepsilon)\cdot\delta\varepsilon dV$$

$$f(\delta u) = \int_V f\cdot\delta u dV + \int_{St} t^{\circ}\cdot\delta u dS$$

and St is the stress boundary, D the elasticity tensor, f the body force vector. The finite element approximation of Eq.(3.1) takes the following form

$$a(u_h,\delta u_h) = f(\delta u_h) \quad (3.2)$$

where u_h is the finite element approximation of the displacement vector u. From Eqs.(3.1) and (3.2), one gets the following expression,

$$a(u-u_h,\delta u_h) = 0 \quad (3.3)$$

Using the Schwarz's inequality, the above equation becomes

$$\sqrt{a(u-u_h,u-u_h)} \leq \sqrt{a(u-v_h,u-v_h)} \quad (3.4)$$

where v_h is perturbed u_h. The measure of error VOC_i is obtained as

$$VOC_i = \sqrt{a(u-v_h,u-v_h)} \quad (3.5)$$

In the actual numerical scheme, the gradient of strain is used as measure of error which is given explicitly as

$$VOC_i = (\nabla\varepsilon\cdot\nabla\varepsilon)_i \times A \quad (3.6)$$

$$A = A_1 + A_2 + A_3 + A_4$$

3.2 Adaptive mesh generation by using r-method

There are three methods, p-method, h-method, and r-method for optimisation of finite element meshes. The p-method is based on the principle of increasing the order of shape functions, and the h-method is associated with the fine discretisation of elements, and the r-method relocates the nodes of elements (Kikuchi 1985). In this study the r-method is employed as it has the advantages of keeping the initial numbering of nodes and node associations of elements. However, one must take care of keeping the appropriateness of the shape of elements while relocating the nodes. In the present study, the amount of movement of nodes are determined by considering not only the error measure but also the smoothness and orthogonality of elements (Carcaillet et al. 1986) . The optimisation problem, thus, can be written as:

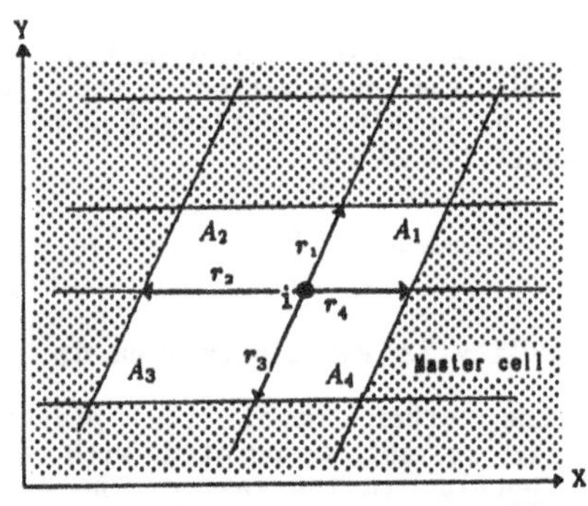

Fig.3.1 A typical master cell in finite element mesh

Searching Min F (3.7)

The objective function F is written explicitly as

$$F = \sum_{i=1}^{N} \theta \times \{\eta ORT_i + (1-\eta) SM_i\} + VOC_i \tag{3.8}$$

where η and θ are the given constants, and the orthogonality ORT_i, smoothness SM_i and VOCi are given respectively as:

$$ORT_i = (\boldsymbol{r}_1 \cdot \boldsymbol{r}_2)^2 + (\boldsymbol{r}_2 \cdot \boldsymbol{r}_3)^2 + (\boldsymbol{r}_3 \cdot \boldsymbol{r}_4)^2 + (\boldsymbol{r}_4 \cdot \boldsymbol{r}_1)^2$$

$$SM_i = (A_1 \cdot A_2)^2 + (A_2 \cdot A_3)^2 + (A_3 \cdot A_4)^2 + (A_4 \cdot A_1)^2$$

$$VOC_i = (\nabla\varepsilon \cdot \nabla\varepsilon)_i \times A \tag{3.9}$$

One should refer to Fig. 3.1 regarding r in the above equation(i denotes an arbitrary node in the finite element mesh, and a master cell consists of the elements associated with that node i). Eq.(3.3) results in the optimised finite element mesh through the minimisation of the linear combination of the smoothness, othogonality and the measure of error evaluated at each master cell on the total region. The amount of movement of nodes.can be controlled by assigning various values to η and θ. η is a weighting parameter in relation with smoothness and orthogonality of elements. This parameter is used to prevent an element from becoming too small and being abnormally distorted. The weighting parameter θ is not only related to the smoothness and the orthogonality but also the measure of error. θ generally ranges inbetween 0 and 1. But, it is desireble to assign a small value to θ in order to reduce the measure of error. In this study, the values of parameters η and θ were 0.2 and 0.03 respectively.

Here, it is shown how to treat pactically the optimisation problem of the finite element mesh. In the actual numerical scheme, this is realised by minimising the evaluation function during the relocation of nodes iteratively.

The flow chart used in the numerical scheme in the case of linear problems is shown in Fig.3.2 and is explained as follows :

1)First, an elastic finite element analysis is carried out using the initial finite element mesh.
2)Strains are evaluated from the calculated displacements, and the measure of error given by Eq.(3.8) at each node is calculated.

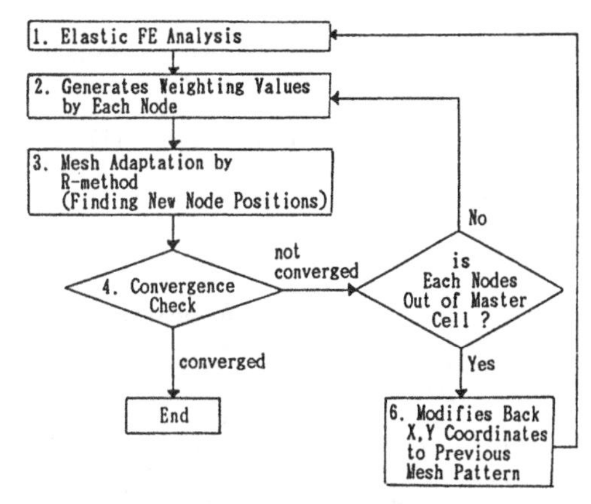

Fig.3.2 Flow chart for adaptive mesh method

3)Calculated smoothness and orthogonality from the finite element mesh by which the measure of error are evaluated and the new location of nodes are determined by using Eq.(3.7).
4)The new location of nodes is checked whether it is within the master cell. This is made due to the reason that if the new location of nodes is out of the master cell, the reliability of the measure of error is greatly reduced.
5)If the new location of the node is within the master cell, then the measure of error is recalculated and mesh adaptation is carried out.
6)If the new location of the node is out of the master cell, a new procedure of finite element analysis is made by using the previous mesh. Procedures are repeated.

4. ANALYSIS OF LOCALISATION CONDITION AND ITS APPLICATIONS

4.1 Analysis of localisation condition

In this section, the global failure of some structures by the localisation theory using an elasto-plastic finite element method is considered. Let us denote $\nu^T D^{ep} \nu$ by the tensor A for the convenience. As will be dealing with only two dimensional problems in this paper, the components of matrix A can be given as follows:

$$\begin{aligned}
A_{11} &= \nu_1 D^{ep}_{11}\nu_1 + \nu_1 D^{ep}_{13}\nu_2 + \nu_2 D^{ep}_{31}\nu_1 + \nu_2 D^{ep}_{33}\nu_2 \\
A_{12} &= \nu_1 D^{ep}_{13}\nu_1 + \nu_1 D^{ep}_{12}\nu_2 + \nu_2 D^{ep}_{33}\nu_1 + \nu_2 D^{ep}_{33}\nu_2 \\
A_{21} &= \nu_1 D^{ep}_{31}\nu_1 + \nu_1 D^{ep}_{33}\nu_2 + \nu_2 D^{ep}_{21}\nu_1 + \nu_2 D^{ep}_{23}\nu_2 \\
A_{22} &= \nu_1 D^{ep}_{33}\nu_1 + \nu_1 D^{ep}_{32}\nu_2 + \nu_2 D^{ep}_{23}\nu_1 + \nu_2 D^{ep}_{22}\nu_2
\end{aligned} \tag{4.1}$$

where $\nu = (\nu_1, \nu_2, 0)$ and it is noted that D^{ep} in Eq.(4.1) is actually a fourth order tensor and contracted to a second order tensor.

The localisation condition (2.11) is rewritten as (refer to Appendix I for the details)

$$det A = A_{11}A_{22} - A_{12}A_{21} \tag{4.2}$$

Finally, the localisation analysis is carried out to find the unit normal to the shear band which satisfies the localisation condition given by Eq.(4.2). An elasto-plastic constitutive equation of non-associative type, employing the Drucker-Prager yield function together with the dilatancy function β is used in the finite element analysis for the plane strain case.

The coefficients α and Hardening function K in Eq.(2.19) determined from laboratory tests carried out on Oya-tuff are given as

$$\alpha = 0.2 \tag{4.3}$$

$$K = 5.80 + a_1(1 - \exp(-\epsilon^p/\tau_1)\exp(-\bar{\epsilon}^p/w_1) + a_2(1 - \exp(-\epsilon^p/\tau_2)\exp(-\bar{\epsilon}^p/w_2)$$

$$a_1 = -0.11 \quad MPa \quad \tau_1 = 7.1 \times 10^{-4}$$
$$a_2 = 2.50 \quad MPa \quad \tau_1 = 1.8 \times 10^{-4}$$
$$w_1 = 6.6 \times 10^{-5}$$
$$w_1 = 2.0 \times 10^{-4} \tag{4.4}$$

and dilatancy funtion β is assumed to be constant in the analyses and has the following value:

$$\beta = 0.10 \tag{4.5}$$

Young's modulus and Poisson's ratio for this material are:

$$E = 2.2 \times 10^3 \tag{4.6}$$

$$\nu = 0.35 \tag{4.7}$$

4.2 Application to bearing capacity problem of foundations

Finite element meshes before and after optimisation for the elasto-plastic analysis of the bearing capacity of a

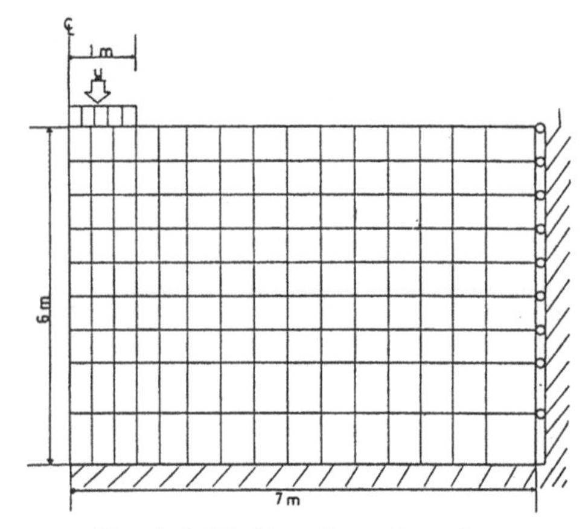

Fig.4.1 Finite element mesh before optimisation

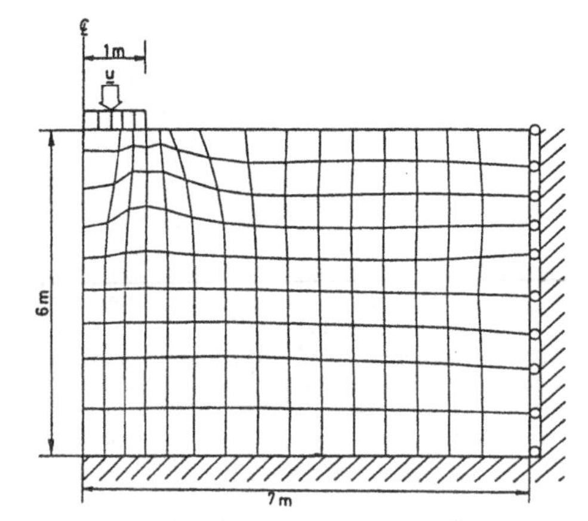

Fig.4.2 Finite element mesh after optimisation

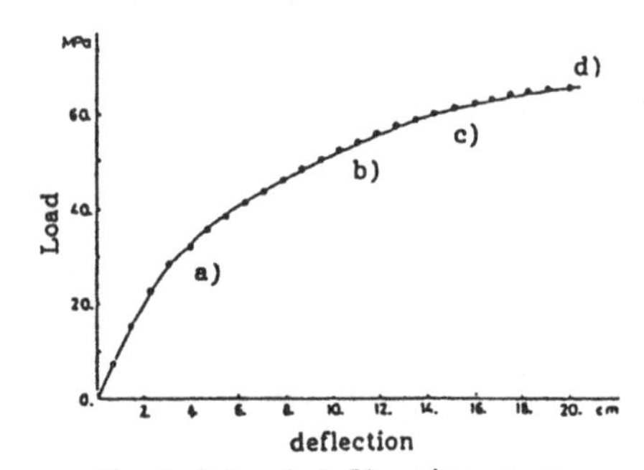

Fig.4.3 Load-deflection curve

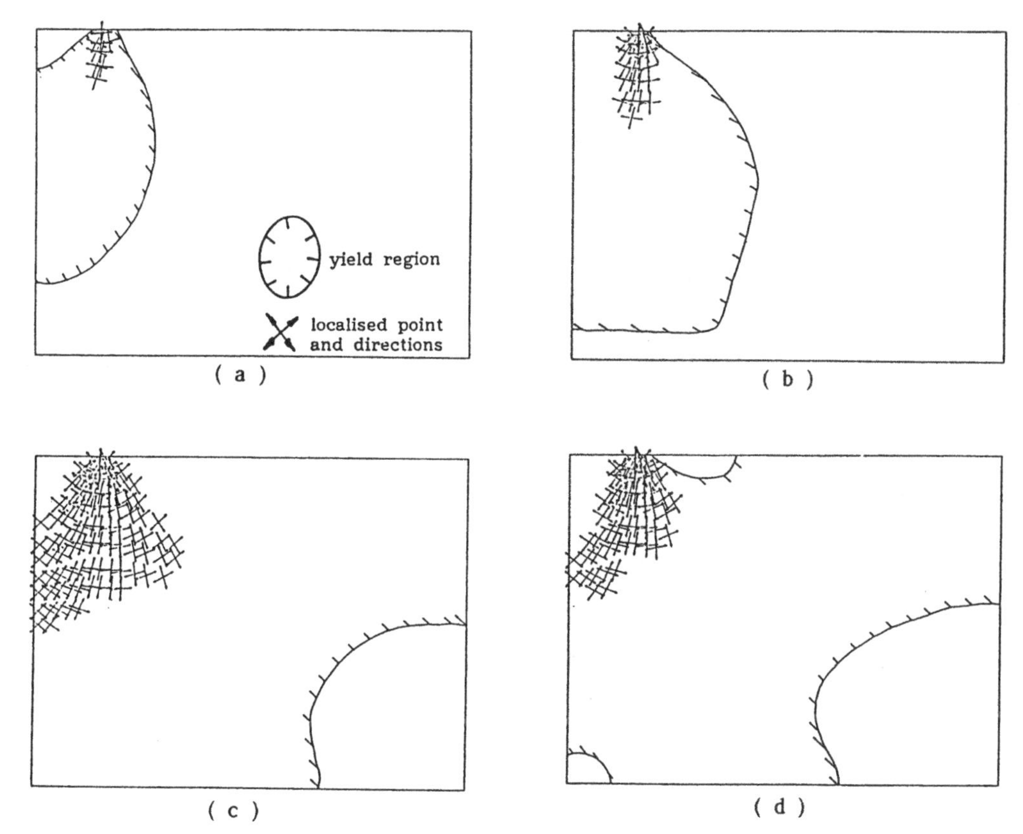

Fig.4.4 Propagation of localisation points and the unit normal vectors and yield region

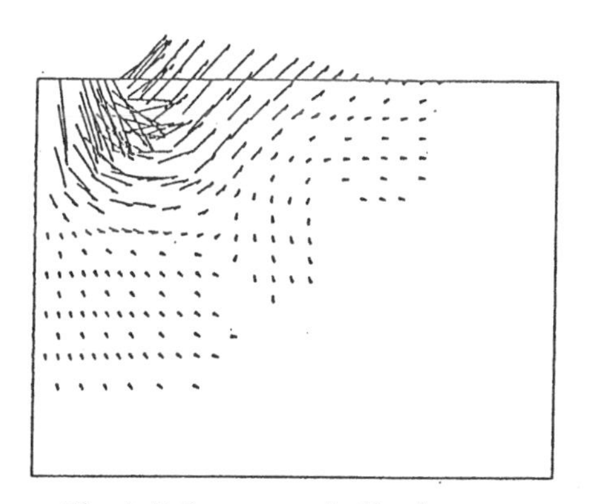

Fig.4.5 Incremental displacement

foundation are shown in Figs. 4.1 and 4.2. The final optimised mesh is used in the localization analysis. The calculated load-deflection curve is shown in Fig. 4.3. Fig. 4.4 shows the plastic zone and the localisation points with associated unit vectors. Although the solution of Eq.(2.11) results in four unit vectors, the unit vectors are associated with only two directions. Regarding this problem, one can easily notice how the localisation occurs and propagates from Fig. 4.4 which illustrates the initiation and propagation of the localisation at various loading steps . The incremental displacement field at the loading step (d) is shown in Fig. 4.5. As noted from this figure, the global failure mode is well illustrated.

Let us discuss the results of this analysis in more details: as expected, the localisation is initiated at the edge of the loading plate and propagates downward along a curved surface. At the loading step (c), this surface coincides with the similar kind of surface emanating from the other edge of the plate. **Thus we suggest that the load bearing capacity of foundations must be taken equal to the load at step (c) since the sliding surface occurs beneath the plate at this load level.**

The bearing capacity of the foundation in the analysed example is found to be 62.1 MPa and corresponding displacement is 160 mm.

4.3 Application to a slope stability problem

Finite element mesh after optimisation for the elasto-plastic analysis of the bearing capacity of a foundation on a spope is shown in Fig. 4.6. The calculated load-deflection curve is shown in Fig. 4.7. Fig. 4.8 shows the plastic zone and the localisation points with associated unit vectors. Regarding this problem, one can easily notice how the localisation occurs and propagates from Fig. 4.8 which illustrates the initiation and propagation of the localisation at various loading steps . The incremental displacement field at the loading step (d) is shown in Fig. 4.9. As noted from this figure, the global failure mode is well illustrated.

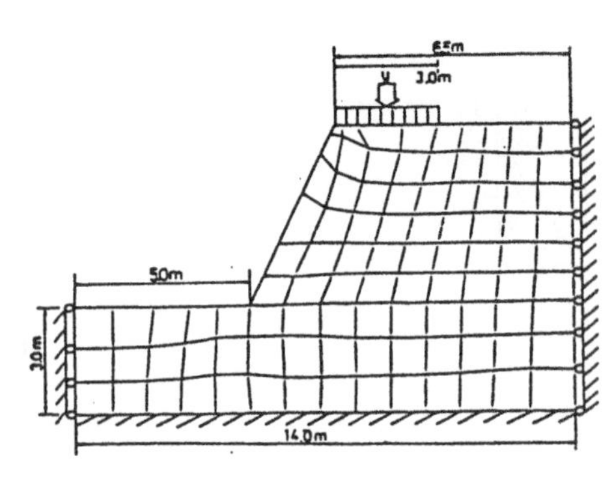

Fig.4.6 The optimised finite element mesh

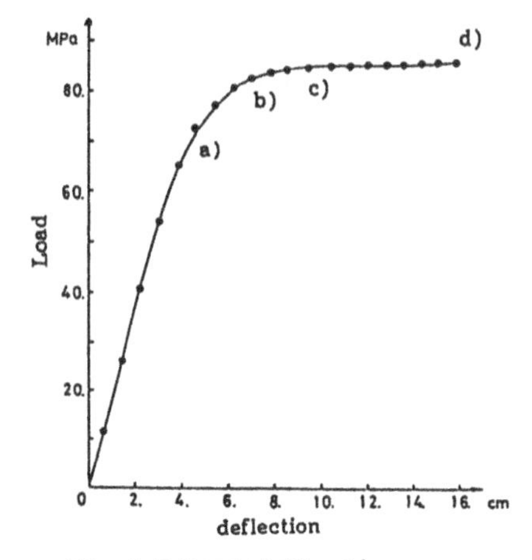

Fig.4.7 Load-deflection curve

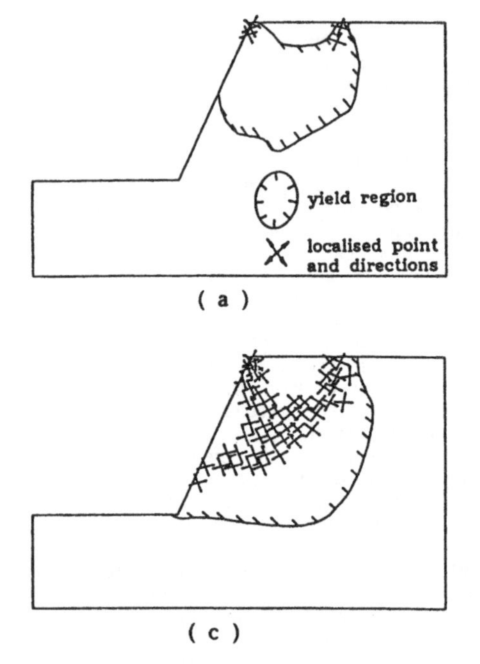

(a)

(c)

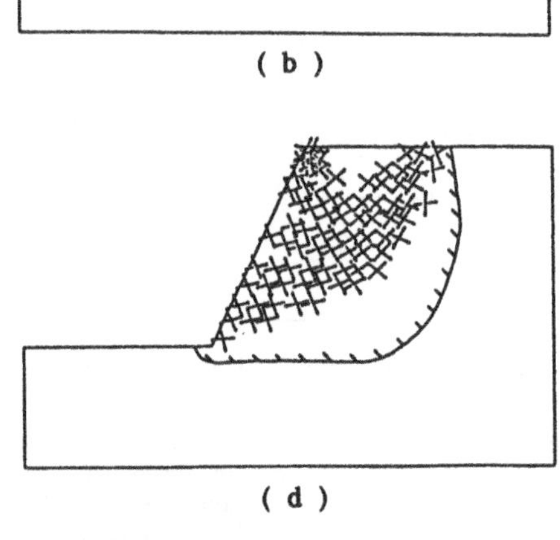
(b)

(d)

Fig.4.8 Propagation of localisation points and the unit normal vectors and yield region

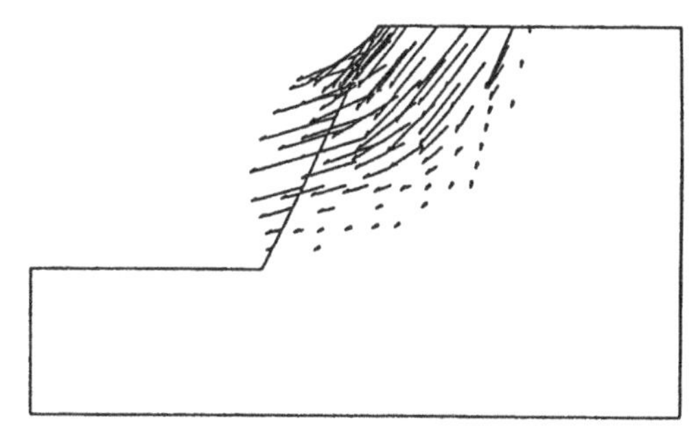

Fig.4.9 Incremental displacement

Let us discuss the results of this analysis in more details: as expected, the localisation is initiated at the edges of the loading plate and propagates downward . At the loading step (c), the localisation surfaces coincides under the plate. Then, the sliding surface propagates towards the slope surface. The sliding surface is almost circular and the initiation of localisation and its propagation is in good agreement with the global failure pattern in such problems.

The bearing capacity of the foundation on a rock slope is found to be 25.1 MPa and corresponding displacement is 94 mm.

5. CONCLUSIONS

The load bearing capacity of foundations and the stability of slopes are estimated by an elasto-plastic finite element method together with the localisation condition. In the analyses, the adaptive mesh method is employed and is found to be very effective for particularly the stress concentration problems such as at the edge of a rigid plate as in this study. The adaptive mesh method is only applied to a linear-elastic scheme, however, it is felt that there is a need to improve the present method for non-linear scheme. By the localisation condition, the slip surfaces can be well expressed in the load bearing capacity and slope stability problems. Especially, the mode of failure in the case of slope stability problem is of great interest: the initial mode failure takes place in the form of wedge and then becomes a circular form as very well experienced in such tests.

ACKNOWLEDGEMENT

The authors wish to express their thanks to Messers. T. Kyoya, Nagoya University and N. Fukuura, Taisei Corporation for their invaluable contributions and usefull discussions and to Mr. Ö. Aydan, Nagoya University for his help in the preparation of this article.

REFERENCES

Carcaillet, R., G.S. Dulikravich & S.R. Kennon 1986. Generation of solution adaptive computational grids using optimization. Comp. Meths. in Appl. Mech. Engng. Vol.57, p.279-295.

Diaz, A.R., N. Kikuchi & J.E. Taylor 1983. A method of grid optimization for finite element methods. Comp. Meths. in Appl. Mech. Engng. Vol.37, p.29-46.

Hill. R. 1962. Acceleration waves in solids. J. Mechs. Phys. Solids, Vol.10 p.1-16.

Ichikawa, Y., T. Kyoya & T. Kawamoto 1985. Incremental theory of plasticity for rock. Procs. 5th Int. Conf. Num. Meths. Geomechanics, Nagoya, Vol.1, p.451-462.

Kikuchi, N. 1985. Adaptive grid design methods for finite element analysis. Procs. 2nd Joint ASCE/ASME Mechanics Conf., Univ. of New Mexico

Lysmer, J. 1970. Limit analysis of plane problems in soil mechanics. Procs. ASCE, SM4, p.1311-1334.

Needleman, A. & J.R. Rice 1978. Limits to ductility set by plastic flow localization. Mechanics of Sheet Metal forming, Plenum Press, New York.

Obara, Y., T. Yamabe, Y. Shimizu, Y. Ichikawa & T. Kawamoto 1982. Elastoplastic analysis by cracked triangular element. Proc. Int. Conf. FEM, Science Press, China.

Rice, J.R. 1976. The localization of plastic deformation. Theoretical and Applied Mechanics, North-Holland Publishing Co.

Rudnicki, J.W. & J.R. Rice 1975. Conditions for the localization of deformation in pressure-sensitive dilatant materials. J. Mechs. Phys. Solids, Vol.23.

Tamura, T. & S. Kobayashi 1984. Limit analysis of soil structure by rigid plastic finite element method. Soils & Found., Vol.21, p.34-42.

APPENDIX

From Eq. (4.2), the localisation condition is rewritten as

$$detA = C_1\nu_1^4 + C_2\nu_1^3\nu_2 + C_3\nu_1^2\nu_2^2 + C_4\nu_1\nu_2^3 + C_5\nu_2^4 = 0 \quad (1)$$

where

$$C_1 = D_{11}D_{33} - D_{13}D_{31}$$

$$C_2 = D_{33}(D_{13} + D_{31}) + D_{11}(D_{32} + D_{23}) - D_{31}(D_{12} + D_{33}) - D_{13}(D_{33} + D_{21})$$

$$C_3 = D_{33}^2 + D_{11}D_{22} + (D_{32} + D_{23})(D_{13} + D_{31}) - (D_{31}D_{32} + D_{13}D_{23}) - (D_{33} + D_{21})(D_{12}D_{33})$$

$$C_4 = D_{33}(D_{32} + D_{23}) + D_{22}(D_{13} + D_{31}) - D_{32}(D_{33} + D_{21}) - D_{23}(D_{12} + D_{33})$$

$$C_5 = D_{22}D_{33} - D_{23}D_{32}$$

The following relationship is valid between the components of the unit normal vector $\nu = (\nu_1, \nu_2, 0)$:

$$\nu_1^2 + \nu_2^2 = 1 \quad (2)$$

Combining Eqs. (1) and (2) yields

$$\nu_2 = -\frac{C_1\nu_1^4 + C_3\nu_1^2(1-\nu_1^2) + C_5(1-\nu_1^2)^2}{C_2\nu_1^3 + C_4\nu_1(1-\nu_1^2)} \quad (3)$$

and inserting Eq. (3) into Eq. (2), one has

$$B_1\nu_1^8 + B_2\nu_1^6 + B_3\nu_1^4 + B_4\nu_1^2 + B_5 = 0 \quad (4)$$

where

$$B_1 = (C_2 - C_4)^2 + (C_1 - C_3 + C_5)^2$$

$$B_2 = 2 \times (C_2 - C_4)C_4 - (C_2 - C_4)^2 + (C_1 - C_3 + C_5)(C_3 - 2 \times C_5)$$

$$B_3 = C_4^2 - 2 \times (C_2 - C_4)C_4 + 2 \times (C_1 - C_3 + C_5)C_5 + (C_2 - 2 \times C_5)^2$$

$$B_4 = 2 \times C_5(C_3 - 2 \times C_5) - C_4^2$$

$$B_5 = C_5^2$$

Eq. (4) is of the fourth degree and its real solution is the unit normal vector ν.

Computer and Physical Modelling in Geotechnical Engineering, Balasubramaniam et al. (eds)
© 1989 Balkema, Rotterdam. ISBN 90 6191 864 2

Stability of cemented tailings mine backfills

R.J.Mitchell
Queen's University, Kingston, Canada

ABSTRACT: Ore bodies, hence mine openings and exposed backfills in large underground metalliferous mines vary in shape from relatively high and narrow to relatively low and wide. Wall rocks adjacent to the backfill may be steeply dipping or may be relatively flat-lying. Fill stability analyses must consider the details of the geometric boundaries of the fill if economical use of stabilizing cement is to be achieved. It is generally prohibitively expensive to use sufficient cement to satisfy a lower bound free standing wall design approach.

This paper considers three typical fill geometries--steeply dipping and gently dipping relatively narrow fills as well as relatively wide fills--and presents the results of both analytical and physical model stability studies. It is shown that efficient use of cement can be achieved by relying on the beneficial effects of soil arching, boundary shear and reinforcing techniques.

1 INTRODUCTION

Cemented tailings sands are being used throughout the world for backfilling large underground mine openings. The purpose of filling these openings is to provide support to adjacent wall rocks as mining progresses, thus avoiding wall slough, caving and overstressing of pillars. The purpose of the cement is to consolidate (stabilize) the sand fill so that exposed fill faces will be self-supporting when adjacent ore pillars are removed. Figure 1 is a schematic showing a room and pillar mining operation and Figure 2 is a section showing an exposed backfill face. In general, self-weight stresses govern the backfill design and a traditional design has been that of a free standing wall, requiring an unconfined compressive strength of

$$UCS = \gamma H \qquad (1)$$

where γ = fill unit weight and H = fill height

In many cases, however, the adjacent rock walls actually help support the fill through boundary shear and arching effects. Thus, the backfill and rock walls are mutually supporting. Recognizing this, Mitchell (1983) developed the expression

$$UCS = \gamma H/(1 + H/L) \qquad (2)$$

where L = strike length of the exposed backfill

As a comparison of equations (1) and (2), consider the typical case where the unconfined compressive strength (UCS) increases with cement as shown on Figure 3 and the backfill has H = 70 m, L = 20 m and γ = 20 kN/m³. Equation (1) gives UCS = 1400 kPa and the cement requirement is 0.115 tonnes per tonne of backfill. Equation 2 gives UCS = 311 KPa and, hence, 0.05 tonnes of cement per tonne of fill. A tonne of fill is required for approximately 2 tonnes of ore removed and roughly 50% of the backfill would need to be cemented (see Figure 1). At a cost of $100 per tonne of cement, the cost difference between the predictions of equation (1) and equation (2) is $1.63 per tonne of ore production. This would amount to over a million dollars per year for a typical bulk scale mining operation. At present day profit margins, a dollar a tonne could make the difference in obtaining

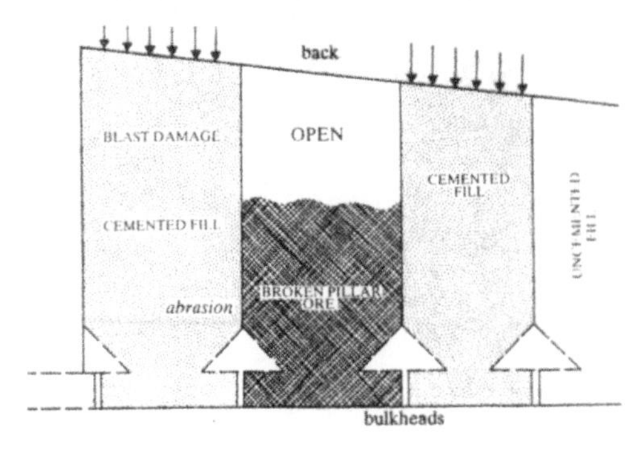

Figure 1 Room and pillar sequence

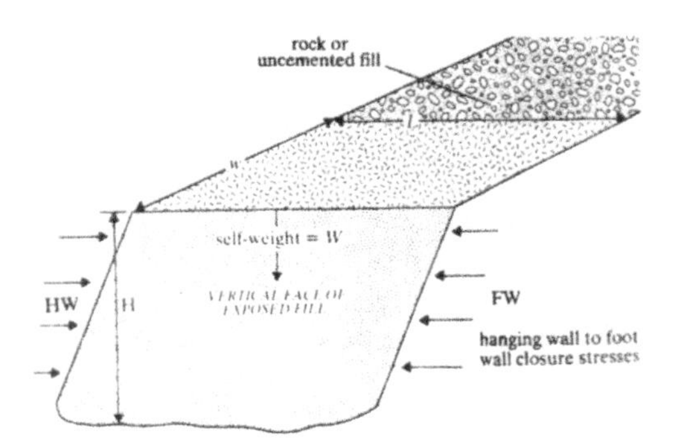

Figure 2 Mine backfill exposure

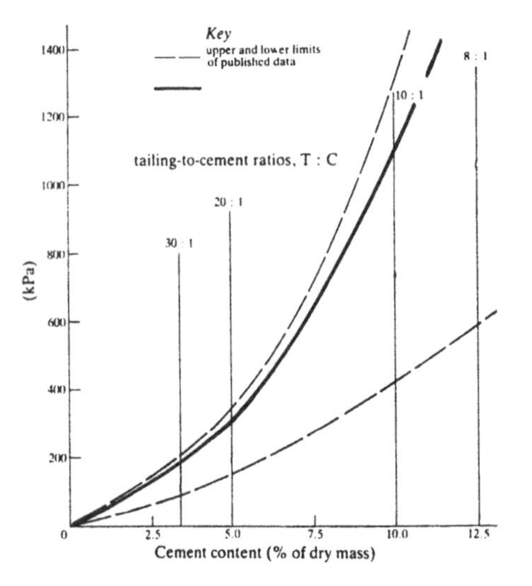

Figure 3 Unconfined compressive strength

financial backing for a low to medium grade ore body.
Further stability analyses and correlating physical model studies are summarized, for three general geometrical cases, in this paper. The three cases are near vertically oriented fills having H>L, sloped fills having H>L, and fills having L>H. In the first two cases, plane cemented fills are considered to be most economical but layered or reinforced cemented backfills are likely to be more economical for the latter case.

1.1 Vertically oriented fills having H>L

The earliest stability model studies carried out on plane cemented fills were models of 2 m height, having $1 \leq (H/L) \leq 4.5$ and have been reported by Mitchell, Olsen and Smith (1982). These authors concluded that the results supported the use of equation (2) for design of vertical backfills having H>L. Figure 4 shows one of the two dozen model failures generated in the study and it is of interest to note that similitude of the dimensionless factor $UCS/\gamma H$ was obtained by limited the cement curing period to a matter of hours.

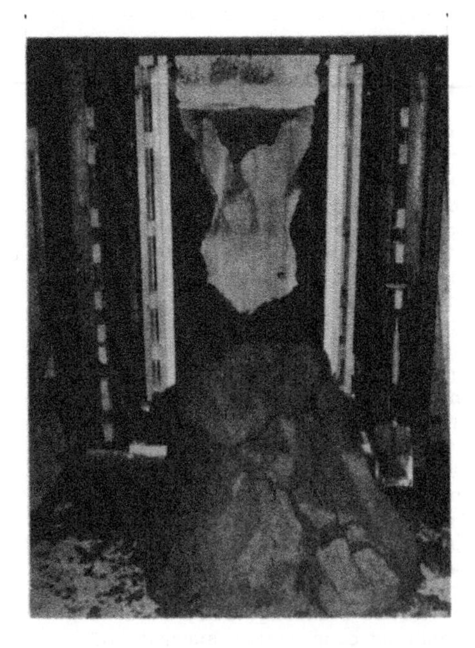

Figure 4 Failure in model fill with H > L

In 1985 a 3 meter radius, 30 g-tonne centrifuge facility was designed and built in Queen's University Civil Engineering Department. Figure 5 shows a schematic of this machine. Figure 6 shows the relation between centrifuge

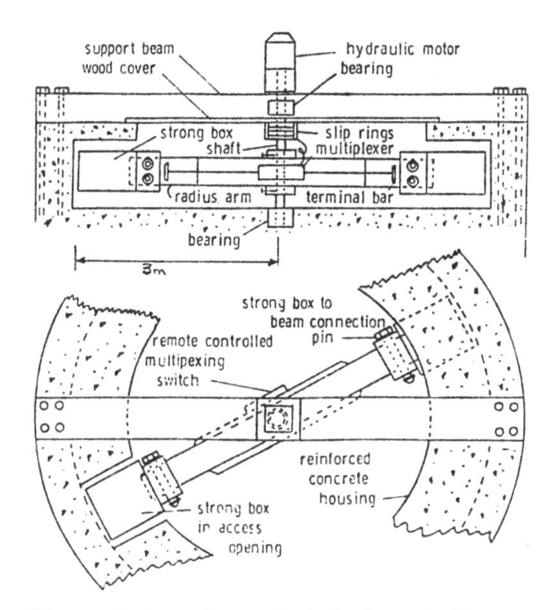

Figure 5 Queen's geotechnical centrifuge

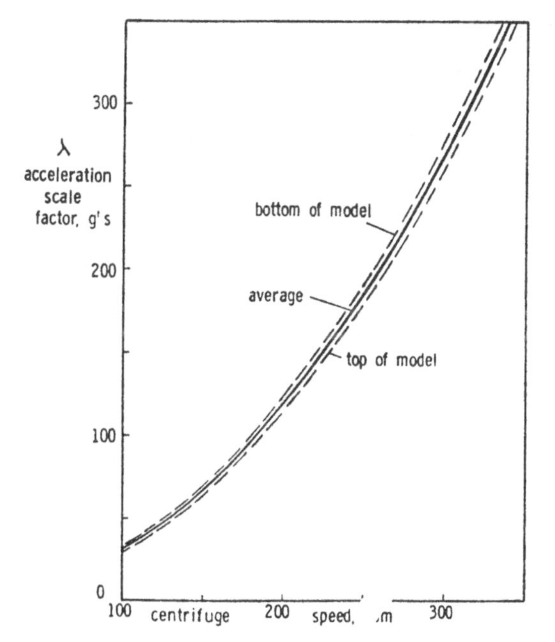

Figure 6 Acceleration scale factor

speed and the model stress scale factor $\lambda = rw^2/g$.
Several model tests were conducted in this centrifuge on 28 day cured, 3.3% by weight cemented backfill (30:1 T:C) and rough wall contacts. Figure 7 shows a typical failure which started at the base of the fill face and caved upwards as the stress levels were increased by increasing the centrifuge speed. Figure 8 shows a centrifuge model with smooth wall conditions, before and after failure. Smooth wall failures occurred, as expected, by the formation of a tensile crack and a slip plane of an angle somewhat steeper than 45°. Mass movement developed at the scale factor when the shear plane was first noted on the centrifuge video monitor.

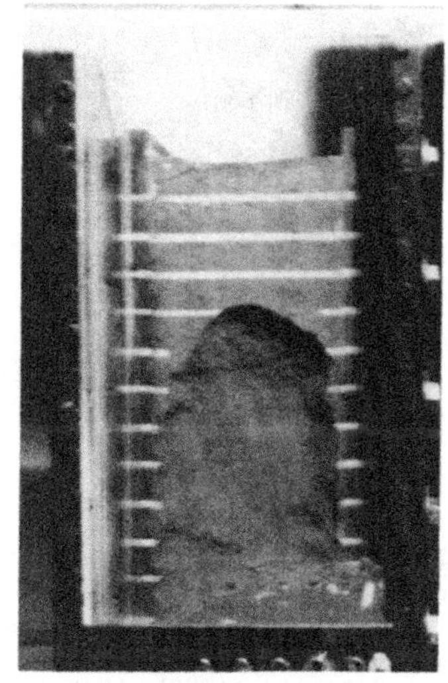

Figure 7 Failure in centrifuge model with rough walls

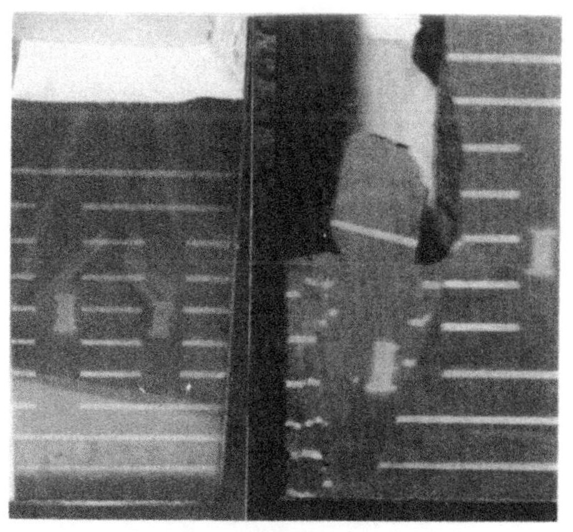

Figure 8 Centrifuge model with smooth walls

A summary of some of the data obtained from testing vertically aligned models having H>L is plotted on Figure 9. This data covers prototype fill heights from 20 m to over 90 m with cement contents from 2.5% (40:1 T:C) to 5% (20:1 T:C). The softer 2 m high models and the smooth wall centrifuge models support the prediction from equation (2) while the centrifuge models having rough wall conditions are significantly more stable

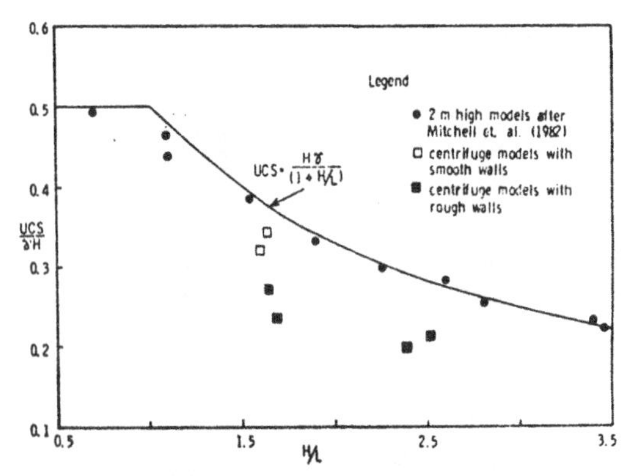

Figure 9 Comparison of model data

than the equation (2) prediction. Thus, the transfer of stress to the rough walls through an arching/shear stress mechanism must be greater than that assumed by Mitchell (1983), that is that the wall shearing resistance is equal to the cohesive strength times the wall area.

1.2 Sloped fills having H>L

When the hanging wall-footwall dip is less than 90° (see Figure 2), a fill failure may be restricted by the vertical distance between hanging wall and footwall and the transfer of stresses to the footwall will be greater than the transfer to the hanging wall. Figure 10 shows a potential failure mechanism for this situation. It is assumed that a parabolic shear surface extends from the edge of the exposed footwall to intersect a tensile plane on the hanging wall side. The location of this intersection varies from a depth, 2d, at the top of the fill to zero at the base of the failure, having an average dept of d. By assuming that the direction of movement was closely parallel to the footwall dip direction, an approximate formulation was derived in the form

$$\frac{UCS}{\alpha H} = \frac{1}{\frac{0.6}{\sin\beta}(\chi + H/L)} \quad (3)$$

where $\chi = \frac{L}{d}$ describes the average depth of failure. This equation predicts a deep critical surface but the maximum practical depth must be restricted, on kinematic considerations, to L/2. Thus the value of χ is restricted to 4 and

$$\frac{UCS}{\alpha H} = \frac{1}{\frac{2.4}{\sin\beta} + \frac{0.6}{\sin\beta}\frac{H}{L}} \quad (4)$$

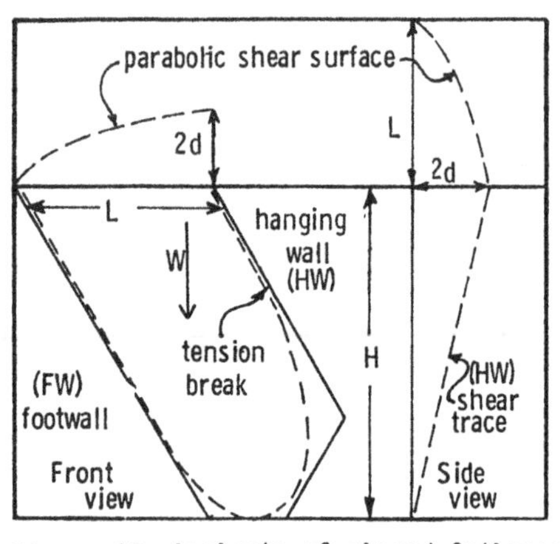

Figure 10 Analysis of sloped failures

Several large scale static models were cast at a mine site into 5 m high test boxes having two different values of H/L, β = 55° and rough walls. Figure 11 shows the failure that developed in one on these models. The models were cured for up to 24 hours to achieve sufficient strength to just attain a failure condition when the facia boards on the test boxes were removed. Further details of this testing has been reported by Smith et. al. (1983). The results of several tests are plotted on Figure 12 and the data points are seen to be scattered about a curve given by equation (4).

It may also be noted, by comparison of Figures 9 and 12, that the vertical rough walls and the sloped rough walls gave much the same stability factors (values of UCS/γH). Further testing of sloped fills is currently being carried out using centrifuge models.

Figure 11 Failure in model with sloped wall

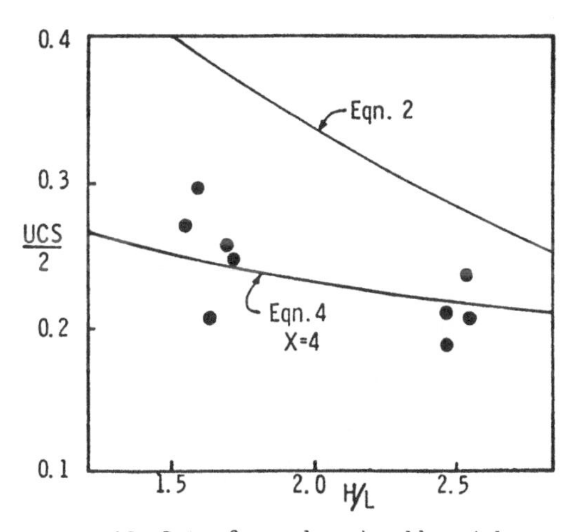

Figure 12 Data from sloped wall models

1.3 Reinforced fills having L>H

When the strike length of the fill exceeds the exposed fill height the effects of arching and wall shear will have diminished to the extent that the common two dimensional vertical slope upper bound stability criterion, UCS = $\gamma H/2$, might be considered appropriate for evaluating the backfill stability. Several recent centrifuge models have been tested with an 'elastic clearance' between the side walls of the centrifuge strong box in order to represent the two dimensional case. The results of most of these models support the two dimensional analysis and, indeed, as the depth of the model decreases below H/2, the stability tends to decrease towards the ultimate lower bound of the free standing wall. Since mines that have L>H cannot take advantage of arching, advantage may be gained through the use of reinforcements.

Cemented backfills may be reinforced by (1) pouring strongly cemented layers, with or without inclusions, at intervals in a low cement content bulk fill, by (2) pouring inclusions in the low cement content bulk fill or by (3) placing geogrid or other continuous reinforcing elements on predetermined spacings as the fill height is increased. Inclusions in bulk pours would increase the UCS of the material and it is simply a matter of finding suitable effective inclusions. The analysis remains the same. Layered systems and reinforced systems can be readily analyzed using existing theories (see, for examples, Vidal (1969), Lee et. al. 1973) and these will not be repeated herein. Rather, the results of some drop-box and centrifuge models will be presented. Before the centrifuge was built at Queen's a drop-box type of linear deceleration device was used to test fully cured plane and reinforced cemented backfill models of 0.4 m height. With a 5.5 m maximum drop height, peak decelerations in the order of 100 gravities could be exerted on a model for about 5 milliseconds. A typical deceleration record is shown on Figure 13 and Figure 14 shows a failed model which contained corrugated metal reinforcing strips at 110 mm vertical spacings. Due to the very short time span in which prototype stresses were simulated, multiple drops were required to fail most models. Significant dynamics also accompanied each drop but it was possible to obtain the relative stability ratios from the data, as shown on Table 1. Details of the analyses are provided by Stone (1985) who concludes that reinforced cemented backfills are likely to provide up to 50% greater stability than plane cemented backfills.

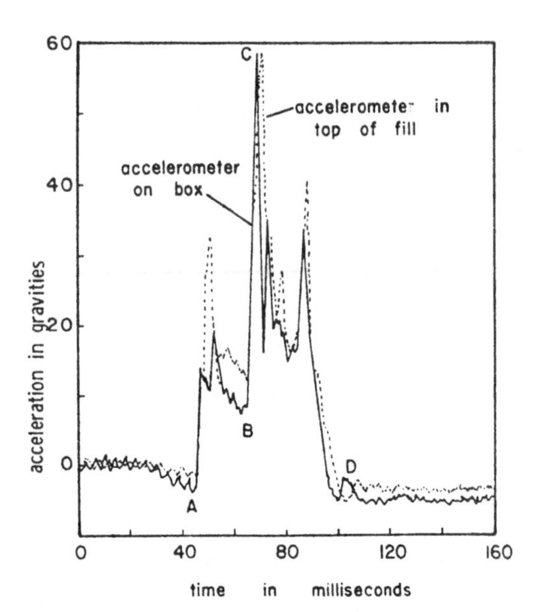

Figure 13 Deceleration record for a drop-box test

Figure 14 Failure in reinforced drop-box model

TABLE 1 - Relative Stability Ratios

Backfill Model Description	Relative Stability
plane backfill 40:1 T:C	1.0
plane 40:1 T:C bulk fill with 8:1 T:C layers, 10 mm thick at 0.1 m spacings	1.2
metal strip reinforced 40:1 T:C fill, strips at 0.11 m spacings	1.5

Several centrifuge tests have recently been carried out on cemented backfill models with smooth walls and relatively small depths (d = 0.25H) but reinforced with model geogrid systems. On the suggestion that it would be easier to install geogrids in a vertical orientation by hanging these from cables into open stopes prior to hydraulic filling, models were tested with this type of alignment of the reinforcements. Figure 15 shows the remains of a failed model after the crushed material in the shear zone was removed. As noted in Table 2, these models were only marginally stronger than the plane cemented backfill models. Figure 16 shows the remains of a failed model with horizontal reinforcement and these models

Figure 15 Model with vertical reinforcements

averaged a 40% higher prototype height. The results in Table 2 support the conclusion from drop-box testing. It may be noted from Figure 16 that the model horizontal reinforcements were sheared, not pulled out, by the forces at failure. Hence, stronger model geo-

Figure 16 Model with horizontal reinforcements

TABLE 2 -- Stability of Reinforced Models

Backfill model description	Centrifuge RPM at model failure (avg. of two	$\lambda = a/g$	Prototype equivalent height at failure, m
0.35 m high plane 22:1 T:C fill	155	73.9	25.8
0.35 m high 22:1 T:C fill with vertically oriented reinforcements on 50 mm c-c	162	80.7	28.3
0.35 m high 22:1 T:C fill with horizontally orientated reinforcements on 50 mm c-c	195	117.0	40.9

grids or closer model geogrid spacings would have increased the stability of this model. Further centrifuge testing of reinforced cemented backfill models is currently underway in order to evaluate the various alternative types of reinforcements. Figure 17 shows centrifuge models with strongly cemented layers. In this case the strong layers resulted in a doubling of the stable prototype height. At this time it can be concluded that it is technically feasible to use various types of reinforcements to increase the stability of a weakly cemented tailings backfill. Further testing is necessary to optimize reinforced designs.

Figure 17 Failure in model with strong layers

Conclusions

The work presented and summarized in this paper is directed towards physical model study evaluations of the stability of cemented tailings mine backfills and the following conclusions result:

(1) Arching and wall shear effects in vertically oriented fills having heights exceeding the exposed strike length cause a reduction in the stresses acting on potential failure planes so that the design strength of such fills can be reduced well below that given by the usual two dimensional analysis. Both wall proximity and wall roughness are significant in determining the stable height of a given cemented backfill. Correlations between proposed analysis and physical model test data are provided in the paper.

(2) Backfills formed between rough sloped hanging wall-footwall geometries are at least as stable as vertically oriented backfills having rough walls and the same height to strike exposure ratio. An analysis of the stability of these fills is presented and correlated

with physical model tests to provide some guidance for the design of prototype fills.

(3) When the strike length of the exposed backfill is greater than the exposed fill height, the major benefits of soil arching and wall shear are lost. For such cases, it has been suggested that reinforcements could be placed in the cemented backfill to increase stability. Physical model test results presented in this paper indicate that the stable height of weakly cemented backfills can be increased by at least 50% with the use of reinforcements.

The results of recent centrifuge model testing have been introduced in this paper and further centrifuge modelling is now underway to evaluate the various alternative backfill designs and design methods in more detail.

Acknowledgements

The author wishes to acknowledge the support of the Natural Sciences and Engineering Research Council of Canada (NSERC) in providing funds for research in earth structures engineering. The centrifuge was constructed from funds provided by NSERC, Westmin Resources Ltd. and Falconbridge Ltd. on a Shared Facilities Grant.

References

Lee, K.L., Adams, B.D. and Vagneron, T.M. (1973) Reinforced Earth Retaining Walls. ASCE J. Soil Mech. Fnd. Eng. 99(10) 745-64.
Mitchell (1983) Earth Structures Engineering. Allen and Unwin.
Smith, J.D., DeJongh, C.L. and Mitchell, R.J. (1983) Large Scale Model Tests to Determine Strength Requirements for Pillar Recovery. Proc. Int. Symposium on Mining with Backfill, Lulea, Sweden, June 1983.
Stone, D.M.R. (1985) Model Studies on the Stability of Reinforced Mine Backfill. Ph.D. Thesis, Queen's University, Kingston, Canada.
Vidal, H. (1969). The Principle of Reinforced Earth. Highway Research Record No. 282, N.A.S., Washington, USA.

Computer and Physical Modelling in Geotechnical Engineering, Balasubramaniam et al. (eds)
© 1989 Balkema, Rotterdam. ISBN 90 6191 864 2

Diffusion-controlled contaminant transport in landfill-clay liner systems

R.Kerry Rowe
University of Western Ontario, Canada
John R.Booker
University of Sydney, Australia

ABSTRACT: A simple, semi-analytic technique for the calculation of contaminant migration through a clay cover, landfill, clay liner and underlying sand, gravel and clay layers is described for two dimensional conditions.

The application of the theory is then illustrated by considering the effect of variability in physical parameters upon both the contamination of groundwater in an aquifer beneath the landfill-liner system and the contamination of surface water passing over the clay cover. Particular attention is given to the effects of variability in the diffusion-dispersion coefficient within the clay cover and landfill together with the effect of the advective velocities in the landfill-liner system and the underlying aquifer.

1 INTRODUCTION

The use of clay liners is an attractive method of inhibiting the migration of pollutants which may contaminate groundwater. There are of course potential difficulties (Anderson 1982; Brown 1983; Daniel 1984, 1985; Day 1984; Fernandez and Quigley 1984). However, as indicated by Daniel (1984), clay liners can be used to control the movement of contaminants from waste disposal sites provided that due consideration is given to the geotechnical design and construction supervision.

It is now recognised that even well designed clay liners are not completely impermeable. Clearly contaminant transport can occur due to advection (seepage). However even in the total absence of advective transport, contaminant migration can occur by molecular diffusion within the pore fluid of the clay. For example, by monitoring the migration of contaminants through a natural clayey till "liner" beneath a domestic landfill in Southwestern Ontario, Canada, Quigley and his co-workers (Goodall and Quigley 1977; Crooks and Quigley 1984; Quigley and Rowe 1985) have shown that over a 12 year period, several soluble contaminants have migrated to a depth of approximately 1 m even though the advective advance was estimated to be only between 0.03 and 0.05 m.

Accepting that some contaminant migration through clay liners is inevitable raises the question as to how much migration will occur and what will be the likely impact in terms of the concentration of contaminants within the general groundwater system? This is particularly critical when the landfill is to be separated from an aquifer by a relatively thin clay liner. Such situations are often encountered in practice. For example in Southern Ontario, Canada, landfills are being constructed by excavation in layers of clayey till or clay which are underlain by thin aquifers (i.e. less than 1 m thick). A similar situation may arise due to the construction of a compacted clay liner over an existing aquifer. In either case an assessment of the potential impact of the landfill upon the groundwater will require contaminant 'migration' analyses for the particular situation.

These contaminant migration analyses could be performed using finite element techniques (Anderson 1979). However to obtain accurate results for both small and large times, this requires a refined finite element mesh (to accommodate the high concentration gradients at low times) and considerable computational effort - particularly if a sensitivity study is required.

Many soil deposits can be idealized as being horizontally layered and in these cases it is not really necessary or efficient to use the finite element method. Rowe and Booker (1985a,b, 1986) have proposed an alternative finite layer procedure which can be used to calculate the concentration of contaminants at specified locations and times without determining the solution at all points or previous times. These analyses can be readily performed on a microcomputer and require minimal data preparation. This previous work has been restricted to one dimensional (1D) contaminant transport in a single or multilayer system and to two dimensional (2D) transport in a single layer system. The present paper generalizes these approaches so that a two dimensional analysis of a fully stratified deposit can be considered. The application of the theory will then be illustrated by considering the effects of variability in physical parameters; such as diffusion-dispersion coefficients and advective velocities, upon both the concentrations of contaminant within an aquifer and the mass of contaminant escaping to the surface by diffusion through the clay cover.

The results of this full 2D analysis will then be compared with results which were obtained assuming 1D transport in the landfill and clay. Finally, the determination of parameters and the implications for design will be discussed.

2 THEORY

2.1 General equations

A typical situation is shown in Fig. 1. Here the landfill has a clay cover and is separated from an underlying aquifer by a clay liner. This situation may be generalized by considering a deposit divided into a number of layers with node planes $z = z_0, z_1, \ldots z_n$ as shown in Fig. 2.

If it is assumed that each layer k ($z_j < z < z_k$ where $j = k-1$) is homogeneous and that the contaminant transport is governed by Fick's law, then for 2D (i.e. plane) conditions, the fluxes f_x, f_z in the x and z directions within layer k, are given by

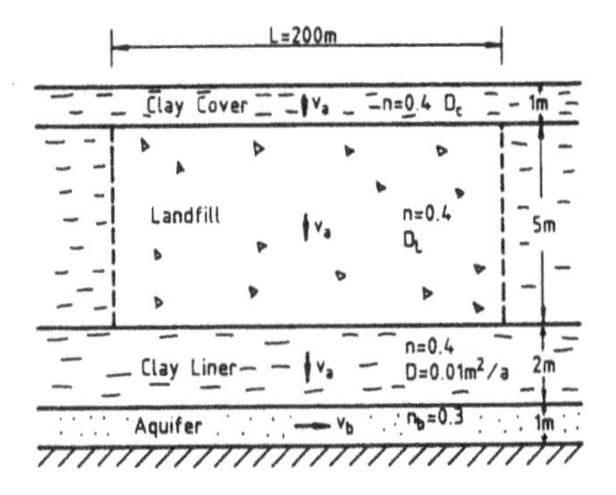

Figure 1. Typical cover-landfill-clay liner system considered.

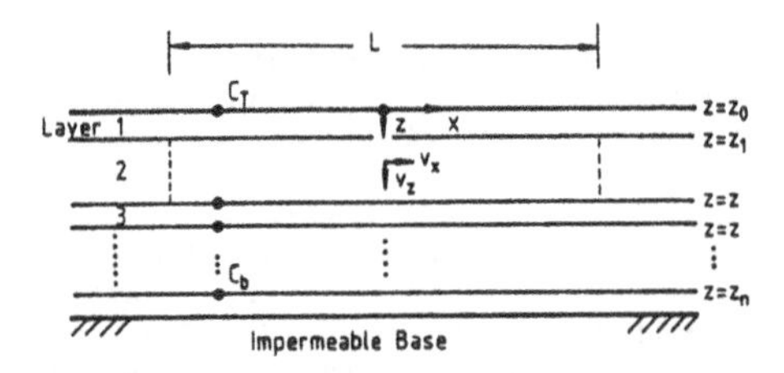

Figure 2. Generalized situation for which the theory is applicable. A region of surface flow, the clay cover, the landfill, the clay liner, and any other horizontally layered geological strata may each be treated as a separate layer.

$$f_x = n\, v_x\, c - n\, D_x \frac{\partial c}{\partial x} \tag{1a}$$

$$f_z = n\, v_z\, c - n\, D_z \frac{\partial c}{\partial z} \tag{1b}$$

in which c = the concentration of the contaminant of interest at some point (x,z) at time t; n = the porosity of the soil; v_x, v_z = the components of the seepage velocity; and D_x, D_z = the coefficients of hydrodynamic dispersion, referred to herein as the diffusion-dispersion coefficients.

For equilibrium controlled ion exchange where the concentration of the exchange ion is relatively low, the absorption of contaminant onto the soil may be approximated by a linear relationship and so consideration of mass balance gives:

$$\frac{1+\rho K}{n} \frac{\partial c}{\partial t} = D_x \frac{\partial^2 c}{\partial x^2} + D_z \frac{\partial^2 c}{\partial z^2} - v_x \frac{\partial c}{\partial x} - v_z \frac{\partial c}{\partial z} \tag{2a}$$

in which ρ = the bulk density of the soil solids, K is a distribution coefficient indicating the level of absorption of this contaminant onto the solids (determined experimentally: Rowe, Caers et al 1985, 1987); and the other quantities are as previously defined. Each of the quantities n, v_x, v_z, D_x, D_z, ρ, and K is assumed to be constant throughout the particular layer under consideration.

2.2 Initial conditions

Equation 2a governs two dimensional contaminant transport subject to the initial conditions

$$c = c_I \quad (z_j \leq z \leq z_k) \text{ at time } t = 0 \qquad (2b)$$

Two cases commonly occur. The concentration c_I may represent an existing background concentration, in which case c_I is uniform throughout the deposit. Alternatively, c_I may represent the concentration in the landfill, c_o, and in that case

$$\begin{aligned} c_I &= c_o \qquad |x| < L/2 \\ c_I &= 0 \qquad |x| > L/2 \end{aligned} \qquad (2c)$$

where L is the width of the landfill in the plane section being considered.

2.3 General Solution

Equations 1 and 2 can be simplified by the introduction of a Laplace transform:

$$(\bar{c},\bar{f}_x,\bar{f}_z) = \int_0^\infty (c,f_x,f_z) \exp(-st)dt \qquad (3)$$

and a Fourier transform

$$(C,F_x,F_z) = \frac{1}{2\pi} \int_{-\infty}^{\infty}(c,f_x,f_z)\exp(-i\xi x)dx \quad (4)$$

It is then possible to establish a relationship between the transformed concentrations $\bar{C}_j$, $\bar{C}_k$ and the transformed flux $\bar{F}_{zj}$, $\bar{F}_{zk}$ at the node planes $z=z_j$, $z=z_k$ having the form

$$\begin{bmatrix} \bar{F}_{zj} \\ -\bar{F}_{zk} \end{bmatrix} = \begin{bmatrix} Q_k & R_k \\ S_k & T_k \end{bmatrix} \begin{bmatrix} \bar{C}_j \\ \bar{C}_k \end{bmatrix} - \begin{bmatrix} U_k \\ V_k \end{bmatrix} \qquad (5)$$

where expressions for Q_k, R_k, S_k, T_k, U_k and V_k are derived in the appendix.

Observing that the transformed flux $\bar{F}_z$ and concentrations $\bar{C}$ must be continuous, the layer matrices defined by Eq. 5 may be assembled for each layer k in the deposit to give

$$\begin{bmatrix} Q_1 & R_1 & & & & \\ S_1 & T_1+Q_2 & R_2 & & & \\ & S_2 & T_2+Q_3 & R_3 & & \\ & & & \vdots & & \\ & & & S_{n-1} & T_{n-1}+Q_n & R_n \\ & & & & S_n & T_n \end{bmatrix} \begin{bmatrix} \bar{C}_T \\ \bar{C}_1 \\ \bar{C}_2 \\ \vdots \\ \bar{C}_{n-1} \\ \bar{C}_b \end{bmatrix} = \begin{bmatrix} \bar{F}_T+U_1 \\ V_1+U_2 \\ V_2+U_3 \\ \vdots \\ V_{n-1}+U_n \\ -\bar{F}_b+V_n \end{bmatrix} \qquad (6)$$

where $\bar{F}_T$ and $\bar{F}_b$ are the transformed fluxes at the top ($z=z_o$) and bottom ($z=z_n$) nodal planes respectively and $\bar{C}_T$ and $\bar{C}_b$ are the corresponding transforms of the concentrations at these points. It should be noted that this matrix will not usually be symmetric but it is tridiagonal and hence solution of these equations is computationally trivial.

2.4 Boundary conditions

Equation 6 must be solved subject to the appropriate boundary conditions. If there is vertical advective transport into the aquifer then, strictly speaking, continuity requires that the base velocity in the aquifer should vary with horizontal position. However if, as is often the case, the vertical velocity is small compared to the horizontal velocity v_b in the aquifer then, as a first approximation, the base velocity may be assumed to be uniform and horizontal. Thus the aquifer can be modelled as a layer of the deposit. If this aquifer layer is underlain by an impermeable boundary, then $F_b = 0$ in the last equation of (6) which becomes:

$$S_n \bar{C}_{n-1} + T_n \bar{C}_b = V_n \qquad (7)$$

Both the overlying clay cover and the landfill can be treated as a separate layer where the landfill has an initial concentration c_I defined by Eq. 2c. In this case the upper boundary (i.e. the surface) may be considered to be washed by surface runoff so that the surface concentration is always zero i.e.

$$C_T = 0 \tag{8}$$

and the first equation in Eq. 6 can be eliminated. Clearly, the surface flux F_T can be found and used to calculate the mass of contaminant which will escape to the surface due to upward diffusion. (The possibility that surface runoff is not sufficient to cause complete washing at the surface can also be modelled by introducing an additional layer above the cover layer and allowing horizontal surface flow through this layer. In this case the upper boundary condition may be specified by setting $F_T = 0$. One could then solve for the unknown surface concentration C_T.)

An alternative approach which was adopted in the authors' previous work (Rowe and Booker 1985b) is to model the landfill by a special boundary condition which allows for the variation in concentration with time in the landfill as contaminant is transported into the soil. The approach implicitly assumes that the concentration of contaminant escapes to the surface. Details of the formulation of this boundary condition are given by Rowe and Booker (1985b) and for the sake of brevity will not be repeated here.

2.5 Solution

The transformed concentrations at all nodal planes $z=z_j$, can be determined by solving Eq. 6 subject to the appropriate boundary conditions (Eq. 7,8). (The concentration at any intermediate depth z, $z_j<z<z_k$ can also be determined using the interpolation formulae given in the appendix.) The concentrations at any particular point and time of interest can then be obtained by numerically inverting the Fourier and Laplace transforms.

The theory described above for 2D conditions has been coded in program MIGRATE (Rowe, Booker et al 1985) and can be implemented on modern microcomputers. The major computational effort involved is associated with the numerical inversion of the Fourier and Laplace transforms for the locations and times of interest. The Fourier transform can be efficiently inverted using 20 point Gauss quadrature. The width and number of integration subintervals which are needed to achieve a reasonable accuracy (say 0.1%) depends somewhat on the geometry and properties of the problem under consideration. The parameters can be determined from a few trial calculations for a representative point and time of interest. This is essentially no different than the normal procedure adopted in finite element analyses where trial calculations are used to determine mesh and time step sensitivity. The Laplace transform can be inverted using Talbot's algorithm (Talbot 1979).

2.6 One Dimensional Contaminant Transport through the Landfill and Liner

The development of theory for one dimensional contaminant transport within the landfill and liner parallels that just described for 2D conditions except that in the 1D case the Fourier transform is omitted. Since the major portion of the computation involved in the 2D analysis is associated with the inversion of the Fourier transform, there is clearly computational advantage in using a 1D formulation where it is appropriate.

In a truly 1D situation, there can be no horizontal flow in a base aquifer. However, the situation where a clay liner which is thin relative to the dimensions of the landfill is underlain by a thin aquifer in which the base velocity is horizontal can be approximately modelled by assuming 1D advective-diffusive transport in the clay cover, the landfill and the clay liner while modelling the horizontal transport in the base aquifer by a boundary condition as previously described by Rowe and Booker (1985a).

3 APPLICATION OF THE THEORY

Major geotechnical uncertainties in the design or evaluation of landfills are the magnitude of the diffusion-dispersion coefficients, particularly in the landfill and clay cover, and the magnitude of the advective velocities v_a and v_b. The theory presented in the previous section provides a convenient means of performing sensitivity studies to indicate the potential effect of these uncertainties upon the mass of contaminant expected to escape through the clay cover and the expected concentrations of contaminant within an underlying aquifer.

To illustrate this application of the theory, consider the hyothetical case of a 5 m thick landfill with a 1 m thick

Table 1. Cases considered ($D=0.01 m^2/a$)

Case (1)	D_L (m^2/a) (2)	D_c (m^2/a) (3)	v_a m/a (4)	v_b m/a (5)	Analysis Type (6)
1	∞	0	0	1	2D
2	0.01	0.01	0	1	2D
3	∞	0	0.005	10	2D
4	0.01	0.002	0.005	10	2D
5	0.01	0.01	0.005	10	2D
6	∞	0	0	1	1D
7	0.01	0.01	0	1	1D
8	0.01	0.002	0	1	1D
9	∞	0	0.005	10	1D
10	0.01	0.01	0.005	10	1D
11	0.01	0.002	0.005	10	1D
12	0.01	0.01	0	*	1D
13	0.01	0.01	0.005	*	1D

* Infinitely deep clay liner.

compacted clay cover which is separated from a 1 m thick aquifer by a 2 m thick layer of clay. In cases such as this, it is often sufficiently accurate to consider migration of the contaminant within a plane section passing through the centre of the landfill and parallel to the direction of the superficial velocity v_b within the aquifer, as shown in Fig. 1. In the following examples, the diffusion-dispersion coefficient (i.e. coefficient of hydrodynamic dispersion) is denoted by D_c in the compacted clay cover, by D_L within the landfill itself, and by D in the saturated natural clay liner. A number of cases will be considered as summarized in Table 1.

3.1 Full 2D analysis

Initially suppose the vertical advective velocity is zero (i.e. vertical contaminant transport is by pure diffusion), that the horizontal advective velocity in the aquifer v_b = 1 m/a, and that the diffusion-dispersion coefficient D for a non-reactive contaminant of interest in the clay liner is 0.01 m^2/a. The effect of uncertainty regarding the diffusion-dispersion coefficient in the landfill and clay cover can then be illustrated by considering the extreme cases:

Case 1: where there is perfect mixing of contaminant within the landfill (i.e. the limit as D_L tends to infinity) and there is no contaminant escape through the cover (i.e. the limit as D_c approaches zero - this corresponds to a perfect cover); and

Case 2: where both diffusion-dispersion coefficients D_c and D_L are equal to that in the clay liner (viz $D_c = D_L = D = 0.01\ m^2/a$).

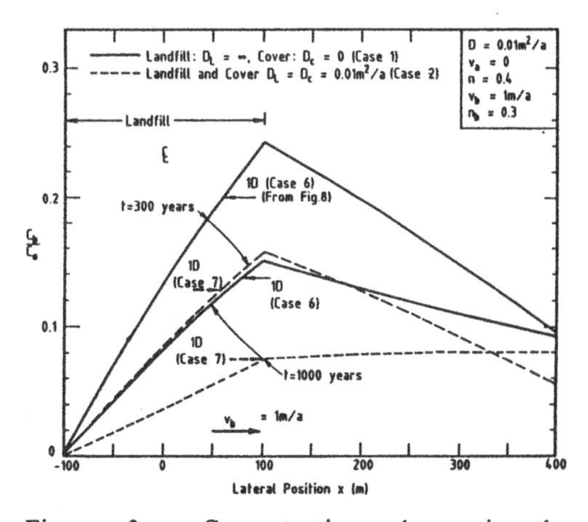

Figure 3. Concentration plume in the aquifer (Geometry as shown in Figure 1) Cases 1 and 2.

Fig. 3 shows the variation in concentration with lateral position in the base aquifer for cases 1 and 2 at two different times. In both cases, the concentration of contaminant increases as one moves from the upstream (in the sense of flow within the aquifer) to the downstream edge ofthe landfill. The maximum concentration beneath the landfill occurs at the downstream edge of the landfill.

In practical situations, the designer would normally be concerned with the concentrations of contaminant at a number of critical monitoring points in the aquifer; for example, beneath the downstream edge of the landfill and at the boundary of the property owned by the landfill operator. Fig. 4 shows the variation in concentration at two such points in the aquifer (x = 100 and 400 m) with time, for the two cases. At both positions, the concentration increases with time until a peak concentration is reached; for later times the concentration decreases. As might be expected, it takes longer to reach the peak concentration at points away from the landfill

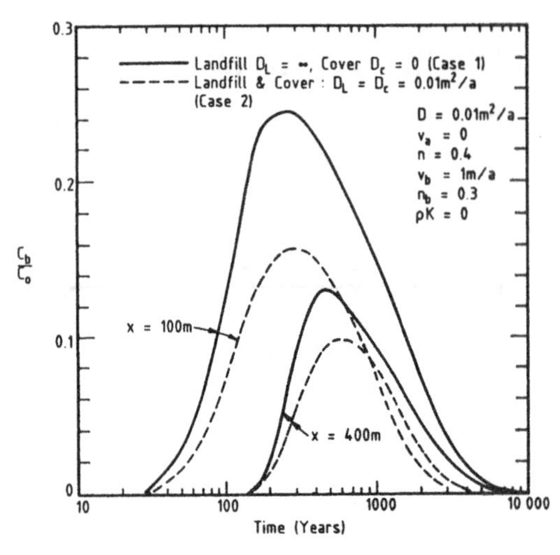

Figure 4. Variation in concentration with time at two points in the aquifer (problem as shown in Figure 1): Cases 1 and 2.

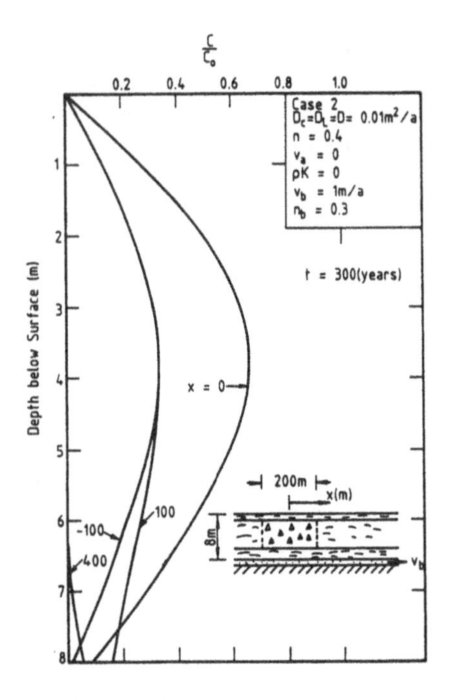

Figure 5. Variation in concentration with depth at 300 years: Case 2

(e.g. x = 400 m) than it does beneath the landfill. A far more interesting observation is that the peak concentrations reached at points outside the landfill are noticeably less than the peak concentrations reached in the aquifer beneath the landfill. This natural attenuation of peak concentration with distance away from the landfill is due to upward diffusion of contaminant from the aquifer into the adjacent clay as indicated by the concentration gradient at x = 400 m in Fig. 5.

Comparing the results for the different values of diffusion-dispersion coefficient in the landfill and cover, it is seen here that the extreme variation in these parameters caused less than a 60% variation in peak concentration at the edge of the landfill and less than a 30% variation in peak concentration at a point, x = 400 m, away from the landfill.

Of the two cases, the limiting case 1 which assumes a perfect cover ($D_c = 0$) and perfect mixing of contaminant within the landfill ($D_c = \infty$), gives the higher (i.e. more conservative) estimate of contaminant concentrations within the aquifer. This is primarily because the assumption of a perfect cover (i.e. $D_c = 0$) precludes contaminant escape to the surface and hence the entire mass of contaminant must escape through the clay liner into the aquifer.

The cover will generally not be perfect (i.e. $D_c > 0$) and there will be potential for upward contaminant transport through a clay cover to the surface by diffusion through the pore fluid of the cover material. Contaminant can then be transported away from the landfill by surface runoff. For example, Fig. 5 shows the distribution of concentration with depth at the upstream (x = -100 m), and downstream (x = 100 m) edges of the landfill, at the centre of the landfill (x = 0) and at a point 300 m downstream of the landfill (x = 400 m) 300 years after construction. At the upstream and downstream edges of the landfill, significant diffusion can occur laterally as well as vertically and hence the concentration in the landfill at these points is less than at the centre (x = 0). Of greater interest though, is the fact that there is significant upward diffusion through the clay cover to the surface as indicated by the concentration gradients above the mid depth of the landfill (z = 3.5 m). Under these circumstances not only are there potential environmental problems associated with contamination of the underlying aquifer but also with the contamination of surface runoff. This will be discussed more fully in later paragraphs.

Groundwater flows will vary considerably from one site to another. In case 2 just considered, it was assumed that there was no advection (seepage) through the cover, landfill and liner and it was also assumed that the diffusion-dispersion coefficient D_c in the compacted clay cover would be similar to that in the natural clay liner. These assumptions may be appropriate for some landfills; however in many cases there will be downward seepage through the clay cover, landfill and clay liner together with a higher horizontal velocity in the aquifer. Furthermore if, as is often the case, the compacted clay cover is only partially saturated, the effective diffusion-dispersion coefficient D_c in the cover will normally be less than that in the saturated natural clay liner.

To illustrate the possible effect of a lower diffusion-dispersion coefficient in the cover combined with higher advective velocities v_a and v_b, analyses were performed for $D_c = 0.002\ m^2/a$, a downward velocity $v_a = 0.005$ m/a and a horizontal velocity in the aquifer v_b = 10 m/a (case 4) and the results at x = 100 m and 400 m are shown as dashed lines in Fig. 6. Compared to case 2, the higher advective velocities v_a, v_b in case 4 decrease the time required to reach the peak concentration while also decreasing the magnitude of the peak because of higher dilution of contaminant associated with higher groundwater flows. Comparing Figs 4 and 6, it is seen that this effect is greater close to the landfill than at points outside the landfill. At x = 100 m the peak concentration for case 4 is 45% less than for case 2, however at x = 400 m there is only a 14% reduction in peak concentration for case 4 compared to case 2. This arises from the fact that at points away from the landfill, the benefits of dilution due to higher groundwater flows are offset by the fact that there is less time for natural attenuation to occur by diffusion into the surrounding clay. Thus for any given situation, there will be a critical base velocity v_b which gives rise to the maximum peak concentration at any specified point in the aquifer outside the landfill. Velocities either greater or less than this critical value will give rise to lower peak concentrations at the point in question. The important practical consequences of this observation are discussed more fully by Rowe and Booker (1985b).

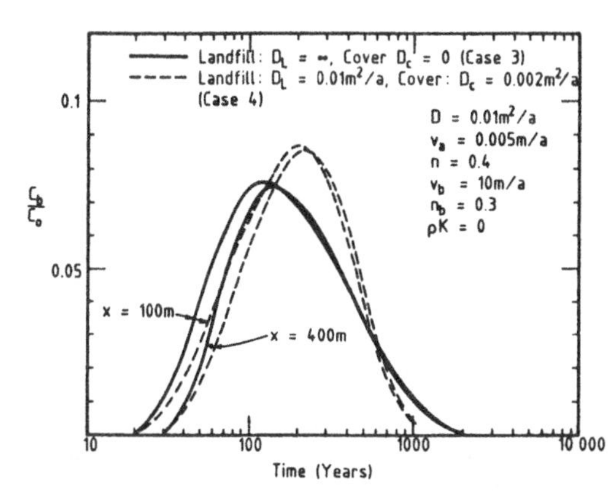

Figure 6. Variation in concentration with time at two points in the aquifer (problem as shown in Figure 1): Cases 3 and 4.

In contrast to the results obtained for case 2, it is found that with the downward advective velocity v_a = 0.005 m/a and $D_c = 0.002\ m^2/a$ in the cover for case 4, there is negligible contaminant escape to the surface. Thus a comparison of these results with those obtained for the extreme case (case 3) of a perfect cover (D_c = 0) and perfect mixing of contaminant in the landfill ($D_L = \infty$) primarily shows the effect of variability in the landfill diffusion-dispersion coefficient D_L. It is found that here, the extreme case (case 3, $D_L = \infty$) does not provide a conservative estimate of the peak concentration although the difference between the peak concentration obtained for cases 3 and 4 is small (less than 10%). The higher peak concentration found for case 4, compared to case 3, is a consequence of the delay of concentration movement through the landfill associated with the relatively small value of diffusion-dispersion coefficient within the landfill adopted in this case. From a practical standpoint, it is fortunate that the variability in peak concentration due to uncertainty regarding the actual value of D_L is not large in these cases. Considering the results for cases 1-4 it is also apparent that contaminant escape to the surface has a much greater effect upon peak concentration in the aquifer than does variability in the landfill diffusion-dispersion coefficient D_L.

The contamination of surface water which can occur due to upward diffusion in the landfill and clay cover will depend on the mass of contaminant reaching the surface together with quantity of surface runoff water. Fig. 7 shows the proportion of the total mass of contaminant within the landfill (at the end of construction) which will have escaped from the surface at a given time for the cases considered. For the situation where there is no downward advective velocity and the diffusion coefficient in the clay cover is similar to that in the clay liner (case 2), the greater part (i.e. approximately 57%) of the contaminant escapes from the surface of the landfill due to upward diffusion. In this case contamination of surface runoff may be as important, if not more important, a consideration as contamination of the underlying aquifer. This will be particularly so if the runoff enters a lake (or other form of impoundment) where the mass of contaminant may accumulate with time.

As previously noted, partial saturation of the clay cover would be expected to reduce the effective diffusion coefficient. In the absence of downward seepage, a five fold reduction in diffusion coefficient in the clay cover (compared to the value for a saturated cover) reduces the proportion of contaminant escaping to the surface to approximately 40% of that originally in the landfill (case 5 - Fig. 7). This is still a significant proportion.

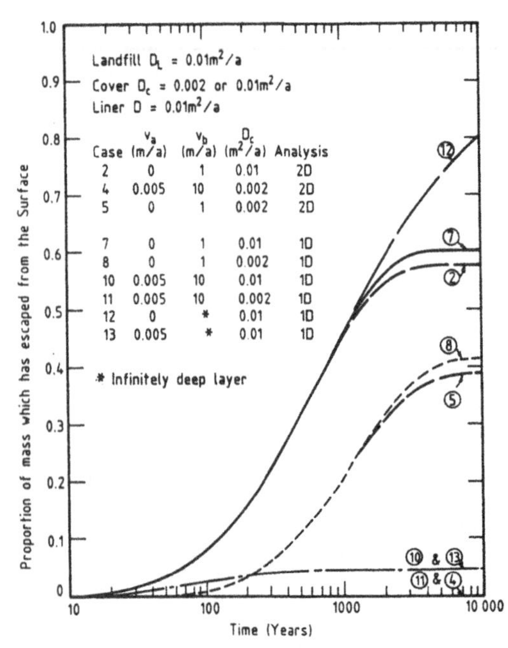

Figure 7. Proportion of the total initial mass of contaminant which has escaped through the clay cover by upward diffusion at a given time: Problem as shown in Figure 1: various cases.

The primary factor affecting contaminant escape to the surface is the advective velocity through the cover. The results shown for case 4 (see Fig. 7) indicated that even a relatively small downward advective velocity v_a = 0.005 m/a would reduce the mass of contaminant escaping to the surface to a negligible amount.

3.2 Modified 1D analyses

The 2D analysis provides a means of examining quite complex problems. Sensitivity studies performed using these techniques can provide insight regarding the impact of uncertainty regarding design parameters upon water quality. However, even though these studies can be performed using a microcomputer, the computation time required to perform a detailed sensitivity study is significant.

An examination of the geometry of the problem shown in Fig. 1 would suggest that for cases such as this, contaminant transport within the clay cover, the landfill and the clay liner would be predominantly vertical. This then raises the question as to what error might occur by adopting the simpler modified one-dimensional analysis, since the computation time involved in performing such analysis is typically less than 1% of that for a 2D analysis.

To allow the effects of horizontal flow within the aquifer to be modelled in a 1D approach, it is necessary to assume that the concentration within the aquifer directly beneath the landfill is spatially homogeneous and that mass transport out of the aquifer directly beneath the landfill is only due to the horizontal advective velocity. This is clearly an oversimplification of the actual situation which involves a variation in concentration with position as already shown in Fig. 3. Nevertheless, a comparison of the results obtained for cases 1 and 2 from the full 2D analysis (and given in Figs 3 and 4) with the corresponding values calculated using the modified 1D analysis (cases 6 and 7), shown by arrows on Fig. 3 and by the curves given on Fig. 8, suggests that

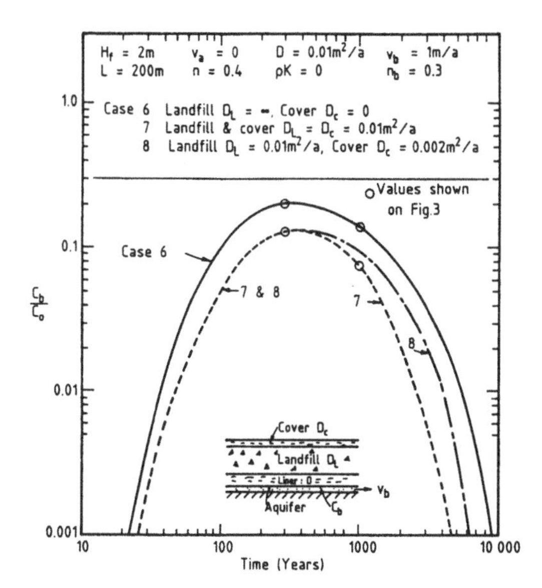

Figure 8. Variation in concentration in the aquifer beneath the landfill with time: Problem as shown in Figure 1 - calculations performed using modified 1D analysis: Cases 6, 7 and 8.

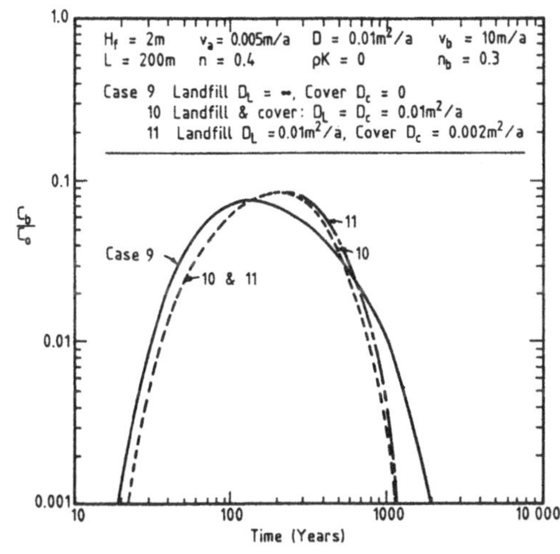

Figure 9. Variation in concentration in the aquifer beneath the landfill with time: Problem as shown in Figure 1 - calculations performed using modified 1D analysis: Cases 9, 10 and 11.

the 1D solution does provide a reasonable estimate of the maximum concentration beneath the landfill. A similar conclusion is reached by comparing the 2D and 1D results given in Figs 6 (cases 3 and 4) and 9 (cases 9 and 11) for v_a = 0.005 m/a, v_b = 10 m/a.

The mass which has escaped from the surface at a given time as calculated from a 1D analysis is shown in Fig. 7 for a number of cases. Comparing these results with the results from corresponding 2D analyses (i.e. compare cases 2 and 7, 4 and 11, 5 and 8), it is seen that they are nearly identical at small to moderate times, and even at very large times (10000 years) the difference between the results given by the two analyses is less than 5% of the total mass of contaminant.

For the problem considered here, the major difference between the 1D and 2D analysis is the more realistic modelling of the base aquifer in the 2D case. Thus the similarity of the results calculated from the 1D and 2D cases indicates that the base aquifer has only a modest effect on the mass escaping from the surface. This is further illustrated by comparing the results for cases 12 and 13, which assume an infinitely deep clay deposit (i.e. no aquifer), with the corresponding cases which consider the aquifer (cases 7 and 10).

These results would suggest that for problems such as that examined here, the modified 1D analysis could be used to perform an initial sensitivity study to identify critical combinations of parameters. The more sophisticated 2D analysis (or its 3D extension) could then be used to obtain more precise results for critical cases, if necessary.

4 DISCUSSION

For landfills constructed with a relatively intact clay liner such that the migration of contaminant will be diffusion controlled, the time required to reach the peak concentration of contaminant within any nearby aquifer may range from hundreds to thousands of years. Clearly in such a situation, the results of calculations performed using parameters based on relatively recent experience should be viewed with caution.

In order that analyses should be meaningful, it is important to be able to estimate values of material parameters. The values of porosity and advective velocity can be bounded with some confidence. Recently techniques have also been developed for determining the diffusion-dispersion coefficient and distribution coefficient of contaminant in undisturbed samples of clay liners (Rowe, Caers et al 1985; Rowe et al 1987). For example, diffusion-dispersion coefficients deduced using these techniques for the "liner" material beneath the Sarnia landfill are consistent with values deduced by backfiguring the diffusion-dispersion coefficients from the observed migration beneath the actual landfill (Rowe et al 1987). Furthermore, these diffusion-dispersion coefficients appear to be reasonably consistent with values deduced by Desaulniers et al (1981) from a study which included examination of diffusion profiles established over a period of more than 10000 years.

The diffusion-dispersion coefficient in a partially saturated clay liner or cover can be estimated by performing a column test on a representative sample. It may be expected that the diffusion coefficient will be a minimum at low levels of saturation and so reasonable bounds can be placed on the likely value by performing tests on the partially and fully saturated samples. Thus although the theory described in the previous section was developed for saturated soils, it could also be used to provide engineering estimates of concentrations for partially saturated systems by performing a sensitivity study using bounding parameters.

The greatest uncertainty regarding parameters is for the landfill itself. Although the actual porosity of landfills may be relatively high, the effective porosity to be used in contaminant analyses may be expected to be somewhat smaller due to the presence of isolated voids which do not provide a path for contaminant transport. The presence of relatively inert and impermeable materials within a landfill may also act to reduce the effective diffusion coefficient within the landfill by increasing the tortuosity. On the other hand, large pores will readily allow diffusion through the leachate and so in some cases the effective diffusion coefficient in the landfill could be higher than in the clay liner. Any advection through the landfill will also result in some dispersion which can further increase the diffusion-dispersion coefficient.

There are serious difficulties associated with even placing close bounds on the expected values of the diffusion-dispersion coefficient D_L within the landfill, and this raises the important practical question as to what effect this uncertainty will have upon the expected concentrations of contaminant within an aquifer beneath the clay liner. The analyses presented in the previous section show that for situations similar to that considered, the peak concentration in the aquifer varied by less than a factor of two for a variation in the diffusion-dispersion coefficient of many orders of magnitude. In these cases, the adoption of even extreme parameters $D_C = 0$ and $D_L = \infty$ for the clay cover and the landfill provided either good (but slightly unconservative) or reasonable (and conservative) estimates of the peak concentrations at points in the aquifer. Clearly, these findings should not be over-extrapolated and in practice, it is desirable that sensitivity analyses similar to those reported here be performed for each particular situation.

The analysis described in this paper provides a simple means of estimating the concentrations of contaminant that are likely to occur in the groundwater due to the construction of a landfill. It is considered that the properties of clay liners should be selected to ensure that with our present state of knowledge the peak concentration of contaminant within aquifers beneath or near landfills will not exceed allowable levels. In situations such as those discussed in this paper, failure to consider diffusion as a contaminant transport mechanism could result in designs which will cause unacceptable levels of contamination within the groundwater for future generations. Similarly, consideration should also be given to the potential impact of diffusion through a clay cover upon the contamination of surface waters.

5 CONCLUSION

A simple, semi-analytic technique for the calculation of contaminant migration through a clay cover, landfill, clay liner and an underlying aquifer has been described. This technique is applicable for soils which can be idealized as being horizontally layered, and the calculations can be performed on a micro-computer.

The application of this theory has been illustrated with reference to a number of typical situations and it has been shown that:

(1) The diffusion coefficient in the clay cover and landfill itself have only a modest effect on the peak concentration of contaminant within an underlying aquifer (provided that the effective diffusion coefficient in the landfill is not less than that in the underlying soil);

(2) Analyses performed using a modified 1D analysis may provide a reasonable indication of the concentrations to be expected in a thin aquifer separated from a landfill by a relatively thin clay liner;

(3) In the absence of downward advective transport through a clay cover and landfill, a significant proportion of the contaminant in the landfill may escape to the surface by a process of molecular diffusion through the pore fluid of the clay cover. For the problem considered, even a relatively small (0.005 m/a) downward advective velocity effectively eliminated contaminant migration to the surface; and

(4) The process of molecular diffusion may result in significant (and possibly unacceptable) levels of contamination in aquifers which are separated from a landfill be a clay liner. It is considered that in any such design, calculations similar to those described in this paper should be performed as a check on the possible implications of advective-diffusive transport.

6 ACKNOWLEDGEMENT

The work described in this paper represents part of a general programme of research into pollutant migration through soil supported by grants No. A1007 and G0921 from the National Sciences and Engineering Research Council of Canada. Additional funding, in the form of sabbatical leave support for Dr. Rowe, was provided by the University of Sydney. The authors also wish to acknowledge the value of many discussions with Dr. R.M. Quigley.

REFERENCES

Anderson, D.C. 1982. Does landfill leachate make clay liners more permeable? Civ. Engrg. Sep.:66-69.

Anderson, M.P. 1979. Using models to simulate the movement of contaminants through groundwater flow systems. CRC Critical Reviews in Environmental Control. 9(2):97-156.

Brown, K.W. 1983. The influence of selected organic liquids on the permeability of clay liners. U.S. EPA 600/9-83-018, 114-125.

Crooks, V.E. & R.M. Quigley 1984. Saline leachate migration through clay: a comparative laboratory and field investigation. Can.Geotech.J. 21(2):349-362.

Daniel, D.E. 1984. Predicting hydraulic conductivity of clay liners. J.Geotech. Engrg. 110(2):285-300.

Daniel, D.E. 1985. Can clay liners work? Civ. Engrg. Apr.:48-49.

Day, R. 1984. A field permeability test for compacted clay liners. M.S. Thesis, Univ. Texas, Austin. p. 105.

Desaulniers, D.D., J.A. Cherry & P. Fritz 1981. Origin, age and movement of pore water in argillaceous quaternary deposits at four sites in Southwestern Ontario. J.Hydrol. 50:231-257.

Fernandez, F. & R.M. Quigley 1984. Hydraulic conductivity of natural clays permeated with simple liquid hydrocarbon. Res. Rep. GEOT-10-84, Fac. Engrg Science, Univ. W. Ontario. p.37.

Goodall, D.E. & R.M. Quigley 1977. Pollutant migration from two sanitary landfill sites near Sarnia, Ontario. Can. Geotech. J. 14:223-236.

Quigley, R.M. & R.K. Rowe 1985. Leachate migration through clay below a domestic waste landfill, Sarnia, Ontario, Canada, chemical interpretation and modelling philosophies. Proc. Int. Symposium on Industrial and Hazardous Waste, Alexandria, see also Hazardous and Industrial Waste Testing and Disposal, 6th Volume, ASTM, STP 933, pp.93-106, 1986.

Rowe, R.K. & J.R. Booker 1985a. 1-D pollutant migration in soils of finite depth. J.Geotech. Engrg, ASCE, 111(GT4):479-499.

Rowe, R.K. & J.R. Booker 1985b. 2D pollutant migration in soils of finite depth. Can. Geotech. J. 4:429-436.

Rowe, R.K. & J.R. Booker 1986. A finite layer technique for calculating 3D pollutant migration in soil. Geotechnique. 36(2):205-214.

Rowe, R.K., J.R. Booker & C.J. Caers 1985. Migrate-2-D pollutant migration through a non-homogeneous soil: user manual. Available through SACDA, Fac. Engrg Science, Univ. W. Ontario.

Rowe, R.K., C.J. Caers & F. Barone 1987. Use of a laboratory column test to determine the diffusion and distribution coefficients of contaminant through undisturbed soil. (Accepted for publication, Canadian Geotechnical Journal, Feb. 1988.)

Rowe, R.K., C.J. Caers, J.R. Booker & V.E. Crooks 1985. Pollutant migration through clayey soils. Proc. XI Int. Conf. Soil Mechanics and Foundation Engrg., San Francisco. 3:1293-1298.

Talbot, A. 1979. The accurate numerical integration of Laplace transforms. J. Inst. Maths. Applics. 23:97-120.

APPENDIX I

Application of repeated Laplace and Fourier transforms (Equations 3 and 4) to the governing equations leads to the set of ordinary differential equations

$$\bar{F}_x = n\, v_x\, \bar{C} - n\, D_x\, i\xi\bar{C} \quad (9a)$$

$$\bar{F}_z = n\, v_z\, \bar{C} - n\, D_z \frac{\partial \bar{C}}{\partial z} \quad (9b)$$

$$(n+\rho K)(s\bar{C}-\bar{C}_I) = -\xi^2 nD_x\,\bar{C} + nD_z \frac{\partial^2 \bar{C}}{\partial z^2} - i\,\xi n v_x - n v_z \frac{\partial \bar{C}}{\partial z} \quad (10)$$

This equation has the solution

$$\bar{C} = E + A\exp(\alpha z) + B\exp(\beta z) \quad (11a)$$

in which $m = \alpha, \beta$ are the roots of the equation

$$nD_z m^2 - nv_z m - [i\xi n v_x + \xi^2 nD_x + (n+\rho K)s] = 0 \quad (11b)$$

and

$$E = \frac{(n+\rho K)\, C_I}{\xi^2 nD_x + i\xi n v_x + (n+\rho K)s} = -\frac{(n+\rho K)\, C_I}{nD_z\ \alpha\beta} \quad (11c)$$

Evaluating Eq. 11 in terms of the concentration of the nodal planes z_j, z_k yields the interpolation formulae

$$\bar{C} = (\bar{C}_j - E)\left\{\frac{\exp[\alpha(z-z_k)]-\exp[\beta(z-z_k)]}{\exp[\alpha(z_j-z_k)]-\exp[\beta(z_j-z_k)]}\right\} + (\bar{C}_k - E)\left\{\frac{\exp[\alpha(z-z_j)]-\exp[\beta(z-z_j)]}{\exp[\alpha(z_k-z_j)]-\exp[\beta(z_k-z_j)]}\right\} + E \quad (12)$$

and thus

$$\frac{F_z}{nD_z} = (\bar{C}_j - E)\left\{\frac{\beta\exp[\alpha(z-z_k)]-\alpha\exp[\beta(z-z_k)]}{\exp[\alpha(z_j-z_k)]-\exp[\beta(z_j-z_k)]}\right\} + (\bar{C}_k - E)\left\{\frac{\beta\exp[\alpha(z-z_j)]-\alpha\exp[\beta(z-z_j)]}{\exp[\alpha(z_k-z_j)]-\exp[\beta(z_k-z_j)]}\right\} + (\alpha+\beta)E \quad (13)$$

If we denote the thickness of the sublayer by $\lambda = z_k - z_j$, this then gives the layer matrix relating nodal concentrations and nodal fluxes as given in Eq. 5 viz:

$$\begin{bmatrix} \bar{F}_{zj} \\ -\bar{F}_{zk} \end{bmatrix} = \begin{bmatrix} Q_k & R_k \\ S_k & T_k \end{bmatrix} \begin{bmatrix} \bar{C}_j \\ \bar{C}_k \end{bmatrix} - \begin{bmatrix} U_k \\ V_k \end{bmatrix}$$

in which

$$\frac{Q_k}{nD_z} = \frac{\beta\exp(-\alpha\lambda)-\alpha\exp(-\beta\lambda)}{\exp(-\alpha\lambda)-\exp(-\beta\lambda)}$$

$$\frac{R_k}{nD_z} = \frac{(\beta-\alpha)}{\exp(\alpha\lambda)-\exp(\beta\lambda)}$$

$$\frac{S_k}{nD_z} = \frac{-(\beta-\alpha)}{\exp(-\alpha\lambda)-\exp(-\beta\lambda)}$$

$$\frac{T_k}{nD_z} = \frac{-[\beta\exp(\alpha\lambda)-\alpha\exp(\beta\lambda)]}{\exp(\alpha\lambda)-\exp(\beta\lambda)}$$

and

$$\frac{U_k}{E_n D_z} = \frac{Q_k}{nD_z} + \frac{R_k}{nD_z} - (\alpha+\beta)$$

$$\frac{V_k}{E_n D_z} = \frac{S_k}{nD_z} + \frac{T_k}{nD_z} + (\alpha+\beta)$$

Computer and Physical Modelling in Geotechnical Engineering, Balasubramaniam et al. (eds)
© 1989 Balkema, Rotterdam. ISBN 90 6191 864 2

Soil-structure interaction – A rational design-orientated approach

L.A.Wood
Department of Civil & Structural Engineering, South Bank Polytechnic, London, UK
W.J.Larnach
Department of Civil Engineering, University of Bristol, Bristol, UK

ABSTRACT: The importance of accounting for real structural behaviour by using a combined structure/soil analysis is emphasised and an economic analysis method using simplified soil and structure models is presented. The method is validated by comparison with more exact methods. It is then used to demonstrate its effectiveness in dealing with real situations by making predictions for comparison with the results from several case histories. From these it becomes clear that careful account must be taken of construction sequence as well as making well-considered choices of soil and structural parameters.

1 INTRODUCTION

An unyielding foundation (the simple hypothesis used in much structural analysis) does not exist. Similarly completely flexible or unconnected foundations (the simple hypothesis used to obtain estimates of settlement) are very rare in practice. The actual behaviour of a foundation (and to a lesser degree the superstructure) is dependent upon the combined action of it and the supporting soil continuum. The effects of this interaction between the structure and the soil are characterised by the relative stiffness of these two components. This may be illustrated from consideration of the same structural foundation resting on two entirely different soils. First, a very stiff soil, in which case the foundation will tend towards the completely flexible situation; and second, a very soft soil, giving rise to a much more rigid mode of deformation. It must be realised,therefore,that it is not sufficient to talk in terms of the rigidity of a structure without regard to the underlying soil conditions.

In so far as the structure itself is concerned it is the differential movements (between adjacent columns, say) that give rise to induced stress resultants and to any structural damage; whilst the total movement is of importance in relation to the surroundings.

In order to facilitate an interactive analysis several idealisations have to be made. These are summarised below.

a) The soil geometry must be idealised on the basis of very limited data and judgements have to be made concerning the continuity and thickness of the various strata.

b) The determination of the soil properties (stiffness, shear strength, permeability, etc.) presents a formidable task. This is especially so when consideration is given to local variations and differences in the vertical and horizontal planes.

c) The resultant foundation loads are usually reasonably well defined although often their distribution is less so. The greatest uncertainty is the order in which the loads are applied, determined by the method of excavation or sequence of construction.

d) The finished structural geometry is usually accurately specified, although that at any time during construction is not.

e) The properties of the structural materials are probably more precisely defined than that of the soil, but are nevertheless variable due to the effects of creep etc. It is extremely difficult to assess the overall stiffness of a structure, although in many instances this may exert more control over the deformations than the stiffness of the foundation members themselves.

The engineer is faced with a situation where no matter how sophisticated the analytical techniques at his disposal the most he can hope to achieve is the imposition of bounds on the overall behaviour, together with an assessment of the effects of various construction features.

Thus in the context of normal foundation design for building structures the available information is rarely sufficient to warrant the use of sophisticated non-linear stress-strain laws in modelling the action of the soil following the application of external loads. Furthermore, in general, the often used reduction of the problem to a two-dimensional axi-symmetric, plane stress or plane strain model, in order to facilitate economic analysis, is inappropriate. Finite element programs which will handle the modelling of three-dimensional continua exist, but for semi-infinite domains, such as that required when representing soils, the computation involved may become excessive and the introduction of non-linear stress-strain laws may present problems in the numerical stability of the solution algorithms. Yet the structural engineer requires a realistic rationale to base his design. The method presented here provides just such an approach which remains compatible with the lack of precision of the field data, takes full account of the interaction between the soil and the structure, and is economic in operation.

The basis of this approach is by no means new and is in essence similar to that proposed by Chamecki (1956) and utilised with respect to a homogeneous, semi-infinite continuum by Cheung and Zienkiewicz (1965). However, the technique has been developed to include transversely isotropic, elastic, layered continua, the effects of local shear failure of the soil and the time dependent deformation characteristics of the consolidation of clay soils (Wood and Larnach, 1975 a, b, Wood, 1977, 1978a). Loss of contact between foundation and soil, uplift due to stress relief and the effect of new construction upon neighbouring structures are also catered for within the method (Wood, 1978a). The analysis yields a complete solution including soil settlements, structural displacements, interfacial contact pressures and stress resultants within the structure (Wood, 1978b).

The method of analysis is described below and is then validated with respect to other numerical and analytical techniques; and, finally, with observed independent field measurements.

2. THE INTERACTIVE ANALYSIS

2.1 The soil model

For the majority of foundations the soil strains that develop in the field are in general small and the assumption of linear elastic behaviour is valid. This assumption may be used to exploit the potential offered by boundary element (boundary integral) methods in providing an economic means of modelling the soil continua in three dimensions.

A simple soil boundary element has been developed to allow the engineer to take full account of spatial variations in elastic moduli and is based upon the assumption that the stress distribution within the inhomogeneous soil strata, due to the initially unknown interfacial contact forces, is identical to that occurring within a homogeneous, semi-infinte continuum. The acceptability of this assumption has been demonstrated (Wood, 1977) with respect to more rigorous solutions for a wide variety of inhomogeneous situations and elastic layers of finite thickness. Indeed the use of more sophisticated soil models with non-linear stress-strain relationships, may not be justified in the context of the normal design situation, and may lead to a false sense of accuracy in the predicted performance of a structure. Geotechnics is one area where the use of sensitivity analysis is of prime importance in determining confidence limits and it is considered that realistic and economic, methods of analysis, such as the boundary element approach described here, are best suited to fulfill this need.

The particular element adopted enables the soil to be represented as a transversely isotropic continuum; that is, a continuum within which the stiffness in the vertical direction differs from that in the horizontal plane. Thus the elastic properties of the continuum are defined by five independent constants:

E_v, E_h, ν_{vh}, ν_{vh}, G_{vh}

where E and ν are Young's modulus and Poisson's ratio respectively, G is the shear modulus and the suffixes v and h refer to the vertical direction and horizontal plane respectively. Although these five constants are in general independent strain energy requirements place limits upon the acceptable combinations (Wood, 1978a). For example, if the continuum is considered to deform at constant volume (such as in the immediate, undrained behaviour of a saturated clay) the number of independent constants is reduced to three. Namely,

E_v, E_h and G_{vh};

with $\nu_{vh} = 0.5$ and $\nu_{hh} = 1 - E_h/2E_v$.

Over-soncolidated soils in particular, are known to exhibit transversely isotropic characteristics due to the nature of their deposition and subsequent weathering. The effect of such anisotropy is to reduce the total and differential settlements occurring within the loaded area.

2.2 Extension of the basic soil model

The use of a purely elastic analysis with regard to the design of raft foundations is somewhat unrealistic due to the concentration of large contact stresses that may occur adjacent to the edges of a stiff raft. Although the adoption of a nonlinear soil model (Wood, 1981) will produce a more realistic assessment of the raft behaviour, the increase in computational effort with respect to a three-dimensional analysis is rarely justified on economic grounds. Hence, the inclusion within the model of an iterative procedure, similar in operation to that for the elimination of tensile contact forces, which allows the imposition of an upper limit on the magnitude of the developed contact forces in order to incorporate and assess the effect of local shear failure of the soil. This approach is considered compatible with the likely level of soils knowledge and fits neatly into the design office process in that the overall bearing capacity of the soil must be assessed prior to more detailed analysis of the chosen foundation type. Hence, the assessment of the limiting bearing pressures will have been largely undertaken already, as will the initial estimate of the soil stiffness from preliminary settlement calculations.

2.3 The structural model

The finite element method lends itself to the formulation for the structural model. Again, the assumption of elastic behaviour for the structural materials is felt to be sufficient in the approach adopted. In principle the complexity of the structural model need not be limited but in practice the quantification of all of the stiffening elements is unlikely to be possible. Hence, it is the main structural elements at foundation level that will be employed in the interactive analysis. However, in order to obtain realistic estimates of differential movements and therefore of the induced stress resultants in the foundation members, some consideration should be given to the additional stiffening effect of the superstructure components.

In the examples quoted the structural model has been limited to assemblages of beam and/or thin plate bending finite elements.

2.4 The combined numerical model

Foundations do not normally lend themselves to two-dimensional plane strain analysis, unless they can be reasonably modelled as strips and even then such an analysis gives very little information about the behaviour in the longitudinal direction. The reduction of a rectangular raft to an equivalent circular form in order to facilitate axi-symmetric finite element analysis has been found in some instances to produce reasonable estimates of total and maximum differential settlements. However, such an idealisation of the structural geometry also requires corresponding idealisation of the loading pattern and is unlikely to provide results of direct value to the structural engineer in the design of the foundation.

The use of a three-dimensional finite element model is rarely justified on economic grounds and in any case requires the solution of a very large number of equations the majority of which will of necessity, in order to achieve remote boundaries, apply to a large area outside the foundation itself.

A much more economical approach is the use of a boundary element model of the soil continuum coupled to a finite element representation of the structure. Several investigators have used this method with encouraging results (Fraser and Wardle 1973, Wood 1979b).

The simplest idealisation of the soil continuum is that of a homogeneous, half-space, facilitating the direct use of the Boussinesq equations in order to form a stiffness matrix for the soil. However, this is of rather limited practical application. A more realistic idealisation, which has been found to produce very acceptable results, follows from the simple assumption that the stress distribution within the soil deposit due to the initially unknown interfacial contact forces is identical to that within a homogeneous halfspace. Thus the stresses are obtained from the classical equations of elasticity and the surface displacements from the numerical integration of the vertical strains taking full account of the inhomogeneous nature of the layered deposit.

2.5 Summary of the method of analysis

From consideration of the foundation, the net force acting is $\underline{Q} - \underline{P}$, where $\underline{P}$ is the vector of soil reactions and $\underline{Q}$ is the applied force vector. Hence

$$\underline{Q} - \underline{P} = \left[K_F\right] \cdot \omega_F$$

where ω_F is the displacement vector and K_F the foundation (structural) stiffness matrix. Note that ω_F is the full displacement vector of rotations and vertical displacements and similarly $\underline{Q}$ is the full applied force vector of moments and vertical forces.

On the assumption that the interface between the foundation and the soil is smooth, compatibility of vertical displacement only is maintained at the interface. (In finite element analysis full adhesion is assumed, but the difference in results obtained using these assumptions is small and the smooth condition is probably more realistic in that the tractions generated for full adhesion could not be sustained by a real soil).

Thus, the equations must be condensed. This is easily achieved if they are rewritten in the form

$$\begin{vmatrix}\underline{V}\\ \underline{M}\end{vmatrix} - \begin{vmatrix}\underline{P}\\ \underline{O}\end{vmatrix} = \begin{bmatrix}K_{11} & K_{12}\\ K_{21} & K_{22}\end{bmatrix} \begin{vmatrix}\underline{U}\\ \underline{\Theta}\end{vmatrix}$$

where M is the vector of applied moments and $\underline{\Theta}$ the vector of rotations. Hence,

$$\underline{V} - \underline{P} = K_{11}\,\underline{U} + K_{12}\,\underline{\Theta}$$

and

$$\underline{M} - \underline{O} = K_{21}\,\underline{U} + K_{22}\,\underline{\Theta}.$$

Therefore

$$\underline{\Theta} = \left[K_{22}^{-1}\,\underline{M} - K_{21}\,\underline{U}\right]. \tag{1}$$

On substitution,

$$\underline{V} - \underline{P} = K_{11}\,\underline{U} + K_{12}K_{22}^{-1}\underline{M} - K_{12}K_{22}^{-1}K_{21}\underline{U},$$

giving

$$\left[\underline{V} - K_{12}K_{22}^{-1}\underline{M}\right] - \underline{P} = \left[K_{11} - K_{12}K_{22}^{-1}K_{12}^{T}\right]\underline{U} \tag{2}$$

since $K_{21} = K_{12}^{T}$ from symmetry.

Equation (2) may be simplified as

$$\underline{Q}' - \underline{P} = K_F'\underline{U} \tag{3}$$

where $\underline{Q}'$ and K_F' are the condensed applied force vector and structural stiffness matrix respectively.

From consideration of the soil continuum, this approach leads to a direct relationship between the soil settlement (vertical displacements), $\underline{U}_s$ and the contact forces,

$$\underline{U}_s = f.\underline{P},$$

where f is a flexibility matrix for the soil, formed as a boundary element. Therefore,

$$\underline{P} = K_s\,\underline{U}_s \tag{4}$$

where $K_s = f^{-1}$ and is the stiffness matrix for the soil.

Substituting for $\underline{P}$ from equation (4) in equation (3) gives

$$\underline{Q}' - K_s\,\underline{U}_s = K_F'\,\underline{U}.$$

Noting that for compatibility

$$\underline{U}_s = \underline{U} \quad \text{then}$$

$$\underline{Q}' = \left[K_F' + K_s\right]\underline{U}, \text{ and hence}$$

$$\underline{U} = \left[K_F' + K_s\right]^{-1}\underline{Q}'.$$

Once $\underline{U}$ has been solved for, $\underline{P}$ may be obtained from equation (4) and $\underline{\Theta}$ from equation (1).

Having obtained $\underline{U}$ and $\underline{\Theta}$ to form ω_F the stress resultants within each element may be determined.

Thus, a complete solution of displacements, contact forces (pressures) and stress resultants is determined from a single analysis.

3 VALIDATION

Several validations have been undertaken and reported in the literature (Wood 1978, a, b; 1981). Other researchers have concentrated their attention upon "bare" raft foundations in which the governing relative stiffness, K, between the soil and structure may be readily quantified.

3.1 Circular raft

For a circular raft of thickness t and radius, a

$$K = \frac{E_R(1 - \nu_s^2)}{E_s(1 - \nu_R^2)}\left(\frac{t}{a}\right)^3$$

where E_R and E_s denote Young's modulus of the raft and soil respectively, and ν_R and ν_s are the respective Poisson's ratios.

For a raft carrying a uniform load/unit area of q and resting on a homogeneous half space (represented by a total soil layer thickness h equal to 10a) close agreement has been obtained for maximum differential settlement and bending moments (in the range K = 0.0 (flexible) to K = 1000.0 (rigid)) computed by the method described and others.

a) Raft resting on a homogeneous layer of finite depth

For poisson's ratio of the soil equal to 0.5 and a relative stiffness K = 0.1 the variation of raft bending moment with the thickness of the soil layer is shown in Fig. 1. Results obtained by Brown (1969)

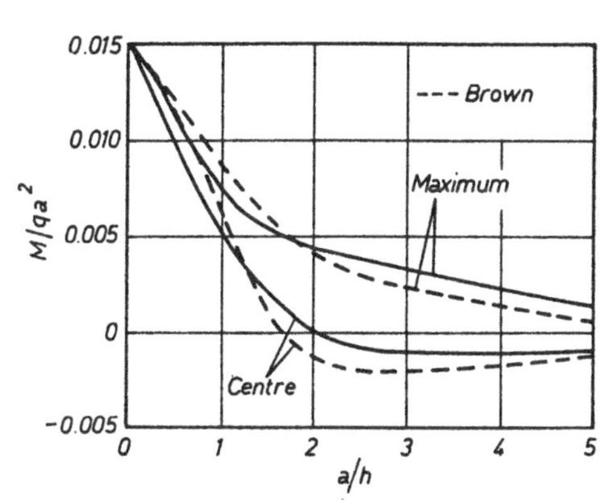

Fig. 1 Bending moments, K = 0.1, ν_s = 0.5, ν_R = 0.3. Homogeneous layer

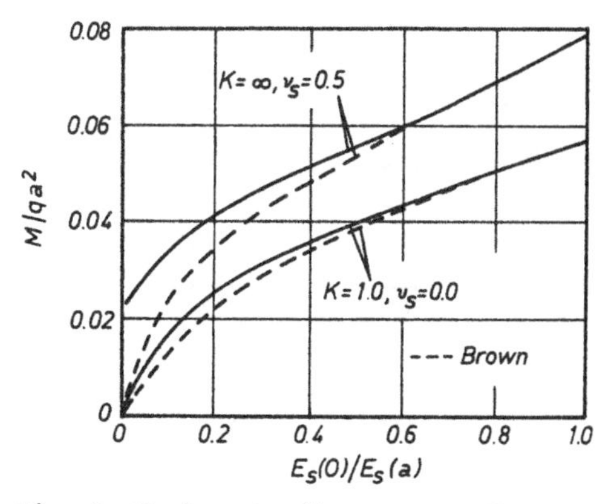

Fig. 2 Maximum bending moment, M. Heterogeneous half-space ν_R = 0.3

are shown for comparison and the agreement is satisfactory. It should be noted that as a/h increases above 0.5 the position at which the maximum moment occurs moves away from the centre, and the central moment experiences a reversal of sign.

b) Raft resting on a heterogeneous half-space

Most soil deposits exhibit a degree of heterogeneity and the validity of the numerical model in this case is of much more interest to the practising Engineer. Results obtained are compared in Fig. 2 with those reported by Brown (1974) for a half-space within which Young's modulus increases linearly with depth. The degree of heterogeneity of the soil is expressed as the ratio of the Young's modulus, $E_s(0)$ at the surface and that at a depth equal to the raft radius $E_s(a)$. The relative raft-soil stiffness is defined, as for the homogeneous case, but in terms of the soil modulus at the surface $E_s(0)$.

The variation of maximum moment with degree of heterogeneity is shown in Fig. 2. Again for K = 1.0 and ν_s = 0.0 agreement with results obtained by Brown is very close. However for the rigid raft and ν_s = 0.5 the maximum moment appears to be overestimated as the degree of heterogeneity increases.

3.2 Rectangular raft

In order to establish the accuracy of the solution the simple example of a rigid rectangular raft of length to width ratio

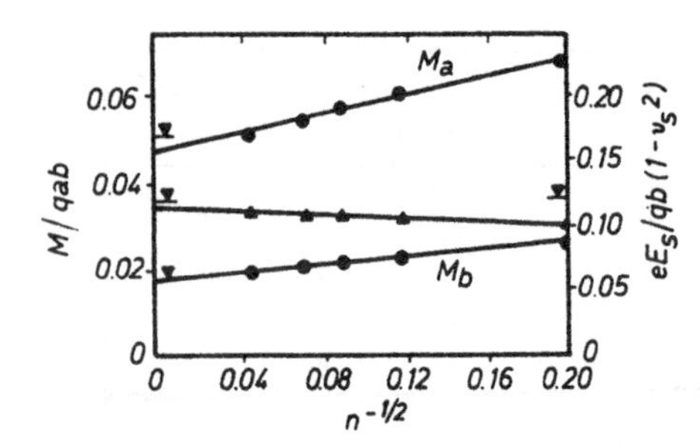

Fig. 3 Effect of element discretization on maximum moment and settlement of rigid raft. a/b = 2, d/b = 2. Fraser and Wardle values at left.

a/b = 2 resting on a 10m thick homogeneous soil mass has been considered. The results obtained for a uniform load, q/unit area are shown in Fig. 3 with respect to different element discretizations of the raft, where n is the total number of square plate bending finite elements. The extrapolated results are in reasonable agreement with those obtained by Fraser and Wardle (1976).

Further confirmation of the achievable precision of the numerical procedure is given in Tables 1 and 2 where results obtained for the same raft resting on a thin soil deposit with a thickness to width ratio d/b = 0.5 are compared again with those of Fraser and Wardle (1976).

Table 1 Raft settlements $E_s/qb(1-\nu_s^2)$, a/b = 2, d/b = 0.5

	Computed	Exact*	% Error	ν_s
A	0.37	0.37	-	0.0
	0.31	0.33	- 6	0.3
	0.20	0.18	+11	0.5
B	0.47	0.47	-	0.0
	0.39	0.42	- 7	0.3
	0.25	0.23	+ 9	0.5
C	0.35	0.35	-	0.0
	0.30	0.30	-	0.3
	0.21	0.19	+11	0.5

*Fraser and Wardle values
A, Settlement of rigid raft;
B, settlement at centre of perfectly flexible raft;
C, maximum differential settlement of perfectly flexible raft

Table 2 Maximum moments (M/qab), rigid raft, a/b = 2, d/b = 0.5

	Computed	Exact*	% Error	ν_s
M_b	0.011	0.012	- 8	0.0
M_a	0.025	0.027	- 7	
M_b	0.010	0.011	- 9	0.3
M_a	0.021	0.023	-9	
M_b	0.009	0.009	-	0.5
M_a	0.012	0.013	-8	

*Fraser and Wardle values

4 CASE HISTORIES

The real value of any method of analysis must be judged in the context of its ability to reflect physical behaviour at full scale. The series of case histories presented here illustrates the power of the numerical method developed in mimicking the behaviour of real structures. At full scale the paucity of soils data is such that any agreement between theory and practice may be somewhat fortuitous. In the examples below the soil parameters used in the "after the event" analyses have all been determined in the light of the information that would have been available at the design stage. They are not therefore back-analysed in which the result is obtained using a set of soil parameters determined by forcing a match of the "predicted" and "actual" behaviour (the class C1 prediction type, Lambe 1973). They are a measure of the ability of the method to model correctly the behaviour at full-scale working from design data.

4.1 Grain silos on chalk

Burland and Davidson (1976) published a case study of the performance of a group of four grain silos. The silos were supported on independent 1.22 m thick circular raft foundations, 23 m in diameter bearing directly on to chalk. The silo floor was supported independently of the walls, on 24 circular columns providing access for lorries with the walls

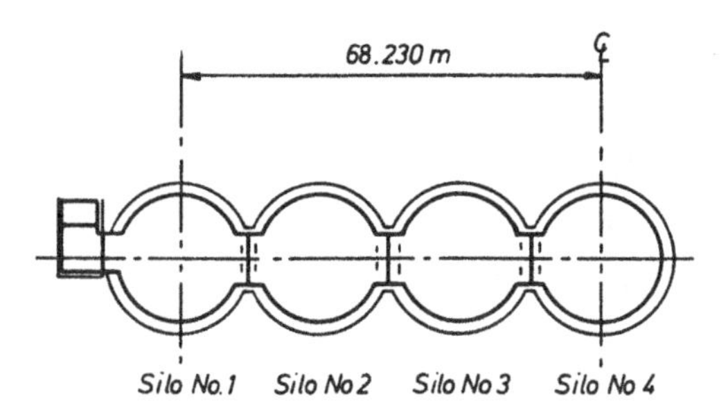

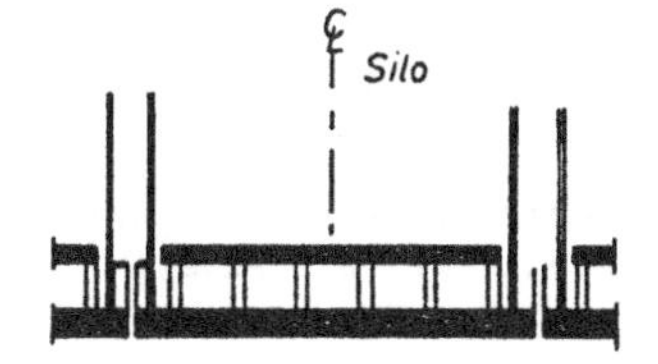

Fig. 4 Silo complex

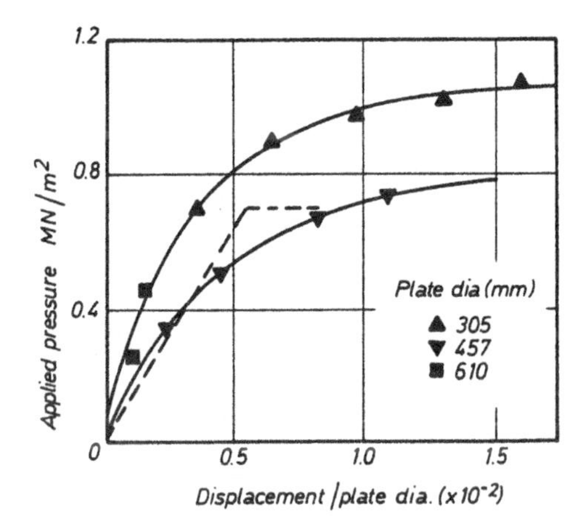

Fig. 5 Results of plate bearing tests

supported around the rim of the raft foundation, as shown in Fig. 4. The results of plate bearing tests carried out on the chalk are shown in Fig. 5. In the analysis the chalk has been characterised with a Young's modulus of 120 MPa and a Poisson's ratio of 0.25, together with a maximum bearing pressure of 700 MPa.

The measured and computed settlement for all four silos, using the method proposed here together with values computed by Majid and Rahman (1982) using a non-linear material model for the chalk are shown in Fig. 6. Both sets of computed settlements take account of the loading sequence for the silos and both are in tolerable accord with the actual performance.

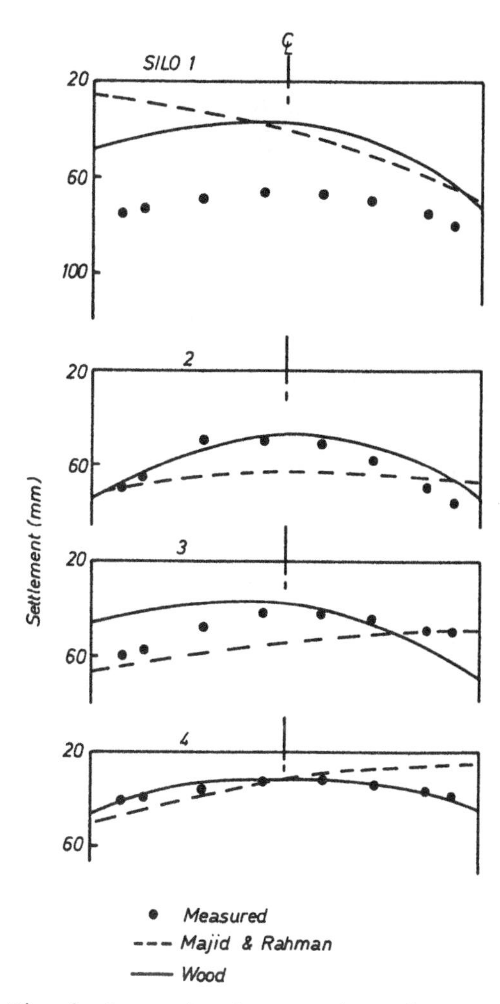

Fig. 6 Computed and measured settlements on longitudinal axis of silos

4.2 Dungeness 'B' Nuclear Power Station, Kent

Recorded movements of the raft foundation for the reactors of Dungeness 'B' Nuclear Power Station have been presented by Dunn (1975). The 3.4 m thick raft was founded at a depth of 9 m below ground level on 30 m of beach sand overlying silty clay mudstone. The layout of the reactor building is shown in Fig. 7 together with the results (N-values) obtained from standard penetration tests (SPT) and values of C_{kd} from static cone penetration tests. Also shown in Fig. 7 is the assumed variation of Young's modulus, E'_v adopted in the present analysis which indicates average ratios of $E'_v/C_{kd} \simeq 4.5$. Thomas (1968) suggests $E = 3.0$ to $12.0\ C_{kd}$.

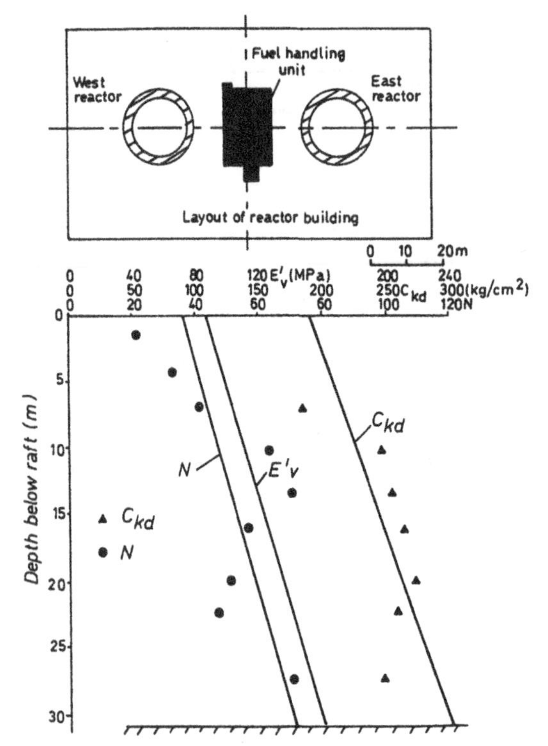

Fig. 7 Raft layout and soil properties Dungeness 'B'

Also from the Fig. 7 $E'_v/N \approx 2.1$ (E'_v in MPa). Parry (1971) suggests $E'_v = 2.5$ N for design. In this case the soil is normally consolidated and has been taken as exhibiting isotropic elastic characteristics with $\nu' = 0.1$.

The major construction sequence involved dewatering of the excavation in order to construct the raft, followed by construction of the fuel handling unit and then the two nuclear reactors. It has been assumed that any movements due to stress relief would have occurred prior to the start of the main construction. The datum for recorded settlements was established after the construction of the raft and therefore these correspond to the imposition of the load from the superstructure of 500 MN for the reactors and 115 MN for the central fuel handling unit. In the analysis the raft concrete has been assigned a Young's modulus of 28 GN/m^2 and Poissons's ratio of 0.15 and the raft has been taken as symmetrical about both axes. Computed and measured settlements along the longitudinal axis of the raft are shown in Fig. 8, exhibiting extremely good agreement. It should be noted that the

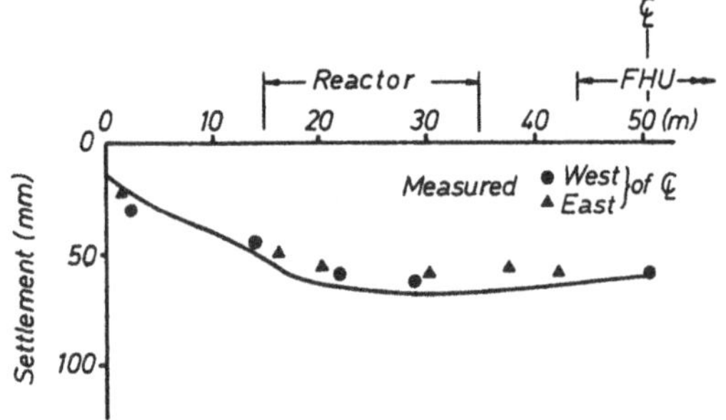

Fig. 8 Computed and measured settlement along longitudinal axis, Dungeness 'B'

measured settlements also exhibited symmetry about the raft axes.

4.3 Royal Free Hospital, London

Settlements of the 'T'-shaped 2.2 m thick raft foundation for the Royal Free Hospital in the London Borough of Camden have been published by Morton and Au(1975). The site has a slope of approximately 6° and is underlain by about 60 m of London Clay, 19 m (assumed) of Woolwich and Reading Beds, Thanet Sand and Chalk strata. The Thanet Sand and Chalk have been taken as incompressible.

The raft is founded at a maximum depth below the ground surface of 12 m and the gross applied load is equivalent to a uniformly distributed load of 270 kN/m^2 giving rise to an average net increase in applied pressure of 55 kN/m^2. The in-situ reinforced concrete frame superstructure is relatively flexible and its contribution to the stiffness of the raft has been ignored. The Young's modulus and Poisson's ratio of the raft concrete have been taken as 15 GN/m^2 and 0.15 respectively.

Two analyses have been undertaken in order to assess both the immediate undrained behaviour of the underlying London Clay and its long term, fully drained performance. The Woolwich and Reading Beds consisting of interbedded sand and clay layers have been assumed to behave at all times in a fully drained manner. The assumed transversely isotropic soil properties are given below:

London Clay, at depths z below the raft:

Undrained $E_v = 40 + 3.3z$ MN/m^2, $E_h/E_v = 1.8$, $\nu_{vh} = 0.5$, $\nu_{hh} = 0.1$ and $G_{vh}/E_v = 0.4$.

Drained $E'_v = 24 + 2z$ MN/m^2, $E'_h/E'_v = 2.3$, $\nu'_{vh} = 0.1$, $\nu'_{hh} = -0.15$ and $G_{vh}/E'_v = 0.661$.

Woolwich and Reading Beds

$E'_v = 200$ MN/m^2, $E'_h/E'_v = 2.3$, $\nu'_{vh} = 0.1$, $\nu'_{hh} = -0.15$ and $G/E'_v = 0.661$.

The relationship of $E_v = 1.67 E'_v$ for the London Clay is based upon elastic theory. The results of one-dimensional consolidation tests showed that m_v, the coefficient of compressibility, varied from 0.03 to 0.11 m^2/MN. (From elastic theory $E'_v = 0.96/m_v$ giving a variation in E'_v of between 8.7 and 32 MN/m^2, approximately one-half of the values used in the present analysis.)

Comparison between the computed time/settlement and measured time/settlement for two points on the raft is shown in Fig. 9. In determining the computed settlements it has been assumed that the undrained movements due to the stress relief caused by the excavation would have occurred prior to the construction of the raft itself. Hence the computed raft settlements are given by the summation of the drained settlements due to the applied load and the drained heave minus the undrained heave due to the stress relief. In addition the computed values have been adjusted in order that the datum coincides with that for the measured settlements (i.e. after construction of the raft). The agreement obtained is encouraging and would suggest that this approach provides a realistic assessment of raft behaviour.

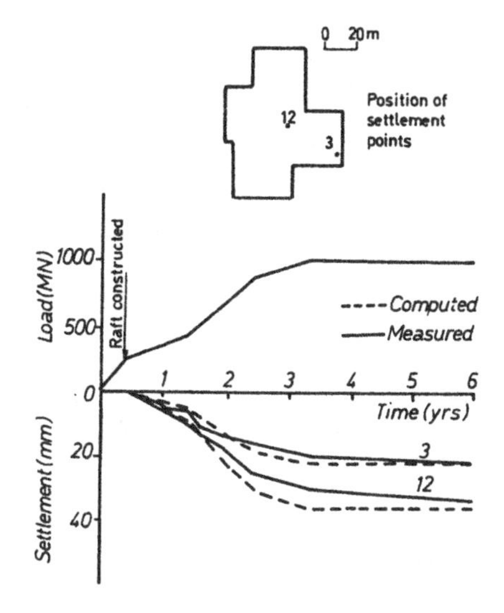

Fig. 9 Computed and measured settlements, Royal Free Hospital

4.4 A two-storey framed structure

Consideration has been given to the behaviour of a two-storey reinforced concrete framed structure supported on isolated footings situated on reclaimed land. Total settlements in excess of 300 mm were reported by Webb (1975) and excessive differential settlement gave rise to cracking in one of the main members. In order to contain these differential movements two of the columns were raised periodically by jacking. The relevant structural details, soil data and measured settlements are summarised below. In the analysis the structure has been modelled as a three-dimensional assemblage of beam elements, and settlements have been computed using parameters obtained from the results of the site investigation (Wood et al 1977b). The ground reactions, stress resultants and corresponding movements induced within the structure have also been determined.

The two-storey block adjoins a large single-storey warehouse and comprises offices on the first floor over truck loading bays. The general arrangement of the block which consists of a reinforced concrete frame and ribbed floor construction is shown in Fig. 10. (The idealised model used for the transverse frames is also shown and the sizes of members used in the analysis are indicated). It measures approximately 12.0 m wide by 61.0 m long by 8.0 m high. In the analysis the reinforced concrete members have been assigned a Young's modulus of 15 GN/m^2 and a Poisson's ratio of 0.15.

The structure is situated on reclaimed land and the individual footings are supported on stone columns (installed using a Vibroflot technique). The soil profile is shown in Fig. 11 and the soil properties are given in Table 3.

A buried ditch (see Fig. 10) ran under columns 010 and P8A and in the analysis the soil immediately beneath these columns was assigned a much reduced stiffness. The assumed soil succession below the stone columns was:

0.0 to 1.4 m Sand, E' = 100.0 MPa;
footings 02A to 08A and P10.
Clay, E' = 1.67 MPa; footings P8A and 010.

1.4 m to 2.3 Silt, E' = 1.43 MPa.

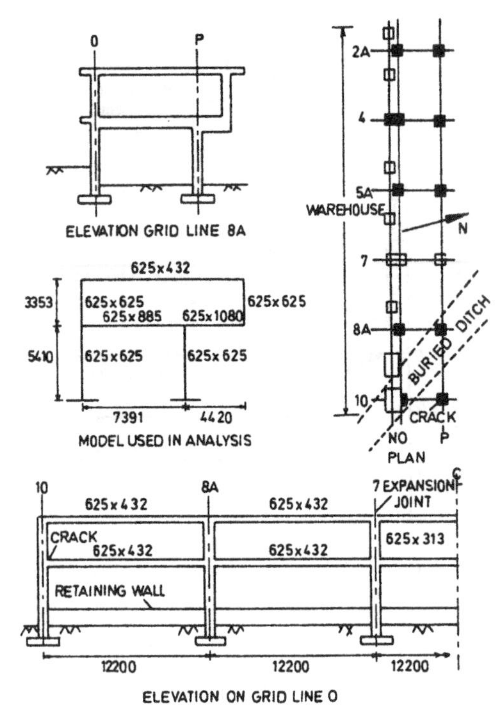

Fig. 10. General arrangement (dimensions in mm)

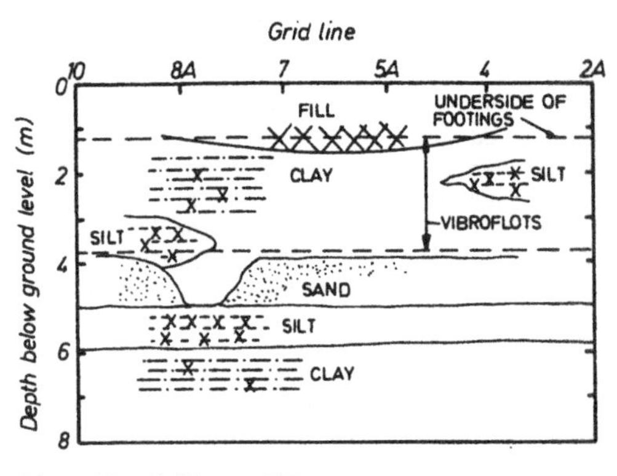

Fig. 11 Soil profile

2.3 m to 22.3 Clay, E' = 1.25 MPa.

Poisson's ratio was taken as zero for all the layers. The sand was assumed to be always in the drained condition and the undrained Young's modulus for the silt and clay layers was taken as E = 1.5 E'.

Periodic jacking of columns O8A and P8A was carried out in order to control the differential settlements. In computing

Table 3 Soil properties

Soil	m_v (m^2/MN)		$E' = 1/m_v$ (MN/m^2)	
	range	av.	range	av.
Black silty clay	0.2-0.6	0.4	1.67-5.0	2.5
Black organic clayey silt	0.6-1.2	0.8	0.83-1.67	1.25
Dark grey silty clay	0.8-1.4	1.0	0.71-1.25	1.0

the final settlement of the structure allowance has been made for this procedure by subjecting the lower ends of columns O8A and P8A to upward movements of 25 mm and 46 mm respectively, corresponding to the movements measured at the time, with the other footings fixed. The application of these enforced movements induced corresponding forces in all the columns and in turn the additional settlement due to these new column loads has been computed. The final settlements have been evaluated by superimposing these movements on the settlements computed from the original column loads.

The total computed and measured settlements, together with the initial applied column loads and the ground reactions are given in Table 4. It is apparent that good agreement has been obtained between the computed settlements taking into account the structural stiffness (column b) and the measured settlements. The settlements computed without including the structural stiffness (column a) are on the whole in tolerable agreement with the measured settlements at the western end of the building but diverge in the area of excessive recorded settlement at the eastern end. This is reflected in the pattern of computed ground reactions where footings P8A and O10 have shed load to the neighbouring footings.

Fig. 12 shows the agreement obtained between the computed and measured time-settlement curves for footing P8A. The early portion of the settlement curve is in reasonable agreement with that obtained from consideration of the undrained behaviour but would, perhaps, suggest that the relationship used to obtain E has produced an under estimate of the undrained stiffness of the clay and silt layers.

Table 4 Principal results

Footing	Initial Column Load (MN)	Computed (a) (mm)	Settlement (b) (mm)	Measured Settlement Feb. '76 (mm)	Computed Ground Reaction (MN)	Footing size (m)
O2A	0.85	192	198	212	0.88	2.44 x 2.44
P2A	0.91	202	210	210	0.96	2.44 x 2.44
O4 + N4	1.27 + 0.20	270	269	261	1.47	4.67 x 2.44
P4	1.37	306	286	280	1.26	2.44 x 2.44
O5A	1.18	276	279	269	1.19	2.44 x 2.44
P5A	1.26	289	295	279	1.29	2.44 x 2.44
O7 + N7	1.18 + 0.20	262	267	275	1.41	4.67 x 2.44
P7	1.26	291	301	296	1.31	2.44 x 2.44
O8A	1.27	263*	276	284	1.37	2.44 x 2.44
P8A	1.37	423*	338	332	1.08	2.44 x 2.44
O10 + N10	0.85 + 0.20	227	212	234	0.94	4.67 x 3.51**
P10	0.91	205	240	255	1.13	2.44 x 2.44

* Including deduction of 25 mm and 46 mm due to raising of columns
** Equivalent rectangle

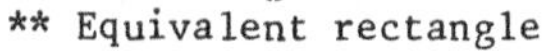

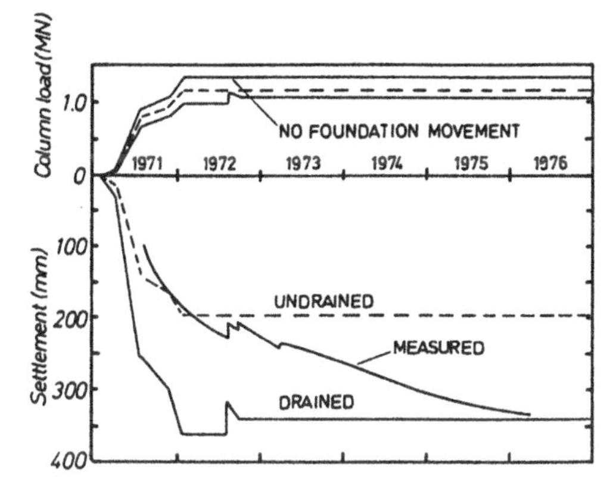

Fig. 12 Time-settlement curves for Column P8A

Estimates of the primary bending moments obtained from the distribution of dead and live loads with no foundation movement are given in Fig.13(a) for the typical transverse frame on grid line 4. The transverse distortion of this frame and the induced secondary bending moments are shown in Fig. 13(b). The sign convention adopted is that of clockwise rotation giving rise to positive moments. It should be noted that joint equilibrium is maintained by the torsional moment contributions from the members framing in at right angles to the frame under discussion. The values obtained are typical of those which would have been induced in all of the transverse frames but for the presence of the irrigation ditch at the eastern end of the building. For comparison the computed induced bending moments in the transverse frame on grid line 8A just prior to jacking of column P8A are given in Fig. 14.

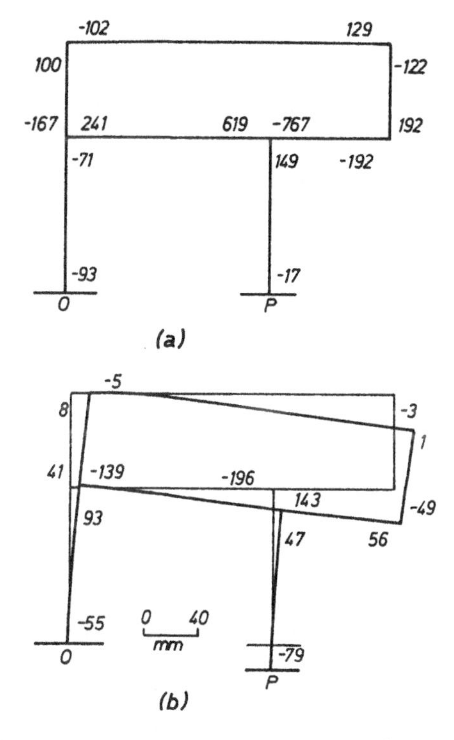

Fig. 13 Computed bending moments (kNm) on Grid Line 4.

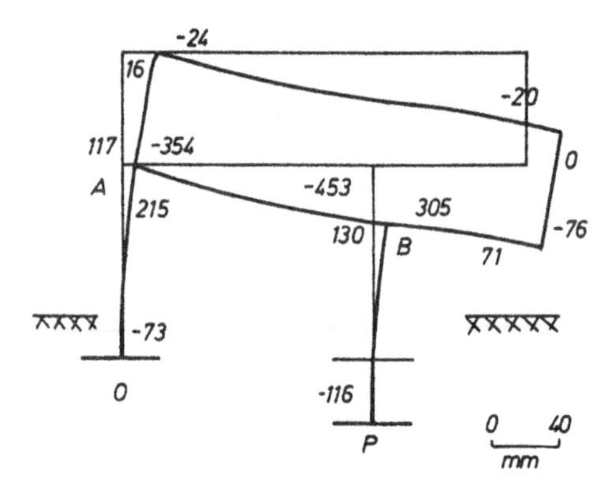

Fig. 14 Computed bending moments (k Nm) on Grid Line 8A, prior to jacking

CONCLUSION

A cost-effective method, in keeping with normal design office practice, has been developed in order to provide the practising engineer with a means of taking account of the interaction between a structure and the supporting soil continuum. The method has been validated with respect to the results obtained from more rigorous analytical and numerical techniques. Finally, the ability to mimic full-scale behaviour has been demonstrated.

The importance of including the effects of soil-structure interaction in any predictive assessment of structural behaviour has been clearly illustrated. Design methods which do not incorporate such effects at the analysis stage will not in any way reflect real structural behaviour and will be liable to produce erroneous predictions. These in turn could lead at the design stage to alterations in foundation or superstructure type and layout which would be unnecessarily conservative or uneconomic. The method described has been shown to be well validated and useful. It is clear from the case histories described that its use must be coupled with the exercise of careful judgement in the choice of appropriate models and parameter values and in the accounting for constructing sequence.

REFERENCES

Brown, P.T. 1969. Numerical analysis of uniformly loaded circular rafts on elastic layers of finite depth. Geotechnique 19.

Brown, P.T. 1974. Influence of soil inhomogeneity on raft behaviour. Soils and Foundations 14.

Burland, J.B. & Davidson, W.A. 1976. A case study of cracking of columns supporting a silo due to differential foundation movement. Proc. Conf. on Performance of Building Structures, London. Pentech Press.

Chamecki, S. 1956. Structural rigidity in calculating settlements. J. S.M. and F. Div. Am. Soc. C.E. 82. SM1

Cheung, Y.K. & Zienkiewicz, O.C. 1965. Plates and tanks on elastic foundations - an application of the finite element method. Int. Jl. Solid Struct. 1.

Dunn, C.S. 1975. Settlement of a large raft foundation on sand. Proc. Conf.on Settlement of Struct., London. Pentech Press.

Fraser, R.A. & Wardle, L.J. 1973. A rational analysis of shallow foundations considering soil structure interaction. Austr.Geomech. J. 1. G5.

Fraser, R.A. & Wardle, L.J. 1976. Numerical analysis of rectangular rafts on layered foundations. Geotechnique 26.

Lambe, T.W. 1973. Predictions in soil engineering, Geotechnique 23.

Majid, K.I. & Rahman, M.A. 1982. Nonlinear analysis of structure-soil systems. Proc. I.C.E.73, Part 2.

Morton, K. & Au, E. 1975. Settlement observations on eight structures in London. Prof. Conf. on Settlement of Structures, London. Pentech Press.

Parry, R.H.G. 1971. A direct method of estimating settlements in sand from SPT values. Proc. Symp. Interaction of Structure and Foundation, Birmingham. Midland Geotech. Soc.

Thomas, D. 1968. Deep sounding test results and the settlement of spread footings on normally consolidated sands. Geotechnique 18.

Webb, D.L. 1975. Observed settlement and cracking of a reinforced concrete structure founded on clay. Proc. Conf. on Settlement of Structures, London. Pentech Press.

Wood, L.A. & Larnach, W.J. 1975a. The effects of soil-structure interaction on raft foundations. Proc. Conf. on Settlement of Structures, London. Pentech Press.

Wood, L.A. & Larnach, W.J. 1975b. The interactive behaviour of a soil-structure system and its effect on settlements. Proc. Symp. Anal. Soil Behaviour and Applic. Geotech.Struct. Univ. N.S.W., Australia.

Wood, L.A., Larnach, W.J. & Woodman, N.J. 1977a. Observed and computed settlements of two buildings. Proc. Int. Symp.

on Soil Struct. Interaction. Univ. of Roorkee, India.
Wood, L.A., Larnach, W.J. & Webb, D.L. 1977b. Observed and computed behaviour of a framed structure. Large Ground Movements and Structures, London. Pentech Press.
Wood, L.A. 1977. The economic analysis of raft foundations. Int. Jl. Nvm. Methods in Geomech. 1.
Wood, L.A. 1978a. A simple boundary element approach to the prediction of the settlement of structures. Recent Advances in Boundary Element Methods. C.A. Brebbia (ed.), London. Pentech Press.
Wood, L.A. 1978b. RAFTS - a program for the analysis of soil-structure interaction. Advanced in Eng. Soft. 1.
Wood, L.A. 1979a. A rational approach to the analysis of building structures taking full account of foundation movements. Engineering Software. R.A.Adey (ed.), London. Pentech Press.
Wood, L.A. 1979b. Simple boundary elements in soil-structure interaction applications. Proc. Conf. Computer Application in Civil Eng. Univ. of Roorkee, India.
Wood, L.A. 1980. Time dependent settlement of structures. New Developments in Boundary Element Methods, C.A.Brebbia (ed.), London. CML Publications.
Wood, L.A. 1981. The application of "RAFTS" to soil-structure interaction problems. Implementation of Computer Procedures and Stress-Strain Laws in Geotechnical Engineering. C.S. Desai, S.K. Saxena (eds.), London, Acorn Press.

Computer and Physical Modelling in Geotechnical Engineering, Balasubramaniam et al. (eds)
© 1989 Balkema, Rotterdam. ISBN 90 6191 864 2

Behaviour of flexible earth retaining walls at full-scale

L.A.Wood & C.J.Forbes King
Department of Civil & Structural Engineering, South Bank Polytechnic, London, UK

ABSTRACT: The importance of the determination of the in-service behaviour of earth retaining structures in order to validate design methodology is emphasised. Field measurements of ground displacements and wall deflections for a temporary king-post wall in London clay, a diaphragm wall in boulder clay and a permanent sheetpile cofferdam are presented. These case histories also include information on the soil and wall material properties and the construction programme. An economic method of analysis, in which account is taken of soil-structure interaction effects, for the design of diaphragm and sheetpile type earth retaining walls is also presented. Comparison between the measured and predicted wall deflections are given. The latter, in which the sequence of construction of the walls was modelled, show reasonable agreement with the field measurements.

1 INTRODUCTION

Within a long standing programme of research in the field of soil-structure interaction undertaken by the senior author (see Wood and Larnach, 1988) the emphasis in recent years has been placed upon correlation of computer models with full-scale physical performance. The field data flowing from a programme of instrumentation and long term performance monitoring, which commenced in 1981 (Wood, 1984) and has to date encompassed nine geotechnical structures, is now such that effective comparison may be made with the original design predictions. Furthermore, these case histories will serve as valuable bench marks for the evaluation and validation of future mathematical and computer models. These case histories include both foundations (Wood and Perrin, 1985) and flexible earth retaining walls (Wood and Perrin, 1984). The emphasis here is on the latter and three previously unpublished case histories are presented. Namely, a temporary king-post wall constructed in London clay with a single level of struts, a permanent cantilever reinforced concrete diaphragm wall in boulder clay and two permanent tied steel sheetpile walls forming the abutments of a flood defence barrier.

All of the walls were analysed at the design stage using the LAWWALL computer program (Wood, 1979, 1980, 1984). The basis of the method of analysis used is outlined below. In addition to the presentation of the field measurements comparison is made with the best design stage predictions.

2 LAWWALL COMPUTER PROGRAM

The method of analysis used in LAWWELL is very similar in concept to that pioneered in a sister program LAWRAFTS (Wood, 1978).

The original version of the program, known as LAWPILE, was developed to provide design engineers with a realistic and economic method of analysing pile groups subjected to lateral loads. The program was subsequently extended, utilizing the same basic soil and structural model, to include the analysis of multipropped sheetpile and diaphragm walls (Wood, 1979). Conceived initially as a purely analytical tool it was expanded to incorporate a design capability and renamed LAWWALL (Wood, 1984). The design facility allows the user to specify the desired factor of safety in terms of net pressure, passive earth pressure or partial factors on the soil shear strength parameters. The depth of penetration having been determined the wall is then analysed without the factors present. In order to take account of the construction programme the analysis is incremental with respect to each excavation stage.

2.1 Method of analysis

An outline of the basic method of analysis adopted is given. All the computations are centred on a unit length of wall, and the wall itself is modelled as an assemblage of beam elements exhibiting two degrees of freedom, namely horizontal translation and rotation. The stiffness matrix $[K_w]$ is condensed to give a relationship between the net force Q_{net} acting on the wall and the horizontal displacement ω only;

$$Q_{net} = [K_w] \quad (1)$$

where K_w may also include, where appropriate, the additional stiffness of props. The soil is modelled as an inhomogeneous elastic continuum utilizing an approximate extension of Mindlin's (1936) solution, based on the assumption that the displacement is a function of the local values of the elastic parameters. Hence, a matrix of flexibility coefficients may be established for the soil, the inverse of which is the soil stiffness matrix $[K_s]$.

The soil reactions P are related to the wall displacements (assumed equal to the soil displacements) by

$$P = [K_s]\omega \quad (2)$$

Combining equations (1) and (2), and noting that

$$Q_{net} = Q - P \quad (3)$$

where Q is the external applied force,

$$\omega = [K_s + K_w]^{-1}Q \quad (4)$$

For the wall Q, the initial disturbing force, is determined from a consideration of the coefficient of earth pressure at rest (K_o) and the out-of-balance earth pressures on either side of the wall. Hence, the wall displacements are obtained from equation (4). The compatible earth pressures are derived and compared with the active and passive envelopes, obtained from a consideration of classical theory. An iterative process is adapted to ensure that the final soil reactions always lie within the active-passive range (no account is taken of arching). At each excavation stage the incremental movements etc. are computed and the summation result given.

The program provides the user with two distinct soil models to choose from: first the continuum model in which the soil stiffness matrix is full, and second an uncoupled Winkler spring analogy in which the off-diagonal terms in the matrix are zero. The results presented in the subsequent case histories were all produced using the uncoupled soil model.

3 CASE HISTORIES

The three case histories presented cover three distinct types of flexible retaining wall in dissimilar ground conditions.

The first was situated in an urban environment in which control of the movement of adjacent structures was of paramount concern. The second is a permanent reinforced concrete diaphragm retaining one side of a cutting for a new highway. The third encompasses the performance of two cofferdams situated at the mouth of a creek forming part of a flood defence barrier.

3.1 King-post wall

Redevelopment of Beaver House in the City of London provided an opportunity to monitor the performance of an unusual temporary retaining wall to a new 7.5m deep basement. As with many new developments in an inner city location neighbouring buildings were situated very close to the excavation. Therefore, a major priority during construction was maintenance of the integrity of the adjoining properties. With this in mind a simple instrumentation and monitoring programme was assembled to provide information about wall and ground displacements. These would in turn provide confidence in the method of excavation support adopted and an early warning should anything unusual begin to happen.

The site is situated immediately south of Mansion House LRT station in the City of London and is flanked on the east and west respectively by Garlick Hill and Trinity Lane. A plan of the site is shown in Fig.1. Also shown are the position of the eight inclinometer ducts attached to the 'I'-section steel king-posts, and the layout of the single level of struts.

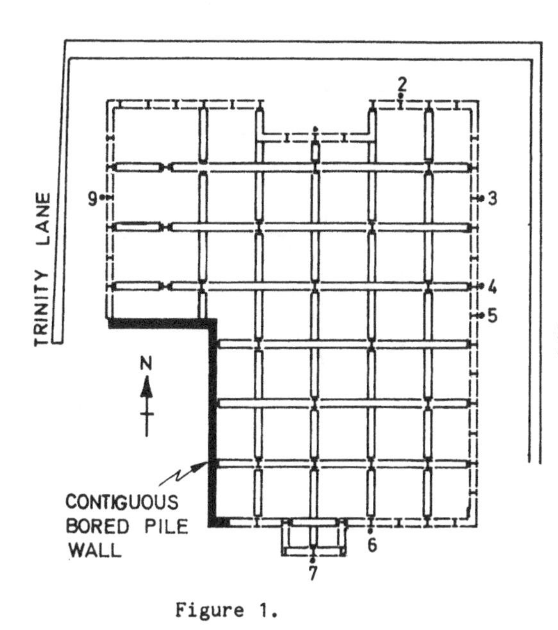

Figure 1.

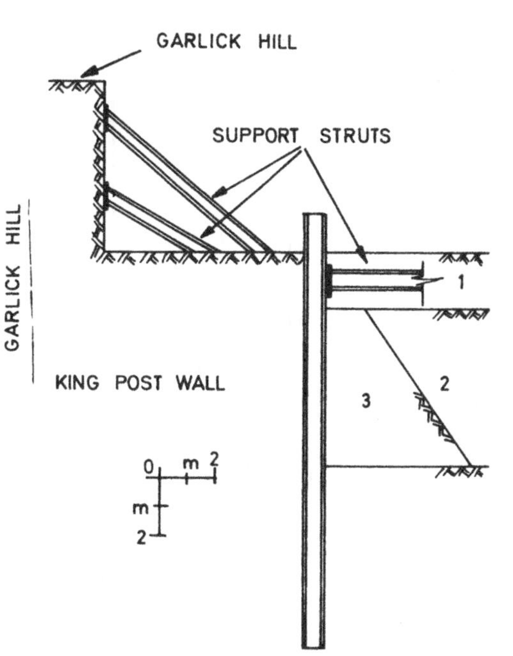

EXCAVATION SEQUENCE

1. EXCAVATE 2m & STRUT WALL
2. GENERAL EXCAVATION
3. REMOVE BERM

Figure 2.

Alternate king-posts were propped by a reinforced concrete waling and. the rectangular grid of struts supported on soldier piles. Both the king-posts and soldier piles were placed in open bores formed prior to the bulk excavation, and backfilled with concrete to final excavation level with granular material above. As excavation proceeded in-situ reinforced concrete poling boards were constructed between the king-posts. The excavation sequence is shown in Fig.2. where the position of the new basement within an existing basement of variable depth, due to the sloping nature of the site, is also indicated.

The ground conditions are very simple with London clay overlying the Woolwich and Reading beds. The whole of the retaining wall was in the London clay. In the design the soil was characterised with following parameters;

$$c' = 20 \text{ kN/m}^2,$$

$$\phi' = 20^\circ,$$

$$\text{Unit weight}, \gamma = 20 \text{ kN/m}^3,$$

Poisson's ratio, ν = 0.1,

Young's modulus, E = 10.0+3.0z MN/m^2, where z is the depth below the top of the wall, all taken as representative of the long term, effective stress state of the London clay.

The ground water table was 5m below the top of the wall. Wall friction and adhesion values of 10 degrees and 10kN/m^2 respectively were assumed to act on the back face of the wall. Although in practice the wall was discontinuous, with only the king-posts at 3m centres penetrating below excavation level it was considered to behave as a continuous wall in the analysis.

Furthermore, in the analysis, the wall was taken as 400mm wide with a Young's modulus of 45 GN/m^2, and the stiffness of the struts as equivalent to 87.5 MN/m/m. The predicted wall deformation at final excavation is shown in Fig.4.

3.1.1 Field results

Typical combined results obtained from the inclinometer measurements and line survey of the king-posts are shown in Fig.3 for inclinometer duct No.5. It is readily apparent that the major proportion of the final movement had occurred at the second stage with only the berm in place.

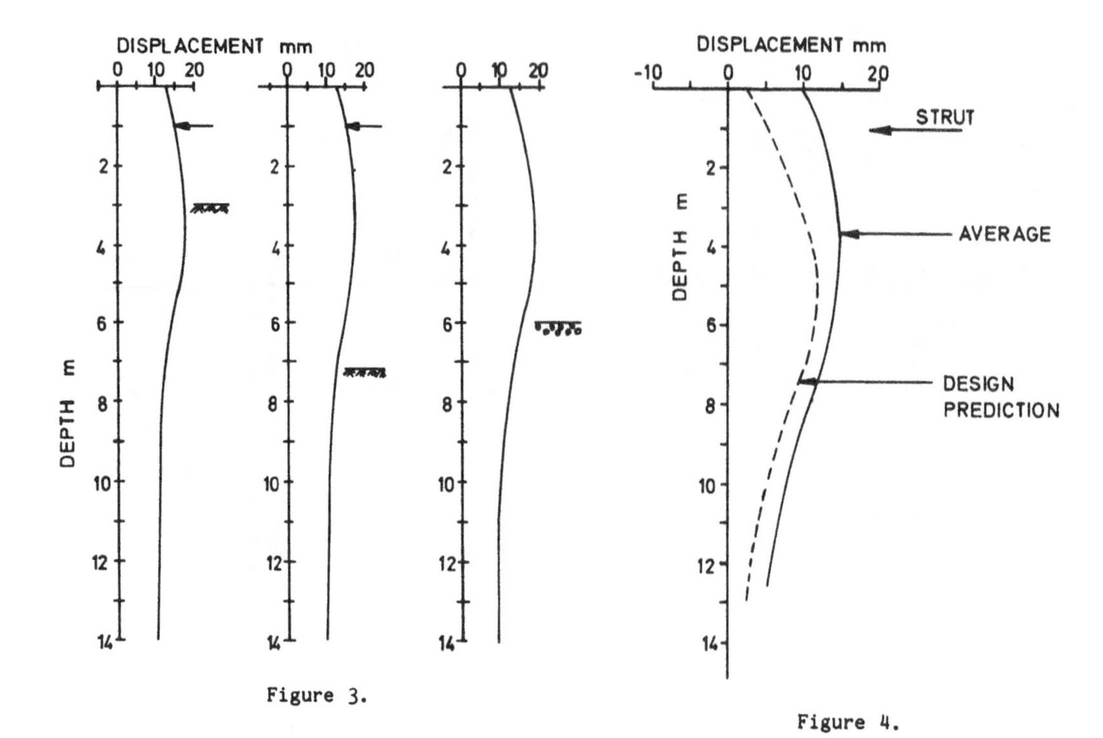

Figure 3.

Figure 4.

Another important feature is the movement of the toe of the wall which is of the same order of magnitude as that recorded at the top.

The average wall deformation at final excavation obtained from consideration of measurements from all eight instrumented king-posts is shown in Fig. 4 together with the design prediction. It would appear that the strut stiffness may have been over-estimated in the design prediction.

The horizontal movement towards the excavation of Garlick Hill was also monitored and the results for a point midway along the excavation side are shown in Fig. 5. This point is located approximately 8m back from the edge of the excavation and some 6m higher than the top of the wall, see Fig. 2.

The recorded movement in May 1985 of around 10mm is similar to that of the top of the king-post wall. The increase in movement in July corresponds to the removal of the raking props shown in Fig. 2.

Although movements of between 10 and 20mm were measured no distress to adjacent buildings was recorded.

3.2 Diaphragm wall

Construction of the Western Distributor Road by Leicestershire County Council, provided an ideal opportunity to monitor an 800mm thick in-situ cantilever diaphragm wall and compare the results with computer prediction. The relevant section of road is shown in Fig. 6, with the location of the diaphragm wall marked. The diaphragm wall extends north and south of a new underbridge, with the maximum retained height of 8m just south of the underbridge. The wall is structurally separate from the underbridge. Two vertical inclinometer ducts were cast into the wall and survey stations established on the wall at this location in order to facilitate the measurement of the deformation of the wall.

The geology of the area is a glacial boulder clay deposit overlying red marl. The ground conditions adjacent to the instrumented section of wall as revealed by a borehole are shown in Fig. 7. The results obtained from consolidated, undrained with pore-water pressure measurement, triaxial tests on the clay are shown in Fig. 8. The values of the relevant soil parameters and are given in Fig. 9 together with a section through the 18.5m deep wall.

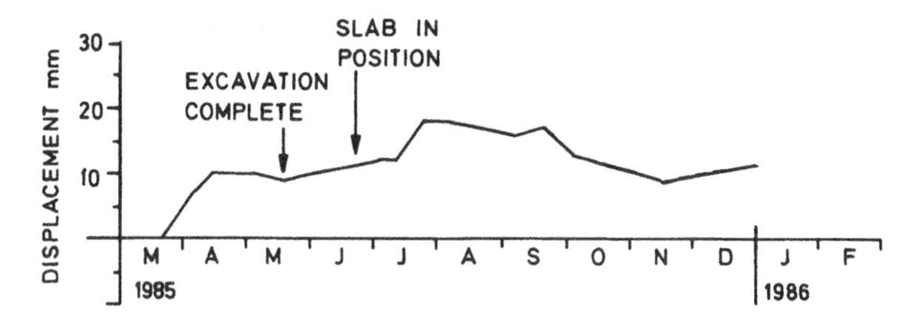

Figure 5.

It should be noted that excavation has taken place in three stages and that in the analysis undertaken using LAWWALL some account has been taken of the long term change in the effective Young's modulus of the wall concrete. It has been assumed that due to the timescale for the excavation the effects of creep etc. would lead to a reduction in the Young's modulus of the concrete from 30 GN/m^2 to 15 GN/m^2.

The computed displacements are shown in Fig. 10, and the corresponding shear force and bending moment diagrams are given in Fig. 11.

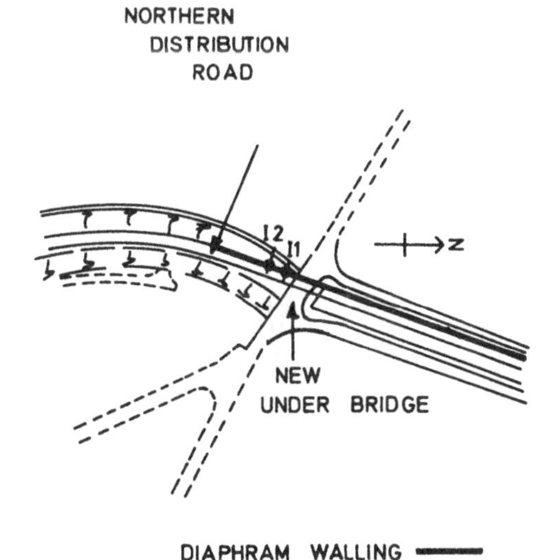

Figure 6.

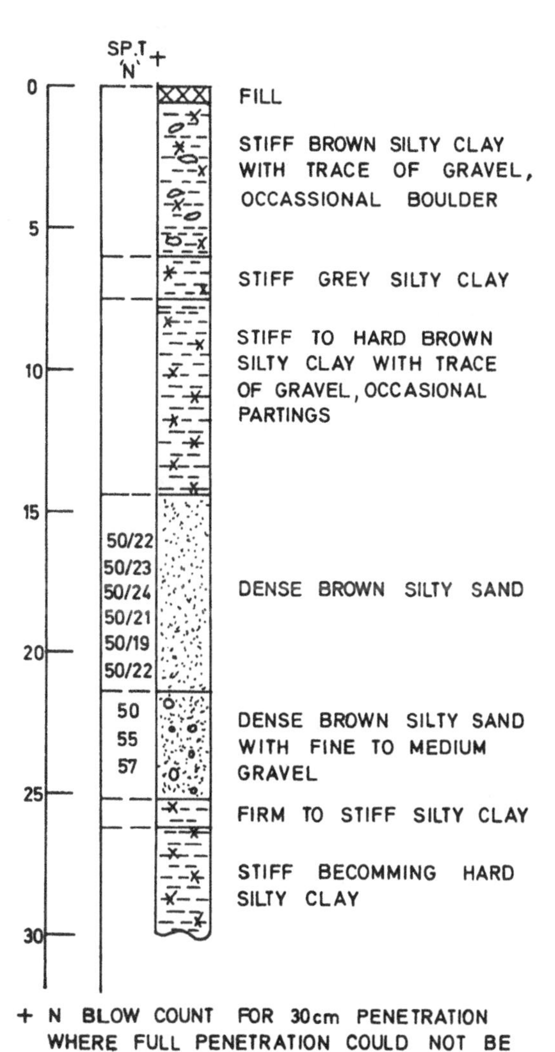

Figure 7.

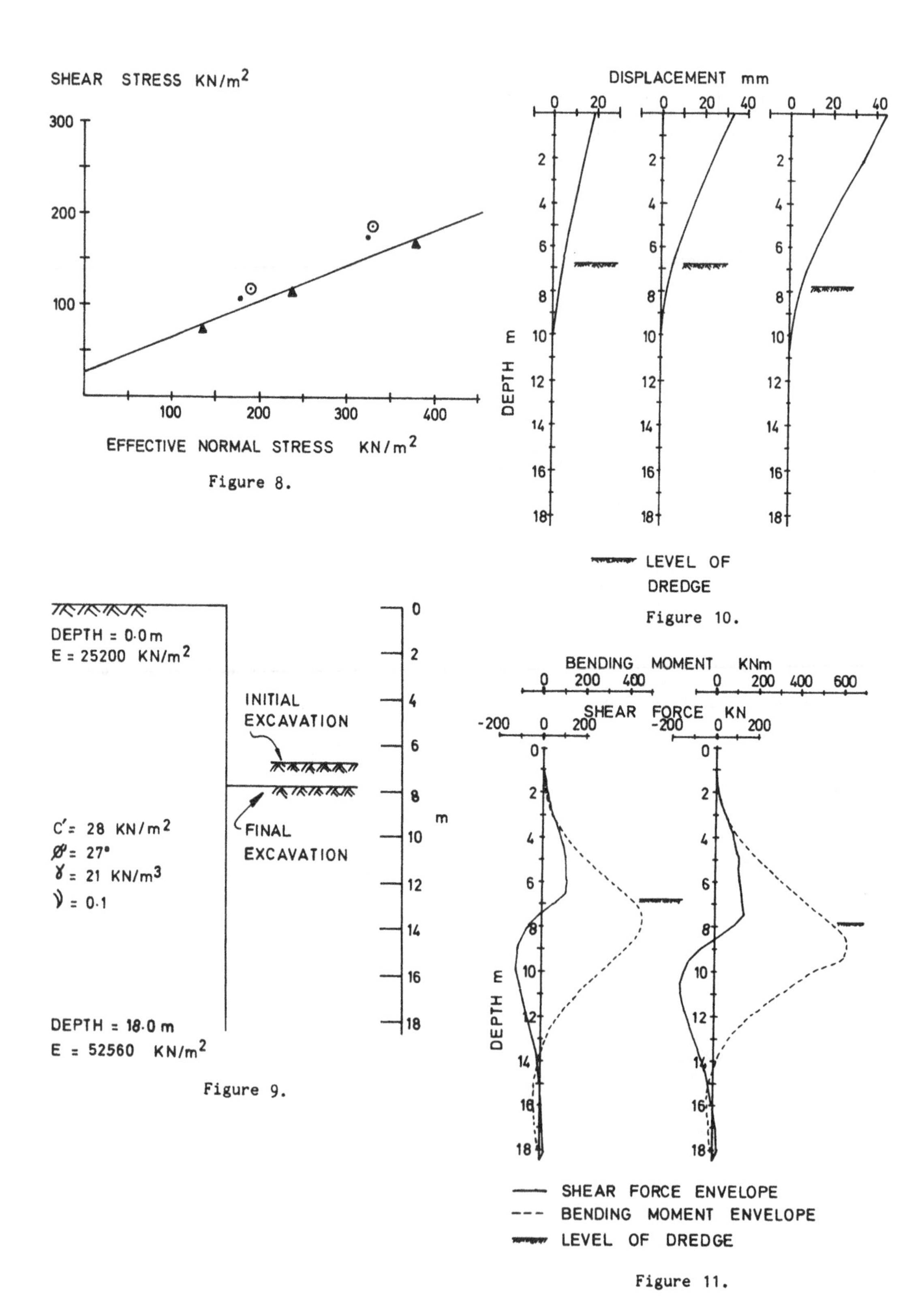

Figure 8.

Figure 9.

Figure 10.

Figure 11.

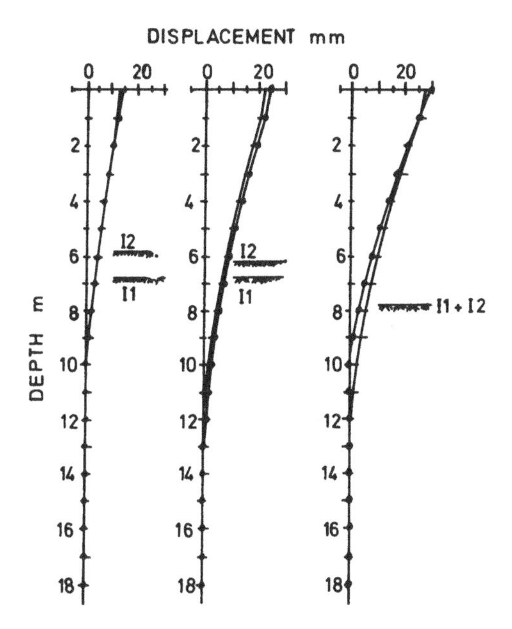

Figure 12.

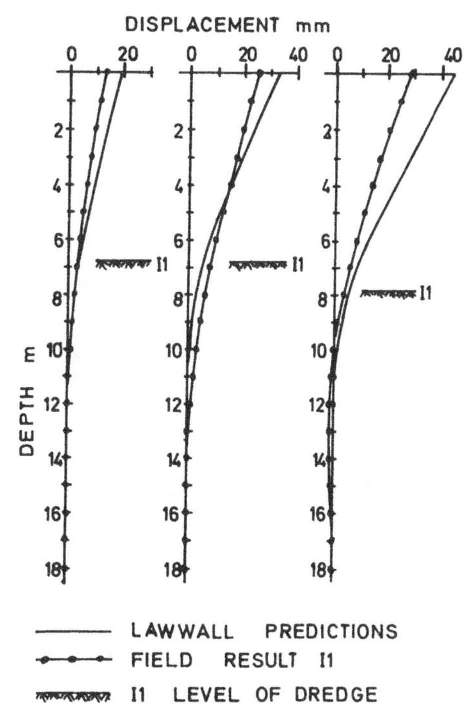

Figure 13.

The increased deflection at constant excavation level in Fig. 10 is entirely due to a reduction in wall stiffness due to creep.

3.2.1 Field results

The results obtained from the two inclinometer ducts are shown in Fig. 12 for the three excavation stages. The absence of any movement at the toe of the wall has been confirmed by the results obtained from an independent survey of the top of the wall.

The computed and observed wall deformations are compared in Fig. 13, where the consistent over-estimation of computed wall curvature is most evident. For a cantilever wall the deflected shape is very dependent upon the stiffness of the wall section.

3.3 Tied sheetpile wall

The construction of a new flood defence barrier in Queenborough Creek on the Isle of Sheppey in Kent provided the opportunity to monitor the performance of two tied sheetpile walls. The permanent sheetpile walls enclose the abutments to a flood-gate located in the main navigational channel. The general arrangement of the two abutments and the instrumented sections are shown on the plan in Fig. 14.

The site was dredged to remove the river silt and expose the surface of the underlying London clay. The sheetpiles were driven into the London clay creating two cofferdams which were subsequently backfilled with single size free draining 40mm aggregate. The parallel lines of sheetpiles forming the north and south walls of the cofferdams are connected together by two levels of passive ties. At the outer end of each cofferdam the flood-gate is supported on a reinforced concrete structure which bears onto 'H' piles driven into the underlying clay.

The top of the sheetpile wall was the subject of a line survey and inclinometer ducts were attached to a number of the sheetpile sections. In addition two sheetpiles in each abutment, one in the north wall and one in the south wall, were instrumented with a vertical array of

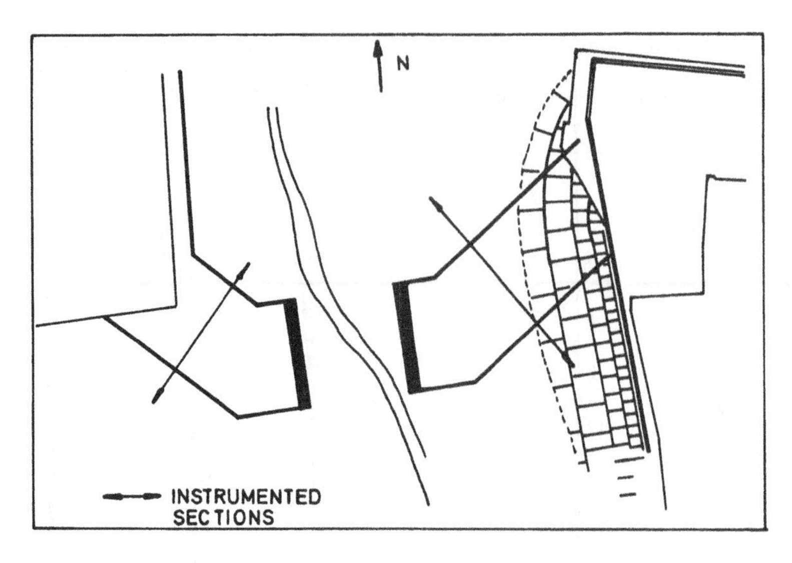

Figure 14.

vibrating wire strain gauges. Similar gauges were also used to measure the force developed in several of the tie rods.

The long term effective stress parameters used in the computation of the design stage predictions were;

aggregate backfill: $c' = 0$
$\phi' = 35^{o}$,
unit weight $\gamma = 18\ kN/m^3$
$\nu = 0.25$
$E = 50\ MN/m^2$;

London clay: $c' = 20\ kN/m^2$,
$\phi' = 16^{o}$,
unit weight $\gamma = 19\ kN/m^3$,
$\nu = 0.1$
$E = 3.0\ MN/m^2$.

Two conditions were analysed. First, with the ties in position and tight before the commencement of filling of the cofferdams. This was an early design exercise and the position of the ties changed subsequently. Second, the positioning and tightening of the ties as filling of the cofferdams proceeded. That is, the aggregate to be placed in order to provide a working platform for the placement of the ties.

3.3.1 Results

Comparison is made between the computed and observed behaviour for the east and west abutments in Figs. 15 and 16 respectively. Although the measured displacements of the sheetpiles on the two abutments exhibit different characteristics agreement with the movements predicted on the basis of the ties being placed as filling proceeded is reasonable. This result concurs with the construction practises adopted on site.

Bending moments have been deduced from the readings obtained from the strain gauges located on the sheetpiles. Agreement between these and the predicted values is poor. Work is continuing in the examination of these results in order to gain a clearer insight into the reasons for this disparity.

The measured force in the lower level of ties on the east and west abutments was 280 and 130 kN/m respectively. The corresponding predictions were 110 and 180 kN/m. For the upper level of ties both the field measurements and the predictions indicated the absence of any significant force. It should be noted that comparison has been made with the predictions based upon the ties being placed as filling proceeded.

4 CONCLUSIONS

Comprehensive case histories of three dissimilar flexible earth retaining walls have been presented. It is to be hoped that these will be utilised by other researchers and practitioners to validate computer based numerical modelling techniques.

The observed behaviour of the walls has been compared in all cases with design predictions obtained using the LAWWALL

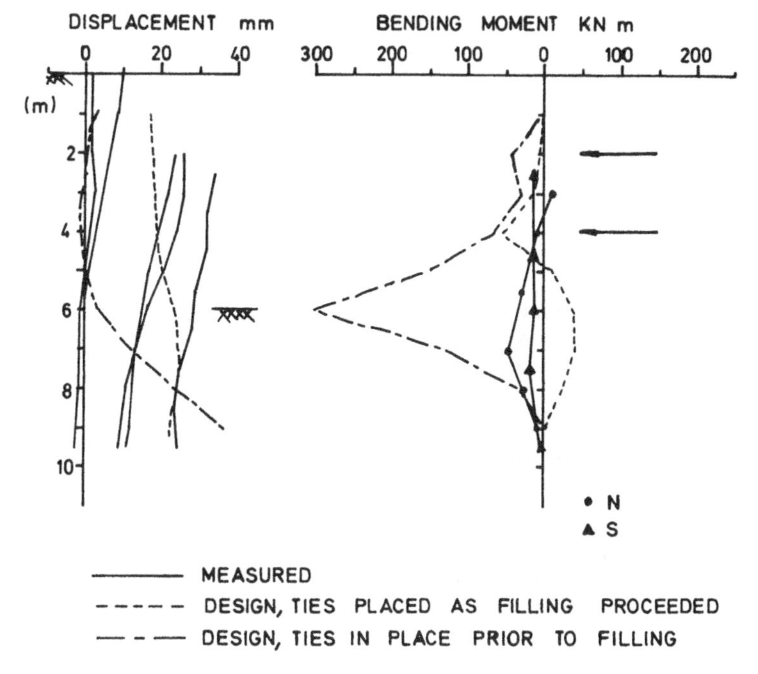

Figure 15. East Abutment

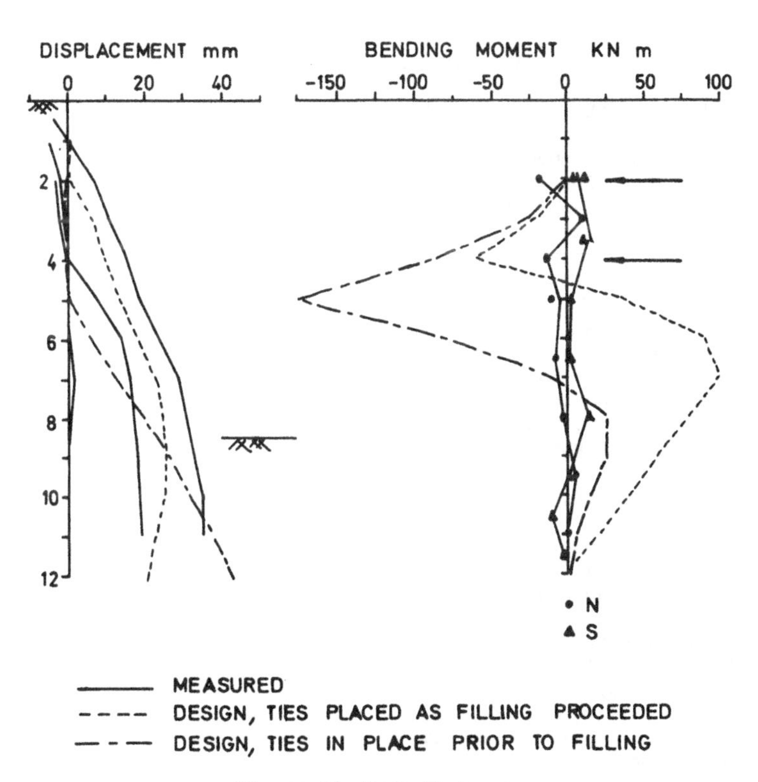

Figure 16. West Abutment.

computer program. The agreement is satisfactory when account is taken of all the imponderables associated with design and construction. Further work is in progress on improvements to this model which it is felt provides a level of sophistication equitable with the design stage of most projects.

5 ACKNOWLEDGEMENTS

A major portion of the funding for the programme of field work has been provided by the Science and Engineering Research Council of Great Britain.

The authors also wish to thank Clarke Nicholls and Marcel, consulting engineers; Mowlem Management Ltd; Leicestershire County Council; and the Southern Water Authority for their support and encouragement in the field work. Thanks are also due to A. Duff for his assistance with the installation and reading of the instruments.

6 REFERENCES

Mindlin, R.A. (1936) Force at a point in the interior of a semi-infinite solid. J. Physics, 77.

Wood, L.A. (1978) RAFTS - a program for soil-structure interaction. Adv. Engng. Soft. 1,1.

Wood, L.A. (1979) LAWPILE - a program for the analysis of laterally loaded pile groups and propped sheetpile and diaphragm walls. Adv. Engng. Soft. 1,4.

Wood, L.A. (1980) The analysis of piles and walls subject to lateral forces. Ground Engineering, 13, 1.

Wood, L.A. (1984a) A new departure for Ground Engineering 17, 7.

Wood, L.A. (1984b) LAWWALL: analysis of cantilever and multi-braced sheetpile and diaphragm walls. User manual. London: SIA Computer Services.

Wood, L.A. and Perrin, A.J. (1984) Observations of a strutted diaphragm wall in London clay: preliminary assessment. Geotechnique, 34, 4.

Wood, L.A. and Perrin, A.J. (1985) The performance of a deep foundation in London clay. Proc. XI Int. Conf. Soil Mech. Fdn. Engng., San Francisco, A.A. Balkema, Rotterdam.

Wood, L.A. and Larnach, W.J. (1988) Soil-Structure Interaction - a rational design orientated approach. Computer and Physical Modelling. ed. A.S. Bolasubramaniam, A.A. Balkema, Rotterdam, Netherlands.

For Product Safety Concerns and Information please contact our EU
representative GPSR@taylorandfrancis.com Taylor & Francis Verlag GmbH,
Kaufingerstraße 24, 80331 München, Germany

Printed and bound by CPI Group (UK) Ltd, Croydon, CR0 4YY
01/05/2025
01858472-0002